8 .P47 1993

Performance Evaluation of High Speed Switching Fabrics and Networks

Performance Evaluation of High Speed Switching Fabrics and Networks

ATM, Broadband ISDN, and MAN Technology

Edited by

Thomas G. Robertazzi
Department of Electrical Engineering
State University of New York at Stony Brook

A Selected Reprint Volume
IEEE Communications Society, *Sponsor*

The Institute of Electrical and Electronics Engineers, Inc., New York

This book may be purchased at a discount from the publisher
when ordered in bulk quantities. For more information contact:

IEEE PRESS Marketing
Attn: Special Sales
P.O. Box 1331
445 Hoes Lane
Piscataway, NJ 08855-1331

Printed in the United States of America

10 9 8 7 6 5 4 3 2 1

ISBN 0-7803-0436-5

IEEE Order Number: PC03335

Library of Congress Cataloging-in-Publication Data
Performance evaluation of high-speed switching fabrics and networks :
  ATM, broadband ISDN, and MAN technology/edited
  by Thomas Robertazzi.
       p.   cm.
  "A selected reprint volume."
  "Communications Society, sponsor."
  Includes bibliographical references and indexes.
  ISBN 0-7803-0436-5 :
    1. Telecommunication—Switching systems. 2. Packet switching
(Data transmission) 3. Computer networks. I. Robertazzi, Thomas
G. II. IEEE Communications Society.
TK5103.8.P47   1993
621.382—dc20   92-44950
                                             CIP

Contents

Preface

This volume was created because of the emergence of a new communication technology with vast potential. Specifically I refer to the use of high speed switching fabrics for packet switching. These switching fabrics are patterned collections of simple switching building blocks. Used as switching hubs, such fabrics could serve as the basis of future local area networks, metropolitan area networks and wide area networks. They make possible asynchronous transfer mode (ATM) technology to deliver broadband ISDN services. This is a topic of much interest to both the communication and the computer industry.

This volume focuses on the performance evaluation of high speed switching fabrics and networks. One of the reasons for putting together this collection is that amazing progress has been made over the past decade by a large number of researchers working on this problem. By focusing on the performance evaluation of such fabrics the reader gains a broad systems level look at the behavior of individual architectures and the tradeoffs between different architectures. Moreover this volume is a rich source of design data for high speed fabrics and systems.

The intended audience for this work are engineers and computer scientists interested in the research and development of high speed switching fabrics and networks. As such it should appeal to communication and computer people in industry, academia and government. Moreover it could be used for university graduate level seminars on high speed switching and as a reference in communication courses.

How is this volume organized? The first section contains a tutorial paper that the switching novice will find extremely useful. The papers that follow are grouped into sections according to common architectural characteristics. Each section is preceded by an introduction which puts the section's papers into perspective and provides extended summaries of each paper. Each introduction also includes additional references. There are a good many excellent papers at this date in the literature on telecommunications switching. Papers were chosen for this volume based on three criteria. One was that they deal with fabrics, be they centralized or distributed. A second was that they contain some form of performance evaluation, preferably of an analytic type. Finally, papers were selected that would help bring about a complete picture, in this volume, of our current knowledge of switching fabric/network performance. While I had the advice of a number of well qualified manuscript reviewers, the final selections were my responsibility. A result of using these criteria was that this volume is extremely focused and can give an in depth look at its topics.

I would like to close by commending the work of the authors of this volume's papers to you, the reader. The progress that they have made in these short years is remarkable and sets the stage for tomorrow's networking technology.

T. Robertazzi
Stony Brook N.Y.

Acknowledgments: This volume would never have seen the light of day without the help of a number of people. I gratefully acknowledge the sponsorship of this volume by the Communications Society and the assistance of Communications Society liaisons Tom Plevyak and Jack Holtzman. For me as an editor, the folks at IEEE Press have been a joy to work with. In particular, thanks are due to Dudley Kay, Executive Editor; Valerie Zaborski, Production Editor; and Marybeth Hunter, Marketing Manager.

I would like to acknowledge the assistance of the reviewers of this reprint book—several valuable suggestions were incorporated into this volume. My understanding of many of these papers was aided by discussions with a number of Stony Brook graduate students who work with me. The support of the SDIO/IST under ONR grant N00014-91-J4063 in the course of my research on this subject is gratefully acknowledged. Finally, the support and encouragement of my wife and daughters contributed significantly to the success of this project.

This book is about the performance evaluation of high-speed switching fabrics and networks. More simply, it concerns how the next generation of telecommunications technology can be designed, analyzed, and understood. But this idea requires some explanation.

The Technology

The roots of today's high-speed switching technology go back to the advent of the digital computer in the 1940s. For the first time, machines were built containing information with a binary, or digital, representation. It was only a matter of time before interest grew in transporting this digital information from place to place. At first the digital information was simply transformed into analog form for transport over analog telephone circuits. During the 1960s, though, the concept of a *packet-switching network* was developed.

The idea behind a packet-switching network was to create a network of dedicated leased lines whose sole function would be to transport digital data traffic. At the source, data would be divided into groups of bits called *packets*. An actual packet has two parts. The first part is the header field, which is a group of bits containing such information as a packet's destination, its origination point, its priority and its error codes. The second part is the actual information field. This part is the group of information bits to be transported over the network. Algorithms running in the switching nodes read a packet's destination address and route the packet over the next successive link on its way to its destination.

A great advantage of packet-switching technology was its inherent *statistical multiplexing* [6]—that is, a leased line is shared by the data traffic between many users. This lowers the cost of transmitting data. Moreover, statistical multiplexing allows the inherently bursty data traffic to be combined into aggregate flows that can be accommodated economically by long-lasting leased-line connections.

Packet-switching technology was first used in early networks such as the U.S. ARPANET. Today packet-switching technology is widely implemented in such general user networks as the Internet and in such specialized applications as the control network for telephone networks, Common Channel Signaling System 7.

Through the 1980s packet-switching technology was used exclusively for digital data transport. As such it filled an important function in the telecommunications world. To put this in perspective, though, the total bandwidth of all the world's packet-switching networks was still a tiny fraction of the total bandwidth of the world's telephone networks. However, this could change with the introduction of the proposed asynchronous transfer mode (ATM) technology [1, 2, 3, 6].

ATM is a proposal to create a broadband (high-speed) packet switching network capable of transporting a wide variety of services in an integrated manner. These services would include voice, data, and video communications.

To understand ATM it is worthwhile to consider an alternative called synchronous transfer mode (STM). Under STM integrated services networks would be built from 64-kbps digital channel building blocks. However this leads to a number of problems discussed in [2], some of which are as follows:

- A rigidly structured hierarchy is involved.
- A complicated time slot assignment (mapping) problem occurs.
- A potential need for separate switches at each data rate exists.
- Although multirate switching is plausible, it is not simple to construct out of 64-kbps building blocks.
- STM is more appropriate to fixed-demand services rather than bursty services.

Under ATM, on the other hand, a single-class high-performance packet switching network would carry voice, data and video traffic. Currently the use of short, fixed-size packets called "cells" is envisioned. The short size reduces packetization delay, reduces potential queueing delay, and simplifies engineering [6].

An advantage of having all traffic treated in the same way (single traffic class) is that capacity is dynamically assigned and the statistical multiplexing makes efficient use of network resources. Put another way, there is inherent rate adaptation [6]. The traffic rate depends on packet arrival rates (which can be agreed upon), and unused bandwidth can be used for other traffic. Moreover, switches can be designed that are independent of service [6]. This simplifies their engineering since the main concern is simply to design the highest-throughput switch for a given bandwidth. A side benefit is that a separate common-channel signaling network may not be necessary [6].

However, as great as the engineering advantages of ATM technology are, perhaps its greatest attraction is economic. It is hard—and sometimes impossible—to predict accurately the future demand for new services. Thus it is hard to justify economically a technology like STM that involves separate investments for specific service classes. ATM, on the other hand, can be justified on the basis of the total traffic load [6]. New services can simply use the same switching and transmission equipment that carries current traffic. This considerably reduces planning uncertainty, an important problem considering the large investments involved.

Performance Evaluation

At the heart of an ATM network are the switching fabrics. *Switching fabrics* are patterned collections of simple switching building blocks that serve to move cells on their way to their ultimate destinations. These switching fabrics are centralized and implemented in VLSI chips. There have also been proposals for distributed patterned collections of switching elements, or distributed fabrics, for such applications as metropolitan area networks (MANs).

In either case an engineer would like to predict the performance of a given switching fabric without actually first having to build it. This prediction is what is referred to as *performance evaluation*. In order to accomplish it the performance analyst has two approaches. One approach is discrete event simulation. The other approach is to make use of statistical models of switch behavior to determine such essential performance measures as throughput, time delay, utilization, blocking, loss probability and buffer occupancy. High-speed switching fabrics, be they centralized or distributed, are highly structured networks with extremely large numbers of packets flowing through them according to well-defined and usually simple rules. Thus it is not too surprising that a statistical approach will often be successful.

The statistical approach to evaluating the performance of switching fabrics uses many of the tools of queueing theory. However, there are some crucial new features to the switching fabric problem, including the following:

- Time is slotted so discrete-time models are usually more appropriate than continuous-time models.
- Rather than studying a single queue, a typical switching fabric will consist of a structured network of hundreds or thousands of buffers.
- For many of these fabrics, blocking is an essential part of their operation.

This book contains the fruit of ten years of work by almost a hundred authors on this problem. A great deal of thought has gone into devising solvable models of switching fabric performance. Why are these important? Compared to a discrete-event simulation, statistical models—once developed—are usually solved more quickly. However, the more important reason is that statistical models give *insight* into the operation of switches and networks. They allow the interaction of thousands of switching nodes and millions of packets to be made accessible to the human mind. Thus one is seeking not only to produce a numerical evaluation but to understand the reasons behind it. In this, the authors of this book's papers have been successful to an extent that would have been considered surprising 10 years ago.

Overview

This reprint book is organized as follows:

- Part 1 *Introduction* Part 1 consists of a tutorial paper by H. Ahmadi and W. E. Denzel on high-speed switching techniques. It is written at a very accessible level, and the reader who is unfamiliar with proposed switching technologies will find it to be extremely helpful.

- Part 2 *Foundations* Part 2 consists of four papers from the early 1980s that contain early approaches to devising statistical models for high-performance switching fabrics. Interestingly, much of this early work was motivated by computer multiprocessor communications; three of these four papers appeared in the *IEEE Transactions on Computers*.

- Part 3 *Nonblocking Switch Analysis* Part 3 deals with building models of nonblocking switches, or switches where packets (cells) are guaranteed to move through the switch unimpeded.

- Part 4 *Blocking Switch Analysis* In Part 4 switches with blocking are studied. "Blocking" arises when a switch is designed so that not every output may always be immediately accessible by every input. The authors of these papers have made excellent progress in spite of the usual difficulty of analyzing blocking [4]. In spite of the blocking phenomena, such switches may be more efficient in terms of hardware complexity than nonblocking switches.

- Part 5 *Knockout-Based Switching* Part 5 contains three papers on a specific architecture known as the Knockout switch. The name comes from the clever use of a knockoutlike tournament to select among competing requests for outputs.

- Part 6 *Multicast Switches* The usual switching model deals with data transport from one input to one output. In multicasting, data may be transferred from one input to multiple outputs. The study of such a multicasting function is motivated by a desire for new broadcasting services.

- Part 7 *Toroidal Switching Networks* Part 7 concerns a specific distributed switch fabric topology, that of the torus. Toroidal networks have received an increasing amount of attention in recent years [5]. It has been proposed for use as a metropolitan area network (MAN)— that is, as a backbone interconnection for local area networks.

- Part 8 *Shuffle Network Switching Networks* Part 8 discusses another type of switch fabric, shuffle networks. Both papers in this section concern the use of deflection routing with shuffle networks. Deflection routing, which is also discussed in the chapter on toroidal fabrics, involves a simplified routing strategy that takes advantage of regularly structured networks.

- Part 9 *Hot Spots, Packet Trains, and Other Extensions* Part 9 contains a potpourri of papers on specialized aspects of high-performance switching. It includes papers on hot spots, packet trains, integrated traffic, hypercubes, and very large switches.

REFERENCES

[1] M. de Prycker, *Asynchronous Transfer Mode: Solution for Broadband ISDN*, Ellis Horwood, New York: Simon and Schuster, 1991.

[2] S. Mintzer, "Broadband ISDN and asynchronous transfer mode," *IEEE Commun. Magazine*, pp. 17–24, 57, Sept. 1989.

[3] J. J. Johnston, "ATM: A dream come true," *Data Commun.*, March 21, 1992, pp. 53–55.

[4] H. G. Perros and T. Altiok, ed., *Proceedings of the First International Workshop on Queueing Networks with Blocking*, North Carolina State University, Raleigh, N.C., May 1988, Elsevier North-Holland, 1989.

[5] T. G. Robertazzi, "Toroidal networks," *IEEE Commun. Magazine*, vol. 26, no. 6, pp. 45–50, June 1988.

[6] M. Littlewood, I. D. Gallagher, and J. L. Adams, "Evolution toward an ATD multi-service network," *British Telecom Tech. J.*, vol. 5, no. 2, pp. 52–62, April 1987.

A Survey of Modern High-Performance Switching Techniques

HAMID AHMADI, MEMBER, IEEE, AND WOLFGANG E. DENZEL

Abstract—The rapid evolution in the field of telecommunications has led to the emergence of new switching technologies to support a variety of communication services with a wide range of transmission rates in a common, unified integrated services network. At the same time, the progress in the field of VSLI technology has brought up new design principles of high-performance, high-capacity switching fabrics to be used in the integrated networks of the future. Most of the recent proposals for such high-performance switching fabrics have been based on a principle known as fast packet switching. This principle employs a high degree of parallelism, distributed control, and routing performed at the hardware level. In this paper, we present a survey of high-performance switch fabric architectures which incorporate fast packet switching as their underlying switching technique to handle various traffic types. Our intention is to give a descriptive overview of the major activities in this rapidly evolving field of telecommunications.

INTRODUCTION

THE existing telecommunications networks, whether circuit-switched or packet-switched, are oriented towards particular applications. Thus, we have different networks for voice, signaling, video, and data applications operating in parallel and independently. While each of these networks is suitable for the application it is designed for, they are not very efficient for supporting other applications. The advantages of an integrated communication system which can accommodate a variety of diverse services with different bandwidth requirements has been recognized for some time [1]–[5]. Reasons of economy and flexibility motivated its development. The objective of having a unified integrated network is the flexibility to cover existing as well as future services with good performance and economical resource utilization, and with a unified network management, operation, and maintenance [6], [7].

The rapid evolution in the field of telecommunications, especially in the area of transmission systems and fiber optics, has led to new switching technologies for integrated networks. Advances in the field of VLSI technology while reducing the cost of circuit fabrication have brought about completely new principles in the design and architecture of high-performance switching fabrics which can accommodate a wide range of bandwidths. The intention of this paper is to give a survey of the state of the art and characteristics of these switching technologies.

The traditional circuit switching concept evolved as a way to handle and transport stream-type traffic such as voice and video. A circuit-switched connection is set up with a fixed bandwidth for the duration of a connection which provides fixed throughput and constant delay. The classical switching techniques employed for this type of switching are *space division multiplexing* (SDM), *time division multiplexing* (TDM), and combinations of both techniques. In pure SDM switching arrangements, there is no multiplexing of connections on the internal data paths or space segments of the switch. In systems which employ the TDM technique, multiple connections are time multiplexed on the internal data paths of the switch. A controller schedules the allocation of time and/or space segments in advance in order to avoid any conflict. While this technology has been very attractive and familiar, it is less efficient in supporting different bit rates that are needed for various services. Even in the multirate circuit-switching systems which allow the allocation of bandwidth equal to integer multiples of some basic rate, as Kulzer and Montgomery point out [8], the choice of a basic rate is a difficult engineering design decision. In order to accommodate the relatively low basic ISDN rate, many parallel low-rate channels must be established for high-rate services. This will imply extra control overhead needed to set up these parallel channels. In addition, all channels that comprise a single connection must be synchronized with no differential delay within each channel.

Packet switching evolved primarily as an efficient way to carry data communications traffic. Its characteristics are access with buffering, statistical multiplexing, and variable throughput and delay, which is a consequence of the dynamic sharing of communication resources to improve utilization. Today, it is used exclusively for data applications, and it employs both virtual circuits and datagram techniques. Very sophisticated protocols have been developed which incorporate complex functions, such as error recovery, flow control, network routing, and session control. The switching capacity of current conventional packet switches ranges from 1 to 4 thousand packets per second [9], [10] with average nodal delays of 20–50 ms.

Manuscript received July 13, 1988.

H. Ahmadi is with the IBM Research Division, T. J. Watson Research Center, Yorktown Heights, NY 10598.

W. E. Denzel is with the IBM Research Division, Zurich Research Laboratory, 8803 Rüschlikon, Switzerland.

IEEE Log Number 8929565.

Reprinted from *IEEE J. Selected Areas Commun.*, vol. 7, no. 7, pp. 1091–1103, September 1989.

The switching functions are typically performed by means of software processing on a general-purpose computer or a set of special-purpose processors. The capabilities of these packet switches are very attractive for applications requiring low throughput and low delay such as inquiry/response-type traffic and those requiring high throughput, but which can tolerate higher delays such as file transfer. However, these capabilities are not sufficient for real-time types of traffic such as voice, video, or computer-to-computer data transfer.

The rapid pace of technological change has brought about new switching system concepts in order to satisfy the high-performance requirement for future systems. Many different switch fabric designs have been proposed and developed at various research organizations around the world over the past few years. All current approaches of high-performance switching fabrics employ a high degree of parallelism, distributed control, and the routing function is performed at the hardware level. In this paper, we will survey many of these architectures. Our intention is to give a descriptive overview of the major recent activities in this rapidly evolving field of telecommunications.

Some recent switching approaches have been based on the *Fast Circuit Switching* (FCS) concept. This idea relies on the fast setting up and taking down of connections such that the system does not allocate any circuit to a user during his idle time. An architecture for integrated data/voice with fast circuit switching was proposed by Ross *et al.* [11]. A form of a distributed fast circuit switching which is called "burst switching" has been reported in [12]–[14].

Our emphasis in this paper is on those switch fabrics which incorporate the *Fast Packet Switching* (FPS) concept [6], [7], [15], [16] as their underlying switching technique to support a wide range of services with different bit rates. We try to categorize and classify them according to their internal fabric structures. In this regard, we will classify them into the following categories:

- Banyan and buffered banyan-based fabrics
- Sort-banyan-based fabrics
- Fabrics with disjoint-path topology and output queueing
- Crossbar-based fabrics
- Time division fabrics with common packet memory
- Fabrics with shared medium.

In the following sections, we give a detailed descriptive overview of most switch architectures within each class.

Banyan and Buffered Banyan-Based Fabrics

The very early theoretical work on *multistage interconnection networks* (MIN) was done in the context of circuit-switched telephone networks [17], [18]. The aim was to design a nonblocking multistage switch with the number of crosspoints less than a single-stage crossbar matrix. Later on, many multistage interconnection networks were realized and studied for the purpose of interconnecting multiple processors and memories in parallel computer systems. Subsequently, several types of these networks such as *banyan* and *delta* networks have been proposed [6]–[8] as a switching fabric for integrated telecommunication switching nodes.

Banyan networks, which were originally defined by Goke and Lipovski [19], belong to a class of multistage interconnection networks with the property that there is exactly one path from any input to any output. Banyan networks are subdivided into several classes [20]. Of practical interest are the regular, rectangular SW-banyans which are constructed from identical switching elements with the same number of inputs as outputs.

Delta networks as defined by Patel [21], [22] are a subclass of banyan networks which have the digit-controlled routing or self-routing property. A rectangular $N \times N$ delta network, called *delta-b*, is constructed from identical $b \times b$ switching elements or nodes in k stages where $N = b^k$.

Many of the well-known interconnection networks such as omega, flip, cube, shuffle-exchange, and baseline belong to the class of delta networks [22]. These networks have been considered for packet-switching techniques to obtain high throughput because several packets can be switched simultaneously and in parallel and the switching function is implemented in hardware. Although these networks have different interconnection patterns, they have the same performance in a packet-switching environment.

The principal characteristics' of these networks are: 1) they consist of $\log_b N$ stages and N/b nodes per stage, 2) they have the self-routing property for packet movements from any input to any output by using a unique k digit, base b destination address, 3) they can be constructed in a modular way from smaller subswitches, 4) they can be operated in a synchronous or asynchronous mode, and 5) their regularity and interconnection pattern are very attractive for VLSI implementation. Fig. 1 shows an example of an 8×8 delta-2 network with the bold lines indicating the routing paths of two packets.

While these networks are capable of switching packets simultaneously and in parallel, they are blocking networks in the sense that packets can collide with each other and get lost. There are two forms of blocking: internal link blocking and output port blocking. The internal link blocking refers to a case where packets are lost due to contention for a particular link inside the network. The output port blocking, however, occurs when two or more packets are contending for the same output port. Fig. 2 shows these effects in an example of an 8×8 delta-2 network. These two effects result in the reduction of the maximum throughput of the switch. The performance of these switches has been studied extensively [22]–[27]. Reference [25] gives a survey on the performance of these networks in a packet-switching environment. This reference also gives detailed attention to how the performance of these networks can be improved.

There are several ways to reduce the blocking or to increase the throughput of banyan-type switches:

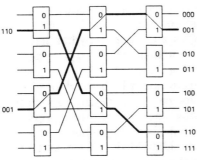

Fig. 1. An 8 × 8 delta-2 network.

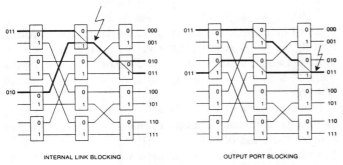

INTERNAL LINK BLOCKING OUTPUT PORT BLOCKING

Fig. 2. Internal link blocking and output port blocking in an 8 × 8 delta-2 network.

• increasing the internal link speeds relative to the external speeds,
• placing buffers in every switching node,
• using a handshaking mechanism between stages or a back-pressure mechanism to delay the transfer of blocked packets,
• using multiple networks in parallel to provide multiple paths from any input to any output or multiple links for each switch connection [26], [27],
• using a distribution network in front of the banyan network to distribute the load evenly.

The idea of using a *buffered* delta network for switching in multiprocessor systems was first introduced and analyzed by Dias [23]. This idea was further developed and considered for a general-purpose packet switching system by Turner [6], [28], [29]. He proposed a buffered banyan switch architecture which led to the dawn of research activities on fast packet switching [6], [28]. He further developed and advanced the concept to the *Integrated Services Packet Network* (ISPN) architecture [29]. The ISPN architecture is based on a large high-performance packet switch structure as illustrated in Fig. 3. The switch interfaces up to 1000 high-speed digital transmission facilities (e.g. 1.5 Mbits/s) via packet processors (PP). The packet processor provides input buffering, adds the routing header, and performs the link level protocol functions. A control processor (CP) performs all connection control functions. The switch fabric is based on a ten-stage self-routing buffered banyan network with 1024 ports made up of 5120 binary switching elements. Each 2 × 2 switching element has a buffer at each input port capable of storing one packet. The internal links joining the nodes are bit

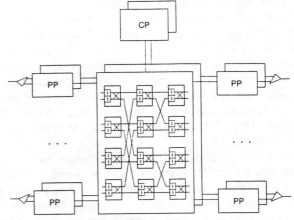

Fig. 3. Turner's ISPN packet switch structure [7].

serial and operate at eight times the speed of the external links. This means that a 100 percent load on the external links yields a 12.5 percent load on the internal switch fabric links. The switch fabric also uses a back-pressure flow control mechanism between stages which prevents buffer overflow or packet loss within the fabric. The buffering technique proposed is based on the *virtual cut-through* concept described by Kermani and Kleinrock [30]. When a packet arrives at a switching element and the desired port is free, it is directly sent through, bypassing the buffer. This way, the delay through the switching elements is reduced to a minimum.

A *Wideband Packet Technology* network with packet switches based on Turner's original buffered banyan development and capable of switching packetized voice, data, and video has been reported by AT&T [31], [32]. A field trial was carried out between several AT&T locations in California in order to demonstrate the feasibility of a fully integrated packet network working in conjunction with existing transmission facilities. The experimental 16 × 16 buffered banyan fabric has a clock rate of 8 Mbits/s and the external transmission line speed is 1.5 Mbits/s. Each binary switching element has a buffer for one complete packet per input. Recently, the same concept has also been proposed for use at much higher speeds, realized in GaAs technology [33].

A high-performance *Broadcast Packet Switching Network* has been proposed recently by Turner [34]. The principle is based on the ISPN architecture described above, but with higher bandwidth and broadcasting capability. The switch architecture consists of a 64 × 64 buffered banyan fabric as a routing network (RN), preceded by a copy network (CN) and a distribution network (DN); see Fig. 4. The distribution network is itself a buffered banyan network and its purpose is to reduce internal link blocking by distributing the traffic offered to the routing network evenly across all input ports. The nodes of the distribution network ignore the destination addresses of the packets and route them alternatively to both node outputs. The copy network also has a banyan structure. Each node, when it receives a broadcast packet (which

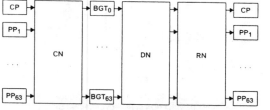

Fig. 4. Basic structure of Turner's Broadcast Packet Switching Network. From [7], © 1986 IEEE.

has a header field indicating the number of copies), copies the packet, and sends it out on both outgoing links and modifies the "number of copies" field. Between the outputs of the copy network and inputs of the distribution network are the broadcast and group translators (BGT). The function of the BGT's is to translate the destination address of the packet copies which emanate from the copy network in order to be routed to different destinations. Thus, they translate the group number and logical channel number of the broadcast packets into an output port number. For details of the copy network routing algorithm, the reader is referred to [29].

The performance of buffered banyan networks has been studied extensively [23]–[27], [35]–[38]. The analytical model presented by Jenq [35] is for a packet switch based on a delta-2 network with single buffers in each node. The result of this model shows that, for a balanced and uniform traffic pattern, the maximum throughput of the switch is about 40 percent. Extensive simulation studies of the buffered banyan network, specifically in the context of the ISPN switch architecture described above, is presented in [36]. This study is different from others in several respects. It considers the performance of the routing network with emphasis on the effect of cut-through switching, different buffer sizes within each node, and the various node sizes. In addition, it also investigates the effect of nonuniform traffic on the routing network and the improvements achieved by the distribution network in front of the routing network. The performance of the copy network for broadcasting packets is also given in this report. In a recent paper [37], an approximate analytic approach based on a Markov chain model is presented for a general delta-b network with multiple buffers within each node. Numerical results in this report as well as the results in [36] indicate that, by increasing the size of the nodes and the buffer size, respectively, the maximum throughput of the switch can be increased to about 60 percent. However, this improvement is not linear. In fact, for a buffer size of more than four packets per port or a node size greater than 4×4, the amount of improvement in throughput is not very significant.

In these studies of buffered banyan networks, the position of the buffers within each switching element has been at the inputs. Recently, buffered banyan structures have also been considered which use switching elements of larger size and have buffers at the module outputs [38], [54]. We will defer the discussion about this architecture until Section IV.

SORT-BANYAN-BASED FABRICS

As mentioned before, one drawback of the banyan networks is that they are internally blocking in the sense that two packets destined for two different outputs may collide in one of the intermediate nodes. However, if packets are first sorted based on their destination addresses and then routed through the banyan network, the internal blocking problem can be avoided completely. This is the basic idea behind the *sort-banyan* type networks. Fig. 5 shows the basic structure of a sort-banyan network. The first segment consists of a sorting network (in this case, a Batcher bitonic sort network [39]) which sorts the packets according to their destination address, followed by a shuffle exchange and a banyan network which routes the packets. It can be shown that packets will not block within the banyan network. Note that Fig. 5 shows two connection patterns which would have caused internal link blocking in a pure banyan network (see Fig. 2). Each node of the sort network as shown in Fig. 5 is a 2×2 switching element which sorts the incoming packets in the order as indicated by the arrow shown. The number of elements in the sorting segment is $N/4((\log_2 N)^2 + \log_2 N)$. The network operates in a synchronous manner, and packets are processed in each stage in parallel. While this self-routing interconnection network is internally nonblocking, blocking still may occur if destination addresses are not distinct, i.e., if there are packets with identical addresses. Hence, simultaneous packets destined for the same output will collide within the banyan part of the network.

The very first switch fabric implementation which proposed and employed the sort-banyan self-routing structure is the *Starlite* switch [40]; see Fig. 6. At the switch interface, various services are converted into constant length "switch packets" with a packet header indicating the routing information which is the destination address. To overcome the output port contention problem, the Starlite approach uses a trap network between the sort and the banyan (called the expander here) segment which detects packets with equal destination addresses at the output of the sort segment. Thus, the packets with repeated addresses are separated from the ones with distinct addresses. The packets with repeated addresses are fed back to the input side of the sort network for reentry within the next cycle. These packets can only use the idle input ports. Since the number of recycled packets at any time can be larger than the free input ports, a buffering stage must be used for the recycled packets if packet loss is not acceptable. The trap network consists of a single-stage comparator followed by a banyan network of the same type as the routing network. Note that if packets were simply recycled through the switch, they would be delivered out of sequence. This problem is solved by "aging" packets as they recirculate and by using the sort network to give old packets priority over new ones. The Starlite switch has also been considered for optical implementation [41].

Recently, Hui [42], [43] proposed a switch architecture

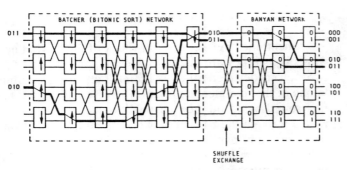

Fig. 5. Basic structure of a sort-banyan network.

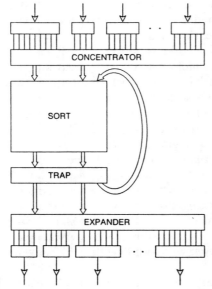

Fig. 6. Basic structure of the Starlite switch. From [40], © 1984 IEEE.

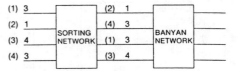

PHASE I : SEND AND RESOLVE REQUEST

- SEND SOURCE-DESTINATION PAIR THROUGH SORTING NETWORK
- SORT DESTINATION IN NON-DECREASING ORDER
- PURGE ADJACENT REQUESTS WITH SAME DESTINATION

(a)

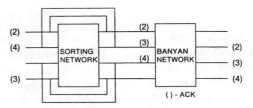

PHASE II : ACKNOWLEDGE WINNING PORT

- SEND ACK WITH DESTINATION TO PORT WINNING CONTENTION
- ROUTE ACK THROUGH BATCHER-BANYAN NETWORK

(b)

Fig. 7. Phases I and II in the switch proposed in [42], [43]. From [43], © 1987 IEEE.

based on the sort-banyan network structure, but with a different scheme than Starlite to overcome the output port contention problem. The advantage of this approach is that the switch fabric delivers exactly one packet to each output port from one of the input ports which request a packet delivery to the same output port. Hence, packets are never lost within the fabric and they are delivered in sequence.

The basic structure of this fabric, like the Starlite switch, is a Batcher sorting network followed by a banyan routing network. As mentioned before, the combined sort-banyan network is internally nonblocking. To resolve the output port conflict, a three-phase algorithm in conjunction with input queueing at each input port is employed. All packets are of a constant length and the switch operates in a synchronous manner. In the first phase [Fig. 7(a)], which is the arbitration phase, each input port which has a packet in its queue for transmission sends a short request packet. The request packet carries just the source/destination address pair. The requests are sorted in ascending order by the Batcher sorting network according to the destination address. All but one request for the same destination are purged. The second phase [Fig. 7(b)] is to acknowledge the input ports which have won the arbitra-

tion since the input ports which made the requests do not know the result of the arbitration. In this phase, the winning requests are looped back to the input of the Batcher network, and the Batcher-banyan network is again used to acknowledge the input ports. Note that, in this phase, the source address is used by the sorting and the self-routing mechanism for sending the acknowledgment packets. The input ports receiving the acknowledgment for their request then transmit the full packet through the same Batcher-banyan network. This process constitutes the third and final phase. Obviously, there will be no conflict at the output ports in this phase. Input ports which fail to receive an acknowledgment retain the packet in their input queue and retry in the next cycle when the three-phase operation is repeated again. The queueing of the unacknowledged packet at the head of the queue will cause the subsequent packets which might have been intended for a free output port to be delayed. This phenomenon is called head of the line (HOL) blocking, and in principle will reduce the switch throughput.

The switch supports a port speed of 150 Mbits/s. Of course, phases one and two constitute overhead processing for the switch fabric. Therefore, the switch fabric would have to be speeded up by a fraction which depends on the size of the switch fabric and the size of the information field of the packet. For example, as shown in [43], for a 1000×1000 switch and the packet size of 1000 bits, the overhead is about 14 percent. This means that the switch has to operate at 170 Mbits/s in order to handle a 150 Mbit/s input port speed.

The performance study of this switch for random traffic indicates that the maximum throughput is about 58 percent. Actually, this is a theoretical limit which can be shown for any nonblocking space switch operating in a packet-switched mode which employs input queueing with random traffic (see [44]). Results in [43] also indicate that, with a reasonable input buffer size, an acceptable buffer overflow probability can be achieved.

The performance behavior and the maximum throughput of this switch would be different if the traffic behavior were nonrandom and specifically periodic. In [43], there is a discussion on how to improve the throughput of the switch in the presence of periodic traffic. In a theoretical paper, Li [45] studies the performance of a generic nonblocking packet switch with input queues and periodic packet traffic streams. He determines the necessary and sufficient switch-internal clock rate as a function of the input stream rate in order for the switch to be nonblocking.

The use of a Batcher-banyan network was also proposed in the *Synchronous Wideband Switch* in [46], [47] which is a time–space–time circuit switching system. The function of the Batcher-banyan fabric is to provide a nonblocking self-routing network in the space segment of the time–space–time switch. The time slot interchangers at the inputs and the outputs of the Batcher-banyan network rearrange the time slots to avoid port contention.

The technological feasibility of the Batcher-banyan concept is described in [33]. A Batcher chip of size 32 × 32 has been built in CMOS technology with each port running at 140 Mbits/s. Furthermore, about 100 of those chips are packaged in a three-dimensional way to form a 256 × 256 Batcher-banyan switch fabric with a total throughput of 35 Gbits/s.

Fabrics with Disjoint-Path Topology and Output Queueing

The switch fabric architectures described so far were based on multistage interconnection networks comprised of small switching elements. The switch fabrics described in this section are based on a fully interconnected topology in the sense that every input has a nonoverlapping direct path to every output so that no blocking or contention may occur internally. They employ output queueing in order to resolve the output port contention. Intuitively speaking, in any nonblocking space-division packet switch, a higher throughput can be achieved with output queueing as compared to input queueing. This is because the head of the line blocking effect which is the limiting factor in a switch with input queues disappears with output queueing. This result has been shown analytically by Karol and Hluchyj in [44], [49]. In fact, one can show that a nonblocking space-division switch with an output queueing of infinite capacity would give the best delay/throughput performance.

The first switch fabric we describe in this class is the *Knockout* switch [50], [51]. The Knockout switch is designed for a pure packet-switched environment and can handle either fixed-length (called Knockout I [50]) or variable-length packets (called Knockout II [51]). The Knockout switch uses one broadcast input bus from every input port to all output ports as shown in Fig. 8. Each output port has a bus interface which can receive packets

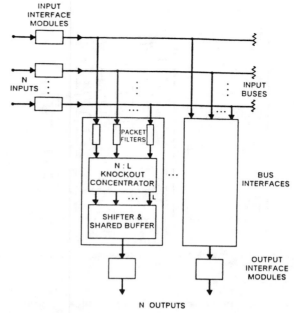

Fig. 8. Basic structure of the Knockout switch. From [75]. © 1988 IEEE.

from each input bus line or input port. Hence, no contention occurs between packets destined for different outputs. In addition, simultaneous packets from several inputs can be transmitted to the same output. In Fig. 8, one of the output bus interfaces is shown in more detail. It has three major components. The first component is the set of N packet filters, each interfacing a bus line. The packet filters, which implement the self-routing function, detect the address of each packet on the broadcast bus and pass those destined to that output on to the next component which is the concentrator. The concentrator uses a novel algorithm to select a fixed number of packets, say L, from the N incoming lines to the concentrator. The L selected packets are stored in the order of their arrivals into a shared buffer which constitutes the third component. The main philosophy behind the N to L concentration mechanism is that the probability of packet loss due to output congestion can be kept below the loss expected from other sources such as channel errors. This means, for example, that with N large and $L = 8$, a packet loss rate of less than 10^{-6} can be achieved. Taking advantage of this observation, the number of separate buffers needed to receive simultaneously arriving packets is reduced from N to L. This result holds under the assumption of uniform traffic patterns. If the traffic pattern is nonuniform, L has to be higher (e.g., up to 20) for the same packet loss rate, depending on the nonuniformity of traffic [52].

The basic principle of selecting L packets out of N possible contenders in the concentrator stage is an algorithm implemented in the hardware analogous to a knockout tournament. For a concentrator with N inputs and L outputs, there are L rounds of competition. The basic building block of the concentrator is a 2 × 2 contention switching element, with one output being the winner and one the loser. When two packets arrive, one is selected ran-

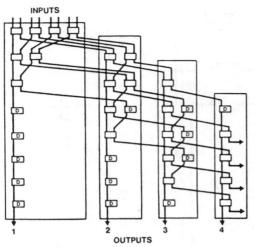

Fig. 9. Example for an 8:4 Knockout concentrator. From [50], © 1987 IEEE.

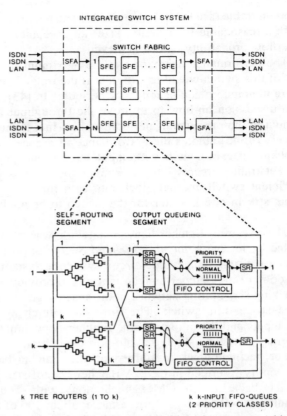

Fig. 10. Basic switch structure and switch fabric element proposed in [53].

domly as a winner. Fig. 9 shows a block diagram of an 8:4 concentrator made of these 2×2 contention switches. The boxes marked "D" indicate a 1-bit delay line to keep the competition synchronous. The first round of the tournament starts with N contenders. The $N/2$ winners from the first round advance to the second round. The winners in the second round advance to the third round, and so on. Note that the final winner at the output number one has won all rounds, the winner at output number two has won all but one round, and so on.

The next segment after the concentrator is a shared buffer structure. The output buffer must be capable of storing up to L packets within one packet time slot. This is achieved by using L separate buffers preceded by a shifter function. This logically implements an L-input single-output FIFO queueing discipline for all packets arriving at the output.

The design of the Knockout switch is based on possible VLSI realization with input/output and internal hardware operating at 50 Mbits/s. Also, a solution is proposed for modular growth which can grow from 32×32 to 1024×1024.

Another switch fabric, the design of which is based on an interconnection structure with no internal blocking and queueing capability at the outputs of it modules, is the *Integrated Switch Fabric* proposed in [53]. This switch fabric is designed to handle both circuit-switched and packet-switched traffic in a unified manner. It is self-routing, and uses uniform fixed-length *minipackets* within the switching fabric for all types of connections. Circuit-switched connections can be provided for various speeds and for constant delay with full transparency to the terminating ports. Its design concept emphasizes modularity; it is based on VLSI technology, aiming for a single chip as the basic building block such that a very large range of switch fabrics can be configured, covering sizes from 16 to more than 1000 input/output ports. This translates into a throughput capability from 500 Mbits/s to 30 Gbits/s, assuming a speed of 32 Mbits/s per port.

Fig. 10 shows the basic structure of the fabric which consists of two major components: the switch fabric adapters (SFA) and the switch fabric elements (SFE). The function of the switch fabric adapter is to convert the user information from packet-switching and circuit-switching interfaces into uniform fixed-length minipackets. Fig. 10 also shows the basic structure of a $k \times k$ switch fabric element which consists of a self-routing segment of an output queueing segment. The self-routing segment of an SFE performs the minipacket routing function. It is a decoder with a tree structure per input which can simultaneously route minipackets from every input to every output. Each node of the tree decoder segment has one input and two outputs and works like a simple banyan node with only one used input. Therefore, within each module, up to k minipackets can be transferred to the same output at the same time. The interface between the terminating points of all trees belonging to a certain SFE output and the output line consists of k small shift registers (SR) for intermediate storage of minipackets and a pair of FIFO queues which constitute the output queueing segment. The importance of the shift registers is to store the minipackets momentarily to allow sequential access to the output FIFO queue pair. The contents of the shift registers are transferred sequentially into the associated output FIFO at a speed k times higher which is realized by parallelism. The combination of shift registers and their corresponding output FIFO is a realization of a multiinput single-output FIFO queue. One major difference between this switch fabric and most others is that it uses a priority scheme for different classes of traffic. Two priority classes are proposed. One is used for the high-priority or time-critical type traffic, and the other for the low-priority traffic. Circuit-switched connections are supported with minipackets which have high priority. Another feature of this switch architecture is its simple modular growth capability. That is, larger switch sizes can be configured by combining $k \times k$ switch fabric elements via stage-expanding and/or multistage arrangements.

A single-stage $N \times N$ switch fabric can be built from the basic $k \times k$ element. Fig. 11 shows an example of a $2k \times 2k$ switch fabric configured from four $k \times k$ switch fabric elements. Each basic module has a selection logic which can be set such that only the appropriate module accepts the incoming minipackets and routes them internally to the destined output. Furthermore, in each module, there are provisions for each output to ensure that only one output FIFO queue is feeding the output line at a time. In the same manner, a $4k \times 4k$ fabric can be realized from a $2k \times 2k$ fabric, and so on. In general, to build a single-stage $N \times N$ fabric from a basic $k \times k$ element, $(N/k)^2$ modules are required. It should be noted that a single-stage switch fabric which is realized this way still preserves its disjoint-path topology with output queueing. However, each output of the $N \times N$ single-stage configuration has one logical queueing segment which is physically realized in different switch fabric elements as parallel queues. For a moderate-size switch, say up to 128×128, the single-stage approach seems feasible and reasonable. But, for a large-size switch, a three-stage realization is proposed which yields a very cost-effective solution.

Another multistage switch fabric architecture which actually considers the buffered banyan structure but with larger modules has been proposed in [54]–[56]. We have included it in this section because each module or switching element has a structure with disjoint-path topology and a substantial output buffering at each of the module outputs. The switching element has eight inputs and eight outputs. A TDM bus interconnecting all input ports to all output ports runs at the sum of all input bit rates and hence is nonblocking. The switch runs at 560 Mbits/s per port. The bus speed within each module is therefore 8×560 Mbits/s. The output queues of each module can be filled at the speed of the TDM bus.

The overall network [56] functionally constitutes a randomization (distribution) network preceding a banyan routing network which actually maps into a Benes topology. In reality, the randomization network and routing network are folded and combined into one network with input/output pairs at one side and a mirror line at the other side of the network. Hence, it is used in one direction for the randomization function and in the other direction for the routing function. Reference [56] presents simulation results of this switch with emphasis on video application.

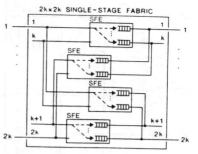

Fig. 11. Single-stage expansion in the switch proposed in [53], © 1988 IEEE.

CROSSBAR-BASED SWITCH FABRICS

Crossbar switches have always been attractive to switch designers because they are internally nonblocking and they are simple. In addition to circuit-switch applications, they have also been considered as a base for switches which operate in a packet mode. But unfortunately, the simple crossbar matrix has the property of square growth and is not economical for large switches. Nearly all known approaches are either designed for relatively small applications or just for a building block which is used in larger multistage arrangements.

Even though the crossbar matrix is internally nonblocking, in packet mode operation, the probability of output port contention remains. Hence, a queueing function has to be added to the pure crosspoint matrix in order to overcome that problem. The location of this queueing function allows the approaches of crossbar-based switches to be categorized. In principle, there are three possibilities, namely, crossbar matrices with input queueing, matrices with queueing within the crosspoints themselves, and matrices with output queueing.

If the queueing function is located at the inputs of the crossbar matrix, then a switch control is necessary which arbitrates all packets waiting at the heads of the different input queues and which are destined for a certain output port.

One solution is a centralized control. This control gets requests over separate control paths from all switch input ports which have packets waiting. It schedules these requests, sets up the necessary crosspoints in the matrix, and grants the requests as soon as the packets can be transmitted. This way, the matrix itself remains as simple as possible and can be realized economically with high-speed technology. By use of several matrices in parallel, even higher speeds can be achieved. Certainly, the central control is the bottleneck of this approach. The control overhead drastically increases with the size of the switch. Nevertheless, with suitable technology and intelligent techniques (e.g., pipelining), very high-speed, centrally controlled packet switches can be built, as long as they are small.

One way to reduce the control overhead is to distribute the control function. This leads to an approach as pro-

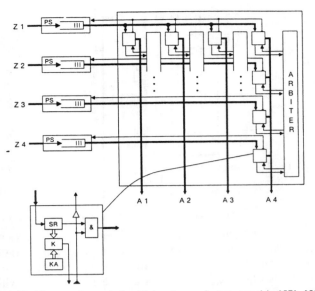

Fig. 12. The crossbar switch with input queueing proposed in [57], [58]. From [57], © 1987 vde-verlag gmbh.

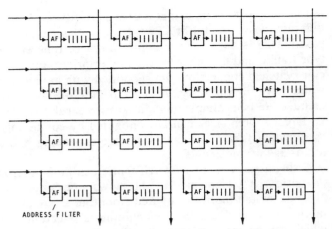

Fig. 13. A crossbar switch with queueing in the crosspoints.

posed, e.g., in [57], [58]; see Fig. 12. There, each output port has its individual control, called arbiter. This arbiter allows only one of the input ports to be switched through at the same time to the corresponding output port by a fair algorithm. This is accomplished by a separate control signal (back pressure) to each input queue which practically stops all but one of those queues which compete for the same output port. Since only one of the arbiters generates a back-pressure signal to a specific input queue at the same time, all of these signals of a corresponding matrix line can be interconnected very simply by a wired OR connection. In addition, this switch proposal does not need separate request lines from the input queues to the arbiters because it has the address decoder function distributed in the crosspoints of the matrix. Hence, this approach represents a self-routing crossbar switch.

The performance of crossbar matrices with input queueing depends very much on the control overhead, and especially on the way in which the input queues are organized. If the input queue is a simple, single FIFO queue, then the throughput of the switch saturates even at 58 percent (see [44], [49]) or less, depending on the control overhead and the traffic distribution. This is due to the head of the line (HOL) blocking phenomenon, already mentioned earlier. This means that the waiting packet at the head of the queue eventually blocks subsequent packets which might be destined for momentarily free output ports. But, if the input queue is organized as multiple queues, i.e., one per destination port, then we come closer to the concept of output queueing, even though these queues are physically located at the inputs. As mentioned earlier, output queueing provides ideal performance. The extent to which the performance of input queueing with multiple queues per input measures up to that of output queueing depends largely on how the input queues are

controlled and scheduled and what the control overhead is.

A second class of crossbar switches has been proposed which implements the queueing function within the crosspoints themselves. The simple on/off switch of a crosspoint is replaced now by a FIFO queue preceded by an address decoder function (packet filter). So, the packets can be sent directly into the matrix. A packet can only pass through the filter whose address matches the packet's destination address. So, the self-routing concept is realized. The queues of a matrix column which belong to one specific output port have to be emptied in a fair way, e.g., in round-robin fashion. This is also done by one very simple arbiter per output port.

The concept is shown in Fig. 13. It has been proposed, e.g., in [59], in the *Bus Matrix Switch* described in [60], or in the experimental broadband ATM switching system described in [61]. The architecture is motivated by the fact that the concept provides almost ideal performance, at least as long as the queues are dimensioned long enough. On the other side, because of the square growth, it is a very expensive approach for large matrices and can only be a good solution either for small switches or as basic architecture for the building blocks of modular multistage arrangements, as proposed in [60] and [61].

Finally, looking at the third class of crossbar switches, namely, matrices with output queueing, it has to be mentioned that they only work if the matrix runs at N times higher speed, assuming a matrix size of $N \times N$ (see [49]). Under this assumption, they are not attractive for very high-speed switches. So, parallelism has to be introduced instead of speed. This leads to an expanded $N \times N^2$ matrix, and finally to the class of switches with disjoint-path topology and output queueing which we already covered separately in a previous section of this paper.

TIME DIVISION FABRICS WITH COMMON PACKET MEMORY

A switch architecture which uses a common memory for all connection paths is the *Prelude* switch [62]–[66];

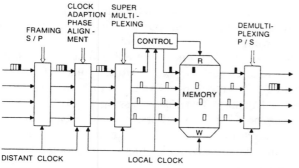

Fig. 14. Basic structure of the Prelude switch. From [66], © 1987 IEEE.

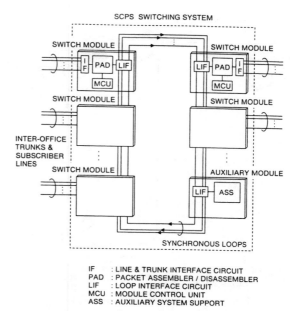

IF : LINE & TRUNK INTERFACE CIRCUIT
PAD : PACKET ASSEMBLER / DISASSEMBLER
LIF : LOOP INTERFACE CIRCUIT
MCU : MODULE CONTROL UNIT
ASS : AUXILIARY SYSTEM SUPPORT

Fig. 15. Basic structure of the SCPS switch. From [70], © 1987 IEEE.

see Fig. 14. Even though this system works in a packet mode, it operates very similarly to a conventional TDM circuit switch. This mode is called asynchronous time division (ATD) and is characterized by a slotted operation with packet queueing. The incoming packets, one per frame, are synchronized first and converted from serial to parallel so they can be written cyclically to the common memory. In advance, the control information is extracted from the packet headers and fed to the common control. Since the control has to handle all packets sequentially, the multiplexing at the input is done such that the headers of all packets arrive at the control sequentially. The control compares the header information with a routing table which delivers a new header for the packet. Additionally, a control memory provides the addressing information for the memory read process. Unlike a TDM switch where the control memory is operated cyclically, here it is operated as FIFO queues. Finally, the outgoing information is demultiplexed again and converted back from parallel to serial.

It has been demonstrated in a Prelude prototype [62] that a module of size 16×16 can be built for a line speed of 280 Mbits/s with a reasonable amount of hardware. This is due to the use of high internal parallelism and fast technology (ECL) at the same time. Since the basic architecture is not suited for large switches, the Prelude authors also suggest that large switches be constructed in a modular way by interconnecting smaller building blocks.

Fabrics with Shared Medium

In this section, we classify switch fabric architectures which are based on a shared medium as switching kernel. We concentrate especially on fast packet switch approaches which employ a bus or ring network as switching medium. But we do not want to cover here all the typical random access local-area network architectures.

Ring and bus networks are used as switching media in many of today's packet switches. Their technology is well understood and advanced. They provide flexibility in terms of the access protocol and distribution of traffic. However, their bandwidth capacity and throughput are limited compared to multipath switch architectures. One way to increase the capacity limitation is to use multiple

rings or multiple buses in a single or multiple hierarchical structure.

An experimental *Synchronous Composite Packet Switching* (SCPS) architecture [67]–[70] was proposed for integrated circuit- and packet-switching functions which uses multiple rings to interconnect the switching modules. The switching system architecture as shown in Fig. 15 comprises switch modules interfacing externally to circuit-switched and packet-switched lines. The switch modules are interconnected by multiple fiber optical rings where each module has a direct interface to each ring. The switching operations are accomplished by transmitting the circuit- and packet-switched channel messages between switch modules via the intermodule network. Intermodule signaling and command messages for system operation and control are transmitted between the modules as well.

The circuit, packet, and signaling information are all transmitted within the switching subsystem in similarly structured packets which are generated in the following way. Each switch module generates a so-called *composite packet* from several circuit-switched channel samples which have to be transmitted simultaneously and have the same destination module. Conventional data user packets are converted into switch packets, called *noncomposite packets*, in which the original packet with its header is encapsulated by an SCPS header and trailer. The signaling and control packets have similar headers and trailers as the other two types of packets. Thus, all three kinds of packets have almost the same structure.

The intermodule network consists of 8 (or 16) independent synchronous optical loops with loop speeds of 393 Mbits/s (98.3 Mbits/s), connecting 32 modules. The SCPS system has a total throughput of 4 Gbits/s (2 Gbits/s). Note that the intramodule switching is done locally; hence, the maximum system throughput is higher

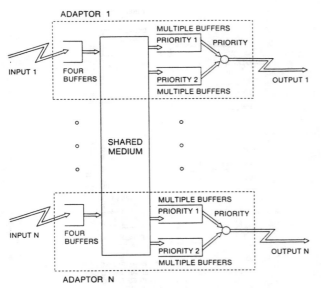

Fig. 16. Basic structure of the PARIS switch.

than the maximum throughput supported by multiple loops. The access method to loops is similar to the slotted ring method [70]. The individual loops operate synchronously. Having a frame structure with a duration of 125 μs, complete time transparency for circuit-switched channels is maintained. Each frame consists of 384 (96) time slots of 128 bits each. This switch architecture can integrate $n \times 64$ kbit/s circuit switching with packet switching in any combination while maintaining its compatibility with existing networks. By using composite packets for circuit-switched channels, the delay for packet assembling and disassembling can be reduced to a value comparable to the conventional TDM switching systems.

A similar switch architecture which also uses multiple rings to provide the fast-packet switching function was also reported in [71].

Several switch architectures have used a shared bus structure as their transport or switching mechanism (see, e.g., [16]). In fact, the switch fabric proposed in [16] is one of the very early switching concepts that used the concept of packet switching as the only means to transport a wide range of services such as voice, data, and video.

A recent experimental high-speed packet switching system which was designed to transport voice, data, and video all in packetized form and uses a bus architecture is the *Packetized Automated Routing Integrated System* (PARIS) [72]. Its design philosophy is to use a very simple protocol to achieve low packet delay with an architecture simple enough that it can be implemented even with off-the-shelf components.

The basic structure of the PARIS switch is shown in Fig. 16. It uses a high-speed bus as shared medium to interconnect input ports to the output ports [73]. The maximum bandwidth of the shared bus is taken to be greater than the aggregate capacity of all input lines. The switch can handle variably sized packets ranging from 32 bits to

a maximum of 8 kbits. Each input port has a buffer that can hold no more than four packets of maximum size. It is shown that this buffering is adequate to ensure no packet loss, provided a round-robin exhaustive service policy is used to arbitrate the access to the bus. Therefore, the switch is nonblocking. The arbitration protocol is implemented using a fast token-passing algorithm [73]. Each output port has two buffers, one for each priority class. The output buffers are sized such that the packet loss due to a momentary overload situation of an output port is less than 10^{-8}.

SUMMARY

We have presented an overview of the major high-performance telecommunications switching fabrics which have been proposed and experimentally developed within the past few years. Most of the architectures we covered in this paper use the principle of fast packet switching as a unified switching architecture for the transport of a wide range of services with different bandwidth requirements. Our intention was to give a descriptive overview by attempting to classify the approaches into six categories. The classification has been chosen in order to emphasize the basic principles which differentiate these architectures.

Some of these basic principles inherently allow covering a wide range of fabric sizes while maintaining their basic architecture in a clean way. Other basic principles are limited, such that they can only be considered for small fabrics. Some principles are also applicable as the basic architecture for modules in large multistage fabrics. The way modules are interconnected could be viewed as a basic principle itself.

Considering only small fabrics, one can make the general observation that almost every known principle allows building small fabrics of approximately the same capabilities and performance. Large multistage fabrics, however, are more sensitive to the architecture of their modules. Several module designs, if they are considered isolated from the whole fabric, may show only minor differences in their module performance. But, when placed within a multistage environment, these minor differences could escalate drastically. In general, multistage fabrics perform better if they employ queueing within the stages or the switch modules, respectively. Furthermore, the location of the queues within the modules is important [38]. Hence, it seems to be worth directing some effort to optimizing the design of a buffered, self-routing, large-fabric switch module.

Finally, we have to mention that in our overview, we only concentrated on the point-to-point connection aspects of these switching architectures. Many of the switching architectures described here have also been enhanced to support broadcasting and multicasting connections using fast packet-switching principles (see, e.g., [29], [74], [75]) which we have not covered other than just mentioning them.

REFERENCES

[1] H. Frank et al., "Issues in the design of networks with integrated voice and data" in Proc. ICC'77, June 1977, pp. 38.1.36–38.1.43.

[2] M. J. Ross et al., "Design approaches and performance criteria for integrated voice/data switching," Proc. IEEE, vol. 65, pp. 1283–1295, Sept. 1977.

[3] H. Frank and I. Gitman, "Study shows packet switching is best for voice traffic, too," Data Commun., pp. 43–62, Mar. 1979.

[4] I. Gitman and H. Frank, "Economic analysis of integrated voice and data networks: A case study," Proc. IEEE, vol. 66, pp. 1549–1570, Nov. 1978.

[5] H. Frank, "Plan today for tomorrow data/voice nets," Data Commun., pp. 51–62, Sept. 1978.

[6] J. S. Turner and L. F. Wyatt, "A packet network architecture for integrated services," in Proc. GLOBECOM'83, San Diego, CA, Nov. 1983, pp. 2.1.1–2.1.6.

[7] J. S. Turner, "New directions in communications (or which way to the information age?)," IEEE Commun. Mag., vol. 24, pp. 8–15, Oct. 1986. Also, in Proc. Int. Zurich Seminar Digital Commun., Zurich, Switzerland, Mar. 1986, pp. A3.1–A3.8.

[8] J. J. Kulzer and W. A. Montgomery, "Statistical switching architecture for future services," in Proc. ISS'84, Florence, Italy, May 1984, pp. 43A.1.1–43A.1.6.

[9] R. E. Cardwell and J. H. Campbell, "Packet switching of data with the 3B-20D computer system," in Proc. ISS'84, Florence, Italy, May 1984, pp. 22A.4.1.–22A.4.6.

[10] J. F. Huber and E. Mair, "A flexible architecture for small and large packet switching networks," in Proc. ISS'87, Phoenix, AZ, Mar. 1987, pp. B10.4.1–B10.4.6.

[11] M. J. Ross et al., "An architecture for a flexible integrated voice/data network," in Proc. ICC'80, June 1980, pp. 21.6.1–21.6.5.

[12] J. D. Morse and S. J. Kopec, Jr., "Performance evaluation of a distributed burst-switching communications system," in Proc. Phoenix Conf. Comput. Commun., Mar. 1983.

[13] S. R. Amstutz, "Burst switching—A method for dispersed and integrated voice and data switching," IEEE Commun. Mag., vol. 21, pp. 36–42, Nov. 1983. Also, in Proc. ICC'83, June 1983, pp. 288–292.

[14] E. F. Haselton, "A PCM frame switching concept leading to burst switching network architecture," IEEE Commun. Mag., vol. 21, pp. 13–19, Sept. 1983. Also, in Proc. ICC'83, Boston, MA, June 1983, pp. 1401–1406.

[15] R. Rettberg et al., "Development of voice funnel system: Design report," Bolt, Beranek and Newman Inc., Rep. 4098, Aug. 1979.

[16] E. H. Rothauser, P. A. Janson, and H. R. Mueller, "Meshed-star networks for local communication systems," in Local Networks for Computer Communications, A. West and P. Janson, Eds. Amsterdam: North-Holland, 1981, pp. 25–41.

[17] C. Clos, "A study of non-blocking switching network," Bell Syst. Tech. J., vol. 32, pp. 406–424, Mar. 1953.

[18] V. E. Benes, "Optimal rearrangeable multistage connecting networks," Bell Syst. Tech. J., vol. 43, pp. 1641–1656, July 1964.

[19] L. R. Goke and G. J. Lipovski, "Banyan networks for partitioning multiprocessor systems," in Proc. 1st Annu. Int. Symp. Comput. Architecture, Dec. 1973, pp. 21–28.

[20] R. J. McMillan, "A survey of interconnection networks," in Proc. GLOBECOM'84, Atlanta, GA, Dec. 1984, pp. 105–113.

[21] J. H. Patel, "Processor-memory interconnections for multiprocessors," in Proc. 6th Annu. Int. Symp. Comput. Architecture, Apr. 1979, pp. 168–177.

[22] ——, "Performance of processor-memory interconnections for multiprocessors," IEEE Trans. Comput., vol. C-30, pp. 771–780, Oct. 1981.

[23] D. M. Dias and J. R. Jump, "Analysis and simulation of buffered delta networks," IEEE Trans. Comput., vol. C-30, pp. 273–282, Apr. 1981.

[24] ——, "Packet switching interconnection networks for modular systems," IEEE Comput. Mag., vol. 14, pp. 43–53, Dec. 1981.

[25] D. M. Dias and M. Kumar, "Packet switching in N log N multistage networks," in Proc. GLOBECOM'84, Atlanta, GA, Dec. 1984, pp. 114–120.

[26] C. P. Kruskal and M. Snir, "The performance of multistage interconnection networks for multiprocessors," IEEE Trans. Comput., vol. C-32, pp. 1091–1098, Dec. 1983.

[27] M. Kumar and J. R. Jump, "Performance of unbuffered shuffle-exchange networks," IEEE Trans. Comput., vol. C-35, pp. 573–577, June 1986.

[28] J. S. Turner, "Fast packet switch," U. S. Patent 4 491 945, Jan. 1, 1985.

[29] ——, "Design of an integrated services packet network," in Proc. 9th ACM Data Commun. Symp., Sept. 1985, pp. 124–133.

[30] P. Kermani and L. Kleinrock, "Virtual cut-through: A new computer communication switching technique," Comput. Networks, vol. 3, pp. 267–286, 1979.

[31] R. W. Muise, T. J. Shonfeld, and G. H. Zimmerman, "Digital communications experiments in wideband packet technology," in Proc. Int. Zurich Seminar Digital Commun., Zurich, Switzerland, Mar. 1986, pp. 135–140.

[32] G. W. R. Luderer et al., "Wideband packet technology for switching systems," in Proc. ISS'87, Phoenix, AZ, Mar. 1987, pp. 448–454.

[33] A. K. Vaidya and M. A. Pashan, "Technology advances in wideband packet switching," in Proc. GLOBECOM'88, Hollywood, FL, Nov. 1988, pp. 668–671.

[34] J. S. Turner, "Design of a broadcast packet switching network," in Proc. INFOCOM'86, Apr. 1986, pp. 667–675. Also, Dep. Comput. Sci., Washington Univ., St. Louis, MO, Tech. Rep. WUCS-84-4, Mar. 1985.

[35] Y.-C. Jenq, "Performance analysis of a packet switch based on a single-buffered banyan network," IEEE J. Select. Areas Commun., vol. SAC-1, pp. 1014–1021, Dec. 1983.

[36] R. G. Bubenik and J. S. Turner, "Performance of a broadcast packet switch," Dep. Comput. Sci., Washington Univ., St. Louis, MO, Tech. Rep. WUCS-86-10, Mar. 1986.

[37] M. N. Huber, E. P. Rathgeb, and T. H. Theimer, "Self routing banyan networks in an ATM-environment," in Proc. ICCC'88, Tel Aviv, Israel, Oct. 1988, pp. 167–174.

[38] E. P. Rathgeb, T. H. Theimer, and M. N. Huber, "Buffering concepts for ATM switching networks," in Proc. GLOBECOM'88, Hollywood, FL, Nov. 1988, pp. 1277–1281.

[39] K. E. Batcher, "Sorting networks and their application," in Proc. Spring Joint Comput. Conf., AFIPS, 1968, pp. 307–314.

[40] A. Huang and S. Knauer, "Starlite: A wideband digital switch," in Proc. GLOBECOM'84, Atlanta, GA, Dec. 1984, pp. 121–125.

[41] A. Huang, "The relationship between STARLITE, A wideband digital switch and optics," in Proc. ICC'86, Toronto, Canada, June 1986, pp. 1725–1729.

[42] J. Y. Hui, "A broadband packet switch for multi-rate services," in Proc. ICC'87, Seattle, WA, June 1987, pp. 782–788.

[43] J. Y. Hui and E. Arthurs, "A broadband packet switch for integrated transport," IEEE J. Select. Areas Commun., vol. SAC-5, pp. 1264–1273, Oct. 1987.

[44] M. J. Karol, M. G. Hluchyj, and S. P. Morgan, "Input versus output queueing on a space-division packet switch," IEEE Trans. Commun., vol. COM-35, pp. 1347–1356, Dec. 1987.

[45] S.-Y. R. Li, "Theory of periodic contention and its application to packet switching," in Proc. INFOCOM'88, New Orleans, LA, Mar. 1988, pp. 320–325.

[46] L. T. Wu and N. C. Huang, "Synchronous wideband network—An interoffice facility hubbing network," in Proc. Int. Zurich Seminar Digital Commun., Zurich, Switzerland, Mar. 1986, pp. 33–39.

[47] C. M. Day, J. N. Giacopelli, and J. Hickey, "Applications of self-routing switching to LATA fiber optic networks," in Proc. ISS'87, Phoenix, AZ, Mar. 1987, pp. 519–523.

[48] W. D. Sincoskie, "Frontiers in switching technology; Part two: Broadband packet switching," Bellcore Exchange, pp. 22–27, Nov./Dec. 1987.

[49] M. G. Hluchyj and M. Karol, "Queueing in space-division packet switching," in Proc. INFOCOM'88, New Orleans, LA, Mar. 1988, pp. 334–343.

[50] Y. S. Yeh, M. G. Hluchyj, and A. S. Acampora, "The Knockout switch: A simple, modular architecture for high-performance packet switching," IEEE J. Select. Areas Commun., vol. SAC-5, pp. 1274–1283, Oct. 1987.

[51] K. Y. Eng, M. G. Hluchyj, and Y. S. Yeh, "A Knockout switch for variable-length packets," IEEE J. Select. Areas Commun., vol. SAC-5, pp. 1426–1435, Dec. 1987.

[52] H. Yoon, M. T. Liu, and K. Y. Lee, "The Knockout switch under nonuniform traffic," in Proc. GLOBECOM'88, Hollywood, FL, Nov. 1988, pp. 1628–1634.

[53] H. Ahmadi et al., "A high-performance switch fabric for integrated circuit and packet switching," in Proc. INFOCOM'88, New Orleans, LA, Mar. 1988, pp. 9–18.

[54] P. Debuysscher, J. Bauwens, and M. De Somer, "Système de commutation," Belgian Patent BE 904100, Jan. 1986.

[55] M. De Prycker and J. Bauwens, "A switching exchange for an asynchronous time division based network," in *Proc. ICC'87*, Seattle, WA, June 1987, pp. 774-781.

[56] M. De Prycker and M. De Somer, "Performance of a service independent switching network with distributed control," *IEEE J. Select. Areas Commun.*, vol. SAC-5, pp. 1293-1301, Oct. 1987.

[57] U. Killat, "Asynchrone Zeitvielfachübermittlung für Breitbandnetze," *Nachrichtentech. Z.*, vol. 40, no. 8, pp. 572-577, 1987.

[58] W. Jasmer, U. Killat, and J. Krüger, German Patent Appl. P 37 14 385.9.

[59] R. Bakka and M. Dieudonne, "Switching circuit for digital packet switching network," U.S. Patent 4 314 367, Feb. 2, 1982.

[60] S. Nojima *et al.*, "Integrated services packet network using bus matrix switch," *IEEE J. Select. Areas Commun.*, vol. SAC-5, pp. 1284-1292, Oct. 1987.

[61] Y. Kato *et al.*, "Experimental broadband ATM switching system," in *Proc. GLOBECOM'88*, Hollywood, FL, Nov, 1988, pp. 1288-1292.

[62] A. Thomas, J. P. Coudreuse, and M. Servel, "Asynchronous time division techniques: An experimental packet network integrating video communication," in *Proc. ISS'84*, Florence, Italy, May 1984, paper 32C2.

[63] P. Gonet, "Fast packet approach to integrated broadband networks," *Comput. Commun.*, pp. 292-298, Dec 1986.

[64] P. Gonet, P. Adams, and J. P. Coudreuse, "Asynchronous time-division switching: The way to flexible broadband communication networks," in *Proc. Int. Zurich Seminar Digital Commun.*, Zurich, Switzerland, Mar. 1986, pp. 141-148.

[65] M. Dieudonne and M. Quinquis, "Switching techniques for asynchronous time division multiplexing (or fast packet switching)," in *Proc. ISS'87*, Phoenix, AZ, Mar. 1987, pp. 367-371.

[66] J. P. Coudreuse and M. Servel, "Prelude: An asynchronous time-division switched network," in *Proc. ICC'87*, Seattle, WA, June 1987, pp. 769-773.

[67] T. Takeuchi and T. Yamaguchi, "Synchronous composite packet switching for ISDN switching system architecture," in *Proc. ISS'84*, Florence, Italy, May 1984, p 42B3.

[68] T. Takeuchi *et al.*, "An experimental synchronous composite packet switching system," in *Proc. Int. Zurich Seminar Digital Commun.*, Zurich, Switzerland, Mar. 1986, pp. 149-153.

[69] H. Suzuki, T. Takeuchi, and T. Yamaguchi, "Very high speed and high capacity packet switching for broadband ISDN," in *Proc. ICC'86*, Toronto, Canada, June 1986, pp. 749-754.

[70] T. Takeuchi *et al.*, "Synchronous composite packet switching—A switching architecture for broadband ISDN," *IEEE J. Select. Areas Commun.*, vol. SAC-5, pp. 1365-1376, Oct. 1987.

[71] J. R. Pierce, "Network for block switching of data," *Bell Syst. Tech. J.*, July-Aug. 1972.

[72] I. S. Gopal, I. Cidon, and H. Meleis, "Paris: An approach to integrated private networks," in *Proc. ICC'87*, Seattle, WA, June 1987, pp. 764-773.

[73] I. Cidon, I. S. Gopal, and S. Kutten, "New models and algorithms for future networks," in *Proc. ACM Principles Distributed Comput.*, Toronto, Canada, 1988.

[74] T. T. Lee, R. Boorstyne, and E. Arthurs, "The architecture of a multicast broadband packet switch," in *Proc. INFOCOM'88*, New Orleans, LA, Mar. 1988, pp. 1-8.

[75] K. Y. Eng, M. G. Hluchyj, and Y. S. Yeh, "Multicast and broadcast services in a knockout packet switch," in *Proc. INFOCOM'88*, New Orleans, LA, Mar. 1988, pp. 29-34.

INTRODUCTION

This part contains four papers from the early 1980s presenting statistical performance models for multistage switching fabrics. Important contributions to our knowledge of such fabrics occurred as early as the 1950s; a representative list of references appears at the end of this introduction [1–10]. However, the papers in this chapter represent the beginnings of our understanding of the statistical performance evaluation of such models.

The paper by D. M. Dias and J. R. Jump is entitled "Analysis and Simulation of Buffered Delta Networks." It specifically mentions packet communication. This paper deals with the general class of delta-switching networks. These networks possess the "self-routing" property; that is, a routing decision can be made at each switching element in each stage based on a single bit in the packet's address. This property greatly simplifies the routing of packets through the switch and, in fact, allows the routing to be performed distributively.

This paper consists of both analytical and simulation results. For the case of unbuffered delta networks, a recursive expression for throughput is presented. A closed-form solution for the throughput of crossbar switches is also presented. For buffered delta networks a solution technique involving the iterative solution of a statistical model is outlined. This is significant, since most statistical models of multistage switching fabrics will involve sets of nonlinear probabilistic equations. These equations cannot usually be solved in closed form but are often susceptible to an iterative solution.

An iterative solution involves starting with an initial guess for the unknown performance values and substituting them into the nonlinear equations to produce a new estimate of the performance values. One continually resubstitutes the latest estimate (or a weighted sum of recent estimates) to produce a slightly better estimate. Eventually the estimates will converge to the true performance values.

The statistical models for switching fabrics can be simplified by noting the presence of independence or by assuming independence. The paper of Dias and Jump points out that "The arrival of a packet at an input link of a switch (element) is independent of the arrival of a packet at the other input link of the same switch (element)." This point finds its way into later work. This paper also makes what will become a common assumption of a uniformly random traffic pattern—that is, each packet at an input is directed to each output independently and with equal probability.

Finally, Dias and Jump propose that all forms of unbuffered delta networks, under certain common assumptions, have the same throughput (packets/second delivered). Simulation results at the end of the paper indicate that this may be true for buffered delta networks.

The paper by J. H. Patel, "Performance of Processor-Memory Interconnection for Multiprocessors," appeared in journal form during the same year (1981) as the Dias and Jump paper (Dias and Jump refer to earlier published work by Patel in their paper). In terms of applications, its perspective comes from the area of processor-to-memory communication in multiprocessors. Patel's paper also deals with delta networks—more specifically, unbuffered delta networks. The paper begins with a construction technique to generate delta networks.

Patel develops a recursion to calculate the throughput of unbuffered delta networks (for the first time) and also an expression for crossbar throughput. The recursions for delta network throughput and the expressions for crossbar throughput of Patel and of Dias and Jump are arrived at by different means but quantify the same phenomena. Patel also assumes a uniform traffic loading and that blocked requests are lost.

One attractive feature of a delta network, compared to the crossbar, is a reduced gate count. Dias and Jump mention that "the number of gates in an $(N \times N)$ delta network is $O(N \log N)$ as opposed to $O(N^2)$ gates in the same size crossbar switch." Patel presents a cost-effectiveness graph (Fig. 14) that shows the delta network's superiority in terms of gate count for medium- to large-scale switches.

The next paper is "The Performance of Multistage Interconnection Networks for Multiprocessors," by C. P. Kruskal and M. Snir. This paper is motivated by the multiprocessor interconnection problem but specifically mentions packet switching. In this paper Kruskal and Snir provide a closed-form expression for unbuffered Banyan network throughput (which is accurate to $O(1/m)$, where m is the number of stages).

They go on to establish expressions for asymptotic throughput of unbuffered Banyan networks. Also discussed are unbuffered dilated Banyan networks (where each link in an unbuffered Banyan network is replaced by d distinct links) and unbuffered replicated Banyan networks (where d parallel Banyan networks replace the original one). Finally, delay formulas for buffered Banyan networks are described.

Last but not least is "Performance Analysis of a Packet Switch Based on Single-Buffered Banyan Network," by Y.-C. Jenq. Chronologically, this is the first paper in this collection with the phrase "packet switched" in the title. During the first half of the paper, Jenq constructs a recursive statistical model of a Banyan network with single buffers for a uniform loading. Jenq makes two independence assumptions that are now quite widely used:

- Under the balanced load the state of each switching element in stage k is statistically equivalent.
- The states of the two buffers in each 2×2 switching element are statistically independent.

Jenq describes models with and without the last assumption (in fact, they differ little in terms of numerical results). The models are based on three quantities: the probability that a switching element buffer at stage k is empty; the probability that packet is ready to enter a switching element buffer at stage k, and the probability that a packet in a switching element buffer at stage k is able to move forward into the next stage. Such recursive models will be widely used by researchers in years to come.

The paper concludes with a study of systems with input buffer controllers with and without infinite buffers. The key feature here is the addition of input buffers to the front end of the Banyan networks. Algorithms are presented to calculate delay for an infinite-input buffer model and blocking probability for a finite input buffer model.

REFERENCES

[1] C. Clos, "A study of nonblocking switching networks," *Bell System Tech. J.*, vol. 32, pp. 406–424, March 1953.

[2] V. E. Benes, "On rearrangeable three stage connecting networks," *Bell System Tech. J.*, vol. 41, no. 5, pp. 1481–1492, 1962.

[3] K. E. Batcher, "Sorting networks and their applications," *Proc. Spring Joint Computer Conf.*, 1968, AFIPS Press, pp. 307–314. This paper appears in the reprint collection by Wu and Feng [10].

[4] D. G. Cantor, "On non-blocking switching networks," *Networks*, vol. 1, pp. 367–377, 1971.

[5] D. P. Bandarkar and S. H. Fuller, "A survey of techniques for analyzing memory interference in multiprocessor systems," *Tech. Report, Carnegie-Mellon University No. AD-762-524*, April 1973.

[6] D. P. Bandarkar, "Analysis of memory interference in multiprocessors," *IEEE Trans. on Computers*, vol. C-24, no. 9, pp. 897–908, Sept. 1975.

[7] F. Baskett and A. J. Smith, "Interference in multiprocessor computer systems with interleaved memory," *Commun. Assoc. for Computing Machinery*, vol. 19, no. 6, pp. 327–334, June 1976.

[8] C. D. Thompson, "Generalized connection networks for parallel processor interconnection," *IEEE Trans. Computers*, vol. C-27, no. 12, pp. 1119–1125, Dec. 1978.

[9] G. M. Masson, G. C. Gingher, and S. Nakamura, "A sampler of circuit switching networks," *IEEE Computer Magazine*, vol. 12, no. 6, pp. 32–48, 1979.

[10] C.-L. Wu and T.-Y. Feng, *Tutorial: Interconnection Networks for Parallel and Distributed Processing*, Los Alamitos, Calif.: IEEE Computer Society Press, 1984.

Analysis and Simulation of Buffered Delta Networks

DANIEL M. DIAS AND J. ROBERT JUMP, MEMBER, IEEE

Abstract—Delta networks are a class of multistage interconnection networks with gate complexity less than crossbar switches that are easy to control, and which include several networks that have been proposed in the literature as special cases. Buffered delta networks have queues of packets between the stages of the network. This paper presents analytic and simulation results for the performance of delta networks in a packet communication environment. The performance of buffered delta networks is compared with unbuffered delta networks and crossbar switches. It is demonstrated that buffering produces considerable improvement in the performance of these networks, making their performance comparable to that of crossbar switches.

Index Terms—Crossbar switches, delta networks, multistage interconnection networks, packet communication architecture, performance analysis, simulation.

I. INTRODUCTION

IN recent years there has been considerable interest in systems consisting of a large number of modules and the important problem of interconnecting these modules. In this paper we present analytic and simulation results for the performance of a class of multistage interconnection networks, called delta networks [4], in a packet communication environment [1]. These networks can be used to interconnect a large number of modules. By a packet communication environment we mean that the modules in the system communicate asynchronously over this interconnection network by sending fixed size packets. The class of delta networks includes the omega network [3], the indirect binary n-cube network [5], and the cube network [7], among others (excluding a permutation at either the input or at the output of these networks). The number of gates in an $(N \times N)$ delta network is $0(N \log N)$ as opposed to the $0(N^2)$ gates in the same size crossbar switch. Packets can be routed through a delta network by examining a certain bit of the packet destination address at each stage of the network. Thus, packet routing is very simple and can be accomplished under local control at each switch. Applications include interprocessor communication and memory-processor interconnections, particularly for some proposed data flow computers [2]. Patel [4] examines the performance

Manuscript received June 2, 1980; revised November 10, 1980. This work was supported by the National Science Foundation under Grant DCR74-14283.

The authors are with the Department of Electrical Engineering, Rice University, Houston, TX 77001.

of these networks for an unbuffered case. In this paper we allow buffering between stages in the network. For this case, we present an approximate analysis that is substantiated and extended to analytically intractable cases by simulation. We compare the resulting performance with that of unbuffered networks and crossbar switches. We show that with buffering, the performance of delta networks is quite comparable to the crossbar switch.

II. THE MODEL

This section defines delta networks, states some of their properties, and specifies the environment that we will consider. The properties of the networks presented here form the basis for the analysis of Section III.

A. Delta Networks

The name "delta networks" was introduced by Patel [4]. We consider $(2^n \times 2^n)$ delta networks. These networks have 2^n network input links and 2^n network output links each labeled by n-bit binary numbers (Fig. 1). They are composed of n stages $S_0, S_1, \cdots, S_{n-1}$. Each stage has 2^{n-1} (2×2) crossbar switches labeled $0, 1, \cdots, (2^{n-1} - 1)$. There are permutations $P_0, P_1, \cdots, P_{n-2}$ between stages. The outputs of the (2×2) crossbar switches are labeled "0" and "1." These switches can connect any of the two input links to any of the two output links with the restriction that both input links cannot be connected to the same output link at the same time. The permutations between stages must satisfy the requirement that there exists a transmission path (hereafter referred to as a path) between any network input link $i_{n-1} \cdots i_0$ and any network output link $d_{n-1} \cdots d_0$ obtained by connecting input links of switches in stage j, $0 \le j < n$, to output d_{c_j}, $0 \le c_j < n$. This property of delta networks is called the *bit-controlled property*. We refer to d_{c_j} as the *control bit* at stage j, while c_j is called the *control index* at stage j.

A procedure to construct delta networks is given by Patel [4]. We note that the omega network [3] (excluding the permutation at the input to the network) and the indirect binary n-cube network [5] (without the permutation at the output of the network), among other networks, belong to the class of delta networks. An example of a $(2^3 \times 2^3)$ delta network with a perfect shuffle permutation [8] between stages is shown in Fig. 2.

We now present some useful properties of delta networks.

Lemma 1: In a $(2^n \times 2^n)$ delta network there is a unique path from each network input link to each network output link.

Reprinted from *IEEE Trans. Comput.*, vol. C-30, no. 4, pp. 273–282, April 1981.

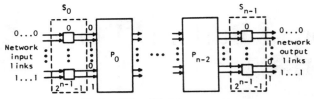

Fig. 1. $(2^n \times 2^n)$ delta networks.

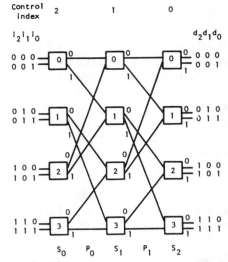

Fig. 2. A $(2^3 \times 2^3)$ delta network. (Omega network without input permutation.)

Proof: There are 2^n paths from each network input link to network output links (since there are n stages and two positions for switch connections in each stage). But in a $(2^n \times 2^n)$ delta network, there must be at least one path to each of the 2^n output links and hence the paths must be unique.

Theorem 1: Consider any switch "m" in stage S_k of a $(2^n \times 2^n)$ delta network with $0 \le k < n$. Let i_x (o_x) and i_y (o_y) denote the input (output) links of m. Let I_x (0_x) be the set of network input (output) links with paths to (from) i_x (o_x) and let I_y (o_y) be the corresponding set with paths to (from) i_y (o_y). Then

1) I_x (0_x) and I_y (0_y) are disjoint.
2) $|I_x| = |I_y| = 2^k$ and $|0_x| = |0_y| = 2^{n-1-k}$.
3) A path from any network input link in $(I_x \cup I_y)$ to any network output link in $(0_x \cup 0_y)$ must pass through switch m.
4) Exactly 2^{n+1} paths from network input links to network output links pass through switch m.

Proof: Follows easily from Lemma 1.

B. The Environment

We make the following assumptions about the environment and the operation of the networks.

1) Packets arriving at network input links contain both the data to be transferred and the label of the network output link to which the data are to be passed.

2) We compare the performance of the networks in an environment of "maximum loading." For this case, we assume a single buffer at network input links which is filled by an input packet whenever it is emptied by the network. It is assumed that buffers at network output links are emptied instantaneously.

3) All input packets are assumed to be independently and equiprobably directed to each network output link.

4) The delay at a (2×2) switch is modeled as consisting of two time intervals. Time "t_select" to select a switch output link and time "t_pass" to pass the data to the selected output link. The minimum delay "min_d" of data through a $(2^n \times 2^n)$ delta network is thus $n \times (t\_select + t\_pass)$.

5) Networks without buffering: For this case output data are delivered in time intervals of length "min_d" referred to as min_d time slots. An attempt is made to pass all input packets along the unique path from the network input link to the destination network output link (Lemma 1). A *conflict* is said to occur if two packets must pass through the same link. In this case, one of the packets is equiprobably selected and passed while the other is rejected. A rejected packet is assumed to be lost. This assumption is made to simplify the analysis. In a realistic environment rejected packets would be resubmitted in the next min_d time slot. Patel [4] has done simulation experiments that indicate that the throughput obtained by resubmitting rejected packets is only slightly lower than that obtained otherwise.

6) Networks with buffering: The operation of buffered networks is modeled using timed Petri nets [6]. Shown in Fig. 3 is a Petri net model for a (2×2) switch with a single buffer between stages. This model is easily extended to bigger switches and longer buffers. The operation is essentially as follows. A fixed maximum queue length of waiting packets is allowed between stages. These queues are modeled by the places labeled as "input buffers" in Fig. 3. A switch can handle a packet at each input queue simultaneously. The model assumes that it takes time "t_select" to determine the output link to which the packet is to be passed. This time interval is modeled by transitions with a firing time of "t_select" as in Fig. 3. If the selected output is in use (i.e., another packet is in the process of being passed to that output link) the packet waits its turn for the use of that output link with equiprobable selection from several packets waiting to be passed to the same output link. When the selected output link becomes available, the packet waits until a buffer at the successor input queue is empty. The packet is then delayed for another time interval "t_pass" which models transmission delay. After this time the data are available in the successor buffer. This process is modeled by transitions with a firing time of "t_pass," and places labeled as "output request" and "buffer empty flags" in Fig. 3. We note that, for analytical purposes, the backward propagation delay in informing a predecessor switch of an empty buffer can be included in t_pass.

III. ANALYSIS

This section outlines an analysis of delta networks for the buffered and the unbuffered cases and of crossbar switches, for the environment and assumptions described in Section II.

The performance criteria used are those of throughput and turn-around-time and normalized versions of these criteria. The throughput of a network for a particular environment is the average number of packets put out in unit time. The turn-around-time is the average time interval between the time

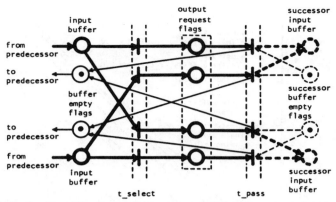

Fig. 3. Timed Petri net model for the operation of a (2 × 2) switch with a single buffer (shown in the initial state).

a packet is placed in a buffer at a network input link and the time at which it is placed in a buffer at a network output link. The maximum throughput and minimum turn-around-time are those for the ideal case in which no conflicts occur in the network. The normalized throughput and turn-around-time are the ratios of the throughput and turn-around-time to the maximum throughput and the minimum turn-around-time, respectively.

Lemma 2: Consider an unbuffered $(2^n \times 2^n)$ delta network with the environment and assumptions of Section II-B. Let p_k denote the probability of a packet passing through an input link of a switch in stage S_k, $0 \leq k < n$, in a min_d time slot. Let p_n denote the probability of a packet arriving at a network output link. Then

1) p_k, $0 \leq k < n$, is the same for all switch input links in the same stage, and p_n is the same for all network output links.

2) The arrival of a packet at an input link of a switch is independent of the arrival of a packet at the other input link of the same switch.

3) p_k, $0 \leq k \leq n$, is the same for all $(2^n \times 2^n)$ delta networks.

Proof: By an induction on k, $0 \leq k \leq n$. The statements are trivially true for stage S_0 from assumptions 2), 3), and 5) of Section II-B. Assume that they are true for stage S_k. Then from Theorem 1 and assumption 3) of Section II-B it directly follows that they are true for stage S_{k+1}.

Theorem 2: All $(2^n \times 2^n)$ unbuffered delta networks, with the environment described in Section II-B, have the same throughput.

Proof: With the above definition of the throughput and the notation of Lemma 2, the throughput is $(p_n \times 2^n/\text{min}\_d)$. Thus, it directly follows from Lemma 2 that the throughput is the same for all $(2^n \times 2^n)$ delta networks.

We have not been able to establish whether or not the entire class of $(2^n \times 2^n)$ buffered delta networks (for any constant n) has the same performance for the environment described in Section II-B. However, the analysis that we will present in this section suggests that the performance of all these networks is approximately the same. This claim is supported by the results of simulation experiments that are described in a later section.

A. Unbuffered Delta Networks

This case has been analyzed by Patel [4]. We obtain the

results of Patel [4] by a simpler analysis. Let p_k be defined as in Lemma 2. Let "b" denote the number of input and output links of crossbar switches in each stage of the network. Thus, $b = 2$ for this case. From Lemma 2 it follows that the number of packets that arrive at the input links of a switch, in a min_d time slot, have a binomial distribution. Thus, the probability of j packet arrivals at a switch m in stage S_k is $\left(\binom{b}{j} p_k^j (1 - p_k)^{b-j}\right)$. The probability of a packet being passed to a particular output link of m given that there are j packets at the input links of m is $\left(1 - \left(\frac{b-1}{b}\right)^j\right)$. Thus,

$$p_{k+1} = \sum_{j=0}^{b} \binom{b}{j} p_k^j (1 - p_k)^{b-j} \left(1 - \left(\frac{b-1}{b}\right)^j\right)$$

which simplifies to

$$p_{k+1} = 1 - \left(1 - \frac{p_k}{b}\right)^b. \tag{1}$$

For the environment we consider, $(p_0 = 1.0)$. The normalized throughput is p_n, while the absolute throughput is $(p_n \times 2^n/\text{min}\_d)$ and these are shown in Figs. 4 and 5.

B. Crossbar Switches

For a $(2^n \times 2^n)$ crossbar switch equation (1) holds with $b = 2^n$. Thus, $p_1 = 1 - \left(1 - \frac{p_0}{2^n}\right)^{2^n}$. For large networks with $p_0 = 1$, $p_1 = 1 - e^{-p_0} = 0.632$. The normalized and absolute throughputs that result are shown in Figs. 4 and 5.

C. Buffered Delta Networks

The analysis we present assumes that $t\_\text{select} = 0$ or that $t\_\text{pass} = 0$ and $(t\_\text{select} + t\_\text{pass}) = t\_\text{delay}$. In Section IV we compare the performance obtained with these assumptions with that for arbitrary values of these parameters.

We assume that at time = 0 the system is in the state with no packet in any buffer. With the above assumptions and the

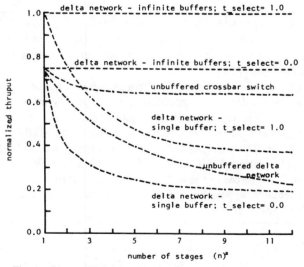

Fig. 4. Normalized throughput versus number of stages (n).

assumed operation of these networks, as described in Section II-B, it immediately follows that packets are placed in buffers only at times that are integer multiples of t_delay. Let $t_k = k \times t\_delay$, $k = 0, 1, 2, \cdots$. The time interval $(t_k, t_{k+1}]$ is denoted by τ_k.

We first consider the case of a single buffer between the stages of the delta network. The system can be modeled by a discrete time, aperiodic, Markov chain with a finite state space and with states representing the network states at times t_k, for $k = 0, 1, \cdots$. However, an exact Markov chain model has an enormously large number of states making it analytically intractable. We develop an approximate model based on the properties of delta networks that have been developed. This approximate model is substantiated by comparison with simulation results.

Theorem 1 asserts that the set of network input links with a path to one of the links of a switch is disjoint from the corresponding set with a path to the other switch input link. We therefore approximate that the event that a packet is placed in a buffer at an input link of a switch is independent of the corresponding event at the other input link of the same switch, both at the same time t_k. We also assume that the event that a packet at the output of a switch is passed to a successor switch in time interval τ_k is independent of the passage of a packet at the other output of that switch. The distinguishable states of a (2×2) switch with a single buffer between stages are shown in Fig. 6. Motivated by Theorem 1 we assume that all switches in the same stage have the same probabilities of being in each of these states at t_k, for $k = 0, 1, \cdots$. We note that Theorem 1 is valid for all delta networks. Thus, the following approximate analysis holds for the entire class of delta networks.

Consider the case of $t\_pass = 0$. For this case the last stage in the network delays all packets for a time of t_delay even if two packets are directed to the same output in time interval τ_k. This is because of the assumption that packets at the output of the network are removed instantaneously. Thus, for this case the last stage of the network does not restrict the performance of the network and can be eliminated from the analysis. Fur-

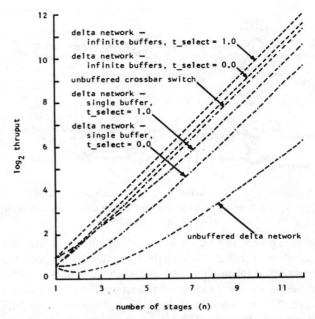

Fig. 5. Throughput versus number of stages (assuming a unit delay per stage).

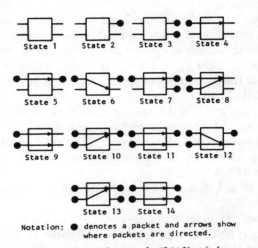

Notation: ● denotes a packet and arrows show where packets are directed.

Fig. 6. Internal states of a (2×2) switch.

$$\approx \frac{(\frac{1}{2}p1^k_{4,j+1} + \frac{1}{2}p1^k_{6,j+1} + \frac{1}{2}p1^k_{8,j+1} + p1^k_{9,j+1} + \frac{1}{2}p1^k_{11,j+1} + \frac{1}{2}p1^k_{12,j+1})}{(\frac{1}{2}p1^k_{4,j+1} + \frac{1}{2}p1^k_{5,j+1} + \frac{1}{2}p1^k_{6,j+1} + \frac{1}{2}p1^k_{7,j+1} + p1^k_{8,j+1} + p1^k_{9,j+1} + p1^k_{10,j+1} + p1^k_{11,j+1} + p1^k_{12,j+1} + p1^k_{13,j+1} + p1^k_{14,j+1})} \quad (2)$$

ther, a packet at an output buffer of stage $(n - 2)$ at t_k, $k = 0, 1, \cdots$, is always advanced to an output buffer of stage $(n - 1)$ in the interval τ_k.

Consider advancing a packet from an input buffer of a stage to an input buffer of the succeeding stage in a τ_k interval, $k = 0, 1, \cdots$. The selection of the desired switch output link in the t_select time interval occurs regardless of the availability of a buffer at that output link. For the case of $t\_pass = 0$, the packet is advanced if it is selected at that output link and if a buffer is available at the end of the t_delay interval. This in turn depends on whether a packet in that buffer was removed in the τ interval.

The analysis for this case is facilitated by modeling the operation of the network in each τ_k interval as consisting of three "logical" steps. In step 1 packet movement at the output of switches occurs. In step 2 packets move from input links of switches to output links of the corresponding switches. In step 3 packets move into buffers at switch input links from preceding stages. Note that these "logical" steps are artificial and the separation has been made to facilitate computation. Note also that from the definition of the states of a switch (Fig. 6), step 1 at stage j, $0 \leq j < n - 1$, corresponds to step 2 at stage $(j + 1)$, and step 3 at stage j, $0 < j \leq n - 1$, corresponds to step 2 at stage $(j - 1$. As discussed above, packets at the output of

$$\approx \frac{(\frac{1}{2}p1^k_{4,j-1} + \frac{1}{2}p1^k_{6,j-1} + \frac{1}{2}p1^k_{8,j-1} + p1^k_{9,j-1} + \frac{1}{2}p1^k_{11,j-1} + \frac{1}{2}p1^k_{12,j-1})}{(p1^k_{1,j-1} + \frac{1}{2}p1^k_{2,j-1} + p1^k_{4,j-1} + \frac{1}{2}p1^k_{5,j-1} + \frac{1}{2}p1^k_{6,j-1} + p1^k_{8,j-1} + p1^k_{9,j-1} + \frac{1}{2}p1^k_{10,j-1} + \frac{1}{2}p1^k_{11,j-1} + \frac{1}{2}p1^k_{12,j-1})} \quad (3)$$

stage $(n - 2)$ are always removed in step 1. Thus, given the state probabilities of switches at t_k, $k = 0, 1, \cdots$, one can compute the state probabilities of switches in stage $(n - 2)$ after step 1 in time interval τ_k. Similar computations at stage j, $0 \leq j < n - 2$, can be made after they have been made at stage $(j + 1)$. Similar comments apply to steps 2 and 3. Thus, the state probabilities after each of these steps can be computed by starting at stage $n - 2$ and proceeding down to stage 0. We use the following notation to formalize this procedure. In the following, if not otherwise stated, $k = 0, 1, \cdots, 0 \leq j < (n - 1)$, $1 \leq i \leq 14$. Let

$p0_{i,j}^k$ = Probability {switch in stage

j is in state i at time $= t_k$}

$p1_{i,j}^k$ = Probability {switch in stage

j is in state i in time interval τ_k after step 1}

$p2_{i,j}^k$ = Probability {switch in stage

j is in state i in time interval τ_k after steps 1 and 2}

$p3_{i,j}^k = p0_{i,j}^{k+1}$ = Probability {switch in stage

j is in state i in time interval τ_k after steps 1, 2 and 3}

$\bar{p}_j$ = Probability {a packet at a switch output link in stage

j is passed in time interval τ_k}

for $0 \leq j < n - 2$.

The denominator in Equation 2 is the probability that a packet exists at a switch input link in stage $j + 1$ after step 1 in time interval τ_k and the numerator is the probability that such a packet is passed in that time interval. The boundary condition is

$$\bar{p}_{n-2} = 1.0$$

$$\bar{q}_j = 1.0 - \bar{p}_j$$

p_j = Probability {a packet from an input link at stage $j - 1$ is placed in a buffer at in input link of stage j in time interval τ_k}

for $0 < j \leq n - 2$.

The denominator in Equation 3 is the probability that an empty buffer exists at a switch output link in stage $j - 1$ after step 1 in time interval τ_k and the numerator is the probability that such a buffer is filled in that time interval. The boundary condition is

$$\overline{p}_0 = 1.0$$

$$\overline{q}_j = 1.0 - \overline{p}_j$$

p_j = Probability {a packet exists at an output link

of a switch in stage j at time t_k}

$$\approx (\tfrac{1}{2}(p0_{2,j}^k + p0_{5,j}^k + p0_{6,j}^k + p0_{10,j}^k + p0_{11,j}^k + p0_{12,j}^k) + p0_{3,j}^k + p0_{7,j}^k + p0_{13,j}^k + p0_{14,j}^k). \quad (4)$$

State transition tables in terms of the above definitions are shown in Fig. 7. The method employed is to start at the network state with no packets in any buffer and to iterate using the above equations and the transition tables of Fig. 7 until a steady state is reached. This steady state is detected by the condition that $(p_j \cdot \bar{p}_j)$ is (relatively) constant for $0 \leq j < n$ (i.e., all stages have the same average number of packets passing through them) and when this value is (relatively) constant over a large number of iterations (typically 100). The normalized and absolute throughputs are estimated as p_{n-2} and $(p_{n-2} \times 2^n/t$ delay), respectively, after this steady state is achieved and are shown in Figs. 4 and 5. An approximate estimate for the turn-around-time is obtained as follows. Observe from the definition of $\bar{p}_j$ and (2) that $\bar{p}_j$ is the probability that a packet at an input buffer of stage $(j + 1)$ is passed in interval τ_k. Thus, the average delay that a packet encounters at stage j is estimated as

$$t\_\text{delay} \cdot \sum_{k=1}^{\infty} k \, \bar{p}_{j-1} (1 - \bar{p}_{j-1})^{k-1} = \frac{t \text{ delay}}{\bar{p}_{j-1}}.$$

The turn-around-time is thus estimated as

$$t\_\text{delay} \cdot \sum_{j=0}^{n-1} \frac{1}{\bar{p}_{j-1}} \quad (5)$$

and the normalized turn-around-time as

$$\frac{1}{n} \cdot \sum_{j=0}^{n-1} \frac{1}{\bar{p}_{j-1}} \quad (6)$$

The normalized turn-around-time is shown in Fig. 9, where it is compared with simulation results.

We now consider the case of $t\_\text{select} = 0$ and a single buffer between stages. Again consider advancing a packet from an input buffer of a stage to a buffer at the input of the succeeding stage in a τ_k interval, $k = 0, 1, \cdots$. Recall that the t_pass phase is not entered unless a buffer is available at the corresponding output link. Thus, with $t\_\text{select} = 0$, the packet is advanced if a buffer is available at the selected output link at the beginning of $\tilde{t}_k$ and if it is selected for passage at that output. (This is opposed to the preceding case of $t\_\text{pass} = 0$ where buffer availability at the end of $\tilde{t}_k$ mattered.) This means that state transitions can be computed in a single step.

The method employed is the same as for the preceding case except that there is a single step at each iteration. The definitions for $p0_{i,j}^k$, $\bar{p}_j$, $\overline{p}_j$, and p_j are the same as before. Equation (2) is valid with $p1$ replaced by $p0$ and for $0 \leq j < n - 1$. The boundary condition is changed to $\bar{p}_{n-1} = 1.0$. Equation (3) holds with $p1$ replaced by $p0$ and for $0 \leq j < n$, while (4) is valid as is. State transition tables are shown in Fig. 8. The normalized and absolute throughputs are estimated as p_{n-1} and $(p_{n-1} \times 2^n/t\_\text{delay})$ at steady state and are shown in Figs. 4 and 5. The normalized turn-around-time is estimated using (6) and is shown in Fig. 9, where it is also compared with simulation results.

Finally, consider the case of infinite buffers between the stages of the delta networks. For $t\_\text{select} = 0$, $p_{n-1} = 0.75$ and for $t\_\text{pass} = 0$, $p_{n-1} = 1.0$. These values follow directly by contradiction. Assuming them to be false leads to infinite length queues between stages for which case they are also true. The resulting normalized and absolute throughputs are shown in Figs. 4 and 5.

D. Comparison

Fig. 4 shows the throughputs of the various networks nor-

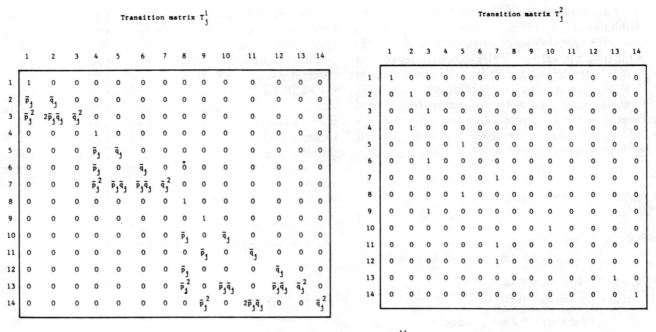

$$p1_{i,j}^{k} = \sum_{m=1}^{14} p0_{m,j}^{k}\, T_j^1(m,i) \quad \text{for } 1 \le i \le 14 \,;\, 0 \le j < n \,;\, k = 0, 1, \ldots .$$

(a)

$$p2_{i,j}^{k} = \sum_{m=1}^{14} p1_{m,j}^{k}\, T_j^2(m,i) \quad \text{for } 1 \le i \le 14 \,;\, 0 \le j < n \,;\, k = 0, 1, \ldots .$$

(b)

Transition matrix T_j^3

	1	2	3	4	5	6	7	8	9	10	11	12	13	14
1	q_j^2	0	0	$2p_jq_j$	0	0	0	$\tfrac{1}{2}p_j$	$\tfrac{1}{2}p_j^2$	0	0	0	0	0
2	0	q_j^2	0	0	p_jq_j	p_jq_j	0	0	0	$\tfrac{1}{4}p_j^2$	$\tfrac{1}{2}p_j^2$	$\tfrac{1}{4}p_j^2$	0	0
3	0	0	q_j^2	0	0	0	$2p_jq_j$	0	0	0	0	$\tfrac{1}{2}p_j^2$	$\tfrac{1}{2}p_j^2$	
4	0	0	0	q_j	0	0	0	$\tfrac{1}{2}p_j$	$\tfrac{1}{2}p_j$	0	0	0	0	0
5	0	0	0	0	q_j	0	0	0	0	$\tfrac{1}{2}p_j$	$\tfrac{1}{2}p_j$	0	0	0
6	0	0	0	0	0	q_j	0	0	0	$\tfrac{1}{2}p_j$	$\tfrac{1}{2}p_j$	0	0	0
7	0	0	0	0	0	0	q_j	0	0	0	0	0	$\tfrac{1}{2}p_j$	$\tfrac{1}{2}p_j$
8	0	0	0	0	0	0	0	1	0	0	0	0	0	0
9	0	0	0	0	0	0	0	0	1	0	0	0	0	0
10	0	0	0	0	0	0	0	0	0	1	0	0	0	0
11	0	0	0	0	0	0	0	0	0	0	1	0	0	0
12	0	0	0	0	0	0	0	0	0	0	0	1	0	0
13	0	0	0	0	0	0	0	0	0	0	0	0	1	0
14	0	0	0	0	0	0	0	0	0	0	0	0	0	1

$$p3_{i,j}^{k} = p0_{i,j}^{k+1} = \sum_{m=1}^{14} p2_{m,j}^{k}\, T_j^3(m,i) \quad \text{for } 1 \le i \le 14 \,;\, 0 \le j < n \,;\, k = 0, 1, \ldots .$$

(c)

Fig. 7.   State transition probabilities for t_pass = 0. (a) State transition probabilities due to packet movement at switch output links. (b) State transition probabilities due to internal packet flow within a switch. (c) State transition probabilities due to packet movement into switches.

malized with respect to the maximum possible throughput that would be obtained if there were no clashes in the network. For large crossbar switches the normalized throughput converges to a constant. For buffered delta networks the normalized throughput decreases very slowly as the size of the networks become large. The rate of fall of the normalized throughput is much faster for unbuffered delta networks. The maximum (conflict free) throughput can be expressed as $2^n/t$_slot. Let t_delay be the minimum delay at a stage of the networks. Then t_slot = min_d = $n \times t$_delay for the unbuffered delta networks, while it is t_delay for the buffered delta network and the crossbar switch. T_delay for an $(N \times N)$ crossbar switch is $0(N)$, while for a single (2×2) switch it is a constant. Thus, the buffered delta network has a better throughput than the simple crossbar switch for large networks. However, the throughput of the crossbar can be improved by internal buffering in the switch. The main asset of the delta networks is that for an $(N \times N)$ network, the number of gates required is $0(N \log N)$, while $(N \times N)$ crossbar switches require $0(N^2)$ gates.

24

Transition matrix T_0^4

	1	2	3	4	5	6	7	8	9	10	11	12	13	14
1	0	0	$\frac{1}{2}$	0	$\frac{1}{2}$	0	0	0	0	0	0	0	0	0
2	0	0	0	0	$\frac{1}{4}\bar{p}_j$	$\frac{1}{2}\bar{p}_j$	$\frac{3}{4}\bar{q}_j$	$\frac{1}{4}\bar{p}_j$	0	$\frac{1}{4}\bar{q}_j$	0	0	0	0
3	0	0	0	0	0	0	0	$\frac{1}{2}\bar{p}_j^2$	$\frac{1}{2}\bar{p}_j^2$	$\frac{1}{2}\bar{p}_j\bar{q}_j$	$\bar{p}_j\bar{q}_j$	$\frac{1}{2}\bar{p}_j\bar{q}_j$	$\frac{1}{2}\bar{q}_j^2$	$\frac{1}{2}\bar{q}_j^2$
4	0	0	$\frac{1}{2}$	0	$\frac{1}{2}$	0	0	0	0	0	0	0	0	0
5	0	0	0	0	0	$\frac{1}{2}\bar{p}_j$	$\frac{1}{2}\bar{q}_j$	$\frac{1}{2}\bar{p}_j$	0	$\frac{1}{2}\bar{q}_j$	0	0	0	0
6	0	0	0	0	$\frac{1}{2}\bar{p}_j$	$\frac{1}{2}\bar{p}_j$	$\bar{q}_j$	0	0	0	0	0	0	0
7	0	0	0	0	0	0	0	$\frac{1}{2}\bar{p}_j^2$	$\frac{1}{2}\bar{p}_j^2$	$\frac{1}{2}\bar{p}_j\bar{q}_j$	$\bar{p}_j\bar{q}_j$	$\frac{1}{2}\bar{p}_j\bar{q}_j$	$\frac{1}{2}\bar{q}_j^2$	$\frac{1}{2}\bar{q}_j^2$
8	0	0	0	0	1	0	0	0	0	0	0	0	0	0
9	0	0	1	0	0	0	0	0	0	0	0	0	0	0
10	0	0	0	0	0	0	0	$\bar{p}_j$	0	$\bar{q}_j$	0	0	0	0
11	0	0	0	0	0	$\bar{p}_j$	$\bar{q}_j$	0	0	0	0	0	0	0
12	0	0	0	0	$\bar{p}_j$	0	$\bar{q}_j$	0	0	0	0	0	0	0
13	0	0	0	0	0	0	0	$\bar{p}_j^2$	0	$\bar{p}_j\bar{q}_j$	0	$\bar{p}_j\bar{q}_j$	$\bar{q}_j^2$	0
14	0	0	0	0	0	0	0	0	$\bar{p}_j^2$	0	$2\bar{p}_j\bar{q}_j$	0	0	$\bar{q}_j^2$

$$po_{1,0}^{k+1} = \sum_{m=1}^{14} po_{m,0}^k\, T_0^4(m,i) \quad \text{for } 1 \le i \le 14 \; ; \; k = 0, 1, \dots .$$

(a)

Transition matrix T_{n-1}^6

	1	2	3	4	5	6	7	8	9	10	11	12	13	14
1	q_j^2	0	0	$2p_jq_j$	0	0	0	$\frac{1}{2}p_j^2$	$\frac{1}{2}p_j^2$	0	0	0	0	0
2	q_j^2	0	0	$2p_jq_j$	0	0	0	$\frac{1}{2}p_j^2$	$\frac{1}{2}p_j^2$	0	0	0	0	0
3	q_j^2	0	0	$2p_jq_j$	0	0	0	$\frac{1}{2}p_j^2$	$\frac{1}{2}p_j^2$	0	0	0	0	0
4	0	q_j	0	0	$\frac{1}{2}p_j$	$\frac{1}{2}p_j$	0	0	0	0	0	0	0	0
5	0	q_j	0	0	$\frac{1}{2}p_j$	$\frac{1}{2}p_j$	0	0	0	0	0	0	0	0
6	0	q_j	0	0	$\frac{1}{2}p_j$	$\frac{1}{2}p_j$	0	0	0	0	0	0	0	0
7	0	q_j	0	0	$\frac{1}{2}p_j$	$\frac{1}{2}p_j$	0	0	0	0	0	0	0	0
8	0	0	0	0	1	0	0	0	0	0	0	0	0	0
9	0	0	1	0	0	0	0	0	0	0	0	0	0	0
10	0	0	0	0	1	0	0	0	0	0	0	0	0	0
11	0	0	1	0	0	0	0	0	0	0	0	0	0	0
12	0	0	0	0	1	0	0	0	0	0	0	0	0	0
13	0	0	0	0	1	0	0	0	0	0	0	0	0	0
14	0	0	1	0	0	0	0	0	0	0	0	0	0	0

$$po_{1,n-1}^{k+1} = \sum_{m=1}^{14} po_{m,n-1}^k\, T_{n-1}^6(m,i) \quad \text{for } 1 \le i \le 14 \; ; \; k = 0, 1, \dots$$

(b)

Transition matrix T_j^5

	1	2	3	4	5	6	7	8	9	10	11	12	13	14
1	q_j^2	0	0	$2p_jq_j$	0	0	0	$\frac{1}{2}p_j^2$	$\frac{1}{2}p_j^2$	0	0	0	0	0
2	$q_j^2\bar{p}_j$	$q_j^2\bar{q}_j$	0	$2p_jq_j\bar{p}_j$	$p_jq_j\bar{q}_j$	$p_jq_j\bar{q}_j$	0	$\frac{1}{2}p_j^2\bar{p}_j$	$\frac{1}{2}p_j^2\bar{p}_j$	$\frac{1}{4}p_j^2\bar{q}_j$	$\frac{1}{2}p_j^2\bar{q}_j$	$\frac{1}{4}p_j^2\bar{q}_j$	0	0
3	$q_j^2\bar{p}_j^2$	$2q_j^2\bar{p}_j\bar{q}_j$	$q_j^2\bar{q}_j^2$	$2p_jq_j\bar{p}_j^2$	$2p_jq_j\bar{p}_j\bar{q}_j$	$2p_jq_j\bar{p}_j\bar{q}_j$	$2p_jq_j\bar{q}_j^2$	$\frac{1}{2}p_j^2\bar{p}_j^2$	$p_j^2\bar{p}_j\bar{q}_j$	$\frac{1}{2}p_j^2\bar{p}_j\bar{q}_j$	$p_j^2\bar{p}_j\bar{q}_j$	$\frac{1}{2}p_j^2\bar{p}_j\bar{q}_j$	$\frac{1}{2}p_j^2\bar{q}_j^2$	$\frac{1}{2}p_j^2\bar{q}_j^2$
4	0	q_j	0	0	$\frac{1}{2}p_j$	$\frac{1}{2}p_j$	0	0	0	0	0	0	0	0
5	0	0	0	$q_j\bar{p}_j$	$q_j\bar{q}_j$	0	0	$\frac{1}{2}p_j\bar{p}_j$	$\frac{1}{2}p_j\bar{p}_j$	$\frac{1}{2}p_j\bar{q}_j$	$\frac{1}{2}p_j\bar{q}_j$	0	0	0
6	0	$q_j\bar{p}_j$	$q_j\bar{q}_j$	0	$\frac{1}{2}p_j\bar{p}_j$	$\frac{1}{2}p_j\bar{p}_j$	$p_j\bar{q}_j$	0	0	0	0	0	0	0
7	0	0	0	$q_j\bar{p}_j^2$	$q_j\bar{p}_j\bar{q}_j$	$q_j\bar{p}_j\bar{q}_j$	$q_j\bar{q}_j^2$	$\frac{1}{2}p_j\bar{p}_j^2$	$\frac{1}{2}p_j\bar{p}_j^2$	$\frac{1}{2}p_j\bar{p}_j\bar{q}_j$	$p_j\bar{p}_j\bar{q}_j$	$\frac{1}{2}p_j\bar{p}_j\bar{q}_j$	$\frac{1}{2}p_j\bar{q}_j^2$	$\frac{1}{2}p_j\bar{q}_j^2$
8	0	0	0	0	1	0	0	0	0	0	0	0	0	0
9	0	0	1	0	0	0	0	0	0	0	0	0	0	0
10	0	0	0	0	0	0	0	$\bar{p}_j$	0	$\bar{q}_j$	0	0	0	0
11	0	0	0	0	0	$\bar{p}_j$	$\bar{q}_j$	0	0	0	0	0	0	0
12	0	0	0	0	$\bar{p}_j$	0	$\bar{q}_j$	0	0	0	0	0	0	0
13	0	0	0	0	0	0	0	$\bar{p}_j^2$	0	$\bar{p}_j\bar{q}_j$	0	$\bar{p}_j\bar{q}_j$	$\bar{q}_j^2$	0
14	0	0	0	0	0	0	0	0	$\bar{p}_j^2$	0	$2\bar{p}_j\bar{q}_j$	0	0	$\bar{q}_j^2$

$$po_{1,j}^{k+1} = \sum_{m=1}^{14} po_{m,j}^k\, T_j^5(m,i) \quad \text{for } 1 \le i \le 14 \; ; \; 0 < j < (n-1) \; ; \; k = 0, 1, \dots .$$

(c)

Fig. 8. State transition probabilities for $t\_select = 0$ (a) State transition probabilities at stage 0. (b) State transition probabilities at stage j, $0 < j < (n-1)$. (c) State transition probabilities at stage $(n-1)$.

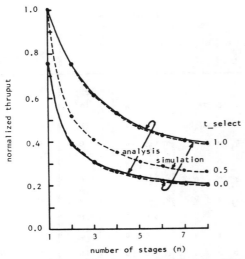

Fig. 9. Normalized throughput versus number of stages. (Single buffer between stages).

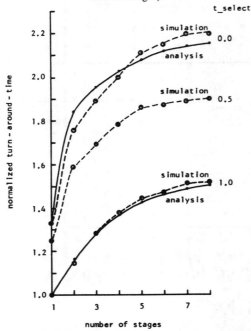

Fig. 10. Normalized turn-around-time versus number of stages. (Single buffer between stages.)

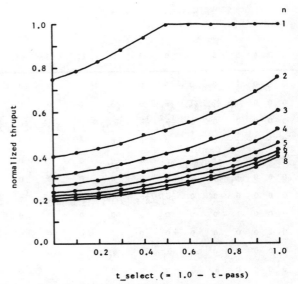

Fig. 11. Normalized throughput versus t_select (=$1.0 - t\_pass$). (Single buffer between stages.)

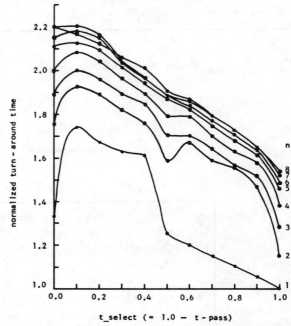

Fig. 12. Normalized turn-around-time versus t_select (=$1.0 - t\_pass$). (Single buffer between stages.)

IV. SIMULATION

This section outlines a simulator and presents simulation results.

A. A Simulator

We have developed a general simulator for multistage interconnection networks in a packet communication environment. The simulator can handle several switch types, interstage interconnection patterns, loading conditions, and switch operation policies. The performance measures used are those of average throughput and turn-around-time and these measures normalized with respect to their maximum possible values. Regenerative simulation is not being used currently because of the large number of distinguishable states making a regeneration point a rare event that is hard to detect. The point estimates reported are obtained by gathering statistics after

a specifiable settling time and averaging over a large number of packets (typically 5000 to 20 000) passing through the network.

B. Simulation Results

There are a large number of topologically distinguishable $(2^n \times 2^n)$ delta networks, for any constant n. As discussed earlier, the approximate analysis of Section III-C suggests that all these networks have the same performance. Simulation experiments were carried out to test this hypothesis. Several topologically different $(2^n \times 2^n)$ buffered delta networks 3 $\leq n \leq 6$ were selected and simulated. It was found that within simulation accuracy the performance of a $(2^n \times 2^n)$ delta network was indistinguishable from the performance of topologically different networks with the same value of n.

These experiments were limited to only a few selected values of the parameters listed below. Extensive simulation was carried out for omega networks [3] and these results are reported in this section.

The following parameters were varied in simulation runs:
1) the size of the network;
2) t_select and t_pass with $(t\_select + t\_pass) = 1$;
3) the number of buffers between stages.

The maximum throughput is that for an ideal network with no conflicts and is estimated as $max\_t = (2^n/(t\_select + t\_pass)$. The minimum delay is $min\_d = n \cdot (t\_select + t\_pass)$. The throughput and turn-around-time are normalized with respect to these bounds.

Figs. 9–12 show the variation of the normalized throughput and turn-around-time with the number of stages and with t_select and t_pass. Also shown are the estimates of Section III-C which compare quite well with the simulation estimates. We note that for large networks, the extreme cases of $t\_select = 0$ and $t\_pass = 0$ provide good bounds on the throughput and turn-around-times obtained by variation of these parameters.

The variation of normalized throughput and turn-around-time with buffer size and with the number of stages as parameter is shown in Figs. 13 and 14. The average throughput approaches the predicted asymptotes for large buffer sizes. However, the turn-around-time for large buffer lengths is very poor.

V. CONCLUSIONS

We have considered the problem of estimating the performance of delta networks with buffers between stages. These networks have the same asymptotic gate complexity as the buffered case, but they have throughputs comparable to crossbar switches which have a worse gate complexity. The turn-around-time with a single buffer between stages is less than three times the minimum possible delay at the maximum load for network sizes up to $(2^8 \times 2^8)$. As the number of buffers between stages is increased, the throughput converges to a constant while the turn-around-time increases almost linearly (Figs. 13 and 14). The rate of increase in throughput with the number of buffers is largest for a change from one to two buffers and this rate of increase falls rapidly as the buffer size increases. It thus seems that, for most applications, the number of buffers between stages should be limited to one or two.

Possible avenues for further work in this area include the estimation of the performance of these networks in different environments, consideration of networks with a differing number of inputs and outputs, and schemes for improving the reliability of these networks. Analysis and simulation indicate that the entire class of delta networks has approximately the same performance for the environment considered. However, this hypothesis has not been proven and it is an open question whether there are any delta networks that perform significantly better than other delta networks of the same size.

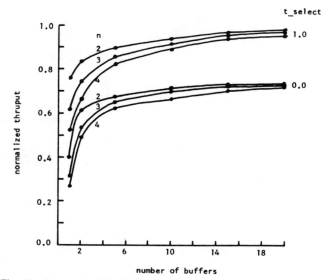

Fig. 13. Normalized throughput versus number of buffers. ($t\_select + t\_pass = 1.0$.)

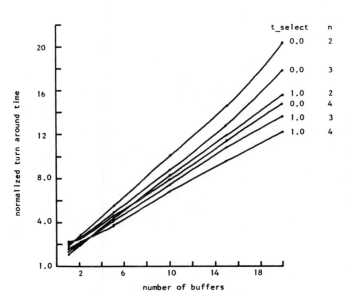

Fig. 14. Normalized turn-around-time versus number of buffers. ($t\_select + t\_pass = 1.0$.)

REFERENCES

[1] J. B. Dennis, "Packet communication architecture," in *Proc. IEEE 1975 Sagamore Comput. Conf. on Parallel Processing,* New York, NY, Aug. 1975, pp. 224–229.

[2] ——, "First version of a data flow procedure language," Mass. Inst. Technol., Cambridge, MA, MAC TM-61, Project MAC, May 1975.

[3] D. H. Lawrie, "Access and alignment of data in an array processor," *IEEE Trans. Comput.,* vol. C-24, pp. 1145–1155, Dec. 1975.

[4] J. H. Patel, "Processor-memory interconnections for multiprocessors," in *Proc. 6th Annu. Symp. on Comput. Arch.,* New York, N.Y, Apr. 1979, pp. 168–177.

[5] M. C. Pease, "The indirect binary *n*-cube microprocessor array," *IEEE Trans. Comput.,* vol. C-24, pp. 458–473, Dec. 1975.

[6] C. Ramachandani, "Analysis of asynchronous concurrent systems by Petri nets," Mass. Inst. Technol., Cambridge, MA, MAC-TR-120, Project MAC, Feb. 1974.

[7] H. J. Seigel and D. S. Smith, "Study of multistage SIMD interconnection networks," in *Proc. 5th Annu. Symp. Comput. Arch.*, New York, NY, Apr. 1978, pp. 223–229.

[8] H. S. Stone, "Parallel processing with the perfect shuffle," *IEEE Trans. Comput.*, vol. C-20, pp. 153–161, Feb. 1971.

Performance of Processor-Memory Interconnections for Multiprocessors

JANAK H. PATEL, MEMBER, IEEE

Abstract—A class of interconnection networks based on some existing permutation networks is described with applications to processor to memory communication in multiprocessing systems. These networks, termed delta networks, allow a direct link between any processor to any memory module. The delta networks and full crossbars are analyzed with respect to their effective bandwidth and cost. The analysis shows that delta networks have a far better performance per cost than crossbars in large multiprocessing systems.

Index Terms—Crossbar, interconnection networks, memory bandwidth, multiprocessor memories.

I. INTRODUCTION

WITH the advent of low cost microprocessors, the architectures involving multiple processors are becoming very attractive. Several organizations have been implemented or proposed. Principally among these are parallel (SIMD) type processors [1], computer networking, and multiprocessor organizations. In this paper we focus our attention on multiprocessors.

The principal characteristics of a multiprocessor system is the ability of each processor to share a single main memory. This sharing capability is provided through an interconnection network between the processor and the memory modules, which logically looks like Fig. 1. The function of the switch is to provide a logical link between any processor and any

memory module. There are several different physical forms available for the processor-memory switch; the least expensive of which is the time-shared bus. However, a time-shared bus has a very limited transfer rate, which is inadequate for even a small number of processors. At the other end of the bandwidth spectrum is the full crossbar switch, which is also the most expensive switch. In fact, considering the current low costs of microprocessors and memories, a crossbar would probably cost more than the rest of the system components combined. Therefore, it is very difficult to justify the use of a crossbar for large multiprocessing systems. It is the absence of a switch with reasonable cost and performance which has prevented the growth of large multimicroprocessor systems. To circumvent the high cost of switch, some "loosely coupled" systems can be defined. In these systems sharing of main memory is somewhat restricted, for example, some memory accesses may be fast and direct while many other references may be slow, indirect, and may even involve operating system intervention. There is considerable research on the permutation networks for parallel (SIMD) processors but very little research on processor-memory interconnections requiring random access capabilities.

In this paper we study a specific class of interconnection networks, termed delta networks, which are far less expensive than full crossbars. Moreover, delta networks are modular and easy to control. The delta networks form a subset of a very broad class of networks called banyan networks, initially defined in graph theoretic terms by Goke and Lipovski [2]. Informally, in terms of Fig. 1 the switch is a banyan network if and only if there exists a unique path from every processor to every memory module. The control of delta networks resembles the control of permutation networks of Lawrie [3] and Pease [4]. However, these networks were not applied or analyzed for random access of memories which invariably involves path conflicts in the network, as well as memory conflicts.

In Section II we present the basic principle involved in the design and control of delta networks. Then we describe the generalized delta networks with some examples. Following this we give the detailed logic of a 2×2 delta module which is the basic building block of $2^n \times 2^n$ delta networks. In Section V we analyze full crossbar networks for effective bandwidth and probability of conflicts, given the rate of memory requests from processors. The results of crossbar are used in turn to analyze arbitrary delta networks. Finally, we present quantitative comparison of crossbars and delta networks.

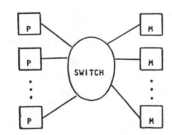

Fig. 1. Logical organization of a multiprocessor.

Manuscript received March 24, 1980; revised February 17, 1981. This work was supported in part by the Joint Services Electronics Program under Contract N00016-79-C-0424.

The author is with the Coordinated Science Laboratory and the Department of Electrical Engineering, University of Illinois, Urbana, IL 61801.

Reprinted from *IEEE Trans. Comput.*, vol. C-30, no. 10, pp. 771–780, October 1981.

II. PRINCIPLE OF OPERATION

Before we define delta networks, let us study the basic principle involved in the construction and control of delta networks. Consider a 2×2 crossbar switch (Fig. 2). This 2×2 switch has the capability of connecting the input A to either the output labeled 0 or output labeled 1, depending on the value of some control bit of the input A. If the control bit is 0, then the input is connected to the upper output and if 1, then it is connected to the lower output. The same description applies to terminal B, but for the time being ignore the existence of B. It is straightforward to construct a 1×2^n demultiplexer using the above described 2×2 module. This is done by making a binary tree of this module. For example, Fig. 3 shows a 1×8 demultiplexer tree. The destinations are marked in binary. If the source A requires to connect to destination $(d_2 d_1 d_0)_2$, then the root node is controlled by bit d_2, the second stage modules are controlled by bit d_1, and the last stage modules are controlled by bit d_0. It is clear that A can be connected to any one of the eight output terminals. It is also obvious that the lower input terminal of the root-node also can be switched to any one of the 8 outputs.

At this point we add another capability to the basic 2×2 module, the capability to arbitrate between conflicting requests. If both inputs require the same output terminal, then only one of them will be connected and the other will be blocked or rejected. The probability of blocking and the logic for arbitration is treated later on.

Now consider constructing an 8×8 network using 2×2 switches; the principle used is the same as that of Fig. 3. Every additional input must also have its own demultiplexer tree to connect to any one of the eight outputs. Basically, the construction works as follows. Start with a demultiplexer tree, then for each additional input superimpose a demultiplexer tree on the partially constructed network. One may use the already existing links as part of the new tree or add extra links and modules if needed. We have redrawn the tree of Fig. 3 as Fig. 4(a). The addition of next tree is shown with heavy lines in Fig. 4(b). This procedure is continued until the final 8×8 network of Fig. 4(d) results. The only restriction which must be strictly followed during this construction is that if a 2×2 module has its inputs coming from other modules, then both inputs must come from upper terminals of other modules or both must be lower terminals of other modules. (All upper output terminals are understood to have label 0 and the lower terminals, label 1.) Other than this there is considerable freedom in establishing links between the stages of the network. In the construction of the above 8×8 network we had the benefit of some hindsight that 12 modules are necessary and sufficient to build this network. If more modules are used then some inputs of some modules will remain unconnected. We could have stopped in the middle of the construction to obtain a 4×8 or a 6×8 network, such as Fig. 4(b) and (c), however in each case some inputs of the 2×2 modules will remain unutilized.

We term the networks constructed in the above manner *digit* controlled or simply *delta* networks, since each module is controlled by a single digit from the destination address. Furthermore, no external or global control is required. Digit controlled networks are not new; Lawrie's omega networks [3] and Pease's indirect binary n-cube [4] are subsets of delta networks.

Under the formal definition of delta networks presented in the next section, Fig. 4(d) is a delta network, but not Fig. 4(a), (b), or (c). Note that the network of Fig. 4(d) does not allow an identity permutation, that is, the connection 0 to 0, 1 to 1, $\cdots$, 7 to 7 at the same time. An identity permutation is useful if say memory module 0 is a "favorite" module of processor 0, and module 1 that of processor 1, and so on. Thus, identity permutation allows most of the memory references to be made without conflict. A simple renaming of the inputs of Fig. 4(d) will allow an identity permutation. This is shown in Fig. 5; in here if all 2×2 switches were in the straight ($=$) position, then an identity permutation is generated. As a matter of fact, since one and only one path is available from any source to any destination, every different setting of the form $\times$ and $=$ generates a different permutation. Thus, the network of Fig. 5 generates 2^{12} distinct permutations. This brings us to another somewhat unrelated topic of permutation networks. The procedure to construct a delta network can be used to generate different permutation networks. For example, we could have started with a tree in which the first stage was controlled by bit d_1, the second by d_0, and third by d_2. This would of course require relabeling the outputs. But does this really produce a "different" network? Siegel [5] has shown that by a simple address transformation the networks of Lawrie [3] and that of Pease [4] can be made equivalent, i.e., they produce the

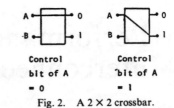

Fig. 2. A 2×2 crossbar.

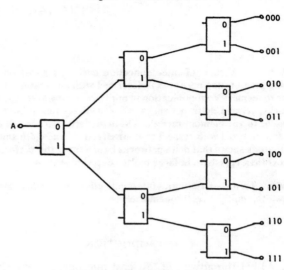

Fig. 3. 1×8 demultiplexer.

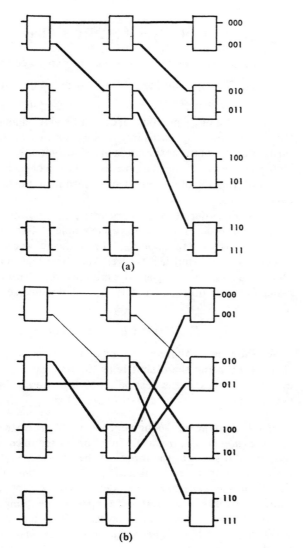

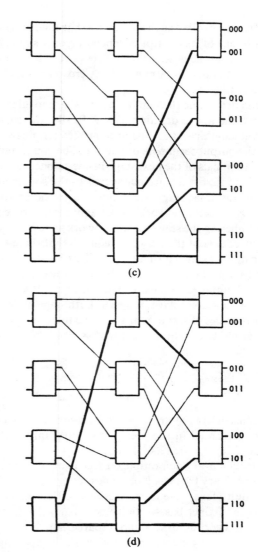

Fig. 4. Construction of an 8 × 8 delta network.

same set of permutations. As far as we know, the network of Fig. 5 cannot be made equivalent to either Lawrie's or Pease's network by a simple address transformation. It is quite possible that there are only two nonequivalent 8 × 8 delta networks, namely Lawrie's omega network and the network of Fig. 5. We shall not pursue this subject any further, as our primary interest lies in the random access capabilities of these networks and not permutations.

III. DESIGN AND DESCRIPTION OF DELTA NETWORKS

So far we have not defined the delta networks in a formal and rigorous manner. For the purpose of this paper we define them as follows.

Let an $a \times b$ crossbar module have the capability to connect any one of its a inputs to any one of the b outputs. Let the outputs be labeled $0, 1, \cdots, b - 1$. An input terminal is connected to the output labeled d if the control digit supplied by the input is d, where d is a base-b digit. Moreover, an $a \times b$ module also arbitrates between conflicting requests by accepting some and rejecting others.

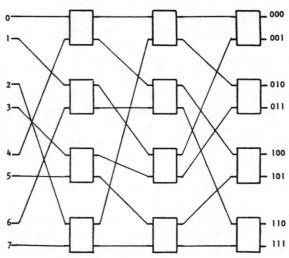

Fig. 5. An 8 × 8 delta network to allow identity permutation.

A *delta network* is an $a^n \times b^n$ switching network with n stages, consisting of $a \times b$ crossbar modules. The link pattern between stages is such that there exists a *unique* path of constant length from any source to any destination. Moreover, the

31

path is digit controlled such that a crossbar module connects an input to one of its b outputs depending on a single base-b digit taken from the destination address. Also, in a delta network no input or output terminal of any crossbar module is left unconnected.

One can determine the number of crossbar modules in each stage from the above definition. Specifically, the conditions of constant length paths and no unconnected terminals require that all the output terminals of one stage be connected to all the input terminals of the next stage in one-to-one fashion. The inputs of the first stage are connected to the source and the outputs of the final stage are connected to the destination. Thus, $a^n \times b^n$ delta network has a^n sources and b^n destinations. Numbering the stages of the network as $1, 2, \cdots$ starting at the source side of the network requires that there be a^{n-1} crossbar modules in the first stage. The first stage then has $a^{n-1}b$ output terminals. This implies that the stage two must have $a^{n-1}b$ input terminals, which requires $a^{n-2}b$ crossbar modules in the second stage. In general, ith stage has $a^{n-i}b^{i-1}$ crossbar modules of size $a \times b$. Thus, the total number of $a \times b$ crossbar modules required in an $a^n \times b^n$ delta network is

$$\sum_{1 \leq i \leq n} a^{n-i}b^{i-1} = \frac{a^n - b^n}{a - b} \qquad a \neq b$$

$$= nb^{n-1} \qquad a = b.$$

The construction of an $a^n \times b^n$ delta network follows the principle presented in the previous section. Informally, the procedure can be stated as follows.

Construct a b-ary demultiplexer tree using $a \times b$ crossbar switches. A b-ary tree has b branches for every node. For a $1 \times b^n$ demultiplexer, the tree has n levels. Each level is controlled by a distinct base-b digit taken from the destination address. For every additional input source, superimpose a new tree on the partially completed network. Each superimposition must satisfy the condition that an $a \times b$ module which receives inputs from other $a \times b$ modules must have all its inputs connected to identically labeled outputs, where the outputs of each $a \times b$ module are labeled $0, 1, \cdots, b - 1$, as was described earlier.

As one can see from the above construction procedure, there is a large number of link patterns available for an $a^n \times b^n$ delta network. It is natural to wonder if one topology is better than the others. We shall see later that, as far as probability of acceptance or blocking for random access is concerned, all delta networks are identical. However, different topologies may have different permutation capabilities in $b^n \times b^n$ delta networks.

Since the link pattern between stages of a delta network is of no particular concern to us, we may ask if there is some regular link pattern, which can be used between all stages and thus avoid the cumbersome construction procedure for every different delta network. There is indeed such a pattern which we describe below.

Let a q-shuffle of qr objects, denoted S_{q*r}, where q and r are some positive integers, be a permutation of qr indices $\langle 0, 1, 2, \cdots, (qr - 1) \rangle$, defined as

$$S_{q*r}(i) = \left(qi + \left\lfloor \frac{i}{r} \right\rfloor \right) \bmod qr \qquad 0 \leq i \leq qr - 1.$$

Alternately, the same function can be expressed as

$$S_{q*r}(i) = qi \bmod (qr - 1) \qquad 0 \leq i < qr - 1$$

$$= i \qquad i = qr - 1.$$

A q-shuffle of qr playing cards can be viewed as follows. Divide the deck of qr cards into q piles of r cards each; top r cards in the first pile, next r cards in the second pile, and so on. Now pick the cards, one at a time from the top of each pile; the first card from top of pile one, second card from the top of pile two, and so on in a circular fashion until all cards are picked up. This new order of cards represents a S_{q*r} permutation of the previous order. Fig. 6 shows an example of 4-shuffle of 12 indices, namely the function S_{4*3}. From the above description it is clear that a 2-shuffle is the well-known perfect shuffle [6]. One can also show that q-shuffle is an inverse permutation of r-shuffle of qr objects. That is, S_{q*r} is an inverse of S_{r*q}. Thus,

$$S_{q*r}(S_{r*q}(i)) = i \qquad 0 \leq i \leq qr - 1.$$

We show below that an $a^n \times b^n$ delta network can be constructed by using the a-shuffle as the link pattern between every two stages. If the destination D is expressed in base-b system as $(d_{n-1}d_{n-2} \cdots d_1 d_0)_b$, where $D = \sum_{0 \leq i < n} d_i b^i$ and $0 \leq d_i < b$, then the base-b digit d_i controls the crossbar modules of stage $(n - i)$. The a-shuffle function is used to connect the outputs of a stage to the inputs of the next stage where the inputs and outputs are numbered $0, 1, 2, \cdots$ starting at the top. Fig. 7 shows a general $a^n \times b^n$ delta network. Note that the shuffle networks between stages are passive, i.e., simply wires, and not active like the stages themselves. Two delta networks, one $4^2 \times 3^2$ and $2^3 \times 2^3$, derived from Fig. 7 are shown in Figs. 8 and 9; the interstage patterns are, respectively, 4-shuffle and 2-shuffle. The destinations are labeled in base 3 in Fig. 8 and in base 2 in Fig. 9. Now we prove that the a-shuffle link pattern indeed allows a source to connect to any destination using the destination-digit control of each $a \times b$ crossbar module.

Theorem: An $a^n \times b^n$ delta network which uses a-shuffles as interstage link pattern, can connect any source to any destination $D = (d_{n-1}d_{n-2} \cdots d_1 d_0)_b$ by switching the source to the output terminal d_{n-i} at each stage i, where the outputs of each $a \times b$ crossbar are assumed to have labels $0, 1, \cdots, (b - 1)$ from top to bottom.

Proof: Let the source S need a connection to destination D, where

$$S = (s_{n-1}s_{n-2} \cdots s_1 s_0)_a \text{ in base-}a \text{ system}$$

and

$$D = (d_{n-1}d_{n-2} \cdots d_1 d_0)_b \text{ in base-}b \text{ system}.$$

In other words

$$S = s_{n-1}a^{n-1} + s_{n-2}a^{n-2} + \cdots + s_1 a + s_0$$
$$\text{and } 0 \leq s_i \leq a - 1$$

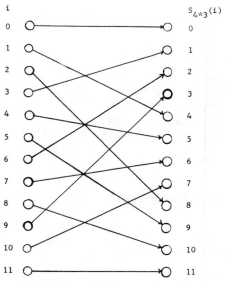

Fig. 6.   4-shuffle of 12 objects.

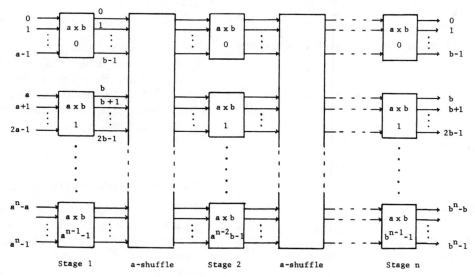

Fig. 7.   An $a^n \times b^n$ delta network.

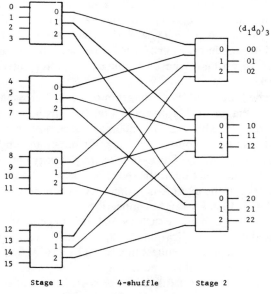

Fig. 8.   A $4^2 \times 3^2$ delta network.

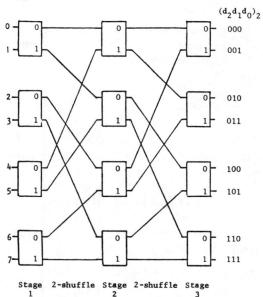

Fig. 9.   A $2^3 \times 2^3$ delta network.

33

$$D = d_{n-1}b^{n-1} + d_{n-2}b^{n-2} + \cdots + d_1 b + d_0$$
$$\text{and } 0 \le d_i \le b - 1.$$

If the $a \times b$ modules are numbered $0, 1, 2, \cdots$ from top to bottom (see Fig. 7), then S is connected to module number $\lfloor S/a \rfloor = s_{n-1}a^{n-2} + \cdots + s_2 a + s_1$. By the destination-digit control algorithm, the source S is switched to the output terminal d_{n-1} of the module number $\lfloor S/a \rfloor$. Assuming all output lines of stage 1 are numbered $0, 1, 2, \cdots$ from top, the source S is switched by stage 1 to its output line number $L_1 = \lfloor S/a \rfloor b + d_{n-1}$, that is

$$L_1 = (s_{n-1}a^{n-2} + \cdots + s_2 a + s_1)b + d_{n-1}.$$

This line, when the a-shuffle of $a^{n-1}b$ output lines of stage 1 is done, becomes

$$L_1' = \left(L_1 a + \left\lfloor \frac{L_1}{a^{n-2}b} \right\rfloor \right) \bmod a^{n-1}b.$$

Let us first evaluate the floor function.

$$\frac{L_1}{a^{n-2}b} = s_{n-1} + \frac{s_{n-2}a^{n-3} + \cdots + s_1 + d_{n-1}/b}{a^{n-2}}.$$

Since $d_{n-1}/b < 1$ the numerator of the second term is less than a^{n-2}. Therefore

$$\left\lfloor \frac{L_1}{a^{n-2}b} \right\rfloor = \lfloor s_{n-1} + \text{fraction} \rfloor = s_{n-1}.$$

Therefore

$$
\begin{aligned}
L_1' &= (L_1 a + s_{n-1}) \bmod a^{n-1}b \\
&= s_{n-1}a^{n-1}b + (s_{n-2}a^{n-2} + \cdots + s_1 a)b \\
&\quad + d_{n-1}a + s_{n-1} \bmod a^{n-1}b \\
&= \left(s_{n-2}a^{n-2} + \cdots + s_1 a + \frac{d_{n-1}a + s_{n-1}}{b} \right)b \bmod a^{n-1}b
\end{aligned}
$$

since $d_{n-1} < b$ and $s_{n-1} < a$ the quantity $(d_{n-1}a + s_{n-1})/b$ is less than a. Thus, the expression in the parentheses is less than a^{n-1}. Therefore

$$L_1' = (s_{n-2}a^{n-2} + \cdots + s_1 a)b + d_{n-1}a + s_{n-1}$$

which is the input to stage 2 of the network. This line is connected to module $\lfloor L_1'/a \rfloor$ of stage 2 and is switched according to digit d_{n-2} and becomes the output line number L_2 of stage 2, where

$$
\begin{aligned}
L_2 &= \left\lfloor \frac{L_1'}{a} \right\rfloor b + d_{n-2} \\
&= (s_{n-2}a^{n-3} + \cdots + s_2 a + s_1)b^2 + d_{n-1}b + d_{n-2}.
\end{aligned}
$$

After the a-shuffle of $a^{n-2}b^2$ lines, the line number L_2 becomes

$$L_2' = \left(L_2 a + \left\lfloor \frac{L_2}{a^{n-3}b^2} \right\rfloor \right) \bmod a^{n-2}b^2.$$

After simplifying the right-hand side as before, the input to stage 3 is line number

$$
\begin{aligned}
L_2' = (s_{n-3}a^{n-3} + \cdots + s_1 a)b^2 \\
+ (d_{n-1}b + d_{n-2})a + s_{n-2}.
\end{aligned}
$$

This line when switched by stage 3 becomes the output of stage 3 at line number

$$
\begin{aligned}
L_3 &= \left\lfloor \frac{L_2'}{a} \right\rfloor b + d_{n-3} \\
&= (s_{n-3}a^{n-4} + \cdots + s_2 a + s_1)b^3 \\
&\quad + (d_{n-1}b^2 + d_{n-2}b + d_{n-3}).
\end{aligned}
$$

By finite induction it can be shown that the source is connected to an output terminal of stage i at line number

$$
\begin{aligned}
L_i = (s_{n-i}a^{n-i-1} + \cdots + s_2 a + s_1)b^i \\
+ (d_{n-1}b^{i-1} + d_{n-2}b^{i-2} + \cdots + d_{n-i})
\end{aligned}
$$

which after the a-shuffle of $a^{n-i}b^i$ lines becomes the input of stage $i + 1$ at line number

$$
\begin{aligned}
L_i' = (s_{n-i-1}a^{n-i-1} + \cdots + s_2 a^2 + s_1 a)b^i \\
+ (d_{n-1}b^{i-1} + \cdots + d_{n-i})a + s_{n-i}.
\end{aligned}
$$

And after the final stage n, the source is connected to the output line number

$$L_n = d_{n-1}b^{n-1} + \cdots + d_1 b + d_0$$

which is the desired destination D. $\qquad\Box$

When $a = b$, the above proof can be simplified because the b-shuffle of b^n lines has a simple form. Take, for example, any integer i, $0 \le i < b^n$ in its base-b representation. Let $i = (c_{n-1}c_{n-2}\cdots c_1 c_0)_b$, then the b-shuffle changes it to $(c_{n-2}c_{n-3} \cdots c_1 c_0 c_{n-1})_b$, which is simply a left rotation of the digits of i. This property was used by Lawrie [3] in proving a statement similar to the theorem above, when $a = b = 2$.

IV. IMPLEMENTATION OF DELTA NETWORKS

Within the current technological limitations it is uneconomical to encode base b digits, where b is not a power of 2. Thus, in practice an $a \times b$ crossbar module for a delta network is more cost-effective if b is a power of 2, since our modules require base b digits for control. Again, due to the cost and technological limitations modules of size 8×8 or greater are not very practical at this time. This leaves $a \times 2$ and $a \times 4$ modules, where a is 1–4, as the most likely candidates for implementation of delta networks. Here we give the functional and logical description of 2×2 modules. This in turn will be used to estimate the cost and delay factors of delta networks.

The functional block diagram of a 2×2 crossbar module of a delta network appears in Fig. 10. All single lines in the figure are one bit lines. The double lines on INFO box represent address lines, incoming and outgoing data lines, and a Read/Write control line. The data lines may or may not be bidirectional. The function of the INFO box is that of a simple 2×2 crossbar; if the input X is 1, then a cross connection exists and if X is 0, then a straight connection exists.

The function of the CONTROL box is to generate the signal X and provide arbitration. A request exists at an input port if the corresponding request line is 1. The destination digit provides the nature of the request; a 0 for the connection to upper output port and a 1 for the lower port. In case of conflict the

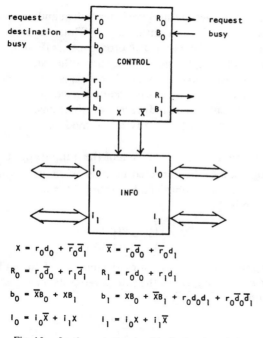

$$X = r_0 d_0 + \overline{r}_0 \overline{d}_1 \qquad \overline{X} = r_0 \overline{d}_0 + \overline{r}_0 d_1$$

$$R_0 = r_0 \overline{d}_0 + r_1 \overline{d}_1 \qquad R_1 = r_0 d_0 + r_1 d_1$$

$$b_0 = \overline{X} B_0 + X B_1 \qquad b_1 = X B_0 + \overline{X} B_1 + r_0 d_0 d_1 + r_0 \overline{d}_0 \overline{d}_1$$

$$I_0 = i_0 \overline{X} + i_1 X \qquad I_1 = i_0 X + i_1 \overline{X}$$

Fig. 10. Implementation details of a 2 × 2 module.

request r_0 is given the priority and a busy signal $b_1 = 1$ is supplied to the lower input port. A busy signal is eventually transmitted to the source which originated the blocked request.

The priority among processors can be randomized by connecting the two outputs of each 2 × 2 module to two different priority input ports at the next stage. For example, output port 0 can be connected to a high priority input port and the port 1 to a low priority input of the next stage. The logic equations for all the labeled signals are given with the block diagram. For INFO box the equations are given for left to right direction. The parallel generation of X and $\overline{X}$ reduces one gate level. Signal X and $\overline{X}$ are valid after 3 gate delays. Assuming that one level of buffer gates for X and $\overline{X}$ in the INFO box exists due to fanout limitations of X and $\overline{X}$, the total delay to establish the connections of INFO box is 6 gate delays, of which 4 gate delays are due to X and $\overline{X}$. Thus, after the initial setup time, the data transfer requires only 2 gate delays per stage of the network.

The operation of a $2^n \times 2^n$ delta network using the above described 2 × 2 modules is as follows; recall that there are n stages in this network.

All processors requiring memory access must submit their requests at the same time by placing a 1 on the respective request lines. After $8n$ gate delays the busy signals are valid. If the busy line is 1, then the processor must resubmit its request. This can be accomplished simply by doing nothing, i.e., continue to hold the request line high. The Read data is valid after $8n$ gate delays plus the memory access time if the busy signal is 0. Thus, the operation of the implementation described here is synchronous, that is, the requests are issued at fixed intervals at the same time. An asynchronous implementation is preferable if the network has many stages. However, such an implementation would require storage buffers for addresses, data, and control in every module and also a complex control mod-

ule. Thus, the cost of such an implementation might well be excessive. We have analyzed only the synchronous networks in this paper.

V. ANALYSIS OF CROSSBARS AND DELTA NETWORKS

In this section first we analyze $M \times N$ crossbar networks and then delta networks. Both networks are analyzed under identical assumptions for the purpose of comparison. We analyze the networks for finding the expected bandwidth given the rate of memory requests. Bandwidth is expressed in number of memory requests accepted per cycle. A cycle is defined to be the time for a request to propagate through the logic of the network plus the time to access a memory word plus the time to return through the network to the source. We shall not distinguish the read or write cycles in this analysis. The analysis is based on the following assumptions.

1) Each processor generates random and independent requests; the requests are uniformly distributed over all memory modules.

2) At the beginning of every cycle each processor generates a new request with a probability m. Thus, m is also the average number of requests generated per cycle by each processor.

3) The requests which are blocked (not accepted) are ignored; that is, the requests issued at the next cycle are independent of the requests blocked.

The last assumption is there to simplify the analysis. In practice, of course, the rejected requests must be resubmitted during the next cycle; thus the independent request assumption will not hold. However, to assume otherwise would make the analysis if not impossible, certainly very difficult. Moreover, simulation studies done by us and by others and more complex analyses reported [7]–[10] for similar problems have shown that the probability of acceptance is only slightly different if the third assumption above is omitted. Thus, the results of the analysis are fairly reliable and they provide a good measure for comparing different networks.

Analysis of Crossbars: Assume a crossbar of size $M \times N$, that is, M processors (sources) and N memory modules (destinations). In a full crossbar two requests are in conflict if and only if the requests are to the same memory module. Therefore, in essence we are analyzing memory conflicts rather than network conflicts. Recall that m is the probability that a processor generates a request during a cycle. Let $q(i)$ be the probability that i requests arrive during one cycle. Then

$$q(i) = \binom{M}{i} m^i (1 - m)^{M-i} \qquad (1)$$

where $\binom{M}{i}$ is the binomial coefficient.

Let $E(i)$ be the expected number of requests accepted by the $M \times N$ crossbar during a cycle; given that i requests arrived in the cycle. To evaluate $E(i)$, consider the number of ways that i random requests can map to N distinct memory modules, which is N^i. Suppose now that a particular memory module is not requested. Then the number of ways to map i requests to the remaining $(N-1)$ modules is $(N-1)^i$. Thus, $N^i - (N-1)^i$ is the number of maps in which a particular module is always requested. Thus, the probability that a particular module is requested is $[N^i - (N-1)^i]/N^i$. For every memory module, if it is requested, it means one request is accepted by the network for that module. Therefore, the expected number of acceptances, given i requests, is

$$E(i) = \frac{N^i - (N-1)^i}{N^i} \cdot N$$

$$= \left[1 - \left(\frac{N-1}{N}\right)^i\right] N. \tag{2}$$

Thus, the expected bandwidth BW, that is, requests accepted per cycle is

$$\text{BW} = \sum_{0 \le i \le M} E(i) \cdot q(i)$$

which simplifies to

$$\text{BW} = N - N\left(1 - \frac{m}{N}\right)^M. \tag{3}$$

Let us define the ratio of expected bandwidth to the expected number of requests generated per cycle as the probability of acceptance P_A, that is, the probability that an arbitrary request will be accepted. Thus, P_A is a measure of the wait time of blocked requests. A higher P_A indicates a lower wait time and a lower P_A indicates higher wait time. The expected wait time of a request is $(1/P_A - 1)$.

$$P_A = \frac{\text{BW}}{mM} = \frac{N}{mM} - \frac{N}{mM}\left(1 - \frac{m}{N}\right)^M. \tag{4}$$

It is interesting to note the limiting values of BW and P_A as M and N grow very large. Let $k = M/N$, then

$$\lim_{N \to \infty} \left(1 - \frac{m}{N}\right)^{kN} = e^{-mk}.$$

Thus, for very large values of M and N

$$\text{BW} \simeq N(1 - e^{-mM/N}) \tag{5}$$

$$P_A \simeq \frac{N}{mM}(1 - e^{-mM/N}). \tag{6}$$

The above approximations are good within 1 percent of actual value when M and N are greater than 30 and within 5 percent when $M, N \ge 8$. Note that for a fixed ratio M/N the bandwidth of (5) increases linearly with N.

Analysis of Delta Networks: Assume a delta network of size $a^n \times b^n$ constructed from $a \times b$ crossbar modules. Thus, there are a^n processors connected to b^n memory modules. We apply the result of (3) for an $M \times N$ crossbar to an $a \times b$ crossbar and then extend the analysis for the complete delta network. However, to apply (3) to any $a \times b$ crossbar module we must first satisfy the assumptions of the analysis. We show below that the independent request assumption holds for every $a \times b$ module in a delta network.

Each stage of the delta network is controlled by a distinct destination digit (in base b) for setting of individual $a \times b$

switches. Since the destinations are independent and uniformly distributed, so are the destination digits. Thus, for example, in some arbitrary stage i an $a \times b$ crossbar uses digit d_{n-i} of each request; this digit is not used by any other stage in the network. Moreover, no digit other than d_{n-i} is used by stage i. Thus, the requests at any $a \times b$ module are independent and uniformly distributed over b different destinations. Thus, we can apply the result of (3) to any $a \times b$ module in the delta network.

Given the request rate m at each of the a inputs of an $a \times b$ crossbar module, the expected number of requests that it passes per time unit is obtained by setting $M = a$ and $N = b$ in (3), which is

$$b - b\left(1 - \frac{m}{b}\right)^a.$$

Dividing the above expression by the number of output lines of the $a \times b$ module gives us the rate of requests on any one of b output lines

$$1 - \left(1 - \frac{m}{b}\right)^a.$$

Thus, for any stage of a delta network the output rate of requests m_{out} is a function of its input rate and is given by

$$m_{\text{out}} = 1 - \left(1 - \frac{m_{\text{in}}}{b}\right)^a.$$

Since the output rate of a stage is the input rate of the next stage, one can recursively evaluate the output rate of any stage starting at stage 1. In particular, the output rate of the final stage n determines the bandwidth of a delta network, that is, the number of requests accepted per cycle.

Let us define m_i to be the rate of requests on an output line of stage i. Then the following equations determine the bandwidth, BW of an $a^n \times b^n$ delta network, given m, the rate of requests generated by each processor.

$$\text{BW} = b^n m_n \tag{7}$$

where

$$m_i = 1 - \left(1 - \frac{m_{i-1}}{b}\right)^a \text{ and } m_0 = m$$

and the probability that a request will be accepted is

$$P_A = \frac{b^n m_n}{a^n m}. \tag{8}$$

VI. Effectiveness of Delta and Crossbar Networks

Since we do not have a closed form solution for the bandwidth of delta networks (7), we cannot directly compare the bandwidths of crossbar (3) and delta networks. We have computed the values of the expected bandwidth and the probability of acceptance for several networks. These results are plotted in Figs. 11, 12, and 13.

Fig. 11 shows the probability of acceptance P_A, for $2^n \times 2^n$

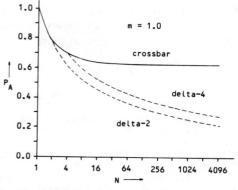

Fig. 11. Probability of acceptance of $N \times N$ networks.

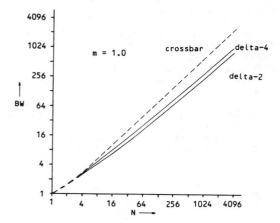

Fig. 12. Expected bandwidth of $N \times N$ networks.

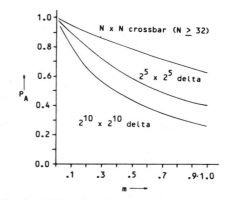

Fig. 13. Probability of acceptance versus mean request rate.

and $4^n \times 4^n$ delta networks and $N \times N$ crossbar, when the request generation rate of each processor is $m = 1$. The curve marked delta-2 is for delta networks using 2×2 switches and delta-4 for delta networks using 4×4 switches. The graphs are drawn as smooth curves in this and other figures only for visual convenience, in actuality the values are valid only at specific discrete points. In particular, an $N \times N$ crossbar is defined for all integers $N \geq 1$, a delta-2 is defined only for $N = 2^n$, $n \geq 1$, and delta-4 is defined only for $N = 4^n$, $n \geq 1$.

Notice in Fig. 11 that P_A for crossbar approaches a constant value as was predicted by (6) of the previous section. P_A for delta networks continues to fall as N grows. Fig. 12 shows the expected bandwidth BW as a function of N. The bandwidth is measured in number of requests accepted per cycle. In all

fairness we must point out that a cycle for a crossbar could be smaller than a cycle for a large delta network. Taking into account fan-in and fan-out constraints, the decoder and arbiter for a $N \times N$ crossbar has a delay of $0(\log_2 N)$ gate delays. A $2^n \times 2^n$ delta network also has $0(\log_2 N)$ gate delays, from the analysis of Section IV. However, the delay of a delta network is approximately twice the delay of a crossbar. If the delay is small compared to the memory access time, then the cycle time (the sum of network delay and memory access time) of a crossbar is not too different from that of a delta network. Thus, the curves for bandwidth provide a good comparison between networks.

Fig. 13 shows P_A as a function of the request generation rate m. The curve for the crossbar is the limiting value of P_A as N grows to infinity. Curves for $N \geq 32$ are not distinguishable with the scale used in that graph.

Finally, the graph of Fig. 14 is an indication of cost-effectiveness of delta networks. The cost of a $N \times N$ crossbar or delta network is assumed to be proportional to the number of gates required. The constant of proportionality should be the same in both cases because the degree of integration, modularity, and wiring complexity in both cases is more or less the same. For the $N \times N$ crossbar the minimum number of gates required is one per crosspoint per data line. Depending on the assumptions used on fan-in, fan-out, the complexity of the decoder and the arbiter, one can estimate the gate complexity of a crossbar anywhere from one gate to six gates per crosspoint. Let us assume the lowest cost figure of one gate per crosspoint.

The cost of $2^n \times 2^n$ delta network is estimated from the Boolean equations of the 2×2 module of Fig. 10. The number of gates in a 2×2 module is 23 gates for the control plus 6 gates per information line. Assuming the number of information lines to be large, the gates for control can be ignored. Thus, the gate count of a $2^n \times 2^n$ delta network is $6n2^{n-1}$ per information line because the network has $n2^{n-1}$ modules.

Thus, the cost of $N \times N$ networks are kN^2 for crossbar and $3kN\log_2 N$ for delta ($N = 2^n$), where k is the constant of proportionality. We have used these cost expressions in the computation of performance-cost ratio for Fig. 14; the ratio is that of expected bandwidth over cost. Taking this ratio for a $1 \times$

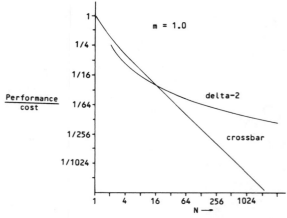

Fig. 14. Cost-effectiveness of $N \times N$ networks.

1 crossbar as unity, the Y-axis of Fig. 14 represents the performance over cost relative to a 1×1 crossbar. The Y-axis may also be interpreted as bandwidth per gate per information line. Notice that delta network is more cost-effective for network size N greater than 16. If the cost of the crossbar was assumed as 2 or more gates per crosspoint, then the curve for the crossbar would shift downward and the effectiveness of delta becomes even more pronounced. However, if one assumed the cycle time of a crossbar half as much as that of a delta network, then the curve for crossbar would shift upwards relative to the curve for delta, thus shifting the crossover point of the two curves towards right. Thus, depending on the assumptions, the crossover point may move slightly left or right; but in any case the curves clearly show the effectiveness of delta networks for medium and large scale multiprocessors.

VIII. CONCLUDING REMARKS

We have presented in this paper a class of processor-memory interconnection networks, called delta networks, which are easy to control and design, and are very cost-effective. We also presented the combinatorial analysis of delta networks and full crossbars. It is seen that delta networks bridge the gap between a single time-shared bus and a full crossbar. The cost of an $N \times N$ delta network varies as $N\log_2 N$, while that of crossbar varies as N^2. Thus, delta networks are very suitable for relatively low cost multimicroprocessor systems.

REFERENCES

[1] M. J. Flynn, "Very high speed computing systems," *Proc. IEEE*, vol. 54, pp. 1901–1909, Dec. 1966.
[2] L. R. Goke and G. J. Lipovski, "Banyan networks for partitioning multiprocessor systems," in *Proc. 1st Annu. Symp. Comput. Arch.*, Dec. 1973, pp. 21–28.
[3] D. H. Lawrie, "Access and alignment of data in an array processor," *IEEE Trans. Comput.*, vol. C-24, pp. 1145–1155, Dec. 1975.
[4] M. C. Pease, "The indirect binary *n*-cube microprocessor array," *IEEE Trans. Comput.*, vol. C-26, pp. 458–473, May 1977.
[5] H. J. Siegel, "Study of multistage SIMD interconnection networks," in *Proc. 5th Annu. Symp. Comput. Arch.*, Apr. 1978, pp. 223–229.
[6] H. S. Stone, "Parallel processing with the perfect shuffle," *IEEE Trans. Comput.*, vol. C-20, pp. 153–161, Feb. 1971.
[7] D. Y. Chang, D. J. Kuck, and D. H. Lawrie, "On the effective bandwidth of parallel memories," *IEEE Trans. Comput.*, vol. C-26, pp. 480–490, May 1977.
[8] W. D. Strecker, "Analysis of the instruction execution rate in certain computer structures," Ph.D. dissertation, Carnegie-Mellon Univ., Pittsburgh, PA, 1970.
[9] F. Baskett and A. J. Smith, "Interference in multiprocessor computer systems with interleaved memory," *Commun. Ass. Comput. Mach.*, vol. 19, pp. 327–334, June 1976.
[10] D. P. Bhandarkar, "Analysis of memory interference in multiprocessors," *IEEE Trans. Comput.*, vol. C-24, pp. 897–908, Sept. 1975.

The Performance of Multistage Interconnection Networks for Multiprocessors

CLYDE P. KRUSKAL, MEMBER, IEEE, AND MARC SNIR

Abstract—This paper studies the performance of unbuffered and buffered, packet-switching, multistage interconnection networks. We begin by reviewing the definition of banyan networks and introducing some generalizations of them. We then present an asymptotic analysis of the performance of unbuffered banyan networks, thereby solving a problem left open by Patel. We analyze the performance of the unbuffered generalized banyan networks, and compare networks with approximately equivalent hardware complexity. Finally, we analyze the performance of buffered banyan networks and again compare networks with approximately equivalent hardware complexity.

Index Terms—Bandwidth, banyan network, bidelta network, buffered network, circuit-switching network, crossbar network, delta network, dilated network, multistage interconnection network, packet-switching network, performance analysis, replicated network, simulation, square network, throughput, unbuffered network, uniform network.

I. INTRODUCTION

WITHIN the last decade interest has increased in large scale multiprocessors composed of thousands of processors sharing a common memory, and several such systems have been proposed [5], [11]. A typical configuration for such a system is illustrated in Fig. 1: many identical processors are connected via an interconnection network to identical memory modules. The interconnection network is an essential component, so it is important to have a solid understanding of its performance. In some proposed designs it supports dynamic access from each processor to each memory module, and the traffic through the network consists of short items (requests to memory and replies), with requests being dynamically generated independently at each processor. The pattern of requests is essentially random and varies rapidly.

In this paper we study the performance of unbuffered and buffered, packet-switching, multistage interconnection networks. Section II reviews the definitions necessary for reading this paper, including specifically the definition of banyan

Manuscript received May 21, 1982; revised October 29, 1982, March 2, 1983, and May 25, 1983. This work was supported in part by the Applied Mathematical Sciences Program of the U.S. Department of Energy under Contract DE-AC02-76ER03077, in part by the National Science Foundation under Grants NSF-MCS79-21258 and NSF-MCS81-05896, and in part by a grant from IBM. A preliminary version of this material appeared in the *Proceedings of the 1982 Conference on Information Sciences and Systems*, Princeton University, March 1982.

C. P. Kruskal is with the Department of Computer Science, University of Illinois, Urbana, IL 61801.

M. Snir was with Courant Institute, New York University, New York, NY 10012. He is now with the Department of Computer Science, Hebrew University of Jerusalem, Jerusalem, Israel.

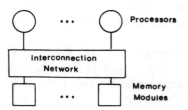

Fig. 1.   Multiprocessor organization.

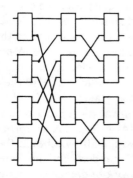

Fig. 2.   3-stage square banyan network of degree 2.

networks and some generalizations of them. Sections III and V analyze the performance of banyan networks. Delta networks [9], Omega networks [8], indirect binary cube networks [11]—among others (see [7])—are all banyan networks, so our analyses apply to these networks as well. Patel [10] presents a recurrence relation for the performance of unbuffered networks, but leaves open the question of how they perform asymptotically. Section III provides such an analysis; the analysis is also applicable to circuit-switching networks. Section IV analyzes the performance of unbuffered generalized banyan networks, and compares networks with approximately equivalent hardware complexity. Section V analyzes the performance of buffered banyan networks, and as in the unbuffered case compares networks with approximately equivalent hardware complexity. Section VI summarizes the paper and contains some concluding remarks.

II. NETWORKS

We consider packet-switching networks built of switches connected by unidirectional lines. A p input, q output ($p \times q$) switch can receive packets at each of its p input ports, and send them through each of its q output ports. A *network* is a directed graph where nodes are of the following three types:

Reprinted from *IEEE Trans. Comput.*, vol. C-32, no. 12, pp. 1091–1098, December 1983.

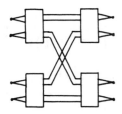

Fig. 3.   2-dilation of banyan network with 2 stages and degree 2.

i) *Source nodes* which have indegree 0;

ii) *Sink nodes* which have outdegree 0;

iii) *Switches* which have positive indegree and outdegree.

Each edge represents one of more lines going from a node to successor. Throughout this paper we assume that for a given network every edge represents the same number of lines. The edges in two different networks, however, may represent different numbers of lines.[1]

A *banyan network* is defined by Goke and Lipovsky [4] to be a network with a unique path from each source node to each sink node. This condition implies that the set of paths leading to a node in the network forms a tree and that the set of paths leading from a node also forms a tree. A *multistage network* is a network in which the nodes can be arranged in stages, with all the source nodes at stage 0, and all the outputs at stage i connected to inputs at stage $i + 1$. If all the sink nodes of a multistage network are at stage $n + 1$ then we have an *n-stage network* (called in [4] L-level). A *uniform network* is a multistage network in which all switches at the same stage have the same number of input ports and the same number of output ports. A *square network of degree k* is network built of $k \times k$ switches. Fig. 2 shows a 3-stage square banyan network of degree 2.

We shall analyze the performance of *n*-stage square banyan networks under the assumptions usually used in the literature [10]: packets are generated at each source node by independent, identically distributed random processes. Each processor generates with probability p at each cycle a packet, and sends a generated packet with equal probability to any sink node. We

---

[1] This definition allows us to study the performance of unidirectional interconnection networks. However, in actual parallel computers of the form described in Fig. 1, it is necessary to send replies back through the network. There are at least four ways to reinterpret or modify our definition to allow this. Assume that the parallel processor has P processors and M memory modules.

i) Each edge really represents a bidirectional line. There are P source nodes and M sink nodes; the orientation of the edges distinguishes processors from memory modules.

ii) Each edge really represents two unidirectional lines—one in each direction. Again, there are P source nodes and M sink nodes, and the orientation of the edges distinguishes processors from memory modules.

iii) There is one network for sending messages from processors to memory modules and a different network for sending replies back. In one network each of the P source nodes represents a processor, and each of the M sink nodes represents a memory module; in the other network each of the P sink node represents a processor and each of the M source node a memory module.

iv) The same (unidirectional) network is used for sending messages and replies. There are $P + M$ source nodes and $P + M$ sink nodes. Each source node is identified with a distinct sink node, and the pair represents either a processor or a memory module.

In each of these cases it is not hard to at least approximate the behavior the full network given our analyses below of unidirectional networks.

---

assume that the network is synchronous, so that packets can be sent only at time $t_c, 2t_c, 3t_c, \cdots$, where t_c is the *network cycle time*.

The uniqueness of paths in banyan networks implies the following result which is implicitly used in all the performance analyses of these networks.

Lemma: Let packets be generated at the source nodes of a banyan network by independent, identically distributed random processes, that uniformly distribute the packets over all of the sink nodes. Assume that the routing logic at each switch is "fair," i.e., conflicts are randomly resolved. Then

i) The patterns of packet arrivals at the inputs of the same switch are independent.

ii) Packets arriving at an input of a switch are uniformly distributed over the outputs of that switch.

Moreover, if the network is uniform, then for each stage in the network, the pattern of packet arrivals at the inputs of that stage have the same distribution

While banyan networks are very attractive in their simplicity, other considerations such as performance or reliability sometimes dictate the use of more complicated networks. Two strategies can be used to augment a network G, without sacrificing much of its structure.

i) The *d-dilation* of G is defined to be the network obtained from G by replacing each edge by d distinct edges (see Fig. 3). A packet entering a switch may exit using any of the d edges going to the desired successor switch at the next stage.

ii) The *d-replication* of G is defined to be the network consisting of d identical distinct copies of G, with the d corresponding source (sink) nodes in each copy identified (see Fig. 4).

In [12] *d*-dilated networks are introduced without name. Replicated banyan networks are called layered banyan networks in [4].

III. UNBUFFERED BANYAN NETWORKS

We first consider packet-switching networks built of $k \times k$ unbuffered switches with the topology of an *n*-stage square banyan network. When several packets at the same switch require the same output, a randomly chosen one is forwarded and the remaining packets are deleted. The relevant figure of merit for such networks is the probability p_m that there is some packet on any particular input at the mth stage of the network. By the above lemma, it is easily seen that p_m is well defined and satisfies the recurrence relation

$$p_{m+1} = 1 - (1 - p_m/k)^k \qquad (1)$$

with boundary condition $p_0 = p$, where p is the probability of packet creation at a source node. Patel [9] leaves open the question of how p_m behaves asymptotically. It turns out that for any fixed initial value $p_0 > 0$, and any fixed k

$$p_m = \frac{2k}{(k-1)m}\left[1 - \frac{(k+1)}{3(k-1)}\frac{\ln m}{m} + O\left(\frac{1}{m}\right)\right], \qquad (2)$$

where $\ln x$ denotes $\log_e x$. A proof of this result is given in Appendix A. It is interesting to note that the first and second order terms in this expansion are independent of p_0. Thus, the probability that a message is not deleted is asymptotically inversely proportional to the number of stages in the network.

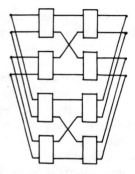

Fig. 4. 2-replication of banyan network with 2 stages and degree 2.

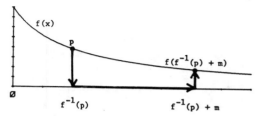

Fig. 5.   Schema for finding asymptotic equation approximating p_m. The arrows show the mapping from the point $(f^{-1}(p), p)$ to the approximation of p_m. The left axis is nominally a probability, but since f only approximates p_m the range of the graph can contain values greater than 1.

In particular, in a square banyan network with N source nodes built of 2×2 switches the *bandwidth* (or *throughput*) of the network, i.e., the average number of packets arriving at the other end of the network per cycle, is

$$N \cdot p_{\lg N} = \frac{4N}{\lg N}\left[1 - \frac{\ln \lg N}{\lg N} + O\left(\frac{1}{\lg N}\right)\right],$$

where $\lg x$ denotes $\log_2 x$.

We can use (2) to obtain for any given $p = p_0$ an approximate formula for p_m. Although we do this only for $k = 2$, the technique is valid in general. For $k = 2$ the formula for p_m reduces to

$$\frac{4}{m}\left[1 - \frac{\ln m}{m} + O\left(\frac{1}{m}\right)\right].$$

Assuming the last term does not oscillate (which can be shown formally by generating the next term in the asymptotic expansion), p_m can be approximated by

$$\frac{4}{m}\left[1 - \frac{\ln m}{m} + \frac{c}{m}\right],$$

where c is some constant to be determined. We can determine a value for c by taking a "large" value for m, finding p_m by brute force using difference equation (1) m times, and solving for c. For example, consider $p = 1$. For $m = 10$, $p_0 = 1$, $p_1 = 0.75$, $p_2 \simeq 0.61, \cdots, p_{10} \simeq 0.26$, so $c \simeq -1.2$. Thus, for m large, p_m is closely approximated by

$$\frac{4}{m}\left[1 - \frac{\ln m}{m} - \frac{1.2}{m}\right]. \tag{3}$$

Unfortunately, for each initial value p we need a different constant c. Moreover, for small p the approximation will not be consistently good until m is very large, no matter what value we choose for c. Using an alternative approach we can derive an asymptotic formula which has p as a parameter.

Let p_m satisfy the recurrence relation with initial value $p_0 = p$, and let $f(x)$ (defined on the real numbers) approximate p_m for some initial value greater than or equal to p. Then f, starting at the point where it has value p, approximates p_m; that point is by definition $(f^{-1}(p), p)$. Thus, $f(f^{-1}(p) + m)$ approximates p_m (see Fig. 5). For our f we use only the first term $4/m$ of the above asymptotic formula for two reasons: first of all, we can invert it in closed form; and second, its range contains the entire half open interval $(0, 1]$. Thus $f^{-1}(p) =$

$\dfrac{4}{p}$, and the approximation for p_m is $f(f^{-1}(p) + m) = f\left(\dfrac{4}{p} + m\right)$

$$= \frac{4}{m + \dfrac{4}{p}}. \tag{4}$$

This is an excellent approximation of p_m for all p and m, except when both p is large and m is small where it is still very good. (See Fig. 6.) In general, for square networks composed of $k \times k$ switches p_m is approximated by

$$\frac{2k}{(k-1)m + \dfrac{2k}{p}}.$$

The above analysis also applies to a circuit-switching network with the topology of a square banyan. We assume that each source node attempts simultaneously to establish a communication path with a randomly chosen sink node (each mapping is equally likely). Then p_m is the probability that a communication is not blocked after the first m stages of the network, and $N \cdot p_{\lg N}$ is the average number of communication paths that will be established in a network with N source nodes built of 2×2 switches. Our asymptotic analysis indicates that the probability that a path will be established is asymptotically inversely proportional to the number of stages in the network.

IV. Unbuffered Dilated and Replicated Networks

We now analyze the performance of d-dilated square banyan networks under the following two assumptions: 1) Every source node issues one message at each cycle. 2) If $m < d$ packets are competing for the d edges leading from a switch to its successor switch at the next stage, all of the packets are forwarded; if $m > d$ packets are competing at a switch for the d edges, then d of them are chosen at random and forwarded, and the remaining ones are deleted. We give the following recurrence for q_m, the probability that the d edges will contain some message after the first m stages of a d-dilated square banyan network of degree 2.

Let $R(m, j)$ be the probability that j packets are transmitted through d identical edges leaving a switch at stage m. R is initialized as

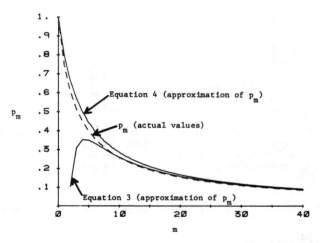

Fig. 6.   Graph of p_m, for $p_0 = 1$, and two asymptotic equations for finding approximate values.

$$R(0, j) = 0, \text{ for } j \neq 1,$$
$$R(0, j) = 1, \text{ for } j = 1.$$

The probability of having i messages entering a switch at stage $m + 1$ is given by $\sum_{r+s=i} R(m, r) \cdot R(m, s)$. The probability that j of these i messages are directed to a fixed output port is $\binom{i}{j} 2^{-i}$. Thus, $R(m + 1, j)$ is given by

$$R(m + 1, j) = \sum_{i=j}^{2d} \left(\sum_{r+s=i} R(m, r) \cdot R(m, s) \right) \binom{i}{j} 2^{-i},$$
$$\text{for } j < d,$$

and

$$R(m + 1, j) = \sum_{i=d}^{2d} \left(\sum_{r+s=i} R(m, r) \cdot R(m, s) \right) \sum_{t=d}^{i} \binom{i}{t} 2^{-i},$$
$$\text{for } j = d.$$

Finally, the probability that at least one of the d edges contains a message is

$$q_m = 1 - R(m, 0).$$

The performance of d-replicated square banyan networks can be easily approximated. A d-replicated network consists of d copies of a square banyan networks, and the results of Section III apply to each of these copies. We assume that every source issues one message at each cycle and randomly sends it to one of the d copies. On each copy the probability p_m that a message survives m stages of the network satisfies the recurrence from equation (1). Assuming the d copies are independent, which is slightly optimistic, the probability q_m that the ith edge of one of the d copies has a surviving request is $1 - (1 - p_m)^d$. Our numerical results assume this model of d-replication. Note, however, that if k divides d, a d-replicated square banyan network of degree k could be organized so that there will be no conflicts until after $\log_k d$ stages of switches. The analysis of each copy is equivalent to assuming each banyan network has $\log_k N - \log_k d$ stages, where the prob-

ability is still $1/d$ of issuing a request at each input port. Now the d copies are independent, so this organization is easy to analyze; it performs better than even the above optimistic approximation assumes.

A d-dilation of a square banyan network of degree 2 with N source nodes has $\lg N$ stages, $(N \lg N)/2$ $2d \times 2d$ switches, and $dN(\lg N + 1)$ edges. A $d(\lg d + 1)$-replication of a square banyan network of degree $2d$ has $\log_{2d} N$ stages, $(N \lg N)/2$ $2d \times 2d$ switches, and $dN(\lg N + \lg d + 1)$ edges. Thus two such networks have the same number and size of switches, and have roughly the same number of edges. Figs. 7(a)–(c) compares the performance—as measured by q_m—of dilated banyan networks with replicated banyan networks of comparable complexity for switches of size 4×4, 8×8, and 16×16. We have also included the crossbar network for comparison purposes. (In a crossbar network of size N the probability that a message is not blocked is $1 - (1 - 1/N)^N$, which asymptotically approaches $1 - 1/e$.) As we see dilated banyan networks have asymptotically better performance than replicated banyan networks, although for practical values of N their performances are approximately the same.

We have also calculated the bandwidth of these networks. To do so we assume that a message is issued at each input edge at each cycle. For d-dilated networks the recurrence is the same as above except for different boundary conditions and a different formula for calculating q_m. In this case

$$R(O, j) = 0, \text{ for } j \neq d,$$
$$R(O, j) = 1, \text{ for } j = d,$$

and

$$q_m = \frac{1}{d} \sum j \cdot R(m, j).$$

For d-replicated networks the bandwidth is exactly d times the bandwidth of a single network. Once again dilated networks have asymptotically better performance: This occurs, however, only for impractically large values of N: for $N < 2^{60}$ replicated networks built of 4×4, 8×8, or 16×16 switches are better than comparable dilated networks.

V. Buffered Banyan Networks

The bandwidth of packet-switching networks can be improved by using buffers to queue conflicting packets. An accurate analysis of the performance of buffered square banyan networks does not seem tractable. Several authors have analyzed the performance with buffers of length one and performed simulations for larger buffers (see [3] and references therein, and more recently [1]). We present here a formula that seems to yield a good approximation for the performance with large buffers and also indicate some tradeoffs suggested by that formula.

Consider a buffered $k \times k$ switch. An ideal switch consists of k infinite queues, one associated with each output port, where each queue can accept at each cycle up to k distinct packets coming from distinct input ports. In practice a switch can be built in the following way. To prevent blocking each such queue is actually implemented by k FIFO buffers, one

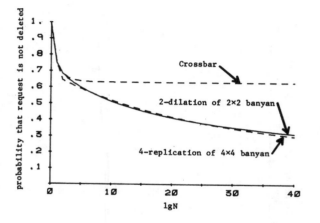

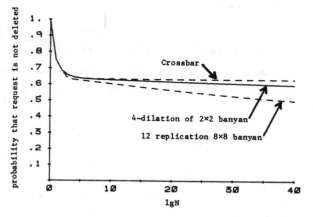

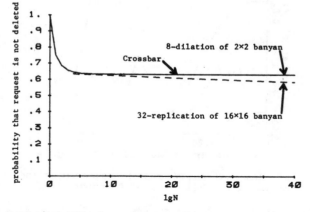

Fig. 7. (a) Probability of a request being satisfied in a network with N source nodes using a crossbar, 2-dilation of 2×2 network, and 4-replication of 4×4 network. (b) Probability of a request being satisfied in a network with N source nodes using a crossbar, 4-dilation of 2×2 network, and 12-replication of 8×8 network. (c) Probability of a request being satisfied in a network with N source nodes using a crossbar, 8-dilation of 2×2 network, and 32-replication of 16×16 network.

at each input port. Incoming packets are therefore directly entered into a buffer associated with the destination output port, with no conflicts arising. An arbitration mechanism is used at each output port to remove packets from the k associated buffers in the order of their arrivals (see [13] for an implementation of such switch). (In the terminology of Dias and Jump [3] we have $t\_$pass $= 0$.) Assuming that the buffers are infinite, each switch gives the same performance as an ideal switch.

We now analyze the performance of networks composed of ideal switches. Let t_c be the *cycle time* of the switch, i.e., the interval between successive packet arrivals; let t_τ be the *transit time* of a packet from one switch to the next one when the buffers on its path are empty ($t_\tau \leq t_c$). In general, the transit time of a packet P through a switch is $t_\tau + bt_c$, where b is the number of packets with the same destination that arrived before P, or arrived at the same time as P, but were transmitted before P (we assume that the order of transmission of packets arriving at the same cycle is random). If at each cycle a packet arrives on each input with probability p, then the average number of queueing cycles for a packet is

$$b_{av} = \frac{(1 - 1/k)p}{2(1 - p)}.$$

A proof of this formula is given in Appendix B. (Chen [2] has independently derived this formula for the case $k = 2$.) Thus, the average transit time of a packet through a $k \times k$ switch is

$$t_k = t_\tau + t_c \frac{(1 - 1/k)p}{2(1 - p)}.$$

This suggests a formula of the form

$$T = \log_k N \cdot t_k$$

$$= \log_k N \cdot \left(t_\tau + t_c \frac{(1 - 1/k)p}{2(1 - p)} \right)$$

for the average transit time through the $\log_k N$ stages of a square banyan network of degree k with N source nodes.

Simulations were run on six stages of a network built of 2×2 switches. The assumption of infinite buffers was dropped: each switch was given a buffer of size eight at each output port. Table I compares the average number of queueing cycles predicted by this analysis with values obtained by the simulation. (The "packets per cycle" row of the simulation section shows the average number of packets actually generated per cycle; only the last column differed from the expected number.)

As we see, the predicted delays are in good agreement with the simulations. Besides statistical error, there are two reasons for the discrepancies between the two. First, buffers have only finite size, and a full switch will not accept a new packet. However, the close agreement between the number of transmissions predicted and simulated indicates this does not occur frequently, so that limited buffer size does not seem to be a significant factor for the loads considered. Second, the distribution of the packet arrivals at the second and the subsequent stages is not time independent anymore. A clustering effect occurs, which tends to increase the average delays. Indeed, delays can be seen to increase from the first stage to the successive stages. It is interesting to note that for each p after the second stage there is no discernible difference between the delay times. This is probably because the distribution of packet arrivals has by that time pretty much settled down to its limiting distribution.

Sometimes each packet will not represent a full message, but a message will be composed of several packets, say m. In this case each message is said to be *time multiplexed* by a

TABLE I

	probability of transmission	0.2	0.4	0.6	0.8
analysis	packets per cycle	0.2	0.4	0.6	0.8
	waiting per stage	0.063	0.167	0.375	1.265
simulation	packets per cycle	0.200	0.400	0.600	0.795
	waiting time — 1st stage	0.066	0.167	0.367	1.082
	2nd stage	0.065	0.175	0.434	1.275
	3rd stage	0.069	0.201	0.457	1.328
	4th stage	0.069	0.195	0.456	1.316
	5th stage	0.070	0.202	0.431	1.298
	6th stage	0.066	0.196	0.459	1.289

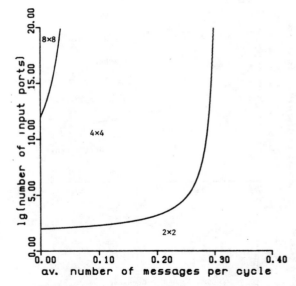

Fig. 8. Domain of best performance for a square banyan network of degree 2^k, $k = 1, 2, \cdots$.

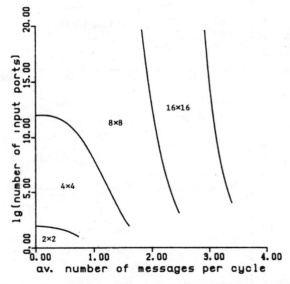

Fig. 9. Domain of best performance for replicated square banyan networks of degree 2^k, $k = 1, 2, \cdots$, of comparable complexity.

factor m. If the cycle time for to process a single packet is t_c and the transit time is t_τ, the cycle time to process a message that is time multiplexed by a factor m will be $T_c = mt_c$, whereas the transit time will be unchanged. The average number of packets per cycle will now be $P = mp$. Using the previous formula, one obtains that the average transit time through a network (built of $k \times k$ switches in which messages are time multiplexed by a factor m) is approximately

$$T = \log_k N \cdot \left(t_\tau + T_c \frac{(1 - 1/k)P}{2(1 - P)} \right) + (T_c - t_c)$$

$$= \log_k N \cdot \left(t_\tau + t_c \frac{m^2(1 - 1/k)p}{2(1 - mp)} \right) + (m - 1)t_c.$$

The last term accounts for the pipe setting delay.

Furthermore, we can consider d-replicated square banyan networks. Recall that a d-replicated banyan network consists of d copies of a banyan network. When a processor issues a message it sends the message to one of the d copies with equal probability. Thus, the average number of packets per cycle on each copy is now p/d. So, ignoring the overhead in connecting the d copies together, the average transit time through a d-replicated square banyan network is approximately

$$T = \log_k N \cdot \left(t_\tau + t_c \frac{m^2(1 - 1/k)p}{2(d - mp)} \right) + (m - 1)t_c.$$

It is possible to build square banyan networks with different performances by increasing the number of ports on each switch while proportionally decreasing the bandwidth of each port (i.e., the number of lines per edge). For VLSI chips if the computation time is small enough relative to the I/O requirements then the chip is said to be pin limited (i.e., the time to compute the desired function mainly depends on how many pins are allocated to its I/O, and is inversely proportional to this number). If each switch is implemented on one chip and the chip performance is pin limited, then without changing the transit time the number of logical lines per switch can be increased by a factor of l while increasing the multiplexing factor of each message by the by the same factor l. Assuming that a 2×2 switch can process each message as one packet, a $k \times k$ switch will require $k/2$ packets per message. Thus the average transit time through a $k \times k$ square banyan network will be approximately

$$T = \log_k N \cdot \left(t_\tau + t_c \frac{k(k - 1)p}{8(1 - kp/2)} \right) + \left(\frac{k}{2} - 1 \right)t_c.$$

We have used this formula to compare the bandwidth of networks built of $k \times k$ switches, $k = 2, 4, 8, \cdots$, assuming that $t_c = t_\tau$. Fig. 8 indicates the domain where each type of switch yields the best performance. Larger switches perform better for low traffic intensities and large networks.

Note, however, that a network built of 2×2 switches has the same number of switches as a $(k \lg k)/2$-replication of a network built of $k \times k$ switches. Furthermore, assuming each edge of a 2×2 switch is given $k/2$ times as many lines per edge as a $k \times k$ switch, the different networks will have the same number of lines per switch, and networks with identical numbers of switches will have roughly the same total number of lines. So two such networks will have comparable complexity. Comparing the performance of networks of comparable complexity gives another picture, which is illustrated in Fig. 9. Networks with larger switches are capable of supporting higher traffic intensities. In particular, four networks built of 4×4 switches have the same number of switches, and always outperform one network built of 2×2 switches.

VI. Summary and Conclusion

We have studied the performance of unbuffered and buffered multistage interconnection networks. For unbuffered banyan networks we presented an asymptotic equation for the probability that a request issued at a source node arrives at its sink node. From the analysis we saw that asymptotically this probability is inversely proportional to the number of stages in the network. Furthermore, using the analysis we derived very simple equations which closely approximate the actual probabilities.

We then analyzed the performance of unbuffered dilated networks and unbuffered replicated networks, and compared networks of comparable hardware complexity. Although dilated networks provide asymptotically better performance, for practical numbers of processors dilated and replicated networks have similar performance. Thus, other considerations such as delay, fault tolerance, and layout would likely decide which is preferable for a given situation.

Finally, we derived an equation for the performance of buffered banyan networks. While the analysis is valid only for the first stage, we have presented evidence that it is a reasonable approximation in general. Using the analysis, we compared buffered banyan networks built of different sized switches but comparable hardware complexity, and determined where each switch size is most effective.

All of our analyses were done in the framework of the banyan network since it is the most general class for which the analyses apply. In practice, however, for ease of communication between sources and sinks one would use delta networks [9] where the paths from the source nodes to any specific sink node have the same descriptor, or more likely bidelta networks [7] where also the paths from the sink nodes to any specific source node have the same descriptor.

Appendix A

We derive an asymptotic estimate for the sequence defined by the recurrence $p_{n+1} = 1 - (1 - p_n/k)^k$, where $0 < p_0 \le 1$.

It is easy to check that the sequence p_n is monotonically decreasing to zero. Let $r_n = 1/p_n$. The sequence r_n is monotonically increasing to infinity, and fulfills the recurrence

$$\frac{1}{r_{n+1}} = 1 - \left(1 - \frac{1}{kr_n}\right)^k$$

$$= \frac{1}{r_n} - \frac{k-1}{2kr_n^2} + \frac{(k-1)(k-2)}{6k^2 r_n^3} + O\left(\frac{1}{r_n^4}\right).$$

It follows that

$$r_{n+1} = r_n + \frac{k-1}{2k} + \frac{k^2-1}{12k^2 r_n} + O\left(\frac{1}{r_n^2}\right). \tag{A.1}$$

Thus, for n large enough,

$$r_{n+1} > r_n + \frac{k-1}{2k}.$$

Summing for n we obtain the inequality

$$r_n \ge \frac{k-1}{2k} \cdot n + O(1).$$

Substituting back into (A.1) we obtain the inequality

$$r_{n+1} \le r_n + \frac{k-1}{2k} + \frac{k^2-1}{12k^2\left(\dfrac{k-1}{2k} \cdot n + O(1)\right)} + O\left(\frac{1}{n^2}\right)$$

$$= r_n + \frac{k-1}{2k} + \frac{k+1}{6kn} + O\left(\frac{1}{n^2}\right).$$

Summing again we obtain

$$r_n \le \frac{k-1}{2k} \cdot n + \frac{k+1}{6k} \cdot \ln(n) + O(1). \tag{A.2}$$

Substituting again into (A.1) we obtain

$$r_{n+1} \ge r_n + \frac{k-1}{2k} + \frac{k^2-1}{12k^2\left(\dfrac{k-1}{2k} \cdot n + \dfrac{k+1}{6k} \cdot \ln(n) + O(1)\right)} + O\left(\frac{1}{n^2}\right)$$

$$= r_n + \frac{k-1}{2k} + \frac{k+1}{6kn} + O\left(\frac{\ln(n)}{n^2}\right).$$

Summing a final time we obtain

$$r_n \ge \frac{k-1}{2k} \cdot n + \frac{k+1}{6k} \cdot \ln(n) + O(1). \tag{A.3}$$

It follows from inequalities (A.2) and (A.3) that

$$r_n = \frac{k-1}{2k} \cdot n + \frac{k+1}{6k} \cdot \ln(n) + O(1),$$

so that

$$p_n = \frac{1}{r_n}$$

$$= \frac{1}{\dfrac{k-1}{2k} \cdot n + \dfrac{k+1}{6k} \cdot \ln(n) + O(1)}$$

$$= \frac{2k}{(k-1)n} \cdot \left(1 - \frac{k+1}{3(k-1)} \cdot \frac{\ln(n)}{n} + O\left(\frac{1}{n}\right)\right).$$

Appendix B

We derive the average waiting time in a $k \times k$ switch with infinite buffers. We assume that a queue of unbounded capacity is associated with each output port, and that at each cycle each such queue can accept up to k packets coming through k distinct input ports. At each cycle a packet arrives at each input port with a fixed probability p, and each packet is equally likely to join each output queue. We take the cycle time t_c to be equal to 1.

Let v_n be the number of packets joining a fixed output queue at cycle n. Then, $v_1, v_2, \cdots$, are independent random variables with a $b(\cdot; k; p/k)$ Bernoulli distribution. The expected number of arrivals is $E = k \cdot (p/k) = p$ and the variance is $V = k \cdot (p/k) \cdot (1 - p/k) = p(1 - p/k)$. Let q_n be the number of packets in the queue at the end of cycle n. We have for q_n the recurrence relation

$$q_{n+1} = q_n + v_{n+1} - 1 \quad \text{if } q_n > 0,$$
$$q_{n+1} = v_{n+1} \quad \text{if } q_n = 0.$$

Note that this recurrence is formally identical to that describing the number of customers at departure times in a $M/G/1$ queueing system [6, eq. (5.33)]. We can therefore use the derivation that leads to the Pollaczek-Khinchine mean-value formula to obtain the expected number of packets in the queue to be

$$\bar{q} = \frac{E}{2} + \frac{V}{2(1 - E)}.$$

Using Little's identity we get that the average system time for a packet is

$$\bar{s} = \frac{\bar{q}}{E} = \frac{1}{2} + \frac{V}{2E(1 - E)},$$

and the average waiting time is

$$\bar{w} = \bar{s} - 1 = \frac{V}{2E(1 - E)} - \frac{1}{2}.$$

Substituting for E and V we get that the average number of queueing cycles at a $k \times k$ switch is equal to

$$\bar{w} = \frac{p(1 - p/k)}{2p(1 - p)} - \frac{1}{2}$$
$$= \frac{(1 - 1/k)p}{2(1 - p)}.$$

ACKNOWLEDGMENT

We would like to thank the referees for carefully reading earlier versions of this paper. Their suggestions improved both the technical contents and presentation.

REFERENCES

[1] S. Cheemalavagu and M. Malek, "Analysis and simulation of banyan interconnection networks with 2 × 2, 4 × 4, and 8 × 8 switching elements," in *Proc. Real-Time Syst. Symp.*, Dec. 1982, pp. 83-89.
[2] P.-Y. Chen, "Multiprocessor systems: interconnection networks, memory hierarchy, modeling, and simulations," Ph.D. dissertation, Univ. Illinois, Urbana, IL, Jan. 1982.
[3] D. M. Dias and J. R. Jump, "Packet switching interconnection networks for modular systems," *Computer*, vol. 14, pp. 43-54, Dec. 1981.
[4] G. R. Goke and G. J. Lipovski, "Banyan networks for partitioning multiprocessor systems," in *Proc. 1st Annu. Symp. Comput. Arch.*, 1973, pp. 21-28.
[5] A. Gottlieb, R. Grishman, C. P. Kruskal, K. P. McAuliffe, L. Rudolph, and M. Snir, "The NYU Ultracomputer-Designing an MIMD, shared-memory parallel machine," *IEEE Trans. Comput.*, vol. C-32, pp. 175-189, 1983.
[6] L. Kleinrock, *Queueing Systems, Vol 1: Theory.* New York: Wiley, 1975.
[7] C. P. Kruskal and M. Snir, "The structure of multistage interconnection networks for multiprocessors," manuscript; see also, "Some results on multistage interconnection networks for multiprocessors," NYU Ultracomputer Note 41, in *Proc. 1982 Conf. Informat. Sci., Syst.*, Princeton Univ., Princeton, NJ, Mar. 1982.
[8] D. H. Lawrie, "Access and alignment of data in an array processor," *IEEE Trans. Comput.*, vol. C-24, pp. 1145-1155, 1975.
[9] J. A. Patel, "Processor-memory interconnections for multiprocessors," in *Proc. 6th Annu. Symp. Comput. Arch.*, pp. 168-177, Apr. 1979.
[10] ——, "Performance of processor-memory interconnections for multiprocessors," *IEEE Trans. Comput.*, vol. C-30, pp. 771-780, 1981.
[11] M. C. Pease, "The indirect binary n-cube microprocessor array," *IEEE Trans. Comput.*, vol. C-26, pp. 458-473, 1977.
[12] J. T. Schwartz, "The Burroughs FMP machine," Ultracomputer Note 5, Courant Institute, NYU, New York, NY, 1980.
[13] M. Snir and J. Solworth, "The Ultraswitch—A VLSI network node for parallel processing," Ultracomputer Note 39, Courant Institute, NYU, New York, NY, 1982.

Performance Analysis of a Packet Switch Based on Single-Buffered Banyan Network

YIH-CHYUN JENQ, MEMBER, IEEE

Abstract —Banyan networks are being proposed for interconnecting memory and processor modules in multiprocessor systems as well as for packet switching in communication networks. This paper describes an analysis of the performance of a packet switch based on a single-buffered Banyan network. A model of a single-buffered Banyan network provides results on the throughput, delay, and internal blocking. Results of this model are combined with models of the buffer controller (finite and infinite buffers). It is shown that for balanced loads, the switching delay is low for loads below maximum throughput (about 45 percent per input link) and the blocking at the input buffer controller is low for reasonable buffer sizes.

I. Introduction

RECENTLY, there has been considerable interest in Banyan networks among computer and communication technology researchers. The Banyan network has been considered as a candidate for the interconnecting network linking a large number of memory and processor modules together in a multiprocessor system and as a switching fabric for packet-switched interconnection of computers (see, e.g., [1]-[3]).

In this paper, we present analytical results for the performance of a packet switch based on a single-buffered Banyan network. Results show reasonably low blocking probabilities and low delays for balanced internal loads. In Section II we describe the switch and its operation. In Section III we present a simple model of the switch and obtain the throughput and the internal delay of the switch. In Section IV we refine the model by removing one independence assumption used in the simple model of Section III and show that the results from two models are very close. In Section V we add a queue with unlimited waiting rooms as a front-end buffer to the switch and obtain the switching delay. The case of a finite waiting rooms front-end buffer is analyzed in Section VI. Conclusions are given in Section VII.

II. The Packet Switch

The packet switch consists of k input buffer controllers, k output buffer controllers, and a $k \times k$ square switch. A Banyan network is considered for the switching fabric

Manuscript received February 4, 1983; revised June 22, 1983.
The author is with Bell Laboratories, Holmdel, NJ 07733.

because of its economy, potential VLSI implementation, and easiness of routing, A 4-stage (16×16) Banyan network is shown in Fig. 1. Each switching element (box) is a 2×2 crossbar switch with one buffer on each of its two input links. The packet routing is done by the hardware as follows. Each packet has an n-bit header in an n-stage switch. The switching element (a 2×2 crossbar) at stage 1 routes the packet up or down according to the first bit of the header ("zero" or "one" indicate up or down routing, respectively) and then removes the first bit from the header. The succeeding switching elements will perform the same routing function by removing one bit from the header and routing the packet to the next stage until the packet reaches its destination output link. It is easy to see that the header is simply the binary address of the output link at the last stage and is independent of the input link at the first stage.

If both buffers at a switching element have a packet and both packets are going to the same output link of the switching element, then a conflict occurs. In this case, we assume that one of the packets will be chosen, randomly, to go to the next stage and the other one will stay in the buffer. Of course, in order for a packet to be able to move forward, either the buffer at the next stage is empty or there is a packet in the buffer and that packet is able to move forward. Thus, the ability for a packet to move forward depends on the state of the entire portion of the network succeeding the current stage.

For simplicity, we can consider the switch operating in a clocked format. In the first portion τ_1 of a clock period $\tau = \tau_1 + \tau_2$, control signals are passed across the network from the last stage toward the first stage, so that every packet knows whether it should move forward one stage or stay in the same buffer. Then, in the second portion τ_2 of a clock period τ, packets move in accordance with control signals and the clock period ends. The whole process repeats in every clock period, and packets continue to be shifted in and out of the switch.

III. Model 1

We consider a Banyan switch with n stages. Considering the clocked operation described in the previous section, it

Reprinted from *IEEE J. Selected Areas Commun.*, vol. SAC-1, no. 6, pp. 1014–1021, December 1983.

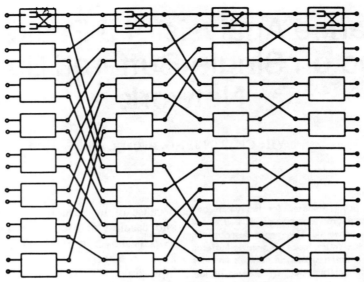

Fig. 1. A 4-stage Banyan network.

is clear that the whole switch can be modeled as a Markov chain. However, the number of the states in the chain grows exponentially with the number of stages n; the number of states in an n-stage network is $k^{n \cdot 2^n}$, where k is the number of states in a node. With such a rapidly growing number, it is almost impossible to even simulate it for large n (say $n > 10$). Therefore, some simplifying assumptions will be used.

We assume that packets arriving at each input link at stage 1 are destined uniformly (or randomly) for all output links at stage n. We also assume a uniform load for each input link at stage 1. Thus, we assume that the loads are balanced in the whole switching network. Under these assumptions, the state of a switching element (2×2 crossbar) at stage k is statistically not distinguishable from that of another switching element of the same stage. Hence, the state of a "stage" can be characterized by that of a "switching element." If we further assume that the two buffers in the same switching element are statistically independent, then the state of a "stage" can be reduced further to that of a single buffer. This last independence assumption is motivated by the fact that input packets arriving at each of the two input links of a switching element are from disjoint sets of input links at stage 1 and, hence, are independent. However, packets in the two buffers of a switching element do interfere with each other, and thus it is not clear that this independence assumption is reasonable. In this section we will assume independence to obtain a very simple model. In the next section we will remove this independence assumption to check if the removal of the independence assumption makes any noticeable difference.

We first introduce the following notation.

n = number of switching stages.

$p_0(k, t)$ = probability that a buffer of a switching element at stage k is empty at the beginning of the tth clock period.

$p_1(k, t) = 1 - p_0(k, t)$.

$q(k, t)$ = probability that a packet is ready to come to a buffer of a switching element at stage k during the tth clock period.

$r(k, t)$ = probability that a packet in a buffer of a switching element at stage k is able to move forward during the tth clock period, given that there is a packet in that buffer.

With these definitions, we can state the equations governing the relationships among these variables:

$$q(k, t) = 0.75 p_1(k-1, t) p_1(k-1, t)$$
$$+ 0.5 p_0(k-1, t) p_1(k-1, t)$$
$$+ 0.5 p_1(k-1, t) p_0(k-1, t),$$
$$k = 2, 3, \cdots, n \quad (1)$$

$$r(k, t) = [p_0(k, t) + 0.75 p_1(k, t)][p_0(k+1, t)$$
$$+ p_1(k+1, t) r(k+1, t)],$$
$$k = 1, 2, 3, \cdots, n-1 \quad (2)$$

$$r(n, t) = p_0(n, t) + 0.75 p_1(n, t)$$

$$p_0(k, t+1) = [1 - q(k, t)][p_0(k, t) + p_1(k, t) r(k, t)]$$
$$(3)$$

$$p_1(k, t+1) = 1 - p_0(k, t+1). \quad (4)$$

Equations (1)–(4) describe the dynamics of the state transition of the system. If this system has a steady state, then these quantities should converge to time-independent quantities $q(k)$, $r(k)$, $p_0(k)$, and $p_1(k)$. Then the normalized throughput S (the number of output packets per output link per clock period) and the normalized mean delay d are given by

$$S = p_1(k) r(k) \qquad \text{for any } k \quad (5)$$

and

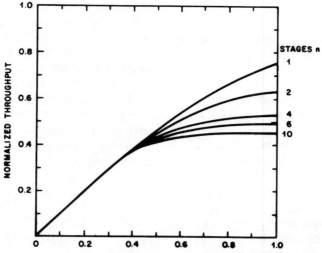

Fig. 2.   Normalized throughput of Banyan networks.

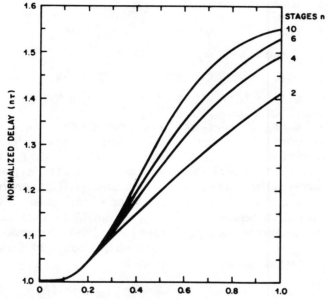

Fig. 3.   Normalized internal delay of Banyan networks.

$$d = \frac{1}{n} \sum_{k=1}^{n} 1/r(k). \qquad (6)$$

The quantity $q(1)$ is determined by the load applied to the switch. In this section it is considered as an independent input parameter. The determination of $q(1)$ for a given input load is deferred to Section V.

In Fig. 2 we plot the normalized throughput S as a function of $q(1)$ for different number of stages n. This was obtained by iteratively solving for the steady state (1)–(5). There are two interesting observations. First, the throughput grows almost linearly with respect to the offered load in the range where the offered load is less than 0.4. This agrees with the intuition that under low load, all input traffic will pass through the switch with little interference. Second, with the maximum load at 1, the normalized throughput seems to converge (monotonically decreasingly) to a quantity of approximately 0.45 as the total number of

stages n in a switch increases to 10. Normalized delay d [(6)] is displayed in Fig. 3. It is interesting to see that the normalized delay also seems to converge to a quantity approximately 1.55 as n increases. A computer run is performed for $n = 20$ to check this convergence, and the result shows that the normalized delay stands at 1.55. Thus, for an n stage switch the average delay is in the range between $n\tau$ and $1.55n\tau$, depending on the load and the number of stages n.

IV.   MODEL 2

In the previous section we assumed that the state of a buffer in a switching element is independent of that of the other buffer in the same switching element. In this section we present a refined model by removing this independence assumption. Numerical results show little difference from the previous model. Therefore, we conclude that the above independence assumption is a good approximation.

This model uses the following notation.

n = number of switching stages.

$p_{ij}(k, t)$ = probability that a switching element at stage k has i packets in the upper buffer and j packets in the lower buffer at the beginning of the tth clock period ($i = 0, 1; j = 0, 1$).

$q(k, t)$ = probability that a packet is ready to come to a buffer of a switching element at stage k during the tth clock period.

It is noted that "arrivals" coming to each of the two buffers in a switching element are independent of each other.

$r_{01}(k, t)$ = probability that the packet in the lower buffer of a switching element at stage k is able to move forward during the tth clock period, given that the upper buffer is empty and the lower buffer has a packet.

$r_{10}(k, t)$ = probability that the packet in the upper buffer of a switching element at stage k is able to move forward during the tth clock period, given that the lower buffer is empty and the upper buffer has a packet.

$r_{11}^{ij}(k, t)$ = probability that i packets in the upper buffer and j packets in the lower buffer of a switching element at stage k are able to move forward during the tth clock period, given that both buffers have packets ($i = 0, 1, j = 0, 1$).

The equations governing the relationships among these variables are as follows:

$$q(k, t) = 0.5 p_{01}(k-1, t) + 0.5 p_{10}(k-1, t)$$
$$+ 0.75 p_{11}(k-1, t) \quad k = 2, 3, 4, \ldots, n \quad (7)$$

$$r_{01}(k, t) = p_{00}(k+1, t)$$
$$+ 0.5 p_{01}(k+1, t)[1 + r_{01}(k+1, t)]$$
$$+ 0.5 p_{10}(k+1, t)[1 + r_{10}(k+1, t)]$$
$$+ p_{11}(k+1, t)[r_{11}^{11}(k+1, t) + 0.5 r_{11}^{01}(k+1, t)$$
$$+ 0.5 r_{11}^{10}(k+1, t)] \quad k = 1, 2, 3, \ldots, n-1 \quad (8)$$

$$r_{01}(n, t) = 1.0$$

$$r_{10}(k, t) = r_{01}(k, t), \quad k = 1, 2, \ldots, n$$

$$r_{11}^{11}(k, t) = 0.25\{[r_{01}(k+1, t)]^2 + [r_{10}(k+1, t)]^2\} \quad (9)$$

$$= 0.5[r_{01}(k+1, t)]^2 \quad (10)$$

$$r_{11}^{00}(k, t) = 0.5[1 - r_{01}(k+1, t)]$$
$$+ 0.5[1 - r_{01}(k+1, t)]^2 \quad (11)$$

$$r_{11}^{01}(k, t) = r_{11}^{10}(k, t)$$
$$= 0.5[1 - r_{11}^{11}(k, t) - r_{11}^{00}(k, t)]. \quad (12)$$

Equation (9) results from the fact that the load is assumed to be balanced and the switch is symmetric. Equation (10) results from the fact that both output links of a switching element in a Banyan network are connected either to two upper input links or to two lower input links of two different switching elements of the next stage. (See Fig. 1.) Equation (11) follows from the same reasoning. Equation (12) results from the symmetry property of the switch and the conservation law of probability.

Equations governing the state transitions are given by the following matrix equation:

$$\begin{pmatrix} p_{00}(k, t+1) \\ p_{01}(k, t+1) \\ p_{10}(k, t+1) \\ p_{11}(k, t+1) \end{pmatrix} = Q \begin{pmatrix} p_{00}(k, t) \\ p_{01}(k, t) \\ p_{10}(k, t) \\ p_{11}(k, t) \end{pmatrix} \quad (13)$$

where

$$Q = \begin{pmatrix} \bar{q}\bar{q} & \bar{q}r_{01}\bar{q} & \bar{q}r_{10}\bar{q} & \bar{q}r_{11}^{11}\bar{q} \\ \bar{q}q & \bar{q}(r_{01}q + \bar{r}_{01}) & \bar{q}r_{10}q & \bar{q}(r_{11}^{01} + r_{11}^{11}q) \\ q\bar{q} & qr_{01}\bar{q} & \bar{q}(r_{10}q + \bar{r}_{10}) & \bar{q}(r_{11}^{10} + r_{11}^{11}q) \\ qq & q(r_{01}q + \bar{r}_{01}) & q(r_{10}q + \bar{r}_{10}) & r_{11}^{00} + r_{11}^{01}q + r_{11}^{10}q + t_{11}^{11}q^2 \end{pmatrix}$$

where the arguments (k, t) of the variables in the coefficient matrix are left out for the simplicity of presentation. In (13) we also use the shorthand notations $\bar{q}$, $\bar{r}_{01}$, and $\bar{r}_{10}$ to denote $(1 - q)$, $(1 - r_{01})$, and $(1 - r_{10})$, respectively.

Equations (7)–(13) govern the dynamics of the state transition of the system. If the system has a steady state, then all variables should converge to some corresponding time-independent quantities (after some iteration). The normalized throughput S and the normalized delay d are given by

$$S = 0.5\{ p_{01}(k)r_{01}(k) + p_{10}(k)r_{10}(k)$$
$$+ p_{11}(k)[r_{11}^{10}(k) + r_{11}^{01}(k) + 2r_{11}^{11}(k)]\} \quad (14)$$

and

$$d = \frac{1}{n} \sum_{k=1}^{n} 1/p(k) \quad (15)$$

where

$$p(k) = \{ p_{10}(k)r_{10}(k)$$
$$+ p_{11}(k)[r_{11}^{11}(k) + r_{11}^{10}(k)]\}/[p_{10}(k) + p_{11}(k)]. \quad (16)$$

It is noted that $p(k)$ is the probability that a packet in the upper buffer of a switching element is able to move forward, given that there is a packet in that buffer. Because of the symmetry of the problem, $p(k)$ is also applicable to the lower buffer. Therefore, (15) gives the correct expression for normalized delay d.

We have used (7)–(16) to compute the normalized throughput and average delay for varying loads and switch-

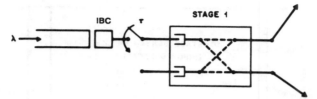

Fig. 4. Packet switch with infinite waiting rooms at its input buffer controllers.

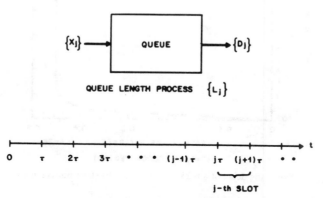

Fig. 5. A discrete time queueing model.

ing network sizes, and have found that the results of this refined model differ very little from those obtained in the previous model (differences are less than 0.1 percent). Thus, we conclude that the independence assumption is reasonable, and model 1 of Section III gives good approximate results with greatly reduce computation effort.

V. SWITCHING DELAY—IBC WITH INFINITE WAITING ROOM

In this section we study a system with input buffer controllers (IBC's) added to the front end of a Banyan switch. The model is shown in Fig. 4. Packets arrive at an input buffer controller at a rate of λ packets per clock period τ. At clock ticks, if the corresponding buffer of the switching element at stage 1 of the switching network is able to accept a packet from the input buffer controller, the IBC will place a packet into that buffer (if there is any packet in the IBC queue). Arriving packets are placed in the IBC queue according to the order of their arrivals, and they are delivered to the switching network in the first-in-first-out (FIFO) fashion. We are interested in quantifying the switching delay, which is the time a packet spends in the IBC queue and in the switch before it leaves the switch. The size of the IBC waiting room is assumed to be infinite. In the next section we will address the problem with a finite IBC waiting room.

Let us consider a discrete-time queueing model for automatic repeat request (ARQ) used in data communications as shown in Fig. 5 and previously analyzed in [4]. The time axis is slotted into uniformly spaced time epochs. The random variable X_i is the number of arriving packets

during the ith slot, D_i is the number of packets removed from the queue during the ith slot, and L_i is the queue length (number of waiting packets) at the beginning of the ith slot. Usually, a slot time is the time required to transmit a packet. If the transmission is successful, D_i is 1, otherwise, D_i is zero. The probability that a transmission is successful is p. This model can be characterized by the following two equations:

$$L_{i+1} = L_i + X_i - D_i \tag{17}$$

and

$$D_i = \begin{cases} 0, & \text{if } L_i = 0 \\ \left\{\begin{array}{ll} 0, & \text{with probability } 1 - p, \\ 1, & \text{with probability } p, \end{array}\right\} & \text{if } L_i > 0. \end{cases} \tag{18}$$

If X_i is independent of i (i.e., stationary), then the steady state exists and we have $E(X) = E(D) = p \cdot \text{prob}\{L \neq 0\}$ from (17) and (18).

If we compare this model to our switching model of Fig. 4, we see that two models are mathematically equivalent and the corresponding variables and parameters are

$$E(X) = \lambda \tag{19}$$

$$p = p_0(1) + p_1(1)r(1) \tag{20}$$

and

$$\text{prob}\{L \neq 0\} = q(1) \tag{21}$$

where $p_0(1)$, $p_1(1)$, $r(1)$, and $q(1)$ are steady-state variables of our Model 1 of Section III.

Now we are ready to describe an algorithmic solution to our switching delay problem.

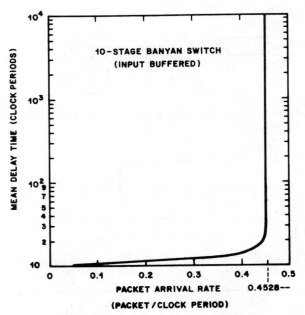

Fig. 6.   Switching delay of a packet switch with 10-stage Banyan network.

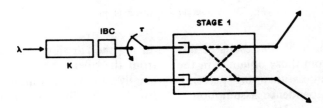

FAST PACKET SWITCH WITH FINITE WAITING ROOM

Fig. 7.   Packet switch with finite waiting rooms at its input buffer controllers.

Step 1) Compute p:

a) Guess an initial $q(1)$.

b) Perform the Banyan network computation using Model 1 to obtain $p_0(1)$, $p_1(1)$, and $r(1)$.

c) Let $p = p_0(1) + p_1(1)r(1)$.

d) Let $q(1) = \lambda/p$.

e) Go to b) and iterate.

After the procedure converges, we obtain p.

Step 2) Solve the discrete-time queuing model.

With the p obtained in step 1), we then solve the well-known discrete-time queuing model. This model has previously been solved [4], and the mean switching delay is given by

$$D_S = \left| \frac{\sigma_x^2 + \lambda(1-\lambda)}{2(p-\lambda)\lambda} - \frac{1}{2} + \sum_{k=1}^{n} 1/r(k) \right| \tau \quad (22)$$

where σ_x^2 is the variance of the arrival X.

A plot of mean switching delay for a 10-stage Banyan switch with front-end IBC queue is shown in Fig. 6. It is seen that this delay is relatively small for loads ($q(1)$) less than 0.4, and increases rapidly to infinity at about 0.45 load. Notice that 0.45 is the maximum achievable throughput for a 10-stage Banyan network as shown in Fig. 2.

VI.   BLOCKING PROBABILITY—IBC WITH FINITE WAITING ROOM

In this section we study the case where the IBC waiting room is of finite size K and the quantity of interest is the blocking probability P_b. The blocking probability is the probability that an arriving packet will find the IBC waiting room is full. Blocked packets are dropped and considered to be lost.

The model is shown in Fig. 7. This model is similar to the one shown in Fig. 4, except that the IBC has only K waiting rooms. Packets arrive at the buffer controller at a rate of λ packets per clock period. Let λ' be the "carried load" of the system, that is, the average number of packets accepted by the switch; then the blocking probability P_b is given by

$$P_b = (\lambda - \lambda')/\lambda. \quad (23)$$

Therefore, if for every arrival rate λ we can compute λ', then we can obtain P_b readily by (23).

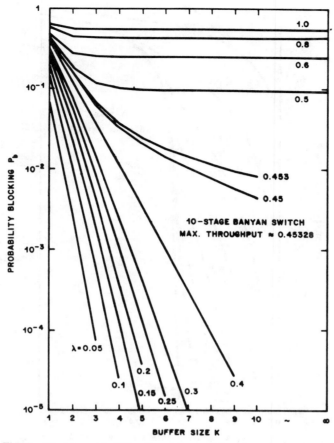

Fig. 8. Blocking probability of a packet switch with a 10-stage Banyan network.

Recall that λ' is simply the normalized throughput of the Banyan switch described in Section III. To obtain λ', we need the quantity $q(1)$. If we consider the Banyan switch as a server for the finite queue, then the service time for a customer (a packet) is $n\tau$ (recall that τ is the clock period) with probability $p(1-p)^{n-1}$ where p is the probability that an input link at the first stage of the Banyan network is able to accept a new packet at clock ticks. This variable p can be expressed in terms of the parameters of the Banyan network:

$$p = p_0(1) + p_1(1)r(1). \qquad (24)$$

Because the switch operates in clocked mode, the server will not return until next clock tick if it finds an empty queue at the clock tick. Hence, assuming Poisson arrivals, the system can be modeled as an $M/G/1/K$ queue with vacation [5] (vacation time is the clock period τ). The quantity $q(1)$ is simply the probability that the server will find a nonempty queue at clock ticks. It is given by $B/(V+B)$ where B and V are, respectively, the mean busy and idle periods of a $M/G/1/K$ queue. Now we are ready to describe the algorithm for computing the blocking probability as follows.

Step 1) Guess an initial $q(1)$.
Step 2) Perform Banyan network computation using Model 1 of Section III.

Step 3) $p = p_0(1) + p_1(1)r(1)$.
Step 4) Do $M/G/1/K$ computation.
Step 5) $q(1) = B/(B+V)$.
Step 6) Go to Step 2).

After the procedure converges, the carried load λ' is given by the normalized throughput of the switch, and the blocking probability is given by (23).

In Fig. 8, we plot the blocking probability P_b against the buffer size K with load λ as a parameter for the 10-stage network. It is noted that, for $\lambda < 0.45$, P_b decreases sharply as K increases, while for $\lambda > 0.45$, the P_b curves behave dramatically differently. This is due to the fact that a 10-stage Banyan switch has the maximum achievable normalized throughput 0.45. If the input load λ is greater than 0.5, the blocking probability P_b is about $(\lambda - 0.45)/\lambda$ for $K > 3$. In Fig. 9, we plot the blocking probability P_b against the load λ with the buffer size K as a parameter. Again, we see the breakdown of the smoothness of the curves happens near $\lambda = 0.45$.

VII. Conclusions

In this paper, we analyzed the performance of the packet switch based on the single-buffered Banyan network. First, we derived a model for computing the normalized throughput and the average internal delay of the Banyan

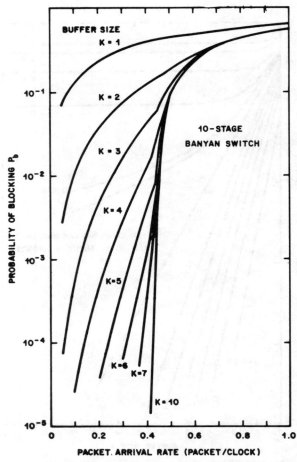

Fig. 9. Blocking probability of a packet switch with a 10-stage Banyan network.

switch. Then we used a discrete-time queuing model to compute the switch delay of the switch with infinite waiting rooms at its input buffer controllers. It was found that the switch delay is not significant until the input load approaches the maximum achievable throughput of the Banyan switch. An $M/G/1/K$ queue with vacation was used to compute the blocking probability of the packet switch with finite waiting rooms at its input buffer controllers.

References

[1] "Interconnection Networks" Special Issue of *IEEE Computer*, Dec., 1981.

[2] D. M. Dias and R. Jump, "Analysis and simulation of buffered delta network," *IEEE Trans. Comput.*, vol. C-30, Apr. 1981.

[3] S. Cheemalavagu and M. Malek, "Analysis and simulation of Banyan interconnection networks with 2×2, 4×4, and 8×8 switching elements," in *Proc. 1982 Real-Time Syst. Symp.*, Los Angeles, CA, Dec. 1982.

[4] Y. C. Jenq, "On calculations of transient statistics of a discrete queuing system with independent general arrivals and geometric departures," *IEEE Trans. Commun.*, vol. COM-28, pp. 908–911, June 1980.

[5] T. T. Lee, "$M/G/1/N$ queue with vacation time and exhaustive service discipline," *J. Oper. Res.*, submitted for publication.

INTRODUCTION

To understand nonblocking switching, one needs to understand the role of blocking in switching. Blocking occurs in a switch when a packet can not immediately access an idle output line or port to which it nominally would have access. This problem may occur because a buffer in the succeeding stage is full or because a packet at the head of the line of a queue can not be immediately switched. Switches are often built with inherent blocking for the following reasons:

- They can be simpler to implement than nonblocking switches.
- The blocking may produce appreciable effects only under heavy loads that are not likely to occur in normal practice.

In a nonblocking switch, packets can always immediately access a desired idle output. Thus blocking in the switch fabric is eliminated. Nonblocking switches are not always used in practice because of the potential difficulty of implementation. A nonblocking $N \times N$ crossbar switch, for instance, requires $O(N^2)$ crosspoints, a traditional measure of switch complexity. A blocking network, such as a delta network, may be implemented with $O(N \log N)$ crosspoints. This is a much smaller number of crosspoints. Still, studies have shown that the amount of chip area required for (blocking) Banyan and (nonblocking) crossbar networks is both $O(N^2)$ [19, 20]. There has been a great deal of work concerning nonblocking switches. In one sense, nonblocking switches represent an ideal approach against which other switching strategies can be compared.

Even with a nonblocking switching fabric, such as a crossbar, there can be queueing. Queueing can occur in several ways. If the fabric switches at the same speed as the input-cell arrival rate, then queues will be needed at each input (*input queueing*) because not every packet will be immediately able to pass through the fabric. If an $N \times N$ fabric switches N times faster than the input-cell arrival rate, then queues will be needed at each output (*output queueing*). This is because the output lines, which are usually assumed to operate at the same speed as the input lines, will not have the necessary capacity for the potential number of arriving packets. Queues may also be placed inside the switch fabric.

The first two papers in this chapter, by M. J. Karol, M. G. Hluchyj, and S. P. Morgan and by J. Y. Hui and E. Arthurs, contain significant contributions. Both show that the normalized saturation throughput for input queueing for a large switch with fixed-length cells obeying a uniform traffic pattern is 0.586 $(2 - \sqrt{2})$. In comparison, output queueing has a 1.00 normalized saturation throughput under the same conditions. Of course, input queueing can be simpler to implement be-cause the switch runs at the same speed as the input-cell arrival rate. Some recent ideas for improving input queueing performance appear in [21–25].

The first of these two papers is "Input Versus Output Queueing on a Space Division Packet Switch," by Karol, Hluchyj, and Morgan. Published originally at the 1986 Global Telecommunications Conference, it presents a discrete-time queueing analysis for both input and output queueing on a nonblocking crossbar type switch. This is based on simple difference equations to describe the behavior of each situation. The authors show that mean queue lengths and mean waiting times are larger for input queueing than for output queueing.

A simple recursive algorithm for calculating the steady-state probabilities of the queue length of the under output queueing is described. This algorithm is helpful because it avoids having to deal with sophisticated transform techniques. Finally, showing that the 0.586 figure is not an immutable constant, the authors point out that if interfering packets are dropped at the end of each time slot, then output trunk utilization (throughput) can be increased to 0.632 $(1 - e^{-1})$. Note that this is only possible if the input trunk utilization exceeds 0.881. As Karol, Hluchyj, and Morgan point out: "Consequently, if the objective is maximum output utilization, rather than 100 percent packet delivery, then below the second threshold (0.881), it is better to queue packets until they are successful, whereas above the second threshold, it is better to reduce input queue blocking by dropping packets whenever there are conflicts."

The second of the two papers mentioned is "A Broadband Packet Switch for Integrated Transport" by Hui and Arthurs. In it they set out to design a switch with the following system objectives:

- Support all services except entertainment video broadcast on a switch fabric.
- Create 1000–4000 ports serving a maximum of 200,000 subscriber lines.
- Employ CMOS VLSI for the switch fabric and support a port speed of 150 Mbps.

For this purpose Hui and Arthurs present a three-phase algorithm running over a sorting network followed by a Banyan network. A basic problem that their design seeks to solve is the arbitration of conflicting requests at multiple inputs for the same output. In the first phase, requests for output ports are sorted in order using the sorting network. In the second phase, one acknowledgement for each output for which there was at least one request is sent back to the appropriate input. In the third phase, the packets receiving permission move through the network *without* conflict.

This paper is noteworthy for a discussion of the hardware implementation of the three-phase algorithm. There is also a discussion of several variations on the basic algorithm. An analysis is then presented to show that the normalized throughput of the switch under random traffic is 0.586. A technique different from that of Karol, Hluchyj, and Morgan is used. It involves taking the expectation of the square of a difference equation. This technique was made popular by D. G. Kendall in his work on the M/G/1 queue [1]. A closed-form expression for mean queue length (and thus delay) as a function of arrival rate is also presented.

In the last section of Hui and Arthurs' paper, the ability of a packet-switched network to carry circuit switched traffic is discussed. The emphasis is on maintaining the periodicity of such traffic as it transits the packet switch.

The next paper in this chapter is "Nonblocking Networks for Fast Packet Switching" by R. Melen and J. S. Turner. In this paper Melen and Turner extend the classical theory of nonblocking switching networks to multirate and ATM switching systems. Here a *multirate* system is one handling traffic with a multiplicity of data rates. The authors point out that one method for operating a multirate ATM style system is, for each connection, to select a path through the switch to be used by all packets in that connection. However, one must be sure that sufficient bandwidth is available on each link carrying the connection. This results in a logical generalization of the theory of nonblocking networks, which Melen and Turner go on to examine in this paper. The conditions under which the Clos, Cantor, and Beněs network are strictly nonblocking are presented. In a *rearrangeable nonblocking network* all permutations can be made, but some existing connections will have to be rearranged when a new connection is made. A *strictly nonblocking network* is one where an algorithm exists for setting up new connections without having to rearrange existing connections [2].

An interesting point made at the end of the paper is that an IC package count tends to be more informative and accurate than the traditional measure of complexity, the crosspoint count.

The next paper in this collection proposes an improved type of switching network that possesses a wide variety of attractive features. It is entitled "Multi-Log$_2$ N Networks and Their Applications in High-Speed Electronic and Photonic Switching Systems," by C.-T. Lea. The new class of networks, multi-log$_2$ networks, is constructed by the "vertical stacking" of self-routing networks (i.e., banyan or baseline or shuffle). Vertical stacking refers to placing a number of self-routing networks in parallel with connections from each input and output to each self-routing network.

The use of self-routing networks is significant as their use avoids time (crosspoint update) and space (control wiring) constraints associated with conventional time-multiplexed systems. However Lea's goal in this paper was to construct a nonblocking network using vertical stacking. Earlier work by

Lea and N.-H. Huang had resulted in a blocking network using vertical stacking [17].

In this paper Lea shows that vertical stacking can be used to develop strictly and rearrangeable nonblocking networks. Vertically stacked networks have the same advantages as self-routing networks: self-routing, simple path hunt, simple fault diagnosis, and increased fault tolerance. Lea goes on to show that the question of the number of self-routing network copies needed to produce nonblocking can be solved using graph theory.

The next paper studies the influence of two priority classes on a nonblocking packet switch with input queueing. Authored by J. S.-C. Chen and R. Guérin, it is titled "Performance Study of an Input Queueing Packet Switch with Two Priority Classes." Classes of traffic with different priorities are of interest because in an integrated network carrying multiple types of traffic, a natural way to manage network resources is in terms of a priority structure. There has been a good deal of work on queues with priority [8–16], but most of this work has involved single queues. A significant feature of the Chen and Guérin paper is that priorities are studied in a network- (fabric-) wide setting.

Chen and Guérin consider two priority policies. Under the first policy, packets of both classes are queued in the input queues. Under the second policy, only low-priority packets are queued; high-priority packets that can not be immediately delivered are dropped (lost). The dependencies in service make an exact analysis infeasible, so an approximate analysis is provided. The approximate analysis uses independence assumptions and an equivalent queueing system. The authors develop a transform expression for the input-queue-length distribution.

Interestingly, it is shown that the maximum switch throughput under two priority classes and the first policy can slightly exceed that under a single class (0.586; see the previous papers by Karol, Hluchyj, and Morgan and by Hui and Arthurs).

It was mentioned before that for a nonblocking fabric with output queueing, 100 percent throughput will be achieved at the price of speeding up the $N \times N$ switch speed to N times the cell arrival rate. In "Effect of Speedup in Nonblocking Packet Switch," Y. Oie, M. Murata, K. Kubota, and H. Miyahara examine a switch operating at $L < N$ times the cell arrival rate. This necessitates the inclusion of input and output buffers but is a significant relaxation of a key design specification.

Oie and his coauthors determine loss probability and average transmission delay for two architectures. In one architecture there are buffers only at the outputs. Analytical results are presented for the infinite buffer case, and a numerical procedure is presented for the finite buffer case. In the other architecture there are buffers at both the inputs and the outputs. An analysis is presented for the infinite buffer case and simulation results are given for the finite buffer case. It is found that a speedup of 3 can achieve 99% throughput. A more recent paper by Gupta and Georganas [3] also examines this problem. They find that modest switch speedups of 2 or 3 can achieve near-maximal throughput.

The performance of a nonblocking space division packet switch under various switching speeds is also examined in "Throughput Analysis, Optimal Buffer Allocation, and Traffic Imbalance Study of a Generic Nonblocking Packet Switch" by J. S.-C. Chen and T. E. Stern. Using a model with exponential packet lengths, Chen and Stern also come to the conclusion that modest speedups of the switch fabric ($L = 4$) can achieve close to maximal throughput. The authors also study policies for deciding on the order in which contending packets receive service. It is found that the choice of policy affects the delay distribution but not throughput.

Chen and Stern determine the optimal allocation of buffers between input and output buffers when the total number of buffers is fixed and the objective is to minimize blocking probability. Finally, Chen and Stern study the switch under two types of traffic nonuniformity. Under one the traffic intensity at the inputs varies, whereas under the other the addressing of packets to destinations is nonuniform. They find that the input imbalance affects throughput more seriously than output imbalance.

Studies of input and output queueing illustrate the importance of architectural considerations for switch design. A somewhat different phrasing of the architectural problem appears in "Comparison of Buffering Strategies for Asymmetric Packet Switch Modules," by S. C. Liew and K. W. Lu. The emphasis here is on constructing large switches out of a basic asymmetric packet switch module. That module consists of n inputs and m outputs. The outputs are segmented into g groups of r output ports ($m = gr$). The basic idea is that a packet destined for each of g addresses can access any of the r output ports; that is, the r ports in a group effectively have the same address. An advantage of this arrangement is that the channel-grouping switch modules are less complex than typical switch modules of the same dimensionality. Here complexity is measured in terms of switch element counts. Moreover, the switch modules can serve as the building blocks of many large multistage switches. It should be mentioned that two papers in this part [5, 6] discuss the idea of grouping channels or outputs and additional work is available in the literature [18].

In this paper Liew and Lu study switch modules under input and output queueing. Using a variety of techniques, they find the saturation throughput of modules with input buffers. They also find the mean delay of switch modules with input and output buffers. An important insight found in this study is that the performance of a switch will improve as the number of output ports per address is increased. They also find that the usual performance superiority of output queueing over input queueing decreases for asymmetric switches.

The paper by Chen and Stern had an example of the consideration of nonuniform traffic. During the first pass at analyzing a switching architecture, it is natural to assume uniform traffic; that is, arriving packets to any input are independently addressed and equally likely to be destined for any output. However actual traffic patterns may depart from the uniform ideal and can lower the performance of switches. Thus as the study of switch performance has developed, there has been an increased interest in the study of nonuniform traffic. One paper devoted to this topic is "Nonuniform Traffic Analysis on a Nonblocking Space-Division Packet Switch" by S.-Q. Li.

This paper considers an infinite-sized $N \times N$ nonblocking switch, with input buffers, under nonuniform traffic. Li finds a formulation for the maximal throughput of a saturated switch given the input traffic intensities and their output distribution. This work also indicates when saturation will occur.

Li also demonstrates that the contention for outputs by packets at the head of the input queues has a phase-type distribution [4]. Each input queue can be modeled as an independent GEOM/PH/1 queueing system. This leads to a procedure for finding the queue-length distribution.

A paper that looks at grouping outputs in a Batcher-Banyan network is "Performance Evaluation of a Batcher-Banyan Interconnection Network with Output Pooling," by A. Pattavina. This paper is actually a successor to a 1988 paper by Pattavina [5] that examined the idea of grouping channels. A channel group is ". . . a set of parallel packet channels that is seen as a single data link connection between two cooperating routing entities." The goal is to take advantage of the "statistical smoothing" of a multitude of sources by a group of channels.

In the 1991 paper in this collection, Pattavina evaluates a Batcher-Banyan interconnection network through a simulation model. It is well known [7] that by using a Batcher network to presort packet addresses and selecting only one packet per output prior to the Banyan network, an internally nonblocking switch can be created. Three different policies for the allocation of output channels within a "pool" to users are presented. Both the 1988 and 1991 papers are noteworthy for detailed discussions of implementation.

Modeling realistic traffic inputs to nonblocking switches is an important problem. In "Performance of a Nonblocking Space-Division Packet Switch with Correlated Input Traffic," S.-Q. Li examines a nonblocking switch with correlated input traffic. Each input in this paper carries traffic from multiple sources. Each source is modeled by a two-state Markov chain. In the ON state, each packet generates fixed-length packets periodically. This is a good model of such traffic as voice traffic.

Li finds lower and upper bounds on maximum mean throughput under a heavy traffic load. For pure input buffering and an infinite sized switch, he shows that

$$0.500 \leq \lambda_{\text{MAX}} \leq 0.586$$

The lower bound corresponds to heavy correlation and the upper bound to no correlation (matching the value of the first two papers of this chapter). For a switch with a speedup of 4 and both input and output buffers,

$$0.993 \leq \lambda_{MAX} \leq 0.996$$

This is consistent with the result of other papers using independent traffic. Li shows that if the source access rate is much less than the link speed (as is typical), then input traffic correlation has little effect on output contention. A consequence of this is that input buffers can be analyzed, as Li does, separately from the switch. Finally, in this paper the model is extended to include nonuniform traffic.

REFERENCES

[1] D. G. Kendall, "Some problems in the theory of queues," *J. Royal Statistical Soc.*, Series B, vol. 13, no. 2, pp. 151–185, 1951.

[2] V. E. Benes, *Mathematical Theory of Connecting Networks*, New York: Academic Press, 1965.

[3] A. K. Gupta and N. D. Georganas, "Analysis of a packet switch with input and output buffers and speed constraints," *Proc. IEEE INFOCOM '91*, Bal Harbour, Fla., pp. 694–700, April 1991.

[4] L. Kleinrock, *Queueing Systems, Vol. I: Theory*, New York: John Wiley, 1975.

[5] A. Pattavina, "Multichannel bandwidth allocation in a broadband packet switch," *IEEE J. on Selected Areas in Commun.*, vol. 6, no. 9, pp. 1489–1499, Dec. 1988.

[6] J. Turner, "Design of a broadcast packet network," *Proc. IEEE INFOCOM '86*, Miami, Fla., pp. 667–675, April 1986.

[7] J. Y. Hui, *Switching and Traffic Theory for Integrated Broadband Networks*, Norwell, Mass.: Kluwer Academic Publishers, 1990.

[8] A. Cobham, "Priority assignment in waiting line problems," *Operations Res.*, vol. 2, pp. 70–76, 1959.

[9] L. Kleinrock, "A conversion law for a wide class of queueing systems," *Naval Res. Logistics Quarterly*, vol. 12, pp. 181–192, 1965.

[10] Y. Lim and J. Kobza, "Analysis of a delay-dependent priority discipline in a multi-class traffic packet switching node," *Proc. IEEE INFOCOM '88*, New Orleans, La., pp. 889–898, April 1988.

[11] R. Chipalkatti, J. F. Kurose, and D. Towsley, "Scheduling policies for real-time and non-real time traffic in a statistical multiplexer," *Proc. IEEE INFOCOM '89*, Ottawa, Canada, pp. 774–783, April 1989.

[12] T. M. Chen, J. Walrand, and D. G. Messerschmitt, "Dynamic priority protocols for packet voice," *IEEE J. Selected Areas in Commun.*, vol. 7, no. 5, pp. 632–643, June 1989.

[13] A. M. Viterbi, "Mean delay in synchronous packet networks with priority queueing disciplines," *IEEE Trans. Commun.*, vol. 39, no. 4, pp. 469–473, April 1991.

[14] J. M. Hyman, A. A. Lazar, and G. Pacifici, "Real-time scheduling with quality of service constraints," *IEEE J. Selected Areas in Commun.*, Nov. 1991.

[15] J. M. Hyman, A. A. Lazar, and G. Pacifici, "Joint scheduling and admission control for ATS-based switching nodes," *IEEE J. Selected Areas in Commun.*, Jan. 1992.

[16] A. Y.-M. Lin and J. A. Silvester, "Priority queueing strategies and buffer allocation protocols for traffic control at an ATM integrated broadband switching system," *IEEE J. on Selected Areas in Commun.*, vol. 9, no. 9, pp. 1524–1536, Dec. 1991.

[17] N.-H. Huang and C.-T. Lea, "Architecture of a time-multiplexed switch," *Proc. IEEE Globecom '86*, Nov. 1986.

[18] S.-Q. Li, "Performance of trunk grouping in packet switch design," *Proc. of IEEE INFOCOM '91*, Bal Harbour, Fla., pp. 688–693, April 1991.

[19] M. A. Franklin, "A VLSI performance comparison of Banyan and crossbar communications networks," *IEEE Trans. Computers*, vol. C-30, no. 4, pp. 283–290, April 1981.

[20] T. H. Szymanski, "A VLSI Comparison between crossbar and switch-recursive Banyan interconnection networks, *Proc. Int. Conf. on Parallel Processing*, pp. 192–199, Aug. 1986.

[21] H. Obara and T. Yasushi, "High-speed transport processor for broadband burst transport system," *Proc. IEEE Int. Conf. Commun. '88*, pp. 922–927, 1988.

[22] H. Obara and T. Yasushi, "An efficient contention resolution algorithm for input queueing ATM cross-connect switches," *Int. J. Digital and Analog Cabled Systems*, vol. 2, no. 4, pp. 261–267, Oct.–Dec. 1989.

[23] H. Obara, M. Sasagawa, and I. Tokizawa, "An ATM cross-connect system for broadband transport networks based on virtual path concept," *Proc. IEEE Int. Conf. Commun. '90*, pp. 839–843, April 1990.

[24] H. Obara, "Optimum architecture for input queueing," *Electronic Letters*, pp. 555–557, March 28, 1991.

[25] M. J. Karol, K. Y. Eng, and H. Obara, "Improving the performance of input-queued ATM packet switches," *Proc. IEEE INFOCOM '92*, Florence, Italy, pp. 110–115, May 1992.

Input Versus Output Queueing on a Space-Division Packet Switch

MARK J. KAROL, MEMBER, IEEE, MICHAEL G. HLUCHYJ, MEMBER, IEEE, AND SAMUEL P. MORGAN, FELLOW, IEEE

Abstract—Two simple models of queueing on an $N \times N$ space-division packet switch are examined. The switch operates synchronously with fixed-length packets; during each time slot, packets may arrive on any inputs addressed to any outputs. Because packet arrivals to the switch are unscheduled, more than one packet may arrive for the same output during the same time slot, making queueing unavoidable. Mean queue lengths are always greater for queueing on inputs than for queueing on outputs, and the output queues saturate only as the utilization approaches unity. Input queues, on the other hand, saturate at a utilization that depends on N, but is approximately $(2 - \sqrt{2}) = 0.586$ when N is large. If output trunk utilization is the primary consideration, it is possible to slightly increase utilization of the output trunks—up to $(1 - e^{-1}) = 0.632$ as $N \to \infty$—by dropping interfering packets at the end of each time slot, rather than storing them in the input queues. This improvement is possible, however, only when the utilization of the input trunks exceeds a second critical threshold—approximately $\ln(1 + \sqrt{2}) = 0.881$ for large N.

I. INTRODUCTION

SPACE-DIVISION packet switching is emerging as a key component in the trend toward high-performance integrated communication networks for data, voice, image, and video [1], [2] and multiprocessor interconnects for building highly parallel computer systems [3], [4]. Unlike present-day packet switch architectures with throughputs measured in 1's or at most 10's of Mbits/s, a space-division packet switch can have throughputs measured in 1's, 10's, or even 100's of Gbits/s. These capacities are attained through the use of a highly parallel switch fabric coupled with simple per packet processing distributed among many high-speed VLSI circuits.

Conceptually, a space-division packet switch is a box with N inputs and N outputs that routes the packets arriving on its inputs to the appropriate outputs. At any given time, internal switch points can be set to establish certain paths from inputs to outputs; the routing information used to establish input–output paths is often contained in the header of each arriving packet. Packets may have to be buffered within the switch until appropriate connections are available; the location of the buffers and the amount of buffering required depend on the switch architecture and the statistics of the offered traffic.

Clearly, congestion can occur if the switch is a blocking network, that is, if there are not enough switch points to provide simultaneous, independent paths between arbitrary pairs of inputs and outputs. A Banyan switch [3]–[5], for example, is a blocking network. In a Banyan switch, even when every input is assigned to a different output, as many as

Paper approved by the Editor for Local Area Networks of the IEEE Communications Society. Manuscript received August 8, 1986; revised May 14, 1987. This paper was presented at GLOBECOM'86, Houston, TX, December 1986.

M. J. Karol is with AT&T Bell Laboratories, Holmdel, NJ 07733.

M. G. Hluchyj was with AT&T Bell Laboratories, Holmdel, NJ 07733. He is now with the Codex Corporation, Canton, MA 02021.

S. P. Morgan is with AT&T Bell Laboratories, Murray Hill, NJ 07974.

IEEE Log Number 8717486.

$\sqrt{N}$ connections may be contending for use of the same center link. The use of a blocking network as a packet switch is feasible only under light loads or, alternatively, if it is possible to run the switch substantially faster than the input and output trunks.

In this paper, we consider only nonblocking networks. A simple example of a nonblocking switch fabric is the crossbar interconnect with N^2 switch points (Fig. 1). Here it is always possible to establish a connection between any idle input–output pair. Examples of other nonblocking switch fabrics are given in [3]. Even with a nonblocking interconnect, some queueing in a packet switch is unavoidable, simply because the switch acts as a statistical multiplexor; that is, packet arrivals to the switch are unscheduled. If more than one packet arrives for the same output at a given time, queueing is required. Depending on the speed of the switch fabric and its particular architecture, there may be a choice as to where the queueing is done: for example, on the input trunk, on the output trunk, or at an internal node.

We assume that the switch operates synchronously with fixed-length packets, and that during each time slot, packets may arrive on any inputs addressed to any outputs (Fig. 2). If the switch fabric runs N times as fast as the input and output trunks, all the packets that arrive during a particular input time slot can traverse the switch before the next input slot, but there will still be queueing at the outputs [Fig. 1(a)]. This queueing really has nothing to do with the switch architecture, but is due to the simultaneous arrival of more than one input packet for the same output. If, on the other hand, the switch fabric runs at the same speed as the inputs and outputs, only one packet can be accepted by any given output line during a time slot, and other packets addressed to the same output must queue on the input lines [Fig. 1(b)]. For simplicity, we do not consider the intermediate case where some packets can be queued at internal nodes, as in the Banyan topology.

It seems intuitively reasonable that the mean queue lengths, and hence the mean waiting times, will be greater for queueing on inputs than for queueing on outputs. When queueing is done on inputs, a packet that could traverse the switch to an idle output during the current time slot may have to wait in queue behind a packet whose output is currently busy. The intuition that, if possible, it is better to queue on the outputs than the inputs of a space-division packet switch also pertains to the following situation. Consider a single road leading to both a sports arena and a store [Fig. 3(a)]. Even if there are no customers waiting for service in the store, some shoppers might be stuck in stadium traffic. A simple bypass road around the stadium is the obvious solution [Fig. 3(b)].

This paper quantifies the performance improvements provided by output queueing for the following simple model. Independent, statistically identical traffic arrives on each input trunk. In any given time slot, the probability that a packet will arrive on a particular input is p. Thus, p represents the average utilization of each input. Each packet has equal probability $1/N$ of being addressed to any given output, and successive packets are independent.

With output queueing, all arriving packets in a time slot are

Reprinted from *IEEE Trans. Commun.*, vol. COM-35, no. 12, pp. 1347–1356, December 1987.

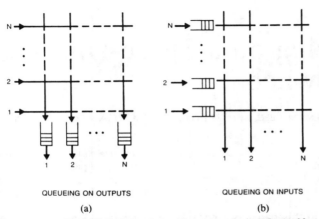

Fig. 1. (a) An $N \times N$ crossbar switch with output queueing. (b) An $N \times N$ crossbar switch with input queueing.

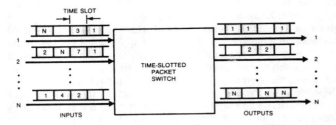

Fig. 2. Fixed-length packets arrive synchronously to a time-slotted packet switch.

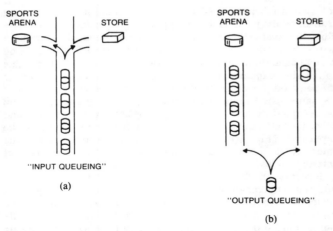

Fig. 3. "Output queueing" (b) is superior to "input queueing" (a). In (a), even if there are no customers waiting for service in the store, some shoppers might be stuck in stadium traffic. In (b), a bypass road around the stadium serves cars traveling to the store.

cleared before the beginning of the next time slot. For example, a crossbar switch fabric that runs N times as fast as the inputs and outputs can queue all packet arrivals according to their output addresses, even if all N inputs have packets destined for the same output [Fig. 1(a)]. If k packets arrive for one output during the current time slot, however, only one can be transmitted over the output trunk. The remaining $k - 1$ packets go into an output FIFO (first-in, first-out queue) for transmission during subsequent time slots. Since the average utilization of each output trunk is the same as the utilization of each input trunk, namely p, the system is stable and the mean queue lengths will be finite for $p < 1$, but they will be greater than zero if $p > 0$.

A crossbar interconnect with the switch fabric running at the same speed as the inputs and outputs exemplifies input queueing [Fig. 1(b)]. Each arriving packet goes, at least momentarily, into a FIFO on its input trunk. At the beginning of every time slot, the switch controller looks at the first packet in each FIFO. If every packet is addressed to a different output, the controller closes the proper crosspoints and all the packets go through. If k packets are addressed to a particular output, the controller picks one to send; the others wait until the next time slot, when a new selection is made among the packets that are then waiting. Three selection policies are discussed in Section III: one of the k packets is chosen at *random*, each selected with equal probability $1/k$, *longest queue* selection, in which the controller sends the packet from the longest input queue,[1] and *fixed priority* selection where the N inputs have fixed priority levels, and of the k packets, the controller sends the one with highest priority.

Solutions of these two queueing problems are given in Sections II and III. Curves showing mean waiting time as a function of p are plotted for various values of N. As expected, the mean waiting times are greater for queueing on inputs than for queueing on outputs. Furthermore, the output queues saturate only as $p \rightarrow 1$. Input queues, on the other hand, saturate at a value of p less than unity, depending weakly on N; for large N, the critical value of p is approximately $(2 - \sqrt{2}) = 0.586$ with the random selection policy. When the utilization p of the input trunks exceeds the critical value, the steady-state queue sizes are infinite, packets experience infinite waiting times, and the output trunk utilization is limited to approximately 0.586 (for large N). In the saturation region, however, it is possible to increase utilization of the output trunks—up to $(1 - e^{-1}) = 0.632$ as $N \rightarrow \infty$—by dropping packets, rather than storing them in the input queues. This improvement is possible, however, only when the utilization of the input trunks exceeds a second critical threshold—approximately $\ln (1 + \sqrt{2}) = 0.881$ for large N. Consequently, if the objective is maximum output utilization, rather than 100 percent packet delivery, then below the second threshold, it is better to queue packets until they are successful, whereas above the second threshold, it is better to reduce input queue blocking by dropping packets whenever there are conflicts. With high probability, new packets (with new destinations) will quickly arrive to replace the dropped packets.

Comparing the random and longest queue selection policies of input queueing, the mean waiting times are greater with random selection. This is expected because the longest queue selection policy reduces the expected number of packets blocked (behind other packets) from traversing the switch to idle outputs. For fairness, the fixed priority discipline should be avoided because the lowest priority input queue suffers large delays and is sometimes unstable, even when the other two selection policies guarantee stability.

II. QUEUES ON OUTPUTS

Much of the following analysis of the output queueing scheme involves well-known results for discrete-time queueing systems [6]. Communication systems have been modeled by discrete-time queues in the past (e.g., [7]); we sketch our analysis and present results for later comparison to the input queueing analysis.

We assume that packet arrivals on the N input trunks are governed by independent and identical Bernoulli processes. Specifically, in any given time slot, the probability that a packet will arrive on a particular input is p. Each packet has equal probability $1/N$ of being addressed to any given output, and successive packets are independent.

Fixing our attention on a particular output queue (the

---

[1] A random selection is made if, of the k input queues with packets addressed to a particular output, several queues have the same maximum length.

"tagged" queue), we define the random variable A as the number of packet arrivals at the tagged queue during a given time slot.[2] It follows that A has the binomial probabilities

$$a_i \triangleq \Pr [A = i] = \binom{N}{i} (p/N)^i (1 - p/N)^{N-i}$$

$$i = 0, 1, \cdots, N \quad (1)$$

with probability generating function (PGF)

$$A(z) \triangleq \sum_{i=0}^{N} z^i \Pr [A = i] = \left(1 - \frac{p}{N} + z \frac{p}{N}\right)^N. \quad (2)$$

As $N \to \infty$, the number of packet arrivals at the tagged queue during each time slot has the Poisson probabilities

$$a_i \triangleq \Pr [A = i] = \frac{p^i e^{-p}}{i!} \quad i = 0, 1, 2, \cdots \quad (3)$$

with probability generating function (PGF)

$$A(z) \triangleq \sum_{i=0}^{N} z^i \Pr [A = i] = e^{-p(1-z)}. \quad (4)$$

Letting Q_m denote the number of packets in the tagged queue at the end of the mth time slot, and A_m denote the number of packet arrivals during the mth time slot, we have

$$Q_m = \max (0, Q_{m-1} + A_m - 1). \quad (5)$$

When $Q_{m-1} = 0$ and $A_m > 0$, one of the new packets is immediately transmitted during the mth time slot; that is, a packet flows through the switch without suffering any delay. The queue size Q_m is modeled by a discrete-time Markov chain; Fig. 4 illustrates the state transition diagram. Using (5) and following a standard approach in queueing analysis (see, for example, [8, sect. 5.6]), we obtain the PGF for the steady-state queue size:

$$Q(z) = \frac{(1-p)(1-z)}{A(z) - z}. \quad (6)$$

Finally, substituting the right-hand side of (2) into (6), we obtain

$$Q(z) = \frac{(1-p)(1-z)}{\left(1 - \frac{p}{N} + z \frac{p}{N}\right)^N - z}. \quad (7)$$

Now, differentiating (7) with respect to z and taking the limit as $z \to 1$, we obtain the mean steady-state queue size $\bar{Q}$ given by

$$\bar{Q} = \frac{(N-1)}{N} \cdot \frac{p^2}{2(1-p)} = \frac{(N-1)}{N} \bar{Q}_{M/D/1} \quad (8)$$

where $\bar{Q}_{M/D/1}$ denotes the mean queue size for an $M/D/1$ queue. Hence, as $N \to \infty$, $\bar{Q} \to \bar{Q}_{M/D/1}$.

We can make the even stronger statement that the steady-state probabilities for the queue size converge to those of an $M/D/1$ queue. Taking the limit as $N \to \infty$ on both sides of (7) yields

$$\lim_{N \to \infty} Q(z) = \frac{(1-p)(1-z)}{e^{-p(1-z)} - z} \quad (9)$$

STATE TRANSITION PROBABILITIES

$$a_i \equiv \Pr(A = i) \qquad i = 0, 1, 2, \ldots$$

$$= \begin{cases} \dfrac{p^i e^{-p}}{i!} & \text{IF } N = \infty \\[2ex] \dbinom{N}{i}\left(\dfrac{p}{N}\right)^i \left(1 - \dfrac{p}{N}\right)^{N-i} & \text{IF } N < \infty \end{cases}$$

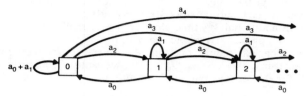

Fig. 4. The discrete-time Markov chain state transition diagram for the output queue size.

which corresponds to the PGF for the steady-state queue size of an $M/D/1$ queue. Expanding (9) in a Maclaurin series [9] yields the asymptotic (as $N \to \infty$) queue size probabilities[3]

$$\Pr (Q = 0) = (1-p)e^p \quad (10)$$

$$\Pr (Q = 1) = (1-p)e^p(e^p - 1 - p) \quad (11)$$

$$\vdots$$

$$\Pr (Q = n) = (1-p) \sum_{j=1}^{n+1} (-1)^{n+1-j} e^{jp}$$

$$\cdot \left[\frac{(jp)^{n+1-j}}{(n+1-j)!} + \frac{(jp)^{n-j}}{(n-j)!}\right] \quad \text{for } n \geq 2 \quad (12)$$

where the second factor in (12) is ignored for $j = (n + 1)$.

Although it is mathematically pleasing to have closed-form expressions, directly using (12) to compute the steady-state probabilities leads to inaccurate results for large n. When n is large, the alternating series (12) expresses small steady-state probabilities as the difference between very large positive numbers. Accurate values are required if one is interested in the tail of the distribution; for example, to compute the probability that the queue size exceeds some value M. Numerically, a more accurate algorithm is obtained directly from the Markov chain (Fig. 4) balance equations. Equations (13)–(15) numerically provide the steady-state queue size probabilities.

$$q_0 \triangleq \Pr (Q = 0) = \frac{(1-p)}{a_0} \quad (13)$$

$$q_1 \triangleq \Pr (Q = 1) = \frac{(1 - a_0 - a_1)}{a_0} \cdot q_0 \quad (14)$$

$$\vdots$$

$$q_n \triangleq \Pr (Q = n) = \frac{(1 - a_1)}{a_0} \cdot q_{n-1}$$

$$- \sum_{i=2}^{n} \frac{a_i}{a_0} \cdot q_{n-i} \qquad n \geq 2 \quad (15)$$

---

[2] We use the phrase "arrivals at the tagged queue *during* a given time slot" to indicate that packets do not arrive instantaneously, in their entirety, at the output. Packets have a nonzero transmission time.

[3] The steady-state probabilities in [9, sect. 5.1.5] are for the *total number* of packets in an $M/D/1$ system. We are interested in *queue size;* hence, the modification to (10)–(12).

where the a_i are given by (1) and (3) for $N < \infty$ and $N = \infty$, respectively.

We are now interested in the waiting time for an arbitrary (tagged) packet that arrives at the tagged output FIFO during the mth time slot. We assume that packet arrivals to the output queue in the mth time slot are transmitted over the output trunk in random order. All packets arriving in earlier time slots, however, must be transmitted first.

The tagged packet's waiting time W has two components. First, the packet must wait W_1 time slots while packets that arrived in earlier time slots are transmitted. Second, it must wait an additional W_2 time slots until it is randomly selected out of the packet arrivals in the mth time slot.

Since packets require one time slot for transmission over the output trunk, W_1 equals Q_{m-1}. Consequently, from (6), the PGF for the steady-state value of W_1 is

$$W_1(z) = \frac{(1-p)(1-z)}{A(z) - z}. \tag{16}$$

We must be careful when we compute W_2, the delay due to the transmission of other packet arrivals in the mth time slot. Burke points out in [10] that many standard works on queueing theory are in error when they compute the delay of single-server queues with batch input. Instead of working with the size of the batch to which the tagged packet belongs, it is tempting to work with the size of an arbitrary batch. Errors result when the batches are not of constant size. The probability that our tagged packet arrives in a batch of size i is given by $ia_i/\bar{A}$; hence, the random variable W_2 has the probabilities

$$\Pr[W_2 = k] = \sum_{i=k+1}^{\infty} \frac{1}{i} ia_i/\bar{A} \qquad k = 0, 1, 2, \cdots$$

$$= \frac{1}{p} \sum_{i=k+1}^{\infty} a_i \tag{17}$$

where $\bar{A} \ (= p)$ is the expected number of packet arrivals at the tagged output during each time slot, and the a_i are given by (1) and (3) for $N < \infty$ and $N = \infty$, respectively. The PGF for the steady-state value of W_2 follows directly from (17).

$$W_2(z) = \frac{1 - A(z)}{p(1-z)}. \tag{18}$$

Finally, since W is the sum of the independent random variables W_1 and W_2, the PGF for the steady-state waiting time is

$$W(z) = Q(z) \cdot \frac{1 - A(z)}{p(1-z)}. \tag{19}$$

$A(z)$ is given by (2) and (4) for $N < \infty$ and $N = \infty$, respectively.

Differentiating (19) with respect to z and taking the limit as $z \to 1$, we obtain the mean steady-state waiting time given by

$$\bar{W} = \bar{Q} + \frac{1}{2p} [\bar{A}^2 - \bar{A}]. \tag{20}$$

Since $\bar{A} = p$ and $\bar{A}^2 = p^2 + p(1 - p/N)$, substituting the right-hand side of (8) into (20) yields

$$\bar{W} = \frac{(N-1)}{N} \cdot \frac{p}{2(1-p)} = \frac{(N-1)}{N} \bar{W}_{M/D/1} \tag{21}$$

where $\bar{W}_{M/D/1}$ denotes the mean waiting time for an $M/D/1$ queue. The mean waiting time $\bar{W}$, as a function of p, is shown

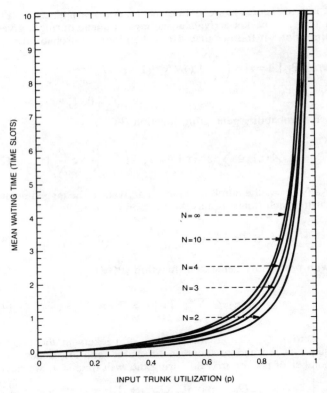

Fig. 5. The mean waiting time for several switch sizes N with output queueing.

in Fig. 5 for several values of N. Notice that Little's result and (8) generate the same formula for $\bar{W}$.

Rather than take the inverse transform of $W(z)$, it is easier to compute the steady-state waiting time probabilities from

$$\Pr[W = k] = \Pr[W_1 + W_2 = k]$$

$$= \frac{1}{p} \sum_{n=0}^{k} q_n \cdot \sum_{i=k+1-n}^{\infty} a_i \qquad k = 0, 1, \cdots$$

$$= \frac{1}{p} \sum_{n=0}^{k} q_n \cdot \left[1 - \sum_{i=0}^{k-n} a_i \right] \tag{22}$$

where the q_n are given by (13)–(15) and the a_i are given by (1) and (3) for $N < \infty$ and $N = \infty$, respectively.

III. Queues on Inputs

The interesting analysis occurs when the switch fabric runs at the same speed as the input and output trunks, and packets are queued at the inputs. How much traffic can the switch accommodate before it saturates, and how much does the mean waiting time increase when we queue packets at the inputs rather than at the outputs? As in the previous section, we assume that packet arrivals on the N input trunks are governed by independent and identical Bernoulli processes. In any given time slot, the probability that a packet will arrive on a particular input is p; each packet has equal probability $1/N$ of being addressed to any given output. Each arriving packet goes, at least momentarily, into a FIFO on its input trunk. At the beginning of every time slot, the switch controller looks at the first packet in each FIFO. If every packet is addressed to a different output, the controller closes the proper crosspoints and all the packets go through. If k packets are addressed to a particular output, one of the k packets is chosen at *random*, each selected with equal probability $1/k$. The others wait until the next time slot when a new selection is made among the packets that are then waiting.

A. Saturation Analysis—Random Selection Policy

Suppose the input queues are saturated so that packets are always waiting in every input queue. Whenever a packet is transmitted through the switch, a new packet immediately replaces it at the head of the input queue. We define B_m^i as the number of packets at the heads of input queues that are "blocked" for output i at the end of the mth time slot. In other words, B_m^i is the number of packets destined for output i, but not selected by the controller during the mth time slot. We also define A_m^i as the number of packets moving to the head of "free" input queues during the mth time slot and destined for output i. An input queue is "free" during the mth time slot if, and only if, a packet from it was transmitted during the $(m - 1)$st time slot. The new packet "arrival" at the head of the queue has equal probability $1/N$ of being addressed to any given output. It follows that

$$B_m^i = \max (0, B_{m-1}^i + A_m^i - 1). \tag{23}$$

Although B_m^i does not represent the occupancy of any physical queue, notice that (23) has the same mathematical form as (5).

A_m^i, the number of packet arrivals during the mth time slot to free input queues and destined for output i, has the binomial probabilities

$$\Pr [A_m^i = k] = \binom{F_{m-1}}{k} (1/N)^k (1 - 1/N)^{F_{m-1}-k}$$

$$k = 0, 1, \cdots, F_{m-1} \tag{24}$$

where

$$F_{m-1} \triangleq N - \sum_{i=1}^{N} B_{m-1}^i. \tag{25}$$

F_{m-1} is the number of free input queues at the end of the $(m - 1)$st time slot, representing the total number of packets transmitted through the switching during the $(m - 1)$st time slot. Therefore, F_{m-1} is also the total number of input queues with new packets at their heads during the mth time slot. That is,

$$F_{m-1} = \sum_{i=1}^{N} A_m^i. \tag{26}$$

Notice that $\bar{F}/N = \rho_0$ where $\bar{F}$ is the mean steady-state number of free input queues and ρ_0 is the utilization of the output trunks (i.e., the switch throughput). As $N \to \infty$, the steady-state number of packets moving to the head of free input queues each time slot, and destined for output i, (A^i) becomes Poisson at rate ρ_0 (see Appendix A). These observations and (23) together imply that we can use the results of Section II to obtain an expression for the mean steady-state value of B^i as $N \to \infty$. Modifying (8), we have

$$\bar{B}^i = \frac{\rho_0^2}{2(1 - \rho_0)}. \tag{27}$$

However, using (25) and $\bar{F}/N = \rho_0$, we also have

$$\bar{B}^i = 1 - \rho_0. \tag{28}$$

It follows from (27) and (28) that $\rho_0 = (2 - \sqrt{2}) = 0.586$ when the switch is saturated and $N = \infty$.

It is interesting to note that this same asymptotic saturation throughput has also been obtained in an entirely different context. Consider the problem of memory interference in synchronous multiprocessor systems [11], [12] in which M memories are shared by N processors. Memory requests are presented at the beginning of memory cycles; a conflict occurs if more than one simultaneous request is made to a particular memory. In the event of a conflict, one request is accepted, and the other requests are held for future memory cycles. If $M = N$ and processors always make a new memory request in the cycle immediately following their own satisfied request, then our saturation model for input queueing is identical to the multiprocessor model. As $N \to \infty$, the expected number of busy memories (per cycle) is $(2 - \sqrt{2}) \cdot N$ [11].

When the input queues are saturated and $N < \infty$, the switch throughput is found by analyzing a Markov chain model. Under saturation, the model is identical to the Markov chain analysis of memory interference in [12]. Unfortunately, the number of states grows exponentially with N, making the model useful only for small N. The results presented in Table I,[4] however, illustrate the rapid convergence to the asymptotic throughput of 0.586. In addition, saturation throughputs obtained by simulation[5] (Fig. 6) agree with the analysis.

B. Increasing the Switch Throughput by Dropping Packets

Whenever k packets are addressed for a particular output in a time slot, only one can be transmitted over the output trunk. We have been assuming that the remaining $k - 1$ packets wait in their input queues until the next time slot when a new selection is made among the packets that are then waiting. Unfortunately, a packet that could traverse the switch to an idle output during the current time slot may have to wait in queue behind a packet whose output is currently busy. As shown in Section III-A, input queue blocking limits the switch throughput to approximately 0.586 for large N.

Instead of storing the remaining $k - 1$ packets in input queues, suppose we just drop them from the switch (i.e., we eliminate the input queues). Dropping packets obviously reduces the switch throughput when the input trunk utilization p is small; more time slots on the output trunks are empty because new packets do not arrive fast enough to replace dropped packets. Although dropping a significant number of packets (say, more than 1 out of 1000) may not be realistic for a packet switch, it is interesting to note that as the input utilization p increases, the reduction in input queue blocking when packets are dropped eventually outweighs the loss associated with dropping the packets.

We define A_m^i as the number of packet arrivals during the mth time slot that are addressed for output i. A_m^i has the binomial probabilities

$$\Pr [A_m^i = k] = \binom{N}{k} (p/N)^k (1 - p/N)^{N-k}$$

$$k = 0, 1, \cdots, N. \tag{29}$$

We also define the indicator function I_m^i as follows:

$$I_m^i = \begin{cases} 1 & \text{if output trunk } i \text{ transmits a packet} \\ & \quad \text{during the } m\text{th time slot} \\ 0 & \text{otherwise.} \end{cases} \tag{30}$$

When we drop packets, only those that arrive during the mth time slot have a chance to be transmitted over output trunks during the mth time slot. If they are not selected in the slot in which they arrive, they are dropped. Consequently, for each output trunk i, the random variables I_r^i and $I_s^i (r \neq s)$ are independent and identically distributed, with probabilities

$$\Pr [I_m^i = 1] = \Pr [A_m^i > 0]$$

$$= 1 - (1 - p/N)^N. \tag{31}$$

---

[4] The entries in Table I were obtained by normalizing (dividing by N) the values from [12, Table III].

[5] Rather than plot the simulation results as discrete points, the saturation throughputs obtained for N between 2 and 100 are simply connected by straight line segments. No smoothing is done on the data.

TABLE I
THE MAXIMUM THROUGHPUT ACHIEVABLE WITH INPUT QUEUEING

N	Saturation Throughput
1	1.0000
2	0.7500
3	0.6825
4	0.6553
5	0.6399
6	0.6302
7	0.6234
8	0.6184
∞	0.5858

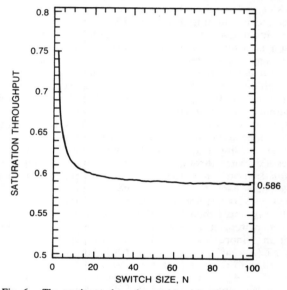

Fig. 6.   The maximum throughput achievable with input queueing.

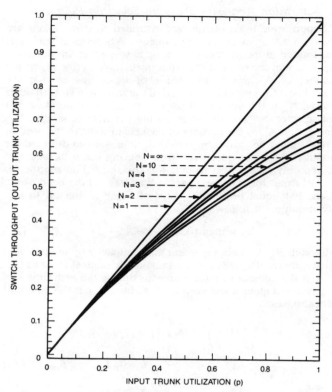

Fig. 7.   The switch throughput when packets are dropped, rather than queued at the inputs.

TABLE II
THE STRATEGY (AS A FUNCTION OF p), INPUT QUEUEING, OR PACKET DROPPING THAT YIELDS THE LARGER SWITCH THROUGHPUT

N	Queues On Inputs - Finite Queue Sizes	Queues On Inputs - Saturated Queues	Drop Packets
1	$0 \leq p < 1$	$p = 1$	----------
2	$0 \leq p < 0.750$	$0.750 \leq p \leq 1$	----------
3	$0 \leq p \leq 0.682$	$0.683 \leq p \leq 0.953$	$0.954 \leq p \leq 1$
4	$0 \leq p \leq 0.655$	$0.656 \leq p \leq 0.935$	$0.936 \leq p \leq 1$
5	$0 \leq p \leq 0.639$	$0.640 \leq p \leq 0.923$	$0.924 \leq p \leq 1$
6	$0 \leq p \leq 0.630$	$0.631 \leq p \leq 0.916$	$0.917 \leq p \leq 1$
7	$0 \leq p \leq 0.623$	$0.624 \leq p \leq 0.911$	$0.912 \leq p \leq 1$
8	$0 \leq p \leq 0.618$	$0.619 \leq p \leq 0.907$	$0.908 \leq p \leq 1$
∞	$0 \leq p \leq 0.585$	$0.586 \leq p \leq 0.881$	$0.882 \leq p \leq 1$

By symmetry, $1 - (1 - p/N)^N$ is the utilization of each output trunk; the switch throughput ρ_0 is given by

$$\rho_0 = 1 - (1 - p/N)^N. \tag{32}$$

As $N \to \infty$,

$$\rho_0 = 1 - e^{-p}. \tag{33}$$

The probability that an arbitrary packet will be dropped from the switch is simply $1 - \rho_0/p$.

The switch throughput ρ_0, as a function of p, is shown in Fig. 7 for several values of N. When the utilization p of the input trunks exceeds a critical threshold, the switch throughput ρ_0 is larger when we drop packets [(32) and (33)] than when we queue them on the input trunks (Table I). For example, when $N = \infty$ and $p > \ln (1 + \sqrt{2})$, the switch throughput when we drop packets is greater than $(2 - \sqrt{2})$ – the throughput with input queues. Table II lists, as a function of p, which of the two strategies yields the larger switch throughput.

C. Waiting Time—Random Selection Policy

Below saturation, packet waiting time is a function of the service discipline the switch uses when two or more input queues are waiting to transmit packets to the same output. In this section, we derive an exact formula for the mean waiting time under the random selection policy for the limiting case of $N = \infty$. The waiting time is obtained by simulation for finite values of N. In Section III-D, numerical results are compared to the mean waiting time under the longest queue and fixed priority selection policies.

When the input queues are not saturated, there is a significant difference between our analysis of a packet switch with input queues and the analysis of memory interference in synchronous multiprocessor systems. The multiprocessor application assumes that new memory requests are generated only after a previous request has been satisfied. A processor never has more than one memory request waiting at any time. In our problem, however, packet queueing on the input trunks impacts the switch performance.

A discrete-time Geom/G/1 queueing model (Fig. 8) is used

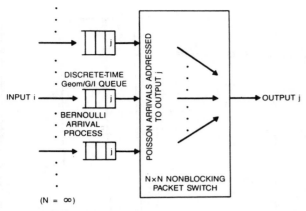

Fig. 8. The discrete-time Geom/G/1 input queueing model used to derive an exact formula for the mean waiting time for the limiting case of $N = \infty$.

to determine the expected packet waiting time for the limiting case of $N = \infty$. The arrival process is Bernoulli: in any given time slot, the probability that a packet will arrive on a particular input is p where $0 < p < 2 - \sqrt{2}$. Each packet has equal probability $1/N$ of being addressed to any given output, and successive packets are independent. To obtain the service distribution, suppose the packet at the head of input queue i is addressed for output j. The "service time" for the packet consists of the wait until it is randomly selected by the switch controller, plus one time slot for its transmission through the switch and onto output trunk j. As $N \rightarrow \infty$, successive packets in input queue i experience the same service distribution because their destination addresses are independent and distributed uniformly over all N outputs. Furthermore, the steady-state number of packet "arrivals" to the heads of input queues and addressed for output j becomes Poisson at rate p.[6] Consequently, the service distribution for the discrete-time Geom/G/1 model is itself the packet delay distribution of another queueing system: a discrete-time M/D/1 queue with customers served in random order. Analysis of the discrete-time M/D/1 queue, with packets served in random order, is given in Appendix B.

Using [6, eq. (39)], the mean packet delay for a discrete-time Geom/G/1 queue is

$$\bar{D} = \frac{p \overline{S(S-1)}}{2(1 - p\bar{S})} + \bar{S} \qquad (34)$$

where S is a random variable with the service time distribution given in Appendix B and mean value $\bar{S}$. The mean waiting time is $\overline{W} = \bar{D} - 1$

$$\overline{W} = \frac{p \overline{S(S-1)}}{2(1 - p\bar{S})} + \bar{S} - 1. \qquad (35)$$

$\overline{S(S-1)}$ and $\bar{S}$ are determined numerically using the method in Appendix B.

The mean waiting time $\overline{W}$, as a function of p, is shown in Fig. 9 for both input queueing and output queueing—in the limit as $N \rightarrow \infty$. As expected, waiting times are always greater for queueing on inputs than for queueing on outputs. Packet waiting times for input queueing and finite values of N, obtained by simulation,[7] agree with the asymptotic analytic results (Fig. 10).

---

[6] This follows from the proof in Appendix A.

[7] Rather than plot the simulation results as discrete points, the simulation results are simply connected by straight line segments; no smoothing is done on the data. The same comment applies to Figs. 11, 12, and 13.

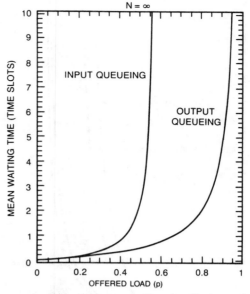

Fig. 9. A comparison of the mean waiting time for input queueing and output queueing for the limiting case of $N = \infty$.

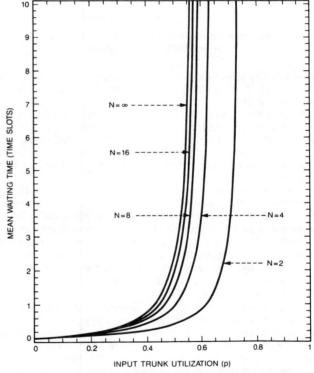

Fig. 10. The mean waiting time for several switch sizes N with input queueing.

D. Longest Queue and Fixed Priority Selection Policies

Until now, we have assumed that if k packets are addressed to a particular output, one of the k packets is chosen at random, each selected with equal probability $1/k$. In this section, we consider two other selection policies: longest queue selection, and fixed priority selection. Under the longest queue selection policy, the controller sends the packet from the longest input queue. A random selection is made if, of the k input queues with packets addressed to a particular output, several queues have the same maximum length. Under the

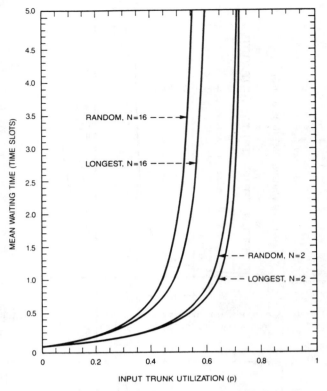

Fig. 11. The mean waiting time for input queueing with the random and longest queue selection policies.

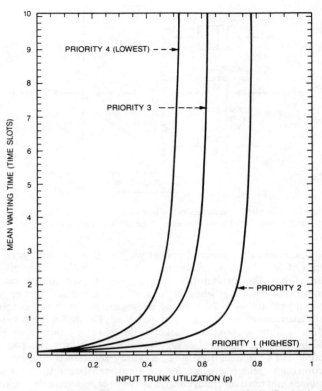

Fig. 12. The mean waiting time for input queueing with the fixed priority service discipline and $N = 4$.

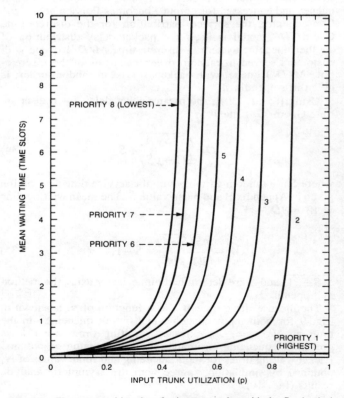

Fig. 13. The mean waiting time for input queueing with the fixed priority service discipline and $N = 8$.

fixed priority selection policy, the N inputs have fixed priority levels, and of the k packets, the controller sends the one with highest priority.

Simulation results for the longest queue policy indicate smaller packet waiting times than those expected with random service (Fig. 11). This is anticipated because the longest queue selection policy reduces the expected number of packets blocked (behind other packets) from traversing the switch to idle outputs.

For the fixed priority service discipline, our simulation results show that the lowest priority input queue suffers large delays and is sometimes saturated, even when the other two service disciplines guarantee stability. Although the saturation throughput is 0.6553 under the random selection policy when $N = 4$ (see Table I), it is shown in Fig. 12 that the lowest priority input queue saturates at approximately 0.55 under the fixed priority discipline. Fig. 13 illustrates the family of waiting time curves for $N = 8$.

These results are interesting because imposing a priority scheme on a single server queueing system usually does not affect its stability; the system remains work conserving. For the $N \times N$ packet switch, however, packet blocking at the low priority input queues does impact stability. More work remains to characterize the stability region.

IV. CONCLUSION

Using Markov chain models, queueing theory, and simulation, we have presented a thorough comparison of input versus output queueing on an $N \times N$ nonblocking space-division packet switch. What the present exercise has done, for a particular solvable example, is to quantify the intuition that better performance results with output queueing than with input queueing. Besides performance, of course, there are other issues, such as switch implementation, that must be considered in designing a space-division packet switch. The Knockout Switch [13] is an example of a space-division packet switch that places all buffers for queueing packets at the outputs of the switch, thus enjoying the performance advantages of output queueing. Furthermore, the switch fabric runs at the same speed as the input and output trunks.

APPENDIX A

POISSON LIMIT OF PACKETS MOVING TO THE HEAD OF FREE INPUT QUEUES

For the input queueing saturation analysis presented in Section III-A, as $N \rightarrow \infty$, we show that the steady-state number of packets moving to the head of free input queues each time slot, and destined for output i, (A^i) becomes Poisson at rate $\rho_0 = \bar{F}/N$. To make clear the dependency of F and ρ_0 on the number of inputs N, we define $\bar{F}(N)$ as the steady-state number of free input queues and $\rho_0(N)$ as the output trunk utilization for a given N where $\rho_0 N = \bar{F}(N)/N$.

We can write

$$\text{Var} \left\{ \frac{F(N)}{N} \right\} = \frac{1}{N} \text{Pr [input queue } r \text{ is free]}$$

$$+ \left(1 - \frac{1}{N} \right) \text{Pr [input queue } r \text{ is free,}$$

$$\text{input queue } s(s \neq r) \text{ is free]}$$

$$- (\text{Pr [input queue } r \text{ is free]})^2. \quad \text{(A1)}$$

As $N \rightarrow \infty$, the events {input queue r is free} and {input queue s is free} become independent for $s \neq r$. Therefore, from (A1),

$$\lim_{N \rightarrow \infty} \text{Var} \left\{ \frac{F(N)}{N} \right\} = 0. \quad \text{(A2)}$$

Given $\epsilon > 0$, we define the set S_N by

$$S_N \triangleq \{ L, L+1, \cdots, U-1, U \} \quad \text{(A3)}$$

where

$$L \triangleq \max \{1, \lfloor \bar{F}(N) - \epsilon N \rfloor \}, \quad \text{(A4)}$$

$$U \triangleq \min \{N, \lceil \bar{F}(N) + \epsilon N \rceil \}, \quad \text{(A5)}$$

and $\lfloor x \rfloor$ ($\lceil x \rceil$) denotes the greatest (smallest) integer less than (greater than) or equal to x. By the Chebyshev inequality,

$$\text{Pr} [F(N) \in S_N] \geq 1 - \frac{\text{Var} \left\{ \frac{F(N)}{N} \right\}}{\epsilon^2}. \quad \text{(A6)}$$

Therefore,

$$\text{Pr} [A^i = a] = \sum_{f = \max(a,1)}^{N} \text{Pr} [F(N) = f]$$

$$\cdot \binom{f}{a} (1/N)^a (1 - 1/N)^{f-a}$$

$$\leq \sum_{f \in S_N} \text{Pr} [F(N) = f]$$

$$\cdot \binom{f}{a} (1/N)^a (1 - 1/N)^{f-a}$$

$$+ \frac{\text{Var} \left\{ \frac{F(N)}{N} \right\}}{\epsilon^2}. \quad \text{(A7)}$$

Case I (a = 0): If $f \in S_N$, then

$$(1 - 1/N)^U \leq \binom{f}{0} (1/N)^0 (1 - 1/N)^{f-0} \leq (1 - 1/N)^L. \quad \text{(A8)}$$

Therefore, from (A6), (A7), and (A8), we obtain

$$\left[1 - \frac{\text{Var} \left\{ \frac{F(N)}{N} \right\}}{\epsilon^2} \right] \cdot (1 - 1/N)^U \leq \text{Pr} [A^i = 0]$$

$$\leq (1 - 1/N)^L + \frac{\text{Var} \left\{ \frac{F(N)}{N} \right\}}{\epsilon^2}. \quad \text{(A9)}$$

As $N \rightarrow \infty$,

$$e^{-(\rho_0 + \epsilon)} \leq \text{Pr} [A^i = 0] \leq e^{-(\rho_0 - \epsilon)}. \quad \text{(A10)}$$

Since this holds for arbitrarily small $\epsilon > 0$,

$$\lim_{N \rightarrow \infty} \text{Pr} [A^i = 0] = e^{-\rho_0}. \quad \text{(A11)}$$

Case II (a > 0): Since $\binom{f}{a}(1/N)^a(1 - 1/N)^{f-a}$ is a nondecreasing function of f for $1 \leq f \leq N$ and $a > 0$, for $f \in S_N$ we have

$$\binom{L}{a} (1/N)^a (1 - 1/N)^{L-a} \leq \binom{f}{a} (1/N)^a (1 - 1/N)^{f-a}$$

$$\leq \binom{U}{a} (1/N)^a (1 - 1/N)^{U-a}. \quad \text{(A12)}$$

Therefore, from (A6), (A7), and (A12), we obtain

$$\left[1 - \frac{\text{Var} \left\{ \frac{F(N)}{N} \right\}}{\epsilon^2} \right] \cdot \binom{L}{a} (1/N)^a (1 - 1/N)^{L-a}$$

$$\leq \text{Pr} [A^i = a] \leq \binom{U}{a} (1/N)^a (1 - 1/N)^{U-a}$$

$$+ \frac{\text{Var} \left\{ \frac{F(N)}{N} \right\}}{\epsilon^2}. \quad \text{(A13)}$$

As $N \rightarrow \infty$,

$$e^{-(\rho_0 - \epsilon)} \frac{(\rho_0 - \epsilon)^a}{a!} \leq \text{Pr} [A^i = a] \leq e^{-(\rho_0 + \epsilon)} \frac{(\rho_0 + \epsilon)^a}{a!}. \quad \text{(A14)}$$

Since this holds for arbitrarily small $\epsilon > 0$,

$$\lim_{N \rightarrow \infty} \text{Pr} [A^i = a] = e^{-\rho_0} \frac{\rho_0^a}{a!}. \quad \text{(A15)}$$

APPENDIX B

DISCRETE-TIME $M/D/1$ QUEUE—PACKETS SERVED IN RANDOM ORDER

In this Appendix, we present a simple numerical method for computing the delay distribution of a discrete-time $M/D/1$

queue, with packets served in random order. The number of packet arrivals at the beginning of each time slot is Poisson distributed with rate λ, and each packet requires one time slot for service. We fix our attention on a particular "tagged" packet in the system, during a given time slot. Let $p_{m,k}$ denote the probability, conditioned on there being a total of k packets in the system during the given time slot, that the remaining delay is m time slots until the tagged packet completes service. It is easy to obtain $p_{m,k}$ by recursion on m.

$$p_{1,1} = 1 \tag{B1}$$

$$p_{m,1} = 0 \qquad m \neq 1 \tag{B2}$$

$$p_{1,k} = \frac{1}{k} \qquad k \geq 1 \tag{B3}$$

$$p_{m,k} = \frac{k-1}{k} \cdot \sum_{j=0}^{\infty} p_{m-1,k-1+j} \cdot \frac{e^{-\lambda}\lambda^j}{j!} \qquad m > 1, \; k > 1. \tag{B4}$$

Averaging over k, the packet delay D has the probabilities

$$\Pr[D = m] = \sum_{k=1}^{\infty} p_{m,k}$$

$$\cdot \Pr[k \text{ packets in system immediately after the tagged packet arrives}]$$

$$= \sum_{k=1}^{\infty} p_{m,k} \cdot \sum_{n=0}^{k-1} q_n \cdot \frac{e^{-\lambda}\lambda^{k-n-1}}{(k-n-1)!} \tag{B5}$$

where the q_n are the steady-state queue size probabilities given by (13)–(15).

The variance and mean of the packet delay distribution are determined numerically from the delay probabilities in (B5).

REFERENCES

[1] J. S. Turner and L. F. Wyatt, "A packet network architecture for integrated services," in *GLOBECOM'83 Conf. Rec.,* Nov. 1983, pp. 45–50.

[2] J. J. Kulzer and W. A. Montgomery, "Statistical switching architectures for future services," in *Proc. Int. Switching Symp.,* May 1984.

[3] T.-Y. Feng, "A survey of interconnection networks," *Computer,* vol. 14, pp. 12–27, Dec. 1981.

[4] D. M. Dias and M. Kumar, "Packet switching in *N* log *N* multistage networks," in *GLOBECOM'84 Conf. Rec.,* Nov. 1984, pp. 114–120.

[5] Y.-C. Jenq, "Performance analysis of a packet switch based on a single-buffered Banyan network," *IEEE J. Select. Areas Commun.,* vol. SAC-1, pp. 1014–1021, Dec. 1983.

[6] T. Meisling, "Discrete-time queueing theory," *Oper. Res.,* vol. 6, pp. 96–105, Jan.–Feb. 1958.

[7] I. Rubin, "Access-control disciplines for multiaccess communication channels: Reservation and TDMA schemes," *IEEE Trans. Inform. Theory,* vol. IT-25, pp. 516–536, Sept. 1979.

[8] L. Kleinrock, *Queueing Systems, Vol. 1: Theory.* New York: Wiley, 1975.

[9] D. Gross and C. M. Harris, *Fundamentals of Queueing Theory.* New York: Wiley, 1974.

[10] P. J. Burke, "Delays in single-server queues with batch input," *Oper. Res.,* vol. 23, pp. 830–833, July–Aug. 1975.

[11] F. Baskett and A. J. Smith, "Interference in multiprocessor computer systems with interleaved memory," *Commun. ACM,* vol. 19, pp. 327–334, June 1976.

[12] D. P. Bhandarkar, "Analysis of memory interference in multiprocessors," *IEEE Trans. Comput.,* vol. C-24, pp. 897–908, Sept. 1975.

[13] Y.-S. Yeh, M. G. Hluchyj, and A. S. Acampora, "The knockout switch: A simple, modular architecture for high-performance packet switching," in *Proc. Int. Switching Symp.,* Phoenix, AZ, Mar. 1987, pp. 801–808.

A Broadband Packet Switch for Integrated Transport

JOSEPH Y. HUI, MEMBER, IEEE, AND EDWARD ARTHURS

Abstract—This paper gives a broadband (total throughput approaching 1 terabit/s) self-routing packet switch design for providing flexible multiple bit-rate broadband services for an end-to-end fiber network. The switch fabric for the slotted broadband packet switch delivers exactly one packet to each output port from one of the input ports which request packet delivery to that output port. The denied requests would try again during the next slot. We discover an effective scheme, implemented by CMOS VLSI with manageable complexity, for performing this function. First, each input port sends a request for a port destination through a Batcher sorting network, which sorts the request destinations in ascending order so that we may easily purge all but one request for the same destination. The winning request acknowledges its originating port from the output of the Batcher network, with the acknowledgment routed through a Batcher-banyan self-routing switch. The acknowledged input port then sends the full packet through the same Batcher-banyan switch without any conflict. Unacknowledged ports buffer the blocked packet for reentry in the next cycle. We also give several variations for significantly improved performance.

We then study switch performance based on some rudimentary protocols for traffic control. For the basic scheme, we analyze the throughput-delay characteristics for random traffic, modeled by random output port requests and a binomial distribution of packet arrival. We demonstrate with a buffer size of around 20 packets, we can achieve a 50 percent loading with almost no buffer overflow. Maximum throughput of switch is 58 percent. Next, we investigate the performance of the switch in the presence of periodic broadband traffic. We then apply circuit switching techniques and packet priority for high bit-rate services in our packet switch environment. We improve the throughput per port to close to 100 percent by means of parallel switch fabric, while maintaining the periodic nature of the traffic.

I. INTRODUCTION

WE believe that broadband services, supported by a broadband fiber network for end-to-end transport, shall be widely accepted [1]. One key element necessary for providing flexible services is a high capacity packet switch for interconnecting a large number of subscriber lines at the central office [2], [3]. In this paper, we design the packet switch with the following system objectives:

- support all services except entertainment video broadcast on a packet switch fabric;
- 1000–4000 ports serving a maximum of 200 000 subscriber lines;
- employ CMOS VLSI for the switch fabric and support a port speed of 150 Mbits/s.

The packet switch we design should preserve the packet sequencing for a service, have minimal variable delay, and be robust for all patterns of load on the ports. The

Manuscript received November 1, 1986; revised March 15, 1987. This paper was presented at ICC'87, Seattle, WA, June 8–11, 1987.

The authors are with Bell Communications Research, Morristown, NJ 07960-1961.

IEEE Log Number 8716300.

switch design described in this paper shall meet all these criteria.

We want to build a switch fabric performing the function of the box in Fig. 1. The fabric has input ports i and output ports j, $1 \leq i, j \leq N$. In the figure, we put a pair of parentheses around the source address i to distinguish it from the destination address j. At the beginning of a time slot, each port i may request delivery of a packet to a certain port j, which we denote by j_i. A switch fabric is internally nonblocking if it can deliver all packets to the requested output ports given that the j_i are distinct. However, an internally nonblocking switch fabric can still block at the output ports due to conflicting requests, namely, the occurrence of $j_i = j_{i'}$, for some $i \neq i'$. We call such an event an output port conflict, or external blocking, for the output port j. When this event occurs, the switch should deliver exactly one request among the requests for the output port. An example of output port conflict for output port 3 is shown in Fig. 2, in which the request from input port 1 conflicts with the request from input port 4, and loses the contention when the request from port 4 is delivered instead. Thus, a mechanism is needed for resolving output port conflict. Section II describes a simple mechanism for this purpose. If we maintain a first come first serve (FCFS) discipline for the arrival of requests at an input port, the denied request would try again during the next time slot. A buffer at each input port is required to hold incoming traffic while the request at the head of the queue is occasionally blocked.

The FCFS system may cause a phenomenon called head of the line (HOL) blocking, illustrated in Fig. 2. Notice that no request at the head of the line seeks delivery to output port 2, while a request for output port 2 is blocked at input port 1 because the HOL request at input port 1 is blocked. HOL blocking reduces throughput of the switch as shown in Section V, unless some mechanism is provided to arbitrate both HOL and non-HOL requests (a solution is given in Section IV), or the HOL requests are deliberately arranged to cause less contention (such as shifting the phase of periodic traffic destined for the same output port, described in Section VI). We shall show how these two techniques significantly improve the throughput of the switch. Furthermore, we shall show how the use of parallel switch planes can achieve up to a 100 percent efficiency in Section V.

The paper is organized as follows. Section II describes a basic 3-phase algorithm for the Batcher-banyan fabric, which resolves conflict through sorting and acknowledges the winning port before actual packet delivery. Section III

Reprinted from *IEEE J. Selected Areas Commun.*, vol. SAC-5, no. 8, pp. 1264–1273, October 1987.

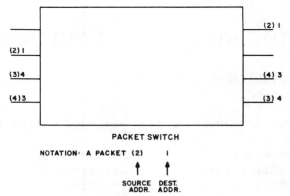

Fig. 1. A nonblocking packet switch.

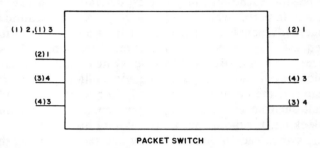

Fig. 2. Output conflict and head of line blocking.

describes the hardware implementation of the switching fabric. Section IV describes variations of the algorithm, some with improved performance. Section V analyzes the queue lengths and throughput of the switch assuming FCFS, uniform, and independent distribution of the destination addresses for packets, and binomial distribution for packet arrivals at each port. The analysis shows reasonable buffer requirement for loading around 50 percent for the single switch plane. Section VI demonstrates that with the use of circuit setup and priority for periodic traffic, significant improvements in throughput can be achieved, while the periodic nature of the traffic is maintained. Section VII summarizes the paper with a switch proposal.

II. A Basic 3-Phase Algorithm for the Batcher-Banyan Fabric

What sort of self-routing interconnection network may be used to deliver packets from input ports which make distinct requests for output ports? Most solutions involve a sorting network [4], [5], [7]. One implementation [7] consists of an internally nonblocking fabric which first sorts the packets according to the destination address by a sorting network (say a Batcher network [6]), subsequently routes the packets through a banyan network preceded by a shuffle exchange. It can be shown that the packets will not block within the banyan network. Unfortunately, blocking may occur if the requests are not distinct, or if we purge packets from the output of the sorter.

To resolve contention for output ports, Huang and Knauer [7] proposed for their Starlite switch the use of

reentry for packets losing the contention as shown in Fig. 3. At the output of the sorting network, a packet loses the contention if the packet on the preceding line in the sorted order has the same destination address. The packets which lose the contention are then concentrated by a concentration network, to be fed back to the front end of the Batcher network for reentry. The output of the sorting network has holes in the sequence due to purged packets. Consequently, another concentration network is required in front of the banyan network for skewing all packets to the top lines, filling all holes left behind by purged packets.

The Starlite approach has several drawbacks. First, packets can be lost due to blocking within the reentry network, and packets may be delivered out of sequence. Second, at least half of the input ports are dedicated for reentry. Third, the two extra concentration networks require more chip sets and subsystem designs. In this paper, we shall provide an alternative solution which is simpler and more flexible.

How may one arbitrate conflicting requests from a number of ports that are spatially separated? Such arbitration would require by itself an interconnection network. One solution is to bring conflicting requests together through sorting the requests. After sorting, the conflicting requests are adjacent to each other (Fig. 4), and a request decides that it wins the contention if the request above it in the sorted order is not making the same request. Thus, in the arbitration phase (Phase I) of the algorithm, each input port i sends a short request packet, which is just a source-destination pair (i, j_i). The requests are sorted in nondecreasing order according to the destination address j_i, and the request is granted only if j_i is different from the one above it in the sorted list.

However, the input port which made the request does not know the result of the arbitration. Consequently, the request packet (i, j_i) which won the arbitration must send an acknowledgment packet to input port i via an interconnection network. This process constitutes the acknowledgment phase (Phase II). By bringing a fixed connection from the kth output of the Batcher network to the kth input of the Batcher network (Fig. 5), the request packet (i, j_i) may send an acknowledgment packet from input port k, the position of the request packet in the sorted order, to input port i. The acknowledgment packets are sent to distinct output ports since each input port can send at most 1 request in Phase I. Consequently, the Batcher-banyan network described at the beginning of this section suffices for nonblocking delivery of the acknowledgments.

Input ports receiving acknowledgments for their request then transmit the full packet in the final Phase III (Fig. 6) through the same Batcher-banyan network, without conflict at the output port. Input ports which fail to receive an acknowledgment retain the packet in a buffer for retry in the next slot, when the 3-phase cycle is repeated.

III. Hardware Implementation of Switch Fabric

The Batcher-banyan network with the 3-phase control algorithm has to be partitioned into subsystems for imple-

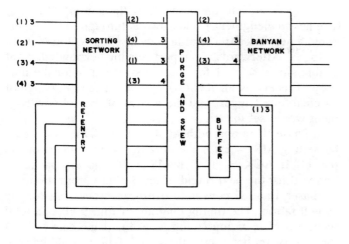

STARLITE RE-ENTRY NETWORK

Fig. 3. Starlite reentry network.

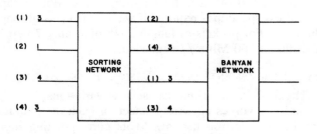

PHASE I: SEND AND RESOLVE REQUEST
• SEND SOURCE-DESTINATION PAIR THROUGH SORTING NETWORK
• SORT DESTINATION IN NON-DECREASING ORDER
• PURGE ADJACENT REQUESTS WITH SAME DESTINATION

Fig. 4. Phase I of 3-phase algorithm.

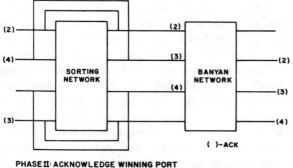

( )-ACK

PHASE II: ACKNOWLEDGE WINNING PORT
• SEND ACK WITH DESTINATION TO PORT WINNING CONTENTION
• ROUTE ACK THROUGH BATCHER-BANYAN NETWORK

Fig. 5. Phase II of 3-phase algorithm.

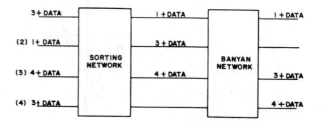

PHASE III: SEND PACKET
• ACKNOWLEDGED PORT SEND PACKET THROUGH BATCHER-BANYAN NETWORK
  BUFFERS AT PORT CONTROLLER

Fig. 6. Phase III of 3-phase algorithm.

the peripheral of the fabric. The reader may read [8] concerning the network partitioning and VLSI chip design.

Given the network implementation, we shall describe the implementation of the port controller, as shown in Fig. 8. One function of the port controller is to provide the 3-phase arbitration mechanism. Besides arbitration, the port controller runs higher level protocols such as address translation, flow control and call setup procedures, etc. The implementation of these protocols is subject to further studies. We expect most of the switch cost results from implementing these protocols and switch interfaces rather than the switch fabric itself, hence concentrating switch functions at the ports helps to modularize switch cost according to number of ports needed.

The hardware implementation of the 3-phase algorithm is shown in Figs. 7, 8, and 9. Fig. 7 indicates the input/output port interface connections to the Batcher-banyan fabric. (We show a 32 × 32 fabric for the sake of illustration.) The port interface i transmits to input i of the Batcher network by BTI_i, and receives from outputs i and $i - 1$ of the Batcher network by BTO_i and BTO_{i-1}, as well as from output i of the banyan network BNO_i. Fig. 8 shows the hardware for the port interface. The 4 : 1 multiplexer controls the phases of the algorithm through selecting one of its inputs by the MUX SELECT (MS) signal. The value of MS selects the phase of the algorithm.

In Phase I with MS = 1, a request packet, consisting of the addresses DESTINATION and SOURCE, is allowed to enter the Batcher network via BTI_i. MS is set to 1 for the duration of the request packet, which equals 2 log N bit times for a switch with N ports (All logarithms are base 2 unless stated otherwise.) Afterwards, MS = 0, consequently blocking any signal into the Batcher network. The request routes through the Batcher network, and its first bit enters another port after a time $\frac{1}{2}$ log N(log $N + 1$), the latency of the Batcher network. (See timing diagram in Fig. 9.) Just before the request enters the port, the comparison enable signal ϕ_1 clears the trigger for arbitration by its rising edge to initiate Phase II of the algorithm.

In Phase II, the two requests entering through BTO_{i-1} and BTO_i are compared by an exclusive OR gate bit-by-bit. If the requested DESTINATION's are different, the

mentation on CMOS VLSI chips. For high speed implementation at 150 Mbits/s, the network must be partitioned with the objective of minimizing the interconnection distance between chips, in order to avoid extensive use of power consuming ECL drivers. Realizing a way of interconnecting subnetworks in a 3-dimensional configuration, [8] proposes a compact mean of implementing the switch fabric, using ECL drivers only at

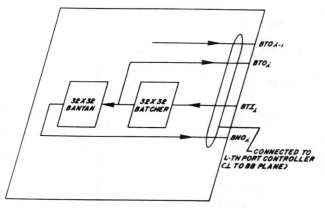

Fig. 7. Interconnections between port and switch fabric.

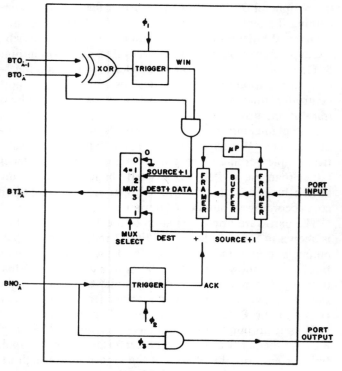

Fig. 8. The port controller.

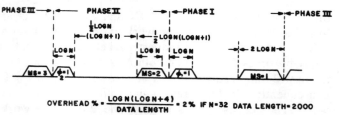

OVERHEAD % = $\dfrac{\text{LOG } N (\text{LOG } N + 4)}{\text{DATA LENGTH}}$ = 2% IF N=32 DATA LENGTH=2000

Fig. 9. Timing diagram for 3-phase algorithm.

exclusive OR gate would give a rising edge to set the trigger WIN = 1, signaling that the request coming in on line BTO_i is granted. Thus, $\phi_1 = 1$ should last only for the duration of DESTINATION, namely, log N. Afterwards, Phase II is started by multiplexing the acknowledgment packet, which is the SOURCE part of the request packet

gated by WIN, into the network by the control MS = 2. The acknowledgment packet travels through the Batcher-banyan network for a total latency of $\frac{1}{2}$ log $N(\log N + 1)$ + log N. Afterwards, each port sets the acknowledgment enable single $\phi_2 = 1$ for a duration of log N for the purpose of receiving an acknowledgment. The rising edge of ϕ_2 clears the trigger, and the arrival of the acknowledgment would set the trigger ACK = 1.

In Phase III, a port with ACK = 1 sends the full packet by setting MS = 3 for the duration of the length of the packet. If ACK = 0, then MS = 0, and the packet is buffered for the next round of the 3-phase arbitration.

Phases I and II constitute overhead processing for the switch fabric. The timing diagram of Fig. 9 gives a total overhead of $h_t = \log N(\log N + 4)$. If the data field of the full packet has l_p bits, the switch fabric would have to be speeded up by a fraction (h_t/l_p) of the speed of the input port. For $N = 1024$ and $l_p = 1024$, the speedup required is 14 percent; consequently, the switch has to operate at 170 Mbits/s for handling a 150 Mbit/s input port. Less speedup is required if we have packets of longer duration. For packets of length 2048 bits, only 7 percent speedup at 160 Mbits/s is needed.

IV. Improvements of the Basic 3-Phase Algorithm

The 3-phase algorithm has several variations. The first variation involves no wire connection from the output of the Batcher sorting network to its corresponding input. The basic idea (Fig. 10) is to allow the requests surviving the contention of Phase I to route to their destinations through the banyan routing network, after going through a concentration network which eliminates all the holes after purging. (A Batcher network would serve that function.) Phases II and III remain the same, except that the acknowledgment is sent from the destination j_i instead. This modified 3-phase scheme can be reduced to a 2-phase scheme as follows. In Phase I, each input port with a packet sends the full packet through the network, and allows those packets winning the contention at the output of the Batcher sorting network to go all the way through the concentration network and then the banyan routing network to their destinations. In Phase II, the output ports receiving these packets acknowledge the origination of the packet by sending an acknowledgment packet to the origin. Viewed alternatively, this 2-phase scheme sends out a packet without the assurance of delivery, but retains a copy of the sent packet in a buffer, and flushes the packet only if an acknowledgment is received.

The second variation involves the designation of packet priority. Priority of packets for the contention resolution process is easily achieved by the built-in sorting function of the network. Packets of higher priority can be distinguished from those of lower priority by a priority field, appended at the end of the destination address of the request packet. The appended address is sorted by the Batcher network, consequently placing the higher priority packet at a favorable position when packets with the same request for an output port are purged.

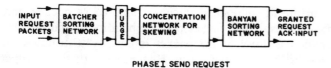

Fig. 10. Modification of 3-phase algorithm.

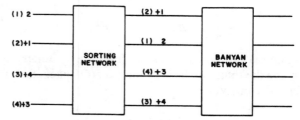

Fig. 11. Use of packet priority.

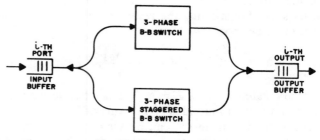

Fig. 12. The duplex switch.

This priority feature can be very profitably used for reducing *HOL* blocking. We may grant a request for the second packet in the line blocked by an *HOL* packet which lost the contention through the following process. After Phases I and II, those *HOL* packets which lost the contention would allow the second packet in the line to make requests in a second round of Phases I and II as shown in Fig. 11. These second attempts must not interfere with the requests granted during the first round. This is achieved by allowing the packets which won the first round to make requests during the second round with a higher priority, thus guaranteeing them to win in the second round. The same procedure can be repeated for a third round for the third packet in the queue. However, the diminishing return for throughput may not justify the speed overhead required for accommodating the extra phases. We shall show in Section VI how packet priority may also be used for slot assignment to avoid *HOL* blocking. Thus, packet priority allows circuit switching in a packet switch environment, in which bursty traffic can be switched by the excess capacity after circuit switching.

The third variation involves the use of 2 switch planes in parallel as shown in Fig. 12. We call this design the duplex switch. Each switch plane operates the 3-phase basic algorithm, with the second switch plane staggering by half a slot behind the first switch plane. The *HOL* packets present themselves to the first switch plane for contention resolution. Since the first 2 phases of the algorithm finish in well less than half the slot time, packets may move up the queue. The resulting *HOL* packets (blocked or fresh) would contend for the second switch plane. This process alternates between the two switch planes. A buffer is needed at the output ports for storing simultaneous arrivals from the two switch planes for the same output port. The output buffer and input buffer at the port controller may reside in the same buffer space.

More generally, we may have more than two switch planes operating in parallel, provided adjacent switch planes are staggered by a period of time longer than that required by Phases I and II of the basic algorithm. Effectively, the throughput capacity of the switch with S planes is equal to S times the single plane switch. The parallel plane switch may be used for the following advantages. First, it serves as a mean to match the speed of the incoming link when the switch fabric cannot run at the same speed as the incoming link. Second, increasing the throughput of the switch increases throughput on the output trunk beyond the 58 percent capacity for random traffic through 1 switch plane to full loading (to be derived in Section V). Third, and most important of all, this duplex

switch provides redundancy of switch fabric for enhancing system reliability.

V. Switch Performance for Random Traffic

In this section, we assume that a packet arrives at each port with probability λ per slot, destined with equal probability $(1/N)$ to any one of the output ports. Each input port has a buffer for incoming packets, which are served on an FCFS basis at the begining of each slot. We are primarily interested in the throughput-delay characteristics of the switch. Since an exact performance analysis for the N-dimensional Markov chain ($N \approx 1000$) is out of the question, an approximate engineering approach will be used. The switch is modeled as N independent single server queues (one for each input port), each queue with an effective service time distribution which accounts for the *HOL* contention. More specifically, we assume for each input port that there is a probability q of winning a switch arbitration, where q is the function of λ, the traffic carried per port. The analysis is then partitioned into two parts: q is determined as a function of λ, and the individual input queues are analyzed.

There are two types of *HOL* packets at an input queue just before the arbitration phase—the blocked *HOL* packets which lost the previous arbitration, and the fresh *HOL* packets which just moved to *HOL*, which could be a new arrival in the previous slot. The fresh *HOL* packets have independent destination addresses, whereas the destination addresses of the blocked *HOL* packets are not independent because they were involved in the previous arbitration. The state of *HOL* blocking is characterized by the number of *HOL* requests for the same output port j, namely, N_j, for $1 \leq j \leq N$. The number of successful deliveries during a slot equals $\Sigma_{j=1}^{N} \epsilon(N_j)$, in which the indicator function $\epsilon(x) = 1$ if $x > 0$ and $\epsilon(x) = 0$ if $x \leq 0$. The throughput per port is defined as the steady-state expected value

$$T = \frac{1}{N} \sum_{j=1}^{N} E[\epsilon(N_j)] = E[\epsilon(N_j)] \qquad (1)$$

in which the last equality results from the symmetry of j. Let N_b be the total number of blocked HOL packets during a slot. Obviously,

$$N_b = \sum_{j=1}^{N} N_j - \sum_{j=1}^{N} \epsilon(N_j). \qquad (2)$$

We then divide (2) by N and take expectation during steady state. Substituting the result into (1), we have the throughput per port

$$T = E[N_j] - \frac{E[N_b]}{N}. \qquad (3)$$

In steady state, we have $T = \lambda$. The last two terms in (3) will be computed individually in terms of λ.

The second term in (3) can be viewed as the steady-state fraction of blocked HOL packets. Let ρ be the steady-state probability that a queue has a fresh HOL packet, given that the input port is not blocked during the previous cycle. (There are $M = N - N_b$ such ports.) Since every packet will eventually become a fresh HOL packet at some point, the following conservation relationship must hold:

$$E[M]\rho = N\lambda. \qquad (4)$$

Consequently, the second term in (3) can be expressed as

$$\frac{E[N_b]}{N} = 1 - \frac{E[M]}{N} = 1 - \frac{\lambda}{\rho}. \qquad (5)$$

The first term in (3) is the steady-state expected number of requests for output port j. Let N_j' be the value of N_j for the next slot, given by

$$N_j' = N_j - \epsilon(N_j) + A_j \qquad (6)$$

for which the random variable A_j is the number of fresh HOL arrivals for output port j. Taking expectations for both sides of (6) during steady state when $E(N_j') = E(N_j)$, we have

$$E[A_j] = E[\epsilon(N_j)] = \lambda. \qquad (7)$$

Squaring both sides of (6), we have

$$(N_j')^2 = (N_j)^2 + \epsilon(N_j) + A_j^2 - 2N_j$$
$$+ 2N_j A_j - 2\epsilon(N_j)A_j. \qquad (8)$$

We now take expectation during steady state for (8). We also notice that A_j is independent of N_j for large N. After applying the substitution in (7), we obtain the following expression:

$$E[N_j] = E[A_j] + \frac{E[A_j(A_j - 1)]}{2(1 - E[A_j])}. \qquad (9)$$

It remains to compute $E[A_j(A_j - 1)]$ for evaluating $E[N_j]$ in (9). For given M, A_j is the fresh HOL arrivals

from the M unblocked queues. Each of M queues may have a fresh HOL packet with probability ρ, and the destination address of the packet is equally distributed for each of the N output ports. Hence, each of the M input port independently contributes, with probability (ρ/N), an arrival for the total arrival A_j. Thus, the moment generating function of A_j for given M is

$$F_M(S) = E[S^{A_j} | M] = \left[1 - \frac{\rho}{N} + \frac{\rho}{N} S \right]^M. \qquad (10)$$

In equilibrium, let π_m denote the probability that $M = m$, $0 \leq m \leq N$. Then the moment generating function of A_j is given by

$$F(S) = \sum_{m=0}^{N} \pi_m F_m(S). \qquad (11)$$

Differentiating $F(S)$ twice with respect to S, and letting $S = 1$, we have

$$E[A_j(A_j - 1)] = \rho^2 \frac{E[M(M - 1)]}{N^2}. \qquad (12)$$

Substituting N_b in (2) into $M = N - N_b$, we can express M as the sum of N random variables X_j, $1 \leq j \leq N$, given by

$$X_j = 1 + \epsilon(N_j) - N_j. \qquad (13)$$

In equilibrium, the X_j are rearrangeable random variables by symmetry. As N goes to infinity, the covariance of any two X_j will go to zero. By the law of large numbers for sums of rearrangeable random variables, M/N becomes a constant (almost surely). Therefore, $N \gg 1$,

$$E[A_j(A_j - 1)] \approx \frac{E^2[M]\rho^2}{N^2} = \lambda^2, \qquad (14)$$

in which the last equality follows from (4). Thus, we obtain the second term of (3) by substituting (14) into (9):

$$E[N_j] = \lambda + \frac{\lambda^2}{2(1 - \lambda)}. \qquad (15)$$

Adding the two terms of (3) from (5) and (15), we have

$$\lambda = T = \lambda + \frac{\lambda^2}{2(1 - \lambda)} - \left(1 - \frac{\lambda}{\rho} \right) \qquad (16)$$

from which we obtain the equation

$$(2 - \rho)\lambda^2 - 2(1 + \rho)\lambda + 2\rho = 0. \qquad (17)$$

Thus, we have a relation between the input traffic rate λ and the degree of saturation ρ, which measures traffic arrival rate conditioned on a successful removal of the HOL packet. In particular, the maximum λ is obtained by setting $\rho = 1$, resulting in $\lambda = 2 - \sqrt{2} = 0.58$. Therefore, the maximum throughput per port is 58 percent for randon traffic. More generally, we shall use (17) to derive a relationship between λ and q, the probability that the HOL packet is served.

In the remainder of this section, we shall focus on a queueing model for each input port i. Let the random variable K be the number of packets in an input queue just before the arbitration phase. We shall consider a buffer of infinite size for the moment. The queue length K' for the next slot is modeled by

$$K' = K - \gamma\epsilon(K) + \alpha \qquad (18)$$

in which γ is a 0–1 random variable with $E[\gamma] = q$, the probability that the HOL packet is served, and α is a 0–1 random variable with $E[\alpha] = \lambda < q$. The value of q as a function of λ can be computed as follows. The mean number of slots before a fresh HOL packet is served is $(1/q)$. After a packet is served, the mean number of slots before an arrival of a fresh HOL packet is $((1 - \rho)/\rho)$. In steady state, the sum of these two terms equals the interarrival time of packets at a switch port, namely, $(1/\lambda)$. This relationship, together with the equation of ρ in terms of λ in (14), gives

$$\frac{1}{q} = 1 + \frac{\lambda}{2(1 - \lambda)}. \qquad (19)$$

From (18), the moment generating function for the steady-state probabilities p_k is given by

$$G(S) = E[S^K] = E[S^\alpha]E[S^{K-\gamma\epsilon(K)}]. \qquad (20)$$

The first term on the right-hand side equal $[1 - \lambda + \lambda S]$. The second term equals

$$E_\gamma\left[\sum_{k=0}^\infty p_k S^{k-\gamma\epsilon(k)} \right]$$

$$= p_0 + E_\gamma\left[\sum_{k=1}^\infty p_k S^{k-\gamma} \right]$$

$$= p_0[1 - E_\gamma[S^{-\gamma}]] + E_\gamma\left[\sum_{k=0}^\infty p_k S^{k-\gamma} \right]$$

$$= p_0[1 - E_\gamma[S^{-\gamma}]] + G(S) E_\gamma[S^{-\gamma}]. \qquad (21)$$

Substituting (21), and $E_\gamma[S^{-\gamma}] = [(1 - q) + qS^{-1}]$ into (20), we have

$$G(S) = \frac{qp_0(S - 1)(1 - \lambda + \lambda S)}{S - (1 - \lambda + \lambda S)((1 - q)S + q)}$$

$$= \frac{qp_0(1 - \lambda + \lambda S)}{((1 - \lambda)q - \lambda(1 - q)S)} \qquad (22)$$

where the last equality results from cancelling the common factor $(S - 1)$. After evaluating p_0 by setting $G(1) = 1$, and substituting $\omega_\lambda = (\lambda(1 - q)/(1 - \lambda)q) = (\lambda^2/2(1 - \lambda)^2)$, we obtain finally

$$\sum_{k=0}^\infty p_k S^k = G(S) = (1 - \omega_\lambda)\frac{1 - \lambda + \lambda S}{1 - S\omega_\lambda}$$

$$= (1 - \omega_\lambda)(1 - \lambda + \lambda S)\sum_{k=0}^\infty \omega_\lambda^k S^k. \qquad (23)$$

Consequently, we have

$$p_0 = (1 - \omega_\lambda)(1 - \lambda) \qquad (24)$$

$$p_k = (1 - \omega_\lambda)((1 - \lambda)\omega_\lambda + \lambda)\omega_\lambda^{k-1}$$

$$= \frac{\lambda(2 - \lambda)}{2(1 - \lambda)}(1 - \omega_\lambda)\omega_\lambda^{k-1}; \qquad k > 0 \quad (25)$$

from which the expected value of K is given by

$$\overline{K} = \frac{\lambda(2 - \lambda)(1 - \lambda)}{(2 - \sqrt{2} - \lambda)(2 + \sqrt{2} - \lambda)}. \qquad (26)$$

Fig. 13 plots the average queue length $\bar{k}$ and average delay $D = (\overline{K}/\lambda)$ (from Little's theorem) as a function of λ.

The buffer overflow probability for a finite buffer of size B is upper bonded by the probability of $K > B$ for the case of infinite buffer size. Therefore,

$$P(\text{loss}) < P(K > B) = \frac{\lambda(2 - \lambda)}{2(1 - \lambda)}\omega_\lambda^B. \qquad (27)$$

Fig. 14 plots this upper bound, which is acceptably low for λ as high as 50 percent, provided $B \approx 16$. Beyond $B = 20$, the packet loss rate is negligible.

VI. CIRCUIT SWITCHED TRAFFIC IN A PACKET SWITCH ENVIRONMENT

Using the results of the previous section, we shall show how periodic traffic is affected by variable packet delay. We first examine a low bit-rate periodic traffic (such as voice), and then full motion video traffic. For periodic traffic occupying a significant fraction of the speed of the port, we show how we may avoid the HOL contention by shifting the relative phase of services destined for the same output port. Finally, we illustrate how we may use traffic priority for maintaining the periodicity of multirate services. We also discuss distributed circuit setup algorithms as well as how the duplex switch can be used for strict sense nonblocking circuit switching.

The use of a packet switch for circuit switching was proposed in the synchronous wideband switch [9], which is a time-space-time switch with nonblocking Batcher-banyan fabric as the space division switch. (Strictly speaking, the synchronous wideband switch provides interconnection at the 1.5 Mbits/s, or DS1 level, instead of 64 kbits/s or DS0 level. However, in this discussion, we are only interested in its time-space-time feature for illustration.) The space division switch is essentially a broadband packet switch, without any output port contention resolution mechanism. Time division switching is performed at the input and output ports of the space division switch by time slot interchangers (TSI), for rearranging the time slots to avoid port contention. At most half of the slots for the input and output of the space division switch are loaded for strict sense nonblocking circuit switching. In this section, we want to examine how the periodicity of mixed services would be affected if we eliminate all TSI in the time-space-time architecture, and

AVERAGE QUEUE LENGTH AND DELAY

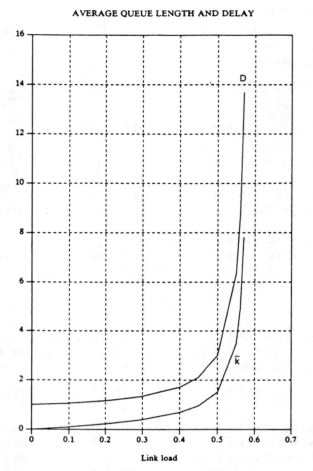

Fig. 13. Delay throughput analysis.

UPPER BOUND ON PACKET LOSS

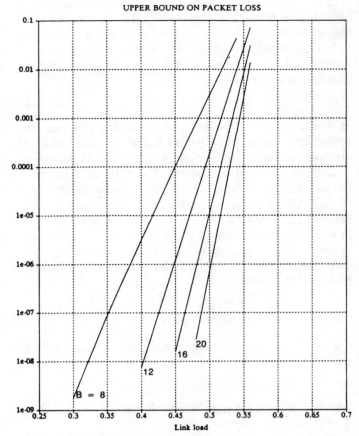

Fig. 14. Buffer overflow probability versus loading.

replace the Batcher-banyan fabric with the 3-phase packet switch described in this paper.

A signal is strictly periodic if the packet interarrival time is a constant. given flexible bit-rates and packet multiplexing of services, maintaining strict periodicity is impossible. However, for reasons discussed later, we would prefer the signal to be fairly periodic per link within the network. Thus, we need a measure of burstiness [10] of the traffic for each service, which is defined by the packet jitter ratio.

$$J = \frac{\text{Var}\,[\text{Packet interarrival time}]}{\text{E}\,[\text{Packet interarrival time}]}.$$

We shall see how J is affected for the different services. It should be noted that the packet loss rate derivations in the previous section also suffice for computing the probability that the packet jitter exceeds a certain tolerable duration, assuming random output addresses.

Consider loading the packet switch with voice traffic only. At a port speed of 150 Mbits/s, voice packet interarrival time for a call is 2000 slots. Subsequently, consecutive hundreds of packets on the channel belong to different calls with essentially random output address. From the analysis of the previous section, we know that we can achieve a throughput of 0.5 provided buffer size is more

than 16. Thus, we can achieve the same efficiency of the synchronous wideband switch without time slot interchanging 1000 packets. From Fig. 13, the variance of delay, which is upper bounded by the mean delay, is less than 2 slots for loading = 0.5. Therefore, J is less than $(2/2000) = 0.001$ per switch for voice. On a 150 Mbits/s link, 2 1000-bit packets has a duration of 28 μs, which is substantially less than the speech sampling interval of 125 μs or the packetization delay of 16 ms for 64 kbit/s speech. We conclude that for low bit-rate services such as speech, signal tranversing several switches and multiplexers in tandem is almost periodic. Therefore, time stamping of packets and buffering of more than one packet may not be required for playout of the signal. Thus, a packet switching network is easy to interface with the existing circuit switched network for low rate services. Furthermore, the random output port address assumption of Section V remains valid for switches in tandem.

For periodic traffic occupying a significant fraction of the capacity of a switch port, both the periodicity of traffic and the property of random output port addresses for consecutive packets are no longer preserved. Consider a 45 Mbit/s video phone service on a 150 Mbit/s packet channel. One out of three packets comes from the same service, suggesting $J \approx (2/3)$. Furthermore, a delay of 2 packets would cause 2 packets of the same service to bundle up. Without proper control of variable delay for the packet switch, this bundling continues to snowball for switches in tandem. This bundling not only requires buffering and time stamping for playout of the signal, but may also create buffer overflow problems and reduce throughput compared to the random traffic case. Thus, it is important that the periodicity of the traffic be preserved, in the sense of maintaining a small J.

There are several mechanisms for maintaining a small J for high bit-rate services. First, we may assign high priority for high bit-rate services, thus reducing their variable delay at the expense of low bit-rate services. However, the high bit-rate services still have to contend with each other. The effect of this contention on variable delay depends on the number of high bit-rate services. For example, given 2 contending services, it can be easily seen that periodicity is hardly disturbed, since each service automatically acquires a phase different from the other.

The second mechanism involves shifting the phase (packet arrival time with respect to a frame reference). Consider each input port and output port to be loaded with at most r services, and each service has a period of n packets, $n \geq r$. The Slepian–Duguid theorem states that there exists a time slot rearrangement of r packets in a frame of n slots such that there will be no output port conflict for the packet switch. Unfortunately, the computation for finding the arrangement, as well as the amount of rearrangement for adding a service to existing arrangements, becomes intractable for large n and $(r/n) > \frac{1}{2}$. For small n, the arrangement is relatively straightforward. Consider $n = 3$ and $r = 2$. It can be readily seen that adding a service to a port with at most 1 service requires no rearrangement of slot assignments. One scenario of arrangements is shown in Fig. 15. Notice the vacant slots at the input are blocking for all but one output port, hence making these slots quite useless for bursty traffic. Rearranging the 2 circuits by circuit packing algorithms would make these vacant slots less blocking. By utilizing these vacant slots at 30 percent efficiency, an output loaded with two 45 Mbit/s services can achieve an efficiency of over 70 percent.

The circuit setup and packing algorithms can be implemented in a distributed fashion. During the call setup procedure, an input port sends a packet to the destination output, requesting a multiple slot assignment within a frame. The packet also contains a vector registering the state of occupancy of the frame at the input. The output then checks the state of occupancy of the output frame, and assigns the earliest available slots common to both input and output frames. Subsequently, the output communicates the assignment (or denial) by sending a packet to the input. During call tear down, the newly vacated slots can be reassigned to services holding slots which are later than the vacated slots, thus performing packing of circuits. This reassignment has to be confirmed by the input. Circuit rearrangements involving other ports can be implemented similarly by having distributed occupancy tables and distributed computation via communication through the switch.

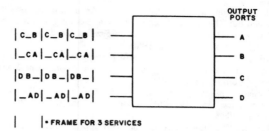

ONE 150 Mbps CHANNEL • THREE 45 Mbps VIDEO SERVICES

Fig. 15. Circuit switched traffic.

The use of the duplex switch can provide strict sense nonblocking circuit setup. The background bursty traffic can be carried by the excess capacity with low priority.

VII. CONCLUSION

We summarize this paper with a proposal for switch design. We advocate using an internally nonblocking switch fabric such as the Batcher-banyan network because of its ability to transport all patterns of loading with tight delay control. An unpredictable loading pattern is inevitable if we want a truly integrated transport network for all services. Such an integrated network has the advantages of being bit-rate flexible, as well as allowing easy bridging of multimedia communications. In contrast, most internally blocking networks, such as the buffered banyan network, perform rather poorly for uneven traffic [11]. To date, we have very little understanding of mixing traffic for these blocking switches in tandem.

In a hybrid transmission environment [12], a 150 Mbit/s link is framed at the speech sampling rate of 8 kHz. By dividing a frame into 18 slots, each slot can transport up to 130 bytes of information. Allowing an up to 10 byte header, each packet can carry an information payload of 120 bytes, or approximately 1000 bits. We allow high bit-rate (>1 Mbit/s) periodic traffic to take up multiple slots within a frame of 18 slots. Strict sense nonblocking slots are assigned through the duplex switch. Excess capacity assigned to a service can be used for random traffic, which has lower priority than slot assigned traffic. Services occupying a bandwidth substantially less than 1/18th of the channel may be treated as random traffic. Thus, the use of packet priority allows us to have a tight control on delay and packet loss rate of individual services. The flexible assignment of slots also allows a broad spectrum of bit-rates.

We have demonstrated switching mechanisms for a broad spectrum of bit-rates and tight control of delay for multirate services. Such switches may provide a bridge for a smooth transition from the predominantly circuit switched services at present to the highly heterogeneous traffic of future broadband services.

REFERENCES

[1] J. S. Turner, "New directions in communications (or which way to the information age)," *IEEE Commun. Mag.*, vol. 24, Oct. 1986.

[2] D. R. Spears, "Broadband ISDN switching capabilities from a services perspective," this issue, pp. 1222-1230.

[3] P. Gonet, P. Adams, and J. P. Coudreuse, "Asynchronous time-division switching: The way to flexible broadband communication networks," in *Proc. IEEE 1986 Int. Zurich Seminar on Digital Commun.*, Zurich, Switzerland, Mar. 1986.

[4] A. Borodin and J. E. Hopcroft, "Routing, merging, and sorting on parallel models of computation," in *Proc. 14th Annu. ACM Symp. on Theory of Comput.*, pp. 65-71.

[5] J. D. Ullman, *Computational Aspects of VLSI.* Rockville, MD: Computer Science Press, p. 230, theorem 6.7, and p. 242, bibliographic notes, 1985.

[6] K. E. Batcher, "Sorting networks and their applications," in *Proc. 1968 Spring Joint Computer Conf.*, pp. 307-314.

[7] A. Huang and S. Knauer, "Starlite: A wideband digital switch," in *Proc. 1984 Globecom Conf.*

[8] C. M. Day, J. N. Giacopelli, and J. Hickey, "Applications of self-routing switching to LATA fiber optic networks," in *Conf. Proc. Int. Symp. Switching*, Phoenix, AZ, Mar. 1987.

[9] L. T. Wu and N. C. Huang, "Synchronous wideband network—An interoffice facility hubbing network," in *Proc. IEEE 1986 Int. Zurich Seminar on Digital Commun.*, Mar. 1986.

[10] K. Sriram and W. Whitt, "Characterizing superposition arrival processes in packet multiplexers for voice and data," *IEEE J. Select. Areas Commun.*, vol. SAC-3, Sept. 1986.

[11] L. T. Wu, "Mixing traffic in a buffered banyan network," in *Proc. 9th Data Commun. Symp.*, Whistler Mountain, B.C., Canada, Sept. 1985.

[12] L. T. Wu, S. H. Lee, and T. T. Lee, "Dynamic TDM—A packet approach to broadband networking," in *Conf. Proc. Int. Conf. Commun.*, Seattle, WA, June 1987.

Nonblocking Networks for Fast Packet Switching

Riccardo Melen and Jonathan S. Turner
Computer and Communications Research Center
Washington University, St. Louis

ABSTRACT

We define and study an extension of the classical theory of nonblocking networks that is applicable to multirate circuit and fast packet/ATM switching systems. We determine conditions under which the Clos, Cantor and Beneš networks are strictly nonblocking. We also determine conditions under which the Beneš network and variants of the Cantor and Clos networks are rearrangeable. We find that strictly nonblocking operation can be obtained for multirate traffic with essentially the same complexity as in the classical context.

1. INTRODUCTION

In this paper we introduce a generalization of the classical theory of nonblocking switching networks to model communication systems designed to carry connections with a multiplicity of data rates. The theory of nonblocking networks was motivated by the problem of designing telephone switching systems capable of connecting any pair of idle terminals, under arbitrary traffic conditions. From the start, it was recognized that crossbar switches with n terminals and n^2 crosspoints could achieve nonblocking behavior, only at a prohibitive cost in large systems. In 1953, Charles Clos [4] published a seminal paper giving constructions for a class of nonblocking networks with far fewer crosspoints, providing much of the initial impetus for the theory that has since been developed by Beneš [1], Pippenger [11] and many others [3,6,8].

Riccardo Melen is with Centro Stude E Laboratori Telecomunicazioni (CSELT), Torino, Italy and his work has been supported in part by Associazione Elettrotecnica ed Elettronica Italiana, Milano, Italy. This work was done while on leave at Washington University.

Jonathan Turner is with the Computer Science Department, Washington University, St. Louis, MO 63130. His work has been supported by the National Science Foundation (grant DCI 8600947), Bell Communications Research, Bell Northern Research, Italtel SIT and NEC.

The original theory was developed to model electromechanical switching systems in which both the external links connecting switches and the internal links within them were at any one time dedicated to a single telephone conversation. During the 1960's and 1970's technological advances led to digital switching systems in which information was carried in a multiplexed format, with many conversations time-sharing a single link. While this was a major technological change, its impact on the theory of nonblocking networks was slight, because the new systems could be readily cast in the existing model. The primary impact was that the the traditional complexity measure of crosspoint count had a less direct relation to cost than in the older technology.

During the last ten years, there has been growing interest in communication systems that are capable of serving applications with widely varying characteristics. In particular, such systems are being to designed to support connections with arbitrary data rates, over a range from a few bits per second to hundreds of megabits per second [5,7,14]. These systems also carry information in multiplexed format, but in contrast to earlier systems, each connection can consume an arbitrary fraction of the bandwidth of the link carrying it. Typically, the information is carried in the form of independent blocks, called *packets* or *cells* which contain control information, identifying which of many connections sharing a given link, the packet belongs to. One way to operate such systems is to select for each connection, a path through the switching system to be used by all packets belonging to that connection. When selecting a path it is important to ensure that the available bandwidth on all selected links is sufficient to carry the connection. This leads to a natural generalization of the classical theory of nonblocking networks, which we explore in this paper. Note that such networks can also be operated with packets from a given connection taking different paths; reference [15] analyzes the worst-case loading in networks operated

Reprinted from *IEEE INFOCOM '89*, pp. 548–557, April 1989.

in this fashion. The drawback of this approach is that it makes it possible for packets in a given connection to pass one another, causing them to arrive at their destination out of sequence.

In Section 2, we define our model of nonblocking multirate networks in detail. Section 3 contains results on strictly nonblocking networks, in particular showing the conditions that must be placed on the networks of Clos and Cantor in order to obtain nonblocking operation in the presence of multirate traffic. We also describe two variants on the Clos and Cantor network that are wide-sense nonblocking in the multirate environment. Section 4 gives results on rearrangeably nonblocking networks, in particular deriving conditions for which the networks of Beneš and Cantor are rearrangeable.

2. PRELIMINARIES

We denote a network N by a quadruple (S, L, I, O), where S is a set of vertices, called *switches*, L is a set of arcs called *links*, I is a set of *input terminals* and O a set of *output terminals*. Each link is an ordered pair (x, y) where $x \in I \cup S$ and $y \in O \cup S$. We require that each input and output terminal appear in exactly one link. Links that include an input terminal are called input links or simply inputs. Those including output terminals are called outputs. The remaining links are called *internal* links. A network with n inputs and m outputs is referred to as an (n, m)-network. An (n, n)-network is also called an n-network.

We consider only networks that can be divided into a sequence of *stages*. We say that the input vertices are in stage 0 and for $i > 0$, a vertex v is in stage i if for all links (u, v), u is in stage $i - 1$. An link (u, v) is said to be in stage i if u is in stage i. In the networks we consider, all output terminals are in the same stage, and no other vertices are in this stage. When we refer to a k stage network, we generally neglect the stages containing the input and output vertices.

There are two basic components from which we construct networks. The first is the m input n output crossbar, denoted $X_{m,n}$. If σ is a permutation on $\{0 \ldots, n - 1\}$, we also let σ denote the network (S, L, I, O) where $I = \{u_0, \ldots, u_{n-1}\}$, $O = \{v_0, \ldots, v_{n-1}\}$, $S = \emptyset$ and $L = \{(u_i, v_{\sigma(i)}) \mid 0 \leq i \leq n - 1\}$. If d_1 and d_2 are positive integers, we define τ_{d_1, d_2} to be the permutation on $\{0, \ldots, d_1 d_2 - 1\}$ satisfying $\tau_{d_1, d_2}(j d_1 + i) = i d_2 + j$ for $0 \leq i \leq d_1 - 1$ and $0 \leq j \leq d_2 - 1$.

Networks are constructed using several basic operations. The *concatenation* of two networks N_1 and N_2 is denoted $N_1; N_2$ and is obtained by identifying output link i of N_1 with input link i of N_2. This operation

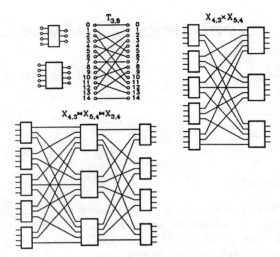

Figure 1: Network Construction Operations

effectively deletes the output terminals of N_1 and the input terminals of N_2. We require of course that the number of outputs of N_1 match the number of inputs of N_2.

The *reverse* of a network N is the network obtained by exchanging inputs and outputs and reversing the directions of all links and is denoted by N'.

If i is a positive integer and N is an (n, m)-network, then $i \cdot N$ denotes the network obtained by taking i copies of N, without interconnecting them. Inputs and outputs to N are numbered in the obvious way, with the first copy receiving inputs $0, \ldots, n - 1$ and outputs $0, \ldots, m - 1$ and so forth.

Let N_1 be a network with n_1 outputs and N_2 be a network having n_2 inputs. The *product* of N_1 and N_2 is denoted $N_1 \times N_2$ and is defined as

$$(n_2 \cdot N_1); \tau_{n_1, n_2}; (n_1 \cdot N_2).$$

Informally, the product is obtained by taking n_2 copies of N_1 and connecting them to n_1 copies of N_2 with a single link joining each pair of subnetworks.

We also define a *three-fold product* which we denote with the symbol $\bowtie$. If N_1 has n_1 outputs, N_2 has n_2 inputs and n_3 outputs and N_3 has n_1 inputs, the product $N_1 \bowtie N_2 \bowtie N_3$ is defined as

$$(n_2 \cdot N_1); \tau_{n_1, n_2}; (n_1 \cdot N_2); \tau_{n_3, n_1}; (n_3 \cdot N_3)$$

These definitions are illustrated in Figure 1.

A *connection request* for a network N is a pair (x, y, ω) where x is an input, y an output and $0 \leq \omega \leq 1$. We refer to ω as the *weight* of the connection and it represents the bandwidth required by the connection. A *route* is a path joining an input to an output together with a weight. A route *satisfies* a request (x, y, ω) if it connects x to y and has weight ω.

A set of connection requests is said to be *compatible* if for all inputs and outputs x, the sum of the weights of all connections involving x is ≤ 1. A set of routes is *compatible* if for all links ℓ the sum of the weights of all routes involving ℓ is ≤ 1. A *state* of a network is a set of mutually compatible routes. If we are attempting to add a connection (x, y, ω) to a network in a given state, we say that a vertex u is *accessible* from x if there is path from x to u, all of whose links have a weight of no more than $1 - \omega$.

A network is said to be *rearrangeably nonblocking* (or simply *rearrangeable*) if for every set C of compatible connections, there exists a state that realizes C. A network is *strictly nonblocking* if for every state S, realizing a set of connections C, and every connection c compatible with C, there exists a route r that realizes c and is compatible with S. For strictly nonblocking networks, one can choose routes arbitrarily and always be guaranteed that any new connections can be satisfied without rearrangements. We say that a network is *wide-sense nonblocking* if there exists a routing algorithm, for which the network never blocks; that is, for an arbitrary sequence of connection and disconnection requests, we can avoid blocking if routes are selected using the appropriate routing algorithm and disconnection requests are performed by deleting routes.

Sometimes, improved performance can be obtained by placing constraints on the traffic imposed on a network. We will consider two such constraints. First, we restrict the weights of connections to the the interval $[b, B]$. We also limit the sum of the weights of connections involving an input or output x to β. Note that $0 \leq b \leq B \leq \beta \leq 1$. We say a network is strictly nonblocking for particular values of b, B and β if for all sets of connections for which the connection weights are in $[b, B]$ and the input/output weight is β, the network cannot block. The definitions of rearrangeably nonblocking and wide-sense nonblocking networks are extended similarly. The practical effect of a restriction on β is to require that a network's internal data paths operate at a higher speed than the external transmission facilities connecting switching systems, a common technique in the design of high speed systems. The reciprocal of β is commonly referred to as the *speed advantage* for a system.

Two particular choices of parameters are of special interest. We refer to the traffic condition characterized by $B = \beta$, $b = 0$ as unrestricted packet switching (UPS), and the condition $B = b = \beta = 1$ as pure circuit switching (CS). Since the CS case is a special case of the multirate case, we can expect solutions to the general problem to be at least as costly as the CS case and that theorems for the multirate case should include known results for the CS case.

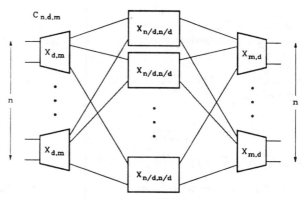

Figure 2: Clos Network

The classical complexity measure for switching networks is the crosspoint count n_C. In our graph model, this can be taken as the sum of the products of the number incoming links and outgoing links for all switches. While the crosspoint count is an appropriate measure for electromechanical switching systems and remains useful, it doesn't give an adequate indication of cost when switching systems are constructed from custom integrated circuits in which input/output constraints at the chip level limit the amount of circuitry that can be placed on a given package. Consequently, we also find it useful to include the *package count* n_P as an additional complexity measure with the understanding that the number of inputs and outputs per package is limited to 2δ. Typical values of δ would be in the range 30–50.

When comparing multirate networks, we also need to take into account the effect of different values of β that may be required by the different networks in order to allow them to achieve comparable performance. We do this by assuming that the speed advantage implied by a given value of β is obtained by providing parallelism in the data paths. This makes the complexity of networks inversely proportional to β.

3. STRICTLY NONBLOCKING NETWORKS

A three stage Clos [4] network with n input and output vertices is denoted by $C_{n,d,m}$, where d and m are parameters, and is defined by $C_{n,d,m} = X_{d,m} \bowtie X_{n/d,n/d} \bowtie X_{m,d}$ (see Figure 2). Note that $n_C = (mn/\beta)(2 + n/d^2)$. To determine the package count, we must partition the large crossbars in the network into smaller portions that meet the pin constraints. Note that at most δ^2 crosspoints can be placed in a single package with 2δ signal pins, so we take n_P to be n_C/δ^2, effectively assuming an ideal situation in which d, m and n/d are multiples of δ so that no fragmentation occurs.

The standard reasoning to determine the nonblocking condition for the Clos network (see [4]) can be extended in a straightforward manner, yielding the following theorem.

THEOREM 3.1. *The Clos network $C_{n,d,m}$ is strictly nonblocking if*

$$m > 2 \max_{b \le \omega \le B} \left\lfloor \frac{\beta d - \omega}{s(\omega)} \right\rfloor$$

where $s(\omega) = \max\{1 - \omega, b\}$.

Proof. Suppose we wish to add a connection (x, y, γ) to an arbitrary state. Let u be the stage 1 vertex adjacent to x and note that the sum of the weights on all links out of u is at most $\beta(d-1) + (\beta - \gamma) = \beta d - \gamma$. Consequently, the number of links out of u that carry a weight of more than $(1-\gamma)$ is $\le \lfloor (\beta d - \gamma)/s(\gamma) \rfloor$, and hence the number of inaccessible middle stage vertices is

$$\le \left\lfloor \frac{\beta d - \gamma}{s(\gamma)} \right\rfloor \le \max_{b \le \omega \le B} \left\lfloor \frac{\beta d - \omega}{s(\omega)} \right\rfloor < m/2$$

That is, less than half the middle stage vertices are inaccessible from x. By a similar argument, less than half the middle stage vertices are inaccessible from y, implying that there is at least one middle stage vertex accessible to both. □

Let us examine some special cases of interest. If we let $b = B = \beta = 1$, the effect is to operate the network in CS mode and the theorem states that we get nonblocking operation when $m \ge 2d - 1$, as is well-known. In the UPS case, the condition on m becomes $m > 2(\beta/(1-\beta))(d-1)$. So $m = 2d - 1$ is sufficient here also if $\beta = 1/2$.

For the UPS case, if we choose $d = \sqrt{n/2}$ and $m = 1 + 2(\beta/(1-\beta))(d-1)$ the crosspoint count of the Clos network becomes

$$\frac{4}{1-\beta} \left[\sqrt{2} n^{3/2} - 2n \right] + 4n/\beta$$

Notice that the complexity becomes unbounded if β is either too close to 0 or too close to 1. Our next result provides a lower bound on the complexity of strictly nonblocking networks when β is unrestricted.

THEOREM 3.2. *Any (m, n)-network that is strictly nonblocking for traffic with $b = 0$ and $B = \beta = 1$ must have at least mn crosspoints.*

Proof. Consider any pair of inputs and outputs x and y. If for each path in the network from x to y there is some link ℓ that is on a path from u to v where $u \ne x$ and $v \ne y$, then the network is not strictly nonblocking, since in this case every path from x to y may contain a link with nonzero weight, which is

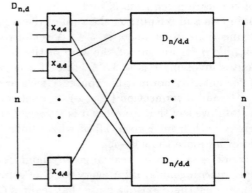

Figure 3: Delta Network

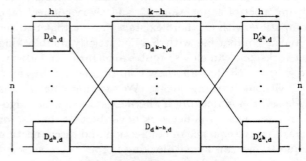

Figure 4: Extended Delta Network

sufficient to block a connection $(x, y, 1)$. Consequently, there must be at least one crosspoint that can be used only to connect x to y and hence there are at least mn crosspoints. □

Theorem 3.2 tells us that we can obtain subquadratic complexity and strictly non-blocking operation, only if we restrict the traffic. Note that Theorem 3.2 leaves open the possibility of rearrangeable or wide-sense nonblocking networks of less than quadratic complexity. In fact, using Theorem 3.1, we can construct a wide-sense nonblocking network for unrestricted traffic by placing two Clos networks in parallel and segregating connections in the two networks based on weight. In particular if we let $m = 4d - 1$, the network $X_{1,2} \bowtie C_{n,d,m} \bowtie X_{2,1}$ is wide-sense nonblocking if all connections with weight $\le 1/2$ are routed through one of the Clos subnetworks and all the connections with weight $> 1/2$ are routed through the other. The complexity of this network is $16\sqrt{2} n^{3/2} - 4n$ or roughly four times that of the strictly nonblocking network for the circuit switching case.

The *delta network* [10] $D_{n,d}$ is defined by

$$D_{d,d} = X_{d,d} \qquad D_{n,d} = X_{d,d} \times D_{n/d,d}$$

and illustrated in Figure 3. Note that $k = \log_d n$ must be an integer. The delta network has k stages and provides exactly one path between each input/output pair. More flexible networks can be obtained by adding additional stages of switching. We define the *extended delta network* $D^*_{n,d,h}$ by

$$D^*_{n,d,h} = D_{d^h,d} \bowtie D_{d^{k-h},d} \bowtie D'_{d^h,d}$$

(see Figure 4). An equivalent definition is

$$D^*_{n,d,0} = D_{n,d} \qquad D^*_{n,d,h} = X_{d,d} \bowtie D^*_{n/d,d,h-1} \bowtie X_{d,d}$$

Between each input/output pair there are d^h different paths, giving greater routing flexibility than the ordinary delta networks. A Beneš network [1], $B_{n,d}$ is equivalent to $D^*_{n,d,k-1}$ where $k = \log_d n$. For the extended delta network, $n_C = dn(h + k)$ and we take $n_P = n_C/\delta^2$ for $d \geq \delta$ and $n_P = n_C/d\delta \log_d \delta$ for $d \leq \delta$, again assuming an ideal situation in which no fragmentation occurs.

THEOREM 3.3. *The extended delta network $D_{n,d,h}$ is strictly nonblocking if*

$$\frac{\beta}{s(B)} \leq \left[\frac{d^{h-1}}{\lceil d^h/2 \rceil} \left(1 + (d-1)h + d^{\lceil (k-h+1)/2 \rceil} - d \right) \right]^{-1}$$

Proof. Let $r = \lfloor (k+h)/2 \rfloor$ and suppose we wish to add a connection (x, y, ω) to an arbitrary state. Note that there are d^h links in stage r that lie on paths from x to y. We will show that at most $\lceil d^h/2 \rceil$ of these links are inaccessible from x if the inequality in the statement is satisfied. By a symmetric argument, at most $\lceil d^h/2 \rceil$ of the links in stage $h + k - r$ that lie on x-y paths are inaccessible from y. Consequently, there must be at least one available path from x to y.

Define W_i to be the set of all links (u, v) in stage i, for which u is accessible from x, but v is not. Define λ_i to be the sum of the weights on all links in W_i and note that $\lambda_i \geq |W_i| s(\omega)$. The number of links in stage r that are not accessible from x is given by

$$\sum_{i=1}^{h} d^{h-i} |W_i| + \sum_{i=h+1}^{r} |W_i|$$

$$\leq \frac{1}{s(\omega)} \left[\sum_{i=1}^{h} d^{h-i} \lambda_i + \sum_{i=h+1}^{r} \lambda_i \right]$$

$$< \frac{1}{s(B)} \left[\beta d^{h-1} + \sum_{i=1}^{h} d^{h-i} (d^i - d^{i-1}) \beta \right.$$

$$\left. + \sum_{i=h+1}^{r} (d^i - d^{i-1}) \beta \right]$$

$$= \frac{\beta}{s(B)} d^{h-1} \left[1 + (d-1)h + d^{r-h+1} - d \right]$$

$$= \frac{\beta}{s(B)} d^{h-1} \left[1 + (d-1)h + d^{\lceil (k-h+1)/2 \rceil} - d \right]$$

$$\leq \lceil d^h/2 \rceil$$

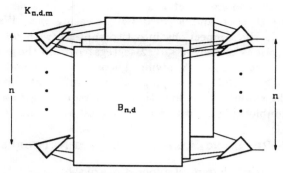

Figure 5: Cantor Network

By the argument above, the theorem follows. $\square$

The following corollaries follow easily from the theorem by substituting the appropriate values of h.

COROLLARY 3.1. *The delta network $D_{n,d}$ is strictly nonblocking if*

$$\frac{\beta}{s(B)} \leq d^{-\lfloor k/2 \rfloor}$$

COROLLARY 3.2. *The Beneš network $B_{n,d}$ is strictly nonblocking if*

$$\frac{\beta}{s(B)} \leq \left[\frac{2}{d} (1 + (d-1) \log_d(n/d)) \right]^{-1}$$

From this corollary it follows that for networks with $b = 0$ and $B = \beta$, an r stage Beneš network is strictly nonblocking if it has a speed advantage of $r - (2/d) \log_d(n/d)$. So for example, the five stage Beneš network with $d = 32$ and $n = 2^{15}$ is strictly nonblocking if it has speed advantage of 4.875.

The Cantor network $K_{n,d,m} = X_{1,m} \bowtie B_{n,d} \bowtie X_{m,1}$ and is shown in Figure 5 [3]. The next theorem captures the condition on m required to make the Cantor network strictly nonblocking.

THEOREM 3.4. *The Cantor network $K_{n,d,m}$ is strictly nonblocking if*

$$m \geq \frac{2\beta}{d \, s(B)} (1 + (d-1) \log_d(n/d))$$

The proof of this theorem is similar to the one for the previous theorem. When we apply it to the CS case for $d = 2$, we find that the condition on m reduces to $m \geq \log_2 n$, as is well known. For the UPS case with $d = 2$, we have $m \geq 2(\beta/(1 - \beta)) \log_2 n$; that is, we again need a speed advantage of two to match the value of m needed in the CS case.

We can construct wide-sense nonblocking networks for $\beta = 1$ and $b = 0$ by increasing m. We divide the connections into two subsets, with all connections of weight $\leq 1/2$ segregated from those with weight $> 1/2$. Applying Theorem 3.4 we find that $m \geq (8/d)(1 + (d-1)\log_d(n/d))$ is sufficient. That is, the complexity is four times that required for strictly nonblocking operation in the circuit switching case.

4. REARRANGEABLY NONBLOCKING NETWORKS

Although in most applications of switching networks it is not practical to operate networks rearrangeably, the property of rearrangeability is important nonetheless, because it implies a topological richness that leads to low blocking probabilities even when the network is not operated in a rearrangeable fashion. In this section, we determine conditions under which the Beneš, Cantor and Clos networks are rearrangeable for multirate traffic.

A d-ary Beneš network [1], can be defined recursively as follows: $B_{d,d} = X_{d,d}$ and $B_{n,d} = X_{d,d} \bowtie B_{n/d,d} \bowtie X_{d,d}$. The Beneš network is rearrangeable in the CS case [1] and efficient algorithms exist to reconfigure it [9]. We start by reviewing a proof of rearrangeability for the CS case, as we will be extending the technique for this case to the multirate situation.

Consider a set of connections $C = \{c_1, \ldots, c_r\}$ for $B_{n,d}$, where $c_i = \{x_i, y_i, 1\}$ and there is at most one connection for each input and output port. The recursive structure of the network allows us to decompose the routing problem into a set of subproblems, corresponding to each of the stages in the recursion. The top level problem consists of selecting, for each connection, one of the d subnetworks $B_{n/d,d}$ to route through. Given a solution to the top level problem, we can solve the routing problems for the d subnetworks independently. We can solve the top level problem most readily by reformulating it as a graph coloring problem. To do this, we define the connection graph $G_C = (V_C, E_C)$ for C as follows.

$$V_C = \{u_j, v_j \mid 0 \leq j < n/d\}$$
$$E_C = \{\{u_{\lfloor x_i/d \rfloor}, v_{\lfloor y_i/d \rfloor}\} \mid 1 \leq i \leq r\}$$

To solve the top level routing problem, we color the edges of G_C with colors $\{0, \ldots, d-1\}$ so that no two edges with a common endpoint share the same color. The colors assigned to the edges correspond to the subnetwork through which the connection must be routed. Because G_C is a bipartite multigraph with maximum vertex degree d, it is always possible to find an appropriate coloring [2]. In brief, given a partial coloring of G_C, we can color an uncolored edge $\{u, v\}$ as follows. If there is a color $i \in \{0, \ldots, d-1\}$ that is not already

in use at both u and v, we use it. Otherwise, we let i be any unused color at u and j be any unused color at v. We then find a maximal *alternating path* from v; that is a longest path with edges colored i or j and v as one of its endpoints. Because the graph is bipartite, the alternating path must end at some vertex other than u or v. Then, we interchange the colors i and j for all edges on the path and use i to color the edge $\{u, v\}$.

To prove results for rearrangeability in the presence of multirate traffic, we must generalize the graph coloring methods used in the CS case. We define a connection graph G_C for a set of connections C as previously, with the addition that each edge is assigned a weight equal to that of the corresponding connection. We say that a connection graph is (β, d)-permissible if the edges incident to each vertex can be partitioned into d groups whose weights sum to no more than β. A legal (β, m)-coloring of a connection graph is an assignment of colors in $\{0, \ldots, m-1\}$ to each edge so that at each vertex u, the sum of the weights of the edges of any given color is no more than β.

Now, suppose we let $Y = Y_1 \bowtie Y_2 \bowtie Y_3$, where Y_1 is a (d, m)-network, Y_2 is an $(n/d, n/d)$-network and Y_3 is an (m, d)-network and also let $0 \leq \beta_1 \leq \beta_2 \leq 1$. Then if Y_1, Y_2, Y_3 are rearrangeable for connection sets with $\beta \leq \beta_2$ and every (β_1, d)-permissible connection graph for Y has a legal (β_2, m) coloring then Y is rearrangeable for connection sets with $\beta \leq \beta_1$,

Our first use of the coloring method is in the analysis of $B_{n,d}$. We apply it in a recursive fashion. At each stage of the recursion, the value of β may be slightly larger than at the preceding stage. The key to limiting the growth of β is the algorithm used for coloring the connection graph at each stage. We describe that algorithm next.

Let $G_C = (V_C, E_C)$ be an arbitrary connection graph. We construct a new graph G'_C by splitting each vertex u with with $x > d$ edges into $r = \lceil x/d \rceil$ vertices $u_0, \ldots, u_{r-1}$ with the d "heaviest" edges assigned to u_0, the next d heaviest edges assigned to u_1 and so forth. When this operation is complete, we are left with a bipartite graph in which every vertex has at most d edges and we can d-color G'_C as before and then color the edges of G_C in the same way that the corresponding edges are colored in G'_C. We refer to this as the balanced vertex splitting algorithm (BVS) algorithm. We can route a set of connections through $B_{n,d}$ by applying BVS recursively. Our first theorem gives conditions under which this routing is guaranteed not to exceed the capacity of any link in the network.

THEOREM 4.1. *The BVS algorithm successfully routes*

all sets of connections for $B_{n,d}$ for which

$$\beta \leq \left[1 + \frac{d-1}{d}(B/\beta)\log_d(n/d)\right]^{-1}$$

Proof. Let G_C be any (β_1, d)-permissible connection graph with maximum edge weight B and $\beta_1 \leq 1 - B(d-1)/d$. We start by showing that the BVS algorithm produces a legal (β_2, d)-coloring for some $\beta_2 \leq \beta_1 + B(d-1)/d$.

Let u be any vertex in G_C. The largest weight that can be associated with any color at u is the sum of the weights of the heaviest edges at each of the corresponding u_i in G'_C. Because of the way u's edges were distributed among the u_is this weight is at most $B + (d\beta_1 - B)/d = \beta_1 - B(d-1)/d$.

Given this, if we route a set of connections through $B_{n,d}$ by recursive application of the BVS algorithm, we will succeed if

$$\beta + \left(\frac{d-1}{d}\right)B\log_d(n/d) \leq 1$$

which is implied by the hypothesis of the theorem. $\square$

As an example, if $n = 2^{15}$, $d = 32$ and $B = \beta$, it suffices to have a speed advantage of 3. We can improve on this result by modifying the BVS algorithm. Because the basic algorithm treats each stage in the recursion completely independently, it can in the worst-case concentrate traffic unnecessarily. The algorithm we consider next attempts to balance the traffic between subnetworks when constructing a coloring. We describe the algorithm only for the case of $d = 2$, although extension to larger values is possible.

Let G_C be a connection graph for $B_{n,2}$. G_C comprises vertices $u_0, \ldots, u_{(n/2)-1}$ corresponding to switches in stage one of $B_{n,2}$ and vertices $v_0, \ldots, v_{(n/2)-1}$ corresponding to switches in stage $2(\log_2 n - 1)$. We have an edge from u_i to v_j corresponding to each connection to be routed between the corresponding switches of $B_{n,2}$. We note that for $0 \leq i < n/4$, the switches corresponding to u_{2i} and u_{2i+1} have the same successors in stage two of $B_{n,2}$. Similarly, the switches in $B_{n,2}$ corresponding to v_{2i} and v_{2i+1} have common predecessors. We say such vertex pairs are *related*.

Let a and b be any pair of related vertices in G_C. The idea behind the modified coloring algorithm is to balance the coloring at a and b so that the total weight associated with each color is more balanced, thus limiting the concentration of traffic in one subnetwork. The technique used to balance the coloring is to constrain it so that when appropriate, the edges of largest weight at a and b are assigned different colors, and hence the corresponding connections

are routed through distinct subnetworks. For any vertex v in G_C, let $\omega_0(v) \geq \omega_1(v) \geq \cdots$ be the weights of the edges defined at v, let $W_0(v) = \sum_{i \geq 0}\omega_{2i}$, $W_1(v) = \sum_{i \geq 0}\omega_{2i+1}$ and $W(v) = W_0(v) + W_1(v)$. Also, let $x(v) = W_0(v) - W_1(v)$.

The *modified* BVS *algorithm* proceeds as follows. For each pair of related vertices a and b in G_C, if $x(a) + x(b) > B$, add a dummy vertex z to G_C with edges of weight two connecting it to a and b. We then color this modified graph as in the original BVS algorithm and on completion we simply ignore the added vertices and edges. The effect of adding the dummy vertex is to constrain the coloring at a and b so that the edges of maximum weight are assigned distinct colors. We apply this procedure recursively except that in the last step of the recursion we use the original BVS algorithm.

THEOREM 4.2. *The modified* BVS *algorithm successfully routes all sets of connections for* $B_{n,2}$ *for which*

$$\beta \leq \left[1 + \frac{1}{4}(B/\beta)\log_2 n\right]^{-1}$$

Proof. Let a and b be related vertices with $\omega_0(a) \geq \omega_0(b)$. Let $z_1 = \max\{W(a), W(b)\}$ and let z_2 be the total weight on edges colored 0 at a and b. If $x(a) + x(b) \leq B$, no dummy vertex is added and we have that

$$
\begin{aligned}
z_2 &\leq W_0(a) + W_0(b) \\
&\leq (z_1 + x(a))/2 + (z_1 + x(b))/2 \\
&\leq z_1 + B/2
\end{aligned}
$$

Similarly, if $x(a) + x(b) \geq B$, a dummy vertex is added and we have that

$$
\begin{aligned}
z_2 &\leq \omega_0(a) + W_1(a) + W_1(b) \\
&\leq \omega_0(a) + (z_1 - x(a))/2 + (z_1 - x(b))/2 \\
&\leq z_1 + B/2
\end{aligned}
$$

Thus, the total weight on a vertex in stage i is at most $2\beta + (i-1)B/2$. In particular, this holds for $i = \log_2 n - 1$. Also note that for an edge (u, v) in stage $j \leq \log_2 n - 2$, the maximum weight is at most B plus half the weight on u. For an edge (u, v) in stage $\log_2 n - 1$, the weight is at most $B/2$ plus the maximum weight at u, since in this last step the original BVS algorithm was used. Consequently, no edge carries a weight greater than $\beta + (B/4)\log_2 n$. $\square$

Theorem 4.2 implies for example that if $\beta = B$, a binary Beneš network with 2^{16} input and output vertices is rearrangeable, if it has a speed advantage of 5. Theorem 4.1, on the other hand gives rearrangeability in this case only with a speed advantage of about 8.5. It turns out that we can obtain a still stronger result by exploiting some additional properties of the original BVS algorithm.

THEOREM 4.3. *The* BVS *algorithm successfully routes all sets of connections for $B_{n,d}$ for which*

$$\beta \leq \left[\max\left\{2, \lambda - \ln\lfloor\beta/B\rfloor\right\}\right]^{-1}$$

where $\lambda = 2 + \ln\log_d(n/d)$.

So, for example if $d = 32$, $n = 2^{15}$ and $\beta = B$, a speed advantage of 2.7 will suffice for rearrangeability. The proof of Theorem 4.3 requires the following lemmas.

LEMMA 4.1. *Let r be any positive integer. If a set of connections for $B_{n,d}$ is routed by repeated applications of the* BVS *algorithm, no link will carry more than r connections of weight $> \beta/(r+1)$.*

Proof. By induction; the condition is true by definition for the external links. If the assertion holds at a given level of recursion, the connection graph G_C for the next stage will have at most rd edges of weight greater than $\beta/(r+1)$ at any given vertex u. These edges are all incident to $u_0, u_1, \ldots u_{r-1}$ in G'_C, implying that the BVS algorithm will use a single color for at most r of them. $\square$

If ℓ is a link in $B_{n,d}$, we define S_ℓ^j to be the set of links ℓ' in stage j for which there is a path from ℓ' to ℓ. If a given set of connections uses a link ℓ, we refer to one connection of maximum weight as the *primary connection* on ℓ and all others as *secondary connections*. We note that if the BVS algorithm is used to route a set of connections through $B_{n,d}$, then if there are $r+1$ connections of weight $\geq \omega$ on a link $\ell = (u,v)$, there are at least $1 + dr$ connections of weight $\geq \omega$ on the links entering u.

LEMMA 4.2. *Let $0 \leq i \leq \log_d(n/d)$, let ℓ be a stage i link in $B_{n,d}$ carrying connections routed by the* BVS *algorithm and let the connections weights be $\omega_0 \geq \omega_1 \geq \cdots \geq \omega_h$. For $0 \leq t \leq h$ and $0 \leq s \leq \min\{i,t\}$, there are at least $(t-s+1)d^s + sd^{s-1}$ connections of weight $\geq \omega_t$ on the links in S_ℓ^{i-s}.*

Proof. The proof is by induction on s. When $s = 0$, the lemma asserts that there are $t+1$ connections of weight $\geq \omega_t$ which is trivially true. Assume then that the lemma holds for $s-1$; that is, there exist $(t-s+2)d^{s-1} + (s-1)d^{s-2}$ connections of weight $\geq \omega_t$ on the links in S_ℓ^{i-s+1}. Because $|S_\ell^{i-s+1}| = d^{s-1}$, by the pigeon-hole principle, at least $(t-s+1)d^{s-1} + (s-1)d^{s-2}$ of these are secondary connections. This implies that there are at least

$$d^{s-1} + d\left[(t-s+1)d^{s-1} + (s-1)d^{s-2}\right]$$
$$= (t-s+1)d^s + sd^{s-1}$$

connections of weight $\geq \omega_t$ in S_ℓ^{i-s}. $\square$

Proof of Theorem 4.3. Consider an arbitrary set of connections for $B_{n,d}$ satisfying the bound on β given in the theorem, and assume that the BVS algorithm is used to route the connections. Let ℓ be any link in stage i, where $i \leq \log_d(n/d)$, and let the weights of the connections on ℓ be $\omega_0 \geq \cdots \geq \omega_h$. Let r be the positive integer defined by $\beta/(r+1) < B \leq \beta/r$ (equivalently, $r = \lfloor\beta/B\rfloor$). By Lemma 4.2, S_ℓ^0 carries connections with a total weight of at least

$$\omega_0 + d\omega_1 + d^2\omega_2 + \cdots + d^{i-1}\omega_{i-1} + d^i(\omega_i + \cdots + \omega_h)$$

Since the total weight on S_ℓ^0 is at most βd^i, we have

$$\beta d^i \geq \sum_{j=0}^{i-1} d^j\omega_j + d^i\sum_{j=i}^{h}\omega_j$$

From this and Lemma 4.1, we have that

$$\sum_{j=0}^{r-1}\omega_j + \sum_{j=r}^{i-1}\omega_j + \sum_{j=i}^{h}\omega_j \leq Br + \beta\sum_{j=r}^{i-1}\frac{1}{j+1} + \beta$$
$$\leq 2\beta + \beta\sum_{j=r+1}^{\log_d(n/d)}\frac{1}{j}$$

If $\lfloor\beta/B\rfloor \geq \log_d(n/d)$, the last summation vanishes and we have that the weight on ℓ is $\leq 2\beta$. Otherwise, the weight is bounded by

$$\leq \beta\left(2 + \ln\frac{1}{r}\log_d(n/d)\right) = \beta(\lambda - \ln\lfloor\beta/B\rfloor)$$

So, if β satisfies the bound in the statement of the theorem, the weight on ℓ is no more than one. By a similar argument, the weight on any link in stage j for $j \geq \log_d n$ is at most one. $\square$

The next theorem gives the conditions for rearrangeability for the Cantor network. The proof is omitted.

THEOREM 4.4. *Let $\epsilon > 0$ and $\lfloor\beta/B\rfloor \leq \log_d(n/d)$. $K_{n,d,m}$ is rearrangeable if*

$$m \geq \lceil(1+\epsilon)(\lambda - \ln\lfloor\beta/B\rfloor)\rceil$$
$$+ 2(2 + \log_2\lambda + \log_2(B/c))$$

where $\lambda = 2 + \ln\log_d(n/d)$ and $c = 1 - \beta\lambda/(1+\epsilon)(\lambda - \ln\lfloor\beta/B\rfloor)$.

The graph coloring methods used to route connections for $B_{n,d}$ can also be applied to networks that "expand" at each level of recursion. Let $C^*_{d,d,m} = X_{d,d}$ and for $n = d^i$, $i > 1$, let $C^*_{n,d,m} = X_{d,m} \bowtie C^*_{n/d,n/d} \bowtie X_{m,d}$. The following theorem gives conditions under which $C^*_{n,d,m}$ is rearrangeable.

THEOREM 4.5. $C^*_{n,d,m}$ is rearrangeable if

$$\beta \leq \left[1/\gamma^c + \frac{m-1}{m}\frac{B}{\beta}\frac{1-1/\gamma^c}{1-1/\gamma}\right]^{-1}$$

where $\gamma = m/d$ and $c = \log_d(n/d)$.

Proof. We use the BVS algorithm to route the connections. If we let β_i be the largest resulting weight on a link in stage i for $1 \leq i \leq \log_d(n/d)$, we have

$$\beta_i \leq B + \frac{d\beta_{i-1} - B}{m} = (\beta_{i-1}/\gamma) + \frac{m-1}{m}B$$

$$\leq (\beta_0/\gamma^i) + \frac{m-1}{m}B\frac{1-(1/\gamma)^i}{1-1/\gamma} \leq 1$$

□

So, for example, $C^*_{n,d,2d-1}$ is rearrangeable if $B \leq 1/2$.

5. CONCLUSIONS

Figure 6 compares the complexity of a variety of different networks. The curves give the complexity of the following networks.

- X, an $n \times n$ crossbar.

- C, a three stage Clos network with $\beta = 1/2$ and m just large enough to make it strictly nonblocking.

- K_2 and K_{32}, Cantor networks with $d = 2$ and $d = 32$, $\beta = 1/2$ and m just large enough to make them strictly nonblocking.

- B_2 and B_{32}, Beneš networks with $d = 2$ and $d = 32$ and β chosen to make them strictly nonblocking.

- B'_2 and B'_{32}, Beneš networks with $d = 2$ and $d = 32$ and β chosen to make them rearrangeably nonblocking.

- S, a Batcher sorting network together with a Banyan routing network as used in the Starlite switching system [7].

The first plot gives the number of crosspoints per port, the second gives the number of packages per port when $\delta = 32$ and the third gives the number of packages per port when $\delta = 2$. In the crosspoint comparison, it's interesting to note the fairly modest difference in complexity attributed to the switch size in the Cantor and Beneš networks. There are two opposing effects at work here. For the Cantor networks, larger switches allows reduction in the value of m and in the Beneš network, reduction in the speed advantage. On the

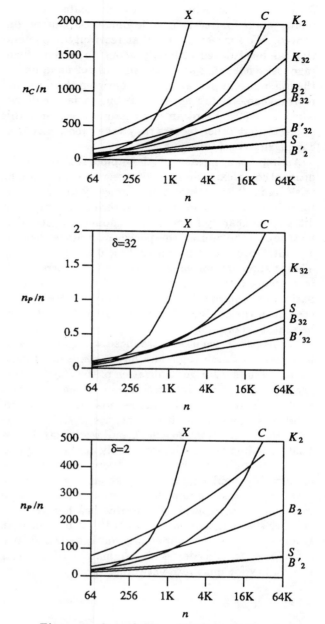

Figure 6: Complexity of Various Networks

other hand, this is partially offset by the larger number of crosspoints in a 32 port switch compared to a five stage network constructed from two port switches. Note that in the package count comparison, the larger switch gives a reduction of more than two orders of magnitude for most of the networks. For the larger package sizes, the Beneš networks appear most attractive, although the sorting network certainly compares favorably. Notice that at switch sizes of 1024 and under there is no difference in the package counts of the strictly nonblocking and rearrangeable Beneš networks for $\delta = 32$. We believe that this indicates that our bound on β for rearrangeable operation can be improved. We suspect in fact that rearrangeable operation can be achieved with only a constant speed advantage. Also note that at switch sizes of 1024 and under, the difference in package counts between the Clos and rearrangeable Beneš networks is only a factor of two when $\delta = 32$. From an engineering viewpoint, this suggests that other factors may take precedence over complexity considerations.

An important message of these plots is that the traditional complexity measure of crosspoint count can be misleading. The package count is clearly the more useful cost measure for making engineering choices and it differs significantly from the crosspoint measure. One final caution regarding the package counts is that the absolute values can be misleading if used carelessly. These values are normalized so that $\beta = 1$ corresponds to systems with bit-serial data paths. If wider data paths are required for reasons outside those considered here, the package counts shown in Figure 6 must be scaled accordingly.

In this paper, we have introduced what we feel is an important research topic and have given some fundamental results. Our generalization of the classical theory is a natural and interesting one, which has direct application to practical systems now under development in various research laboratories [5,7,14]. There are several directions in which our work may be extended. While our results for strictly nonblocking networks are tight, we believe that our results for rearrangeably nonblocking networks can be improved. Another interesting topic is nonblocking networks for multipoint connections. While this has been considered for space-division networks [6,12], it has not been studied for networks supporting multirate traffic. Another area to consider is determination of blocking probability for multirate networks.

References

[1] Beneš V. E. *"Mathematical Theory of Connecting Networks and Telephone Traffic,"* Academic Press, New York, 1965.

[2] Bondy, J. A. and U. S. R. Murty. *Graph Theory with Applications*, North Holland, New York, 1976.

[3] Cantor, D. G. "On Non-Blocking Switching Networks," *Networks*, vol. 1, 1971, pp. 367–377.

[4] Clos, C. "A Study of Non-blocking Switching Networks," *Bell Syst. Tech. J.*, vol. 32, 3/53, pp. 406–424.

[5] Coudreuse, J. P. and M. Servel "Prelude: An Asynchronous Time-Division Switched Network," *International Communications Conference*, 1987.

[6] Feldman, P., J. Friedman and N. Pippenger "Non-Blocking Networks," *Proceedings of STOC 1986* 5/86, pp. 247–254.

[7] Huang, A. and S. Knauer "Starlite: a Wideband Digital Switch," *Proceedings of Globecom 84*, 12/84, pp. 121–125.

[8] Masson, G. M., Gingher, G. C., Nakamura, S. "A Sampler of Circuit Switching Networks," *Computer*, 6/79, pp. 145–161.

[9] Opferman, D. C. and N. T. Tsao-Wu "On a Class of Rearrangeable Switching Networks, Part I: Control Algorithm," *Bell Syst. Tech. J.*, vol. 50, 1971, pp. 1579–1600.

[10] Patel, J.K. "Performance of Processor-Memory Interconnections for Multiprocessors," *IEEE Transactions on Computers*, vol. C-30, 10/81, pp. 301–310.

[11] Pippenger, N. "Telephone Switching Networks," *Proceedings of Symposia in Applied Mathematics*, vol. 26, 1982, pp. 101–133.

[12] Richards, G. and F. K. Hwang "A Two Stage Rearrangeable Broadcast Switching Network," *IEEE Transactions on Communications*, 10/85, 1025–1035.

[13] Shannon, C. E. "Memory Requirements in a Telephone Exchange," *Bell Syst. Tech. J.*, vol. 29, 1950, pp. 343–349.

[14] Turner, J. S. "Design of a Broadcast Packet Network," *IEEE Transactions on Communications*, 6/88, pp. 734–743.

[15] Turner, J. S. "Fluid Flow Loading Analysis of Packet Switching Networks," Washington University Computer Science Department, WUCS-87-16, 7/87.

Multi-Log$_2 N$ Networks and Their Applications in High-Speed Electronic and Photonic Switching Systems

CHIN-TAU LEA, MEMBER, IEEE

Abstract—A new class of switching networks has been proposed to remove the time and space bottlenecks of conventional RAM-controlled switching architectures. The proposed networks possess many desirable characteristics for high-speed electronic and photonic switching systems—such as tolerance of faults, $O(\log_2 N)$ stages between each inlet–outlet pair, self-routing capability, easy path hunt, and easy fault diagnosis.

Graph theory provides the theoretical basis for the proposed networks. It is shown that the key problem in the construction of this new class of networks is related to the coloring problem in graph theory.

I. INTRODUCTION

RECENT advances in lightwave technology force us to rethink the relationship between transmission and switching. Before, the computer-controlled *electronic* switching systems could easily handle the fastest digital transmission facility. Now, the situation is reversed because of the deployment of optical fibers as a transmission medium. The speed of transmission facilities has been improved by several orders of magnitude in recent years. In comparison, the progress made in switching seems to be relatively slow. Facing the performance challenge posed by transmission, switching is now at a transition period. New switching technologies such as fast VLSI, nonlinear optical logic devices, directional couplers, etc., have been developed. They provide new possibilities to tackle the speed mismatch problem between transmission and switching. But technology advances on the device level alone are not enough since each new technology also has its own constraints. Component-by-component replacement of the old architecture simply cannot make the best use of these technologies. New architectures are needed. For example, in directional-coupler-based photonic switching systems, signal regeneration is expensive and the network size is still dominated by the signal-to-noise ratio (SNR). How to reduce SNR value and avoid unwanted signal attenuation are important design issues [1], [2], [16].

A centralized RAM-controlled architecture (see Fig. 1) dominates the conventional time multiplexed switching (TMS) systems. But it possesses some time and space bottlenecks in a high-speed switching environment. For example, the states of all the crosspoints need to be updated at each slot by a central controller. This can become a bottleneck in the *time* domain. Also, vertical wires that connect the crosspoints and the controlling RAM's are needed. As the node-density in VLSI-based electronic switching systems or free-space pho-

Paper approved by the Editor for Communication Switching of the IEEE Communications Society. Manuscript received August 2, 1988; revised February 14, 1989 and June 2, 1989. This work was supported by Bell South under Contract E21-691. This paper was published in part in the INFOCOM '89 Proceedings.

The author is with the School of Electrical Engineering, Georgia Institute of Technology, Atlanta, GA 30332.

IEEE Log Number 9038365.

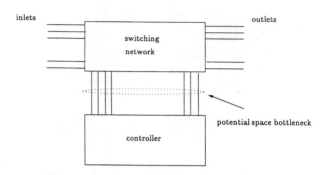

Fig. 1. Conventional TMS switching architecture.

tonic switching systems gets higher and higher, this wiring problem becomes a bottleneck in the *space* domain. Other potential bottlenecks in the time domain are related to call processing. They include path hunt, reconfiguration of the network (if a rearrangeable nonblocking network is used), and database update. Database update is generally independent of the switching network used and is not directly related to our discussion.

A new class of switching networks, called multi-log$_2 N$ networks, is proposed to remove the time and space bottlenecks of conventional switching architectures. The proposed networks, based on multiple log$_2 N$ networks, possess many useful characteristics for high-speed electronic and photonic switching systems—such as tolerance of faults, $O(\log_2 N)$ stages between each inlet–outlet pair, self-routing capability, easy path hunt, and easy fault diagnosis. Their capability to relieve the bottlenecks in the space and time domains will be discussed. Graph theory is used to construct *strictly* nonblocking or *rearrangeable* nonblocking networks that are based on the new architecture.

II. Log$_2 N$ SELF-ROUTING NETWORKS AND THEIR ADVANTAGES

Self-routing networks have been extensively used in parallel processing. Recently, a significant amount of interest has been generated in applying these self-routing networks to the design of the next-generation high-speed electronic and photonic switching systems. Three topologically equivalent self-routing networks are shown in Fig. 2: *banyan* [3], *baseline* [4], and *shuffle* [5] networks. By changing the position sof the nodes, the topology of one network can be transformed into the topology of another network. A unique feature of these networks is that they are all self-routing in the sense that the path between each source–destination pair can be determined by the corresponding bit of the binary representation of the destination address. If the corresponding address bit is zero, the data will be sent to the upper link; otherwise, they will be sent to the lower link. By appending a path header, a data block can be steered to the destination without the involvement of a central controller. Since log$_2 N$ address bits are needed to represent N different

Reprinted from *IEEE Trans. Commun.*, vol. 38, no. 10, pp. 1740–1749, October 1990.

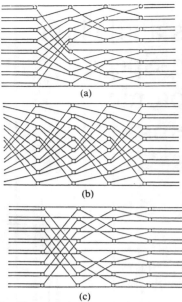

Fig. 2. Banyan, baseline, and shuffle networks are topologically equivalent. (a) A 16-by-16 baseline network. (b) A 16-by-16 shuffle network. (c) A 16-by-16 banyan network.

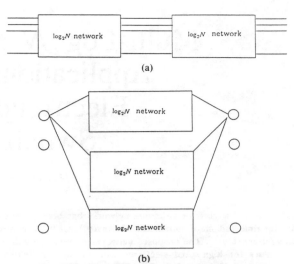

Fig. 3. (a) Horizontal cascading scheme. (b) Vertical stacking scheme.

stations, these self-routing networks normally have $\log_2 N$ stages. Throughput this paper, the term $\log_2 N$ *network* network is used to refer to the self-routing networks in Fig. 2, and the *baseline* network will be used as the representative in the analysis throughput this paper. Although this self-routing feature demands a slight increase in the complexity of each crosspoint owing to the intelligence neded for self-routing, it has a significant payoff: it can remove the time and space bottlenecks of conventional RAM-controlled switching systems. First, the stae of each crosspoint will be set up in a distributed fashion and no central controller is needed. Second, no vertical wires are needed to access the crosspoints; hence, the space bottleneck does not exist in such a system. Third, the path hunt in these self-routing networks is extremely simple. A unique path exists bewteen each inlet–outlet pair and is implied by the destination address. This characteristic also simplifies fault diagnosis. A faulty switching node will be located by two paths which share that node only. By sending data packets with a certain pattern, we can diagnose the network thoroughly. This high-level algorithm-based fault diagnosis scheme makes much more sense in a high-speed switching environment than a low-level hardware-based diagnosis scheme. In the latter, wires are needed to connect the node to the outside world to report a fault. As the discussion on the space bottleneck problem indicates, this can become a serious problem for some switching technologies. Another important feature of these self-routing networks is that there are only $O(\log_2 N)$ crosspoints between each inlet–outlet pair, as compared to $O(N)$ in a crossbar network. This characteristic is important for directional-coupler-based photonic switching systems in which the number of bidirectional couplers a light passes determines the signal attenuation and the SNR value.

But a major obstacle encountered in applying them to switching is that they are *blocking* networks, in the sense that even when both an inlet and an outlet are idle, the connection between them may be blocked by other existing connections. A network is *rearrangeable nonblocking* [14] if all permutations are possible, but some existing connections wil be reconnected when a new call is added. A network is *strictly nonblocking* [14] if an algorithm exists for setting up all new calls without disturbing any existing connection. Traditionally, horizontal cascading of two $\log_2 N$ networks has been used to create a rearrangeable nonblocking network [see Fig. 3(a)]. On example is the Benes network [14] which comprises two baseline networks cascaded back to back. But the simplicity of path hunt in a self-routing network is destroyed by the horizontal cascading

scheme. Another example is a cascade of two shuffle networks [15]. Again, it has the same characteristics of a Benes network. Sometimes these rearrangeable nonblocking networks become impractical due to the amount of reconfiguration activity. Some applications do not allow for reconfiguration of existing circuits. In addition, cascading results in networks with twice as many stages as in a regular self-routing network. For directional-coupler-based photonic switching networks, this means higher signal attenuation and lower SNR value. Still another example was given in [6] where a Batcher sorting network [7] and a banyan network [3] are cascaded together. Although this network can be considered as a rearrangeable network, the rearrangement is done in a totally distributed fashion, and self-routing is its salient characteristic. The major concern about this approach is its vulnerability to faults and difficulty to locate a fault. (Fault tolerance is particularly important for a high-speed high-density switching system, be it electronic or photonic. It is unthinkable that a single fault can destroy tens of thousands of Gb/s transmission facilities.) For example, in a circuit-switching environment when a node in the sorting network fails, an existing call that is routed through the faulty node will be disconnected by the customer. Then another call below it will be sent to the bad switching node. After the second call is disconnected, a third call will be sent to the bad node, and so forth. A single fault in the sorting network has the potential to destroy the whole switching function. Complicating the matter is the fault isolation problem. Since the path between an inlet–outlet pair is random, fault isolation is much harder than in a fixed-path self-routing network. In the case of directional couplers [8] photonic switching technology, this approach is unsuited since the number of stages between an inlet–outlet pair is large. Signal attenuation will be a serious problem.

The focus of this paper is on a different scheme: vertical stacking [see Fig. 3(b)]. Vertical stacking has been used by Huang and the author in [9] to create a self-routing TMS. In that paper, four copies of banyan networks were used to create a large TMS. Although the resulting network is blocking, it was shown that the added blocking rate is negligible in a time-space-time environment. How to create *nonblocking* networks with this vertical stacking scheme was unknown at that time, and it is the focus of this paper. It will be shown that multiple self-routing networks can be vertically stacked together to create strictly and rearrangeable nonblocking networks. A method, based on graph theory, is developed that determines the number of copies required for a strictly or rearrangeable nonblocking network. Unlike the horizontal cascading scheme, vertical stacking can preserve the desirable characteristics of a self-routing network—self-routing, easy path hunt, easy fault diagnosis. Furthermore, multiplicity creates fault tolerance. The effect of a fault in the new networks is a slight increase of call blocking rate. In a TMS environment, even if the network is nonblocking, the time-slot-mis-

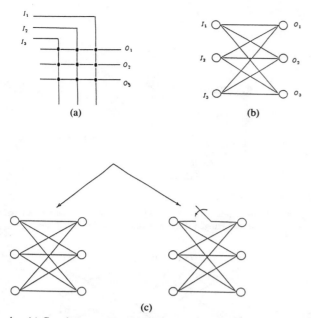

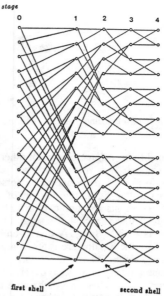

Fig. 4. (a) Crossbar representation. (b) Bipartite representation. (c) Correspondence between an edge and a crosspoint in a bipartite graph.

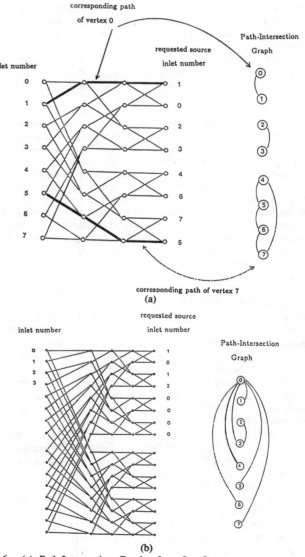

Fig. 5. Bipartite graph representation of a baseline network in Fig. 2(b).

match problem still makes the whole TMS blocking. Thus, a slight increase of the overall blocking rate is not obvious to the users. Hereafter, the term *multi-log₂ N network* is used to refer to this class of self-routing networks created by vertically stacking multiple log₂ N networks together.

III. MULTI-LOG₂ N SELF-ROUTING NETWORKS

A. Bipartite Graph, Path-Intersection Graph, and Coloring Problem

Several important concepts in graph theory are briefly introduced here. They provide the theoretical basis for the discussion later.

In general, there are several graphical representations of a switching network. One is the *crossbar* representation. Inlets and outlets are represented by lines and a crosspoint between them represent a switching element [see Fig. 4(a)]. Another graphical representation is based on bipartite graphs [10] [see Fig. 4(b)]. In a bipartite graph, the inlets and outlets are represented by *vertices* and switching elements by *edges* [see Fig. 4(c)]. As demonstrated in [1], bipartite graphs are a powerful tool for designing a switching system. In a multistage bipartite graph switching network, no two paths are allowed to intersect at a vertex. This characteristic was used in [1] to design a directional-coupler-based photonic switching system with with zero crosstalk. Also, broadcast is an intrinsic property of a bipartite graph, and this property is important in a multiconnection network design [1]. The bipartite graph representation of the baseline network in Fig. 2(a) is given in Fig. 5. Throughout this paper, all the discussion will be based on the bipartite graph representation.

In a $2^n \times 2^n$ self-routing network, there is a unique path between each inlet–outlet pair, and there are at most 2^n paths need to be set up. If the network is used for *one-to-one* connection, these 2^n paths do not have any inlet or outlet in common. If the network is used for *multiconnection* (one-to-many), some paths can share the same inlet (although their outlets are always different). Given a connection pattern (one-to-one or multiconnection) in a $\log_2 N$ network, a Path-Intersection Graph (PIG) can be constructed in the following way. In the PIG, there is a vertex corresponding to each "path." Since the destinations of the paths in the network are always different (regardless of whether the network is one-to-one or multiconnection), the corresponding vertices in the PIG are labeled by the destination of a path in the network. For example, vertex 7 in the PIG in Fig. 6 corresponds to the path in the network with outlet 7 as its destination. Moreover, if two paths intersect, an edge joining their corresponding vertices in the PIG is added. Two

Fig. 6. (a) Path-Intersection Graph of an 8 × 8 one-to-one connection network. (b) Path-Intersection Graph of a 16 × 16 multiconnection network.

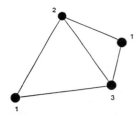

Fig. 7. An example of a coloring problem. The chromatic number of this graph is 3.

vertices are defined to be *adjacent* to each other if there is an edge between them. Thus, if two vertices are adjacent in a PIG, their corresponding paths in the $\log_2 N$ network intersect each other. The *degree* of a vertex in the PIG is the number of edges incident with it. Obviously, it represents the number of paths that intersect the corresonding path of this vertex. Let's study two examples below. The first example is an 8×8 $\log_2 N$ one-to-one switching network [see Fig. 6(a)] where the 8 paths are represented by 8 vertices. The second example is a multiconnection network [see Fig. 6(b)]. Two paths with the same inlet are allowed to overlap in a multiconnection network. Therefore, even when two paths with the same inlet meet at a vertex in the $\log_2 N$ network, there is *no* edge linking their corresponding vertices in the PIG.

In graph theory, *coloring* a graph is to paint the vertices of the graph with one or more distinct colors. A graph is *properly colored* if it is painted in such a way that no two adjacent vertices are painted with the same color. The *chromatic number* of a graph is the least number of distinct colors that can be used to color the graph properly. An example is given in Fig. 7, where the vertices are labeled with three different colors 1, 2, and 3. The chromatic number of this graph is 3 since we cannot use only two colors to properly color this graph.

There is a strong parallel between the problem of coloring a graph and that of determining the number of copies required in a rearrangeable or strictly nonblocking multi-$\log_2 N$ network. If a PIG can be properly colored with n_c colors, we can use n_c copies of a $\log_2 N$ network to construct a multi-$\log_2 N$ network (let each copy correspond to one color), in which no two intersecting paths are assigned to the same $\log_2 N$ network; hence, no path is blocked. From this discussion, it can be seen that the problem of determining the number of copies in a nonblocking multi-$\log_2 N$ network can be transformed to the problem of properly coloring its PIG. The *maximum degree* of the vertices and the *chromatic number* of a PIG have special bearings in determining the number of copies needed for *strictly nonblocking* or *rearrangeable nonblocking* networks.

B. One-to-One Connection Networks

The most important question pertaining to the vertical stacking mechanism is how many copies are required to build a nonblocking network. The answer to this question depends on the nature of the networks, i.e., strictly nonblocking or rearrangeable nonblocking.

1) Strictly Nonblocking Networks: The number of copies of a $\log_2 N$ network needed in the vertical stacking method to make a strictly nonblocking network is determined by Theorem 1, which is given later. But before it is discussed, one lemma has to be introduced first.

Lemma 1: Assume 1) that the vertices in a graph are dynamically added or removed, 2) that the maximal degree of the vertices in the graph never exceeds k, and 3) that there are $k + 1$ colors available. Then when a vertex is added or removed, this graph can always be properly colored without changing the colors of existing vertices.

Proof: It is obvious that the colors of the existing vertices need not be changed when a vertex is removed. According to assumption 2), when a vertex is added to an existing graph, it is connected to at most k vertices in the existing graph. To properly color the new graph, we can simply choose the color which is different from all these k vertices that are connected to the newly added vertex. This

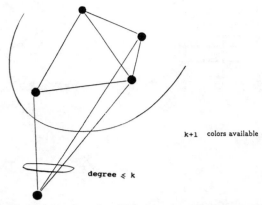

Fig. 8. When a new vertex is added, the colors of existing vertices need not be changed if there are enough colors available.

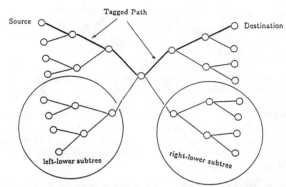

Fig. 9. The inlets from which the traffic originates that intersect the tagged path and the outlets for which the traffic destined, that also intersect the tagged path, form a double-tree.

is always possible since there are $k + 1$ colors are available.
Q.E.D.

Theorem 1: A multi-$\log_2 N$ network comprises n_c copies of a $\log_2 N$ networks. Assume $N = 2^n$, i.e., n is the number of stages in the $\log_2 N$ network. This multi-$\log_2 N$ network is strictly nonblocking if $n_c \geq \left(\frac{3}{2} 2^{n/2} - 1\right)$ when n is even, and $n_c \geq \left(2^{(n+1)/2} - 1\right)$ when n is odd.

Proof: To facilitate the discussion, the baseline network is used as the representative of the $\log_2 N$ networks. The case of an even number n is discussed first (see Fig. 9, in which $n = 6$). To prove this theorem, let's consider an inlet–outlet pair. The path between them is termed the *tagged* path. The inlets from which the traffic originates that can intersect the tagged path, and the outlets for which the traffic is destined that can also intersect the tagged path, form a back-to-back double-tree, with the vertex of center stage of the network as the root. The names *left* and *right* are used to distinguish the two binary trees in this double-tree; and each tree contains $\left\lceil \dfrac{n}{2} \right\rceil$ inlet (or outlets) vertices. Fig. 9 shows a back-to-back double-tree for a 64×64 network. As can be seen, the left (or right) tree contains only eight inlet (or outlet) vertices. The left (right) tree can also be divided into two groups: the *upper* and the *lower* subtrees. Apparently, in terms of the number of paths intersecting the tagged path, the worst case scenario corresponds to the traffic pattern where the traffic originating from the inlets of the left tree, and the traffic destined for the outlets of the right tree, are mutually exclusive. But it is easy to see that in order to intersect the tagged path, the traffic originating from the inlets of the left-lower subtree has to be destined for the outlets of the double-tree, and the traffic destined for the outlets of the right-lower subtree has to originate from the inlets of the double-tree. Therefore, the traffic pattern that creates the maximum number of intersecting paths corresponds to scenario that the traffic originating from the inlets of

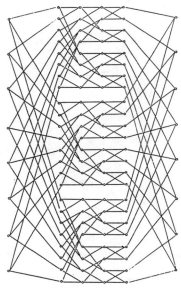

Fig. 10.   Strictly nonblocking 8 × 8 multi-log₂ N network.

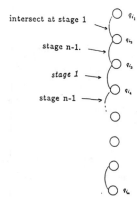

Fig. 11.   The length of a path between two vertices in a first-shell path-intersection graph implies where the two corresponding paths in the network intersect.

the left-lower subtree is destined for the outlets of the right-lower subtree, and the traffic originating from the remaining inlets and the traffic destined for the remaining outlets are mutually exclusive. It is a direct matter to see that under this condition, the maximum number of intersecting paths is $\left(\frac{3}{2}2^{n/2} - 2\right)$. Therefore, the maximum number of intersecting paths is $\left(\frac{3}{2}2^{n/2} - 2\right)$. Therefore, the maximum degree of vertices in any PIG is $\left(\frac{3}{2}2^{n/2} - 2\right)$. According to Lemma 1, it can be concluded that the number of copies of a log₂ N network needed for constructing a strictly nonblocking one-to-one connection network is $\left(\frac{3}{2}2^{n/2} - 1\right)$ when n is even. Following a similar argument, we can prove that when n is odd, the total number of copies needed is $\left(2^{(n+1)/2} - 1\right)$.   Q.E.D.

Theorem 1 gives the sufficient condition for a strictly nonblocking multi-log₂ N network. It is obvious that the condition set in Theorem 1 is also the necessary condition. An example of an 8 × 8 strictly nonblocking multi-log₂ N network is given in Fig. 10. This type of network has $O(3\log_2 N \times N^{3/2})$ crosspoints and $O(\log_2 N)$ stages.

2) Rearrangeable Nonblocking Networks: It is easy to see that in Fig. 9, some of the intersecting paths can be placed in the same network as long as they do not intersect each other. It is also true that the chromatic number of a graph is usually less than the maximum degree of its vertices. In this section, the largest chromatic number of all PIG's is investigated. Its value determines the number of copies needed to construct a rearrangeable multi-log₂ N network.

In graph theory, a *path* with *length k* is defined to be a sequence of k edges $(e_{i_1}, e_{i_2}, \cdots, e_{i_k})$ such that the terminal vertex of e_{i_j} coincides with the initial vertex of $e_{i_{(j+1)}}$. A *circuit* is a path in which the terminal vertex of e_{i_k} coincides with the initial vertex of e_{i_1}. The following lemma is very important to the development of rearrangeable nonblocking multi-log₂ N network. Its proof was given in [10].

Lemma 2: A graph can be properly colored with two colors if and only if it contains no circuits of odd length.

Proof: See [10].

Let's introduce the concept of shell. In the bipartite graph of a log₂ N network, stages are numbered from 0 to n (see Fig. 5). Two paths can intersect each other at the vertices of stages 1, 2, ⋯, n − 1, but not stages 0 and n. Stage 1 and stage n − 1 are collectively called the first shell; stage 2 and stage n − 2 are called the second shell; and so forth. Based on this definition, there are $\left\lceil \dfrac{n}{2} \right\rceil$ shells in total in a n-stage log₂ N network, where $N = 2^n$, and $[x]$ is the integral part of x. Two paths intersecting at stage 1 or stage n − 1 are said to intersect at the first shell. If we focus only on the *first*

shell, then a PIG can be constructed in which two vertices are adjacent if and only if their corresponding paths in the network intersect at the first shell. This PIG is called the First-Shell Path-Intersection Graph (FS-PIG). If a log₂ N network is used for one-to-one connection, then every path can intersect, at most, two other paths at the first shell. The maximum degree of any vertex in an FS-PIG is 2. From Lemma 1, it can be concluded that three colors, at most, are needed to properly color an FS-PIG. But, as shown in the following lemma, two colors are sufficient.

Lemma 3: The FS-PIG of a baseline (or log₂ N) network can be properly colored with two colors.

Proof: Assume the number of stages is n. Then there are, at most, 2^n paths to be set up. Denote these paths by p_0, p_1, $p_2, \cdots, p_{2^n - 1}$. Their corresponding vertices in the FS-PIG are denoted by q_0, q_1, $q_2, \cdots, q_{2^n - 1}$. In the FS-PIG, two paths can intersect only at stage 1 and stage n − 1. The vertices (in the network graph) of stage 1 and stage n − 1 of path p_i form a two-tuple $(v_{i_1}, v_{i_{n-1}})$. If v_{i_1} ($v_{i_{n-1}}$) is the same as v_{j_1} ($v_{j_{n-1}}$), we can conclude that p_i and p_j intersect each other. Two properties in the FS-PIG need to be discussed. First, if two paths intersect at both vertices in stage 1 and n − 1 in the first shell, they cannot intersect any other paths at stages 1 or n − 1. An example can be found in Fig. 6(a) where the first two paths intersect each other at both stages 1 and n − 1. Under this condition, the two corresponding vertices in the FS-PIG are isolated from the rest vertices. Hence, we can conclude that if several vertices constitute a circuit in the FS-PIG, their corresponding paths intersect each other at either stage 1 or stage n − 1, but not both. Second, assume that a path linking vertices q_{i_1}, q_{i_2}, $q_{i_3}, \cdots, q_{i_m}$ exists in the FS-PIG (see Fig. 11). If we further assume that the two corresponding paths of q_{i_1} and q_{i_2} intersect at *stage 1* in the network, then we can conclude that the corresponding paths of q_{i_2} and q_{i_3} must intersect at *stage n − 1*; that the corresponding paths of q_{i_3} and q_{i_4} must intersect at *stage 1*; and so forth. In other words, how the two corresponding paths of q_{i_j} and $q_{i_{j-1}}$ intersect in the network is implied by the *length* of the path between q_1 and q_{i_j} in the FS-PIG. If the length is an even number, then q_{i_j} intersect $q_{i_{j-1}}$ at stage n − 1; if odd, at stage 1.

Now let's assume that a circuit with *odd* length exists in the FS-PIG that comprises the vertices $\{q_{i_1}, q_{i_2}, q_{i_3}, \cdots, q_{i_m}\}$. Then there have to be two paths between q_{i_1} and q_{i_j} in the FS-PIG—one of the paths having odd length and the other having even length. From the previous discussion, it is clear that if we trace the path with even length in the FS-PIG, we will conclude that the two corresponding paths of $q_{i_{j-1}}$ and q_{i_j} intersect at stage n − 1 in the baseline network. But if we trace the path with odd length in the FS-PIG, we will conclude that their corresponding paths intersect at stage 1 in the network. This obvious contradiction leads to the conclusion that no circuits with odd length can exist in the FS-PIG. Based on Lemma 2, we can conclude that an FS-PIG can be properly colored with only two colors.   Q.E.D.

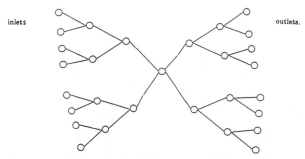

Fig. 12. All the inlets are destined for the outlets. This connection pattern requires at least $\left\lceil \dfrac{n}{2} \right\rceil$ copies.

TABLE I
NUMBER OF COPIES REQUIRED FOR STRICTLY NONBLOCKING AND
REARRANGEABLE NONBLOCKING NETWORKS

Size($N \times N$)	Number of Copies	
N=	Strictly Nonblocking	Rearrangeable Nonblocking
4	2	2
8	3	2
32	7	4
64	11	8
128	15	8
256	23	16

Theorem 2: Assume that a multi-$\log_2 N$ network contains n_c copies of a $\log_2 N$ network. Then it is rearrangeable nonblocking if and only if $n_c \geq 2^{\lceil n/2 \rceil}$, where n is the number of stages of the $\log_2 N$ network.

Proof: To prove this theory, we have to prove that 1) any PIG of a $\log_2 N$ network can be properly colored with $\left\lceil \dfrac{n}{2} \right\rceil$ colors, and 2) we need at least $\left\lceil \dfrac{n}{2} \right\rceil$ colors to properly color all PIG's. Let's prove 1) first. From Lemma 3, we can divide all the paths into two groups, and within each group no two paths intersect at the first shell, i.e., stage 1 and stage $n - 1$. After that, we can divide the connections of each group into two subgroups, and no two paths in a subgroup intersect at the second shell. Repeating the same procedure $\left\lceil \dfrac{n}{2} \right\rceil$ times, we can properly color any PIG. Therefore, $2^{\lceil n/2 \rceil}$ colors are sufficient to color any PIG of a $\log_2 N$ network.

To prove 2), we only need to look at the connection pattern shown in Fig. 12 where all of the inlets are destined for the outlets. Obviously, we need at least $2^{\lceil n/2 \rceil}$ copies to resolve the contention.

Q.E.D.

The numbers of copies required for strictly and rearrangeable nonblocking entworks are tabulated in Table I.

An example of an 8×8 multi-$\log_2 N$ rearrangeable network is shown in Fig. 13, where only two baseline networks are needed—in contrast to 3 for strictly nonblocking networks. Theorem 2 determines the number of copies required to construct a rearrangeable multi-$\log_2 N$ network. We still need a routing algorithm to set up any permutation in the rearrangeable network. Fortunately, a very simple routing algorithm exists which has an $O(N \times \log_2 N)$ computing time. It is based on the PIG. Start with the FS-PIG. We can pick an arbitrary vertex (corresponding to a path) and paint it with one of the two colors (say red). We then paint the vertices that are adjacent to this red vertex with the other color, say blue, and paint the vertices that are adjacent to a blue vertex with the red color. This procedure is repeated until all the vertices are painted. It is guaranteed that, in this way, no vertex is ever painted with both colors, since this will happen only when there are two paths between them: one with even length and the other with odd length. According to Theorem 2, this will never happen. After we successfully divide the permutation into two groups, the same coloring method is repeated for the second shell, and so forth. The comput-

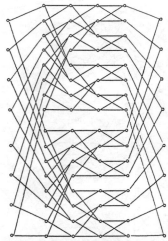

Fig. 13. Rearrangeable nonblocking 8×8 multi-$\log_2 N$ network.

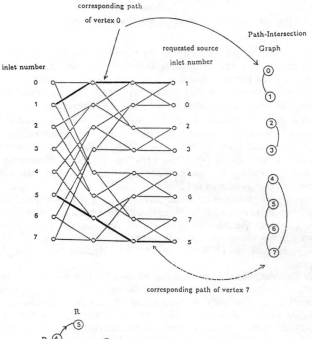

Fig. 14. A simple routing algorithm in a rearrangeable multi-$\log_2 N$ network.

ing time needed to properly color the FS-PIG is $O(N)$, and there are $\dfrac{1}{2} \times \log_2 N$ shells. Thus, the total computing time for the routing algorithm is $O(N \times \log_2 N)$. An example is illustrated in Fig. 14, where two colors—blue (B) and red (R)—are used to properly color the FS-PIG.

The number of crosspoints in this network depends on the technology used. For electronic switching technology, we use the conventional counting method that four crosspoints are needed to implement a 2×2 network. Thus, the number of crosspoints in a $\log_2 N$ self-routing network is $2N \times \log_2 N$. However, if directional couplers are used, we only need one coupler to implement a 2×2 network. Therefore, the total number of crosspoints is only $\dfrac{N}{2} \times \log_2 N$ [11]. The number of "electronic" crosspoints in each network listed in Table I is tabulated in Table II.

TABLE II
NUMBER OF CROSSPOINTS IN MULTI-LOG$_2$ N NETWORKS
AND SQUARE MATRIX

Size($N \times N$)	Number of Electronic Crosspoints		
$N=$	Square Matrix	Multi-log$_2N$ Networks	
		strictly nonblocking	rearrangeable nonblocking
4	16	32	32
8	64	144	96
32	1024	2240	1280
64	4096	8448	6144
128	16384	26880	14336

From Table II, it is clear that multi-log$_2$ N networks do not minimize the number of crosspoints. But crosspoint minimization has never been the motive behind the multi-log$_2$ N proposal. In the VLSI era, crosspoint minimization is no longer important. The main objective of the proposed architectures is to remove the time and space bottlenecks in high-speed electronic and photonic switching systems. For directional-coupler-based photonic switching technology, the proposed architectures can be implemented with much lower signal attenuation and zero crosstalk [1]. Since, at the present time, crosstalk is the most prohibitive factor in constructing a practical size directonal-coupler-based photonic switching system, how to reduce crosstalk on the device and system levels is a very important issue. In addition, many other characteristics of the new networks, such as tolerance of faults and easy path hunt, are extremely valuable in a high-speed environment.

As the network gets larger, the number of copies increases. Whether this scheme is practical for large switching networks depends on the technology used. For electronic switching systems, the wiring complexity of the first and the last stages might be a problem. For free-space-transmission photonic switching systems, this is less of an issue since optics can achieve interconnection more easily. It was demonstrated in [12] that a perfect shuffle interconnection pattern can be implemented with lenses. The topology shown in Figs. 5, 10, and 13 can be transformed into a perfect-shuffle-based topology. Thus, the same technique demonstrated in [12] can also be applied. But there are alternatives to the direct application of a multi-log$_2$ N network for large switching networks. First, we can combine the vertical stacking and horizontal cascading schemes together to create another new class of switching networks (see [17] for details). The number of copies needed in this new class of switching networks is greatly reduced. An example is that when one extra stage is added at the end of a 1024 × 1024 log$_2$ N network, the number of copies needed to make a rearrangeable network is reduced to 16 (from 32); while the rearrangement activities at both networks are virtually zero. The general formula for the number of copies needed to make a rearrangeable nonblocking network and the frequency of rearrangement activities in this new class of switching networks are discussed in [17]; readers are referred to it for details. Second, as shown in [9], even when the number of copies is less than that needed for rearrangeable nonblocking networks, the blocking rate is relatively small compared to other blocking factors inherent in a TMS environment. Third, as will be shown later, small and medium size nonblocking multi-log$_2$ N networks can serve as building blocks of large nonblocking networks. Another point worth mentioning is that although the network based on Theorem 2 is rearrangeable nonblocking, it differs from other rearrangeable nonblocking networks, such as the Benes–Clos network, in many aspects. First, the network in Fig. 13 is self-routing. The path hunt in this network is extremely simple. Second, rearrangement is a rather rare activity in this multi-log$_2$ N network. Although this subject will be treated in detail in a future paper [17], it can be understood through the following intuitive explanation. Assume that a multi-log$_2$ N network contains n_c copies of a log$_2$ N network. Let ρ denote the utilization of each inlet, where $0 \leq \rho \leq 1$, and p_b is the probability that a new call will be blocked in one copy when it arrives. Then p_b is a function of the utilization of the link at each copy, i.e., $\dfrac{\rho}{n_c}$. The probability that the network needs rearrange-

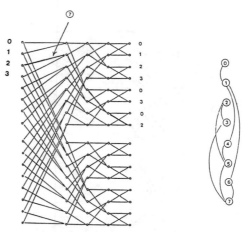

Fig. 15. A circuit with odd length can exist in the first-shell path-intersection graph of a multiconnection log$_2$ N network.

ment is the same as the probability that this call is blocked in all copies, i.e, $p_b^{n_c}$. Let's plug in some numbers. Assume the network's size is 64 × 64. Then n_c is 8. We further assume the slot utilization ρ is 0.9 (much higher than the normal utilization in a TMS environment), then the average utilization of each link at each copy is only 0.15. Therefore, it is safe to say that p_b is small; p_b^6 is almost negligible under this condition. In general, a rearrangeable multi-log$_2$ N network behaves almost like a *strictly* nonblocking network.

C. Multiconnection(Multicasting) Networks

A single inlet can be connected to one or many outlets in a multiconnection network, and two paths with the same inlet are allowed to overlap and share the same vertex in the network. Therefore, even when two paths with the same inlet overlap, their corresponding vertices in the PIG will not be linked by an edge (i.e., not considered adjacent). An example has been shown in Fig. 6(b). There are two major differences between the PIG of a multiconnection network and that of a one-to-one connection network. First, owing to the multiconnection characteristic, the maximum degree in the PIG of a multiconnection network can become very large. Therefore, a strictly nonblocking multiconnection network deduced from Lemma 1 is not practical. We have to focus on rearrangeable nonblocking multiconnection networks. Second, two colors are not sufficient to properly color the FS-PIG of a multiconnection log$_2$ N network. In other words, Lemma 3 developed for one-to-one connection networks does not apply to a multiconnection log$_2$ N network. An example is given in Fig. 15 where the FS-PIG corresponding to this connection pattern contains a circuit with odd length.

The number of copies needed for rearrangeable nonblocking multi-log$_2$ N multiconnection network has not been determined yet. But for 4 × 4, 8 × 8, and 16 × 16 rearrangeable nonblocking multiconnection networks, the numbers of copies needed are the same as rearrangeable nonblocking one-to-one connection networks. It is unknown at this moment whether this result can be extended generally.

Lemma 4: The length of a circuit in a FS-PIG (one-to-one or multiconnection) is always $\leq 2^{n-1}$, where n is the number of stages in the log$_2$ N network.

Proof: Let's use the 16 × 16 baseline network as an example (see Fig. 16). A path in this network is represented by the (inlet–outlet) pair. Without loss of generality, we pick an arbitrary path, say (i_0, o_0), where i_0 (o_0) represents the inlet 0 (outlet 0). This path can intersect other paths only at the first vertex of stage 1 and stage 3. An inlet that can reach the first vertex of stage 3 is from the inlet set $I = \{i_0, i_1, i_2, i_3, i_4, i_5, i_6, i_7\}$, and an outlet that can be reached from the first vertex of stage 1 are from the outlet set $O = \{o_0, o_1, o_2, o_3, o_4, o_5, o_6, o_7\}$. The shaded vertices in Fig.

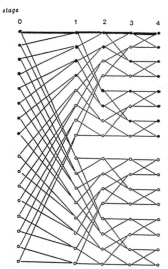

stage

Fig. 16. The corresponding vertices of the paths (including the one indicated by heavy lines) forming a circuit in the path-intersection graph must draw their inlets and outlets from the sets indicated above.

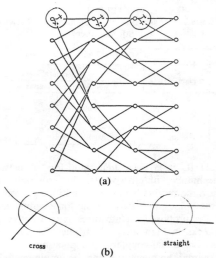

Fig. 17. (a) The crosspoints are moved to the upstream sites in a one-to-one connection network. (b) Each node has only two states: straight and cross.

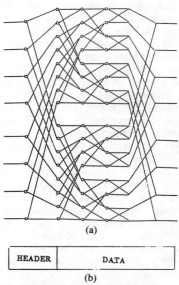

HEADER	DATA

(b)

Fig. 18. (a) A stage of nodes is added to route the data to the selected copy. (b) The header.

16 are those that can be used to link the inlet set and outlet set. Assume a circuit in the FS-PIG exists that contains the vertex corresponding to the path (i_0, o_0). We now trace the circuit and start from the path (i_0, o_0) to see how long it can be. Assume the second path intersects the first path at stage 3. Again, we can pick an arbitrary path with outlet 1 that intersects path (i_0, o_0) at stage 3, say path (i_3, o_1). We then continue from the second path. It is clear that to continue the trace, the inlets and outlets have to be from the sets I and O given above. Otherwise, the new path will not intersect the existing paths, and the trace will stop. In general, $|O| = 2^{n-1}$. Thus, the length of the circuit in an FS-PIG must be $\geq 2^{n-1}$.

Lemma 5: No circuit of length 3 exists in an FS-PIG.

Proof: From Lemma 3, we know that no circuit with odd length can exist in an FS-PIG of a one-to-one connection $\log_2 N$ network. If the $\log_2 N$ network is used as a multiconnection network, then the existence of a circuit with length equal to 3 implies two possible scenarios: a) one corresponding path intersects the other two corresponding paths at stage $n - 1$, and the latter two paths intersect each other at stage 1; or b) one path intersect the other two paths at stage 1, and the other two paths intersect each other at stage $n - 1$. Scenario a) is impossible because it means the destinations of two paths are the same. Scenario b) is also false. If two paths intersect another path at stage 1, their two paths must share the same inlet. According to the definition of a PIG, their corresponding vertices in the PIG cannot be linked by an edge.
Q.E.D.

Theorem 3: The numbers of copies of $\log_2 N$ networks needed for 4×4, 8×8, 16×16 rearrangeable nonblocking multi-$\log_2 N$ multiconnection networks are 2, 2, and 4.

Proof: The case of the 4×4 network is trivial. We now prove the case of the 8×8 network. Since there is only one shell in an 8×8 $\log_2 N$ network, its FS-PIG is the same as its PIG. The possible values for odd length are 3, 5, and 7 in an 8×8 network. From Lemmas 4 and 5, we know none of them is possible in an 8×8 $\log_2 N$ network. Hence, only two colors are needed to properly color any PIG of an 8×8 $\log_2 N$ network.

In the case of a 16×16 network, we can divide all the connections into two groups so that no contention occurs at stage 1. Each group consists of two separate 8×8 connection patterns which can be set up with two copies of $\log_2 N$ networks. Therefore, in total, 4 copies are sufficient to make a 16×16 rearrangeable nonblocking multiconnection network.
Q.E.D.

A multi-$\log_2 N$ rearrangeable nonblocking network with a size of 8×8 is an interesting one. This network almost has the same number of crosspoints as a rearrangeable nonblocking network based on the cascading scheme. But it has only half the number of stages. Another important difference is that this multi-$\log_2 N$ rearrangeable nonblocking network has a multiconnection (i.e., selective broadcast) capability.

IV. Implementation of Multi-$\log_2 N$ Networks

The node design of a multiconnection network is different from that of a one-to-one connection network, and will be discussed in this section. The implementation of a one-to-one connection switching network is discussed first. The 8×8 rearrangeable nonblocking network shown in Fig. 13 is used as an example. To implement this network, the crosspoints in the bipartite graphs are moved to the upstream sites [see Fig. 17(a)]. The switching element for each vertex has two states: cross and straight [see Fig. 17(b)]. To route the data block to the corresponding $\log_2 N$ network that is determined during the call setup time, one stage of the same switching nodes are added in the front. If electronic switching technology is used, the whole network can be implemented as a self-routing network. The destination address is appended to the header of each data block. An example is given in Fig. 18 that shows the implementation and the data format. If directional-coupler photonic switching technology is used (see the discussion later), it can be

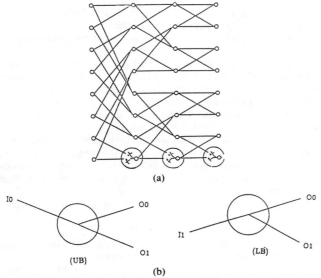

Fig. 19. (a) Crosspoints are moved to the down sites in a broadcast network. (b) Two operation states: upper broadcast (UB) and lower broadcast (LB).

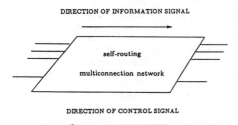

Fig. 20. Directions of control and data signals in a multiconnection multi-log$_2$ N network.

seen that only one directional coupler is needed to implement the switching node in a one-to-one connection network.

For multiconnection networks, the node design is a bit different from that of a one-to-one connection network. First, the crosspoints in the bipartite graphs are moved to the downstream sites. Since no two paths are allowed to intersect, each vertex can be implemented with two states: upper broadcast (UB) or lower broadcast (LB) modes (see Fig. 19). To maintain the self-routing capability, a destination-initiated setup scheme is needed to control the states of the switching nodes. The controlling signals travel from the left to the right—i.e., from the destination to the source (see Fig. 20); and the data travel from the left to the right. Each slot is divided into two periods: the setup period and the data transfer period. During the path setup period, the control packets containing the source addresses are sent from the destinations to the sources to set up the states of the switching nodes. When two control packets arrive at the same node, they must contain the same source address (this is because no two paths with different source inlets are allowed to intersect at a vertex), and only one of the two control packets will be sent to the next stage. The corresponding bit in the source address determines the modes (i.e., UB or LB) of a switching node. After the transmission time of n bits, all the nodes are set up correctly and data transfer can start. In multiconnection multi-log$_2$ N networks, the self-routing characteristic is maintained through this destination-initiated scheme. The overhead (the path setup period) is only about n bits, where n is the total number of stages. Compared to the length of a data block, it is really negligible.

A directional coupler is an integrated optical device [8]. It is created by diffusing titanium at high temperature into a lithium-niobate crystal to create channels in the crystal. A directional coupler can behave like a 2×2 switching node with two states: bar

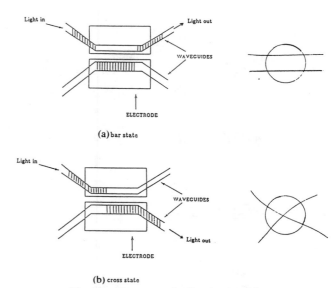

Fig. 21. Two states of a directional coupler.

and cross (see Fig. 21). The state of a directional coupler can be set by controlling the electrodes on top of the directional coupler. Once the state of a directional coupler has been set up properly, signals can be transmitted through it with the rate of several Tb/s. There are some crucial tasks to be tackled before a directional-coupler-based photonic switching becomes a commercial reality (see [13]). One major technology constraint is that the directional coupler does not regeneratively process the optical signal. Thus, signal attenuation and signal-to-noise ratio (SNR) are two subjects which switching architects have to deal with carefully. In terms of signal attenuation, multi-log$_2$ N networks have a big advantage—the number of directional couplers between each inlet–outlet pair has the order of $O(\log_2 N)$; while a square matrix has the order of $O(N)$. In terms of SNR value, it was shown in [1] that any network based on bipartite graphs can be implemented with zero crosstalk. Consequently, the multi-log$_2$ N networks in Figs. 10 and 13, all based on bipartite graphs, also can be implemented with low crosstalk. The node designs of one-to-one and multiconnection directional-coupler-based photonic switching networks have been discussed in [11], and will not be repeated here. Readers are referred to them for details.

V. OTHER APPLICATIONS

In addition to the direct application discussed in the previous sections, multi-log$_2$ N networks can serve as building blocks of other networks. An example is given in this section. A Clos network is modified to become a self-routing network. The intention is to remove its time and space bottlenecks discussed in Section I.

The three-stage Clos decomposition technique is commonly used to design a strictly nonblocking network that has fewer crosspoints than a square matrix. An example of a three-stage 16×16 Clos network is shown in Fig. 22, in which there are four 4×8 switches in the first stage, eight 4×4 switches in the middle stage, and four 8×4 switches in the third stage (although a minimum of 7 middle switches is needed, the number "8" is chosen for easy replacement by a multi-log$_2$ N network). To make the Clos network self-routing, we can replace each switch with a multi-log$_2$ N type of network. But there is a difference: the switches in the first and the last stages are not square. The number of copies required under this circumstance is different from that in preceding sections. A 4×8 degenerate one-to-one log$_2$ N network is shown in Fig. 23. It is clear that if two paths do not intersect at the first three stages, they will not intersect at the last stage. Thus, the number of copies required to make the network strictly nonblocking is the same as that for a 4×4 one-to-one network, i.e., 2. The same is also true of the switches in the last stage. In this self-routing version, the path is

97

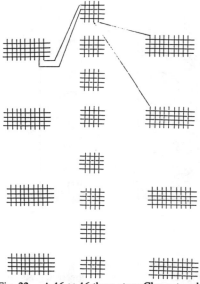

Fig. 22. A 16 × 16 three-stage Clos network.

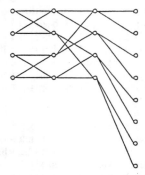

Fig. 23. Degenerate 4 × 8 one-to-one $\log_2 N$ network.

specified by the combination of three addresses. The overall overhead is 8 bits.

If directinal coupler photonic switching technology is used to implement the network, the total number of directional couplers between an inlet–outlet pair in the revised Clos network is only 8, as compared to 32 in the original Clos network. Signal attenuation is greatly improved. In addition, the network can be implemented with zero crosstalk [1].

VI. Concluding Remarks and Future Research

In this paper, a new class of switching networks has been proposed for high-speed electronic and photonic switching systems. They do not have the space and time bottlenecks of conventional RAM-controlled switching architectures. The proposed networks possess many characteristics—tolerance of faults, easy fault diagnosis, $O(\log_2 N)$ stages between each inlet–outlet pair—which are very useful for high-speed photonic and electronic switching systems.

Many interesting topics related to the architectures deserve extensive study in the future. First, Theorem 2 is for one-to-one rearrangeable nonblocking networks. Is the number of copies needed for rearrangeable nonblocking multiconnection network the same as that for a one-to-one network? What are the upper and lower bounds of the number of calls needed to be rearranged in a one-to-one or multiconnection rearrangeable nonblocking multi-$\log_2 N$ network? How often does the network need to be rearranged under a given traffic load? Another important topic is: when the number of copies is less than that required for a rearrangeable nonblocking multi-$\log_2 N$ network, what is the blocking probability? These problems, as well as many other related issues, are left for future research.

References

[1] C.-T. Lea, "Bipartite graph design principle for photonic switching systems," *IEEE Trans. Commun.*, to appear.

[2] K. Padmanabhan and A. Netravali, "Dilated networks for photonic switching," presented at the 1987 Topical Meet. Photonic Switching, Reno, Mar. 1987.

[3] L. R. Goke and G. J. Lipovski, "Banyan networks for partitioning multiprocessor systems," in *Proc. 1st Annu. Symp. Comput. Arch.*, 1973, pp. 21–28.

[4] C.-L. Wu and T.-Y. Feng, "On a class of multistage interconnection networks," *IEEE Trans. Comput.*, vol. C-29, pp. 694–702, 1980.

[5] H. L. Stone, "Parallel processing with the perfect shuffle," *IEEE Trans. Comput.*, vol. C-20, pp. 153–161, Feb. 1971.

[6] S. C. Knauer and A. Huang, "STARLITE: A wideband digital switch," in *Proc. GLOBECOM '84*, Nov. 1984.

[7] K. E. Batcher, "Sorting networks and their applications," in *Proc. Spring Joint Comput. Conf.*, 1968, pp. 307–314.

[8] R. C. Aferness *et al.* "Characteristics of Ti-diffused lithium niobate optical directional couplers," *Appl. Opt.*, vol. 18, no. 23, pp. 4012–4016, Dec. 1979.

[9] N.-H. Huang and C.-T. Lea, "Architecture of a time-multiplexed switch," in *Proc. GLOBECOM '86*, Nov. 1986.

[10] C. L. Liu, *Introduction to Combinational Mathematics.* New York: McGraw-Hill, 1968.

[11] C.-T. Lea, "A new broadcast switching network," *IEEE Trans. Commun.* vol. COM-36, pp. 1128–1137, Oct. 1988.

[12] K.-H. Brenner and A. Huang, "Optical implementation of the perfect shuffle interconnections," *Appl. Opt.*, vol. 27, p. 135, 1988.

[13] W. A. Payne and H. S. Hinton, "System considerations for the lithium niobate photonic switching technology," presented at the 1987 Topical Meet. Photonic Switching, Reno, Mar. 1987.

[14] V. E. Benes, *Mathematical Theory of Connecting Networks.* New York: Academic, 1965.

[15] C.-L. Wu and T.-Y. Feng, "The universatility of the shuffle-exchange network," *IEEE Trans. Comput.*, pp. 324–332, May 1981.

[16] C.-T. Lea, "Cross-over minimization in directional-coupler-based photonic switching systems," *IEEE Trans. Commun.*, vol. COM-36, pp. 355–363, Mar. 1988.

[17] C.-T. Lea and D.-J. Shyy, "Tradeoff of horizontal decomposition versus vertical stacking in rearrangeable networks," *IEEE Trans. Commun.*, submitted for publication.

Performance Study of an Input Queueing Packet Switch with Two Priority Classes

JEANE S.-C. CHEN, MEMBER, IEEE, AND ROCH GUÉRIN, MEMBER, IEEE

Abstract—This paper studies an $N \times N$ nonblocking packet switch with input queues and two priority classes which can be used to support traffic with different requirements. The switch operation is slotted and, at each time slot, fixed size packets arrive at the inputs with distinct Bernoulli distributions for both the high and low priority classes. Two policies are studied. In the first policy, packets of both priority classes are queued when waiting for service. In the second policy, only low priority packets are queued, and high priority packets not delivered at first attempt are dropped from the system. Under both policies, high priority packets prevail over low priority packets at the inputs as well as the outputs. Because of the service dependencies introduced by the switch structure, an exact analysis of this system is intractable. This paper provides an approximate analysis based on some independence assumptions and uses an equivalent queueing system to estimate the service capability seen by each input. Using this approach, an expression for the input queue length distribution is obtained. The maximum system throughput is also derived and shown to exceed that of a single priority switch. Numerical results are compared to simulations and found to be in good agreement.

I. INTRODUCTION

THIS paper is concerned with the performance of a packet switch with input queues and two priority classes. The use of priorities provides the means to give different classes of service to different types of traffic. This can, for example, be useful to accommodate, over a single switch fabric, real-time traffic with stringent delay requirements and data traffic which is less sensitive to delay. More generally, priorities can be used in packet switches to support both synchronous and asynchronous traffic [3], [10], [16]. Performance measures for each individual class of traffic are necessary to effectively dimension the system.

Packet switches have been extensively studied in the literature because of the wide range of environments where they can be used. Good overviews on commonly used structures and available analyses can be found in [1] and [2]. Because of the speed limitation of the software overhead, packet switches were originally restricted to pure data communication applications [4], [19], [25]. However, as technology evolved and made feasible high speed hardware-based packet switches, or rather fast packet switches, the range of potential applications started to increase. A first example suggesting how to exploit such possibilities can be found in [26], and following this early work many papers have been devoted to the analysis of fast packet switches integrating a variety of traffics [3], [7]–[17], [20], [22], [23], [27].

Previous analyses on fast packet switches with input queues provide approximation techniques for switches operating with only a single priority class [11], [12], [14], [16], [24], [25]. In the few cases (e.g., [6], [7], [21]) where more than one priority class was introduced, the influence of the switch structure was not really taken

Paper approved by the Editor for Communication Switching of the IEEE Communications Society. This paper was presented in part at INFOCOM'89 and at the ITG/GI Conference on Distributed Systems, February 1989.

The authors are with the IBM Research Division, Thomas J. Watson Research Center, Yorktown Heights, NY 10598.

IEEE Log Number 9040028.

into account. This paper attempts to analyze the performance of a two priority input queueing switch, while taking into account the influence of both the switch structure and the different classes of service. This is done by extending some of the approximation techniques used for single priority switches.

In this paper we characterize the switch operation in terms of high and low priority traffic loads. In particular, we obtain the amount of switch capacity that remains available to low priority packets for a given high priority traffic load. From this, we determine the total maximum throughput of a two priority input queueing switch and establish that it can exceed that of an equivalent single priority switch. Based on these results, we obtain an expression for the average service rate seen by low priority packets. This quantity is used to define an equivalent queueing system that approximates a low priority input queue. From the study of this system, we obtain performance measures (i.e., average queue length) for low priority packets. Simulation results are compared to analytical ones and found to be in good agreement.

The paper is organized as follows. In Section II, we describe the system structure and operation and briefly outline the method of analysis. Section III presents the derivation of the probability of service seen by low priority packets. Section IV is devoted to the computation of the maximum system throughput and to identifying regions of stable operation. In Section V, we use an equivalent queueing system to obtain performance measures for a low priority input queue. Several system configurations are studied and compared to simulation results. Finally, Section VI provides a brief conclusion and proposes possible extensions.

II. SYSTEM DESCRIPTION

The switch studied is assumed to be an $N \times N$ internally non-blocking packet switch with input queues. Internally nonblocking means that blocking occurs only if two or more inputs attempt to transmit packets simultaneously to the same output. Examples of such switch fabrics include conventional crossbars and Batcher–Banyan Sorting networks [13], [14]. Packets are of fixed length and the switch operation is slotted such that, at each time slot, every input/output can send/receive one packet. The number of priority classes is limited to two and, for every time slot, packets arrive at each input according to independent Bernoulli distributions with λ_H and λ_L giving the probability of packet arrival for high and low priorities, respectively. In addition, we assume balanced traffic and uncorrelated arrivals. This means that destinations of newly arriving packets are uniformly distributed among all outputs and are independent of those of previous packets.

At each time slot, the switch attempts to serve the packets present at the Head-of-the-Line (HOL) position of each input queue. Service consists of routing packets from inputs to the correct outputs and is only completed when a packet successfully reaches its destination. Service completion depends on the packet priority class as well as on the amount of contention for the output. High priority packets preempt low priority ones at the input and move ahead of all low priority packets waiting for service in their input queue. Similarly, high priority packets always prevail over low priority packets contending for the same output. For each output, one among the contending highest priority packets is randomly selected and then transmitted through the switch.

Reprinted from *IEEE Trans. Commun.*, vol. 39, no. 1, pp. 117–126, January 1991.

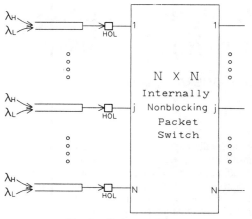

Fig. 1. System model.

Low priority packets that lose the contention phase, either because of preemption by higher priority packets or simply because they are not selected, return to the HOL position and reattempt at the next slot if they are not preempted by a higher priority arrival. For high priority packets, two policies are considered. In the first policy, high priority packets not delivered remain in the queue and keep trying until successfully delivered. In the second policy, high priority packets are not queued and they are dropped from the system if not delivered at first attempt. Herein we refer to the first policy as Policy A and to the second policy as Policy B. Policy B guarantees fixed delay through the system for the high priority packets at the expense of lost packets. This tradeoff can be desirable for some system implementations. Queued packets are, within each class, served on a First-Come-First-Served (FCFS) basis. A description of the system is provided in Fig. 1.

As can be seen from the above discussion, at each time slot high priority packets make a number of inputs and outputs unavailable to low priority packets. This results in a decrease of the effective service rate provided by the switch to low priority packets. On the other hand, the presence of low priority packets is transparent to high priority ones for which the switch behaves as a single priority switch. Performance measures for high priority packets are therefore available from previous works on single priority switches (e.g., [12], [14], [17] for Policy A and [5], [17], [25] for Policy B). This paper is devoted to studying system's performance for low priority packets and, with the exception of one expression needed in Section III, readers are referred to the cited references for details on results for high priority packets.

The switch can be viewed as a system of N parallel queues (the inputs) contending for the service of N servers (the outputs). The state space associated with this system explodes rapidly with N and an exact solution is intractable. The analysis provided in this paper is an approximation method. In this approach, we first study a "virtual" queue formed by the HOL packets, and track how packets are getting served in this queue. Based on information obtained from this virtual queue and the law of flow conservation, we derive, under some independence and limit assumptions, an equivalent service probability for low priority packets. This probability of service defines how often, on the average, HOL packets get served. We then approximate each input port by a Geom/Geom/1 queueing system where the probability of service determines the average service time of a low priority packet. Performance measures are then obtained from this equivalent queueing system. The results are compared to simulations and found to be in good agreement.

III. PROBABILITY OF SERVICE

In this section, we derive an equivalent probability of service, or service rate, for low priority packets. This equivalent service rate is a measure of how many slots it takes a low priority packet from its first time at the HOL position until its successful delivery to its destination. This time can also be viewed as an average virtual service time for low priority packets. Ideally, the switch is capable of delivering one packet from each input every time slot. However, because of the presence of high priority packets at the inputs and the contention of both high and low priority packets for the outputs, low priority packets typically take more than one slot to effectively go through the switch. More specifically, the successful transmission of a low priority packet requires that the three conditions listed below are met.

1) The low priority packet occupies the HOL position of its queue.

2) There are no high priority packets with the same destination at other HOL positions.

3) The low-priority packet is selected (randomly) among all other HOL low priority packets contending for the same output.

From the point of view of low priority packets, an input port is a queueing system where packets are removed from the HOL position at a rate defined by how often the three above conditions are met. We define this equivalent service rate or service probability by P_s, where $1/P_s$ is the expected number of slots it takes a contending low priority packet to go through the switch. An important factor influencing P_s is the effect of memory, because of output contention and busy periods generated by high priority arrivals. In other words, the computation of P_s should attempt to take into account the following phenomena. First, the number of packets contending for the same output in successive slots are correlated since no more than one packet can be delivered to one output in each slot, and those not delivered may retry in the following slot. Second, at each input, high priority packets make the switch unavailable to low priority packets for periods of k consecutive slots, where k is determined by the duration of the busy period generated by the first high priority arrival.

In order to obtain P_s, we first introduce a quantity, denoted as $\hat{P}_s$, representing the conditional probability a low priority packet receives service once it reaches HOL. It follows that $\hat{P}_s \geq P_s$. The relationship between these two quantities will be further explained in Section V. We now proceed with the derivation of $\hat{P}_s$. In Section III-A, the derivation is carried out in detail for Policy A. The derivation for Policy B being similar, only key steps are presented in Section III-B. We denote the conditional service probability for Policy A by $\hat{P}_{s,A}$, and that for Policy B by $\hat{P}_{s,B}$. The resultant expressions for these two measures can be found in (3.22) and (3.26), respectively.

A. Policy A

Before proceeding with the derivation of the probability of service, we simply quote a result from [14] which is needed to obtain $\hat{P}_{s,A}$.

Let K_H (a random variable) be the number of high priority packets waiting for service at a given input. The steady-state distribution of K_H can be approximated by

$$\Pr(K_H = i) \simeq \begin{cases} (1-\omega)(1-\lambda_H), & i = 0 \\ \dfrac{\lambda_H(2-\lambda_H)}{2(1-\lambda_H)}(1-\omega)\omega^{i-1}, & i > 0 \end{cases} \quad (3.1)$$

where $\omega = \lambda_H^2/[2(1-\lambda_H)^2]$. The probability $\Pr(K_H = 0)$ gives the fraction of time during which low priority packets at a given input can access the switch. This quantity is needed to derive $\hat{P}_{s,A}$. The complete queue length distribution of high priority packets is required to compute the distribution of the total queue length at the inputs.

The assumptions of identical switch inputs and packet destinations independent of each other and uniformly distributed over all outputs ensure that all outputs are equivalent. Each time slot consists of three phases: arrival, contention, and departure phases. During the arrival phase, new packets can move into HOL positions, and all HOL packets contend for delivery of the switch during the contention phase. Those packets chosen for delivery departs from HOL during the departure phase. Output j is tagged and we record packet destinations at the HOL positions before the contention phase. Denote $H_j(i)$ as the number of HOL high priority packets destined for output j at the beginning of the contention phase of slot i and H_j as the same measure at steady state. Similarly, $L_j(i)$ and L_j denote the corresponding measures of low priority packets. The definitions of $H_j(i)$, $L_j(i)$, together with $A_j(i)$, $C_j(i)$, $M(i)$, and $R_j(i)$ to be used later, are summarized below.

$M(i)$: number of HOL positions available to take in a "new" low priority packet at beginning of slot i;

$H_j(i)$: number of HOL high priority packets destined for output j before contention phase in slot i;

$L_j(i)$: number of HOL low priority packets destined for output j before contention phase in slot i;

$A_j(i)$: number of "new" HOL low priority packets destined for output j before contention phase in slot i;

$C_j(i)$: number of "old" HOL low priority packets destined for output j before contention phase in slot i;

$R_j(i)$: number of HOL low priority packets destined for output j and not delivered at the end of slot i.

Assuming that the system operates below saturation, the average low priority throughput T_L per output port equals the input arrival rate λ_L and can be expressed in terms of H_j and L_j.

$$T_L = \lambda_L = \frac{1}{N} \sum_{j=1}^{N} E[\epsilon(L_j)\bar{\epsilon}(H_j)] \qquad (3.2)$$

where $\epsilon(x)$ is the indicator function of $\{x \in \mathbb{R}, x > 0\}$ and $\bar{\epsilon}(x) = 1 - \epsilon(x)$. Another quantity of interest is the number R of HOL low priority packets that are left over at the end of a contention phase. Using the fact that all outputs are identical, the expected value of R can be related to the expected value of L_j and the average output throughput T_L.

$$E[R] = NE[L_j] - NT_L. \qquad (3.3)$$

Combining (3.2) and (3.3) gives

$$\lambda_L = E[L_j] - \frac{1}{N} E[R]. \qquad (3.4)$$

We now introduce two additional intermediate variables, M and ρ. At a random slot, M, $0 \le M \le N$, denotes the number of HOL positions available to new[1] low priority packets, and ρ gives the probability that an input has a new low priority packet waiting to move into an available HOL position. Input ports are excluded from M if they have high priority packets or low priority packets that already attempted to go through the switch. The quantity ρ is simply the probability that an input port has a nonempty queue of new low priority packets, given that it is among the M available ones. These two new variables can be related using flow conservation:

$$E[M]\rho = N\lambda_L. \qquad (3.5)$$

Furthermore, it can be seen that $E[M]$ is equal to the difference between $E[N_0^H]$, the expected number of input ports per time slot without high priority packets, and $E[R]$, the expected number of low priority packets that fail to go through the switch in a time slot. All inputs being identical, we have

$$E[N_0^H] = NP_0^H \qquad (3.6)$$

where $P_0^H = \Pr(K_H = 0)$ is given in (3.1). $E[M]$ is then given as

$$E[M] = NP_0^H - E[R]. \qquad (3.7)$$

From (3.5) and (3.7), we have

$$\frac{E[R]}{N} = P_0^H - \frac{\lambda_L}{\rho}. \qquad (3.8)$$

[1]The first time the packet accesses the HOL position.

Equation (3.8) provides an expression for $E[R]$ in terms of ρ and known quantities. As we will see, ρ can be directly related to the quantity of interest $\hat{P}_{s,A}$. Therefore, if $E[L_j]$ can also be expressed in terms of ρ and known quantities, (3.4) can then be used to determine $\hat{P}_{s,A}$. $E[L_j]$ can in fact be expressed solely in terms of λ_L and λ_H, and we now proceed with this derivation.

As a first step, we express $L_j(i)$, the number of low priority packets destined for output j at time slot i, in terms of past and present contributions from all inputs. Denote $C_j(i)$ as the number of "old" low priority packets in the HOL list before the contention phase of slot (i) destined for output j, and $A_j(i)$ as the number of "new" HOL low priority packets destined for output j at the same time instance. By "old" and "new," we mean whether or not this packet has been involved in previous contentions. We have

$$L_j(i) = C_j(i) + A_j(i). \qquad (3.9)$$

$C_j(i)$ in turn can be expressed as follows:

$$C_j(i) = \sum_{q=1}^{N} R_{j,q}(i - B_q) \qquad (3.10)$$

where $R_{j,q}(i)$ is an indicator function for input port q, output port j at time slot i, defined as

$$R_{j,q}(i) = \begin{cases} 1 & \text{there is a low priority packet in input } q \text{ contending} \\ & \text{for output } j \text{ at time } i \text{ and it is not delivered} \\ 0 & \text{otherwise.} \end{cases}$$

$$(3.11)$$

It follows that $R_j(i)$, denoting the total number of HOL low priority packets destined for j and not delivered at the end of time slot i, is

$$R_j(i) = \sum_{q=1}^{N} R_{q,j}(i). \qquad (3.12)$$

Because of preemption by high priority packets, a low priority packet not delivered experiences certain delay before it gets another chance at contention. B_q in (3.10) represents this delay. $R_{j,q}(i - B_q)$ in (3.10) is then an indicator for the event that an "old" packet, which was a left-over packet in input queue q at slot $i - B_q$, gets its first chance to reattempt at slot i. It can be shown that (see Appendix A for intermediate steps) at steady state

$$E[C_j] = E[R_j] \qquad (3.13)$$

and

$$E[C_j^2] = E[R_j^2]. \qquad (3.14)$$

Assuming A_j and C_j are independent, and applying (3.13) and (3.14), we square (3.9) and take expectations to get

$$E[L_j^2] = E[R_j^2] + 2E[A_j]E[R_j] + E[A_j^2]. \qquad (3.15)$$

In order to extract an expression for $E[L_j]$ from (3.15), we first note that $R_j(i)$ and $L_j(i)$ are related in the following way:

$$R_j(i) = L_j(i) - \epsilon(L_j(i))\bar{\epsilon}(H_j(i)). \qquad (3.16)$$

At steady state,

$$E[R_j] = E[L_j] - E[\epsilon(L_j)\bar{\epsilon}(H_j)]$$
$$= E[L_j] - \lambda_L. \qquad (3.17)$$

The last equality in the above equation follows from (3.2). In order to obtain $E[R_j^2]$, an assumption of independence between L_i and $\bar{\epsilon}(H_j)$ has to be applied. With this assumption, it follows that

$$E[R_j^2] = E[L_j^2] - 2E[L_j\epsilon(L_j)\bar{\epsilon}(H_j)] + E[(\epsilon(L_j)\bar{\epsilon}(H_j))^2]$$
$$= E[L_j^2] - 2E[L_j]E[\bar{\epsilon}(H_j)] + E[\epsilon(L_j)\bar{\epsilon}(H_j)]$$
$$= E[L_j^2] - 2E[L_j](1 - \lambda_H) + \lambda_L. \qquad (3.18)$$

It remains to determine the terms involving A_j. We approximate the distribution of A_j by a Poisson distribution with parameter λ_L arrivals/slot. This assumption has already been used with single priority switches [12], [17] for which the distribution of A_j can be shown to in fact tend toward a Poisson distribution as $N \to \infty$. A similar asymptotic result holds for a two priority switch. With the use of this last assumption, and substituting (3.17) and (3.18) into (3.15), an expression for $E[L_j]$ involving only λ_H and λ_L is obtained as

$$E[L_j] = \frac{2\lambda_L - \lambda_L^2}{2(1 - \lambda_H - \lambda_L)}. \quad (3.19)$$

Equations (3.8) and (3.19) are substituted into (3.4) to yield an expression involving ρ and known quantities, which is

$$\lambda_L = \frac{2\lambda_L - \lambda_L^2}{2(1 - \lambda_H - \lambda_L)} - P_0^H + \frac{\lambda_L}{\rho}. \quad (3.20)$$

In order to obtain a desired expression for $\hat{P}_{s,A}$, it now remains to relate it to ρ. This can be achieved using the expected interdeparture time between consecutive low priority packets from the same input where departures correspond to successful transmissions through the switch. Let i and $i + 1$ be the indexes of two consecutive low priority packets at a given input. The expected time between the departure of packet i and the departure of packet $i + 1$ can be obtained from different components. First, it takes on the average $1/(\rho P_0^H) - 1$ slots between the departure of packet i and the time packet $i + 1$ moves for the first time into the HOL position. Then, once packet $i + 1$ has moved into the HOL position, it is served with probability $\hat{P}_{s,A}$ and therefore departs on the average $1/\hat{P}_{s,A}$ slots later. Finally, since from flow conservation the average time between two successfully delivered packets equals the reciprocal of the input throughput $1/\lambda_L$, we get

$$\frac{1}{\lambda_L} = \frac{1}{\hat{P}_{s,A}} + \frac{1}{\rho P_0^H} - 1. \quad (3.21)$$

Using (3.1), (3.20), and (3.21), we now obtain the desired expression for $\hat{P}_{s,A}$:

$$\hat{P}_{s,A} = \frac{(\lambda_H^2 - 4\lambda_H + 1)(1 - \lambda_H - \lambda_L)}{(\lambda_H^2 - 4\lambda_H + 1)(1 - \lambda_H - \lambda_L) + (2\lambda_H + \lambda_L)(1 - \lambda_H)}. \quad (3.22)$$

B. Policy B

Under Policy B, high priority packets not delivered are dropped from the system. The conditional probability of service $\hat{P}_{s,B}$ seen by the low priority packets under this policy can be obtained following the same derivation as in Section III-A, with a few modifications. In this subsection, we only outline the modifications, while intermediate steps, similar to those in Section III-A, can be found in [5].

The main difference between Policy A and B is in the duration of busy periods generated by high priority packets. In Policy A, a high priority packet can disrupt the service of low priority packets for several consecutive time slots since a high priority packet not delivered at first attempt continues to occupy the HOL position until it is successfully delivered. In Policy B, however, a high priority packet disrupts the service of low priority packets for only the time slot of its arrival. At the end of this time slot, it is either delivered to its destination or cleared from the system. Two key equations are modified as a result of this difference. Equation (3.20) becomes

$$\lambda_L = \frac{2\lambda_L - \lambda_L^2}{2\left[\left(1 - \frac{1}{N}\lambda_H\right)^N - \lambda_L\right]} - \left(1 - \lambda_H - \frac{\lambda_L}{\rho}\right), \quad (3.23)$$

and (3.16) becomes

$$\frac{1}{\lambda_L} = \frac{1}{\hat{P}_{s,B}} + \frac{1}{\rho(1 - \lambda_H)} - 1.. \quad (3.24)$$

From (3.23) and (3.24), we have

$$\hat{P}_{s,B} = \frac{2(1 - \lambda_H)\left[\left(1 - \frac{1}{N}\lambda_H\right)^N - \lambda_L\right]}{2 - 2\lambda_H\left[\left(1 - \frac{1}{N}\lambda_H\right)^N - \lambda_L\right] - \lambda_L}. \quad (3.25)$$

For very large systems ($N \to \infty$), using the fact that $(1 - \frac{1}{N}\lambda_H)^N \to e^{-\lambda_H}$, (3.25) can be replaced by

$$\hat{P}_{s,B} = \frac{2(1 - \lambda_H)\left[e^{-\lambda_H} - \lambda_L\right]}{2 - 2\lambda_H\left[e^{-\lambda_H} - \lambda_L\right] - \lambda_L}. \quad (3.26)$$

IV. SYSTEM THROUGHPUT

In this section we investigate the maximum throughput achievable by the switching system under both Policy A and B. We first obtain a stability criterion which gives the maximum high and low priority loads that can be simultaneously supported by the system. This stability criterion then serves as a guideline for obtaining an estimate of the equivalent service probability to be used in the queueing model of Section V.

A. Policy A

We first investigate the maximum throughput of the system under Policy A. This maximum throughput corresponds to the maximum admissible input load under which the system remains stable. In the case of a single priority switch with fixed size packets, the maximum throughput has been shown to be about 0.586 [14], [17], while it is only 0.5 if exponentially distributed packet lengths are assumed [9]. In the case of a two priority switch, contributions to the total arrival rate from high and low priority packets must be distinguished.

Assuming a given arrival rate λ_H for high priority packets, it is possible to find an expression for the maximum low priority arrival rate $\lambda_{\max,A}(\lambda_H)$ such that the system remains stable. This relation can be obtained from the parameter ρ which was defined as the probability that there is a new low priority packet waiting to move into an available HOL position. The value of ρ can never exceed unity. Therefore, setting $\rho = 1$ in (3.20) gives $\lambda_{\max}(\lambda_H)$. Letting λ_L grow beyond this value cannot increase the throughput since ρ is already at its maximum. The resultant value is given by

$$\lambda_{\max,A}(\lambda_H)$$
$$= \frac{\lambda_H^2 - 6\lambda_H + 4 - \sqrt{-3\lambda_H^4 + 12\lambda_H^3 - 16\lambda_H + 8}}{2[1 - \lambda_H]} \quad (4.1)$$

where $0 \leq \lambda_H \leq 2 - \sqrt{2} = 0.586$. Equation (4.1) gives $\lambda_{\max,A}(0) = 0.586$ and $\lambda_{\max,A}(0.586) = 0$ which is in agreement with the result for single priority switches. $\lambda_{\max,A}$ as a function of λ_H is plotted in Fig. 3.

The maximum total system throughput $T_{\max,A}(\lambda_H)$ as a function of the offered high priority load can also be obtained from equation (4.1) as

$$T_{\max,A}(\lambda_H) = \lambda_{\max,A}(\lambda_H) + \lambda_H, \quad 0 \leq \lambda_H \leq 0.586. \quad (4.2)$$

The maximum system throughput is illustrated in Fig. 2 over the range $0 \leq \lambda_H \leq 0.586$. It can be seen that $T_{\max,A}$ starts at 0.586 for $\lambda_H = 0$ and increases to a maximum of about 0.607 for $\lambda_H \simeq 0.45$, after which it decreases to reach again 0.586 for $\lambda_H = 0.586$.

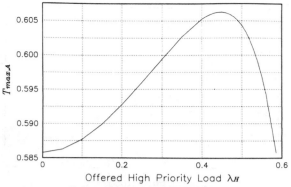

Fig. 2. Maximum system throughput under Policy A.

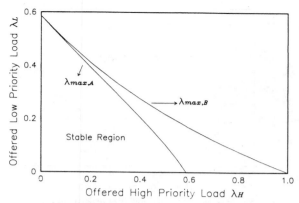

Fig. 3. Stability regions under Policies A and B.

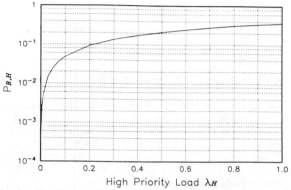

Fig. 4. Blocking probability for high priority traffic under Policy B.

In other words, the presence of two priority classes allows the total system throughput to exceed that of a single priority switch. It should be noted that the Y-axis below 0.585 in Fig. 2 is suppressed to exhibit the result. The result was verified through simulations of a 64×64 switch[2] for which throughputs above the theoretical single priority limit were observed.

This increase in throughput is attributed to the use of two priority classes which reduces the amount of memory in the distribution of packet destinations at the end of each cycle. In a single priority switch with input queues, the distributions of packet destinations in successive cycles are strongly correlated because packets that cannot go through reattempt at the next cycle. On the other hand, a switch with no input queues, where blocked packets are cleared, has independent distributions of packet destinations in successive cycles. For such a system operating under saturation (each input receives a new packet every cycle), it can be shown [25] that, assuming uniform destinations distribution, the asymptotic value of the maximum throughput is $1 - e^{-1} \simeq 0.632$ rather than 0.586. In a two priority switch, the preemption of low priority packets from HOL positions by high priority arrivals modifies somewhat randomly the distribution of packet destinations in successive cycles. It is this reduction in memory which is responsible for the increase, although not up to the level of a blocking switch, in maximum system throughput.

This points to the importance of the service policy not only in selecting between competing inputs, as studied in [17], but also in arbitrating between packets from the same input queue. This point is further illustrated in Section V, where we compare the average input queue length of a single priority switch under both First-In-First-Out (FIFO) and First-In-Random-Out (FIRO) service policies. As expected, the FIRO policy, which minimizes the memory in the distribution of successive packet destinations, outperforms the FIFO policy. Other examples of the improvement in switch throughput available by not using a strict FIFO policy can also be found in [11]. In [11], the contention process is extended to have w phases corresponding to a "window" of w packets at each input. Packets from a window are allowed to contend for the switch sequentially until one is successful. Simulation results demonstrate the increase in switch throughput that results from the use of this policy.

B. Policy B

Under Policy B, the maximum achievable low priority throughput $\lambda_{\max, B}$ given a high priority load of λ_H can similarly be obtained by setting ρ to one in (3.23) and letting $N \to \infty$. This gives

$$
\lambda_{\max, B}(\lambda_H)
$$
$$
= \frac{4 - 2\lambda_H - \sqrt{16 - 16\lambda_H + 4\lambda_H^2 - 8e^{-\lambda_H} + 8e^{-\lambda_H}\lambda_H}}{2}
$$

(4.3)

[2] Such a switch is large enough to have a maximum throughput very close to the asymptotic limit of 0.586 [17].

with $0 \le \lambda_H \le 1$. For $\lambda_H = 0$, (4.3) gives $\lambda_{\max, B}(0) = 0.586$ which is again in agreement with the result for single priority switches. $\lambda_{\max, B}$ is plotted, together with $\lambda_{\max, A}$, in Fig. 3 as a function of λ_H. In Fig. 3, the area to the left of each curve represents the stability region under each policy.

The maximum system throughput $T_{\max, B}$ under policy B is

$$
T_{\max, B}(\lambda_H) = \lambda_{\max, B}(\lambda_H) + \lambda_H[1 - P_{B, H}(\lambda_H)], \quad 0 \le \lambda_H \le 1,
$$

(4.4)

where $\lambda_{\max, B}(\lambda_H)$ is the maximum admissible low priority load given λ_H and $P_{B, H}$ is the blocking probability for high priority packets. $P_{B, H}$ can be obtained following the method used in [25], which gives a closed-form expression as

$$
P_{B, H}(\lambda_H) = \frac{\lambda_H - 1 + \left[1 - \dfrac{\lambda_H}{N}\right]^N}{\lambda_H}.
$$

(4.5)

In the limit, as $N \to \infty$, (4.5) becomes

$$
P_{B, H}(\lambda_H) = \frac{\lambda_H - 1 + e^{-\lambda_H}}{\lambda_H}.
$$

(4.6)

Fig. 4 plots $P_{B, H}$ as a function of the offered high priority load λ_H, while the maximum system throughput under Policy B is plotted in Fig. 5 using (4.3), (4.4), and (4.6).

V. QUEUEING MODEL AND NUMERICAL RESULTS

In order to obtain the average queue length of the low priority class, we use a Geom/Geom/1 queueing model to approximate the queueing behavior in an input port. In this model, low priority packets arrive with probability λ_L at each time slot as is described in Section II, and receive service with a fixed probability P_s per time slot. This queueing model is illustrated in Fig. 6. The differ-

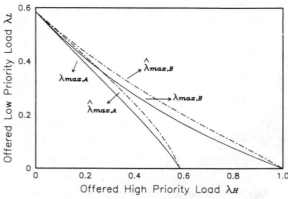

Fig. 5. Maximum system throughput under Policy B.

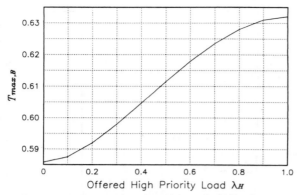

Fig. 7. Capacity overestimation.

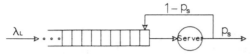

Fig. 6. Queueing model.

ence between P_s and the $\hat{P}_s$ obtained in Section III is explained as follows.

The presence of the high priority class affects the service received by the low priority class in the following two ways. First, the high priority arrivals in the same input port increase the number of time slots needed for a low priority packet to reach the HOL position. Second, once reaching HOL, the contention from other HOL high priority packets increases the number of time slots needed before the low priority packet can be delivered by the switch. The past two terms on the right-hand side in (3.21) and (3.24) account for the first influence up to the point a low priority packet arrives at HOL for the first time. Subsequent preemptions by high priority class along with the second influence mentioned above are accounted for in the quantity $\hat{P}_s$. In the queueing model proposed in this section, however, both influences have to be accounted for in the quantity P_s. From the above arguments, it follows that $P_s \leq \hat{P}_s$, $\forall \lambda_H > 0$. In order to obtain P_s, a heuristic factor, derivable from the maximum throughput results given in Section IV, is used to scale down $\hat{P}_s$.

If $\hat{P}_s$ were to be used as an estimate for the service rate seen by a low priority packet, since part of the effect from high priority preemption is not accounted for in this quantity, it would yield a higher maximum low priority arrival rate when the stability condition is applied. Denote the maximum rate associated with $\hat{P}_s$ as $\hat{\lambda}_{max}$. $\hat{\lambda}_{max, A}$ and $\hat{\lambda}_{max, B}$ for Policies A and B can be obtained by substituting the stability condition $\hat{P}_s = \hat{\lambda}_{max} = \lambda_L$ into (3.22) and (3.26), respectively, and solving for $\hat{\lambda}_{max}$. This yields

$$\hat{\lambda}_{max, A}(\lambda_H) = \frac{-\lambda_H^3 + 4\lambda_H^2 - 8\lambda_H + 4 - \sqrt{\lambda_H^6 - 4\lambda_H^5 + 16\lambda_H^3 - 4\lambda_H^2 - 16\lambda_H + 8}}{2\lambda_H^2 - 6\lambda_H + 2} \quad (5.1)$$

where (5.1) holds for $0 \leq \lambda_H \leq 0.586$ and $\lambda_H \neq (3 - \sqrt{5})/2$. In the case where $\lambda_H = (3 - \sqrt{5})/2$, when the denominator vanishes, we have $\hat{\lambda}_{max, A}((3 - \sqrt{5})/2) = 0.259$. For Policy B, when $\lambda_H \neq 0.5$,

$$\hat{\lambda}_{max, B}(\lambda_H) = \frac{2 - \lambda_H - \lambda_H e^{-\lambda_H} - \sqrt{[e^{-2\lambda_H} - 2e^{-\lambda_H} + 1]\lambda_H^2 - [4 - 2e^{-\lambda_H}][\lambda_H - 1]}}{1 - 2\lambda_H}, \quad (5.2)$$

while for $\lambda_H = 0.5$ we have $\hat{\lambda}_{max, B}(0.5) = 0.253$. Fig. 7 plots $\hat{\lambda}_{max}$ and $\hat{\lambda}_{max}$ for both policies. It is seen that, as to be expected, $\hat{\lambda}_{max} \geq \hat{\lambda}_{max}$ for each policy.

We use the ratio $\lambda_{max}/\hat{\lambda}_{max}$ to scale down $\hat{P}_s$, yielding P_s as

$$P_s = \hat{P}_s \frac{\lambda_{max}}{\hat{\lambda}_{max}}. \quad (5.3)$$

This scaling factor is heuristic. The motivation is to ensure that the correct maximum throughput is obtained when applying the stability condition on P_s.

With P_s determined, the Geom/Geom/1 queueing model can be solved using standard methods (e.g., [18]) and we find (the derivation is provided in Appendix B for completeness) that the generating function $G(z)$ of the steady-state queue length K_L is

$$G(z) = \frac{(P_s - \lambda_L)(1 - \lambda_L + \lambda_L z)}{P_s(1 - \lambda_L) - \lambda_L(1 - P_s)z}. \quad (5.4)$$

From (5.4) we obtain several quantities of interest. The first one is the average queue length $E[K_L]$ of low priority packets,

$$E[K_L] = G'(z)|_{z=1} = \frac{\lambda_L(1 - \lambda_L)}{P_s - \lambda_L}. \quad (5.5)$$

The expected delay $E[D_L]$ of low priority packets can be directly deduced from (5.5) using Little's formula. The values of the mean queue length are plotted in Figs. 8 and 9 for Policies A and B, respectively. Several system configurations are considered and results are compared to simulation outcomes. The simulation assumed a 64 × 64 switch size. As shown in the figures, there is good agreement between the analysis and simulation, indicating that the various approximations used in the analysis are valid for systems of reasonably large size.

We now justify, through simulation, the remark in Section IV regarding the influence of the service discipline at the inputs on the maximum throughput of the system. In Fig. 10 we plot simulation results of a 64 × 64 switch for the average queue length as a function of the offered load (only one class of traffic is assumed) for two queueing disciplines. The first one is the FIFO (First-In-First-Out) policy and the second is the FIRO (First-In-Random-Out) policy. In the FIRO policy, the server randomly chooses for service one packet from the queue at each time slot. As the queue length grows, packet destinations in successive slots become less corre-lated. Thus, under the FIRO policy, the maximum throughput should approach 0.632, the maximum throughput achievable with independent packet destination in successive slots [17], [25]. This is verified in Fig. 10, where the FIRO policy is seen to result in lower average queue lengths than the FIFO policy and to remain stable up to loads of about 0.62 rather than 0.58 for the FIFO policy. It should, however, be pointed out that, in an FIRO policy, packets

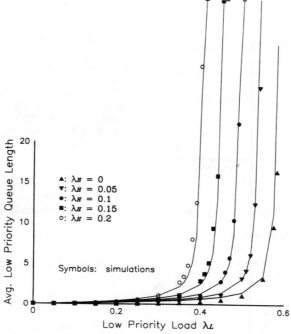

Fig. 8.   Average queue length for low priority traffic under Policy A.

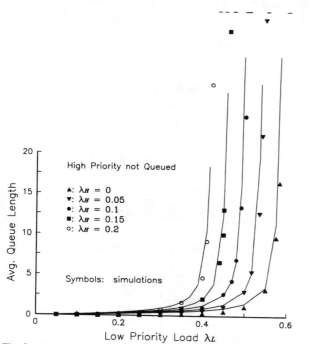

Fig. 9.   Average queue length for low priority traffic under Policy B.

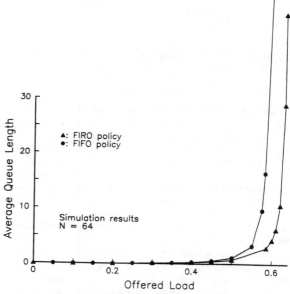

Fig. 10.   Impact of queueing discipline.

are delivered out of sequence. Because of the resequencing problem, it is unlikely that the modest improvement in throughput and average delay could justify its use.

The queue length distribution for low priority packets is obtained by inversion of the generating function in (5.4) giving

$$\Pr\left(K_L = i\right) = \begin{cases} (1-a)(1-\lambda_L), & i = 0 \\ (1-a)(1-\lambda_L)a^i + \lambda_L(1-a)a^{i-1}, & i \geq 1 \end{cases}$$
(5.6)

where

$$a = \frac{\left[\lambda_L(1-P_s)\right]}{\left[P_s(1-\lambda_L)\right]}.$$

Based on this queueing model, the accuracy of the expression we obtain for the queue length distribution is not as good as that of the expression for the mean queue length. This is illustrated in Fig. 11 where we compare the queue length probability density to simulation results for a 64×64 system with $\lambda_H = 0.2$, $\lambda_L = 0.3$, and operating under Policy A. The figure shows that the density function obtained from (5.6) underestimates the actual values for both small and large queue lengths, while it overestimates them for intermediate queue lengths. This lack of accuracy is to be expected in view of the simplifying assumptions associated with the Geom/Geom/1 queueing system used to model a switch input.

This model accurately characterizes the average behavior of the server but ignores the dependencies between consecutive time slots. More specifically, it neglects the fact that, when the server is

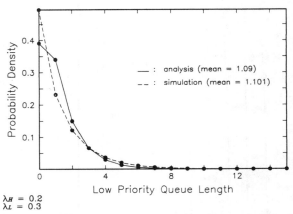

$\lambda_H = 0.2$
$\lambda_L = 0.3$

Fig. 11. Probability density function for low priority queue length.

unavailable at a given time slot because of a nonempty high priority queue, it is more likely to be unavailable in the next time slot. The busy period of the high priority queue gates the server availability, therefore resulting in more "bursty" behavior of the server than is reflected in the "random server" model used here. This implies that the random server system underestimates the probability of large queues and therefore the blocking probability.

In the case of a system with finite rather than infinite buffers, the queue length can be obtained by truncating and scaling the expression given in (5.6). Assuming a finite buffer size B_L for low priority packets, the blocking probability obtained from the equivalent queueing system is equal to

$$P_{B,L} = \frac{\Pr(K_L = B_L)}{\sum_{i=0}^{B_L} \Pr(K_L = i)}. \qquad (5.7)$$

As mentioned above, this value is only an approximation of the actual blocking which typically will be higher.

VI. Conclusion

This paper presents performance results for a two priority packet switch with input queues. The main contribution of the paper is the derivation of performance measures for low priority packets that take into account the influence of both high and other low priority packets. This result is obtained through an expression for the probability of service seen by low priority packets. This probability of service takes into account the memory in the output contention process and in the unavailability of inputs because of high priority packets. In addition to these results, the effect of the two priority classes on the total maximum switch throughput is discussed and "shown" to typically exceed that of a single priority switch.

The probability of service for low priority packets is used to study an equivalent queueing system from which expressions for the average queue length and delay as well as queue length distribution can be obtained. The average queue length is compared to simulation values for various system configurations and found to be in good agreement. The queue length distribution is used to obtain approximation for blocking probabilities with finite buffers. The results are, however, less accurate than for average values. As mentioned in the previous section, this is due to the simplifying assumptions made in using a random server to model the switch availability to low priority traffic.

In order to improve the accuracy of distribution results, it is necessary to use a more accurate queueing model for the inputs. A possible solution is the use of a vacation-type server. The server represents a switch input link which goes on vacation each time the high priority queue leaves the empty state because of an arrival. The server stays on vacation for a period determined by the duration of the busy period generated by the initial high priority arrival. This captures the influence of high priority packets at the switch inputs,

but does not account for the memory in output contention. This second factor is, however, expected to be less significant since the assumption of a random server is probably a more reasonable approximation of the random selection process at the output.

Other interesting directions for future research exist. For example, it might be of interest to change the assumption of Bernoulli arrivals so as to better represent the actual nature of the traffic. Another interesting possibility is to relax the assumption of internally nonblocking switch to include such systems as Banyan and buffered Banyan networks. Finally, a more complete study of the influence of the service policy on the maximum system throughput can potentially help design more efficient systems.

Appendix A

In this appendix, we provide the intermediate steps in (3.13) and (3.14). First, it is seen that for the system to be stable, every left-over packet eventually gets served, thus,

$$\lim_{k \to \infty} \Pr(B_q = k) = 0. \qquad (A.1)$$

Since (A.1) indicates that B_q, a random variable, takes only finite number, it follows that

$$\lim_{i \to \infty} E[R_{j,q}(i - B_q)] = \lim_{i \to \infty} E[R_{j,q}(i)]. \qquad (A.2)$$

We further assume that the $R_{j,q}(j - B_q)$'s are independent random variables for all q's and all B_q's. This is an approximation, but probably reasonable for large N. Using this independent assumption and (3.12) and (A.2), the first two moments of C_j at steady state are obtained as

$$
\begin{aligned}
E[C_j] &= \lim_{i \to \infty} E[C_j(i)] \\
&= \lim_{i \to \infty} E\left[\sum_{q=1}^{N} R_{j,q}(i - B_q)\right] \\
&= \lim_{i \to \infty} \sum_{q=1}^{N} E[R_{j,q}(i - B_q)] \\
&= \sum_{q=1}^{N} \lim_{i \to \infty} E[R_{j,q}(i - B_q)] \\
&= \sum_{q=1}^{N} \lim_{i \to \infty} E[R_{j,q}(i)] \\
&= \lim_{i \to \infty} E\left[\sum_{q=1}^{N} R_{j,q}(i)\right] \\
&= \lim_{i \to \infty} E[R_j(i)] \\
&= E[R_i] \qquad (A.3)
\end{aligned}
$$

and

$$
\begin{aligned}
E[C_j^2] &= \lim_{i \to \infty} E[C_j^2(i)] \\
&= \lim_{i \to \infty} E\left[\left(\sum_{q=1}^{N} R_{j,q}(i - B_q)\right)^2\right] \\
&= \lim_{i \to \infty} E\left[\sum_{q=1}^{N} R_{j,q}^2(i - B_q)\right] \\
&\quad + 2 \lim_{i \to \infty} E\left[\sum_{q=1}^{N} \sum_{p \neq q}^{N} R_{j,q}(i - B_q) R_{j,p}(i - B_p)\right] \\
&= \lim_{i \to \infty} \sum_{q=1}^{N} E[R_{j,q}^2(i - B_q)]
\end{aligned}
$$

$$+ 2 \lim_{i \to \infty} \sum_{q=1}^{N} \sum_{p \neq q}^{N} E[R_{j,q}(i - B_q) R_{j,p}(i - B_p)]$$

$$= \sum_{q=1}^{N} \lim_{i \to \infty} E[R_{j,q}^2(i - B_q)]$$

$$+ 2 \sum_{q=1}^{N} \sum_{p \neq q}^{N} \lim_{i \to \infty} E[R_{j,q}(i - B_q) R_{j,p}(i - B_p)]$$

$$= \sum_{q=1}^{N} \lim_{i \to \infty} E[R_{j,q}^2(i - B_q)]$$

$$+ 2 \sum_{q=1}^{N} \sum_{p \neq q}^{N} \lim_{i \to \infty} E[R_{j,q}(i - B_q)] E[R_{j,p}(i - B_p)]$$

$$= \sum_{q=1}^{N} \lim_{i \to \infty} E[R_{j,q}^2(i)]$$

$$+ 2 \sum_{q=1}^{N} \sum_{p \neq q}^{N} \lim_{i \to \infty} E[R_{j,q}(i)] E[R_{j,p}(i)]$$

$$= \sum_{q=1}^{N} \lim_{i \to \infty} E[R_{j,q}^2(i)]$$

$$+ 2 \sum_{q=1}^{N} \sum_{p \neq q}^{N} \lim_{i \to \infty} E[R_{j,q}(i) R_{j,p}(i)]$$

$$= \lim_{i \to \infty} \left(\sum_{q=1}^{N} E[R_{j,q}^2(i)] \right.$$

$$\left. + 2 \sum_{q=1}^{N} \sum_{p \neq q}^{N} E[R_{j,q}(i) R_{j,p}(i)] \right)$$

$$= \lim_{i \to \infty} E\left[\sum_{q=1}^{N} R_{j,q}^2(i) + 2 \sum_{q=1}^{N} \sum_{p \neq q}^{N} R_{j,q}(i) R_{j,p}(i) \right]$$

$$= \lim_{i \to \infty} E\left[\left(\sum_{q=1}^{N} R_{j,q}(i) \right)^2 \right]$$

$$= \lim_{i \to \infty} E[R_j^2(i)]$$

$$= E[R_j^2]. \tag{A.4}$$

APPENDIX B

In this appendix we provide the computational steps needed in the derivation of (5.4). We start from the expression relating the queue length in consecutive time slots. We have

$$K_L(i + 1) = K_L(i) - \gamma \epsilon(K_L(i)) + \alpha, \tag{B.1}$$

where $K_L(i)$ is the queue length at time i, $\epsilon(x)$ is the indicator function of Section III, and α and γ are binary random variables such that $E[\alpha] = \lambda_L$ and $E[\gamma] = P_s$. From (B.1) we have

$$G(z) = E[Z^{K_L}] = E[z^\alpha] E[z^{K_L - \gamma \epsilon(K_L)}]. \tag{B.2}$$

The individual terms on the right-hand side of (B.2) can be expressed as follows:

$$E[z^\alpha] = P_r(\alpha = 0) z^0 + P_r(\alpha = 1) z^1$$
$$= 1 - \lambda_L + \lambda_L z, \tag{B.3}$$

and

$$E[z^{K_L - \gamma \epsilon(K_L)}] = E_\gamma[E_{K_L}[Z^{K_L - \gamma \epsilon(K_L)}]]$$

$$= E_\gamma \left[\sum_{i=0}^{\infty} P_r(K_L = i) z^{i - \gamma \epsilon(i)} \right]$$

$$= E_\gamma[P_r(K_L = 0)]$$

$$+ E_\gamma \left[\sum_{i=1}^{\infty} P_r(K_L = i) z^{i - \gamma \epsilon(i)} \right]$$

$$= P_r(K_L = 0) - E_\gamma[P_r(K_L = 0) z^{-\gamma}]$$

$$+ E_\gamma \left[\sum_{i=0}^{\infty} P_r(K_L = i) z^{i - \gamma} \right]$$

$$= P_r(K_L = 0)[1 - E_\gamma[z^{-\gamma}]]$$

$$+ G(z) E_\gamma[z^{-\gamma}], \tag{B.4}$$

while

$$E[z^{-\gamma}] = P_r(\gamma = 0) z^0 + P_r(\gamma = 1) z^{-1}$$
$$= 1 - P_s + P_s z^{-1}. \tag{B.5}$$

Combining (B.2), (B.3), (B.4), and (B.5) yields

$$G(z) = (1 - \lambda_L + \lambda_L z) [P_r(K_L = 0)(P_s - P_s z^{-1})$$
$$+ G(z)(1 - P_s + P_s z^{-1})]. \tag{B.6}$$

This gives

$$G(z) = \frac{P_s P_r(K_L = 0)(1 - \lambda_L + \lambda_L z)}{(1 - \lambda_L) P_s - \lambda_L (1 - P_s) z}. \tag{B.7}$$

$P_r(K_L = 0)$ can be obtained using the fact that $G(1) = 1$. Thus,

$$G(1) = \frac{P_s P_r(K_L = 0)}{P_s - \lambda_L} = 1. \tag{B.8}$$

Therefore,

$$P_r(K_L = 0) = \frac{P_s - \lambda_L}{P_s}. \tag{B.9}$$

Substituting (B.9) into (B.7) yields (5.4).

ACKNOWLEDGMENT

The authors would like to thank Prof. T. Stern for reading through the manuscript and for many valuable comments regarding the organization of this paper.

REFERENCES

[1] C.-L. Wu and T.-Y. Feng, Eds., *Tutorial: Interconnection Networks for Parallel and Distributed Processing.* New York: IEEE Computer Society Press, 1984.

[2] H. Ahmadi and W. Denzel, "A survey of modern high-performance switching techniques," *IEEE J. Select. Areas Commun.*, Issue on Architecture and Protocols for Computer Networks: The State-of-the-Art, vol. 7, pp. 1091–1103, Sept. 1989.

[3] H. Ahmadi, W. Denzel, C. A. Murphy, and E. Port, "A high performance switch fabric for integrated circuit and packet switching," in *Proc. 1988 INFOCOM Conf. (INFOCOM'88 Conf. Rec.)*, pp. 9–18.

[4] D. P. Bhandarkar, "Analysis of memory interference in multiprocessors," *IEEE Trans. Comput.*, vol. C-24, pp. 897–908, Sept. 1975.

[5] J. S.-C. Chen and R. Guérin, "Performance study of an integrated packet switch with two priority classes," in *Proc. 1989 ITG/GI-Conf.: Commun. Distributed Syst.*, Stuttgart, F.R.G., Feb. 1989. Also, IBM Res. Div., RC 13774, May 1988.

[6] J. N. Daigle, "Message delays in prioritized packet switching systems," in *Proc. 1986 INFOCOM Conf. (INFOCOM'86 Conf. Rec.)*, 1986, pp. 214–223.

[7] D. M. Dias, M. Kumar, and Y.-C. Lien, "Design and analysis of a multistage voice data switch," in *Proc. 1987 GLOBECOM Conf. (GLOBECOM'87 Conf. Rec.)*, vol. 3, Tokyo, Japan, Nov. 1987, pp. 1856–1860.

[8] K. Y. Eng, M. G. Hluchyj, and Y.-S. Yeh, "A knockout switch for variable-length packets," *IEEE J. Select. Areas Commun.*, vol. SAC-5, pp. 1426–1435, Dec. 1987.

[9] S. W. Fuhrmann, "Performance of a packet switch with crossbar architecture," IBM Res. Div., RZ 1829, June 1989.

[10] G. Hebuterne, "STD switching in an ATD environment," in *Proc. 1988 Infocom Conf. (INFOCOM'88 Conf. Rec.)*, 1988, pp. 449–458.

[11] M. G. Hluchyj and M. J. Karol, "Queueing in high-performance packet switching," *IEEE J. Select. Areas Commun.*, vol. 6, pp. 1587–1597, Dec. 1988.

[12] ——, "Queueing in space-division packet switching," in *Proc. 1988 Infocom Conf. (INFOCOM'88 Conf. Rec.)*, 1988, pp. 334–343.

[13] A. Huang and S. Knauer, "STARLITE: A wideband digital switch," in *Proc. 1984 Globecom Conf. (GLOBECOM'84 Conf. Rec.)*, vol. 1, Atlanta, GA, Nov. 1984, pp. 5.3.1–5.3.5.

[14] J. Y. Hui and E. Arthurs, "A broadband packet switch for integrated transport," *IEEE J. Select. Areas Commun.*, vol. SAC-5, pp. 1264–1273, Oct. 1987.

[15] Y.-C. Jenq, "Performance analysis of a packet switch based on single-buffered Banyan network," *IEEE J. Select. Areas Commun.*, vol. SAC-1, pp. 1014–1021, Dec. 1983.

[16] M. J. Karol and M. G. Hluchyj, "Using a packet switch for circuit-switched traffic: A queueing system with periodic input traffic," *IEEE Trans. Commun.*, vol. 37, pp. 623–625, June 1989. Also in *Proc. ICC'87.*

[17] M. J. Karol, M. G. Hluchyj, and S. P. Morgan, "Input versus output queueing on a space-division packet switch," *IEEE Trans. Commun.*, vol. COM-35, pp. 1347–1356, Dec. 1987. Also in *Proc. GLOBECOM'86.*

[18] H. Kobayashi and A. G. Konheim, "Queueing models for computer communications system analysis," *IEEE Trans. Commun.*, vol. COM-25, pp. 2–29, Jan. 1977.

[19] C. P. Kruskal and M. Snir, "The performance of multistage interconnection networks for multiprocessors," *IEEE Trans. Comput.*, vol. C-32, pp. 1091–1098, Dec. 1983.

[20] S. Li and M. J. Lee, "A study of traffic imbalances in a fast packet switch," in *Proc. 1989 INFOCOM Conf. (INFOCOM'89 Conf. Rec.)*, 1989, pp. 538–547.

[21] Y. Lim and J. Kobza, "Analysis of a delay dependent priority discipline in a multi-class traffic packet switching node," in *Proc. 1988 INFOCOM Conf. (INFOCOM'88 Conf. Rec.)*, 1988, pp. 889–898.

[22] B. N. W. Ma and J. W. Mark, "Performance analysis of burst switching for integrated voice/data networks," *IEEE Trans. Commun.*, vol. 36, pp. 282–297, Mar. 1988.

[23] S. Nojima, E. Tsutsui, H. Fukuda, and M. Hashimoto, "Integrated services packet network using bus matrix switch," *IEEE J. Select. Areas Commun.*, vol. SAC-5, pp. 1284–1292, Oct. 1987.

[24] Y. Oie M. Murata, K. Kubota, and H. Miyahara, "Effect of speedup in nonblocking packet switch," in *Proc. ICC'89*, 1989, pp. 410–414.

[25] J. K. Patel, "Performance of processor-memory interconnections for multiprocessors," *IEEE Trans. Comput.*, vol. C-30, pp. 771–780, Oct. 1981.

[26] J. S. Turner and L. F. Wyatt, "A packet network architecture for integrated services," in *Proc. 1983 GLOBECOM Conf. (GLOBECOM'83 Conf. Rec.)*, Nov. 1983, pp. 45–50.

[27] Y.-S. Yeh, M. G. Hluchyj, and A. S. Acampora, "The knockout switch: A simple, modular architecture for high-performance packet switching," *IEEE J. Select. Areas Commun.*, vol. SAC-5, pp. 1274–1283, Oct. 1987.

Effect of Speedup in Nonblocking Packet Switch

Yuji OIE[1], Masayuki MURATA[2], Koji KUBOTA[2] and Hideo MIYAHARA [2]

(1) Department of Electrical Engineering, Sasebo College of Technology, Sasebo, 857-11 Japan.
(2) Department of Information and Computer Sciences, Faculty of Engineering Science,
Osaka University, Toyonaka, 560 Japan.

Abstract

The nonblocking packet switch to be considered has N inputs and N outputs and operates L times as fast as the input and output trunks. Karol et al. has analyzed special cases of $L = 1, N$. This paper analyzes the effect of speedup (L) on packet loss probability and average transmission delay in the case of an arbitrary number of L such that $1 \leq L \leq N$ in two types of nonblocking packet switch. One has buffers only at outputs. The other has buffers at inputs as well as at outputs. The maximum thoughput is obtained on the assumption of an infinite number of buffers in the latter type; that is 0.8845 and 0.9755 when $L = 2$ and $L = 3$, respectively. Some simulation results are also provided in the case where a finite number of buffers are given at each input.

1 Introduction

A packet switch plays an important part in high speed communication networks. Two classes of packet switching arise in this context; i.e., blocking and nonblocking switches. In a blocking switch, a packet going to some output possibly contend with packets destined for other outputs in an internal switch element. On the other hand, a nonblocking switch (e.g., cross-bar switch, Batcher-banyan network [3], Knockout switch [8]) offers a different path to each packet destined for a different output. A contention is inevitable among packets addressed to a same output even in this case.

The focus of this paper is on the nonblocking switch because of its ability to provide desirable performance. Its performance primarily depends upon (1) switching speed and (2) where and how many buffers are placed. The packet switch to be considered has N inputs and N outputs and operates L times as fast as the input and output trunks, where $1 \leq L \leq N$. Note that in Knockout switch, L refers to the number of buffers in the output concentrator [8]. A packet is assumed to arrive according to a Bernoulli process at each time slot and to go to one of N outputs with equal probability $\frac{1}{N}$.

In one special case where packets get through at the same speed as an input and ouput trunks (i.e., $L = 1$), only one packet can reach a given output. If two or more packets placed on different inputs go to a common output, they have to wait until the following time slot except one successful packet. Thus, input buffers for packets to be queued are required to accomodate them. In [4,5], it has been shown that this switch achieves a throughput of 0.5858 for a large number of N.

In the other special case where $L = N$, packet queueing does not occur at inputs but occurs at outputs. This case offers an optimum performance in the sense that even if N packets arriving at N inputs go to a common output, these N packets can enter the output buffers at once. Karol et al. analyzed the average transmission delay in this switch. The probability of packet being lost in a finite number of output buffers was obtained in [2].

As mentioned above, $L = N$ offers the optimum performance. However, it becomes harder to switch packets at a speed of N as N becomes larger. Thus, the case with L such that $1 < L < N$ is of more practical importance than the case with $L = N$. It is noted that in this case packet queueing occurs at both input and output buffers if the switch has those buffers; otherwise some packets could be lost. Yeh et al. [8] has shown that $L = 8$ achieves a probability of packet loss less than 10^{-6} for an arbitrarily large value of N at 90 percent traffic load if an infinite number of buffers are given only at outputs.

The major objective of this paper is to examine the speedup effect of the switching fabric on the average transmission delay and packet loss probability in the nonblocking packet switch (NPS). Two cases will be dealt with. (is that buffers are provided with only outputs same as in [2,8] although we will treat the case where $1 \leq L \leq N$ and the number of output buffers (B_o) is finite in this paper (see Fig.1). The other is that buffers are given at each input as well as at each output (see Fig.2). Let B_i denote the number of buffers at each input. The input buffers are provided in an attempt to reduce lost packets and to attain the desirable performance even for a relatively small value of L. In Section 2.1, the former case is analyzed with respect to the average transmission delay based on the assumption of an infinite number of output buffers. Packet loss probability will be also obtained analytically when a finite number of output buffers are assumed in Section 2.2. Next, Section 3 provides maximum throughput analysis in the latter case assuming that $B_i = B_o = \infty$ and $N = \infty$, and also gives simulation results for packet loss probability

Reprinted from *Proc. IEEE International Conference on Communications*
'89, pp. 410–414, June 1989.

at a finite number of input and output buffers.

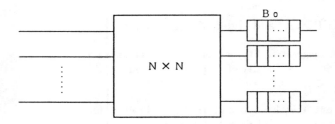

Figure 1 NPS with output buffers

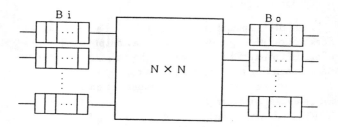

Figure 2 NPS with input and output buffers

2 NPS with Output Buffers

2.1 Average Transmission Delay Analysis

Karol et al. [5] has obtained the average transmission delay in the case where L equals N and an infinite number of buffers are placed in each output (i.e., $B_o = \infty$). On the same assumption, we will analyze the average transmission delay for an arbitrary number of L such that $1 \leq L \leq N$ below.

Let $\tilde{A}$ denote the random variable associated with the number of packet arrivals destined for a particular output (tagged output) during a given slot. We define the random variable $\tilde{V}$ as the number of packets entering the tagged output queue among $\tilde{A}$ packets. In this switch, if more than L packets, say k packets, are addressed to a same output, L packets of the k packets are chosen at random, thereby $k - L$ packets being lost. From the assumption stated in Section 1, the probabilities associated with these random variables are given by

$$a_i \triangleq Pr[\tilde{A} = i] = \begin{cases} \binom{N}{i}(\frac{p}{N})^i(1 - \frac{p}{N})^{N-i} & (N < \infty) \\ \frac{p^i}{i!}e^{-p} & (N = \infty). \end{cases} \quad (1)$$

$$v_i \triangleq Pr[\tilde{V} = i] = \begin{cases} a_i & (i < L) \\ \sum_{n=L}^{N} a_n & (i = L). \end{cases} \quad (2)$$

The packet loss probability at input trunk was obtained in [8] as follows:

$$p_{loss1} = \begin{cases} \frac{1}{p} \sum_{k=L+1}^{N}(k - L)a_k & (N < \infty) \\ (1 - \frac{L}{p})(1 - \sum_{k=0}^{L} \frac{p^k}{k!}e^{-p}) + \frac{p^L}{L!}e^{-p} & (N = \infty). \end{cases} \quad (3)$$

Furthermore, we define the probability generating function (pgf)

$$V(z) \triangleq \sum_{i=0}^{\infty} v_i z^i \quad (4)$$

$$= \sum_{i=0}^{L-1} a_i z^i + \sum_{i=L}^{N} a_i z^L = \sum_{i=0}^{L}(z^i - z^L)a_i + z^L.$$

Let Q_m denote the number of packets in the tagged output queue at the end of the mth time slot, A_m denote the number of packets arrivals addressed to the tagged output during the mth time slot, and V_m denote the number of packets entering the output queue during the mth time slot. Thus, we have

$$Q_m = \max(0, Q_{m-1} + V_m - 1), \quad (5)$$
$$V_m = \min(L, A_m). \quad (6)$$

We can obtain the pgf of the steady–state queue size from eqs.(5) and (6) using the M/G/1 queueing theory [6]:

$$Q(z) = \frac{(1 - p(1 - p_{loss1}))(z - 1)}{z - V(z)}.$$

Differentiating $Q(z)$ with respect to z and taking the limit as z approaches 1, we have

$$\overline{Q} = \dot{Q}(1) = \frac{(1 - p(1 - p_{loss1}))\ddot{V}(1)}{2(\dot{V}(1) - 1)^2}, \quad (7)$$

where $\dot{V}(1)$ and $\ddot{V}(1)$ are derived from eq.(4).

Since our system can be regarded as a single–server system with a batch arrival, a marked packet suffers from two types of delay. One is the waiting time due to transmission times of packets being at the output buffer when the marked packet arrives at the buffer. Noting that each transmission time is one, the average waiting time of this type, say $\overline{W_1}$, is given by $\dot{Q}(1)$ (i.e., eq.(7)). The other is the waiting time due to the order in which the marked packet is transmitted among a group of packets including it. Defining $\tilde{W}_2$ as the random variable of this type of delay time and using the queueing theory associated with a batch arrival [1], we have $Pr[\tilde{W}_2 = k] = \sum_{i=k+1}^{L} \frac{v_i}{\overline{V}}$ $(0 \leq k \leq L - 1)$. Thus,

$$\overline{W_2} = \frac{\overline{V^2} - \overline{V}}{2\overline{V}} = \frac{\ddot{V}(1)}{2\dot{V}(1)}$$

Hence, it follows that the average total waiting time, $\overline{W}$, a packet experiences in the switch is given by

$$\overline{W} = \overline{W_1} + \overline{W_2}$$
$$= \frac{(1 - p(1 - p_{loss1}))\ddot{V}(1)}{2(\dot{V}(1) - 1)^2} + \frac{\ddot{V}(1)}{2\dot{V}(1)}. \quad (8)$$

2.2 Packet Loss Probability in Output Buffer

The plobability, p_{loss2}, that packets are lost at a finite number of output buffers ($B_o < \infty$) will be obtained as follows. This problem was solved in a special case where $L = N$ in [2]. Let B_o denote the number of buffers provided with each output. It follows that Q_m is related to Q_{m-1} and V_m as

$$Q_m = \max(0, \min(B_o, Q_{m-1} + V_m - 1)). \qquad (9)$$

Define the random variables X_m and X_{m-1} as the number of packets being in the output buffer in mth and $(m-1)$st time slot. From eq.(9), the transition probability $p_{ij} \triangleq Pr[X_m = j | X_{m-1} = i]$ being independent of m is given by

$$p_{ij} = \begin{cases} v_0 + v_1 & (i = j = 0) \\ v_{j+1} & (i = 0, 1 \le j < \min(L-1, B_o)) \\ v_{j-i+1} & (i > 0, i-1 \le j < \min(L+i-1, B_o)) \\ \displaystyle\sum_{n=\min(L, B_o-i+1)}^{L} v_n & (j = \min(L+i-1, B_o)) \\ 0 & (\text{otherwise}). \end{cases}$$

Defining the steady state probability q_j of being in state j

$$q_j \triangleq Pr[X = j],$$

we can establish the following recursive equation for calculating q_j:

$$q_j = \sum_{i=0}^{B_o} p_{ij} q_i \qquad (0 \le j \le B_o).$$

Since the normalized switch throughput is given by $1 - q_0 a_0$, we have the probability that a marked packet arriving at output buffer is lost, p_{loss2}, as follows

$$p_{loss2} = 1 - \frac{1 - q_0 a_0}{p(1 - p_{loss1})}, \qquad (10)$$

where p_{loss1} is given by eq.(3).

2.3 Numerical Results

Figure 3 depicts the average transmission delay, which is derived from eq.(8), in the case where $N = \infty$. The average transmission delay times in cases where $L = 2, 3$ are different from others because of a large number of lost packets at inputs in both cases. As shown in this figure, when $L \ge 4$ the average transmission delay is very close to the optimum one, i.e., the case where $L = \infty$. In addition, other numerical results show that the delay performance is not sensitive to N and the case of $L = \infty$ provides a good approximation for the cases with L such that $4 \le L < \infty$.

A packet switch should be designed to gurantee packet loss probability of no more than some value. For example, to achieve p_{loss1} less than 10^{-6} at $p = 0.9$, it follows from eq.(3) that L should be greater than or equal to 8. Figure

4 shows the sensitivity of p_{loss2} given by eq.(10) to L when $N = \infty$. As the case with the delay performance, two cases of $L = 2, 3$ are very different from the others in terms of p_{loss2}. The case with $N = \infty$ is a good approximation for the cases with L such that $L \ge 8$. From this figure, we also see that B_o should be greater than 55 to gurantee p_{loss2} of no more than 10^{-6} at $p = 0.9$ when $L \ge 8$.

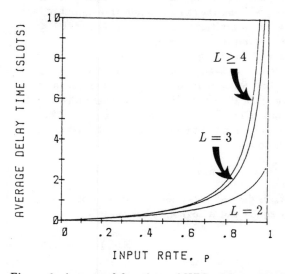

Figure 3 Average delay time of NPS with output buffers ($N = \infty$, $B_o = \infty$)

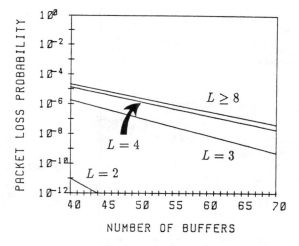

Figure 4 Packet loss probability at output buffers ($N = \infty$)

3 NPS with Input and Output Buffers

This section deals with the switch having some buffers at each input as well as at each output (see Fig.2). Section 3.1 assumes that N is infinite and both B_i and B_o are also infinite while both B_i and B_o are assumed to be finite to investigate the effect of buffer size on performance in Section 3.2.

3.1 Maximum Throughput Analysis in the Case with $N = \infty$

Karol et al. [5] has obtained the maximum throughput of input queueing in the case where L equals 1. In this section, we will analyze the maximum throughput of input queueing for an arbitrary number of L such that $1 \leq L \leq N$ on the assumption of $N = \infty$.

Let us pay our attention to a virtual queue consisting of packets at the head of inputs destined for the tagged output. Let R_m denote the length of the queue at the end of the mth time slot. Since the arrival process at the head of inputs is a Poisson process in the case where $N = \infty$ (refer to [5] for the proof), R_m and R_{m-1} are related according to

$$R_m = \max(0, R_{m-1} + A_m - L),$$

where A_m is defined in Section 2, and its probability a_j is given by eq.(1) and its pgf $A(z)$ becomes

$$A(z) \triangleq \sum_{i=0}^{\infty} a_i z^i = e^{-p(1-z)}.$$

Defining the probability and its pgf:

$$r_i \triangleq Pr[R = i]$$
$$R(z) \triangleq \sum_{i=0}^{\infty} r_i z^i,$$

$R(z)$ becomes

$$
\begin{aligned}
R(z) &= \sum_{\{k,j \mid 0 \leq k+j < L, k \geq 0, j \geq 0\}} r_k a_j z^0 \\
&+ \sum_{\{k,j \mid k+j \geq L, k \geq 0, j \geq 0\}} r_k a_j z^{k+j-L} \\
&= \sum_{k=0}^{L-1} r_k \sum_{j=0}^{L-k-1} a_j (1 - z^{k+j-L}) + z^{-L} R(z) A(z).
\end{aligned}
$$

Hence, we have

$$R(z) = \frac{\sum_{k=0}^{L-1} r_k \sum_{j=0}^{L-k-1} a_j (z^{k+j} - z^L)}{A(z) - z^L}. \quad (11)$$

Using the theory of an L-server queueing system with constant service time, i.e., an M/D/L system, we can analyze the behavior of input queue [7]. The average length of the virtual queue is given as follows (refer to [7] for the derivation).

$$\overline{R} = \dot{R}(1) = \frac{p^2 - L(L-1)}{2(L-p)} + \sum_{i=1}^{L-1} \frac{1}{1 - z_i}, \quad (12)$$

where z_i refers to a root of the equation $A(z) - z^L = 0$.

Suppose such a saturation phenomenon that packets are always waiting at every input buffer. In this case, $\overline{R}$ satisfies the following equation besides eq.(12) (see [5])

$$\overline{R} = 1 - p. \quad (13)$$

Using eqs.(12) and (13), we can obtain the maximum throughput of the nonblocking packet switch with input and output buffers when $N = \infty$. The maximum throughput is given for several values of L in Table 1. Two cases with $L = 1$ and ∞ were already analyzed in [5]. $L = \infty$ offers the optimum performance. It is noted that $L = 2$ can not achieve a throughput of 0.9 when $N = \infty$. In other words, L should be greater than or equal to 3 in order to achieve a throughput of 0.9 with packet loss probability of no more than some value, say 10^{-6}.

Table 1
Maximum throughput of NPS
with input and output buffers ($N = \infty$)

L	Maximum Throughput
1	0.5858
2	0.8845
3	0.9755
4	0.9956
5	0.9993
6	0.9999
∞	1.0000

3.2 Simulation Results for Packet Loss Probability

In Section 3.1, B_i and B_o are assumed to be infinite. This section examines how many buffers are required to gurantee some requirement regarding packet loss probability through simulations. The simulation results are based on statistics of about six million packets when N is fixed at 32. Thus, we can not discuss packet loss probability of less than about 1.7×10^{-7}.

The simulation results show that about 20 buffers at each input and about 27 buffers at each output are needed to achieve packet loss probability of no more than 10^{-6} at $p = 0.8$. The probability does not seem to become less than 10^{-6} at $p = 0.9$ even with a larger number of B_i. This behavior agrees with the saturation analysis that the maximum throughput is 0.8845 when $L = 2$ and $N = \infty$. In addition, it is shown from the simulation that about 8 input buffers and 47 output buffers are needed to meet the same requirement at $p = 0.9$ when $L = 3$, and about 4 input buffers and 49 output buffers when $L = 4$.

Figure 5 gives the packcet loss probability in the case where $L = 4$. A solid line in this figure represents probability of packet loss in the case where $B_i = 0$, which is derived from eq.(3). It follows from eq.(3) that packet loss probability of 10^{-6} is achieved only in a short range of

$p < 0.10653$ when B_i is zero and $L = 4$. Furthermore, the equation also shows that L should be at least 8 to gurantee the same requirement at $p = 0.9$ if no buffers are provided at each input. We see from this fact and Fig.5 that buffering packets at inputs results in drastic improvement in packet loss probability.

The delay performance is depicted in Fig.6. The solid line represents the average transmission delay in an M/D/1 queueing system which is equivalent to the case with $N = L = \infty$. The average transmission delay of the case with $L = 2$ is different from others while the others are very similiar to that of the M/D/1 system.

4 Concluding Remarks

This paper has examined the effect of speedup on packet loss probability and delay performance in two types of non-blocking packet switch. One has buffers only at outputs. The other has buffers at inputs as well as at outputs. In the former, the case with $L = \infty$ has been shown to give a good approximation for the case with $L \geq 8$ when $N = \infty$ in terms of the average transmission delay and packet loss probability at output buffers. In the latter, it has become clear that buffering packets at inputs results in a significant improvement in packet loss probability. Our saturation analysis has shown that rather lower switching speed, $L = 3$, offers a maximum throughput of 0.9755 for an arbitrarily large value of N on the assumption of an infinite number of buffers. Furthermore, we can see from the simulation results that the maximum throughput could be achieved with a relatively small number of buffers at each input; for example, when $L = 3$, 8 buffers at each input and 47 buffers at each output approximately gurantee the packet loss probability of no more than 10^{-6} at $p = 0.9$. We are now investigating the analysis of packet loss probability in NPS with a finite number of input and output buffers.

References

[1] P. J. Burke, "Delays in Single–Server Queues with Batch Input," *Oper. Res.*, vol.23, pp.830–833, July–Aug. 1975.

[2] M. G. Hluchyj and M. J. Karol, "Queueing in Space–Division Packet Switching," *Proc. IEEE INFO-COM'88*, pp.4A.3.1–4A.3.10, Mar. 1988.

[3] A. Huang and S. Knauer, "Starlite: A Wideband Digital Switch," *Proc. GLOBECOM'84*, pp.5.3.1-5.3.5, Nov. 1984.

[4] J. Y. Hui and E. Arthurs, "A Broadband Packet Switch for Integrated Transport," *IEEE J. Select. Areas Commun.*, vol.SAC–5, pp.1264–1273, Oct. 1987.

[5] M. J. Karol, M. G. Hluchyj, and S. P. Morgan, "Input versus Output Queueing on a Space–Division Packet Switch," *IEEE Trans. Commun.*, vol.COM–35, pp.1347–1356, Dec. 1987.

[6] L. Kleinrock, *Queueing Systems, Vol. 1: Theory*, John Wiley and Sons, New York, 1975.

[7] J. Riordan, *Stochastic Service Systems*, John Wiley and Sons, 1962.

[8] Y.-S. Yeh, and M.G. Hluchyj, and A.S. Acampora, "The Knockout Switch:A Simple, Modular Architecture for High–Performance Packet Switching," *IEEE J. Select. Areas Commun.*, vol.SAC–5, pp.1274–1283, Oct. 1987.

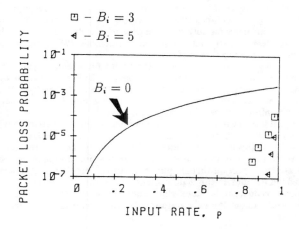

Figure 5 Packet loss probability at input buffers ($N = 32$, $L = 4$)

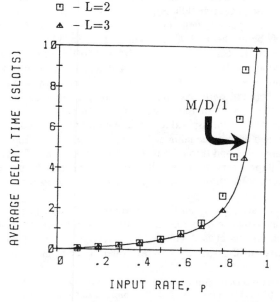

Figure 6 Average delay time of NPS with input and output buffers ($N = 32$, $B_i = 10$, $B_o = \infty$)

Throughput Analysis, Optimal Buffer Allocation, and Traffic Imbalance Study of a Generic Nonblocking Packet Switch

Jeane S.-C. Chen, *Member, IEEE*, and Thomas E. Stern, *Fellow, IEEE*

Abstract—This paper provides a general model to study the performance of a family of space-domain packet switches, implementing both input and output queueing and varying degrees of speedup. Based on this model, the impact of the speedup factor on the switch performance is analyzed. In particular, the maximum switch throughput and the average system delay for any given degree of speedup are obtained. The results demonstrate that the switch can achieve 99% throughput with a modest speedup factor of 4.

Packet blocking probability for systems with finite buffers can also be derived from this model and the impact of buffer allocation on blocking probability is investigated. Given a fixed buffer budget, this analysis obtains an optimal placement of buffers among input and output ports to minimize the blocking probability.

The model is also extended to cover a nonhomogeneous system where traffic intensity at each input varies and destination distribution is not uniform. Using this model, the impact of traffic imbalance on the maximum switch throughput is studied. It is seen that input imbalance has a more adverse effect on throughput than output imbalance. It is also observed that, under the same input imbalance conditions, the throughput reduction is more significant for switches with higher speedup while, under the same output imbalance conditions, this trend is reversed.

I. Introduction

IN this paper, we provide a unified model to study the family of space-domain fast packet switches. A space-domain non-blocking fast packet switch is generically modeled as having both input and output queueing and a speedup factor of C, $1 \leq C \leq N$; N being the size of the switch. This speedup factor of C enables up to C packets to be delivered to the same output port simultaneously. These delivered packets are queued in the output buffers to wait for the service of the output links, while those packets not delivered are queued in the input buffers to wait for the service of the switch. The extreme situations of $C = 1$ and $C = N$ correspond to the conventionally termed input queueing and output queueing switches, respectively. Using this model, the delay and throughput measures are obtained for various speedup factors. A key result of this paper is that the switch can achieve 99% throughput with a modest speedup factor of 4.

The analysis focuses on the performance of large switches, and the analytical model represents the limiting case $N \rightarrow \infty$. A simulation model is developed to investigate the impact of switch size on the validity of the model. It is found that for $C \geq 2$ and loadings below saturation, the model predicts average system delay within 95% confidence intervals of the simulation results for $N \sim 20$. This is also true for $C = 1$ at low load situations. However, for $C = 1$ and loading close to saturation, a larger system is needed ($N \sim 100$) for the analysis to fall within 95% confidence intervals of simulation results. The approximation error for $N = 20$ in this case is still less than 10%.

When multiple packets contend for the same output, only up to C packets can be simultaneously served. The order in which these contending packets receive service represents the arbitration policy of the switch. Various arbitration policies are studied to investigate their impacts on switch performance. It is found that these arbitration policies do not affect the throughput results since maximum throughput is obtained based on the first moment of the service time. However, different arbitration policies do result in different switch delay distributions.

Packet blocking probability for systems with finite buffers can also be derived from this model and the impact of buffer allocation on blocking probability is investigated. Given a fixed buffer budget, this analysis obtains an optimal placement of buffers among input and output ports to minimize the blocking probability.

The model is also extended to cover a nonhomogeneous system where traffic intensity at each input varies and destination distribution is not uniform. Using this model, the impact of traffic imbalance on the maximum switch throughput is studied. It is seen that input imbalance has a more adverse effect on throughput than output imbalance. It is also observed that, under the same input imbalance conditions, the throughput reduction is more significant for switches with higher speedup while, under the same output imbalance conditions, this trend is reversed.

The organization of this paper is as follows. In Section II, we describe the system assumptions and the modeling approach. In Section III, we analyze the model and obtain the switch delay distributions under various arbitration policies, and the input and output delay distributions. A simulation model is presented in Section IV and various switch configurations are simulated to determine the switch size N for which the analysis model is valid. In Section V, we obtain numerical results for the maximum switch throughput as a function of the speedup factor C for both fixed size packets and packets with exponential length distribution. We also present the delay through the system experienced by a packet from the moment it arrives at an input port until it is sent to its distribution output link, and compare these values with simulation results. Packet blocking probability is derived for systems with finite buffer space, and

Manuscript received April 13, 1990; revised November 13, 1990.

J. S.-C. Chen is with the IBM T. J. Watson Research Center, Yorktown Heights, NY 10598.

T. E. Stern is with the Center for Telecommunications Research and Department of Electrical Engineering, Columbia University, New York, NY 10027.

IEEE Log Number 9042046.

Reprinted from *IEEE J. Selected Areas Commun.*, vol. 9, no. 3, pp. 439–449, April 1991.

optimal buffer allocation strategies are investigated in Section VI. In Section VII, we study systems with traffic imbalance and discuss their impact on switch throughput. The findings of this paper are summarized in Section VIII.

II. System Description and Modeling Approach

We consider an $N \times N$ packet switch, as shown in Fig. 1, where both input and output buffers are provided in the system and the switch fabric is operated with a speedup factor of C, $1 \leq C \leq N$.

The switch fabric is nonblocking, which means that contention only occurs when several packets from different input ports request a path to the same output port simultaneously. This output contention is resolved by both input and output queueing. The speedup factor of C enables up to C packets to be delivered to the same output port simultaneously and it dictates the proportion of packets to be queued in either the input or output when contention occurs. This variable speedup factor distinguishes our system from the one studied by Iliadis [10], [11], where the switch is implemented with a full speedup of N (for an $N \times N$ switch). In his system, queueing can also take place at both the inputs and outputs due to the employment of a back-pressure mechanism.

We first consider a homogeneous system where traffic intensity is the same to each input and the traffic is uniformly distributed to each output. The model is extended to include systems with unbalanced traffic in Section VII. Packet arrivals to each input are assumed to be independent and identical Poissson processes with rate λ packet/unit time, and packet has exponentially distributed service time with mean $1/\mu$ unit time. A packet arrives at an input queue Q_i^I and waits until it reaches the head-of-the-line (HOL) position to access the switch. At this point, it contends with possibly other packets at other HOL positions that have the same destination. If it wins the contention, it is transferred by the switch to its destination output port. Otherwise, it continues to occupy the HOL position until it is eventually transferred. After having been transferred to the output port, the packet is sent out on the output link if the link is idle. Otherwise, it is stored in the output buffer until the output link becomes available. The service is considered complete after the packet is sent out on the output link.

We model the system as queues in tandem. We logically divide the packets in the HOL positions into N virtual queues where virtual queue Q_j^V, $1 \leq j \leq N$, is composed of packets in the HOL positions of all input queues Q_i^I, $1 \leq i \leq N$, destined to output port j. Because the destinations are uniformly distributed, the contribution of arrivals to Q_j^V from each input has rate λ/N. The arrival process to Q_j^V is the superposition of N such processes and it tends to a Poisson process with rate λ as $N \to \infty$ [11]. Herein we limit our study to the limiting case where this Poisson assumption holds. The size N required for this limiting assumption to be valid is investigated through simulations in a later section. Each virtual queue has C servers to represent the C-fold speedup of the switch. Thus, Q_j^V, $1 \leq j \leq N$ can be realized as an M/M/C queue with arrival rate λ and mean service time $1/\mu$. Note that the virtual queue has waiting room of size N, which becomes infinite in the limiting case considered here. Fig. 2 shows a virtual queue formed by HOL packets destined to outport j.

A packet, after reaching the HOL position, stays there until it is transferred to an output port. The duration of this stay is determined by the delay it experiences in the virtual queue it

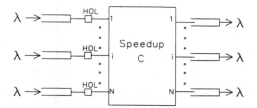

Fig. 1. System model.

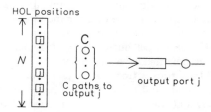

Fig. 2. Virtual queue formed by HOL packets destined to output j.

joins. If we view the HOL position as a service station and the time a packet stays at this position as its service time, we can model an input port as an M/G/1 queue with the service time distribution being the delay distribution of the M/M/C virtual queue. The input queue can then be analyzed by applying results for an M/G/1 queue.

The output process of Q_j^V feeds into output queue Q_j^O. This process is Poisson with rate λ according to Burke's theorem [4]. Service time distribution in Q_j^O is exponential with mean $1/\mu$. However, Q_j^O is not a simple M/M/1 queue. The service requirement of a packet is determined by its length and this remains fixed at any queue it visits. Thus, the interarrival time and the service requirement in Q_j^O are correlated since they correspond to the interdeparture time and the service requirement in Q_j^V, which are correlated. This is a well-recognized obstacle in modeling packet systems as queueing networks when exact analysis is required. Kleinrock studied this phenomenon extensively through simulations [13] and came to the conclusion that, in many packet networks, this dependence is diluted due to the mixing of traffic streams from different directions. He proposed applying independence assumptions to model these packet systems [15]. The dependency in our model derives from the tandem of an M/M/C queue followed by an M/M/1 queue. Boxma studied the case of a tandem of two M/M/1 queues and analytically derived the dependency coefficients [1]-[3]. We investigate the tandem of an M/M/C and an M/M/1 queue through extensive simulations. It is seen from both Boxma's analytical results and our simulation findings that the dependency between interarrival time and service time at the second queue is minimal in the range below the saturation points of the switch. (Maximum throughput for switches with varying speedup factors are obtained in Section V.) We thus adopt Kleinrock's independence assumption and model the output queues as simple M/M/1 queues.

We now have a general model which consists of queues in tandem to describe this family of packet switches. Fig. 3 shows the tandem of queues traversed by a packet originating at input port i and destined to output port j.

III. Model Analysis

In this section, we analyze the model described in the previous section. In Section III-A, we obtain the distribution for the delay of the virtual queue. This delay, denoted as W_s, cor-

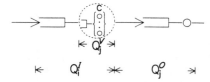

Fig. 3. Tandem queues traveled by packets from input i to output j.

responds to the contention time and switching time of a packet. In Section III-B, we obtain the delay in the input queue by solving an M/G/1 queue with service time distribution $W_s(t)$. In Section III-C, the output queueing delay is obtained from an M/M/1 model. The mean system delay, that is, the total delay experienced by a packet from the instant it arrives at an input port until it is delivered to an output link, is given in Section III-D. Throughout this paper, we use the term "queueing delay" to represent the time spent waiting on queue and the term "delay" to represent the total time spent in system. We assume infinite buffer size for both the input and output queues in this part of the analysis.

A. Delay of the Virtual Queue

Recall that a virtual queue Q^V can be modeled as an M/M/C queue with arrival rate λ. The service discipline of Q^V represents the arbitration policy of the switch in selecting among contending packets for delivery. Since Q^V is a logical queue and packets in the queue physically reside in different input ports, it might not be economical to maintain a global clock and it would not be feasible to distinguish the order of arrivals for packets in Q^V. In such a case, the servers simply choose randomly among packets in the queue for next service. This service policy corresponds to the "random order of service (ROS)" discipline. However, if a global clock is available, service policies based on the order of arrivals can be adopted. Common policies of this type are "first come first serve (FCFS)" and "last come first serve (LCFS)." Different service policies result in different delay distributions. We obtain results for all three policies.

We first consider the FCFS policy. Let $w_s(t)$ be the density function for delay, i.e., queueing time plus service time. In the M/M/C queue under FCFS service discipline, it can be shown that [7]:

$$w_s(t) = \frac{\mu e^{-\mu t}[\lambda - c\mu + \mu W_q(0)] - [1 - W_q(0)][\lambda - c\mu]\mu e^{-(c\mu - \lambda)t}}{\lambda - (c-1)\mu} \tag{3.1}$$

where

$$W_q(0) = 1 - \frac{c(\lambda/\mu)^c}{c!(c - \lambda/\mu)} p_0$$

$$= 1 - E_{2,c}(\lambda/\mu) \tag{3.2}$$

is the probability a packet spends no time waiting in the queue. $E_{2,c}(\lambda/\mu)$ is the well-known Erlang's second formula, and

$$p_0 = \left[\sum_{n=0}^{c-1} \frac{(\lambda/\mu)^n}{n!} + \frac{c(\lambda/\mu)^c}{c!(c - \lambda/\mu)}\right]^{-1} \tag{3.3}$$

is the steady-state probability of zero packets in the system.

Let $W_s(s)$ be the Laplace transform for $w_s(t)$. Then,

$$W_s(s) = \frac{1}{\lambda - (c-1)\mu} \left(\frac{\mu[\lambda - c\mu + \mu W_q(0)]}{(s - \mu)} - \frac{[1 - W_q(0)][\lambda - c\mu]\mu}{(s + c\mu - \lambda)}\right). \tag{3.4}$$

The mean delay is:

$$\overline{W}_s = \frac{\mu(\lambda/\mu)^c}{(c-1)!(c\mu - \lambda)^2} p_0 + \frac{1}{\mu} \tag{3.5}$$

the second moment of the delay is:

$$\overline{W^2_{s,\text{FCFS}}} = \frac{2(\lambda/\mu)^c}{(c-1)!(c\mu - \lambda)^2} \left(\frac{\mu}{c\mu - \lambda} + 1\right) p_0 + \frac{2}{\mu^2} \tag{3.6}$$

and the third moment is:

$$\overline{W^3_{s,\text{FCFS}}} = \frac{(\lambda/\mu)^2 \mu}{(c-1)!(c\mu - \lambda)} p_0 \left(\frac{6}{(c\mu - \lambda)^3} + \frac{6}{\mu(c\mu - \lambda)^2} + \frac{6}{\mu^2(c\mu - \lambda)}\right) + \frac{6}{\mu^3}. \tag{3.7}$$

Both the ROS and LCFS service policies yield the same mean delay as in the FCFS policy, i.e., (3.5). The second and third moments, denoted by $\overline{W^2_{s,\text{ROS}}}$, $\overline{W^3_{s,\text{ROS}}}$ and $\overline{W^2_{s,\text{LCFS}}}$, $\overline{W^3_{s,\text{LCFS}}}$ for ROS and LCFS policies, respectively, can be derived as follows. In [19], given that a packet has to be queued, moments of the conditional queueing delay for M/M/C with ROS policy are found to be:

$$m_{2,\text{ROS}} = \frac{4c\mu}{(c\mu - \lambda)^2(2c\mu - \lambda)} \tag{3.8}$$

$$m_{3,\text{ROS}} = \frac{6c\mu(4c\mu + 2\lambda)}{(c\mu - \lambda)^3(2c\mu - \lambda)^2} \tag{3.9}$$

and for LCFS policy:

$$m_{2,\text{LCFS}} = \frac{2c\mu}{(c\mu - \lambda)^3} \tag{3.10}$$

$$m_{3,\text{LCFS}} = \frac{6c\mu(c\mu + \lambda)}{(c\mu - \lambda)^5}. \tag{3.11}$$

The ith moment of the queueing delay, denoted by $\overline{W^i_q}$, is related to that of the conditional queueing delay by:

$$\overline{W^i_q} = m_i E_{2,c}(\lambda/\mu) \tag{3.12}$$

where $E_{2,c}(\lambda/\mu)$ is given in (3.2). Finally, the moments of the delay (queueing time plus service time) of the virtual queue are obtained as:

$$\overline{W^2_s} = \overline{W^2_q} + \frac{2}{\mu^2} + \frac{2}{\mu}\overline{W_q} \tag{3.13}$$

and

$$\overline{W^3_q} = \overline{W^3_q} + \frac{3}{\mu}\overline{W^2_q} + \frac{6}{\mu^2}\overline{W_q} + \frac{6}{\mu^3}. \tag{3.14}$$

Substituting (3.8), (3.9), (3.10), and (3.11) first into (3.12), and then into (3.13) and (3.14) yield the following:

$$\overline{W^2_{s,\text{ROS}}} = \frac{2(\lambda/\mu)^c}{(c-1)!(c\mu-2)^2} p_0 \left(\frac{2c\mu^2}{(c\mu-\lambda)(2c\mu-\lambda)} + 1 \right)$$
$$+ \frac{2}{\mu^2} \tag{3.15}$$

$$\overline{W^3_{s,\text{ROS}}} = \frac{(\lambda/\mu)^c \mu}{(c-1)!(c\mu-\lambda)} p_0$$
$$\times \left(\frac{6c\mu(4c\mu+2\lambda)}{(c\mu-\lambda)^3(2c\mu-\lambda)^2} \right.$$
$$\left. + \frac{12c\mu}{\mu(c\mu-\lambda)^2(2c\mu-\lambda)} + \frac{6}{\mu^2(c\mu-\lambda)} \right) + \frac{6}{\mu^3} \tag{3.16}$$

$$\overline{W^2_{s,\text{LCFS}}} = \frac{2(\lambda/\mu)^c}{(c-1)!(c\mu-\lambda)^2} p_0 \left(\frac{c\mu^2}{(c\mu-\lambda)^2} + 1 \right) + \frac{2}{\mu^2} \tag{3.17}$$

$$\overline{W^3_{s,\text{LCFS}}} = \frac{(\lambda/\mu)^c \mu}{(c-1)!(c\mu-\lambda)} p_0 \left(\frac{6c\mu(c\mu+\lambda)}{(c\mu-\lambda)^5} \right.$$
$$\left. + \frac{6c\mu}{\mu(c\mu-\lambda)^3} + \frac{6}{\mu^2(c\mu-\lambda)} \right) + \frac{6}{\mu^3}. \tag{3.18}$$

Higher moments can also be derived, and an approximate distribution can be obtained by moment matching. These are not included since, for our purposes, only the first three moments are needed.

B. Input Queue Analysis

The input queue is modeled as an M/G/1 queueing system with arrival rate λ and service time distribution $W_s(t)$ as given in Section III-A. Let $W_i(s)$ denote the Laplace transform of the time spent in the input queueing system. The Pollaczek–Khintchine transform equation gives [14]:

$$W_i(s) = W_s(s) \frac{s(1-\lambda\overline{W}_s)}{s - \lambda + \lambda W_s(s)} \tag{3.19}$$

and the mean input delay $\overline{W}_i$ is:

$$\overline{W}_i = \overline{W}_s + \frac{\lambda\overline{W^2_s}}{2(1-\lambda\overline{W}_s)}. \tag{3.20}$$

C. Output Queue Analysis

The output queue is modeled as an M/M/1 queue with arrival rate λ and mean service time $1/\mu$. Let W_o denote the queueing time at the output queue and $W_o(s)$ be its Laplace transform, then:

$$W_o(s) = 1 - \rho + \frac{\lambda(1-\rho)}{(\mu(1-\rho)+s)} \tag{3.21}$$

where $\rho = \lambda/\mu$ is the utilization, and the mean output queueing delay $\overline{W}_o$ is:

$$\overline{W}_o = \frac{\rho/\mu}{1-\rho}. \tag{3.22}$$

D. System Delay

Let $\overline{D}$ denote the mean system delay experienced by a packet from the time it arrives at an input port until it is delivered to an output link. We have

$$\overline{D} = \overline{W}_i + \overline{W}_o \tag{3.23}$$

with $\overline{W}_i$ and $\overline{W}_o$ given in (3.20) and (3.22), respectively. As can be seen from (3.20), $\overline{W}_i$ depends on the moments of W_s which are determined by the arbitration policy of the switch in selecting contending HOL packets for delivery to their desired destinations, that is, the service policy of the virtual queue. Three service policies, FCFS, ROS, and LCFS, are investigated. $\overline{W}_s$ is the same for all these service policies and is given in (3.5) while $\overline{W^2_{s,\text{FCFS}}}$, $\overline{W^2_{s,\text{ROS}}}$, and $\overline{W^2_{s,\text{LCFS}}}$ are given in (3.6), (3.15), and (3.17), respectively.

IV. SIMULATION MODEL

In order to validate the analysis, a simulation model for this system is constructed as seen in Fig. 4. This model consists of N input queues, N virtual queues, and N output queues. The HOL position for each input port is represented by a token. A packet seizes this token when it reaches the HOL position and does not release it until delivered to an output queue. The dotted line in Fig. 4 indicates the path of a packet from input 1 to output N. The shaded areas in the diagram are queues visited by this packet. When the packet reaches the head of Q'_1, it seizes the HOL token H_1 and contends with other HOL packets destined to output N for the C available paths to Q^O_N. This contention takes place in the virtual queue Q^V_N where various arbitration policies can be implemented. The packet releases H_1 after it receives service from Q^V_N and, at this time, it joins the output queue Q^O_N. The holding time of token H_1 corresponds to W_s in the analysis model. Note that, in this simulation model, we do not impose any independence assumption. The dependencies between interarrival time and service time in Q^V and Q^O are preserved in the simulation and, therefore, the validity of the independence assumption in the analysis can be verified through simulation results.

To check the infinite size approximation in the analytical model, we simulate switches of various configurations under a FCFS arbitration policy. For each simulation run, we obtain mean system delay within 95% confidence interval and compare it with the value predicted by the analytical model. Figs. 5 and 6 plot the mean system delay as a function of switch size for switches with speedup factor $C = 1$ and arrival rates of 0.2 and 0.4, respectively. The range of system delay shown in the graphs is between 50 and 110% of the analysis value. As will be demonstrated in a later section, a load of 0.4 is already very close to the maximum allowable throughput for switches with speedup of 1. At this load, the analysis only falls within the 95% confidence interval of the simulations at $N = 100$ but the error is already less than 10% at $N = 15$. At a load of 0.2 with $C = 1$, the analysis is only 10% from the 95% cofidence interval at $N = 3$, and at $N = 20$ it is within the interval. We also investigate systems with $C = 2, 3, 4$ at loadings that are close to their maximum allowable throughputs. For $C = 2$ and $\lambda = 0.6$, the analysis is only 10% from the 95% confidence interval at $N = 4$, and at $N = 30$ it is within the interval. The analysis is within 5% from the 95% confidence interval for all N and within the interval at $N = 10$ for $C = 3$ and $\lambda = 0.7$ while, for $C = 4$ and $\lambda = 0.7$, the analysis is within 2% from the 95% confidence interval for all N, and at $N = 4$ it is within

117

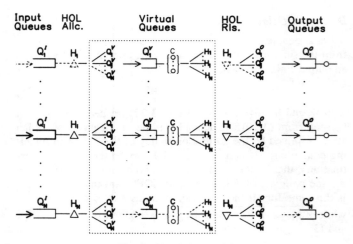

Fig. 4. Simulation model.

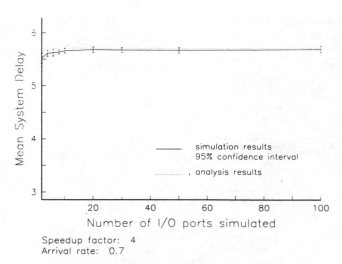

Speedup factor: 4
Arrival rate: 0.7

Fig. 7. Model accuracy w.r.t. switch size. (Case 3).

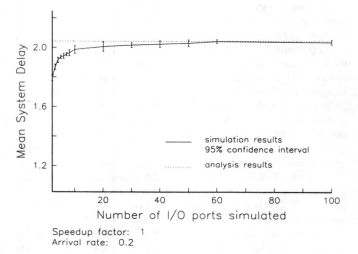

Speedup factor: 1
Arrival rate: 0.2

Fig. 5. Model accuracy w.r.t. switch size. (Case 1).

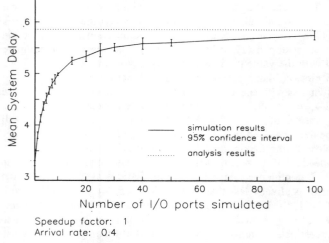

Speedup factor: 1
Arrival rate: 0.4

Fig. 6. Model accuracy w.r.t. switch size. (Case 2).

the interval. We plot only the result for the case of $C = 4$ in Fig. 7.

The discrepancy between simulation and analysis for each system appears to stem from the lack of validity of the Poisson assumption. With $C = 1$ and high loading, packets tend to stay in the HOL positions longer before they can be delivered to the output ports. This HOL blocking perturbs the statistics of the arrivals into virtual queues and results in these arrival processes deviating from Poisson. At low loads, the contention is less and the HOL blocking is significantly reduced, so that the arrival processes to the virtual queues can be well characterized as Poisson. The HOL blocking is also reduced by increasing the speedup factor C because packets can be quickly transferred to the output queues and resolve their contentions there. Hence, the Poisson assumption is also valid for these systems. For large systems, the analysis is valid because aggregation of a large number of processes tends to be Poisson process.[1]

V. NUMERICAL RESULTS

In this section, we present some numerical results obtained from the model. In particular, we present the maximum switch throughput for packets with both exponential and fixed length as functions of the speedup factor. We also obtain the mean system delay under various arbitration policies as functions of the speedup factor and compare them with simulation results.

A. Maximum Switch Throughput

The maximum switch throughput, denoted as λ_{max}, is the minimum of the maximum input throughput λ_{max}^I and the maximum output throughput λ_{max}^O. Since each output queue is served by a dedicated output link with unit service rate, λ_{max}^O is determined from:

$$\rho_{max}^O = \lambda_{max}^O / \mu = 1. \qquad (5.1)$$

Similarly, λ_{max}^I can be obtained from:

$$\rho_{max}^I = \lambda_{max}^I \overline{W}_s = 1 \qquad (5.2)$$

where $\overline{W}_s$ is given in (3.5) and represents the mean packet service time in an input queue. Since $\overline{W}_s \geq 1/\mu$, as can be seen

[1]A theorem, due to C. Palm and A. Y. Khintchine, states that the superposition of a large number of independent equilibrium renewal processes, each with small intensity, is asymptotically a Poisson process [8]. Our statement relates loosely to this theorem without rigorously examining the characteristics of each individual process.

from (3.5), it follows that

$$\lambda_{max} = \min (\lambda_{max}^I, \lambda_{max}^O)$$

$$= \lambda_{max}^I. \qquad (5.3)$$

Letting $\mu = 1$, the maximum throughput as a function of C is shown in Fig. 8.

The maximum throughput of 0.5 for $C = 1$ agrees with the known result for a crossbar switch without speedup [6]. It is also seen that, as expected, $\lambda_{max} \to 1$ as $C \to \infty$. In fact, λ_{max} grows quite rapidly as C increases, and a maximum throughput of more than 99% can be achieved with a modest speedup factor of 4.

Throughput results for packets with fixed size have been obtained by Oie *et al.* [17]. We include the analysis and results of this case for comparison.

With fixed size packets, several parts of the model described in Section II have to be modified. First, instead of an M/M/C queue, the virtual queue is modeled with an M/D/C queue. Second, the arrival process to an output port is no longer Poisson. For the purpose of obtaining switch throughput, we only have to account for the first modification. Let packet size be 1, the mean queueing time $\overline{W_{q,D}}$ for an M/D/C queue with arrival rate λ is found to be [19]:

$$\overline{W_{q,D}} = \frac{1}{\lambda} \left[\sum_{j=1}^{C-1} \frac{1}{1-Z_j} + \frac{\lambda^2 - C^2 + C}{2(C-\lambda)} \right] \qquad (5.4)$$

where the Z_j's are roots of the function:

$$e^{-\lambda(1-Z)} - Z^C = 0. \qquad (5.5)$$

Equation (5.5) is the denominator of the moment generating function for the number of packets in the system. Interested readers are referred to Riordan [19] for the derivation. It can be proved by applying Rouche's theorem that there exist C roots of (5.5), $Z_o = 1$ being the obvious one. The remaining $C - 1$ roots required in (5.4) can be obtained by numerical methods.

The mean delay $\overline{W_{s,D}}$ is simply:

$$\overline{W_{s,D}} = \overline{W_{q,D}} + 1. \qquad (5.6)$$

Again applying the relationship $\lambda_{max} \overline{W_{s,D}} = 1$, the maximum switch throughput for fixed size packets is shown as a function of C in Fig. 9. Again the results for the extreme cases, i.e., $\lambda_{max} = 0.58579$ for $C = 1$ and $\lambda_{max} \to 1$ for $C \to \infty$, agree with known results [9], [12] for pure input queueing and pure output queueing, respectively. We also see that the switch can achieve better throughput with fixed size packets, as is to be expected.

B. Delays in the System

We first examine the delay in virtual queue, W_s, which represents the delay experienced by a packet from the time it first arrives at the HOL of an input queue until it is successfully delivered to an output port. The mean delay in the virtual queue $\overline{W_s}$ is the same for all three arbitration policies considered and a closed-form expression for this measure is given in (3.5). The second moment of the delay in virtual queue, $\overline{W_s^2}$, varies with arbitration policies. Expressions for $\overline{W_s^2}$ for FCFS, ROS, and LCFS policies are given in (3.6), (3.15), and (3.17), respectively. As can be seen from (3.20), a higher second moment of the service time $\overline{W_s^2}$ yields a higher input delay $\overline{W_i}$, thus higher $\overline{D}$. Fig. 10 plots $\overline{D}$ with $C = 1$ as a function of λ for the three policies studied, and Fig. 11 plots $\overline{D}$ with $C = 1, 2, 3, 4$, and

C	Maximum Throughput
1	0.5
2	0.82843
3	0.96114
4	0.99341
5	0.99905
6	0.99988
7	0.99999
8	≈ 1

Fig. 8. Maximum switch throughput for Poisson arrival, exponential packet length ($N = \infty$).

C	Maximum Throughput
1	0.58579
2	0.88454
3	0.9755
4	0.99559
5	0.99931
6	0.99991
7	0.99999
8	≈ 1

Fig. 9. Maximum switch throughput for Poisson arrival, fixed packet length ($N = \infty$).

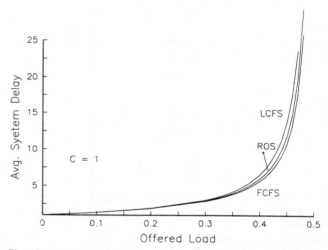

Fig. 10. Average system delay yielded by different arbitration policies.

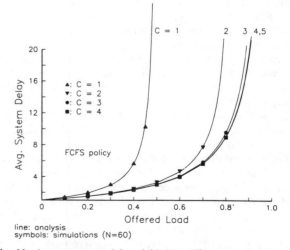

line: analysis
symbols: simulations (N=60)

Fig. 11. Average system delay yielded by different speedup factors.

5 for the FCFS policy. The symbols in the graph indicate results obtained from simulation. We see that the delay performance is not improved by increasing C beyond 4.

VI. BUFFER ALLOCATION

There is only finite buffer space available in a switching system. Packets arriving at a full buffer are blocked and discarded from the system. In this section, we derive the packet blocking probability for systems with input buffer size K_i and output buffer size K_o. We also investigate the impact of buffer allocation on packet blocking probability and obtain optimal placements of buffers among input and output ports to minimize the blocking probability for a given fixed buffer budget.

A. Input Buffer

Input buffers are modeled as M/G/1 queues. For an input queue with infinite buffer, the z-transform for the queue length, denoted by $N_i(z)$, is given by the Pollaczek–Khintchine transform equation [14] as:

$$N_i(z) = W_s(\lambda - \lambda z) \frac{(1 - \lambda \overline{W}_s)(1 - z)}{W_s(\lambda - \lambda z) - z} \qquad (6.1)$$

where W_s is the service time distribution given in Section III. The mean input queue length $\overline{N}_i$ is:

$$\overline{N}_i = \lambda \overline{W}_i \qquad (6.2)$$

with $\overline{W}_i$ given in (3.20). The variance of input queue length, $\sigma^2_{N_i}$, follows from p. 240 of [14] as:

$$\sigma^2_{N_i} = \lambda^2 \sigma^2_{W_i} + \lambda \overline{W}_i \qquad (6.3)$$

with

$$\sigma^2_{W_i} = \frac{\lambda^2 (\overline{W_s^2})^2 + 2\lambda \overline{W}_s \overline{W_s^2} - 2\lambda^2 (\overline{W}_s)^2 \overline{W_s^2}}{2} (1 - \lambda \overline{W}_s)^2$$
$$+ \overline{W_s^2} + \frac{\lambda \overline{W_s^3}}{3(1 - \lambda \overline{W}_s)} - (\overline{W}_i)^2 \qquad (6.4)$$

being the variance of delay in the input queue. $\overline{W}_s$, W_s^2, and W_s^3 are the first three moments of the service time distribution W_s given in Section III.

Mean and variance for the input queue length are plotted in Fig. 12 and Fig. 13 for $C = 1$, 2, 3 and 4. Simulation results for switches with $N = 60$ are also given. Analysis and simulation are seen to be in good agreement. The queue length distribution, Prob $(N_i = k)$, can theoretically be obtained by inverting (6.1) and, with this distribution, the blocking probability for input queues with finite buffer K_i can be computed [5], [7]. In practice, however, this inversion is difficult to obtain.

Instead, we approximate the input blocking probability $P_{B,i}$ with the tail probability of N_i, i.e., Prob $(N_i > K_i)$, which can be computed from (6.1) using an existing algorithm [18]. Recall that the service time distribution W_s is derived based on the amount of traffic contending for the switch. W_s is changed if some packets are blocked, resulting a reduction of the contending traffic. We thus compute $P_{B,i}$ iteratively as follows. Based on arrival rate λ and no input blocking, we first derive $W_s^{(0)}$ and then compute $P_{B,i}^{(0)}$. A new service time, $W_s^{(1)}$, can then be obtained based on arrival rate $\lambda(1 - P_{B,i}^{(0)})$ and yields a new blocking probability $P_{B,i}^{(1)}$. We iterate this procedure until $P_{B,i}^{(i)}$ converges and use this quantity as an estimate for the input blocking probability $P_{B,i}$. In most cases, less than five iterations are required for the calculation to converge.

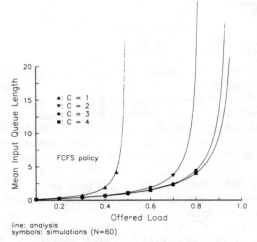

Fig. 12. Mean input queue length.

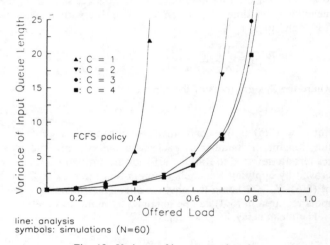

Fig. 13. Variance of input queue length.

B. Output Buffer

Output buffers are described by M/M/1 queues. With infinite buffer, the output queue length, denoted by N_o, has mean and variance given as:

$$\overline{N}_o = \frac{\lambda/\mu}{1 - \lambda/\mu} \qquad (6.5)$$

and

$$\sigma^2_{N_o} = \frac{\lambda/\mu}{(1 - \lambda/\mu)^2}. \qquad (6.6)$$

Notice that the expressions given in the above two equations are independent of C, the speedup factor. $\overline{N}_o$ and $\sigma^2_{N_o}$ are compared with simulation results for $C = 1$, 2, 3, and 4 and plotted in Figs. 14 and 15. It is found that the comparisons are in good agreements for $C = 2$, 3, and 4 while, for $C = 1$, the difference is larger. This is probably attributable to the tandem queue effect discussed in Section II. $P_{B,o}$, the output blocking probability, can be estimated using an M/M/1/K model. With blocking in the input, the arrival rate to an output queue is $\lambda(1 - P_{B,i})$.

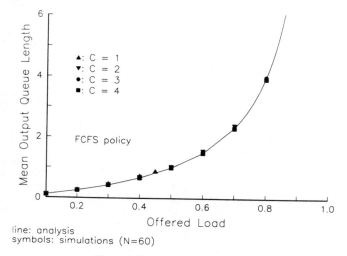

line: analysis
symbols: simulations (N=60)

Fig. 14. Mean output queue length.

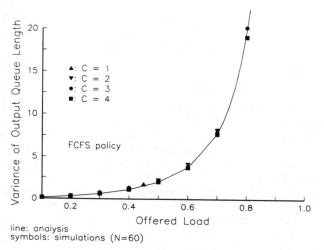

line: analysis
symbols: simulations (N=60)

Fig. 15. Variance of output queue length.

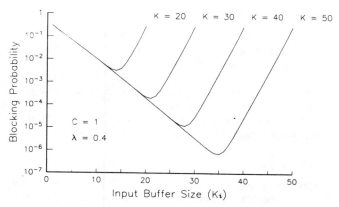

Fig. 16. Blocking probability. (Case 1).

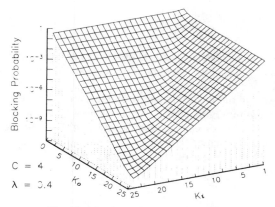

Fig. 17. Blocking probability. (Case 2).

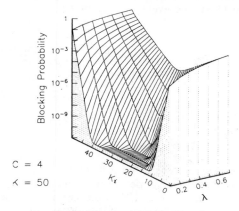

Fig. 18. Blocking probability. (Case 3).

$P_{B,o}$ for an output buffer with size K_o is then:

$$P_{B,o} = \frac{(1 - \rho_o)\rho_o^{K_o}}{1 - \rho_o^{K_o+1}} \qquad (6.7)$$

with

$$\rho_o = \frac{\lambda(1 - P_{B,i})}{\mu}. \qquad (6.8)$$

C. Blocking Probability

Let P_B denote the probability that a packet is blocked somewhere in the system. It is:

$$P_B = 1 - (1 - B_{B,i})(1 - P_{B,o}). \qquad (6.9)$$

P_B is a function of both K_i and K_o. For a fixed total buffer size $K = K_i + K_o$, P_B varies with the division of K_i and K_o and there exists an optimal allocation strategy. This is demonstrated in Fig. 16 where P_B is plotted as a function of K_i for $C = 1$, $\lambda = 0.4$ with $K = 20, 30, 40$, and 50, and $K_o = K - K_i$ for each point in the graph. At $C = 1$, most queueing is done in the inputs. Output buffers are also needed because packets are of variable length and arrival is asynchronous (output queueing is not needed if the switch has synchronous slotted operation with fixed size packets). Thus, at $C = 1$, an optimal strategy tends toward allocating more buffers in the input. As C is increased, more packets are queued in the outputs and the optimal strategies shift accordingly. Fig. 17 shows P_B versus (K_i, K_o) for $C = 4$, $\lambda = 0.4$. The optimal points occur roughly when $K_i = K_o$. Fig. 18 plots P_B versus (λ, K_i) with $C = 4$, $K = 50$, and $K_o = K - K_i$. We see that $K = 50$ is an overdesign at low loads. For example, at $\lambda = 0.1$, $(K_i, K_o) = (10, 10)$ yields about the same P_B as do $(K_i, K_o) = (10, 40)$ and $(K_i, K_o) = (40, 10)$.

VII. TRAFFIC IMBALANCE

The throughput results given in Section V are obtained based on balanced traffic, i.e., arrival intensity is the same to each input, and packet destination to each output has uniform distribution. The switch throughput is adversely affected if traffic is not balanced. Li *et al.* [16] studied the effect of traffic imbalance for an input queueing packet switch. In this section, we investigate the impact of traffic imbalance on the family of space-domain packet switches studied in this paper.

A. General Imbalance

Consider a switch with general traffic condition, as shown in Fig. 19, where arrival rate, i.e., λ_i^I for input port i and probability of choosing output j as destination for each packet is P_j, $\Sigma_{j=1}^N P_j = 1$. Packet arrivals at each input port are still independent Poisson processes. Assume

$$\sum_{i=1}^N \lambda_i^I = N\lambda. \tag{7.1}$$

Then $\lambda_i^I = \lambda$, $\forall i$ corresponds to the case of balanced input traffic while $P_i = 1/N$, $\forall j$, corresponds to the case of balanced output traffic.

Based on the modeling approach presented in Section II, the maximum throughput for a switch fabric with traffic pattern as shown in Fig. 19 can be obtained by first deriving the corresponding W_s then applying (5.2).

Let λ_j^O be the arrival rate at output j, it is:

$$\lambda_j^O = P_j \sum_{i=1}^N \lambda_j^I. \tag{7.2}$$

λ_j^O is also the rate of arrival to Q_j^V, the virtual queue depicting the contention for delivery to output j. Replacing λ with λ_j^O in (3.3) and (3.5), the average delay in Q_j^V, denoted by $\overline{W}_{s_j}$, is given as:

$$\overline{W}_{s_j} = \frac{\mu(\lambda_j^O/\mu)^c}{(c-1)!(c\mu - \lambda_j^O)^2} p_{0_j} + \frac{1}{\mu} \tag{7.3}$$

with

$$p_{0_j} = \left[\sum_{n=0}^{c-1} \frac{(\lambda_j^O/\mu)^n}{n!} + \frac{c(\lambda_j^O/\mu)^c}{c!(c - \lambda_j^O/\mu)} \right]^{-1}. \tag{7.4}$$

Since a packet chooses to go to output j, i.e., join Q_j^V, with probability P_j, the average delay experienced by any packet from the instant it arrives at a HOL position until it is delivered to an output queue, denoted by $\overline{W}_s$, is the weighted sum of the mean delay in Q_j^V, $1 \le j \le N$. Thus,

$$\overline{W}_s = \sum_{j=1}^N P_j \overline{W}_{s_j}. \tag{7.5}$$

This is illustrated in Fig. 20 with $\overline{W}_s$ representing the average duration a packet stays inside the dashed circle. Maximum switch fabric throughput is obtained by solving $\lambda_{\max}\overline{W}_s = 1$. Note that here we are only interested in the throughput supported by the switch fabric. The switch system throughput can be lower, as it can be also limited by the output link throughput. Equation (7.5) is for exponential packets and results for fixed-size packets can be similarly obtained by applying the above arguments on the analysis in Section V.

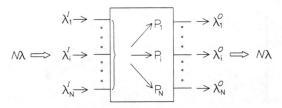

Fig. 19. General imbalance pattern.

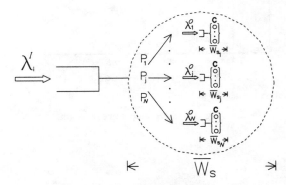

Fig. 20. Service time under traffic imbalance.

B. Bi-Group Imbalance

While the results presented in the above section offer a complete coverage for any imbalance pattern, it is difficult to provide a systematic study of the system since there are too many parameters. We present numerical examples for systems with bi-group traffic imbalance as illustrated in Fig. 21. Based on the traffic patterns, input and output ports are each divided into two groups. N_1^I ports, each with arrival rate λ_1^I, belong to input group 1 while N_2^I ports, each with rate λ_2^I, belong to input group 2. Similarly, there are N_1^o ports in output group 1 and N_2^o ports in output group 2 and ports in each group are chosen as destination with probabilities P_1 and P_2, respectively.

Let R^I and R^O denote the input and output traffic intensity ratios defined as:

$$R^I = \frac{\lambda_1^I}{\lambda_2^I} \tag{7.6}$$

and

$$R^o = \frac{P_1}{P_2}. \tag{7.7}$$

Also, let

$$S_1^I = \frac{N_1^I}{N} \tag{7.8}$$

be the normalized size of input group 1 and

$$S_1^O = \frac{N_1^O}{N} \tag{7.9}$$

be the normalized size of output group 1. These four parameters, R^I, R^O, S_1^I, S_1^O, completely characterize the imbalance pattern of the system. $R^I = 1$ or $S_1^I = 1$ represents systems with balanced input traffic while $R^O = 1$ or $S_1^O = 1$ represents systems with balanced output traffic. Given these imbalance parameters, the maximum switch throughput for the corresponding system can be obtained following the procedures in Section VII-A.

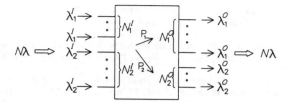

Fig. 21. Bigroup imbalance.

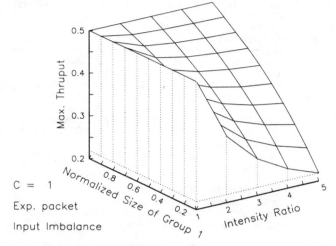

Fig. 22. Maximum throughput with input imbalance, $C = 1$.

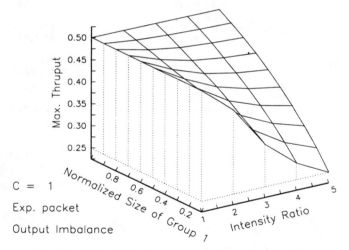

Fig. 23. Maximum throughput with output imbalance, $C = 1$.

For the purpose of illustrating the effects of traffic imbalance, a few graphs were generated for systems with exponential packets by varying the imbalance parameters. Results for systems with fixed-size packets can also be obtained as indicated in Section VII-A. We do not include them here since the trend of changes is similar to that of systems with exponential packets. However, we did compare throughputs for the case $C = 1$ with fixed-size packets with existing results in [16] and find them to be in agreement. Fig. 22 plots maximum throughput versus (R^I, S_1^I) for $C = 1$ with input imbalance only while Fig. 23 plot maximum throughput versus (R^O, S_1^O) for $C = 1$ with output imbalance only. Figs. 24 and 25 plot similar quantities for $C = 4$. As to be expected, large $R^I(R^O)$ or small $S_1^I(S_1^O)$ re-

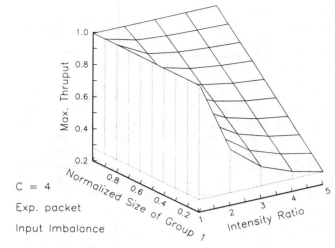

Fig. 24. Maximum throughput with input imbalance, $C = 4$.

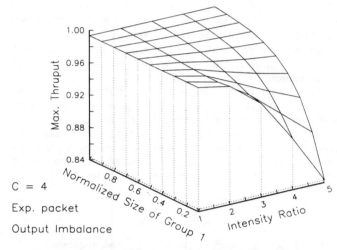

Fig. 25. Maximum throughput with output imbalance, $C = 4$.

duces system throughput. With $C = 1$, either input imbalance only or output imbalance only exhibits about the same rate of throughput reduction w.r.t. the corresponding parameters. For $C = 4$, however, the throughput reduction is much more severe in input imbalance systems than in output imbalance systems. This is because high concentration of traffic in a small group of input ports causes these ports to be quickly saturated, thus making the other input group underutilized and resulting in low total throughput. However, in output imbalance systems, since there are four parallel paths associated with each output port, even if there is a high concentration of traffic for only a small group of output ports, the throughput is not significantly impaired.

VIII. CONCLUSION

A general model was provided to study the family of space-division fast packet switches with input queueing, output queueing, and a speedup factor of C, $1 \leq C \leq N$. The model represents packet flow through the switch as going through queues in tandem. Performance measures are obtained by analyzing this tandem queueing system. The maximum switch throughput given any speedup factor of C was obtained for both packets with exponential length and fixed size. The results obtained for the extreme cases with $C = 1$ and $C \to \infty$ agree with known results for input queueing switches and output queueing switches, respectively. It was also shown that the switch can achieve higher throughput with fixed size packets. Also, for both cases, a modest speedup factor of 4 was found to be sufficient to achieve more than 99% throughput.

Average system delay, i.e., the mean delay a packet experiences from the moment it enters an input queue until it is successfully transmitted on an output link, was also obtained for different values of C. Various switch arbitration policies such as FCFS, ROS, and LCFS were investigated to study their impact on the system delay. It was found that FCFS policy yields the best average delay. ROS delay is the second and LCFS policy gives the worst average delay. However, these differences are not significant. The average delay decreases rapidly as C increases and, with $C = 4$, it approaches its optimal performance. The delay improvement for $C > 4$ is minimal.

Packet blocking probabilities for systems with finite buffer space were also obtained based on this model. The impact of buffer allocation on blocking probability was investigated. Given a fixed buffer budget, the analysis obtains an optimal buffer placement to minimize the blocking probability. It was seen that with small C, the optimal strategy tends toward allocating more buffers in the input while, as C is increased, more packets are queued in the outputs and the allocation strategy shifts accordingly.

The impact of traffic imbalance on switch throughput was also studied. Formulations were given for general traffic patterns with both input and output imbalances. For ease of comparison, numerical examples were given for systems with bi-group imbalance only. It was seen that throughput reduction caused by input imbalance is further amplified for systems with large C while the impact of output imbalance is less severe when C is large.

REFERENCES

[1] O. J. Boxma, "Analysis of models for tandem queues," Ph.D. thesis, Univ. Utrecht, 1977.

[2] ——, "On a tandem queueing model with identical service times at both counters, I," *Adv. Appl. Prob.*, no. 11, pp. 616–643, 1979.

[3] ——, "On a tandem queueing model with identical service times at both counters, II," *Adv. Appl. Prob.*, no. 11, pp. 644–659, 1979.

[4] P. J. Burke, "The output process of a stationary M/M/S queueing system," *Ann. Math. Stat.*, vol. 39, no. 4, pp. 1144–1152, 1968.

[5] R. B. Cooper, *Introduction to Queueing Theory.* New York: Macmillan, 1972.

[6] S. W. Fuhrmann, "Performance of a packet switch with crossbar architecture," *IBM Res. Div.*, RZ 1829, June 1989.

[7] D. Gross and C. M. Harris, *Fundamentals of Queueing Theory.* New York: Wiley, 1985.

[8] D. P. Heyman and M. J. Sobel, *Stochastic Models in Operations Research.* New York: McGraw-Hill, 1982.

[9] M. G. Hluchyj and M. J. Karol, "Queueing in space-division packet switching," in *Proc. 1988 INFOCOM Conf. (INFOCOM'88 Conf. Rec.)*, 1988, pp. 334–343.

[10] I. Iliadis, "Head of the line arbitration of packet switches with input and output queueing," presented at Fourth Int. Conf. Data Commun. Syst. Perform., Barcelona, Spain, June 20–22, 1990.

[11] I. Iliadis and W. Denzel, "Performance of packet switches with input and output queueing," in *Proc. ICC/SUPERCOM'90*, Atlanta, GA, Apr. 1990.

[12] M. J. Karol, M. C. Hluchyj, and S. P. Morgan, "Input versus output queueing on a space-division packet switch," *IEEE Trans. Commun.*, vol. COM-35, no. 12, pp. 1347–1356, Dec. 1987; also in *Proc. GLOBECOM'86*.

[13] L. Kleinrock, *Communication Nets.* New York: McGraw-Hill, 1964. Out of print, reprinted by Dover Publications, 1972.

[14] ——, *Queueing Systems Vol. 1: Theory.* New York: Wiley, 1975.

[15] ——, *Queueing Systems Vol. 2: Computer Applications.* New York: Wiley, 1976.

[16] S. Li and M. J. Lee, "A study of traffic imbalance in a fast packet switch," in *Proc. 1989 INFOCOM Conf. (INFOCOM'89 Conf. Rec.)*, 1989, pp. 538–547.

[17] Y. Oie, M. Murata, K. Kubota, and H. Miyahara, "Effect of speedup in nonblocking packet switch," in *Proc. ICC'89*, 1989, pp. 410–414.

[18] L. K. Platzman, J. C. Ammons, and J. J. Bartholdi III, "A simple and efficient algorithm to compute tail probabilities from transforms," *Operat. Res.*, vol. 36, no. 1, pp. 137–144, Jan.-Feb. 1988.

[19] J. Riordan, *Stochastic Service Systems.* New York: Wiley, 1962.

Comparison of Buffering Strategies for Asymmetric Packet Switch Modules

Soung C. Liew, *Member, IEEE*, and Kevin W. Lu, *Member, IEEE*

Abstract—This paper analyzes the performance of a class of asymmetric packet switch modules with channel grouping. The switch module considered has n inputs and m outputs. A packet destined for a particular output address (out of g) needs to access only one of the r available physical output ports; $m = gr$. The motivation for the study of these switch modules is that they are the key building blocks in many large multistage switch architectures. We concentrate on the performance of input-buffered and output-buffered switch modules under geometrically bursty traffic. A combination of exact derivation, numerical analysis, and simulation yields the saturation throughput of input-buffered switch modules and the mean delay of the input-buffered and output-buffered switch modules. Tables and formulas useful for traffic engineering are presented. Our results show that increasing the number of output ports per output address (r) can significantly improve switch performance, especially when traffic is bursty. An interesting observation is that although output-buffered switch modules have significantly better performance than input-buffered switch modules when there are equal numbers of input and output ports, this performance difference becomes significantly smaller when the switch dimensions are asymmetric.

I. Introduction

RECENT research activities in asynchronous transfer mode (ATM) switching have progressed to the study of large switch architectures constructed of interconnections of smaller switch modules [1]–[4]. In many cases, the underlying switch modules are of asymmetric dimensions in that there are unequal numbers of input and output ports. In addition, channel grouping, the technique of allocating more than one output port to each output address, is often used to improve switch performance. To gain insight into the design of large switch architectures of this type, it is important to understand the performance of the individual switch modules thoroughly.

Toward this goal, this paper considers the performance of a general class of asymmetric packet switch modules illustrated in Fig. 1. There are hs input ports consisting of h input groups of s input ports each, and gr output ports of g output groups of r output ports each. To achieve acceptable performance with the overall switch architecture, it is necessary to choose the various parameters of the basic switch modules properly. The objective of this paper is to quantify the performance of these switch modules as a function of the switch dimensions, buffering strategies, and traffic characteristics.

Before proceeding further, for motivation, we give three examples of switch architectures in Figs. 2–4 which make use of the class of switch modules considered here. Fig. 2 is a modular nonblocking switch architecture proposed by Lee [1]. The first stage consists of Batcher-banyan switch modules of dimensions

Manuscript received April 18, 1990; revised November 3, 1990.
The authors are with Bell Communications Research, Morristown, NJ 07960-1910.
IEEE Log Number 9142935.

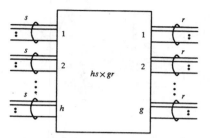

Fig. 1. The asymmetric switch module with channel grouping.

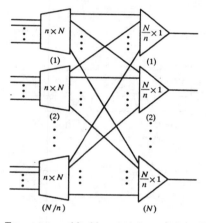

Fig. 2. Two-stage nonblocking modular switch architecture.

$n \times nk$ (i.e., with respect to the switch module in Fig. 1, $s, r \rightarrow 1$, $h \rightarrow n$, and $g \rightarrow nk$). The second-stage switch modules are statistical multiplexers of dimensions $k \times 1$. Fig. 3 is a general three-stage switch architecture proposed by Liew and Lu [2]. The dimensions of the first-stage, second-stage, and third-stage switch modules are $n \times m$ ($m > n$), $l \times l'$, and $m' \times n'$ ($m' > n'$), respectively. Here, a channel group of r (r') channels interconnects switch modules of adjacent stages. The structure is such that if r and r' were to be 1, there would be one and only one path between any input and output. For better performance, however, $r, r' > 1$ (in general) and packets have several alternative paths from their inputs to their destination outputs. Finally, Fig. 4 is a three-stage Clos switch architecture [4] that employs asymmetric switch modules at the two outer stages, and symmetric switch modules at the middle stage. There is no channel grouping internally, however. In all three schemes, asymmetric switch modules at the first stage result in internal line expansion which improves the performance of the overall switch architecture. It is also worth pointing out that the class of asymmetric switches we study here can also be used as

Reprinted from *IEEE J. Selected Areas Commun.*, vol. 9, no. 3, pp. 428–438, April 1991.

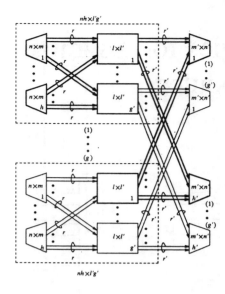

$$N = n g h = n' g' h'$$

$$g' = \frac{m}{r} \qquad g = \frac{m'}{r'}$$

$$h = \frac{l}{r} \qquad h' = \frac{l'}{r'}$$

Fig. 3. General structure of a three-stage switch architecture.

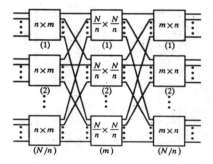

Fig. 4. Three-stage Clos switch architecture.

stand-alone concentrators or expanders rather than components of an overall switch.

Referring to Fig. 1, to the extent that packets at different input ports within the same group are uncorrelated, the switch module reduces to that shown in Fig. 5 in which $hs \rightarrow n$. For simplicity, this paper focuses on the structure shown in Fig. 5, assuming any correlations between packets of different input ports are small and negligible. An output group [2], [5] corresponds to an output address, and a packet can access any of the r output ports of its output address. In any given time slot, at most r packets can be cleared from a particular output group, one on each of the r output ports. Furthermore, we assume packets are destined for a particular output group (address) rather than a particular output port. That is, it does not matter which particular output port a packet accesses as long as the output port belongs to the correct output group. Reference [2] provides several designs of channel-grouping switch modules. It turns out that channel-grouping switch modules have smaller complexity (in terms of switch element counts) than ordinary switch modules of the same dimensions.

We focus on the performance of the input queueing and output queueing buffering strategies under geometric traffic. The

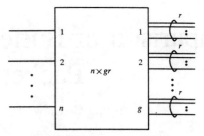

Fig. 5. The asymmetric switch module with uncorrelated inputs.

paper is a generalization of the work reported in [6]–[8] in two respects: the switch dimensions as well as the traffic characteristics have been generalized. With input queueing, an arriving packet enters a FIFO buffer on its input and waits for its turn to access its destination output. With output queueing, a logical FIFO buffer is allocated to each output group, and arriving packets destined for this output group are immediately placed into its FIFO. For simplicity, we assume infinite buffer queues for both input queueing and output queueing in our mean delay analysis.

The organization of this work is outlined as follows. Section II describes the geometric traffic model that we use to model bursty traffic. Section III investigates the maximum throughput and mean delay of input-buffered switch modules, and discusses the maximum throughput degradation due to head-of-line blocking under various settings. Section IV considers the mean delay of output-buffered switch modules. Finally, Section V concludes this work and discusses issues that deserve further attention.

II. TRAFFIC MODEL

We consider ATM transport in which data streams are partitioned and transferred in cells (or packets) of fixed size. On a conceptual level, time is therefore divided into slots corresponding to the cell transmission time. For performance analysis, we assume synchronous switch operation in which cells arrive at the beginning of each time slot, and cells gaining access to their output lines are cleared by the end of each time slot. To quantify the traffic characteristics, we focus on the uniform geometrically bursty traffic model in which an input alternates between active and idle periods of geometrically distributed duration [9]. During the active period of an input, packets destined for the same output arrive at the input continuously in consecutive time slots (see Fig. 6). Termination of the active period is a renewal process, and it occurs with probability p after each active time slot. Thus, the probability that the active period (burst) will last for a duration of i time slots (consists of i packets) is

$$P(i) = p(1 - p)^{i-1}, \quad i \geq 1. \tag{1}$$

Note that we assume there is at least one cell in the burst. This geometric burst-length assumption yields a mean burst length of

$$l = E_B = \sum_{i=1}^{\infty} iP(i) = 1/p. \tag{2}$$

The idle period is also geometrically distributed and is characterized by another parameter q. The probability that an idle period lasts for j time slots is

$$Q(j) = q(1 - q)^j, \quad j \geq 0. \tag{3}$$

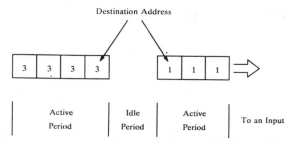

Fig. 6. Geometric packet arrivals to an input.

Unlike the duration of an active period, the duration of an idle period can be 0. The mean idle period is given by

$$E_I = \sum_{i=0}^{\infty} j Q(j) = (1-q)/q. \qquad (4)$$

Given p and q, the offered load ρ can be found by

$$\rho = E_B/(E_I + E_B). \qquad (5)$$

For simplicity, we focus on uniform output destination distribution in which any burst has equal probability of being destined for any output group. In addition, there is no correlation between different bursts. Note that the uniform random traffic model discussed in [6] and [7] is a special case with $p = 1$ and $q = \rho$, i.e., the burst length is deterministic, and it is always one packet long.

III. INPUT QUEUEING

In this section, the maximum throughput of asymmetric input-buffered switch modules with channel grouping is obtained by numerical analysis, and the mean delay by simulation.

In an input-buffered switch, when there are multiple packets at the heads of input queues destined for the same output group, only up to r packets may access the output group. As a result, some inputs idle because the first packets in their queues are blocked. Meanwhile, subsequent packets in the queues, which may be destined for other available output groups, are also blocked because of the FIFO queueing discipline. This is often referred to as head-of-line blocking [6], [7], and it is well known that head-of-line blocking limits the maximum throughput of a symmetric input-buffer switch ($r = 1$, $n = g \to \infty$) to 0.586 under the uniform random traffic condition ($p = 1$). Although the maximum throughput when $r = 1$ can be derived in closed form (including when $p \neq 1$), a general closed-form solution is not possible when $r > 1$. Nevertheless, a similar approach could be taken to a point where the solution could be found by numerical analysis. The same analysis yields the throughput of the switch module as a concentrator ($n > g$) or an expansion network ($n < g$).

To find the maximum throughput, we consider the situation in which the input queues are saturated so that one can always find packets in every queue. In particular, there is always a packet at the head of each queue, waiting to access its destination. Only after this packet is cleared can the next packet move to the head of the queue.

We define the *free input queues* to be input queues with packets transmitted in the previous time slot. The subsequent packet in a free input queue immediately moves to the head, ready to access its output destination in the current time slot. This subsequent packet could be from the same burst as the cleared

packet, in which case the destination remains the same, or it could be from a new burst, in which case it is equally likely to be destined for any output group. For our traffic model, the probability that the subsequent packet belongs to the same burst is given by $1 - p = (l - 1)/l$. Now, strictly speaking, for any finite buffer queue and $l > 1$, the mean burst length of bursts that arrive at the head of queue is not l, even though the mean burst length of the incoming traffic to the queue is l. This is because of the finite packet loss probability due to buffer overflow. For instance, if we overload the switch with a load of 1, then the mean burst length that arrives at the head of queue is actually $l\rho^*$, where ρ^* is the maximum allowable throughput. As far as analysis is concerned, the situation becomes even worse because the burst length is not strictly geometrically distributed after packets are dropped. Nonetheless, the queue would also saturate even if the offered load is just slightly over ρ^*, and in this situation the effective burst length would still be close to l and roughly geometrically distributed. For simplicity, therefore, we assume this to be the case for our saturation analysis. This assumption is further justified by later results which show that the maximum throughput approaches an asymptotic value very quickly as l increases; that is, the maximum throughput is not a strong function of l for moderate l values.

We now set up the framework for derivation of the maximum throughput. Consider a tagged output group i. Let A_j^i be the number of new bursts destined for output group i that move into the heads of free input queues in the beginning of time slot j. Note that under random uniform traffic ($l = 1$), A_j^i is also the number of packets destined for output group i since there is no distinction between bursts and packets. Under bursty traffic ($l > 1$), A_j^i does not include packet arrivals that belong to the same bursts as the packets just cleared. Let D_j^i be the number of bursts that terminate at the end of time slot j. Under bursty traffic, D_j^i is the number of departed packets minus departures which are subsequently replaced by packets of the same bursts. Let C_j^i be the number of head-of-line bursts that are destined for output group i at the beginning of time slot j, and let G_j^i be the number of head-of-line bursts left at the end of time slot j. Then,

$$C_{j+1}^i = G_j^i + A_{j+1}^i \qquad (6)$$

where

$$G_j^i = C_j^i - D_j^i. \qquad (7)$$

Note that C_j^i includes the bursts that are granted output access as well as the bursts that are blocked during time slot j.

By the assumption that packet output destinations are uniformly distributed across all output groups, all output groups face the same situation, and the superscript i can be dropped. The subscript j can also be dropped as the system approaches equilibrium. To simplify analysis, we will assume $n, g \to \infty$ while keeping a fixed value of g/n. This approximation is valid when n is large (e.g., $n \geq 16$). As in [8], it can be shown that $\lim_{n,g \to \infty} \Pr[A = k] = e^{-p\rho_0}(p\rho_0)^k/k!$, where $p\rho_0$ is the average arrival rate of new bursts, and ρ_0 the offered load per output group. For $n \to \infty$, there is no correlation between G and A, and the moment-generating functions of the parameters in (6) are related as follows:

$$C(z) = G(z) A(z) \qquad (8)$$

where

$$A(z) = e^{-p\rho_0(1-z)}. \qquad (9)$$

The key to finding the maximum throughput lies in the observation that the sum of the numbers of backlogged bursts over all output groups at the beginning of each time slot must be n, because there is always a head-of-line burst at each of the n input queues under the saturation condition. Since $C'(1)$ is the expected number of backlogged bursts per output group, we have

$$C'(1) = n/g. \tag{10}$$

The maximum throughput per output group can be found by equating $C'(1)$ obtained from the analysis based on (8) with n/g.

Let us define $P_i = \Pr[C = i]$. Then,

$$
\begin{aligned}
G(z) &= \sum_{i=0}^{\infty} G(z \mid C = i) P_i \\
&= P_0 + [p + (1 - p)z] P_1 \\
&\quad + \left[p^2 + \binom{2}{1} p(1 - p)z + (1 - p)^2 z^2 \right] P_2 \\
&\quad \vdots \\
&\quad + \left[p^{r-1} + \binom{r-1}{1} p^{r-2}(1 - p)z \right. \\
&\quad\quad + \binom{r-1}{2} p^{r-3}(1 - p)^2 z^2 \\
&\quad\quad \left. + \cdots + (1 - p)^{r-1} z^{r-1} \right] P_{r-1} \\
&\quad + \sum_{j=0}^{\infty} z^j \left[p^r + \binom{r}{1} p^{r-1}(1 - p)z \right. \\
&\quad\quad + \binom{r}{2} p^{r-2}(1 - p)z^2 \\
&\quad\quad \left. + \cdots + (1 - p)^r z^r \right] P_{r+j}.
\end{aligned}
$$

The above can be simplified to

$$
\begin{aligned}
G(z) &= \sum_{i=0}^{r-1} [p + (1 - p)z]^i P_i \\
&\quad + z^{-r} [p + (1 - p)z]^r \left[C(z) - \sum_{i=0}^{r-1} P_i z^i \right]. \tag{11}
\end{aligned}
$$

From (8) and (11), we obtain

$$
C(z) = \frac{\displaystyle\sum_{i=0}^{r-1} \left\{ z^r [p + (1 - p)z]^i - z^i [p + (1 - p)z]^r \right\} P_i}{z^r / A(z) - [p + (1 - p)z]^r}. \tag{12}
$$

The equilibrium probabilities P_i, $i = 1, \cdots, r - 1$, can be obtained using a standard method described as follows. It can be shown by Rouche's Theorem [10] that the denominator of $C(z)$ has $r - 1$ zeros, z_k, $k = 1, \cdots, r - 1$, with magnitudes less than 1. Since $C(z)$ is a moment-generating function, it must be analytical for all $|z| < 1$, and therefore, z_k, $k = 1, \cdots, r - 1$ must also be zeros of the numerator of $C(z)$. Thus, given z_k, $k = 1, \cdots, r - 1$, we have $r - 1$ linear equations relating r unknown P_k's. The normalization requirement $C(1) = 1$ gives us the other equation needed: $\Sigma_{i=0}^{r-1} (r - i) P_i = r - \rho_0$.

To summarize the above, the maximum throughput for an output group ρ_0 can be found numerically as follows. Starting with a guess of ρ_0, we first solve for z_k, $k = 1, \cdots, r - 1$, with the following $r - 1$ complex equations

$$
A(z_k)^{1/r} [p + (1 - p)z_k] = z_k \left(\cos \frac{2k\pi}{r} + i \sin \frac{2k\pi}{r} \right)
$$

$$k = 1, \cdots, r - 1. \tag{13}$$

The following r linear equations are then solved to find P_i.

$$
\sum_{i=0}^{r-1} \left\{ z_k^r [p + (1 - p)z_k]^i - z_k^i [p + (1 - p)z_k]^r \right\} P_i = 0
$$

$$k = 1, \cdots r - 1$$

$$
\sum_{i=0}^{r-1} (r - i) P_i = r - \rho_0. \tag{14}
$$

A new ρ_0 is found by

$$
C'(1) = \frac{\rho_0(2r - p\rho_0) - r(r - 1)(2 - p) + \displaystyle\sum_{i=0}^{r-1} [r(r - 1) - i(i - 1)](2 - p)P_i}{2(r - \rho_0)} = \frac{n}{g}. \tag{15}
$$

The three steps are repeated with the new ρ_0, and the process is iterated until the solution converges to the desired accuracy. The maximum throughput per input is related to ρ_0 by

$$
\rho^* = \frac{g}{n} \rho_0. \tag{16}
$$

The above is the general method for finding ρ^*. Various specific cases described below are amenable to simpler analysis, and they are described as follows.

(r arbitrary, p = 1): For uniform random traffic ($p = 1$), the second step of the numerical iterations can be eliminated because it is not necessary to explicitly solve for P_i, $i = 1$, $\cdots$, $r - 1$. In this case, instead of $2r - 1$ zeros, there are only r zeros in the numerator of $C(z)$, and they are all equal to the roots of the denominator, $z = 1$ and z_k, $k = 1, \cdots, r - 1$. We can directly express $C(z)$ in terms of z_k as follows.

$$C(z) = \frac{K(z - 1)(z - z_1) \cdots (z - z_{r-1})}{z^r/A(z) - 1} \quad (17)$$

where $K = p(r - \rho_0)/(1 - z_1) \cdots (1 - z_{r-1})$ is a normalization constant found by setting $C(1) = 1$. Differentiating (17) with respect to z, and setting $z = 1$ yields

$$C'(1) = \frac{\rho_0(2r - \rho_0) - r(r - 1)}{2(r - \rho_0)} + \sum_{k=1}^{r-1} \frac{1}{1 - z_k}. \quad (18)$$

(r arbitrary, p → 0): In the limiting case when the average burst length $l \to \infty$ ($p \to 0$), (12) becomes

$$\lim_{p \to 0} C(z) = \frac{\sum_{i=0}^{r-1} (r - i)z^i P_i}{r - \rho_0 z}. \quad (19)$$

By definition, $C(z) = \sum_{i=0}^{\infty} P_i z^i$. Multiplying both sides of (19) by $(r - \rho_0 z)$, and equating the coefficient of z^i on the left side with that on the right side, we obtain

$$P_i = \begin{cases} \rho_0 P_{i-1}/i & \text{if } i < r \\ \rho_0 P_{i-1}/r & \text{if } i \geq r. \end{cases}$$

This simplifies to

$$P_i = \begin{cases} \rho_0^i P_0/i! & \text{if } i < r \\ \rho_0^i P_0/(r! r^{i-r}) & \text{if } i \geq r \end{cases} \quad (20)$$

where

$$P_0 = \frac{r - \rho_0}{\sum_{i=0}^{r-1} (r - i)\rho_0^i/i!} \quad (21)$$

which is obtained by normalizing $C(1) = 1$. It is remarkable that (20) is the exact result of the $M/M/r$ queue with λ/μ, the ratio of the arrival rate to the service rate, equal to ρ_0. This, however, does conform to the intuitive understanding that as $p \to 0$, the geometrically distributed burst length becomes exponentially distributed.

Differentiating $C(z)$ in (19) and setting $z = 1$, we obtain

$$C'(1) = \frac{\sum_{i=0}^{r-1} i(r - i)P_i + \rho_0}{(r - \rho_0)}. \quad (22)$$

Equating (22) with n/g gives us a polynomial of ρ_0, from which we can obtain ρ_0 numerically. For specific values of r and n/g listed below, $\rho^* = \rho_0 g/n$ can be obtained in closed form by solving for the roots of the corresponding second-order polynomials directly.

$$\rho^* = \begin{cases} \dfrac{g}{n} \bigg/ \left(1 + \dfrac{g}{n}\right) & \text{if } r = 1 \\[3mm] 2\dfrac{g}{n}\left(\sqrt{\left(\dfrac{g}{n}\right)^2 + 1} - \dfrac{g}{n}\right) & \text{if } r = 2 \\[3mm] (\sqrt{162} - 6)/7 \approx 0.961 & \text{if } r = 3 \text{ and } g/n = 1. \end{cases} \quad (23)$$

For the interested readers, it turns out that ρ^* at $r \geq 4$ and $g/n = 1$ can be easily approximated. Consider $r = 4$ for an example. We know the corresponding ρ^* must be very close to 1 since ρ^* at $r = 3$ is already close to 1. Substituting $\rho^* = (1 - \epsilon)$ into $C'(1) = 1$ and ignoring the second and higher order ϵ terms, we get $\epsilon = 0.007$. The resulting $\rho^* = 0.993$ agrees with the exact result to three decimal places. For general g/n and r, however, the numerical root-finding method is needed to find ρ^*.

(r = 1, p arbitrary): Finally, for $r = 1$ and arbitrary p, the numerator of $C(z)$ in (12) has only one root, $z = 1$, and ρ^* can be solved in closed form:

$$\rho^* = \frac{(1 + g/n) - \sqrt{(1 + g/n)^2 - 2pg/n}}{p}. \quad (24)$$

We are now ready to examine results generated by the above analysis. Table I(a) lists the maximum throughput per input port for various values of r and g/n under uniform random traffic. The column in which $g/n = 1$ corresponds to the special cases studied by [8] and [11]. For a given r, the maximum throughput increases with g/n because the load on each output group decreases with g/n. For a given g/n, the maximum throughput increases with r because each output group has more output ports for clearing packets. This is analogous to increasing the number of servers in a queueing system. As shown in the table, when g/n is fairly large (say, $g/n > 4$), there is less incentive to use channel grouping to increase the throughput because the throughput is already close to 1. When g/n is small (say, $g/n < 2$), however, the use of channel grouping can increase the throughput substantially. For concentrators ($g/n < 1$), increasing the number of output ports per output address from 1 to 2 approximately doubles the maximum throughput.

As the average burst length l increases (or p decreases), the maximum throughput degrades. As shown in Fig. 7, the maximum throughput in general approaches an asymptotic value rather quickly as l increases. In particular, the maximum throughput for $l > 5$ is essentially equal to the asymptotic value. Table I(b) lists the asymptotic maximum throughput as $l \to \infty$. As shown, the difference in maximum throughput between the two extreme cases of $p = 1$ and $p \to 0$ is very small. Furthermore, it can be seen that the qualitative results for uniform random traffic described above also hold here. In addition, it can be easily verified that for a fixed r (i.e., for a particular row in the table), the percentage change in maximum throughput by varying p from 1 to 0 is the greatest when $gr/n = 1$, i.e., when the switch dimensions are symmetric. For instance, for $r = 2$, $g/n = 1/2$ yields the greatest percentage change in maximum throughput.

If there were no head-of-line blocking, then the maximum allowable throughput per input would be min $(1, gr/n)$. This

129

TABLE I
MAXIMUM THROUGHPUT FOR AN INPUT QUEUE WITH q/n KEPT CONSTANT WHILE BOTH g AND $n \to \infty$
(a) $p = 1$

r	$\frac{g}{n}$												
	$\frac{1}{32}$	$\frac{1}{16}$	$\frac{1}{8}$	$\frac{1}{4}$	$\frac{1}{3}$	$\frac{1}{2}$	1	2	3	4	8	16	32
1	0.031	0.061	0.117	0.219	0.279	0.382	0.586	0.764	0.838	0.877	0.938	0.969	0.984
2	0.061	0.121	0.233	0.426	0.531	0.686	0.885	0.966	0.984	0.991	0.998	0.999	1.000
3	0.092	0.181	0.346	0.613	0.736	0.875	0.975	0.996	0.999	0.999	1.000	1.000	
4	0.123	0.241	0.457	0.768	0.875	0.959	0.996	1.000	1.000	1.000			
8	0.245	0.476	0.831	0.991	0.998	1.000	1.000						
16	0.487	0.878	0.999	1.000	1.000								
32	0.912	1.000	1.000										

(b) $p = 0$

r	$\frac{g}{n}$												
	$\frac{1}{32}$	$\frac{1}{16}$	$\frac{1}{8}$	$\frac{1}{4}$	$\frac{1}{3}$	$\frac{1}{2}$	1	2	3	4	8	16	32
1	0.030	0.059	0.111	0.200	0.250	0.333	0.500	0.667	0.750	0.800	0.889	0.941	0.970
2	0.061	0.117	0.221	0.390	0.481	0.618	0.828	0.944	0.974	0.985	0.996	0.999	1.000
3	0.091	0.176	0.328	0.565	0.678	0.823	0.961	0.994	0.998	0.999	1.000	1.000	
4	0.121	0.234	0.432	0.715	0.828	0.937	0.993	0.999	1.000	1.000			
8	0.241	0.460	0.791	0.987	0.997	1.000	1.000	1.000					
16	0.477	0.849	0.999	1.000	1.000								
32	0.891	1.000	1.000										

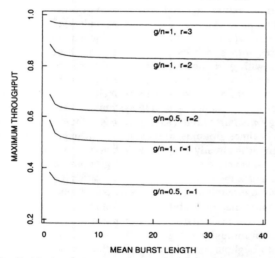

Fig. 7. Maximum throughput as a function of mean burst length.

is because we cannot load each input line with load greater than 1 or each output group with load greater than r. We can therefore define the degradation due to head-of-line blocking as

$$\Delta(r, g/n) = \min(1, gr/n) - \rho^*. \quad (25)$$

Since $\min(1, gr/n)$ is also the maximum throughput of the output-buffered switch module, $\Delta(r, g/n)$ can be interpreted as the throughput advantage of the output-buffered switch module over the input-buffered switch module. Table II(a) and (b) show the $\Delta(r, g/n)$ values for $p = 0$ and $p = 1$, respectively. It can be seen that for either a row (i.e., fixed r) or a column (i.e., fixed g/n), the degradation is the biggest when $gr = n$, and the degradation becomes progressively smaller as we deviate from this point. Thus, whenever the switch dimensions become asymmetric ($gr \neq n$), the throughput advantage of the output-buffered switch module over the input-buffered switch module diminishes. This can be explained intuitively as follows. When $gr < n$, both input queueing and output queueing are limited by the fact that there are fewer number of output ports than input ports, and head-of-line blocking is not the main limiting factor in input-buffered switch modules anymore. When $gr > n$, the maximum throughput of output-buffered switch modules is still limited by 1, while that of input-buffered switch modules improves because the detrimental effect of head-of-line blocking is alleviated by the fact that more head-of-line packets can be cleared now. The table also reveals that for a fixed number of output ports gr, decreasing the number of output addresses g while increasing the channel group size r also alleviates the head-of-line blocking effect and decreases the maximum throughput difference between the two buffering strategies.

As an example of the application of the above results, consider the two-stage switch architecture in Fig. 3. According to our results, the expanded Batcher-Banyan switch modules would have no significant throughput limitations if $N/n \geq 32$.

Analysis of the mean delay of input-buffered switch modules is difficult, so simulation is used here. For input queueing, a contention resolution scheme is needed in order to resolve conflicts when there are more than r packets destined for the same output group. Whereas the maximum throughput of input-buffered switch modules under geometric traffic is insensitive to the particular contention resolution scheme adopted (as long as no head-of-line packets are withheld from clearance when there are free destination output ports), the mean delay does depend on

TABLE II
INCREMENT OF MAXIMUM THROUGHPUT FOR AN OUTPUT QUEUE OVER AN INPUT QUEUE $\Delta(r, g/n) = \min(1, gr/n) - \rho^*$

(a) $p = 1$

r	$\frac{g}{n}$												
	$\frac{1}{32}$	$\frac{1}{16}$	$\frac{1}{8}$	$\frac{1}{4}$	$\frac{1}{3}$	$\frac{1}{2}$	1	2	3	4	8	16	32
1	0.000	0.002	0.008	0.031	0.054	0.118	**0.414**	0.236	0.162	0.123	0.062	0.031	0.016
2	0.002	0.004	0.017	0.074	0.136	**0.314**	0.115	0.034	0.016	0.009	0.002	0.001	0.000
3	0.002	0.007	0.029	0.137	**0.264**	0.125	0.025	0.004	0.001	0.001	0.000	0.000	
4	0.002	0.009	0.043	**0.232**	0.125	0.041	0.004	0.000	0.000	0.000	0.000		
8	0.005	0.024	**0.169**	0.009	0.002	0.000	0.000						
16	0.013	**0.122**	0.001	0.000	0.000								
32	**0.088**	0.000	0.000										

(b) $p = 0$

r	$\frac{g}{n}$												
	$\frac{1}{32}$	$\frac{1}{16}$	$\frac{1}{8}$	$\frac{1}{4}$	$\frac{1}{3}$	$\frac{1}{2}$	1	2	3	4	8	16	32
1	0.001	0.004	0.014	0.050	0.083	0.167	**0.500**	0.333	0.250	0.200	0.111	0.059	0.030
2	0.002	0.008	0.029	0.110	0.186	**0.382**	0.172	0.056	0.026	0.015	0.004	0.001	0.000
3	0.003	0.012	0.047	0.185	**0.322**	0.177	0.039	0.006	0.002	0.001	0.000	0.000	
4	0.004	0.016	0.068	**0.285**	0.172	0.063	0.007	0.001	0.000	0.000			
8	0.009	0.040	**0.209**	0.013	0.003	0.000	0.000	0.000					
16	0.023	**0.151**	0.001	0.000	0.000								
32	**0.109**	0.000	0.000										

the contention scheme. Our simulation experiments assume a random strategy in which a random input port, say Port_{top}, is chosen to have the highest priority at the beginning of each time slot. The priorities of the input ports for that time slot are then ordered in a cyclic manner: Port_{top}, $\text{Port}_{\text{top}+1(\text{mod}\,n)}$, $\cdots$, $\text{Port}_{\text{top}+n-1(\text{mod}\,n)}$. Fig. 8 shows the graphs of the mean delay versus the input offered load for various values of r and g/n, fixing n at 32, and l at 1 and 16. Simulation results show that for a given r and g/n, but $n > 32$, the mean delay is closely approximated by the results of $n = 2$. For all cases shown, enough packet statistics are collected so that the 95% confidence interval is no more than $\pm6\%$ of the collected mean delay value.

As shown in the figure, for uniform random traffic ($l = 1$), the mean delay is rather low except for offered loads close to the maximum allowable throughput. This is, however, not the case for the bursty traffic ($l = 16$), where the maximum delay increases rather quickly as the offered load increases. Comparing cases to $r = 1$, $r = 2$, and $r = 4$, we also see that as traffic becomes bursty, the mean delay does not degrade as much for large r than for small r. To further illustrate this point, we plot in Fig. 9 the mean delay for various cases with $n = gr$ (the dotted lines) as a function of l for an offered load of 0.5 (the asymptotic maximum throughput when $n = g$, $r = 1$ and $p \rightarrow 0$). As shown, the slopes of the mean delay versus l curves decrease quite rapidly as r increases. Thus, in general, channel grouping improves the mean delay, as well as the maximum throughput, under bursty traffic conditions.

IV. OUTPUT QUEUEING

For output queueing switch modules, we assume there is a single FIFO queue for each output group. Arriving packets destined for a given output group are immediately placed on the

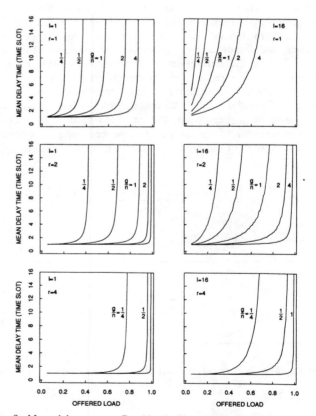

Fig. 8. Mean delay versus offered load of input-buffered switch modules.

corresponding output queue. Unlike input queueing, there is no head-of-line blocking in output queueing, and the maximum throughput per output group is bounded by r. Except for cases

131

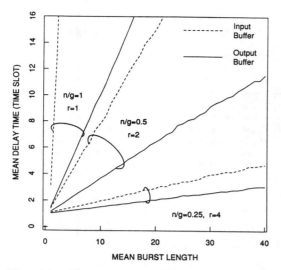

Fig. 9. Mean delay of input- and output-buffered switch modules as a function of the average burst length for an offered load of 0.5.

with $r > 1$ and $l > 1$ (i.e., $p < 1$), the mean delay of output-buffered switches can be obtained using the same framework for deriving the maximum throughput of input-buffered switches. In the following, we consider the three cases ($r \geq 1$, $p = 1$), ($r = 1$, $p < 1$), and ($r > 1$, $p < 1$), separately.

($r \geq 1$, $p = 1$): The situation faced by an output group is described precisely by the framework used to derive the maximum throughput of input-buffered switches, except that arriving packets are immediately presented to the output group for clearance, and do not have to first proceed to the heads of input queues. Unlike the derivation of the maximum throughput, however, ρ_0 is interpreted as the given offered load here.

We shall use the same notation as in the maximum throughput derivation. Consider a particular output group. Given an output offered load of ρ_0 per group, the expected number of backlogged packets for an output group at the beginning of a time slot is given by $C'(1)$ in (18). By Little's Law, the mean delay is

$$\bar{T} = C'(1)/\rho_0 = \frac{\rho_0(2r - \rho_0) - r(r - 1)}{2(r - \rho_0)\rho_0} + \frac{1}{\rho_0} \sum_{k=1}^{r-1} \frac{1}{1 - z_k}$$

(26)

where z_k, $k = 1, \cdots, r - 1$ are the $r - 1$ roots of

$$\exp\left[\frac{-\rho_0(1 - z_k)}{r}\right] = z_k\left[\cos\left(\frac{2k\pi}{r}\right) + i\sin\left(\frac{2k\pi}{r}\right)\right]$$

$$k = 1, \cdots, r - 1. \quad (27)$$

($r = 1$, $p < 1$): When $l > 1$, a complete analysis involves a two-dimensional Markov chain which keeps track of the number of backlogged packets and the number of bursts with pack-

ets still arriving [9]. However, as shown below, simpler analysis is sufficient for deriving the mean delay.

The situation for $r = 1$ is closely related to the $M/G/1$ queue; bursts are analogous to customers and burst lengths to durations of service. As far as the mean delay of a packet is concerned, it does not matter whether we finish serving (clearing) packets of one burst before serving packets of the next burst, or serve packets of the backlogged bursts in an arbitrary order, since the unfinished work, or the number of remaining packets, is the same in either case. Without losing generality, we focus our attention on the former burst-by-burst service discipline.

There is a subtle difference between the $M/G/1$ queue and our situation, however. In the $M/G/1$ queue, when a customer arrives, it arrives in its totality, whereas, in our case, packets in a burst arrive in consecutive time slots. Nonetheless, the waiting time of a packet has the same distribution as that of a burst, and it can be obtained from $M/G/1$ analysis. To see this, consider the jth packet in a burst. Suppose that the waiting time of the burst (or the first packet), or the time the burst spends waiting in the queue before it is served, is W. Although the jth packet arrives $j - 1$ time slots later than the first packet, it is also served $j - 1$ time slots later than the first packet. Thus, the waiting time of the jth packet is also W. Notice that not only are the mean waiting times of the burst and its packets the same, the waiting times are also identically distributed under the burst-by-burst service discipline.[1]

By Little's Law, the mean delay of a burst is $C'(1)/p\rho_0$, where $C'(1)$ is given by (15). Therefore, the mean burst or packet waiting time is $C'(1)/p\rho_0 - 1/p$, and the mean packet delay is

$$\bar{T} = C'(1)/p\rho_0 - 1/p + 1 = \frac{2 - p\rho_0}{2(1 - \rho_0)p} - 1/p + 1.$$

(28)

($r > 1$, $p < 1$): When $r > 1$ and $l > 1$, things become more complicated because the analogy between bursts and customers breaks down. To see this, consider the following. If there are fewer than r customers in an $M/G/r$ queue, then some of the servers are not active. For a switch, however, even if there are fewer than r backlogged bursts in the output queue, as long as there are at least r packets, all the r output lines would be active, and multiple packets from some bursts are served simultaneously. Nevertheless, this observation implies that $M/G/r$ analysis can be used to obtain an upper bound to the actual mean delay. That is, given an offered load ρ_0, the mean packet delay is upper-bounded by $C'(1)/p\rho_0 - 1/p + 1$, where $C'(1)$ is given by (15). This yields

$$\bar{T} < \frac{\rho_0(2r - p\rho_0) - r(r - 1)(2 - p) + \sum_{i=0}^{r-1}\left[r(r - 1) - i(i - 1)\right](2 - p)P_i}{2(r - \rho_0)p\rho_0} - 1/p + 1 \quad (29)$$

where P_i, $i = 0, \cdots, r - 1$, are obtained by solving (13) and (14).

A lower bound to the mean delay can be obtained by considering a modified system in which all packets in a burst are assumed to arrive simultaneously in the beginning of the burst. The basic idea is as follows. In the modified system, the arrival instants of all packets in a burst are shifted to the arrival instant

[1]This implies that the probability $P[W > b]$ obtained from $M/G/1$ analysis can be used as an upper bound for the packet loss probability of a finite buffer queue of b packets deep.

of the first packet. In contrast to the original system, the modified system allows all r output lines to be utilized even when the burst arrives at a queue with fewer than $r - 1$ backlogged packets. Consequently, the departures of some packets are also shifted to earlier instants in the modified system. After the mean delay of the modified system is found, it is easy to compensate for the extra delay due to the shift in arrival instants. It is, however, difficult to compensate for the shift in departure instants or else we would have found an exact solution. Nevertheless, by compensating only for the shift in arrival epochs, a lower bound to the mean delay of the original system is obtained.

In the following lower-bound analysis, instead of focusing on bursts, we focus on packets, and use C and A to denote the number of backlogged packets and the number of packet arrivals, respectively. We essentially have a $G/D/r$ system in which

$$C_{j+1} = \max(0, C_j - r) + A_j \qquad (30)$$

where the moment-generating function of A_j is given by

$$
\begin{aligned}
A(z) &= \sum_{k=0}^{\infty} A(z|k) P[k \text{ bursts arrive}] \\
&= \sum_{k=0}^{\infty} \left[\frac{pz}{1 - (1-p)z}\right]^k \frac{(p\rho_0)^k e^{-p\rho_0}}{k!} \\
&= \exp\left[\frac{-p\rho_0(1-z)}{1 - (1-p)z}\right].
\end{aligned} \qquad (31)
$$

Using an analysis similar to that in the derivation of the maximum input-queueing throughput, we obtain

$$C(z) = A(z) \frac{\sum_{i=0}^{r-1} (z^r - z^i)P_i}{z^r - A(z)}. \qquad (32)$$

This gives

$$
\begin{aligned}
C'(1) &= \frac{\rho_0(2r - \rho_0) + 2\rho_0(1-p)/p - r(r-1)}{2(r - \rho_0)} \\
&+ \sum_{k=1}^{r-1} \frac{1}{1 - z_k}
\end{aligned} \qquad (33)
$$

where z_k, $k = 1, \cdots, r - 1$ are obtained by solving for the roots of

$$\exp\left\{\frac{-p\rho_0(1 - z_k)}{[1 - (1-p)z_k]r}\right\} = z_k\left[\cos\left(\frac{2k\pi}{r}\right) + i \sin\left(\frac{2k\pi}{r}\right)\right]$$
$$k = 1, \cdots, r - 1. \qquad (34)$$

The delay of a packet in this $G/D/r$ queue is then given by $C'(1)/\rho_0$.

We now compensate for the extra waiting time due to the earlier arrival assumption. Given a packet is in a burst of length k, its expected *extra* waiting time in the modified system is $(k - 1)/2$. The probability that a packet is in a burst of length k is $k(1 - p)^{k-1}p^2$. The expected extra waiting time is, therefore,

$$\sum_{k=1}^{\infty} \frac{(k - 1)}{2} k(1 - p)^{k-1}p^2 = (1 - p)/p. \qquad (35)$$

Thus, we have

$$\overline{T} > \frac{C'(1)}{\rho_0} - (1 - p)/p. \qquad (36)$$

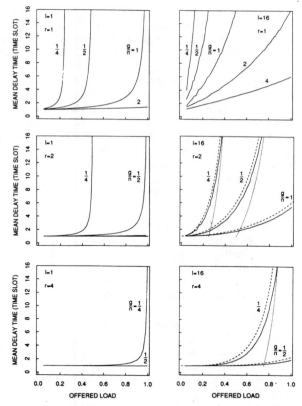

Fig. 10. Mean delay versus offered load of output-buffered switch modules.

It is worth pointing out that both the upper bound and lower bound described above become the exact solution when $p = 1$ or when $r = 1$.

Based on the above analysis, Fig. 10 shows the mean delay versus the offered input load ($g\rho_0/n$) for various values of r and g/n ratio, when $l = 1$ and when $l = 16$. Cases with ($r = 1, 2,$ or $4, l = 1$) and $r = 1, l = 16$) are numerical results, whereas cases with ($r = 2$ or $4, l = 16$) are simulation results. The figure also compares the analytical upper and lower bounds with the simulation results. Although the simulation assumes $n = 32$ and the analytical results assume $n \rightarrow \infty$, this slight discrepancy would not invalidate the following discussion, since the results are not very sensitive to n for $n > 16$.

As expected, comparing Fig. 10 with Fig. 8, the mean delay versus throughput performance of output queueing is uniformly better than that of input queueing for all cases. As in input queueing, however, bursty traffic tends to degrade the performance significantly. Also similar to input queueing is the fact that as the traffic becomes bursty, the mean delay does not degrade as much for large r than for small r. This point is further illustrated in Fig. 9, where we plot the mean delay for various cases with $n = gr$ as a function of l for an offered load of 0.5. As in input queueing, the slopes of the mean delay versus l curves decrease quite rapidly as r increases. It is also interesting to observe that for a fixed number of output ports gr, the difference in mean delay between input queueing and output queueing also decreases as r increases. So, channel grouping tends to decrease the performance gap between the two buffering strategies.

For cases with $r > 1$ and $l > 1$ shown in Fig. 10, the upper bound (dotted lines) is rather close to the exact solution when

$r = 2$, especially at regions of low mean delay. When $r = 4$, the upper bound is not very good at high mean delay. This shows that a switch with channel grouping has significant better delay performance than an $M/G/r$ queue when r is large. In contrast to the upper bound, the lower bound is poor when the mean delay is low. This is not surprising if we recall that the $G/D/r$ queue associated with the lower bound allows all r output lines to be utilized, even when a burst arrives when there are fewer than $r - 1$ packets in the queue. This artificial advantage occurs rather frequently when the offered load is low, but disappears when the buffer occupancy is high. In fact, the lower bound approximates the exact solution better than the upper bound at high mean delay.

V. Conclusions

This paper has quantified the throughput and mean-delay performance of a class of $n \times gr$ asymmetric packet switch modules with channel grouping at the outputs. These switch modules constitute the building blocks of many large switch architectures, and it is important to understand the performance of the switch modules in order to design the large switches properly.

Both input-buffered and output-buffered switch modules have been studied. It is shown that increasing the number of output ports per output address can significantly improve the delay-throughput performance of both buffering strategies, particularly when the ratio of the number of output addresses to the number input ports, g/n, is small. This agrees in principle with the idea originally propounded in the knockout switch [12], [3]. If we fix the line expansion ratio (gr/n), the performance is better for larger r. In other words, decreasing the number of output addresses while fixing the numbers of output and input ports improves the performance. However, reducing the number of output addresses implies reduced switching and, to the extreme that there is only one output address, no switching is performed. Thus, the result simply says one would perform switching to the extent that it is necessary. "Overswitching" not only degrades performance, but also increases switch complexity.

We have also shown that the mean delay performance of both buffering strategies degrades significantly as traffic becomes more bursty, although the maximum allowable throughput of the input-buffered switch module decreases only slightly. In general, however, channel grouping at the outputs tends to decrese the degradation in delay performance due to bursty traffic.

Although output queueing has uniformly better delay-throughput performance than input queueing for all switch dimensions, the advantage of output queueing over input queueing decreases as the switch dimensions become more and more asymmetric (for cases with $n < gr$ as well as $n > gr$). Intuitively, for $gr < n$, the performance limitation is mainly due to line concentration (i.e., fewer output ports than input ports). But this limitation applies to both input and output queueing switch modules. For $gr > n$, the effect of head-of-line blocking on input queueing switch modules is alleviated because of line expansion, and the performance approaches that of output queueing switch modules. In short, $n = gr$ is a special case in which the difference in performance between input and output queueing is the largest. The performance gap between the two buffering strategies also decreases when we increase r and decrease g while keeping gr constant. In fact, the largest performance gap is found in the previously studied case [6], [7] with $n = g$ and $r = 1$.

Finally, some research issues deserve further attention to extend the understanding of input and output queueing strategies in high-speed packet switches.

1) For simplicity, we have assumed that the traffic patterns on different input ports are uncorrelated. Strictly speaking, this is not true when the input ports are also grouped, as in the second stage of the switch architecture shown in Fig. 3. In fact, two packets of the same burst may arrive simultaneously on two input ports of the same group when switch modules with channel grouping are cascaded. It would be interesting to see how the results here need to be modified under this situation.

2) The study of nonuniform traffic distribution in which more packets are destined for some outputs than others also requires further attention. In particular, how would input-buffered and output-buffered switch modules compare with each other under nonuniform but geometrically distributed traffic?

3) When the burst length is not geometric, the maximum throughput of the input-buffered switch module would in general depend on the contention resolution scheme assumed. For instance, when the burst length is deterministic, the optimal strategy is the burst-by-burst service discipline in which we finish serving the packets of one burst before starting on the next burst. In fact, it can be shown that the maximum throughput in this case is the same as that of the uniform random traffic case, for arbitrary burst length l. It is interesting to investigate the sensitivity of our results to the particular bursty traffic model adopted.

Acknowledgment

The authors thank T. Lee for generously sharing his knowledge and expertise with them. This paper has also benefited much from comments by H. Lemberg and the anonymous reviewers.

References

[1] T. Lee, "A modular architecture for very large packet switches," *Conf. Rec., GLOBECOM '89*, vol. 3, pp. 1801–1809, 1989.

[2] S. C. Liew and K. W. Lu, "A 3-stage interconnection structure for very large packet switches," *Conf. Rec., ICC '90*, pp. 316.7.1–316.7.7, 1990.

[3] K. Y. Eng, M. J. Karol, and Y. S. Yeh, "A growable packet (ATM) switch architecture: Design principles and applications," *Conf. Rec., GLOBECOM '89*, pp. 32.2.1–32.2.7, 1990.

[4] H. Suzuki et al., "Output-buffer switch architecture for asynchronous transfer mode," *Conf. Rec., ICC '89*, vol. 1, pp. 99–103, 1989.

[5] A. Pattavina, "Multichannel bandwidth allocation in a broadband packet switch," *IEEE J. Select. Areas Commun.*, vol. 6, no. 9, pp. 1489–1499, Dec. 1988.

[6] M. G. Hluchyj and M. J. Karol, "Queueing in high-performance packet switching," *IEEE J. Select. Areas Commun.*, vol. 6, no. 9, pp. 1587–1597, Dec. 1988.

[7] J. Y. Hui and E. Arthur, "A broadband packet switch for integrated transport," *IEEE J. Select. Areas Commun.*, vol. 5, no. 8, pp. 1264–1273, Oct. 1987.

[8] M. Y. Karol, M. G. Hluchyj, and S. Morgan, "Input versus output queueing on a space-division packet switch," *IEEE Trans. Commun.*, vol. 35, no. 12, pp. 1347–1356, Dec. 1987.

[9] A. Descloux, "Contention probabilities in packet switching networks with strung input processes," in *Proc. ITC 12*, 1988.

[10] L. Kleinrock, *Queueing Systems, Vol. 1: Theory.* New York: Wiley, 1975.

[11] Y. Oie et al., "Effect of speedup in nonblocking packet switch," *Conf. Rec., ICC '89*, vol. 1, pp. 410–415, 1989.

[12] Y. Yeh, M. Hluchyj, and A. Acampora, "The knockout switch: A simple modular architecture for high-performance packet switching," *IEEE J. Select. Areas Commun.*, vol. 5, no. 8, pp. 1274–1283, Oct. 1987.

Nonuniform Traffic Analysis on a Nonblocking Space-Division Packet Switch

SAN-QI LI, MEMBER, IEEE

Abstract—This paper studies the nonuniform traffic performance on a nonblocking space division packet switch. When an output link is simultaneously contended by multiple input packets, only one can succeed and the rest will be buffered in the queues associated with each input link. Given the condition that the traffic on each output is not dominated by individual inputs, our study indicates that the output contention involved by packets at the head of input queues can be viewed as an independent phase type process for a sufficiently large size of the switch. Therefore, each input queue can be modeled by an independent Geom/PH/1 queueing process. Once the relative input traffic intensities and their output address assignment functions are defined, a general formulation can be developed for the maximum throughput of the switch in saturation. The result clearly indicates under what condition the input queue will saturate. We also propose a general solution technique for the evaluation of the queue length distribution. Our numerical study based on this analysis agrees well with simulation results.

I. INTRODUCTION

FUTURE communication services are expected to be provided via packet switching networks [1]. Various space-division packet switching architectures are currently under development [1]–[4]. Such networks will interconnect a large number of subscriber lines at one central office. With fiber optic technology, the bandwidth available on a link has the potential to increase by a few Gbps. One of the most distinctive properties in space division packet switching is that packets need to be buffered whenever there are multiple input packets arriving for the same output. The buffering can be either within the switching fabric, or at the queue of the input/output links.

This paper studies the performance of input queueing architectures. We consider a nonblocking packet switch fabric with N input links and N output links. Examples of this switch include crossbar and Batcher–Banyan [3]. Packets from each input link are simultaneously and independently self-routed to the output links. Each link is logically partitioned into time slots of a duration equal to the packet transmission time through the switch fabric. Assuming that all packets are of equal size, the operation of the switch is synchronized by the time slot, and no packet is queued inside the switch. When an output link is simultaneously requested by multiple input links, only one can be transmitted at a slot interval, and the rest of the contending packets are buffered in the queues associated with each input link. Packets arriving at each input queue in a slot interval are modeled by an independent Bernoulli process, and they are served on the FCFS basis at the beginning of each slot. The output address of the packets from each input link is independently assigned according to certain probability distribution functions.

Recently, the queueing analysis on such switching systems has been done with the assumption of uniform traffic [3], [5]. By uniform, we mean that every link carries the same amount of load and every input

Paper approved by the Editor for Communication Switching of the IEEE Communications Society. Manuscript received September 22, 1988; revised March 28, 1989. This work was supported by the National Science Foundation under Grant CDR-84-21402.

The author is with the Department of Electrical and Computer Engineering, University of Texas at Austin, Austin, TX 78712.

IEEE Log Number 9036334.

packet is equally likely to be addressed to any output link. Under this condition, the system has the maximum throughput equal to 0.586 for sufficiently large N.

This paper analyzes the queueing performance with nonuniform traffic in the limit as $N \to \infty$. The approach taken here is similar to [3], [5], but in a more general form. The only constraint we have on traffic flow is that the traffic on each output comes from a large number of inputs and is not dominated by individual inputs. Mathematically, it means that the arrivals from individual inputs to an output must be asymptotically negligible, in the limit as $N \to \infty$.

Given the input traffic intensities and their output distribution at the switch, we present a general formulation for the maximum throughput of the switch in saturation. The resultant expression clearly indicates under what condition the input queue will saturate.

We further show that each output contention process involved by packets at the head of input queues has a distribution of phase type, and that each input queue can therefore be modeled by an independent Geom/PH/1 queueing process. A solution technique is then developed for the analysis of the queue length distribution at each input.

The outline of this paper is as follows. Section II gives the model of the switch. Section III analyzes the output contention process. Section IV derives the maximum throughput of the switch. Section V presents the input queue analysis. Section VI further refines the results derived in Sections III and IV by taking account of the effect of individual inputs on each output contention process. Examples are shown in Section VII. The paper is summarized in Section VIII.

II. MODELING

The model consists of a nonblocking packet switch fabric with N inputs and N outputs, described in Fig. 1. Assume that the buffer size at each input is infinite. As we will see, the same analytical approach can be applied to the finite buffer case. [A word on notation used hereafter: any notation with subscript $\bigcirc$ ($\bullet$) indicates that notation is associated with input side (output side).] Packets arriving at input i in a slot time interval form an independent Bernoulli process, at rate denoted by λ_i°. Define

$$(\lambda_0^\circ, \lambda_1^\circ, \cdots, \lambda_{N-1}^\circ) \tag{1}$$

The output address of the incoming packets is assigned at random via a matrix T

$$T = \begin{pmatrix} t_{0,0} & t_{0,1} & \cdots & t_{0,N-1} \\ t_{1,0} & t_{1,1} & \cdots & t_{1,N-1} \\ \vdots & \vdots & \ddots & \vdots \\ t_{N-1,0} & t_{N-1,1} & \cdots & t_{N-1,N-1} \end{pmatrix} \tag{2}$$

where $t_{i,k}$ is the possibility for a packet from input i to be destined to output k, given that $\sum_{\forall k} t_{i,k} = 1$. The output traffic intensities are denoted by

$$\Lambda^\bullet = (\lambda_0^\bullet, \lambda_1^\bullet, \cdots, \lambda_{N-1}^\bullet). \tag{3}$$

Reprinted from *IEEE Trans. Commun.*, vol. 38, no. 7, pp. 1085–1096, July 1990.

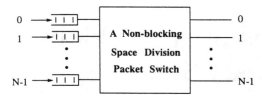

Fig. 1.   Structure of a nonblocking space division packet switch.

For a stable system, we have

$$\Lambda^{\bullet} = \Lambda^{\circ} T. \qquad (4)$$

For uniform traffic, $t_{i,k} = 1/N$ and $\lambda_i^{\circ} = \lambda_k^{\bullet}, \forall i, k$. In practice, the values of Λ° and T are expected to be found by high layer protocols, such as those at service admission and network control/management levels.

III. Output Contention Analysis

Let us focus on a tagged packet in an input queue destined to a particular output (say, from input i to output k). Once this packet moves to the head of the queue, it joins the contention for the transmission with the packets at the head of the other input queues destined to the same output. The contention process to output k is characterized by a queueing model described by

$$C_k' = (C_k - 1)^+ + A_k \qquad (5)$$

where the symbol $(\cdot)^+$ denotes the larger of 0 or its argument. C_k represents the total number of packets at the heads of all input queues destined to output k in the last slot. One of them, which has won the contention in the last slot, will move to the output k in this slot, if $C_k > 0$. A_k consists of those packets which will move to the heads of input queues in this slot, and destined to output k, C_k' then gives the number of packets accumulated for the output contention in this slot. Note that C_k is defined to include the total number of packets in contention as well as in transmission, destined to output k. The output contention time is therefore defined to contain the one slot transmission time through the switch fabric after the contention.

Letting $N \to \infty$, under the condition that the arrivals from individual inputs to an output are asymptotically negligible, the time interval between two adjacent packets moving to output k from an individual input will be sufficiently long. Hence, the output contention process involving every such packet can be viewed as an i.i.d. process. Also, the contention processes from an individual input to different outputs tend to be mutually independent, since the arrival processes to different output contentions A_k become mutually independent as $N \to \infty$. The reason for this is that the number of inputs, which have new packets moved to the heads of the queues at a slot time interval, approaches infinity, and that the output address for packets from each input is independently and randomly assigned. Therefore, one may make the statement that the output contention process involving each packet from input i is mutually independent, in the limit as $N \to \infty$. Moreover, by the same argument, the output contention process involved by the packets from two different inputs can also be viewed as mutually independent in the limit. These observations imply that in the limit, each input queue will form an independent discrete-time Geom/G/1 queue. So far we have considered only a stable system where no input queue is in saturation.

A_k is defined as the newly joined packets at a slot time interval destined to output k. The following shows that as $N \to \infty$, A_k forms an independent Poisson process, provided that the arrivals from individual inputs to output k are asymptotically negligible. There are three types of input queues at the end of a slot: empty, nonempty with a successful contention, and nonempty with a failing contention. During the next slot interval, A_k can only come from the first two types of input queues. Type I is the queues which are empty in the last slot and may possibly receives a new packet arrival in this slot. Type II is the nonempty queues which have won the output contention in the last slot, and a new packet may possibly refill the head of the

queues. Note that, due to the Bernoulli input arrivals, the steady-state distribution of input queues observed at each slot is identical to that observed at the end of each contention (i.e., at the departure instants).

Denote the average service time per packet on the ith input queue by $\overline{s_i^{\circ}}$, and its utilization factor by

$$\rho_i^{\circ} = \lambda_i^{\circ} \overline{s_i^{\circ}}. \qquad (6)$$

If it is a type I queue in the last slot, the probability for a new packet arrival and destined to output k in this slot is given by $\lambda_i^{\circ} t_{i,k}$. Let $x_{i,k}$ denote this arrival process, which forms a Bernoulli process with its generating function given by

$$1 - \lambda_i^{\circ} t_{i,k} + \lambda_i^{\circ} t_{i,k} z. \qquad (7)$$

If it is a type II queue in the last slot, it must be at the departure instant, and therefore the probability in this slot to have a packet at the head of this queue destined to output k is given by $\rho_i^{\circ} t_{i,k}$. Let $y_{i,k}$ denote this arrival process, which is another independent Bernoulli process with its generating function given by

$$1 - \rho_i^{\circ} t_{i,k} + \rho_i^{\circ} t_{i,k} z. \qquad (8)$$

Obviously, the possibility for the ith queue to be type I equals $1 - \rho_i^{\circ}$, which is the probability of the empty queue. The possibility for the ith queue to be type II is λ_i°, since this is also given by the packet department rate. Thus far we have assumed that each input queue has infinite capacity and no packet will be lost.

We then write

$$A_k = \sum_{\forall i} [(1 - \rho_i^{\circ}) x_{i,k} + \lambda_i^{\circ} y_{i,k}]. \qquad (9)$$

On the average,

$$\overline{A_k} = \sum_{\forall i} [(1 - \rho_i^{\circ}) \lambda_i^{\circ} t_{i,k} + \lambda_i^{\circ} \rho_i^{\circ} t_{i,k}] = \sum_{\forall i} \lambda_i^{\circ} t_{i,k} = \lambda_k^{\bullet}. $$
$$(10)$$

The variance is given by

$$\sigma_{A_k}^2 = \sum_{\forall i} [(1 - \rho_i^{\circ}) \lambda_i^{\circ} t_{i,k} (1 - \lambda_i^{\circ} t_{i,k}) + \lambda_i^{\circ} \rho_i^{\circ} t_{i,k} (1 - \rho_i^{\circ} t_{i,k})]$$

$$= \lambda_k^{\bullet} - \sum_{\forall i} [(1 - \rho_i^{\circ}) \lambda_i^{\circ 2} + \lambda_i^{\circ} \rho_i^{\circ 2}] t_{i,k}^2. \qquad (11)$$

Now, at a given time, assume we can identify these two queue types among the N input queues. Denote the set of type I queues by Ω_1 and the set of type II queues by Ω_2. The generating function of A_k, conditioned on Ω_1 and Ω_2, will then be expressed by

$$A_k(z|\Omega_1, \Omega_2) = \prod_{i \in \Omega_1} (1 - \lambda_i^{\circ} t_{i,k} + \lambda_i^{\circ} t_{i,k} z)$$

$$\cdot \prod_{j \in \Omega_2} (1 - \rho_j^{\circ} t_{j,k} + \rho_j^{\circ} t_{j,k} z). \qquad (12)$$

It means that each conditional process A_k consists of a sufficiently large number of independent and nonidentical Bernoulli processes. The following shows that such a conditional A_k actually forms a Poisson process in the limit, provided that the arrivals from each input to output k must be asymptotically negligible as $N \to \infty$. Denote the probability for the conditional A_k equal to j by g_j with $j = 0, 1, \cdots, n$. n is the number of Bernoulli processes in A_k conditioned on Ω_1 and Ω_2, which approaches infinity as $N \to \infty$. Let B_k represent the corresponding Poisson process, and the probability of $B_k = j$ is denoted by h_j. Here, we introduce the concept of *distance function* $d(A_k, B_k)$, defined by

$$d(A_k, B_k) = \sum_{\forall j} |g_j - h_j|. \qquad (13)$$

It is shown in [6, p. 285] that as $n \to \infty$, $d(A_k, B_k) \to 0$ implies proper convergence $A_k \to B_k$. Furthermore, for the process like conditional A_k, it has been proved in [7] that for any size of n

$$d(A_k, B_k) \leq \frac{9}{4} \left[\sum_{i \in \Omega_1} (\lambda_i^\circ t_{i,k})^2 + \sum_{i \in \Omega_2} (\rho_i^\circ t_{i,k})^2 \right]. \quad (14)$$

Define

$$c_{\max} = \max_{i \in \Omega_1, j \in \Omega_2} (\lambda_i^\circ t_{i,k}; \rho_j^\circ t_{j,k}). \quad (15)$$

$c_{\max}$ must be less than or equal to $\max_{\forall i} (\rho_i^\circ t_{i,k})$ since $\rho_i^\circ = \lambda_i^\circ \overline{s_i^\circ}$ $\overline{s_i^\circ} > 1$ and $\rho_i^\circ > \lambda_i^\circ$. We write

$$d(A_k, B_k) \leq \frac{9}{4} c_{\max} \left[\sum_{i \in \Omega_1} \lambda_i^\circ t_{i,k} + \sum_{j \in \Omega_2} \rho_j^\circ t_{j,k} \right]$$

$$< \frac{9}{4} c_{\max} \left[\sum_{\forall i} \lambda_i^\circ \overline{s_i^\circ} t_{i,k} \right]. \quad (16)$$

For a stable system, $\max_{\forall i} \overline{s_i^\circ}$ must be finite, hence

$$d(A_k, B_k) < \frac{9}{4} c_{\max} \max_{\forall i} (\overline{s_i^\circ}) \lambda_k^\bullet$$

$$= \max_{\forall i} (\rho_i^\circ t_{i,k}) \max_{\forall i} (\overline{s_i^\circ}) \lambda_k^\bullet \quad (17)$$

Note that the set $[\rho_i^\circ t_{i,k}, \forall i]$ gives the input throughput distribution on output k and $\sum_{\forall i} \Sigma \rho_i^\circ t_{i,k} < 1$. If we guarantee that the arrivals from every individual input to output k must be asymptotically negligible, we will have $\max (\rho_i^\circ t_{i,k}, \forall i) \to 0$ in the limit as $n \to \infty$. Mathematically, the probability for the traffic from individual inputs to be the same order of magnitude as the overall traffic on that output must tend to zero as $n \to \infty$. Practically, this means that the traffic on each output must come from a sufficiently large number of inputs and is not dominated by individual inputs. Under this constraint, we have $d(A_k, B_k) \to 0$ as $N \to \infty$, so the conditional A_k becomes a Poisson process. Without doubt, one can apply this result to any output under any condition. Also from (11), in the limit as $\max_{\forall i} (\rho_i^\circ t_{i,k}) \to 0$,

$$\sigma_{A_k}^2 = \lambda_k^\bullet \quad (18)$$

which is equal to the mean. One can therefore characterize A_k by a Poisson process.

The output contention process C_k, defined in (5), will be modeled as an $M/D/1$ queue. Letting $N \to \infty$, the average number of packets in contention for output k is then given by [8]

$$\overline{C_k} = \frac{\lambda_k^{\bullet 2}}{2(1 - \lambda_k^\bullet)} + \lambda_k^\bullet. \quad (19)$$

Using Little's result, the average of packet contention time destined to output k, denoted by $s_k^\bullet$, is equal to

$$\overline{s_k^\bullet} = \overline{C_k} \lambda_k^{\bullet -1} = \frac{\lambda_k^\bullet}{2(1 - \lambda_k^\bullet)} + 1 \quad (20)$$

where "1" is for the one slot transmission time through the switch fabric (which is defined as the minimum contention time).

The $M/D/1$ queue is a Markov chain with its state represented by the value of C_k, namely the number of contending packets. Let $a_{k,i} = \Pr(A_k = i)$.

$$a_{k,i} = \begin{cases} \dfrac{\lambda_k^{\bullet i} \exp(-\lambda_k^\bullet)}{i!}, & \text{for } i \geq 0; \\ 0, & \text{for } i < 0; \end{cases} \quad (21)$$

The transition probability from state j to state l in adjacent slots is then given by $a_{k,l-j+1}$. Denote by $E_k^\bullet$ the state transition matrix of this Markov chain, which has the following upper triangular form

$$E_k^\bullet = \begin{pmatrix} a_{k,0} & a_{k,1} & a_{k,2} & a_{k,3} & a_{k,4} & \cdots \\ a_{k,0} & a_{k,1} & a_{k,2} & a_{k,3} & a_{k,4} & \cdots \\ 0 & a_{k,0} & a_{k,1} & a_{k,2} & a_{k,3} & \\ 0 & 0 & a_{k,0} & a_{k,1} & a_{k,2} & \cdots \\ 0 & 0 & 0 & a_{k,0} & a_{k,1} & \cdots \\ 0 & 0 & 0 & 0 & a_{k,0} & \cdots \\ \vdots & \vdots & \vdots & \vdots & \vdots & \end{pmatrix}. \quad (22)$$

Let $p_{k,j} = \Pr(C_k = j)$, or in a row vector form

$$p_k^\bullet = (p_{k,0}, p_{k,1}, p_2, \cdots) \quad (23)$$

which can be derived from

$$p_k^\bullet (I - E_k^\bullet) = 0, \quad p_k^\bullet e^T = 1. \quad (24)$$

e^T is a column unit vector. As an engineering approximation, the size of this matrix can always be truncated. The matrix truncation will be discussed in Section V. Here we may write a numerical solution similar to the one in [5]

$$\begin{cases} p_{k,0} = (1 - \lambda_k^\bullet) \\ p_{k,1} = \dfrac{(1 - a_{k,0})}{a_{k,0}} p_{k,0} \\ p_{k,i} = \dfrac{(1 - a_{k,1})}{a_{k,0}} p_{k,i-1} - \dfrac{a_{k,i-1}}{a_{k,0}} p_{k,0} - \sum_{j=2}^{i-1} \dfrac{a_{k,j}}{a_{k,0}} p_{k,i-j} \\ \quad \text{for } i \geq 2. \end{cases}$$

$$(25)$$

An advanced fast packet switching system is often designed in such a way that the packets from each input are simultaneously and independently self-routed to the output through the switch fabric. Therefore, it is very reasonable to assume that the successful packet will be randomly selected from all contending packets destined to the same output. It means that, given the current value of C_k in a slot, the possibility for any contending packet to succeed is C_k^{-1}.

Let us characterize the contention process based on the random selection policy. By definition, the time spent in contention for each packet is equal to the transition time in the $M/D/1$ queue. Consider the contention process involved by packets from input i to output k. Under the above mentioned traffic condition, the time interval between two adjacent packets from input i to output k will be sufficiently long. It is then reasonable to assume that at the time when a tagged packet moves to the head of the input queue, the contention process will be in equilibrium. Therefore, the probability for this tagged packet to find j packets already in contention is equal to the steady-state probability $p_{k,j}$, which is the possibility for the contention process to start at the initial state $C_k = j$. Note that for $M/D/1$ queues the steady-state distribution of the queue size observed at each slot is identical to the steady-state distribution seen only at the arrival instants.

Once in the contention process, the possibility for the tagged packet to leave the contention while in state j is given by $(j + 1)^{-1}$, since there are $j+1$ contending packets (including the tagged packet itself). Therefore, the contention time will have the distribution of transition time in a transient Markov chain beginning at the initial state until absorption. The state of this transient Markov chain is C_k. Fig. 2 gives the respective state transition diagram. The initial state of this transient Markov chain is assigned according to the probability vector

137

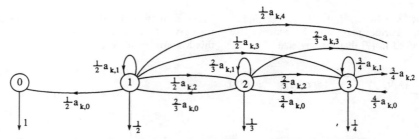

Fig. 2. State transition diagram of contention process for output k.

$p_k^\bullet$, representing the current state of C_k when the tagged packet moves to the head of the input queue. While in state j, the possibility of absorption from this chain is $(j+1)^{-1}$, and the possibility of moving to state l is given by $[1-(j+1)^{-1}]a_{k,l-j+1}$.

Obviously, the contention time is 1 if the initial state is 0, which is the transmission time through the switch in the case when no other packet contends with the tagged one. The state transition matrix for this transient Markov chain $H_k^\bullet$ is expressed by

$$H_k^\bullet = \text{diag}\,[1-(l+1)^{-1}]E_k^\bullet \qquad (26)$$

$\text{diag}\,[1-(l+1)^{-1}]$ denotes a diagonal matrix with its lth diagonal element equal to $1-(l+1)^{-1}$ for $l=0, 1, \cdots$.

From [10, p. 209] we may write the mean and the variance of the contention time by

$$\overline{s_k^\bullet} = p_k^\bullet(I-H_k^\bullet)^{-1}e^T \qquad (27)$$

and

$$\sigma_{s_k^\bullet}^2 = 2p_k^\bullet(I-H_k^\bullet)^{-2}e^T - \overline{s_k^\bullet} - \overline{s_k^\bullet}^2 \qquad (28)$$

where $I-H_k^\bullet$ is nonsingular since the process is transient. Note that in (20) we have already derived $\overline{s_k^\bullet}$ in the closed form. Rewrite (28)

$$\sigma_{s_k^\bullet}^2 = 2p_k^\bullet(I-H_k^\bullet)^{-2}e^T - \left[\frac{\lambda_k^\bullet}{2(1-\lambda_k^\bullet)}+1\right] - \left[\frac{\lambda_k^\bullet}{2(1-\lambda_k^\bullet)}+1\right]^2 \qquad (29)$$

which is in a simpler and more explicit form compared to the numerical solution derived in [5]. One may realize that the distribution of contention time $s_k^\bullet$ has phase-type, defined by $(p_k^\bullet, H_k^\bullet)$, [9].

IV. SATURATION ANALYSIS

By saturation, we mean that some input queues are saturated and packets are transmitted with infinite delay. As we will soon discover, there are two possible saturations: either all the input queues are saturated, or only individual queues are saturated. The saturation analysis will help us to derive the system maximum throughput and study the stability of the switch.

The service time at input i is given by

$$s_i^\circ = \sum_{\forall k} t_{i,k}s_k^\bullet. \qquad (30)$$

Hence,

$$\overline{s_i^\circ} = \sum_{\forall k} t_{i,k}\overline{s_k^\bullet} \qquad (31)$$

and

$$\sigma_{s_i^\circ}^2 = \sum_{\forall k} t_{i,k}\sigma_{s_k^\bullet}^2. \qquad (32)$$

Substituting (20) into (31) gives

$$\overline{s_i^\circ} = \sum_{\forall k} \frac{t_{i,k}\lambda_k^\bullet}{2(1-\lambda_k^\bullet)}+1. \qquad (33)$$

Using (6) and (33), the input utilization factor is expressed as

$$\rho_i^\circ = \lambda_i^\circ \sum_{\forall k} t_{i,k}\left[\frac{\lambda_k^\bullet}{2(1-\lambda_k^\bullet)}+1\right]. \qquad (34)$$

Let us define the system throughput by $\bar{\lambda}$,

$$\bar{\lambda} = \frac{1}{N}\sum_{\forall i}\lambda_i^\circ = \frac{1}{N}\sum_{\forall k}\lambda_k^\bullet. \qquad (35)$$

We further have

$$\lambda_i^\circ = f_i^\circ\bar{\lambda} \qquad (36)$$

and

$$f^\circ = (f_1^\circ, f_2^\circ, \cdots, f_{N-1}^\circ). \qquad (37)$$

The vector f° describes the relative distribution of traffic intensities among all the inputs. Assume f° is given. Similarly, using $\lambda_k^\bullet = \sum_{\forall i}\lambda_i^\circ t_{i,k}$,

$$\lambda_k^\bullet = \bar{\lambda}\sum_{\forall i}f_i^\circ t_{i,k} = \bar{\lambda}f_k^\bullet. \qquad (38)$$

We have

$$f^\bullet = f^\circ T \qquad (39)$$

with $f^\bullet = (f_1^\bullet, f_2^\bullet, \cdots, f_{N-1}^\bullet)$. The vector $f^\bullet$ gives the relative distribution of traffic intensities among all the outputs. Rewrite (34) as

$$\rho_i^\circ = f_i^\circ\bar{\lambda}\sum_{\forall k} t_{i,k}\left[\frac{f_k^\bullet\bar{\lambda}}{2(1-f_k^\bullet\bar{\lambda})}+1\right]. \qquad (40)$$

In order for the system to be stable, we must have $\rho_i^\circ < 1$, $\forall i$. The system saturates as $\max_{\forall i}(\rho_i^\circ) \to 1$. Suppose that the relative distribution of the traffic intensities among inputs and outputs is given, which involves fixing f°, T, and $f^\bullet$. The maximum throughput is achieved as the system approaches saturation, i.e., $\bar{\lambda} \to \bar{\lambda}_{\max}$ as $\max_{\forall i}(\rho_i^\circ) \to 1$. Let $\rho_{\max}^\circ$ be the maximum of ρ_i°, $\forall i$, and denote the input number associated with $\rho_{\max}^\circ$ by $i^\star$. We use $f_k^\bullet\bar{\lambda}/1 - f_k^\bullet\bar{\lambda} = \sum_{n=1}^\infty(f_k^\bullet\bar{\lambda})^n$ to rewrite (40)

$$\rho_i^\circ = \frac{1}{2}\sum_{n=1}^\infty\bar{\lambda}^{n+1}f_i^\circ\sum_{\forall k}t_{i,k}f_k^{\bullet n} + f_i^\circ\bar{\lambda}. \qquad (41)$$

One may find that, given f°, T, and $f^\bullet$, the relative magnitude among $\bar{\lambda}^{n+1}f_i^\circ\sum_{\forall k}t_{i,k}f_k^{\bullet n}$, $\forall i$, will not be changed by $\bar{\lambda}$ for all n's, as long as $0 < \bar{\lambda} < 1$. So does the relative magnitude among ρ_i°, $\forall i$, for $0 < \bar{\lambda} < 1$. In order to identify the input number $i^\star$, one may simply assign $\bar{\lambda} = 1/2$ and find

$$\rho_{i^\star}^\circ = \max_{\forall i}\left[\frac{1}{2}f_i^\circ\sum_{\forall k}t_{i,k}\left(\frac{\frac{1}{2}f_k^\bullet}{2\left(1-\frac{1}{2}f_k^\bullet\right)}+1\right)\right] \qquad (42)$$

which is possibly greater than one. The identification of $i^\star$ through (42) is then only dependent on f°, T, and $f^\bullet$. Once $i^\star$ is identified, the maximum throughput is found by letting $\rho_{i^\star}^\circ \to 1$ as $\bar\lambda \to \bar\lambda_{\max}$, and solving

$$1 - f_{i^\star}^\circ \bar\lambda_{\max} = f_{i^\star}^\circ \bar\lambda_{\max} \sum_{\forall k} \frac{t_{i^\star,k} f_k^\bullet \bar\lambda_{\max}}{2(1 - f_k^\bullet \bar\lambda_{\max})}. \tag{43}$$

Note that the input $i^\star$ may represent a group of inputs with equal utilization, which will all saturate at the maximum throughput.

V. QUEUEING ANALYSIS

This section analyzes the performance of each input queue. In Section III, we have found that the output contention process $s_k^\bullet$ has a phase-type distribution, characterized by $(p_k^\bullet, H_k^\bullet)$ in (23) and (26), and each service process can be viewed as independent. Focus on the ith input queue. Based on (30), we may form the distribution of the input service time s_i° by phase-type as well. Define

$$p_i^\circ = (t_{i,0} p_0^\bullet, t_{i,1} p_1^\bullet, \cdots, t_{i,N-1} p^\bullet\text{N-1}). \tag{44}$$

Using (26), we get

$$H^\circ = \begin{pmatrix} H_0^\bullet & & & & \\ & H_1^\bullet & & & \\ & & H_2^\bullet & & \\ & & & \ddots & \\ & & & & H_{N-1}^\bullet \end{pmatrix}. \tag{45}$$

The service time at input i can then be described by a phase type distribution characterized by (p_i°, H°). Therefore, each input queue is modeled by an independent Geom/PH/1 queue.

We have already derived the solution for the mean and variance of the input service time in (33) and (32). Based on the general solution of Geom/G/1 queue, we may write the solution for the mean queue length at input i, denoted by $\bar q_i$ [5],

$$\bar q_i = \frac{\lambda_i^{\circ 2} \sigma_{s_i^\circ}^2 + \rho_2^{\circ i} - \lambda_i^\circ \rho_i^\circ}{2(1 - \rho_i^\circ)} + \rho_i^\circ. \tag{46}$$

The mean waiting time is

$$\bar W_i = \frac{\lambda_i^\circ \sigma_{s_i^\circ}^2 + \rho_i^\circ \bar s_i^\circ - \rho_i^\circ}{2(1 - \rho_i^\circ)} + \bar s_i^\circ - 1. \tag{47}$$

Neuts has shown in his book [9] how to derive a simple matrix geometric solution for the distribution of $M/PH/1$ queue with finite or infinite buffer size. The extension of the distribution of Geom/PH/1 queue is straightforward. Introduce the absorption probability column vector a for an individual contention process to be absorbed from different phases, given by

$$a = (1^{-1}, 2^{-1}, 3^{-1}, \cdots)^T. \tag{48}$$

The absorption of a contention process means the end of the con-

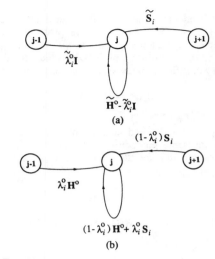

Fig. 3. (a) Transition diagram to level j in $M/PH/1$ queue. (b) Transition diagram to level j in Geom/PH/1 queue.

tention. Define

$$\Delta = \underbrace{(a^T, a^T, \cdots, a^T)^T}_{N} \tag{49}$$

and

$$S_i = \Delta p_i^\circ = \begin{pmatrix} a \\ a \\ \vdots \\ a \end{pmatrix} (t_{i,0} p_0^\bullet, t_{i,1} p_1^\bullet, \cdots, t_{i,N-1} p_{n-1}^\bullet). \tag{50}$$

Δ represents the absorption probability column vector for all the contention processes. p_i° is the initial probability row vector for the new packet to join one of the N output contention processes, starting at different phases. S_i is a rank one matrix.

Consider an $M/PH/1$ queue with its service time distribution described by $(\tilde p_k^\bullet, H^\circ)$. $\bar\lambda_i^\circ$ is the arrival rate. It is modeled as a two-dimensional continuous-time Markov process, constructed by levels and phases. The level indicates the queue size and the phase gives the service state. Fig. 3(a) shows the transition diagram to level j for $j > 1$ where all the phases on each level are superimposed and therefore all the transition rates are in matrix form. The correspondence transition diagram for the Geom/PH/1 queue is given by Fig. 3(b), which is defined in the discrete-time domain. For example, the transition from level j to level j in Fig. 3(b) contains two parts. One is the service time phase transition given no packet arrival in the current slot, which is described by $(1 - \lambda_i^\circ)H^\circ$. The other one is to initiate the next packet service time upon the present packet service time absorption, and meanwhile a new packet arrives at the queue, which is represented by $\lambda_i^\circ S_i$. Similarly, for the transitions from level $j \pm 1$ to level j. Note that both models have the structure of quasi-birth–death type [9]. The Geom/PH/1 queue has the transition matrix in the block-partitioned representation, given by

$$P_i = \begin{pmatrix} 1 - \lambda_i^\circ & \lambda_i^\circ p_i^\circ & 0 & 0 & \cdots \\ (1 - \lambda_i^\circ)\Delta & (1 - \lambda_i^\circ)H^\circ + \lambda_i^\circ S_i & \lambda_i^\circ H^\circ & 0 & \cdots \\ 0 & (1 - \lambda_i^\circ)S_i & (1 - \lambda_i^\circ)H^\circ + \lambda_i^\circ S_i & \lambda_i^\circ H^0 & \cdots \\ 0 & 0 & (1 - \lambda_i^\circ)S_i & (1 - \lambda_i^\circ)H^\circ + \lambda_i^\circ S_i & \cdots \\ 0 & 0 & 0 & (1 - \lambda_i^\circ)S_i & \cdots \\ 0 & 0 & 0 & 0 & \cdots \\ \vdots & \vdots & \vdots & \vdots & \cdots \end{pmatrix} \tag{51}$$

assuming P_i is irreducible and the buffer size is infinite. Let $\Pi_i = (\Pi_{i0}, \Pi_{i1}, \cdots, \Pi_{ij}, \cdots)$ be the steady-state probability row vector where each element Π_{ij} is a row vector except that Π_{i0} is a scale. By definition, Π_{ij} is the steady-state probability vector for the ith input queue equal to j at different phases. The balance equation is

$$\Pi_i(P_i - I) = 0 \tag{52}$$

which is normalized by $\sum_{\forall j} \Pi_{ij} e^T = 1$.

Refer to [9] for the detailed analysis of $M/PH/1$ queue. Following exactly the same procedure as described in [9, p. 84], the matrix-geometric solution for the $Geom/PH/1$ queue is readily derived as

$$\Pi_{i0} = 1 - \rho_i^\circ \tag{53}$$

$$\Pi_{ij} = (1 - \rho_i^\circ) p_i^\circ R_i (H^\circ R_i)^{j-1} \quad \text{for } j \leq 1 \tag{54}$$

The matrix R_i is given by

$$R_i = \lambda_i^\circ [I - \lambda_i^\circ e^T p_i^\circ - (1 - \lambda_i^\circ) H^\circ]^{-1}. \tag{55}$$

Denoting by q_{ij} the probability for the ith input queue equal to j, we get

$$q_{ij} = \Pi_{ij} e^T. \tag{56}$$

Obviously, the numerical analysis using this approach will become intractable if there are too many nonidentical matrices of $H_k^\bullet$. For identical matrices of $H_k^\bullet$, one can always identify them by a single matrix, which will be explained by examples given in Section VII. If one is interested only in the mean queue size performance, (46) will be a more practical solution, since the variance of input service time $\sigma_{s_\circ}^2$ can always be derived by independently calculating the variance of each individual output contention time, $\sigma_{s_k^\bullet}^2$, $\forall k$, from (29) and (32). It should be pointed out that the matrix-geometric solution given in (53) and (54) becomes rather simple with uniform traffic because all the matrices $H_k^\bullet$, $\forall k$, are identical.

Let us now truncate the size of each matrix $H_k^\bullet$ in (45), which is determined by the maximum number of contending packets allowed in C_k. Denote this number by M_k. The original contention process is modeled by a transient Markov process with $M_k \to \infty$, described in Fig. 2. As an engineering approximation, however, M_k must be taken to be finite for the evaluation of $\sigma_{s_\bullet}$ and the queue length distribution. The truncated matrix $H_k^\bullet$ is then expressed by

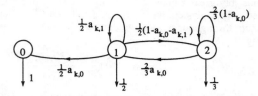

Fig. 4. State transition diagram of contention process at $M_k = 2$.

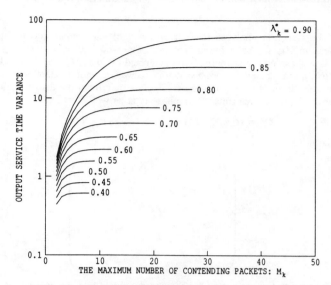

Fig. 5. Output service time variance with state truncation.

$(p_{k,0}, \cdots, p_{k,M_k-1}, p_{k,M_k})$, with $p_{k,M_k} = 1 - \sum_{j=0}^{M_k-1} p_{k,j}$ for normalization. $p_{k,j}$ for $j < M_k$ is given in (25).

From the numerical analysis point of view, it is better to keep M_k small. However, the selection of M_k is dependent of the arrival rate $\lambda_k^\bullet$. The higher the $\lambda_k^\bullet$, the larger the M_k should be. In Fig. 5 we have plotted the results of $\sigma_{s_k^\bullet}^2$ as a function of M_k for different $\lambda_k^\bullet$ values. We let $\lambda_k^\bullet$ vary from 0.4 to 0.9. M_k is initially given by 2. For each given $\lambda_k^\bullet$ in Fig. 5, the maximum of M_k is selected in such a way that the first three valid digits in $\sigma_{s_k^\bullet}^2$ will no longer be affected by the further increase of M_k. One may find that it is sufficient to choose $M_k = 10$ for $\lambda_k^\bullet \leq 0.75$ and $M_k = 20$ for $0.75 < \lambda_k^\bullet < 0.9$.

$$
H_k^\bullet =
\begin{pmatrix}
0 & 0 & \cdots & 0 & 0 \\
\frac{1}{2} a_{k,0} & \frac{1}{2} a_{k,1} & \cdots & \frac{1}{2} a_{k,M_k-1} & \frac{1}{2}\left(1 - \sum_{j=0}^{M_k-1} a_{k,j}\right) \\
0 & \frac{2}{3} a_{k,0} & \cdots & \frac{2}{3} a_{k,M_k-2} & \frac{2}{3}\left(1 - \sum_{j=0}^{M_k-2} a_{k,j}\right) \\
\vdots & \vdots & \ddots & \vdots & \vdots \\
0 & 0 & \cdots & \frac{M_k-1}{M_k} a_{k,1} & \frac{M_k-1}{M_k}\left(1 - \sum_{j=0}^{1} a_{k,j}\right) \\
0 & 0 & \cdots & \frac{M_k}{M_k+1} a_{k,0} & \frac{M_k}{M_k+1}(1 - a_{k,0})
\end{pmatrix}. \tag{57}
$$

The contention time $s_k^\bullet$ is equal to the transition time within the chain before absorptions, characterized by its transition matrix $H_k^\bullet$. The initial state of each contention process is independently assigned by $p_k^\bullet$. Fig. 4 shows the state transition diagram at $M_k = 2$. The mean and the variance of the service time, given the value of M_k, are numerically derived from (27) and (28). Also, the probability vector $p_k^\bullet$ in (27) and (28) needs to be truncated to $p_k^\bullet =$

The results in Fig. 5 exhibit that, as $\lambda_k^\bullet$ varies from 0.4 to 0.9, the value of $\sigma_{s_k^\bullet}^2$ will increase from 0.61 to 62 with sufficiently large M_k. Typically, at $\lambda_k^\bullet = 0.586$, which corresponds to the maximum throughput for the uniform traffic, we have $\sigma_{s_k^\bullet}^2 = 1.59, 1.90, 1.99,$ and 2.01, with respect to $M_k = 4, 6, 8,$ and 10.

The closed-form solution for $\bar{s}_k^\bullet$ has been given in (20). It may also be numerically derived from (27) using the truncated matrix

$H_k^\bullet$. The difference in $\bar{s}_k^\bullet$ between the truncated and the nontruncated results is plotted in Fig. 6, under the same conditions as in Fig. 5. It also displays the fast convergence of the truncated result to the nontruncated one.

Here it should be pointed out that the analysis of input queue distribution based on the Geom/PH/1 model can easily be extended to a finite buffer case. Assume that, the packet will be dropped if it arrives while the buffer is full. As we know, however, the service rate is dependent on the input rates subtracted by the dropping rates, whereas the derivation of the dropping rate is dependent on the respective service rates. One possible method is to use the original input rates to get the first estimation of the service rates, and then use these estimations to find the next dropping rates. In this way, one can always recursively calculate the next dropping rates based on the newly derived service rates, and vice versa, until the result converges.

To improve the system performance, a switch can be designed in such a way that each output will have the capacity to receive multiple packets (up to m) during a slot time interval. The advantage by this technique has been studied in [5]. The extension of the analysis for $m > 1$ is straightforward. The only change is to replace $E_k^\bullet$ and $H_k^\bullet$ in (22) and (26) by

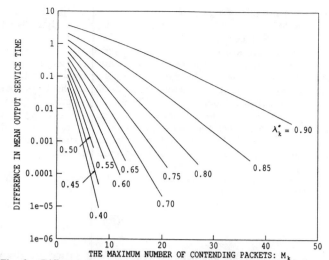

Fig. 6. Difference between truncated and nontruncated mean output service times.

$$
E_k^\bullet =
\begin{array}{c}
0 \\ \\ \vdots \\ \\ m-1 \\ \\ m \\ \\ m+1 \\ \\ \vdots \\ \\ M_k-1 \\ \\ M_k
\end{array}
\left(
\begin{array}{cccccccc}
a_{k,0} & a_{k,1} & \cdots & a_{k,M_k-m-1} & \cdots & a_{k,M_k-1} & \left(1-\sum_{j=0}^{M_k-1} a_{k,j}\right) \\ \\
\vdots & \vdots & \ddots & \vdots & \ddots & \vdots & \vdots \\ \\
a_{k,0} & a_{k,1} & \cdots & a_{k,M_k-m-1} & \cdots & a_{k,M_k-1} & \left(1-\sum_{j=0}^{M_k-1} a_{k,j}\right) \\ \\
a_{k,0} & a_{k,1} & \cdots & a_{k,M_k-m-1} & \cdots & a_{k,M_k-1} & \left(1-\sum_{j=0}^{M_k-1} a_{k,j}\right) \\ \\
0 & a_{k,0} & \cdots & a_{k,M_k-m-2} & \cdots & a_{k,M_k-2} & \left(1-\sum_{j=0}^{M_k-2} a_{k,j}\right) \\ \\
\vdots & \vdots & \ddots & \vdots & \ddots & \vdots & \vdots \\ \\
0 & 0 & \cdots & a_{k,0} & \cdots & a_{k,m} & \left(1-\sum_{j=0}^{m} a_{k,j}\right) \\ \\
0 & 0 & \cdots & 0 & \cdots & a_{k,m-1} & \left(1-\sum_{j=0}^{m-1} a_{k,j}\right)
\end{array}
\right)
\tag{58a}
$$

and

$$
H_k^\bullet = \mathrm{diag}\left[\left(\frac{l+1-m}{l+1}\right)^+\right] E_k^\bullet \tag{58b}
$$

where diag$[(l+1-m/l+1)^+]$ denotes a diagonal matrix with its lth diagonal element equal to $(l+1-m/l+1)^+$ for $l = 0, 1, \cdots, M_k$. Then, one may use (6), (27), and (31) to derive the maximum throughput, and (28), (32), and (46) for the mean queue length performance. The expressions for the queue length distribution remain unchanged.

VI. FURTHER REFINEMENT OF THE RESULTS

This section will further refine the results derived in Sections III and IV, by taking account of the effect of individual inputs on each output contention process. In practice, this means that to some extent we may relax the condition required in the above analysis, which is that the arrivals from individual inputs to each output must be asymptotically negligible as $N \to \infty$. Such a refinement will become necessary in order to get more accurate solutions, especially if the traffic has a high degree of nonuniformity, or if one wants to study a switch where N is not sufficiently large.

In Section III, we formed each contention process as an independent $M/D/1$ queue, by taking the limit of $N \to \infty$. As mentioned there, the arrival process A_k originally consists of a large number of independent and nonidentical Bernoulli processes; each represents the arrivals from an individual input. The variance of A_k is given in (11). As $N \to \infty$, the second part in (11) diminishes and A_k becomes Poisson. The $G/D/1$ queue, with the arrival process A_k defined by (9), is a more precise model for the contention process. This new model takes account of the stochastic property of each individual Bernoulli process.

As discussed in Section III, only two types of Bernoulli processes exist in A_k for the uniform traffic. We have $t_{i,k} = 1/N$, $\rho_i^\circ = \rho^\circ$ and $\lambda_k^\bullet = \bar{\lambda}$. From (11),

$$
\sigma_{A_k}^2 = \bar{\lambda} - \frac{1}{N}\left[\bar{\lambda}^2(1-\rho^\circ) + \bar{\lambda}\rho^{\circ 2}\right] \tag{59}
$$

where the second part diminishes at rate $1/N$, as $N \to \infty$. Therefore, there will be no need to consider the second part for sufficiently large N. We found, for example, that the result obtained by simulation at $N = 64$ is very close to the one derived analytically at $N \to \infty$, which means that taking $N = 64$ will be sufficient.

For the nonuniform traffic, however, all the Bernoulli processes in A_k can be different. The overall stochastic behavior of A_k is more likely to be governed by those processes at high rate only. In practice, this represents the case where the traffic on the kth output comes mainly from a group of inputs. The size of this group is finite, and needs no longer to go to infinity as $N \to \infty$. This is to say that the second part in (11) may not diminish as $N \to \infty$. The size of this group should be large enough in order to guarantee that the contention process can still be modeled as an independent queueing process.

Denote by $\bar{C}'_k$ and $\rho^{\circ}{}'_i$ the respective results derived from the $G/D/1$ model. We get

$$\bar{C}'_k = \bar{C}_k + \frac{\sigma^2_{A_k} - \lambda^{\bullet}_k}{2(1 - \lambda^{\bullet}_k)} \tag{60}$$

where $\bar{C}_k$ denotes the one derived from the $M/D/1$ model, given in (19). Applying this result to (20) and (31), and then to (6),

$$\rho^{\circ}{}'_i = \rho^{\circ}_i + \lambda^{\circ}_i \sum_{\forall k} \frac{t_{i,k}(\sigma^2_{A_k} - \lambda^{\bullet}_k)}{2\lambda^{\bullet}_k(1 - \lambda^{\bullet}_k)} \tag{61}$$

where ρ°_i is given in (34). Using (11), we obtain a more complicated expression

$$\rho^{\circ}{}'_i = \bar{\lambda} f^{\circ}_i \sum_{\forall k} t_{i,k} \left[\frac{\bar{\lambda} f^{\bullet}_k}{2(1 - \bar{\lambda} f^{\bullet}_k)} + 1 \right]$$
$$- \bar{\lambda} f^{\circ}_i \sum_{\forall j} [\bar{\lambda} f^{\circ 2}_j (1 - \rho^{\circ}{}'_j) + f^{\circ}_j \rho^{\circ}{}'^2_j] \sum_{\forall k} \frac{t^2_{j,k} t_{i,k}}{2 f^{\bullet}_k (1 - \bar{\lambda} f^{\bullet}_k)}. \tag{62}$$

For each given $\bar{\lambda}$, the value of $\rho^{\circ}{}'_i$ needs to be recursively calculated from (62). Let $\rho^{\circ}{}'_i(n)$ denote the value of $\rho^{\circ}{}'_i$ obtained by the nth iteration. One may write the following recursive equation:

$$\rho^{\circ}{}'_i(n) = \bar{\lambda} f^{\circ}_i \sum_{\forall k} t_{i,k} \left[\frac{\bar{\lambda} f^{\bullet}_k}{2(1 - \bar{\lambda} f^{\bullet}_k)} + 1 \right]$$
$$- \bar{\lambda} f^{\circ}_i \sum_{\forall j} [\bar{\lambda} f^{\circ 2}_j (1 - \rho^{\circ}{}'_j(n-1))$$
$$+ f^{\circ}_j \rho^{\circ}{}'_j(n-1)^2] \sum_{\forall k} \frac{t^2_{j,k} t_{i,k}}{2 f^{\bullet}_k (1 - \bar{\lambda} f^{\bullet}_k)} \tag{63}$$

starting with $\rho^{\circ}{}'_i(0) = \rho^{\circ}_i$, $\forall i$, given by (40). The iteration terminates once $|\rho^{\circ}{}'_i(n) - \rho^{\circ}{}'_i(n-1)|$ becomes significantly small. Only a small number of iterations are needed, as found by numerical study.

The procedure to derive the maximum throughput is exactly the same as that discussed in Section IV, i.e., $\bar{\lambda} \to \bar{\lambda}_{\max}$ as $\max_{\forall i} (\rho^{\circ}{}'_i) \to 1$. For simplicity, we still use (28) to derive $\sigma^2_{s_k^{\bullet}}$ on the basis of the $M/D/1$ model. The input queue analysis remains unchanged as in Section V. One may replace ρ°_i by $\rho^{\circ}{}'_i$ in (46) and (47) for a more accurate solution of the mean queue length. For the queue length distribution, however, we find by numerical study that it is better not to replace ρ°_i in (54), since the value of $\sum_{\forall j} q_{ij}$ would perhaps be greater than one. The reason for this is that the Geom/PH/1 model for the input queues was developed from the $M/D/1$ model for each contention process. In other words, the matrix $H^{\bullet}_k$ given in (57) is formed from the $M/D/1$ model, which has not yet been refined.

VII. Examples

Consider an $N \times N$ switch with $N = 100$. Both input and output are numbered from 0 to $N - 1$. The output address at each input is assumed to be assigned by a binomial distribution function (except

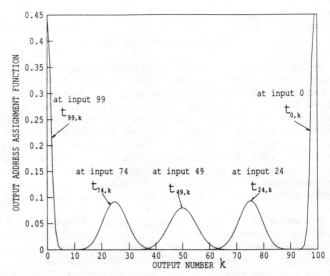

Fig. 7. Output address assignment function at inputs 0, 24, 49, 74, 99.

at inputs 0 and $N - 1$). Define for input i

$$t_{i,k} = \binom{N-1}{k} p^k_i (1 - p_i)^{N-1-k} \quad \text{for } k = 0, 1, \cdots, N-1 \tag{64}$$

except at $i = 0$ and $N - 1$. As a binomial distribution, the maximum probability is always achieved at $k = \mathrm{int}\,[(N - 1)p_i]$ where the function $\mathrm{int}\,[\,\cdot\,]$ takes the integer part of its argument. Therefore, one may select the parameter p_i in such a way that the maximum output for input i is given by $\mathrm{int}\,[(N - 1)p_i]$. The maximum output is defined as the one which carries more traffic from input i than any other output. Here we let the output $N - 1 - i$ be the maximum output for input i, $\forall i$. This leads to

$$p_i = \frac{N - 1 - i}{N - 1}. \tag{65}$$

The binomial distribution function cannot be used for inputs 0 and $N - 1$ to retain their maximum output at $N - 1$ and 0, respectively. Assume that the output assignment function for inputs 0 and $N - 1$ is a normalized Poisson distribution at rate p, defined by

$$t_{0,k} = \frac{p^{N-1-k}/(N-1-k)!}{\sum_{\forall j} p^{N-1-j}/(N-1-j)!} \tag{66}$$

and

$$t_{N-1,k} = \frac{p^k/k!}{\sum_{\forall j} p^j/j!}. \tag{67}$$

Take $p = 0.8$. Fig. 7 shows the distribution function of the output address at input 0, 24, 49, 74, and 99, respectively, which are no longer uniform. The traffic from an individual input goes mainly to 20 outputs, except for inputs 0 and 99.

The input traffic intensity is defined by (36), i.e.,

$$\Lambda^{\circ} = \bar{\lambda} f^{\circ} \tag{68}$$

where $\bar{\lambda}$ is called the system throughput and the vector f° gives the relative distribution for the input traffic intensity. The relative distribution for the output traffic intensity is described by $f^{\bullet}$ from (39).

Three possible input traffic distributions are considered: a) uni-

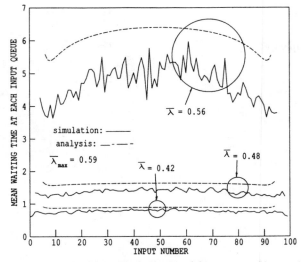

Fig. 8. Mean waiting time performance at each input queue with nonuniform input traffic distribution.

form, b) binomial, and c) normal. The following studies the mean waiting time performance at each input queue for a given input traffic distribution. Both analytical and simulation results will be examined. In simulation studies, each case requires 8 simulation runs, each run lasting 8000 slots. The confidence intervals of the simulation results are found to be generally small. In order to keep the figure clean, these intervals will not be plotted in the following figures. In numerical studies, we choose $M_k = 10$, $\forall k$. The reason for this is that, as we will soon find, the maximum output arrival rate at the system saturation, defined by $\lambda_{max}^{\bullet} = \bar{\lambda}_{max} f_{max}^{\bullet}$, is always less than 0.6, based on the three different input traffic distributions. Therefore, as described in Figs. 5 and 6, taking $M_k = 10$ will be sufficient. All the numerical solutions are refined using the technique developed in Section VI.

The first example is the uniform input traffic distribution, described by $f_i^{\circ} = 1$, $\forall i$. Using (39) one finds that the output distribution is also uniform, except for the outputs 0–4 and 95–99 which are slightly nonuniform, which is caused by the Poisson distribution of output address at inputs 0 and 99 shown in Fig. 7. Letting $f_i^{\circ} = 1$ and $f_k^{\bullet} = 1$, $\forall i, k$, from (43) we get

$$1 - \bar{\lambda}_{max} = \frac{\bar{\lambda}_{max}^2}{2(1 - \bar{\lambda}_{max})} \qquad (69)$$

which gives the maximum throughput $\bar{\lambda}_{max} = 0.586$. The same result has been derived in [3], [5] for the uniform traffic case. The analysis in [3], [5] requires $t_{i,k} = 1/N$, $\forall i, k$, but here we show such a condition is unnecessary. Based on (63), the maximum throughput is further refined to the value of 0.592. The curves plotted in Fig. 8 displays the mean waiting time performance at each input queue for $\bar{\lambda} = 0.56$, 0.48, and 0.42. Both analytical and simulation results are shown in Fig. 8. The inspection shows some degrees of overestimation in the analysis.

For the binomial input traffic distribution,

$$f_i^{\circ} = N \binom{N-1}{i} q^i (1-q)^{N-1-i}, \quad \forall i. \qquad (70)$$

Note that both $\sum_{\forall k} f_k^{\bullet}$ and $\sum_{\forall i} f_i^{\circ}$ need to be normalized to N, according to the definition given in (36) and (39). Suppose that the 39th input carries more traffic than any other input. As with (65), we can get $q = 39/99$. Fig. 9 shows both the relative input and output traffic distributions. The output distribution is somehow like binomial as well, and the 60th output is found to be the maximum output. Both input traffic and output traffic are extremely imbalanced. At input, we have $f_{min}^{\circ} < 10^{-38}$ and $f_{max}^{\circ} = 8.18$, while at output, $f_{min}^{\bullet} < 10^{-21}$

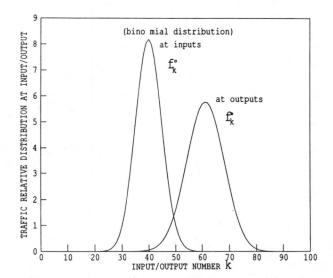

Fig. 9. Traffic relative distribution at input/output with binomial input traffic distribution.

and $f_{max}^{\bullet} = 5.77$. Due to such significant traffic imbalances, the system saturates at $\bar{\lambda}_{max} = 0.0884$, or 0.0903 after the refinement. The results of the mean waiting time are plotted in Figs. 10(a)–(c) at $\bar{\lambda} = 0.086$, 0.080, and 0.075, respectively. Similar behavior is observed for each given value of $\bar{\lambda}$.

For less imbalanced traffic, consider a normal input traffic distribution. The mean and the standard deviation of this distribution are denoted by u and σ_u. Let

$$f_i^{\circ} = \frac{c}{u\sqrt{2\pi}} e^{-(i-u)^2/\sigma_u^2} \ \forall i \qquad (71)$$

where c is the normalization factor to satisfy $\sum_{\forall i} f_i^{\circ} = N$. Taking $u = 39$ and $\sigma_u = 40$, both f° and $f^{\bullet}$ are plotted in Fig. 11. The output distribution $f^{\bullet}$ is somehow like normal distribution as well. We get $f_{min}^{\circ} = 0.42$ and $f_{max}^{\circ} = 1.29$ for input and $f_{min}^{\bullet} = 0.44$ and $f_{max}^{\bullet} = 1.27$ for output. The maximum throughput in this case is equal to $\lambda_{max} = 0.459$, or 0.470 after the refinement. The mean waiting time performance is displayed in Fig. 12 for $\bar{\lambda} = 0.44$, 0.40, and 0.30.

Throughout these examples, one may discover that the analysis gives some degree of overestimation as compared to the simulation result, especially when the system throughput is high. We attribute

143

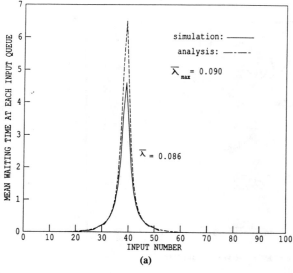

(a)

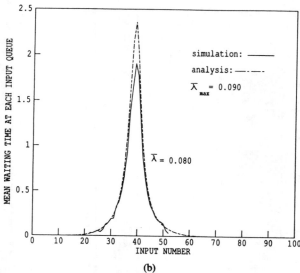

(b)

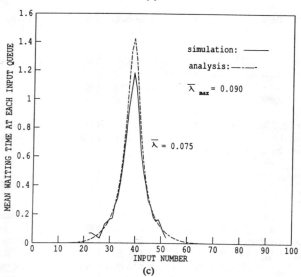

(c)

Fig. 10. (a) Mean waiting time performance at each input queue with binomial input traffic distribution at $\overline{\lambda} = 0.086$. (b) Mean waiting time performance at each input queue with binomial input traffic distribution at $\overline{\lambda} = 0.080$. (c) Mean waiting time performance at each input queue with binomial input traffic distribution at $\overline{\lambda} = 0.075$.

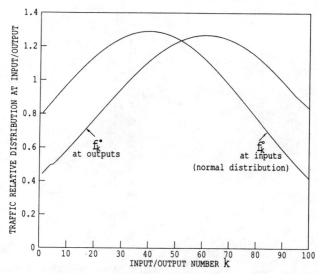

Fig. 11. Traffic relative distribution at input/output with normal input traffic distribution.

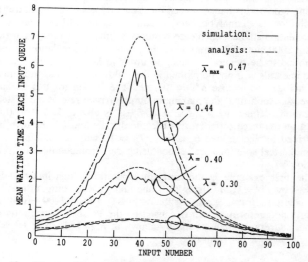

Fig. 12. Mean waiting time performance at each input queue with normal traffic distribution.

this to the fact that the analysis is based on the limit as $N \rightarrow \infty$ while the simulation takes only $N = 100$. As N increases, more packets will be involved in output contentions and therefore the system throughput will be reduced. Such phenomenon can be observed from the examples given in [5].

For the queue length distribution, we consider a simple example where all the inputs are divided into two input groups while all the outputs are divided into two output groups. The size of the groups are denoted by N_1° and N_2° for input and $N_1^\bullet$ and $N_2^\bullet$ for output. The traffic intensities within each group are identical. By proper reordering, we may write

$$f^\circ = [\underbrace{f_1^\circ, \cdots, f_1^\circ}_{N_1^\circ}, \underbrace{f_2^\circ, \cdots, f_2^\circ}_{N_2^\circ}] \tag{72}$$

and

$$f^\bullet = [\underbrace{f_1^\bullet, \cdots, f_1^\bullet}_{N_1^\bullet}, \underbrace{f_2^\bullet, \cdots, f_2^\bullet}_{N_2^\bullet}]. \tag{73}$$

Further assume that the output address assignment probability from any input to any output within a given pair of input/output groups must be identical. Denote such a probability from the ith input group to the kth output group by $t_{i,k}$. We can write

$$T = \begin{pmatrix} t_{1,1}E_{1,1} & t_{1,2}E_{1,2} \\ t_{2,1}E_{2,1} & t_{2,2}E_{2,2} \end{pmatrix} \qquad (74)$$

where $E_{i,k}$ represents an $N_i^\circ \times N_k^\bullet$ matrix with all its elements equal to 1.

By such divisions, all the contention processes within each output group are statistically identical. Similarly, all the queueing processes within each input group are also statistically identical. As discussed in Section III, each output contention time has a distribution of phase type. For the kth output group, the contention time distribution is characterized by $p_k^\bullet$ and $H_k^\bullet$, defined in (24) and (26). From (44) and (45), the service time of the input queue in the ith input group is then characterized by

$$p_1^\circ = (t_{i,1}N_1^\bullet p_1^\bullet, t_{i,2}N_2^\bullet p_2^\bullet) \qquad (75)$$

and

$$H^\circ = \begin{pmatrix} H_1^\bullet & \\ & H_2^\bullet \end{pmatrix}. \qquad (76)$$

Applying these results to (54) and (55) gives the input queue distributions. The input utilization factor for the ith input group is expressed from (40)

$$\rho_i^\circ = \bar{\lambda}f_i^\circ \sum_{k=1}^{2} t_{i,k} N_k^\bullet \left(\frac{\bar{\lambda}f_k^\bullet}{2(1 - \bar{\lambda}f_k^\bullet)} + 1 \right) \qquad (77)$$

Here we take $N_1^\circ = 30$ and $N_2^\circ = 70$ for the input and $N_1^\bullet = 70$ and $N_2^\bullet = 30$ for the output. Let the traffic intensity in input group **1** be twice as much as in input group **2**, i.e., $f_1^\circ = 2f_2^\circ$. Due to normalization, we must have $N_1^\circ f_1^\circ + N_2^\circ f_2^\circ = N$. This leads to $f_2^\circ = N/N_1^\circ + N$, which in our case leads to $f_1^\circ = 1.539$ and $f_2^\circ = 0.769$.

Denote the probability for the packet from input group i to be assigned to output group k by $\psi_{i,k}$. Assume

$$\begin{pmatrix} \psi_{1,1} & \psi_{1,2} \\ \psi_{2,1} & \psi_{2,2} \end{pmatrix} = \begin{pmatrix} 0.6 & 0.4 \\ 0.4 & 0.6 \end{pmatrix} \qquad (78)$$

By definition, $\psi_{i,k} = t_{i,k}N_k^\bullet$. In our case, we get the following output address assignment matrix:

$$\begin{pmatrix} t_{1,1} & t_{1,2} \\ t_{2,1} & t_{2,2} \end{pmatrix} = \begin{pmatrix} 0.00857 & 0.0133 \\ 0.00571 & 0.0200 \end{pmatrix} \qquad (79)$$

One may also get

$$(f_1^\bullet N_1^\bullet, f_2^\bullet N_2^\bullet) = (f_1^\circ N_1^\circ, f_2^\circ N_2^\circ) \begin{pmatrix} \psi_{1,1} & \psi_{1,2} \\ \psi_{2,1} & \psi_{2,2} \end{pmatrix} \qquad (80)$$

Letting $\rho_i^\circ \rightarrow 1$ in (77), we get the maximum throughput $\bar{\lambda}_{\max} = 0.4114$, or 0.4216 after the refinement. Since $f_{\max}^\circ \bar{\lambda}_{\max} = 0.665$, from Figs. 5 and 6 it is sufficient to choose $M_1 = M_2 = 15$ for the matrix size of $H_1^\bullet$ and $H_2^\bullet$ in (76). Fig. 13(a) and (b) show the queue length distribution at the input of the two groups, including both numerical and simulation results. The simulation results are given with 95% confidence intervals. The inspection indicates some degrees of over-estimation in the analysis. Again, this is attributed to the fact that the simulation only takes $N = 100$ while the analysis is done by letting $N \rightarrow \infty$.

Finally, the same example is considered with $m = 4$, i.e., each output has the capacity to receive up to 4 packets in every slot time interval. Using (58) in (6), (27), and (31) gives the maximum throughput $\bar{\lambda}_{\max} = 0.6482$, at which we get $\lambda_1^\circ = 0.9972$ and $\lambda_2^\circ = 0.4986$. We

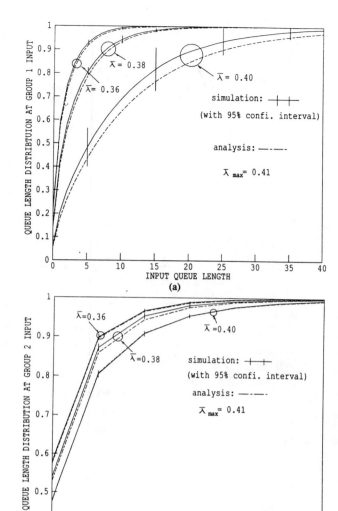

Fig. 13. (a) Queue length distribution at group 1 input for $m = 1$. Queue length distribution at group 2 input for $m = 1$. Queue length distribution at group 1 input for $m = 4$. Queue length distribution at group 2 input for $m = 4$.

must have $\lambda u_i^\circ \leq 1$, $\forall i$, at the maximum throughput. Fig. 14(a) and (b) display the queue length distribution at the input of the two groups $\bar{\lambda} = 0.63, 0.58$, and 0.54, derived by both the analysis and the simulation. A significant improvement is observed as compared to the result for $m = 1$.

VIII. Summary

This paper has analyzed the queueing performance on a nonblocking space division packet switch with nonuniform traffic. Given the condition that the traffic on each output must come from a large number of inputs and that it must not be dominated by individual inputs, our study indicates that in the limit as $N \rightarrow \infty$, the output contention process involved with each packet at the head of input queues can be viewed as an independent phase type process, and therefore each input queue is modeled by an independent Geom/PH/1 queueing process. Once the relative input traffic intensities and their output address distributions at the switch are defined, we have a general formulation for the maximum throughput of the system in saturation. The resultant expressions have clearly indicated under what conditions the input queue will saturate. These results have been further refined by considering the effect of each individual input on an output contention process. A general solution technique for the evaluation of the queueing performance was then proposed. We have also developed an engineering approximation approach for simplifying the numerical

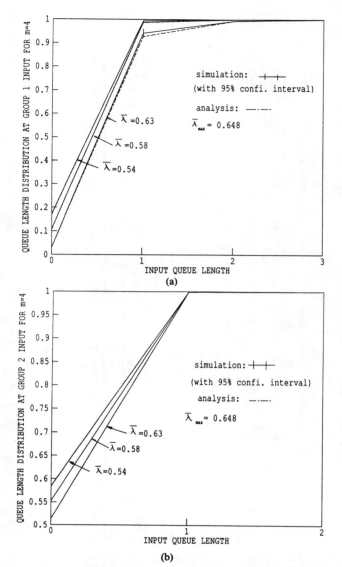

Fig. 14. Queue length distribution at group 1 input for $m = 4$. (b) Queue length distribution at group 2 input for $m = 4$.

analysis of the queue length distribution. The numerical study based on this analysis agrees well with simulation results.

REFERENCES

[1] J. S. Turner, "Design of an integrated services packet network," *IEEE Select. Areas Commun.*, vol. SAC-4, pp. 1373–1380, Nov. 1986.

[2] A. Hwang and S. Knauer, "Starlite: A wideband digital switch," in *Proc. GLOBECOM'84*, Dec. 1984, pp. 659–665.

[3] J. Y. Hui and E. Arthurs, "A broadband packet switch for integrated transport," *IEEE Select. Areas Commun.*, vol. SAC-5, pp. 1264–1273, Oct. 1987.

[4] Y.-S. Yeh, M. G. Hluchyj, and A. S. Acampora, "The knockout switch: A simple, modular architecture for high-performance packet switching," *IEEE Select. Areas Commun.*, vol. SAC-5, pp. 1274–1283, Oct. 1987.

[5] M. J. Karol, M. G. Hluchyj, and S. P. Morgan, "Input vs. output queueing on a space-division packet switch," *IEEE Trans. Commun.*, vol. COM-35, pp. 1247–1356, Dec. 1987; also presented at *GLOBECOM'86*, Dec. 1986.

[6] W. Feller, *An Introduction of Probability Theory and Its Applications*. New York; Wiley, 1971, vol. II.

[7] L. LeCam, "An approximation theorem for the Poisson binomial distribution," *Pacific J. Math.*, vol. 10, pp. 1181–1197, 1960.

[8] L. Kleinrock, *Queing Systems, Volume I: Theory*. New York: Wiley, 1975.

[9] M. F. Neuts, *Matrix-Geometric Solutions in Stochastic Models: An Algorithmic Approach*. Baltimore, MD: The Johns Hopkins University Press, 1981.

[10] R. A. Howard, *Dynamic Probabilistic Systems, Volume I: Markov Models*. New York: Wiley, 1971.

Performance Evaluation of a Batcher-Banyan Interconnection Network with Output Pooling

Achille Pattavina

Abstract—**The basic Batcher–Banyan interconnection network with input queueing is characterized by two intrinsic limitations: the head-of-line blocking severely limits the maximum switch throughput and the available bandwidth is not allocated fairly to the requesting users. A Batcher–Banyan interconnection network that overcomes these two limitations is proposed here and its performance is evaluated. The throughput limitations of the interconnection network are substantially reduced by defining pools of outputs, in which groups of outputs that offer the same service behave each as a set of servers sharing a single waiting list. Through a careful system design and a suitable priority scheme, the servers in a set can be evenly allocated to the users requesting service from the inputs of the interconnection network. For this purpose, three different solutions are proposed for the allocation of the output channels within a pool to those users requesting the same pool. Extensive computer simulation is used to evaluate the overall packet delay performance and the degree of fairness provided by each of these solutions.**

I. INTRODUCTION

SELF-ROUTING multistage interconnection networks represent one of the most interesting solutions for providing an aggregate communication flow between users on the order of hundreds of Gb/s and even more. Researchers in the area of broadband packet networking are looking at these networks with particular interest, owing to the recent progresses in CMOS integration technology that make distributed packet processing feasible [1] on a 150 Mb/s digital pipe with the 48 + 5 bytes/packet format recently agreed in CCITT for broadband networks [2]. Different classes of interconnection networks (IN's) have been proposed (e.g., [3]–[6]) for the design of a broadband packet switch with, say, 1024 inputs and 1024 outputs each carrying 150 Mb/s. Each solution can be characterized as slotted or unslotted (i.e., based on fixed- or variable-size packets), buffered or unbuffered (if conflicts arise between I/O paths sharing internal interstage links), input-, internally-, or output-queued (with reference to the place where user packets are stored when the desired I/O path is not available).

Interconnection networks with output queueing [5], [7] show the best throughput performance, but raise implementation questions related to the broadcast nature of their internal connections and the speed requirements in the output queues. Solutions using a Banyan network to route packets to specific destinations [3], [4], [8] are characterized by lower maximum throughputs, but present less implementation problems. Each switching element outlet drives only one inlet of a switching element in the following stage. It has been shown [9] that the throughput limits in a Batcher–Banyan IN can be substantially reduced if a condition of "output pooling" holds. This condition holds if: 1) the IN outputs are partitioned into disjoint sets of outputs, each output (server) in a set offering the same service as the other elements in the set; 2) each set behaves as a server pool sharing a single waiting list. For example, this condition can hold for applications such as a set of 150 Mb/s channels joining two packet switching nodes, or a set of identical read-only data bases, each accessible through an IN output. Nevertheless, a careful system design is to be carried out so that the servers in a set are evenly allocated to the requesting users. In the limiting case of each output pool including only one output, we are referring to a "classical" Batcher-Banyan IN as that described, e.g., in [8]. The problem of a fair bandwidth allocation for this particular case is examined in [10].

We refer here to the class of input-queued Batcher-Banyan IN's with output pooling and define one of these IN's to be used as the core of a packet switch for broadband networks (see, e.g., [9]). Thus, users are *packets* and servers are *output channels*. The "Batcher-Banyan" interconnection network of our packet switch is described in Section II, where different solutions for achieving fairness are proposed. These solutions differ in the implementation of the sorting function and the allocation of servers. In each solution, users requesting a service have a priority that basically indicates the "age" of the packet in the switch.

The basic solution is represented by the use of a Batcher network, as adopted in [8]. In this case, owing to the conservative order in which service requests with the same priority received on different switch inputs are served by the output servers, the available bandwidth is not fairly distributed among users. Two alternative solutions for the implementation of the sorting network are presented here that aim at providing a fair access to the output servers independent of the specific switch input port requesting the service. The first uses two "complementary" Batcher networks alternating between top-down and bottom-up sorting. The second adopts a preprocessing of the service requests entering the interconnection network, so that each input port is cyclically served first within the set of input ports requesting the same set of output servers with the same priority.

We are interested in evaluating the degree of fairness in bandwidth allocation to the single switch inputs provided by these different solutions, when a given priority range is used. The grade of service measure we adopt is the average switching delay incurred by packets received at each single switch input. It is very unlikely that analytical models can be developed to evaluate this performance measure. Thus, the only viable solution

Manuscript received September 27, 1989; revised March 14, 1990 and July 3, 1990.

A. Pattavina is with the INFOCOM Department, University "La Sapienza" of Rome, 00184, Rome, Italy.

IEEE Log Number 9039395.

Reprinted from *IEEE J. Selected Areas Commun.*, vol. 9, no. 1, pp. 95–103, January 1991.

147

is using extensive computer simulation. The simulation model developed for this problem is described in Section III. The delay figures provided by each of the proposed solutions and the corresponding degree of fairness in bandwidth allocation to the users are given and discussed in Section IV.

II. THE INTERCONNECTION NETWORK

A. Functional Description

The IN architecture of the packet switch for broadband applications [9] considered in this section is based on the self-routing principle typical of Banyan networks. Packets to be switched through the IN are preceded by a routing tag that specifies the (binary) physical address of the IN outlet they are destined for. The IN is slotted, in that all the (fixed-size) packets that are switched concurrently enter the IN at the same time. Moreover, the IN is internally unbuffered, i.e., all the IN switching elements do not include packet buffers and the packets are delayed within each switching element only for the time needed to read the packet self-routing tag. Packet buffers are only placed at the input terminations of the IN.

A set of packets offered to different inlets of a Banyan network can be switched without internal conflicts to their respective outlets if two conditions hold: i) each outlet of the Banyan network is requested by at most one of the packets offered at the inlets of the Banyan network; ii) the set of packets entering the Banyan network satisfies the so-called compactness and address-monotone conditions. Let i and j indicate the addresses of any two inlets in the Banyan network receiving each a packet with $i < j$; let $a(i)$ and $a(j)$ denote the addresses of their requested outlets in the Banyan network. The set of packets is compact if any Banyan network inlet with address k such that $i < k < j$ receives a packet too. The set of packets is increasing (decreasing) address-monotone if $a(i) < a(j)$ $(a(i) > a(j))$. Condition i) is satisfied, adopting an output allocation algorithm based on the three-phase algorithm originally described in [8]. Condition ii) is satisfied by adopting a sorting network that preprocesses the packets to be routed by the Banyan network. The sorting network receives a generic set of packets satisfying only condition i) and switches them so as to generate at its outlets a compact and address-monotone set of packets. More details about the characteristics of a basic Batcher-Banyan IN can be found, e.g., in [8].

The IN adopted here is also capable of handling groups of outputs, or *channel groups*, as output pools. This pooling operation is made possible by a hardware structure designed to operate with the three-phase allocation algorithm. In the first phase, the user requests of transmitting a user packet[1] are concurrently offered to the IN inputs, each request specifying the output pool requested. The IN allocates a specific output in the desired pool to each request, within the limits of the pool service capacity, and, in the second phase, forwards this information back to the requesting user. In the third phase, those users winner of the output channel contention actually transmit their user packet, whose transfer within the IN is conflict-free.

Fig. 1 gives the general structure of our interconnection network with N inputs, $I_0, \cdots, I_{N-1}$, and N outputs, $O_0, \cdots, O_{N-1}$, that support G output pools, numbered 0 through $G - 1$. It differs from the structure described in [9] in that it

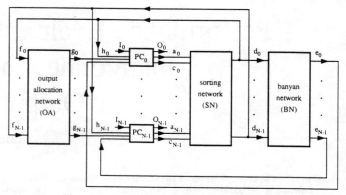

Fig. 1. Structure of the interconnection network.

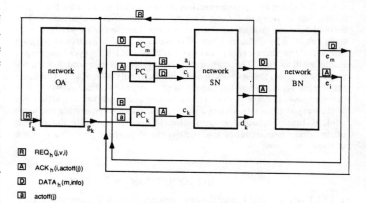

R	$REQ_h(j,v,i)$
A	$ACK_h(i,actoff(j))$
D	$DATA_h(m,info)$
a	$actoff(j)$

Fig. 2. Example of the three-phase algorithm.

adopts a pipelined version of the output allocation algorithm. The IN includes N port controllers $PC_i (i = 0, \cdots, N - 1)$, each terminating an IN input and an IN output, a sorting network SN, a Banyan network BN, and an output allocation network OA. Each port controller is provided with a FIFO packet buffer for the packets received on the associated input and waiting to be switched without conflicts through the IN. Output pool h includes R_h outputs, or packet channels, $O_i, \cdots, O_{i+R_h}$, the smallest address (O_i) representing the output pool address.[2] $R_{\max} = \max\{R_0, \cdots, R_{G-1}\}$ is the maximum output pool capacity.

The packet flow inside the interconnection network concerning an information unit to be transmitted by port controller i to output pool j is represented in Fig. 2. PC_i issues a request packet $REQ_h(j, v, i)$[3] on outlet a_i in slot h for the head-of-line packet stored in its queue. Both *destination* field j and *source* field i consist of $n = \log_2 N$ bits. The *priority* field v represents the packet time stamp and is used to give priority to older packets. If b is the number of bits used to code the priority field, v is given by $2^b - 1$ decreased by the number of slots spent by the user packet in the queue at the input port controller. All port controllers whose older user packet has been queued for more than $2^b - 1$ slots will contend for the outputs with the same priority v.

In contrast with this "global time stamp" scheme, a "local

---

[1] The information unit to be transferred from IN inputs to IN outputs is referred to as a user packet to avoid confusion with the (control) packets used in the output allocation algorithm.

[2] For the sake of simplicity, we purposely omit the adoption of a more flexible logical addressing scheme (see, e.g., [9]), since it does not affect the performance figures studied here.

[3] The fields composing a packet appear in the same order of transmission, e.g., j is the first and i is the last field transmitted in packet $REQ_h(j, v, i)$. The subscript h for a packet indicates the slot in which the reservation cycle corresponding to that packet has started.

time stamp'' approach would consist in field v specifying the user packet head-of-line age; that is, only the number of slots spent in the queue as oldest user packet are considered. The local scheme would require a smaller priority range than the global scheme, as it records delays only starting from the time a user packet becomes head-of-line. However, the global scheme is expected to be more effective in terms of packet delay since it minimizes the packet delay variations, taking into account the overall time spent within the switch by the packet.

After being sorted by network SN, packets $REQ_h(\cdot, \cdot, \cdot)$ enter concurrently network OA and a set of PC's, on the adjacent set of their inlets. Network OA computes an actual offset $actoff(j)$ for packet $REQ_h(j, v, i)$, specifying the output within the requested output pool. This $actoff(j)$ depends on the number of other requests for the same output pool concurrently received by network OA on lower address inlets. This offset is sent back in slot h to PC_i in the acknowledgment packet $ACK_h(i, actoff(j))$, by a port controller, say PC_k, acting as a relayer. Thus, the outlet used to transmit the acknowledgment packet is c_k. To minimize complexity and signal delay in network OA, we guarantee that only the $actoff(\cdot)$ values belonging to the interval $[0, R_{max} - 1]$ are assigned once per slot. Thus, the minimum number of bits coding $actoff(\cdot)$ is $d = \lceil \log_2 R_{max} \rceil$. Larger values of d can be required by the specific implementation of network OA. PC_i, which receives packet $ACK_h(i, actoff(j))$ through networks SN and BN, checks if $actoff(j) < R_j$. If this is true, in slot $h + 1$ it transmits on outlet c_i the head-of-line user packet within the data packet $DATA_h(x, info)$ to the output $x = j + actoff(j)$ ($x = m$ in Fig. 2). Otherwise, it transmits packet $REQ_{h+1}(j, v, i)$ in slot $h + 1$.

With this pipelined output allocation procedure, the transmission of packet $DATA(x, info)$ takes place one slot later than the winning outcome of the output contention. This enables PC_i to transmit in slot h the packet $DATA_{h-1}(x, info)$, which has been granted the transmission by a winning contention in slot $h - 1$, immediately after packet $ACK_h(\cdot, \cdot)$.

B. Implementation

We now explain how the interconnection network described in the preceding section can be implemented. As far as the port controller structure is concerned, we refer to the PC implementation described in [9] for a similar application. Minor modifications to this scheme are required to suite its operations to the PC functions of our IN.

Network BN is an unbuffered delta network [11]. Network OA can be implemented completely in hardware by using the running adder scheme described in [9], in which $d = 1 + \lceil \log_2 R_{max} \rceil$ and R_{max} equals a power of 2. If f_i is an inlet belonging to the complete set of inlets $\{f_k, \cdots, f_{k+l}\}$ receiving packets $REQ_h(j, \cdot, \cdot)$, then the offset on outlet g_i is

$$actoff(j) = \begin{cases} i - k & \text{if } i \leq k + R_{max} - 1 \\ R_{max} & \text{if } i > k + R_{max} - 1. \end{cases} \quad (1)$$

As far as the implementation of network SN is concerned, we describe three different solutions (Fig. 3), whose peformances will be compared in the next section.[4] In network SN, the sort-

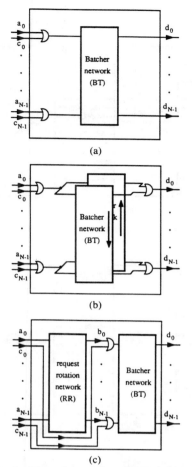

(a)

(b)

(c)

Fig. 3. Alternative solutions for network SN.

ing key is the concatenation of field j and v in packets $REQ_h(j, v, i)$, field i in packet $ACK_h(i, actoff(j))$, and field x in packet $DATA(x, info)$.

With solution (a), only one Batcher network[5] [12] is used adopting top-down sorting (i.e., the packet with the smallest sorting key emerges on the top outlet, d_0). Since network OA conservatively assigns increasing $actoff(j)$ values starting from requests received on lower address inlets f_i [see (1)], the higher address PC's are more likely to lose the output contention, when the requests for an output pool exceed the pool capacity. Thus, this solution is expected to serve PC's with a decreasing priority scale starting from the lower address PC's.

Solution (b) consists of alternating bottom-up sorting (i.e., the packet with the smallest sorting key emerges on the bottom outlet, d_{N-1}) and top-down sorting every Q slots. Logically, this corresponds to adopting two Batcher networks, one performing a top-down sorting and the other a bottom-up sorting. Physically, solutions can be envisioned that use only one Batcher network and some additional hardware to "invert" the set of ouputs every Q slots. Only one Batcher network is active in each slot and the active state is switched between the two networks every Q slots. This solution is conceived to have the output allocation process switch between two "complementary" precedence relationships among PC's every Q slots. To start output allocation from the lower address PC's with top-

---

[4] In the description of network SN, we implicitly assume that an activity bit A set to 0 precedes a packet $REQ_h(\cdot, \cdot, \cdot)$ and an idle port controller always issues an ''idle'' packet $REQ_h(\cdot, \cdot, \cdot)$ preceded by a bit $A = 1$. The activity bit is also used to collect all idle packets at high output addresses.

[5] A Batcher network includes $n(n + 1)/2$ stages of 2×2 sorting elements.

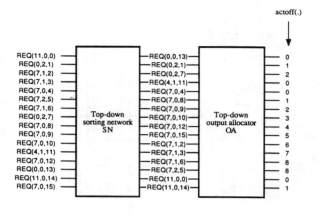

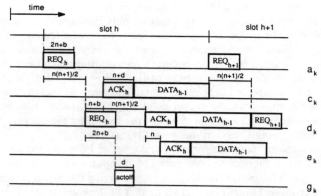

Fig. 5. Transmission periods in the interconnection network for solutions (a) and (b).

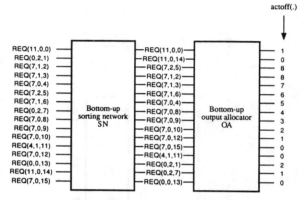

Fig. 4. Example of output allocation.

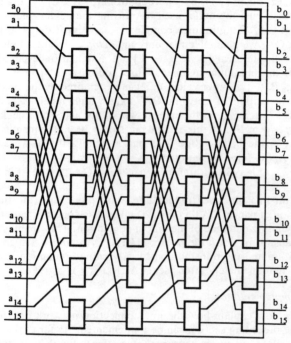

Fig. 6. Network RR.

down sorting and from higher address PC's with bottom-up sorting, the output allocation network must operate in two different modes. When top-down sorting is performed, top-down output allocation [according to (1)] is carried out. Lower address PC's are given priority within the requests for the same output pool with the same priority. When bottom-up sorting is performed, a bottom-up output allocation is performed, in which network OA would assign increasing actual offsets starting from the higher address port f_i receiving a packet $REQ_h(j, \cdot, \cdot)$. That is, if f_i is an inlet belonging to the complete set of inlets $\{f_k, \cdots, f_{k+l}\}$ receiving packets $REQ_h(j, \cdot, \cdot)$, then the offset on outlet g_i with bottom-up output allocation is

$$actoff(j) = \begin{cases} k + l - i & \text{if } i \geq k + l - (R_{max} - 1) \\ R_{max} & \text{if } i < k + l - (R_{max} - 1). \end{cases}$$

(2)

With solution (b), the sorting key of packets $REQ_h(j, v, i)$ has to be limited to the concatenation of fields j and v, otherwise lower address PC's would always be served first in the output allocation procedure within each priority level, independent of the type of sorting operated. Thus, according to (2), higher address PC's are given priority within the requests for the same output pool with the same priority. An example is given in Fig. 4 to show these different output allocation procedures for $N = 16$ and $R_{max} = 8$.

If switching inefficiency is to be avoided, one slot does not have to last more than the time required to receive on input I_i a user packet l_{info} bits long. Fig. 5 shows the transmission periods of the different packets in the outlets of port controllers

and networks SN, BN, and OA for both solutions (a) and (b). From this figure, we can compute the speed-up required to packet channels within the interconnection network compared to the channel rate on input channels I_i, in order to perform the output allocation algorithm. Since the slot length is given by $(n(n + 1)/2) + n + b + n + d + n + l_{info} = l_{over} + l_{info}$ bits (the first n bits in packet $DATA(\cdot, \cdot)$ representing routing overhead), the speed-up within the switch is equal to (l_{over}/l_{info}) %.

In solution (c), request packets cross a request rotation network RR before being sorted by the Batcher network. Packets $ACK_h(\cdot, \cdot)$ and $DATA_{h-1}(\cdot, \cdot)$ enter the Batcher network directly. Network RR performs a cyclic rotation of the inputs so that inlet $a_i(i = 0, \cdots, N - 1)$ is connected to outlet $b_{(i+h) \bmod N}$ in slot h. An omega network [13], which consists of $n = \log_2 N$ stages of 2×2 switching elements, is able to perform this rotation. An example is given in Fig. 6 for $N = 16$. With this solution, each PC accesses cyclically all the input ports of the Batcher network, so that all PC's are expected to

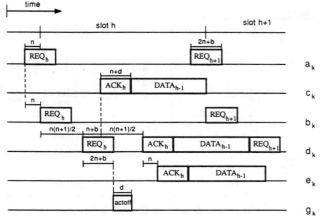

Fig. 7. Transmission periods in the interconnection network for solution (c).

receive an even share of the available bandwidth. Also with solution (c), the sorting key of packets $REQ(j, v, i)$ has to be limited to the concatenation of fields j and v, otherwise lower-address PC's would always be served first in the output allocation procedure within each priority level, independent of the status of network RR. To avoid an increase of the speed-up within the switch owing to the adoption of network RR, packet transmissions are further pipelined by allowing packet $REQ(\cdot, \cdot, \cdot)$ to be transmitted n bits before the end of transmission of packet $DATA(\cdot, \cdot)$ by the port controller (see Fig. 7). This anticipation compensates the additional delay introduced by network RR, which has been assumed equal to n bits, so that all solutions require the same speed-up computed above.

III. THE SIMULATION MODEL

The performance of the interconnection network described in Section II has been evaluated by computer simulation. We did not use existing general-purpose simulation tools, since the particular structure of the system to be modeled suggested us to write a code specifically suited for our problem. The simulation model was written in Fortran, whose drawback of a static memory allocation did not represent a problem as far as the speed of the simulation is concerned. The simulation model required a total amount of memory (program + data) on the order of 2 Mbytes that is easily available in any medium-to-large size computer.

The system to be studied is slotted in that all the events, in particular packet arrival at the input queues and packet transmission through the IN, take place only at discrete time-instants, specifically at the end (or the beginning) of a slot. The slotted nature of the system and the particular traffic type (Bernoulli) offered to the IN (see Section IV), with packet arrivals at an input queue in a slot independent of any other preceding or concurrent packet arrival in the IN, could suggest the adoption of a time-driven simulation model. The simulation clock in this model is incremented by a fixed-length time interval (the slot) and, at each clock time, we check if each of the potential events actually occurs. Using a general-purpose language implies that all the data structures have to be built from scratch. In general, a time-driven model is simpler to be implemented (and sometimes faster to run) than an event-driven simulation model, which requires the handling of a calendar of events.

We are interested in evaluating the IN performance for an offered traffic p per channel in the range $0 < p < 1$. With the time-driven simulation model, the average number of executions of the random number generation routine would be $N(1 + p)$ times per slot for determining the packet arrival event and, in case of arrival, the requested IN output. For this kind of system, an event-driven simulation model can be more efficient than a time-driven model, given that we can reduce the number of calls to the random number generation routine, which can be time-consuming. In our problem, it has been possible to reduce the number of calls to that routine to $2Np$, since the Bernoulli arrival process can be obtained through a geometric arrival process in which the random number of slots between two packet arrivals is computed through standard inverse transformation techniques (see, e.g., [14]). In such a way, the random number generation routine is only executed twice per packet arrival: once for determining the IN output for the arrived packet and once for scheduling the next packet arrival on the same input.

Thus, by disregarding events used for debugging or service, the calendar of events of our event-driven simulation model is dimensioned to contain up to $N + 5$ events: the packet arrival (one per input), the conflict resolution and packet transmission (occurring every slot), the end of the transient state, the end of each subrun, the end of the steady state, and the end of the experiment.

In general, an event-driven simulation model reduces the CPU time required by an experiment with a given total number X of transmitted packets compared to a time-driven model. This is not only due to the smaller number of calls to the random number generation routine in the former model. The latter model, in fact, requires that the set of all possible events is carefully identified *a priori* and their actual occurrence is tested with a given periodicity (the simplest solution is a period of 1 slot). The specific amount of CPU time that can be saved is strictly dependent on the characteristics and software structure of the simulation model, e.g., programming language used, type of calls between software modules, percentage of CPU time spent in tasks independent from the simulation model adopted (the conflict resolution for the output ports, the updating of the current statistics, the analysis of the steady-state results, etc.).

Out of the different methods to obtain the estimates of our grade-of-service figure, which is the average packet delay, we selected the method of batch means [14]. With this approach, only one simulation run is carried out to determine the average packet delay for the IN configuration currently examined. In particular, after discarding the initial data corresponding to the transient state, the run is considered as the sequence of L adjacent subruns of equal length. The packet delay values observed in each single subrun will give L "independent" delay estimates. Through a classical approach based on the t-statistics [14], these estimates provide the confidence interval of the average packet delay observed in the whole run. In our model, we have selected $L = 50$ (smaller values could have also been chosen as well). Thus, two main questions had to be answered: how to determine the end of the transient state and the duration of the steady-state experiment.

In order to obtain steady-state conditions for any offered load p, the end of the transient state criterion could not be chosen based on a fixed amount of slots. In fact, IN's with lower loads tend to reach a steady-state condition in a longer time period than IN's with a higher load. Thus, a reasonable criterion has seemed to be the stability conditions reached by the IN throughput $r(n_p) = n_p/Nt_c$, in which n_p is the current total number of packets transmitted and t_c is the current value (slots) of the simulation clock. If $[|r(n_p - 1) - r(n_p)|/r(n_p)] < k_1$ for k_2 packets transmitted consecutively, the transient state is considered over and the steady state begins. Normalizing with respect to the current throughput in this criterion aims at providing independence of the throughput stability conditions from the particular level of the offered load. The values of the two parameters selected in our experiment are $k_1 = 0.001$ and $k_2 = 50000$ which, for an IN with $N = 256$, resulted in a transient state length of about 2000 slots for $p = 0.10$ and about 250 slots for $p = 0.90$.

We are interested in evaluating delay figures for different offered loads with comparable confidence intervals. Since the confidence intervals width depends on the number of packets transmitted in the steady state, our aim is to run simulation experiments with different offered load so that an equal number of packets are transmitted in the steady state of each experiment. If we set the end of the steady-state condition based on the number of packets transmitted, the data gathered from the transient state become useful in determining the duration in slots of the steady state. In particular, the effective duration of the transient state and the number of packets transmitted in that state enables us to estimate, through a simple proportion, the duration of the steady state. In principle, this duration could also be obtained *a priori* from the offered load p, but this is true only if packet loss does not take place. When packet loss occurs, whose amount is not *a priori* known, our methods gives the expected result. The duration of each subrun is also determined at the end of the transient state and set equal to $1/L$ of the steady-state duration, suitably rounded. For each network configuration and load, about 10^6 packets were transmitted in the steady state. The CPU time for each run was on the order of 1 h for a computer of the SUN 3/60 class.

In general, data from the transient state can also be useful to tune the measurements in the steady state. For example, the evaluation of a given percentile of the packet delay can be done in general by computing the histogram of the packet delay with a given number of delay intervals, say 100, whose length is to be determined. If the maximum packet delay is not known *a priori*, the total span of the x-axis in the histogram should be set large enough to avoid occurrence of a packet delay larger than the estimated maximum that could invalidate the measure. The larger the estimated maximum, the smaller the precision in the percentile evaluation. A reasonable solution could be to set the upper bound of the histogram equal to the maximum delay detected during the transient state multiplied by a proper ''security factor.'' A good rule of thumb in our case is to set the security factor to, say, 2–3, so as to never exceed in the steady state the estimated maximum delay.

IV. PERFORMANCE EVALUATION

We considered an IN with $N = 256$, an input buffer of 100 packets per PC, a global time stamp scheme, and equal output pool capacities with values $R_h = R = 1, 2, 8, 32$. The performance is expressed in terms of packet delay, defined as the number of slots between the packet arrival at the input queue and the slot in which the corresponding request packet is a winner of the output contention. With this definition that makes easier the presentation of the results, the minimum delay is conventionally set to 1 slot, even if the pipelined packet transmission implies a minimum delay of 2 slots. The performance figure that has been evaluated is the averaged packet delay, T (slots).

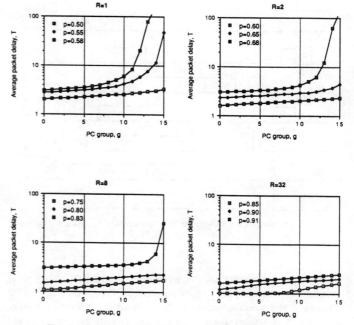

Fig. 8. Average packet delay for solution (a) with $b = 2$.

The averaging is done either on the whole set of PC's or on a group $g(g = 0, 1, \cdots, 15)$ of 16 adjacent PC $\{ PC_{16g}, \cdots, PC_{16g + 15} \}$. For example, group $g = 1$ includes port controllers PC_{16} to PC_{31}. The delay values plotted in the figures of this section have a 95% confidence interval smaller than 20% of the plotted values.

A random traffic modeling was assumed. Let p indicate the probability that a user packet is received by port controller PC_i on input I_i in a slot and the packet receipt events be all independent of one another. A random selection of the destination output pool also takes place for each user packet. Under these traffic assumptions, the maximum channel throughput r_{max} for the output pool capacities considered are: $r_{max} = 0.586, 0.687, 0.831, 0.913$ for $R = 1, 2, 8, 32$ channels, respectively.

Fig. 8 shows the average packet delay measured for each PC group for solution (a) with four priority levels ($b = 2$ b) when the switch is close to saturation conditions. The average packet delay increases slightly with the PC group index for nonasymptotic throughputs. This is due to the conservative way in which contentions of requests with the same priority for the same output pool are solved by the interconnection network. Lower address PC's always win. For small output pool capacities, a much faster increase of the averaged packet delay occurs for higher address port controllers close to asymptotic throughputs. No such increase takes place for $R = 32$ channels. This fast delay increase occurs when the priority range is likely to be always saturated by PC's. In this situation, the higher address PC's will keep losing across consecutive slots, since experiencing more losses in the output contention does not increase their probability of winning the next slots. The priority range required to avoid such a fast increase results to be function of the output pool capacity. In fact, the larger the output pool capacity, the larger the asymptotic channel throughput, and, correspondingly, the smaller the averaged queue length. For the switch under consideration, a priority range coded by $b = 8$ b has been found to be enough to provide for asymptotic throughputs: the light increase with the PC group index typical of low offered loads.

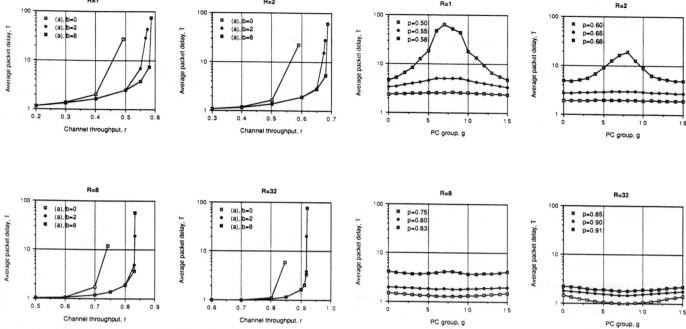

Fig. 9. Average packet delay for solution (a).

Fig. 10. Average packet delay for solution (b) with $b = 2$ and $Q = 1$.

The packet delay averaged over all PC's for solution (a) is given in Fig. 9 as a function of the channel throughput r for the priority ranges corresponding to $b = 0, 2, 8$ b. If the switch supports at least eight outputs per pool, four priority levels ($b = 2$ b) are enough to provide an average delay closest to the minimum obtainable value for almost any channel throughput. If smaller output pools are supported, more bits should be used for field v in order to obtain a comparable delay performance.

For solution (b) with $b = 2$ b and $Q = 1$ slot, Fig. 10 gives the same performance figure as Fig. 8. For very small output pools, e.g., $R = 1, 2$ channels, the port controllers with lowest and highest address ("peripheral" PC's) give the best performance for throughputs close to the asymptotic values, while the PC's with intermediate address ("central" PC's) are the most penalized. Indeed, the maximum throughput r_{max} characterizing these R values enables the peripheral PC's to empty their buffer quickly. Owing to the priority range saturation, they almost always win the output contention in odd (or even) slots, when they are given the highest service priority. Therefore, the probability of winning the contention in even (or odd) slots is larger than the probability of a packet arrival in two consecutive slots.

For larger values of R, r_{max} increases so that in each slot more PC's can be served (e.g., about 90% for $R = 32$ channels). So, the central PC's give the best performance, as they are served with a very high probability both in odd and even slots. The queue length is more likely to grow in the most peripheral PC's, which receive the worst service within the same priority level. Intermediate values of R, e.g., $R = 8$ channels, are characterized by a mix of the two phenomena described above for small and large output pools.

When the priority range is increased by assuming $b = 8$ b in solution (b), a "flat" delay curve is obtained for small R values, as no priority range saturation takes place (see Fig. 11). No significant change occurs for large output pools.

Within solution (b), to decrease the network inversion rate $1/Q$ does not seem to provide significant improvements in terms of fairness. Fig. 12 compares the switch performance for the

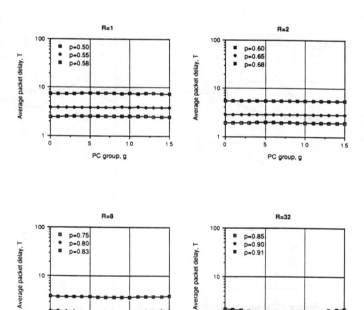

Fig. 11. Average packet delay for solution (b) with $b = 8$ and $Q = 1$.

sorting inversion rates $Q = 1, 4, 8$ slots when $b = 2$ b and the channel throughput is close to the asymptotic value. It shows that a performance improvement of central PC's is obtained by increasing Q. Indeed, maintaining the same priority scale for more consecutive slots results in an increased probability of serving PC's more distant from the peripheral PC's served first. Therefore, differently from peripheral PC's, central PC's can better take advantage of this fact with both top-down and bottom-up sorting.

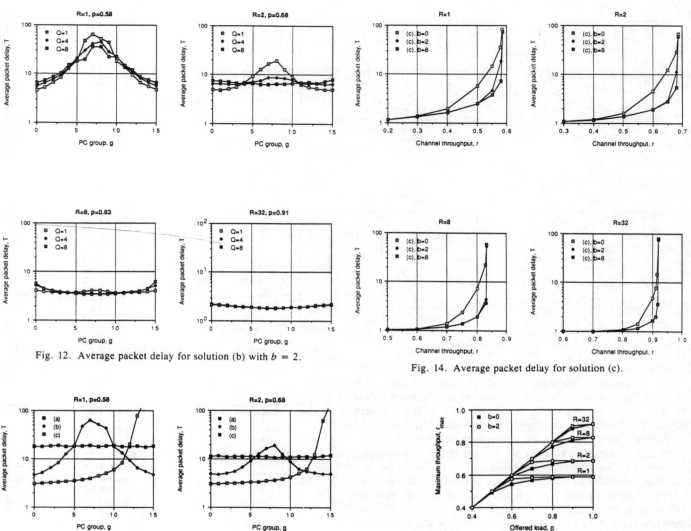

Fig. 12. Average packet delay for solution (b) with $b = 2$.

Fig. 14. Average packet delay for solution (c).

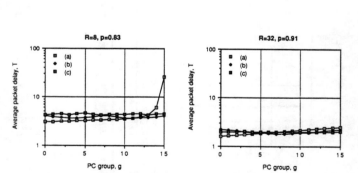

Fig. 13. Average packet delay with $b = 2$.

Fig. 15. Maximum throughput for solution (a).

As one might expect, solution (c) provides, within the precision limits of our simulation tool, equal average packet delays across the different PC groups, independent of the channel throughput. By considering asymptotic throughput values, the average packet delays versus PC group for solutions (a)–(c) are compared in Fig. 13 for $b = 2$ b. Solution (c) provides a "flat" curve for any R, whereas solution (a) shows a monotonically increasing curve whose derivative decreases for larger channel groups, and solution (b) provides a more or less concave curve.

The cases $b = 1, 2, 8$ are compared in Fig. 14 for solution (c). The same comments done for the corresponding curves for solution (a) (Fig. 9) apply in this case. Nevertheless, for an offered load below the asymptotic throughput, the throughput value in solution (c) is independent of the priority range, whereas in solution (a), e.g., only one priority level results in throughputs lower than in the case $b \geq 2$.

This behavior in solution (a) is due to the fact that a lower priority range is more likely to cause packet overflow in some queues (in higher address PC's), whereas others (in the lower address PC's) are almost always empty. Fig. 15 gives a graphical representation of this fact showing the maximum throughput r_{max} for solution (a) with $0.4 \leq p \leq 1$. The asymptotic throughput is reached with offered loads very close to $p = 1$ for $b = 0$, whereas it is approached much faster as p increases for $b = 2$. For any b, the throughput r is independent from the priority range if we offer a load $p = 1$. In particular, the theoretical value $r_{max} = 0.586$ [15] is obtained from our computer simulation for an IN with $R = 1$ and $p = 1$.

Finally, independently from fairness problems, Fig. 16 shows that solutions (a)–(c) provide a comparable average performance, provided that a certain priority range is guaranteed, e.g., with $b = 2$, as assumed in Fig. 16.

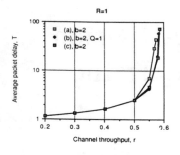
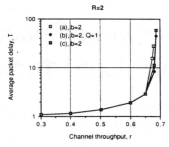

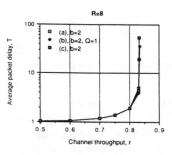
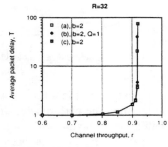

Fig. 16. Average packet delay with $b = 2$.

V. Conclusions

We have examined different solutions for the implementation of the sorting and output allocation function in Batcher-Banyan interconnection networks with output pooling that adopt the three-phase algorithm as a contention resolution scheme. Both solutions of one and two Batcher networks require a minimum range of priority levels to guarantee against uncontrolled delay increase in some PC's, when the average channel throughput approaches the saturation value. If this minimum range is actually provided, the average packet delay versus the PC group, g, is an almost linear increasing function with low derivative in solution (a), while it is either an almost flat (for small R) or a slightly concave (for large R) curve in solution (b). Nevertheless, in both cases the ratio between the largest and the smallest average delay values in any two PC's is always less than 2. If, according to system specifications, we can neither provide the minimum priority range, nor accept such an unfairness degree, the most expensive solution, (c), can be adopted. This solution

guarantees fairness in bandwidth allocation among PC's and the the priority range would be chosen just to set the average packet delay to be expected. It has been shown that 2 b are sufficient in solution (c) to provide an almost ideal throughput/delay performance.

Acknowledgment

The author would like to thank the anonymous reviewers for their constructive criticism and valuable comments.

References

[1] H. J. Chao, T. J. Robe, and L. S. Smoot, "A 140 Mbit/s CMOS LSI framer chip for a broadband ISDN local access system," *IEEE J. Solid State Circuits*, vol. 23, no. 1 pp. 133–141, Feb. 1988.
[2] CCITT, Draft Recommendation I.361, "ATM layer specification for B-ISDN," Geneva, Jan. 1990.
[3] A. Huang and S. Knauer, "Starlite: A wideband digital switch," in *Proc. GLOBECOM*, Atlanta, GA, Nov. 1984, pp. 121–125.
[4] J. Turner, "Design of a broadcast packet network," in *Proc. INFOCOM*, Miami, FL, Apr. 1986, pp. 667–675.
[5] K. Y. Eng, M. G. Hluchyj, and Y. S. Yeh, "A knockout switch for variable length packets," in *Proc. Int. Conf. Commun.*, Seattle, WA, June 1987, pp. 794–799.
[6] P. E. White, J. Y. Hui, M. Decina, R. Yatsuboshi, Eds. Special Issue on Switching Systems for Broadband Networks, *IEEE J. Select. Areas Commun.*, vol. SAC-5, no. 8, Oct. 1987.
[7] Y. S. Yeh, M. G. Hluchyj, and A. S. Acampora, "The knockout switch: A simple, modular architecture for high-performance packet switching," *IEEE J. Select. Areas Commun.*, vol. SAC-5, no. 8, pp. 1274–1283, Oct. 1987.
[8] J. Hui and E. Arthurs, "A broadband packet switch for integrated transport," *IEEE J. Select. Areas Commun.*, vol. SAC-5, no. 8, pp. 1264–1273, Oct. 1987.
[9] A. Pattavina, "Multichannel bandwidth allocation in a broadband packet switch," *IEEE J. Select. Areas Commun.*, vol. 6, no. 9, pp. 1489–1499, Dec. 1988.
[10] A. Pattavina, "Fairness in a broadband packet switch," in *Proc. Int. Conf. Commun.*, Boston, MA, June 1989, pp. 404–409.
[11] R. J. McMillan, "A survey of interconnection networks," in *Proc. GLOBECOM*, Atlanta, GA, Nov. 1984, pp. 105–113.
[12] K. E. Batcher, "Sorting networks and their applications," in *AFIPS Proc. Spring Joint Comput. Conf.*, 1968, pp. 307–314.
[13] D. H. Lawrie, "Access and alignment of data in an array processor," *IEEE Trans. Comput.*, vol. C-24, no. 12, pp. 1145–1155, Dec. 1975.
[14] S. Lavenberg, Ed., *Computer Performance Modeling Handbook*. London: Academic Press, 1983.
[15] M. G. Hluchyj and K. J. Karol, "Queueing in high-performance packet switching," *IEEE J. Select. Areas Commun.*, vol. 6, no. 9, pp. 1587–1597, Dec. 1988.

Performance of a Nonblocking Space-Division Packet Switch with Correlated Input Traffic

San-Qi Li, *Member, IEEE*

Abstract—This paper studies the performance of a nonblocking space-division packet switch in a correlated input traffic environment. In constructing the input traffic model, we consider that each input is a TDM link connecting to multiple sources. Every source on a link supports one call at a time. Each call experiences the alternation of ON and OFF periods, and generates packets periodically while in ON period. The stochastic property of each call does not have to be identical. Packets from each individual call are destined to the same output. The output address of each call is assumed to be uniformly assigned in random. In this multimedia correlated input traffic environment, we derive both upper and lower bounds of the maximum throughput at system saturation. Our study indicates that, if the source access rate is substantially lower than the link transmission rate, the effect of input traffic correlation on the output contentions can generally be ignored. Note that the large disparity between link transmission rate and source access rate always exist in most practical systems. Therefore, not only the input traffic correlation can be neglected in the maximum throughput analysis, but also the analysis of each input queue becomes separable from the rest of the switch. The service time of each input queue in the separate analysis is characterized by an independent renewal process. The same study is carried out with nonuniform call address assignment.

I. INTRODUCTION

VARIOUS advanced, space-division, fast packet switching architectures are currently under development for future broad-band integrated services [1]–[6]. This paper considers a nonblocking space-division packet switching system with N input links and N output links, shown in Fig. 1. Examples of this switch include crossbar, Batcher–banyan [3] and knockout [4]. Packets from each input link are simultaneously and independently self-routed to the output links. Each link is logically partitioned into time slots of duration equal to one packet transmission time through the switch fabric. All packets are of equal size. The operation of the switch is then synchronized in time slots. The capacity of a nonblocking switch fabric is defined by the maximum number of packets say n which can be simultaneously routed from multiple input links to each output link. No packets are queued inside the switch. The input buffering is designed to store packet arrivals which momentarily cannot be routed through the switch fabric. The output buffering is required if more than one packet arrives

Paper approved by the Editor for Wide Area Networks of the IEEE Communications Society. Manuscript received August 6, 1989; revised June 16, 1990 and February 18, 1991. This paper was presented in part at IEEE GLOBECOM '89, Dallas, TX, November 27–30, 1989.

The author is with the Department of Electrical and Computer Engineering, The University of Texas at Austin, Austin, TX 78712.

IEEE Log Number 9105170.

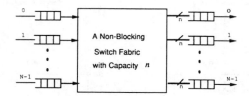

Fig. 1. Structure of a nonblocking space division packet switch.

at the output port in a time slot. The buffer size is assumed to be infinite. The performance of such a switch is measured by both system maximum throughput and queueing statistics.

Analysis, so far, has been done with two assumptions. One is to assume that the packet interarrival process at each input is formed by an independent geometric process [3], [5], [7], [8]. Such a traffic model is used to characterize the interactive behavior of data arrivals in computer networks; there is no time correlation between arrivals at each input. The second assumption is that the output address of each packet is independently assigned. Under these two assumptions, the study in [7] shows that for sufficiently large N's each output contention, involved by packets at the head of input queues, can be characterized by an independent phase type process, and therefore each input queue is modeled by an independent $Geom/PH/1$ queueing process. With this modeling, in [9], [10] we also studied the trunk grouping effect as well as the performance tradeoffs in input/output buffer design, where the packet loss performance with finite input buffer is obtained. The first assumption has recently been loosened to some extent in [11] by using a single underlying Markov chain to modulate all the geometric interarrival processes, which is used to characterize the time variation of input traffic flow.

Future fast packet switching networks, however, are expected to carry much more stream-like traffic, such as voice, video and file transfer. For a call of stream-like traffic, there are high-positive correlations in its adjacent packet arrivals, and all these packets must be destined to the same output link. The above two assumptions are therefore no longer valid. Many studies, focussed on a single transmission link, have shown the significant negative effect of traffic correlated on queueing performance [12], [13]. It is crucial then for system designers to understand the behavior of fast packet switch in such a correlated input traffic environment. This paper attempts to obtain some insight into the behavior of the switch by removing the above two traffic assumptions. To the author's knowledge, no such studies are yet available in the literature.

Reprinted from *IEEE Trans. Commun.*, vol. 40, no. 1, pp. 97–108, January 1992.

In constructing the multimedia correlated input traffic model, we consider that each input is a TDM link connecting to multiple sources. Every source on a link supports one call at a time. Each call is characterized by an independent two-state Markov chain with alternation of ON and OFF periods. Packets from each call will have the same output address. Once a call terminates, the next call will arrive with its output address uniformly assigned. Based on this construction, we are able to obtain both the upper and lower bounds of the maximum throughput at system saturation.

In practice, each link is likely to be shared by a large number of sources. Recognizing the existing large disparity between link transmission rate and source access rate, our study indicates that the effect of input traffic correlation on the output contentions can generally be ignored. Therefore, not only the input traffic correlation can be neglected in the maximum throughput analysis, but also the analysis of each input queue becomes separable from the rest of the switch. By the separate analysis, the service time of each input queue can then be viewed as a renewal process, irrelevant to the traffic correlations. Without loss of the accuracy, one can further simplify the analysis by making the fluid flow approximation to the service time distribution. Finally, the same technique is used to study the correlated input traffic with nonuniform call address distributions.

II. SUMMARY OF THE MAIN RESULTS

We here outline the main results of each following section.

Section II constructs a correlated input source model, which characterizes both the correlation of traffic arrivals and the dependence of packet address assignment. Each source has four parameters:

ϕ_{on}: control the number of successive packets in a burst,

ϕ_{off}: control the number of successive time slots in a silence,

γ: control the frequency of changing to a new output address, and

M: measure the number of slots between successive packets within a burst.

The size of the switch is described by N and the capacity of the switch is defined by n. Two fundamental questions then arise with respect to correlated input traffic: i) What is the maximum throughput λ_{max} of the switch, and ii) what is the input queue performance for throughput $\lambda < \lambda_{max}$?

Section IV deals with throughput. An analytic solution is obtained when $M = 1$ and $N \to \infty$, for arbitrary values of n, ϕ_{on}, ϕ_{off}, and γ. For $n = 1$, which corresponds to pure input buffering, we have

$$0.500 \leq \lambda_{max} \leq 0.586. \tag{1}$$

The lower bound represents a "worst cast" occurred at $M = 1$ where the input correlation is maximized; the upper bound gives a "best case" performance taken as $M \to \infty$ at which the correlation is removed. As n increases, which corresponds to both input and output bufferings, the two bounds move quickly toward 1. For $n = 4$, we obtain

$$0.993 \leq \lambda_{max} \leq 0.996, \tag{2}$$

which indicates that choosing $n = 4$ is sufficient for system to exploit the maximum utilization of a nonblocking switch fabric. As one can see, the correlation effects rarely make much difference in throughput performance. In fact, they become negligible when the ratio of link transmission rate to source access rate, measured by M, is sufficiently large. For instance, one can hardly observe any difference in throughput when $M \geq 40$ for $n = 1$.

Section V studies input queue performance for the case $n = 1$ and $M \geq 40$. The same technique can be used for $n > 1$ with less constraint on M. The constraint on M is made to neglect correlation effect on input packet service time (which is defined through output contentions). Hence, the analysis of each input queue can be separated from rest of inputs as $N \to \infty$. An equivalent model of each input queue is then derived and is analytically solved with fluid-flow approximation. The approximate solution compares favorably with simulations of both the equivalent model and the original switch. The effect of γ on the performance is found to be insignificant.

Section VI extends the model to nonuniform traffic, using the technique developed in [7]. The matrix $\mathbf{T}$ specifies the probability for a call to be addressed from a given input link to a given output link. Assume that traffic on each output is not dominated by individual input links. Our study confirms the general validity of the separate input queue analysis under the nonuniform correlated traffic conditions.

Note that one can ignore the correlation effects on input packet service time which helps us to separate the input queue analysis. Like any other single link queuing analyses, however, the correlation effects on each separate input queue will still be dominant and by no means they can be ignored [12], [13].

III. A CORRELATED INPUT TRAFFIC MODEL

Consider that each input/output link is essentially a TDM link with equal capacity. The input link is logically partitioned into frames and each frame is further decomposed into M slots. As described in Fig. 2, there are M sources on each input link and every one of them is associated with one slot in each frame. Each source supports one call at a time, and so simultaneously there are always M calls transmitting on each input link. The alternation of ON and OFF periods in each call is characterized by an i.i.d. discrete-time two-state Markov chain defined at the transition time interval equal to one frame. The holding time on each state is therefore geometrically distributed with the transition probabilities denoted by ϕ_{on} and ϕ_{off} as in Fig. 3. While in the ON state, a constant-length packet is periodically generated in frames. Note that only the packets generated by the same call are correlated and they must be destined to the same output link. Let γ be the probability for a call to determinate at the end of each OFF period, upon which a new call will arrive with its call address uniformly assigned among the output links. Hence, the traffic in steady state is uniformly distributed among all the input and output links.

A typical example of correlated traffic is packet voice with silence detection [14], [15]. In our case, each input link has the capacity of M voiceband channels, supporting M voice

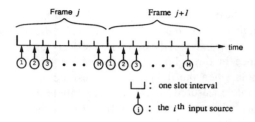

Fig. 2. Time division of M sources on each input link.

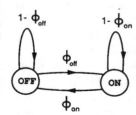

Fig. 3. A two-state Markov chain model.

calls simultaneously. The output address of each call is independently and uniformly assigned. Each call is characterized by a two-state Markov chain alternating between talkspurt and silence. While in talkspurt, packets are periodically generated at intervals of M slots. Therefore, the value of M represents the correlation interval of each source, or the ratio of the input link transmission rate to the source access rate. The average talkspurt and silence are equal to ϕ_{on}^{-1} and ϕ_{off}^{-1} in frame units, respectively. The value of ϕ_{on}^{-1} also gives the average number of packets in each talkspurt period. A call terminates with probability γ at the end of each silence. The holding time of each call is therefore geometrically distributed at talkspurt-silence intervals. The average number of talkspurts in each call is equal to γ^{-1}. The average call holding time is given by $\gamma^{-1}(\phi_{on}^{-1} + \phi_{off}^{-1})$ in frame units. The call activity factor is defined by $\lambda = \phi_{on}^{-1}/(\phi_{on}^{-1} + \phi_{off}^{-1})$, which is also the average input rate of each link. A similar approach has been used to construct a packet video source [16], [17]. For a stable system with uniform address assignment, λ is also the average rate at each output. Let λ_{max} be the maximum throughput, at which the input buffers saturate.

Based on this modeling, the following properties of a correlated input source are characterized:

$\lambda \in [0, \lambda_{max})$: average input/output rate,

$M \in [1, \infty]$: ratio of link transmission rate to source access rate,

$\phi_{on}^{-1} \in [1, \infty]$: average number of packets in each ON period, and

$\gamma^{-1} \in [1, \infty]$: average number of ON periods at each call.

Note that the correlation behavior of each source is measured by both ϕ_{on}^{-1} and M. If ϕ_{on}^{-1} is large, more packets will be consecutively generated at intervals M. On the other hand, if M is small, the traffic intensity of each source will be high during the ON periods.

We need to study how the performance of the switch is affected by the above four parameters. There are two extremes. One is the uncorrelated traffic, characterized by $\phi_{on}^{-1} = 1$ and

$\gamma^{-1} = 1$, or as $M \to \infty$. In this extreme, each call will only generate one packet and so the address of each packet is independently assigned. The other extreme occurs at $M = 1$ and $\phi_{on}^{-1} \to \infty$ where the call on each input link experiences the alternation of two extremely long ON and OFF periods. While in ON period, the call generates the most stream traffic to occupy the entire link capacity.

IV. THE MAXIMUM THROUGHPUT PERFORMANCE

Under the uniform traffic assumption, each input buffer will have the same steady state distribution, and also each output contention process will be statistically identical. Packets on each input queue are assumed to be served on a FIFO basis. We focus on a tagged packet, say, from input i to output k. Once this packet moves to the head of the input queue, it joins the contention with the packets at the head of the rest input queues destined to the same output. The switch fabric has the capacity to rout up to n contending packets to each output in a time slot. The contention process to output k is then characterized by a queueing model, described by

$$C_k' = (C_k - n)^+ + A_k \tag{3}$$

where the symbol $(\cdot)^+$ denotes the larger of 0 and its argument. C_k represents the total number of packets at the head of all input queues destined to output k in a slot. A_k consists of those packets which will move to the head of input queues during this slot interval and destined to the same output k. C_k' then gives the number of packets accumulated for the k-output contention in the next slot. Note that C_k is defined to include the total number of packets in contention as well as in transmission. The output contention time is therefore defined to include the one slot transmission time through the switch fabric at the end of contention.

Let us study the output contention when the system saturates, i.e., as $\lambda \to \lambda_{max}$. With uniform address assignment, all input queues will simultaneously saturate with infinite packet transmission delay. Let $\overline{C}$ be the average number of packets in contention to each output, which is irrelevant to k under the uniform condition. Using Little's result, the average contention time, denoted by μ^{-1}, is equal to $\mu^{-1} = \overline{C}\lambda^{-1}$, which also represents the mean service time at each input. The input link utilization is then defined by $\rho = \lambda\mu^{-1} = \overline{C}$. Therefore,

$$\lambda \to \lambda_{max} \quad \text{as } \overline{C} \to 1. \tag{4}$$

For the uncorrelated traffic, the study in [7] shows the A_k forms an independent Poisson process for sufficiently large N's, given that the traffic on each output is not dominated by individual inputs. In other words, each output contention process tends to be asymptotically independent as $N \to \infty$, and modeled by an $M/D/n$ queue. For instance, at $n = 1$ $\overline{C}$ represents the mean buffer size of $M/D/1$ queue, resulting in $\lambda_{max} = 0.586$ under the uniform condition [3], [5].

The question is what the maximum throughput of the switch will be for the correlated traffic. In principle, due to the high input traffic correlation, the contention arrival process A_k will also tend to be positively correlated at adjacent frames. For instance, once the present A_k becomes high, more packets are expected to join the contention subsequently at approximate frame intervals (each frame has M slots). Such a correlation tends to be strengthened if M is small or ϕ_{on}^{-1} is large. Many queueing analyses indicate that the existence of positive correlation in the arrival process will have adverse effect on the queueing performance (or contention performance in our case) [12], [13]. Hence, the maximum throughput of the switch for the correlated traffic is expected to be less than that of the uncorrelated one. On the other hand, packets are only correlated if they arrive on the same input link and generated by the same call. Through input buffering, these correlated packets never directly contend each other. Hence, the existence of high input traffic correlation may not necessarily cause that much adverse effect on the maximum throughput. Indeed, the following study indicates that the reduction of the maximum throughput caused by the input traffic correlation is insignificant.

A. The Worst Case Maximum Throughput Performance

Similar to the study of uncorrelated traffic given in [3], [5], [7], one may expect that each output contention process tends to be independent asymptotically as $N \to \infty$. Due to the input correlation, however, A_k is no longer characterized by a renewal process, not even by a Markov chain. The output contention process then cannot be generally characterized.

Consider $M = 1$, which is the case where each input link carries one source only. While the call is in ON state, packets are periodically generated in every adjacent slot. All these packets will be accumulated at the input buffer and consecutively move to the head of the queue to join the contention for the same output. The output contention process is therefore most intensified. In other words, the maximum throughput derived at $M = 1$ gives the worst case performance of the switch. Increasing M means to increase the time interval of packet correlation, which reduces the correlation effect on the output contentions. Of course, the correlation effect diminishes as $M \to \infty$.

Although the contention arrival process A_k cannot be characterized in general, for the maximum throughput analysis we only need to characterize A_k when the system saturates. At the saturation point, one can form A_k from two types of input queues. The type I queues are those with their front packets presently destined to output k. The rest of the input queues are type II. Note that the input queues are never empty at the saturation point. There are C_k number of type I queues, up to n of which will transmit a packet to the output k. Upon the transmission, the next packet on the queues will move to the front. The probability for this new packet destined to the same output k is equal to $1 - \phi_{on}\gamma\left(1 - \frac{1}{N}\right)$. By definition, ϕ_{on} is the probability to terminate the present ON period and γ is the probability to terminate the present call. Hence, $\phi_{on}\gamma\left(1 - \frac{1}{N}\right)$ gives the probability for the next packet destined to the rest outputs. It should be mentioned that at $M = 1$ the packets from different calls will never be interleaved on each input queue.

On the other hand, there are $N - C_k$ number of type II queues. The probability for each of these queues to transmit a packet in a slot to the other outputs is equal to its arrival rate λ_{max}. Therefore, the probability for each of these queues to

have its next front packet destined to the output k is given by $\lambda_{max}\phi_{on}\gamma\frac{1}{N}$. As $N \to \infty$, each of these queues is expected to be independent asymptotically. In the limit, the overall contention arrival process from the type II queues forms a Poisson process at rate equal to $\lambda_{max}\phi_{on}\gamma$. The output k contention process at the saturation point can then be modeled by an $M/D/n$ queue with feedback loops, shown in Fig. 4. The new arrival from each type I queue upon a departure is characterized by the feedback loop with probability $1 - \phi_{on}\gamma(1 - 1/N)$, which approaches to $1 - \phi_{on}\gamma$ as $N \to \infty$. The analysis of this queueing model is straightforward. Using the generating function approach, one can readily show that

$$E\left[z^{C_k}\right] =$$

$$\frac{\sum_{j=0}^{n-1} \Pr(C_k = j)\left\{[z - \phi_{on}\gamma(z-1)]^{j-n} - z^{j-n}\right\}}{[z - \phi_{on}\gamma(z-1)]^{-n}e^{\lambda_{max}\phi_{on}\gamma(1-z)} - z^{-n}}. \quad (5)$$

There are n number of distinctive roots in the denominator of $E\left[z^{C_k}\right]$ in $|z| \leq 1$, corresponding to the same number of boundary probabilities $\Pr(C_k = j)$ for $0 \leq j < n$. Denote the ith root by z_i. In the complex domain, one may further decompose the denominator into the following n root equations:

$$z_i = [z_i - \phi_{on}\gamma(z_i - 1)]e^{-\lambda_{max}\phi_{on}\gamma(1-z_i)/n + 2i\pi\sqrt{-1}/n}$$
$$\text{for } i = 0, \cdots, n-1 \quad (6)$$

where each individual root can be recursively derived. Based on the functional analysis, one can set up a boundary linear equation at each root. Especially, we have $z_0 = 1$, which leads to the following boundary equation:

$$\sum_{j=0}^{n-1} \Pr(C_k = j)(n - j) = n - \lambda_{max}. \quad (7)$$

The rest of boundary equations for $0 < i < n$ are expressed by

$$\sum_{j=0}^{n-1} \Pr(C_k = j)\left\{[z_i - \phi_{on}\gamma(z_i - 1)]^{j-n} - z_i^{j-n}\right\} = 0. \quad (8)$$

Once the boundary probabilities are derived from the boundary equations, one can always construct the solution for the first moment function $\overline{C}$ based on (5). Note that the model is only valid at the saturation point, i.e., when $\overline{C} = 1$. In other words, the maximum throughput λ_{max} needs to be numerically obtained at $\overline{C} = 1$. At $n = 1$, we get the closed-form solution

$$\overline{C} = \frac{\lambda_{max}}{2(1 - \lambda_{max})}(2 - \lambda_{max}\phi_{on}\gamma). \quad (9)$$

Rewrite (9) at $\overline{C} = 1$ under the condition of $0 < \lambda_{max} \leq 1$, yielding

$$\lambda_{max} = \frac{2 - \sqrt{4 - 2\phi_{on}\gamma}}{\phi_{on}\gamma}. \quad (10)$$

λ_{max} has two extremes. One occurs at $\phi_{on}\gamma = 1$, which removes the feedback loop of the model. It corresponds to the uncorrelated traffic, where the contention process is modeled by an $M/D/n$ queue. We get $\lambda_{max} = 0.586$, similar to [3], [5]. The value of λ_{max} then decreases as $\phi_{on}\gamma$ reduces, as expected

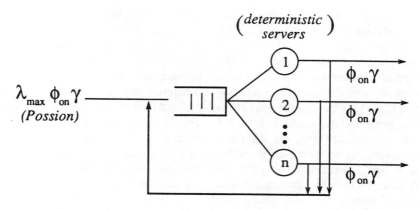

Fig. 4. Output contention process at saturation point for $M = 1$ with multiple output servers.

from the above discussions. The other extreme is given at $\phi_{on}\gamma = 0$. The contention model in Fig. 4 then becomes a closed loop, where the number of packets in the loop is statistically determined by the initial call address assignment on each input link. If $\phi_{on} = 0$, it means that the ON period of each call tends to be infinitely long, representing the most stream traffic type. If $\gamma = 0$, it means that the holding time of each call tends to be infinitely long. From (10) we get $\lambda_{max} = 0.500$ at $\phi_{on}\gamma = 0$. This is the worst case result for the correlated traffic at $M = 1$, which also corresponds to the worst case throughput performance of the switch in general when $n = 1$. The boundary conditions given in (1) are then obtained for $n = 1$. In Table I we list the upper and lower bounds of the maximum throughput at $n = 1, 2, 3, 4$, derived from (4) and (5) in both $\phi_{on}\gamma = 1$ and $\phi_{on}\gamma = 0$ conditions. As one can see, the gap between the two bounds closes rapidly as n increases, and choosing $n = 4$ is sufficient to fully utilize the link capacity.

Although it is instructive to consider each source to be an i.i.d. process, the above bounds can also be met for a switch supporting a mixture of multiple nonidentical sources. We thereby conclude that for the class of nonblocking space-division packet switches the maximum throughput boundary conditions listed in Table I is generally applied in a multimedia traffic environment, under the assumption of uniform call address assignment.

One application of this result is to consider a case where each input link is connected to a data source. Instead of generating equal size messages (which are fixed by the packet size of the switch), we allow each source to generate variable size message. The number of packets in each message is geometrically distributed and on the average is equal to ϕ_{on}^{-1}. Since each input link is a TDM link, the packets of each message are expected to arrive consecutively at the input buffer in adjacent slots, addressing to the same output link. Such a data source is characterized by the above traffic model with $\gamma^{-1} = 1$. The exact maximum throughput can then be derived from (4) and (5) for different ϕ_{on}^{-1}'s.

B. The Maximum Throughput for $M > 1$

The above analysis is based on the observation at $M = 1$. The upper bound performance should not be affected by the value of M, since it is characterized by the uncorrelated traffic.

TABLE I
THE MAXIMUM THROUGHPUT BOUNDARY PERFORMANCE

| | λ_{max} | |
n	lower bound	upper bound
1	0.500	0.586
2	0.828	0.885
3	0.961	0.976
4	0.993	0.996

As M increases, the lower bound eventually converges to the upper bound. The reason is that any type of correlated traffic will become uncorrelated as $M \to \infty$. Note that M gives the ratio of link transmission rate to source access rate. Since the gap between the two bounds at $M = 1$ is already small, one may suggest that for relatively large M's the effect of traffic correlation on the maximum throughput is possibly ignored. Without such a simplification, the output contention process can hardly be characterized even at the saturation point.

At $M = 1$, each input link is connected to a single source, which occupies the entire input bandwidth while in ON period. With advanced electronic switching technologies, however, the capacity of each input link will be in the range of several hundred Mbps. Such a large input bandwidth is more likely to be shared by many sources. Typically, each input link is connected to a time division multiplexer, concentrating traffic from multiple sources. The packet generation rate of each individual source is therefore expected to be substantially smaller than that of the link. It means that M should be relatively large in practice.

Here we take the simulation approach to find out the maximum throughput for different values of M at $n = 1$. The size of the switch is chosen to be $N = 100$. The parameters $(\phi_{on}^{-1}, \gamma^{-1})$ are set at $(110, 10)$, $(11, 10)$, and $(110, 1.0)$, respectively. The results are listed in Table II where the confidence intervals are not given since they are too small. For comparison purposes, we also list the simulation results at $M = 1$, which are shown to be very close to the theoretical limit 0.5. This implies that choosing $N = 100$ and setting $(\phi_{on}^{-1}, \gamma^{-1})$ at the above values are sufficient for measuring the limiting performance. For the uncorrelated traffic, which in our case is represented by $M \to \infty$, one may analytically derive $\lambda_{max} = 0.5920$ at $N = 100$ [7], which reduced to 0.5860 as

TABLE II
THE MAXIMUM TROUGHPUT PERFORMANCE $\lambda_{\max}$ AT $N = 100$

M	$\phi_{on}^{-1} = 10,$ $\gamma^{-1} = 10$	$\phi_{on}^{-1} = 100,$ $\gamma^{-1} = 1$	$\phi_{on}^{-1} = 100,$ $\gamma^{-1} = 10$
1	0.5033	0.5046	0.5087
2	0.5189	0.5168	0.5246
5	0.5528	0.5512	0.5568
8	0.5732	0.5717	0.5782
10	0.5786	0.5784	0.5840
15	0.5858	0.5858	0.5908
20	0.5885	0.5888	0.5932
30	0.5897	0.5904	0.5924
40	0.5900	0.5903	0.5921
∞	0.5920	0.5920	0.5920

$N \to \infty$ [3], [5]. Therefore, one may observe that the effect of input traffic correlation on the throughput performance can generally be ignored for $M \geq 40$ when $n = 1$. This should also be true for a switch carrying multiple nonidentical sources, as long as the access rate of each individual source is less than or equal to 2.5% of the link transmission rate. With the link capacity of 140 Mbps, for example, it means that the maximum bit arrival rate of each source should be less than or equal to 3.5 Mbps. Except for high-speed packet video and HDTV, most sources in multimedia services environment are expected to be in this category.

In fact, the above traffic model may not be suited for the characterization of high-speed video and HDTV sources. In our modeling, we require that the traffic on each output link is not dominated by individual input sources. This is no longer true for the sources that generate very smooth and large volume traffic, such as high speed video and HDTV. Due to the service time constraint as well as the link capacity limitation, some control mechanisms must be implemented at the call admission level to limit the acceptance of such sources. For instance, if a packet video source has its maximum bit arrival rate at 45 Mbps, for the link capacity equal to 140 Mbps the system may not be able to accept more than four or five such sources on each input/output link. Under this condition, the traffic on a output link will be dominated by such individual sources. It means that less number of sources will be involved in the output contention and so a higher throughput can be achieved. The transmission of this traffic type, however, will not be considered here.

With the above limitation on each source access rate, we have shown that the maximum throughput of the switch at $n = 1$ is not affected by the input traffic correlations. Of course, increasing the switching capacity n will further relieve the limit on each source access rate, because of the reduced output contentions. In principle, due to the existing large disparity between link transmission rate and source access rate in real systems, one may generally ignore the effect of input traffic correlation on the maximum throughput performance.

V. INPUT QUEUE PERFORMANCE

This section studies the input queue performance. Here we only consider the case with $n = 1$ and no output buffers are required. For the uncorrelated traffic, each output contention process can be viewed as an independent phase-type process [7]. Also, the contention process involved by each packet on an input queue is expected to be renewable as $N \to \infty$. The analysis of each input queueing process is then separated from that of the rest inputs and is formulated by an independent $Geom/PH/1$ queueing model. For the correlated traffic, however, the output contention process described in (3) cannot be exactly characterized without being able to formulate the arrival process A_k. In addition, due to the input traffic correlation, the contention process involved by the packets of each queue may no longer be viewed as a renewal process. The exact analysis of input queues can then be hardly carried out.

Consider again the output contention process. Under the uniform assumption, we have $\overline{A_k} = \lambda$. For the system to be stable, from (3) and $C = \rho$ one get $\lambda < \overline{C} < 1$. Given $\overline{C}$ in this range, the possibility of having a large number of packets in contention must be significantly small. Indeed, the analysis in [7] for the uncorrelated traffic has shown that the possibility for $C_k > 8$ is negligible under the uniform assumption. For the correlated traffic, packets from each call will sequentially move into the same output contention process at approximate intervals of M slots. If M is small, such a correlation will adversely affect the output contentions and result in the increase of $\overline{C}$ (refer to [12] for the details). However, if M is sufficiently large, one may expect that the contention process will reach a steady state before the next packet of the same call arrives, such that the contention process involved by the packets of each call will be independent and identically distributed. Similar to [7], in the limit as $N \to \infty$, the contention process involved by the packets from different calls can be viewed as mutually independent. Hence, for sufficiently large M's, each output contention process can be modeled by a renewal process, and the analysis of each input queue then becomes separable from that of the rest inputs. Mathematically, such a separation approximates each contention arrival process A_k by an independent Poisson, so that the contention process C_k is modeled by an $M/D/n$ queue as for the uncorrelated traffic.

We know that M cannot be too small in practice. The question is how large the M should be before the separation approximation can be made. For instance, we have found in Section IV that such an approximation is not adequate for the maximum throughput analysis if $M < 40$. Various cases have been simulated to study the effect of M on input queues with different λ, γ^{-1}, and ϕ_{on}^{-1} values. All the simulations are run with $N = 100$. Essentially, we find that for $n = 1$ the separation approximation can generally be made if $M \geq 40$. The following emphasizes the simulation studies for $M \geq 40$.

Our study first indicates that, as an engineering approximation, the effect of γ^{-1} on input queues can generally be ignored. γ^{-1} gives the average number of ON periods in each call. Since all the packets of each call must be destined to the same output, reducing γ^{-1} will reduce the correlation of packets in adjacent ON periods of each source. For instance, the packet address will change at the end of each ON period

TABLE III
EFFECT OF γ^{-1} ON INPUT QUEUES

λ	mean queue length				queue standard deviation			
	$\gamma^{-1} = 100$	$\gamma^{-1} = 25$	$\gamma^{-1} = 2$	$\gamma^{-1} = 1$	$\gamma^{-1} = 100$	$\gamma^{-1} = 25$	$\gamma^{-1} = 2$	$\gamma^{-1} = 1$
0.55	26.1	25.9	24.0	23.9	66.9	63.6	60.1	59.5
0.53	8.40	7.82	7.60	7.78	27.3	25.0	24.1	24.0
0.50	2.54	2.38	2.41	2.37	7.95	7.16	6.77	5.44

if $\gamma^{-1} = 1$. In Table III listed are the results of the mean and the standard deviation of the queues at $M = 40$ and $\phi_{on}^{-1} = 72$ for different γ^{-1}'s. It is observed that the queueing performance is improved as γ^{-1} reduces, which is attributed to less packet correlations in adjacent ON periods of each source. The improvement however is insignificant, especially for the mean queue length. This is because the OFF period of each source, which is at least several intervals of M slots, is statistically much longer than a packet contention time, such that the contention processes involved by the packets in adjacent ON periods are almost mutually independent. Without loss of generality, we fix γ^{-1} at 25 in the following simulation studies.

A. An "Equivalent" Input Queue Model

For the uniform traffic we have $\overline{A}_k = \lambda$ and $\rho = \overline{C}$. Once the correlation effect on the output contentions is eliminated, the work in [7] shows that for a finite N one can form A_k as a renewal process with its variance equal to,

$$\sigma_{A_k}^2 = \lambda - \frac{\lambda}{N}\left[(1 - \rho)\lambda + \rho^2\right]. \tag{11}$$

The contention process C_k is therefore modeled by a $G/D/1$ queue with its mean equal to

$$\overline{C} = \rho = \frac{\lambda^2}{2(1 - \lambda)} + \lambda - \frac{(1 - \rho)\lambda^2 + \lambda\rho^2}{2N(1 - \lambda)}. \tag{12}$$

The average contention time is given by $\lambda^{-1}\overline{C}$, which is also the mean service time of each input queue as defined μ^{-1}. Substituting $\rho = \lambda/\mu$ into (12), we get

$$\mu^{-1} = \frac{-2N(1 - \lambda) + \lambda^2 + \sqrt{4N^2(1 - \lambda)^2 + \lambda^4 + 4N\lambda^2 - 4\lambda^3}}{2\lambda^2}. \tag{13}$$

Especially as $N \to \infty$, A_k becomes a Poisson process and

$$\mu^{-1} = \frac{\lambda}{2(1 - \lambda)} + 1. \tag{14}$$

As shown in [7], the service time of each input queue can then be characterized by an independent phase-type process, where the second moment of the service time has to be numerically derived.

For simplicity, we further approximate the service time of each input by a geometric process with its mean and variance equal to μ^{-1} and $(1 - \mu)\mu^{-2}$, respectively. Such an approximation was also used in [3], [8] for the uncorrelated traffic.

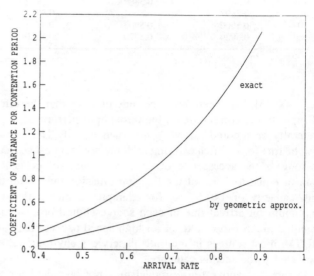

Fig. 5. Coefficient of variance for contention period.

We have plotted in Fig. 5 both solutions for the coefficient of service time variance as a function of λ, assuming $N \to \infty$. The coefficient of variance is defined as the ratio of variance to square mean. The solution of geometric process is equal to $1 - \mu$. The solution of phase-type was derived numerically in [7] based on the assumption that the packets in contention are randomly selected for transmission at each slot. For the uniform traffic, we must have $\lambda < 0.5860$. As one can see, the service time variance is to some extent under-estimated by the geometric approximation. On the other hand, the study in [12], [13] clearly indicates that the queueing performance is mainly dominated by high-input traffic correlations, whereas the service time variance has no significant influence on the queues, given that the service time is renewable.

We have thus far constructed an "equivalent" model for each input queue with an independent geometric server at rate μ given by (13). Due to the input traffic correlation, however, the exact analysis of this model is still difficult. We first examine the accuracy of this "equivalent" model by simulation, and later develop an estimation technique for its analytical evaluation.

Both the exact space-division switch and the "equivalent" single input queue model have been simulated at $N = 100$ under various traffic conditions. We here select some results for demonstration. Fig. 6 shows the mean queue size performance as function of λ at $M = 40$, with respect to ϕ_{on}^{-1} equal to 24, 72, 144. As one can see, the two curves are very close to each other. To keep the figures clean, the 95%

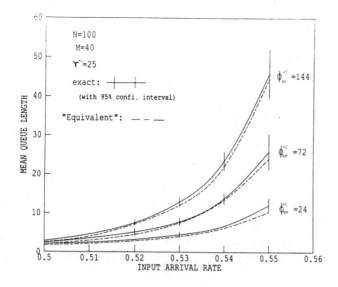

Fig. 6. Mean queue length performance at $M = 40$.

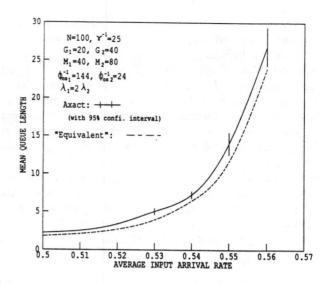

Fig. 8. Mean queue length performance with $G_1 = 20$ and $G_2 = 40$ at $M_1 = 40$ and $M_2 = 80$.

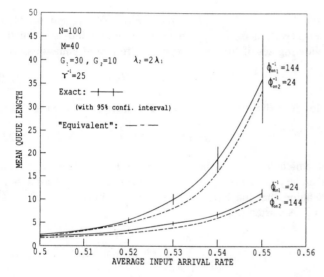

Fig. 7. Mean queue length performance with $G_1 = 30$ and $G_2 = 10$ at $M = 40$.

confidence intervals of the "equivalent" results are not given, which are relatively smaller than that of the exact ones.

So far we have assumed that each call is an i.i.d. process, characterized by a two-state Markov chain. To further verify the validity of the "equivalent" model, one may also consider a mixture of different sources at each input. Without loss of generality, we divide the sources at each input into two groups, G_1 and G_2. The sources access rate in both groups is identical, characterized by a single frame size M. That is, $G_1 + G_2 = M$. The sources in the ith group are also described by λ_i and $\phi_{on\,i}^{-1}$. The overall arrival rate at each input is then given by $\lambda = \lambda_1 G_1/M + \lambda_2 G_2/M$. As an example, we arbitrarily choose $(G_1, G_2) = (30, 10)$ and $\lambda_2 = 2\lambda_1$. The simulation results are presented in Fig. 7 with respect to $\left(\phi_{on\,1}^{-1}, \phi_{on\,2}^{-1}\right)$ equal to $(24, 144)$ and $(144, 24)$. A small degree of under-estimation by the "equivalent" model is observed.

In the next example, we allow the sources in both groups to

have a different access rate, characterized by M_1 and M_2. Hence, $\lambda = \lambda_1 G_1/M_1 + \lambda_2 G_2/M_2$. For convenience, we choose $M_1 = 40$ and $G_1 = 20$ for the first group, and $M_2 = 80$ and $G_2 = 40$ for the second group. The results are plotted in Fig. 8 with $\lambda_1 = 2\lambda_2$ and $\left(\phi_{on\,1}^{-1}, \phi_{on\,2}^{-1}\right) = (144, 24)$. All these examples have shown a good agreement between the exact solutions and the "equivalent" ones. The under-estimation by the "equivalent" model is mainly due to the under-estimation of the service time variance and the neglect of the γ^{-1} effect. One should have no problem in generalizing the "equivalent" model to include multiple heterogeneous sources at each input link, as long as $M_i \geq 40$, $\forall\, i$, for all the sources.

In conjunction with the observation made in Section IV, the above study shows that the analysis of each individual input queue is generally separable from the rest of the switch in a multimedia traffic environment, given that each source access rate is no more than 2.5% of the link transmission rate. The service time of each input queue can be viewed as a renewal process with its mean given by (17) and the coefficient of variance as plotted in Fig. 5. For simplicity, one may approximate the input service time by a geometric process in the separate analysis. A similar statement can be made for $n > 1$ with less constraint on the source access rate.

B. Input Queue Analysis

We now focus on the analysis of each input queue with a geometric server. For analytical purposes, we need to introduce a framing structure for the queue, as described by the following equation

$$q(n + 1) = [q(n) + B(n) - D(n)]^+ \qquad (15)$$

where the symbol $[\cdot]^+$ denotes the larger of 0 or its argument. Here, $q(n)$ is the buffer size at the beginning of the nth frame and $D(n)$ denotes the number of packets which can be transmitted in the nth frame. $B(n)$ represents the number of packet arrivals from all the sources during the nth frame, which is also the number of sources being in active. The formation of this model requires a so-called gated service assumption, that is, all the packets arriving in a frame must be available at the beginning of that frame. Many studies [18], [19] have indicated that, as long as the frame size is small enough compared to the mean ON period of each source (i.e., $\phi_{on}^{-1} \gg 1$), solutions provided by this model will be sufficiently accurate. One reason for this is the dominating effect of input traffic correlation on queues, such that the local dynamics of individual arrivals and departures within a frame can generally be ignored [12].

In our case, $B(n)$ is modeled by a superposition of M independent two-state Markov chains (which are not necessarily to be identical). With the approximation of a geometric server defined at slot intervals, the number of packets which can be transmitted in a frame, denoted by $D(n)$ in the nth frame, is binomial

$$\Pr(D = i) = \binom{M}{i}\mu^i(1-\mu)^{M-i}. \tag{16}$$

The average input link capacity offered by the switch fabric in a frame time is equal to $\overline{D} = \mu M$ in packet units, and M is the maximum number of packets that is possibly transmitted within each frame. The exact analysis of this model can be done using the technique developed in [12], [13]. In particular, it is shown in [12] that the average queueing performance, denoted by $\overline{q}$, can be estimated by the following form:

$$\overline{q} \approx \overline{q^*} + \frac{c}{2(1-\rho)} \tag{17}$$

where c is defined as the coefficient of the service time variance. $\overline{q^*}$ denotes the average queue size of the following system

$$q^*(n+1) = [q^*(n) + B(n) - \overline{D}]^+ \tag{18}$$

where the process $D(n)$ in (15) has been replaced by its mean value $\overline{D}$. Since $\overline{D}$ may no longer be an integer, the discrete queueing analysis technique developed in [12], [13] will not be applied. We here use a continuous flow model developed in [20] for the evaluation of $\overline{q^*}$.

The continuous model characterizes the buffer size at time t by a continuous random variable $q^*(t)$. That is, changing (18) into

$$q^*(t + \Delta) = [q^*(t) + \Delta B(t) - \Delta\overline{D}]^+ \tag{19}$$

where Δ must be taken to be small enough so that $\Delta B(t)$ can be considered as a continuous variable representing the arrival rate. $\Delta\overline{D}$ is the service rate. Given $B(t) = i$, the queue, unless it is empty, will change at the rate Δr_i with $r_i = i - \overline{D}$. Now, taking $\Delta \to 0$, the arrival process $B(t)$ will be characterized by a continuous time Markov process, with its transition rate defined by

$$p_{i,j} = \Pr(B(t + \Delta) = j | B(t) = i)$$
$$\text{for } i, j \in (0, 1, \cdots M). \tag{20}$$

Denoted by P the arrival transition rate matrix $[p_{i,j}]$.

Define the steady-state probability distribution

$$F_i(x) = \Pr(B = i, q^* \leq x) \quad \text{for } 0 \leq i \leq M, x \geq 0 \tag{21}$$

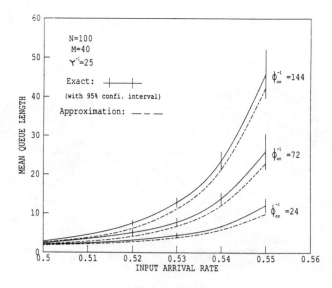

Fig. 9. Mean queue length performance at $M = 40$.

with $F(x) = [F_0(x), F_1(x), \cdots, F_M(x)]^T$. Based on (19), the derivation of $F(x)$ can be transformed into solving the following set of first order system differential equations,

$$F'(x) = R^{-1}PF(x) \tag{22}$$

with $R = \text{diag}[r_0, r_1, \cdots, r_M]$. The solution of (22) is

$$F(x) = e^{R^{-1}Px}b \tag{23}$$

in which b is a constant column vector determined by the boundary condition at $q^* = 0$, which is $F_i(0) = 0$ for $i > \overline{D}$ meaning that the buffer will never be empty when $B > \overline{D}$ [20]. Accordingly,

$$\overline{q^*} = \sum_{i=0}^{M} \int_0^\infty x F_i'(x)\, dx. \tag{24}$$

The solution of (22) can be greatly simplified if the arrival process is a special birth–death process type [20], which in our case is constructed by M i.i.d. two-state Markov chains. Since the multiple transition in these Markov chains can be ignored as $\Delta \to 0$, we get the following birth–death arrival process

$$p_{i,j} = \begin{cases} i\phi_{on} & \text{if } j = i - 1, \\ -i\phi_{on} - (M-i)\phi_{off} & \text{if } j = i, \\ (M-i)\phi_{off} & \text{if } j = i + 1, \\ 0 & \text{else.} \end{cases} \tag{25}$$

The validity of this continuous flow model in its application to voice and video has been examined in [16], [18], [19].

In numerical studies, we use the same examples as used in Fig. 6. For comparison purposes, both exact simulation result and the analytical approximation are plotted in Fig. 9. The analytical approximation is derived based on (17) and (24) where instead of using $c = 1 - \mu$ we have used the exact coefficient of service time variance as plotted in Fig. 5. It conforms well to the simulation result.

VI. Extension to Nonuniform Traffic

So far we have only considered the uniform traffic, i.e., the arrival process at each input link is statistically identical and the output address of each call is uniformly assigned. It has been found that, due to the existing large disparity between the source access rate and the link transmission rate in real systems, the input traffic correlation effect on output contentions can generally be ignored and such that the service time at each input becomes renewable. We now extend to the nonuniform traffic, i.e., each input link will be allowed to carry different sources and the output address of each call can be nonuniformly assigned. Since the output contentions are irrelevant to the input traffic correlation given the above limitation on each source access rate, the characterization of input service time with the correlated traffic should be exactly the same as that of the uncorrelated one. One can therefore use the analytic results, derived in [7] for the uncorrelated traffic, to calculate the input service time statistics for the correlated traffic, under the nonuniform condition. Again, we assume that no output link is dominated by individual input links.

For simplicity, any notation with superscript $\circ(\bullet)$ indicates that notation is associated with input link (output link). Denoted by λ_i° the average arrival rate at input i. Define the input traffic intensity vector by $\Lambda^\circ = (\lambda_1^\circ, \lambda_2^\circ, \cdots, \lambda_N^\circ)$. The output address of each call is assigned at random via a matrix $\mathbf{T} = [t_{i,k}]$ where $t_{i,k}$ is the probability for a call from input i to be destined to output k, given that $\sum_{\forall k} t_{i,k} = 1$. The output traffic intensity vector is denoted by $\Lambda^\bullet = (\lambda_1^\bullet, \lambda_2^\bullet, \cdots, \lambda_N^\bullet)$. For a stable system, we have $\Lambda^\bullet = \Lambda^\circ \mathbf{T}$. Using Little's result, the average contention time to output k, denoted by $\mu_k^{\bullet-1}$, is equal to $\overline{C}_k \lambda_k^{\bullet-1}$ where $\overline{C}_k$ is the average number of packets contending to output k. The average service time of input i is then given by $\mu_i^{\circ-1} = \sum_i t_{i,k} \mu_k^{\bullet-1}$ and so the utilization of input link i is represented by $\rho_i^\circ = \lambda_i^\circ \mu_i^{\circ-1}$, or

$$\rho_i^\circ = \lambda_i^\circ \sum_i t_{i,k} \overline{C}_k \lambda_k^{\bullet-1}. \tag{26}$$

Define the system throughput by $\overline{\lambda} = N^{-1} \sum_i \lambda_i^\circ = N^{-1} \sum_k \lambda_k^\bullet$. The input traffic intensity is described by $\lambda_i^\circ = f_i^\circ \overline{\lambda}$ with $\mathbf{f}^\circ = (f_1^\circ, f_2^\circ, \cdots, f_N^\circ)$ which gives the relative distribution of input traffic intensities. Similarly, the relative output traffic intensities are described by $\mathbf{f}^\bullet = \mathbf{f}^\circ \mathbf{T}$ with $\mathbf{f}^\bullet = (f_1^\bullet, f_2^\bullet, \cdots, f_N^\bullet)$ and $\lambda_k^\bullet = f_k^\bullet \overline{\lambda}$. For the uniform traffic, we have $t_{i,k} = 1/N$ and $\lambda_i^\circ = \lambda_k^\bullet = \overline{\lambda}$, $\forall i, k$.

Define the system saturation by the saturation of individual input queues. Hence, the system saturates as $\max_{\forall i}(\rho_i^\circ) \to 1$, i.e., $\overline{\lambda} \to \overline{\lambda}_{\max}$ as $\max_{\forall i}(\rho_i^\circ) \to 1$. Refer to [7] for detail procedures to evaluate both the system maximum throughput and each input service time variance.

In examples, we choose the one used in [7]. Both input and output links are numbered from 0 to $N - 1$. Except at inputs 0 and $N - 1$, the call output address at each input is assumed to be assigned by a binomial distribution function. For input i,

$$t_{i,k} = \binom{N-1}{k} p_i^k (1 - p_i)^{N-1-k}$$
$$\text{for } k = 0, 1, \cdots, N - 1 \tag{27}$$

except at $i = 0$ and $N - 1$. As a binomial distribution, the maximum probability is always achieved at $k = \text{int}\lfloor (N - 1) p_i \rfloor$ where the function $\text{int}\lfloor . \rfloor$ takes the integer part of its argument. Therefore, one may select the parameter p_i in such a way that the maximum output for input i is given by $\text{int}\lfloor (N - 1) p_i \rfloor$. The maximum output is defined as the one which carries more traffic from input i than any other output. Here we let the output $N - 1 - i$ be the maximum output for input i, $\forall i$. This leads to $p_i(N - 1 - i)/(N - 1)$. As one may see from the curves plotted in [7], with $N = 100$ the traffic from each individual input mainly goes to a group of 20 outputs, which is no longer uniform.

The binomial distribution function cannot be used for inputs 0 and $N - 1$ to retain their maximum output at $N - 1$ and 0, respectively. Assume that the call output address assignment function for inputs 0 and $N - 1$ is a normalized Poisson distribution at rate p, defined by

$$t_{0,k} = \frac{p^{N-1-k}/(N - 1 - k)!}{\sum_{\forall j} p^{N-1-j}/(N - 1 - j)!},$$
$$t_{N-1,k} = \frac{p^k/k!}{\sum_{\forall j} p^j/j!} \tag{28}$$

We here take $p = 0.8$.

Consider a normal distribution for the input traffic intensities. The mean and the standard deviation of this distribution are denoted by u and σ_u. Let

$$f_i^\circ = \frac{x}{u\sqrt{2\pi}} e^{-\frac{(i-u)^2}{\sigma_u^2}} \quad \forall i \tag{29}$$

where x is the normalization factor to satisfy $\sum_{\forall i} f_i^\circ = N$. Taking $u = 39$ and $\sigma_u = 40$, we get $\min_{\forall i} f_i^\circ = 0.42$ and $\max_{\forall i} f_i^\circ = 1.29$ for the input, and $\min_{\forall i} f_k^\bullet = 0.44$ and $\max_{\forall i} f_k^\bullet = 1.27$ for the output, which indicate the nonuniformity of the traffic at both input and output. The maximum throughput in this case is found to be $\overline{\lambda}_{\max} = 0.470$.

In the first example, we choose $M_i = 40$ and $\phi_{\text{on }i}^{-1} = 72$, $\forall i$. Both simulation and analytical results for the mean queue size at each input are plotted in Fig. 10 at $\overline{\lambda} = 0.44$ and 0.40. The simulation results are collected from 8 computer runs. The analytical results are derived using (17). Similar behavior is observed in Fig. 11 as we change $\phi_{\text{on }i}^{-1}$ from 72 to 144, $\forall i$.

In the second example, we choose $\phi_{\text{on }i}^{-1}$ equal to 144 if i is even, and 72 if i is odd, and denote them by $\phi_{\text{on }E}^{-1} = 144$ and $\phi_{\text{on }O}^{-1} = 72$. Again, we let $\lambda_i = 0.44$ and 0.40, $\forall i$, with $M = 40$ and $\gamma^{-1} = 25$. The performance of both even and odd input queues are plotted in Figs. 12 and 13. Since the service time statistics at each input is independent on $\phi_{\text{on }i}^{-1}$, $\forall i$, the "equivalent" model constructed for the approximate analysis of each input queue is identical to the one given in the first example. Hence, the analytical results in Figs. 12 and 13 are the same as those given in Figs. 10 and 11, depending on the respective value of $\phi_{\text{on }i}^{-1}$.

In both examples, we found that the analytical results agree well with simulations. This further confirms the general validity of the separate analysis for each input queue at the switch, under the nonuniform correlated traffic conditions.

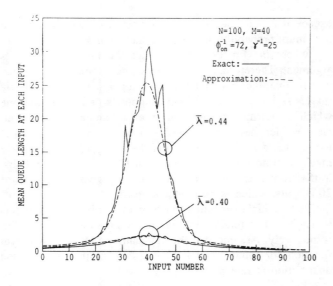

Fig. 10. Mean queue length performance with $M = 40$, $\phi_{on}^{-1} = 72$ and $\gamma^{-1} = 25$.

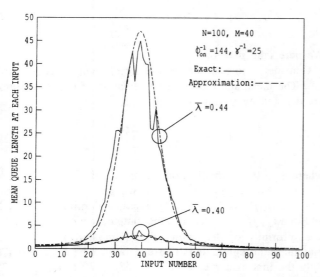

Fig. 11. Mean queue length performance with $M = 40$, $\phi_{on}^{-1} = 144$ and $\gamma^{-1} = 25$.

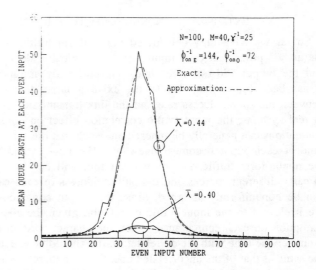

Fig. 12. Mean queue length performance at each even input with $M = 40$, $\varphi_{on\,E}^{-1} = 144$, $\varphi_{on\,O}^{-1} = 72$ and $\gamma^{-1} = 25$.

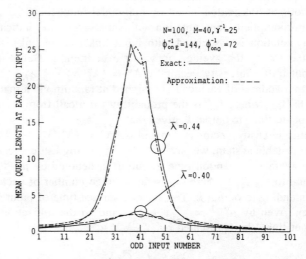

Fig. 13. Mean queue length performance at each odd input with $M = 40$, $\varphi_{on\,E}^{-1} = 144$, $\varphi_{on\,O}^{-1} = 72$ and $\gamma^{-1} = 25$.

VII. Conclusion

This paper has studied the performance of non-blocking space-division packet switches in a correlated input traffic environment. In constructing the input traffic model, we have considered that each input is a TDM link concentrating traffic from multiple sources. Each source supports one call at a time. A call experiences the alternation of ON and OFF periods, and periodically generates packets while in the ON period. The packets generated by each call are destined to the same output. The output address of each call is assumed to be uniformly assigned. The stochastic property of each call is not necessarily identical. In this multimedia correlated input traffic environment, we first have derived the upper and lower bounds of the maximum throughput at system saturation, and found that the gap between the two bounds is insignificant and rapidly diminishes as the switching capacity increases.

Given the fact that in most practical systems the source access rate is substantially smaller than the link transmission rate, our study has further indicated that the effect of input traffic correlation on output contentions can generally be ignored. Therefore, not only one can ignore the input traffic correlation in the maximum throughput analysis, but also the analysis of each input queue becomes separable from the rest of the switch. The service time of each input queue in the separate analysis is characterized by an independent renewal process. The same study has been carried out with nonuniform call address assignment.

Acknowledgment

The author would like to thank the reviewers for their helpful comments.

References

[1] J. S. Turner, "Design of an integrated services packet network," *IEEE J. Select. Areas Commun.*, pp. 1373–1380, Nov. 1986.

[2] A. Hwang and S. Knauer, "Starlite: A wideband digital switch," in *Proc. of GLOBECOM'84*, Dec. 1984, pp. 659–665.

[3] J. Y. Hui and E. Arthurs, "A broadband packet switch for integrated transport," *IEEE J. Select. Areas Commun.*, pp. 1264–1273, Oct. 1987.

[4] Y.-S. Yeh, M. G. Hluchyj, and A. S. Acampora, "The knockout switch: A simple, modular architecture for high-performance packet switching," *IEEE J. Select. Areas Commun.*, pp. 1274–1283, Oct. 1987.

[5] M. J. Karol, M. G. Hluchyj, and S. P. Morgan, "Input vs. output queueing on a space-division packet switch," *IEEE Trans. Commun.*, vol. COM-35, pp. 1247–1356, Dec. 1987; also in *GLOBECOM'86*, Dec. 1986.

[6] H. Ahmadi, W. Denzel, C. Murphy, and E. Port, "A high-performance switch fabric for integrated circuit and packet switching," in *INFOCOM'88*, 1988, pp. 9–18.

[7] S.-Q. Li, "Non-uniform traffic analysis on a non-blocking space division packet switch," *IEEE Trans. Commun.*, vol. 38, pp. 1085–1096, July 1990.

[8] S.-Q. Li and M. J. Lee, "A study of traffic imbalances in a fast packet switch," in *Proc. INFOCOM'89*, pp. 538–547.

[9] S.-Q. Li, "Performance of trunk grouping in packet switch design," *Perform. Eval.*, vol. 12, pp. 207–218, 1991.

[10] M. J. Lee and S.-Q. Li, "Performance trade-offs in input/output buffer design for a non-blocking space-division fast packet switch," *Int. J. Digital Analog Commun. Syst.*, vol. 4, pp. 21–31, 1991.

[11] ——, "Performance of a non-blocking space division packet switch in a time variant non-uniform traffic environment," in *IEEE Trans. Commun.*, vol. 39, pp. 1515–1524, Oct. 1991.

[12] S.-Q. Li and J. W. Mark, "Traffic characterization for integrated services networks," *IEEE Trans. Commun.*, pp. 1231–1243, Aug. 1990.

[13] S.-Q. Li, "A general solution technique for discrete queueing analysis of integrated traffic on ATM," *IEEE Trans. Commun.*, vol. 39, pp. 1115–1132, 1991.

[14] P. T. Brady, "A statistical analysis of on-off patterns in 16 conversation," *Bell Syst. Tech. J.*, pp. 73–91, Jan. 1968.

[15] R. L. Easton, P. T. Hutchinson, R. W. Kolor, R. C. Moncello, and R. W. Muise, "TASI-E communication system," *IEEE Trans. Commun.*, vol. COM-30, pp. 803–807, Apr. 1982.

[16] B. Maglaris, D. Anastassiou, P. Sen, G. Karlsson, and J. Robbins, "Performance analysis of statistical multiplexing for packet video sources," *IEEE Trans. Commun.*, vol. 36, pp. 834–844, July 1988.

[17] W. Verbiest, L. Pinnoo, and B. Voeten, "The impact of the ATM concept on video coding," *IEEE J. Select. Areas Commun.*, vol. 6, pp. 1623–1632, Dec. 1988.

[18] J. N. Daigle and J. D. Langford, "Models for analysis of packet voice communication systems," *IEEE J. Select. Areas Commun.*, vol. SAC-4, pp. 847–855, Sept. 1986.

[19] N. Yin, S.-Q. Li, and T. E. Stern, "Congestion control for packet voice by selective packet discarding," *IEEE Trans. Commun.*, pp. 674–683, May 1990.

[20] D. Anick, D. Mitra, and M. M. Sondhi, "Stochastic theory of data-handling systems with multiple sources," *Bell Syst. Tech. J.*, vol. 61, no. 8, pp. 1871–1893, Oct. 1982.

INTRODUCTION

Blocking occurs in a switch when a packet cannot immediately access an idle output line or port which it would nominally have access. Switches are often built with inherent blocking for the following reasons:

- They can be simpler to implement than nonblocking switches.
- The blocking may produce appreciable effects only under heavy loads that are not likely to occur in normal practice.

All the papers in this section will deal with, to a greater or lesser extent, Banyan networks. These are a popular choice for use as a blocking network. Banyan networks were developed by Goke and Lipovski [1]. They are a class of interconnection networks that include delta networks.

A blocking network, such as a delta network, may be implemented with $O(N \log N)$ crosspoints rather than the $O(N^2)$ crosspoints required by a nonblocking switch. However it should be mentioned that in terms of chip area (blocking), Banyan networks and (nonblocking) crossbar networks have similar requirements [5, 6].

A Banyan network consists of simple 2×2 switching elements interconnected in a structured pattern. Illustrations of Banyan networks usually show a rectilinear placement of switching elements with packets moving from left to right [see Fig. 1 in the paper by Jenq (Part 2)]. Buffers can be placed within each switching element. A vertical column of switching elements is called a *stage*. Packets move through a Banyan network synchronously. That is, during each time slot, stage k's packets attempt to move to stage $k + 1$. A packet in a switching element can move into a switching element in the next stage only if buffer space is available. Blocking occurs when the packet cannot move forward in a given slot. Notice that blocking is internal to the switch and may occur at several switching elements during the same slot.

One reason for the attractiveness of Banyan networks is that they possess a distributed self-routing property. The switch outputs are labeled with binary addresses in ascending order from the top to the bottom of the switch (see Jenq's Fig. 1). Each bit in the destination address is associated with a specific stage of the switch. The decision as to which of the two outputs of a 2×2 switching element to direct a packet is based on the bit in the destination address (contained in the packet header) associated with the stage in which the switching element is located. The simple rule is that if the bit is a 0, the packet is directed to the upper output of the switching element, and if it is a 1, the packet is directed to the lower output of the switching element.

The great advantage of this routing approach is that the routing is performed distributively; that is, each switching element can make decisions independently and simply, based on a single bit in the packet header. There is no need for a centralized control.

The papers in this chapter provide evaluations of Banyan networks (and in some, crossbar networks). That so much progress has been made in understanding Banyan networks is somewhat surprising, since large blocking networks are traditionally difficult to analyze. But substantial progress has been made, as these papers will indicate.

The first paper in this chapter is "Performance Analysis of Multibuffered Packet-Switching Networks in Multiprocessor Systems," by H. Yoon, K. Y. Lee, and M. T. Liu. In it they show how to extend Jenq's model for the single-buffer Banyan case to a finite number of buffers in delta networks. This serves to boost throughput.

Rather than the two-state Markov chain inherent in Jenq's earlier work, Yoon and his coauthors make use of a one-dimensional M-state Markov chain to model buffers of finite length in each switching element. Traffic is assumed to be uniform. They construct a set of nonlinear probabilistic equations for this type of model that can be solved iteratively for the mean delay and the mean throughput of the switch. The results are within 10 to 25% of actual simulation values. The difference is due to a number of independence and randomness assumptions in the model.

The approach taken by Yoon and his coauthors is applicable to networks with switching elements of arbitrary size and type (that is, bus or crossbar). Moreover, the end of the chapter includes a demonstration of their applicability in connection with multibuffered PM2I networks.

The important problem of nonuniform traffic is examined in "Performance of Buffered Banyan Networks under Nonuniform Traffic Patterns," by H. S. Kim and A. Leon-Garcia. In it they present a nonlinear probabilistic model of single buffered Banyan networks that can handle nonuniform input loading. The model can be solved iteratively. In the latter part of the paper they present results for different traffic patterns for single and multibuffered Banyan networks (the modeling of which appears in [2]). They also discuss the effect of traffic patterns on a parallel connection of two Banyan networks.

Kim and Leon-Garcia find that nonuniform traffic lowers the performance of single-buffer Banyan networks. Adding buffers improves performance unless the network is large. The use of parallel Banyan networks results in a substantial improvement of mean throughput and mean delay at the price of additional hardware.

At this point we have seen several papers using probabilistic models to capture the performance of (nonblocking) crossbars and (blocking) Banyan networks. This issue is examined further in *Design and Analysis of Buffered Crossbars and Banyans with Cut-Through Switching* by T. Szymanski and C. Fang. This paper emphasizes virtual cut-through switching where a packet arriving to an empty queue buffer may immediately be transmitted to the next stage. Thus a packet need not spend one clock cycle at each switching stage.

Szymanski and Fang use Markov chains to develop simplified models for analyzing the performance of various switching strategies. One is "input bypass" queueing where several packets near the head of the lines of the input buffers are examined during a slot for possible transmission through the switch. The overall effect of this strategy is to boost throughput. Another strategy is "restricted-output" queueing where a limit is placed on the number of simultaneous arrivals to an output queue. Naturally, accepting up to L packets implies a speedup of L. Szymanski and Fang find that modest speedups of 3 or 4 achieve nearly maximal throughput.

Among other results in this paper are methods for the calculation of the delay density of a Banyan network and a comparison of various Banyan networks with different switching element sizes and equivalent amounts of hardware. Note that a related paper to this one is [3].

Another look at the problem of performance evaluation for buffered Banyan networks appears in "Performance Analysis of Buffered Banyan Networks," by T. H. Theimer, E. P. Rathgeb, and M. N. Huber. The paper begins with a discussion of the classification of Banyan networks. It then goes on to present a performance evaluation model for buffered Delta networks with $b \times b$ switching elements (b inputs and b outputs per switching element). Theimer and his coauthors make use of previous work by Jenq, Dias and Jump, and Patel.

However, this model diverges from simulated values at high loads because of certain independence assumptions that are part of the model. Theimer and his coauthors therefore develop a "refined" model for delta networks with 2×2 switching elements that is more detailed. It makes use of a nine-state Markov chain for each switching element. The model is solved iteratively. It is demonstrated that it produces a closer agreement with simulated values of mean throughput and mean delay.

There are two schools of thought regarding the role of priorities for traffic in broadband networks. One is that networks can be designed without priorities. Essentially the network is engineered to provide a quality of service that can satisfy the most demanding class of traffic. Another school of thought is

that establishing priority classes of traffic allows one to allocate network resources in a fair and flexible manner.

The latter view is taken in "Priority Performance of Banyan-based Broadband ISDN Switches," by S. Tridandapani and J. S. Meditch. In it they provide a performance evaluation for unbuffered Banyan networks carrying several classes of traffic. Link contention in these Banyan networks is handled through direct priorities: Packets with the highest priority succeed at contention and ties are settled randomly. Blocked packets generate negative acknowledgements to the input processors of the Banyan network. Retransmissions then follow.

Tridandapani and Meditch provide a solution for mean throughput for unbuffered Banyan networks with $k^n \times k^n$ switching elements. Work in [4] is used as a starting point for this purpose. The work is then extended to dilated networks (networks with each link replaced by a number of parallel links). Speedups of internal links are also considered. Finally, the mean delay under input queueing is described.

In the numerical results section of this paper the authors point out that in a 64×64 module carrying 10,000 voice channels operating at SONET rates, the voice can occupy only 2.5% of the switch capacity. The extra capacity can be used for data or video traffic. They show that the delay experienced by the voice traffic in such a situation is reasonable.

[1] L. R. Goke and G. J. Lipovski, "Banyan network for partitioning multiprocessor systems," *Proc. 1st Annual Symposium on Computer Architecture*, pp. 21–30, 1973.

[2] H. S. Kim, "Performance of buffered Banyan networks under nonuniform traffic patterns," *M.A.Sc. Thesis, Dept. of Elec. Eng.*, University of Toronto, Toronto, Ontario, Canada, Dec. 1986.

[3] T. H. Szymanski and S. Z. Shaikh, "Markov chain analysis of fast packet switched crossbars and Banyans with input, output and combined input-output queueing," *Columbia University Center for Telecommunications Technical Report CTR-214-90-44*, 1990, submitted for publication in *IEEE Transactions on Communications*.

[4] T. H. Szymanski and V. C. Hamacher, "On the permutation capability of multistage interconnection networks," *IEEE Trans. on Computers*, vol. C-36, no. 7, pp. 810–822, July 1987.

[5] M. A. Franklin, "A VLSI performance comparison of Banyan and crossbar communications networks," *IEEE Trans. on Computers*, vol. C-30, no. 4, pp. 283–290, April 1981.

[6] T. H. Szymanski, "A VLSI comparison between crossbar and switch-recursive Banyan interconnection networks," *Proc. Int. Conf. Parallel Processing*, pp. 192–199, Aug. 1986.

Performance Analysis of Multibuffered Packet-Switching Networks in Multiprocessor Systems

HYUNSOO YOON, KYUNGSOOK Y. LEE, AND MING T. LIU, FELLOW, IEEE

Abstract—We present a new analytic model and analytic results for the performance of multibuffered packet-switching interconnection networks in multiprocessor systems. Previous analyses of buffered interconnection networks in the literature have assumed either single or infinite buffers at each input (or output) port of a switch. As far as multibuffered interconnection networks are concerned, only some simulation results for delta networks have been known [1].

We first model single-buffered delta networks using the state transition diagram of a buffer. We then extend the model to account for multiple buffers.

The analytic results of multibuffered delta networks are compared to simulation results. We also analyze the performance of multibuffered data manipulator networks to demonstrate the generality of the model.

Index Terms—Buffer size, multibuffered delta networks, multibuffered PM2I networks, performance analysis, uniform traffic model.

I. INTRODUCTION

PACKET-SWITCHING multistage interconnection networks (MIN's) can be used for interconnecting a large number of processors and memory modules in multiprocessor systems, or for the resource arbitration and token distribution in the data flow computers [2]. They can also be used as a switching fabric of a packet switch for the high-speed packet-switching computer communications [3]–[5]. Since MIN's play a critical role in the overall performance of such systems, extensive studies have been done on characterizing their performance behavior.

MIN's can be categorized into two groups: unbuffered MIN's and buffered MIN's, depending on whether there are buffer(s) at each switching node. In unbuffered MIN's, whenever a path conflict occurs at a switch among competing packets, only one packet can take the resource and others are discarded. With buffer(s) at each switch, as in buffered networks, the packets, which would be lost otherwise, can be stored as long as buffers are available when a conflict occurs.

Manuscript received August 14, 1987; revised September 22, 1988 and July 3, 1989. H. Yoon and M. T. Liu were supported by the U.S. Army CECOM, Ft. Monmouth, NJ, under Contract DAAB07-88-K-A-003. The views, opinions, and/or findings contained in this paper are those of the authors and should not be construed as an official Department of the Army position, policy, or decision.

H. Yoon is with the Korea Advanced Institute of Science and Technology, Korea.

K. Y. Lee and M. T. Liu are with the Department of Computer and Information Science, The Ohio State University, Columbus, OH 43210.

IEEE Log Number 8932908.

A good body of performance analysis work exists both for unbuffered ([6]–[10] and references therein) and buffered MIN's [1], [11], [3], [7]. For the performance analyses of buffered networks, either a single buffer or infinite buffers at each input (or output) port of a switch have previously been considered. Dias and Jump [1], [11] analyzed the performance of delta networks with single buffers based on the timed Petrinet model. Jenq [3] analyzed the performance of a packet switch based on single-buffered banyan networks, while the performance of banyan networks with infinite buffers was studied by Kruskal and Snir [7].

The performance of multibuffered MIN's (multiple but finite buffers at each input port of a switch) has only partially been known through simulations [1], [11], [12]. In addition to the restriction in the number of buffers (either single or infinite), earlier performance analyses have considered only banyan (delta) networks based on (2×2) crossbar switches. These networks are unique-path networks, which provide a unique path for each input–output connection. Recently, the performance of single-buffered delta networks constructed from switching elements (SE's) of arbitrary sizes and the performance of single-buffered multiple-path networks (plus–minus-2^i networks) have been analytically studied by us [13].

In this paper, we present a simple analytic model, which can be used to analyze multibuffered packet-switching MIN's constructed from SE's of an arbitrary size and type (crossbar or bus). Jenq's elegant analytic model for single-buffered banyan networks [3] laid the ground work for our model.

This paper is organized as follows. Delta networks are briefly introduced in Section II. In Section III, a new analytic model is used to analyze the performance of single-buffered delta networks. The model is generalized for multibuffered delta networks in Section IV. The analytic results are presented and compared to simulation results in Section V. To demonstrate the generality of the model, multibuffered plus–minus-2^i (PM2I) networks are analyzed in Section VI. Conclusions are given in Section VII.

II. DELTA NETWORKS

An $(N \times N)$ delta network consists of n stages of N/a $(a \times a)$ crossbar switches, where $N = a^n$. A packet movement through the network can be controlled locally at each SE by a single base-a digit of the destination address of the packet. Delta networks are chosen in this paper because they are very general, i.e., they are a subclass of banyan net-

Reprinted from *IEEE Trans. Comput.*, vol. 39, no. 3, pp. 319–327, March 1990.

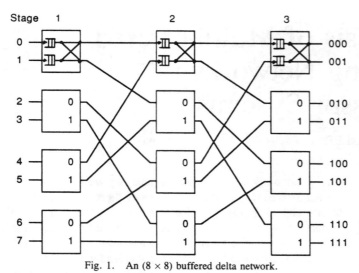

Fig. 1. An (8 × 8) buffered delta network.

works which encompass all the useful unique path MIN's. Delta networks include the cube, omega, indirect binary n-cube, and baseline network [6], [1]. An example of an (8 × 8) buffered delta network is given in Fig. 1. For a detailed description of delta networks, refer to [6].

III. Performance Analysis of Single-Buffered Delta Networks

We analyze the network under the uniform traffic model as in [6], [1], [11], [3], [7], [9], [10], and [13], i.e., the following is assumed.

- Packets are generated at each source node with equal probability.
- Packets are of a fixed size and are directed uniformly over all of the network outputs.
- The routing logic at each SE is *fair*, i.e., conflicts are randomly resolved.

These assumptions imply that, for each switching stage of the network, the pattern of packet distribution is identical and statistically independent for all the SE's. Therefore, each switching stage can be characterized by a single SE, and this fact makes the analysis of the network very simple.

For the operation of single-buffered networks we assume the following as in [1], [11], and [3].

- Each input port of an SE has a single buffer to accommodate a single packet.
- A buffered network operates synchronously at a rate of τ (called *stage cycle*), which consists of two phases.

  — In the first phase, the availability of the buffer space at the subsequent stage along the destined path of a packet in the current buffer is determined; the packet is informed whether it may go to the next stage or should stay in the current buffer.
  — In the second phase, packets may move forward one stage.

- A packet is able to move forward only if it is selected (among competing packets) by the routing logic of the current

SE, *and* either the buffer of the SE it is destined to go to at the next stage is empty *or* the packet in that buffer is able to move forward.

Two important performance measures for buffered networks are throughput and delay. The *throughput* of a network is defined as the average number of packets passed by the network per stage cycle. The *delay* is the average number of stage cycles required for a packet to pass through the network.

We first define the following variables in the same manner as in [3], and derive a set of state equations relating these variables.

n = number of switching stages.

$p_0(k, t)$ = probability that a buffer of an SE at stage k is empty at the beginning of the tth stage cycle.

$p_1(k, t) = 1 - p_0(k, t)$: probability that a buffer of an SE at stage k is full at the beginning of the tth stage cycle.

$q(k, t)$ = probability that a packet is *ready to come* to a buffer of an SE at stage k during the tth stage cycle.

$r(k, t)$ = probability that a packet in a buffer of an SE at stage k is *able to move forward* during the tth stage cycle, *given that there is a packet in that buffer*.

S = normalized throughput (throughput normalized with respect to the network size N; the average number of packets passed by the network per output link).

d = normalized delay (delay normalized with respect to the number of switching stages, i.e., average number of stage cycles taken for a packet to pass a single stage).

Next, we derive relations among these variables. The $q(k, t)$, probability that a packet is ready to come to a buffer of an SE at stage k during the tth stage cycle, is related to the number of packets at the beginning of the tth stage cycle in the buffers of a switch in stage $k - 1$ feeding the buffer under consideration (see Fig. 2), and the destinations of those packets. Derivation of the following three equations for $q(k, t)$, $r(k, t)$, and $r'(k, t)$ assumes the independence of packets in different buffers, i.e., uniformity in destination of these packets, as in [3]. This is an optimistic assumption since these packets may not have been generated during the same stage cycle at the source nodes. As we will see later in Section V, this assumption leads to optimistic analytic results compared to simulation results.

The probability that an $(a \times a)$ SE at stage $k - 1$ has a total of j packets in its "a" single buffers at the beginning of the tth stage cycle is $\binom{a}{j} p_1(k, t)^j p_0(k, t)^{a-j}$. The probability that at least one of j packets is destined to the buffer in stage k under consideration is $[1 - (1 - 1/a)^j]$.[1] Then $q(k, t)$ is the sum over all possible numbers of buffered packets, of the product of these two probabilities.

$$q(k, t) = \sum_{j=0}^{a} \binom{a}{j} p_1(k - 1, t)^j p_0(k - 1, t)^{a-j}$$

$$\cdot [1 - (1 - 1/a)^j] = 1 - [1 - p_1(k - 1, t)/a]^a.$$

[1] This equation was given by Patel [6].

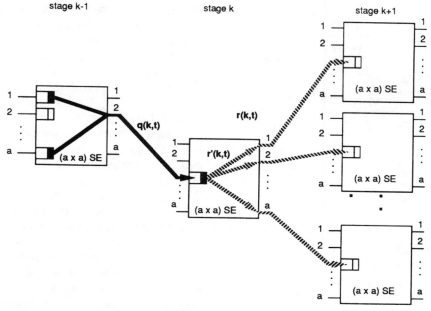

Fig. 2. The illustration $q(k, t)$, $r_1(k, t)$, and $r(k, t)$.

Now, let $r'(k, t)$ be the probability that a packet in a buffer of an SE at stage k is able to pass that SE to the desired output port of that SE during the tth stage cycle by winning the contention among competing packets, given that there is a packet in that buffer. That particular packet in stage k is able to move forward if it can pass the SE and if there is an available buffer at its next stop in stage $k + 1$. The buffer in stage $k + 1$ is available if it is empty or if it is full and the packet in that buffer (in stage $k + 1$) is able to move forward to stage $k + 2$ (see Fig. 2). Thus, $r(k, t)$ can be expressed as following.

$$r(k, t) = r'(k, t)[p_0(k + 1, t) + p_1(k + 1, t)r(k + 1, t)].$$

A packet in a buffer of an SE at stage k is able to pass that SE to the destined output port if it wins the contention among competing packets if any. Since the remaining $a - 1$ buffers may contain j packets, $0 \leq j \leq a - 1$, and the probability that the given packet is chosen out of $j + 1$ packets is $1/(j + 1)$, and the destination of this particular packet can be any one of the "a" output ports, $r'(k, t)$ can be expressed as the following.

$$r'(k, t) = \sum_{j=0}^{a-1} \binom{a-1}{j} p_1(k, t)^j p_0(k, t)^{a-1-j}$$

$$\left[1 - \left(1 - \frac{1}{a}\right)^{j+1}\right] \frac{1}{j+1} \binom{a}{1}$$

Since $\binom{a-1}{j}\binom{a}{1}/(j+1) = \binom{a}{j+1}$, $r'(k, t)$ can be simplified to $q(k+1, t)/p_1(k, t)$ after some manipulations. In summary, the following set of state equations hold for single-buffered delta networks.

$$q(k, t) = 1 - [1 - p_1(k - 1, t)/a]^a, \qquad 2 \leq k \leq n \quad (1)$$

$$r(k, t) = [q(k + 1, t)/p_1(k, t)]$$
$$\cdot [p_0(k + 1, t) + p_1(k + 1, t)r(k + 1, t)],$$
$$1 \leq k \leq n - 1 \quad (2)$$

$$r(n, t) = [1 - \{1 - p_1(n, t)/a\}^a]/p_1(n, t) \quad (3)$$

$$p_0(k, t + 1) = [1 - q(k, t)][p_0(k, t) + p_1(k, t)r(k, t)],$$
$$1 \leq k \leq n \quad (4)$$

$$p_1(k, t + 1) = 1 - p_0(k, t + 1), \qquad 1 \leq k \leq n. \quad (5)$$

Equation (3) assumes that if a packet is available at a network output at the beginning of a stage cycle, it is removed during that stage cycle.

We note that (1)–(5) reduce to Jenq's equations [3] when $a = 2$. To solve this set of equations iteratively, initial conditions need to be defined. We assume that initially at time $t = 0$, all the buffers of the network are empty. We also assume that $q(1, t)$ is the load applied to the network. With these initial conditions, the above set of equations can be solved iteratively to get the time-independent steady-state values.

As an example, Table I shows, for $n = 4$ and $q(1, t) = 1.0$, how $p_1(k, t)$, $q(k, t)$, and $r(k, t)$ change over successive stage cycles and converge to steady-state values. Denoting the $x(k, t)$ in the steady-state as $x(k)$, the normalized throughput (S) and the normalized mean delay (d) can be obtained by the following equations.

$$S = p_1(n)r(n) \quad (6)$$

$$d = \frac{1}{n} \sum_{k=1}^{n} \frac{1}{r(k)}. \quad (7)$$

Equation (7) is due to the fact that the stage delay encountered by a packet is the reciprocal of the probability that the packet is able to pass the stage.

TABLE I
VALUES OF $p_1(k, t)$, $q(k, t)$, AND $r(k, t)$ DURING SUCCESSIVE STAGE CYCLES FOR $n = 4$ AND $q(1, t) = 1.0$

	$t=0$	$t=1$	$t=2$	$t=3$	$t=4$	$t=5$	$t=6$	$t=7$	$t>7$
$p_1(1,t)$	0.00	1.00	1.00	1.00	1.00	1.00	1.00	1.00	1.00
$q(1,t)$	1.00	1.00	1.00	1.00	1.00	1.00	1.00	1.00	1.00
$r(1,t)$	0.00	0.75	0.65	0.59	0.56	0.54	0.54	0.53	0.53
$p_1(2,t)$	0.00	0.00	0.75	0.79	0.80	0.81	0.82	0.82	0.82
$q(2,t)$	0.00	0.75	0.75	0.75	0.75	0.75	0.75	0.75	0.75
$r(2,t)$	0.00	0.00	0.81	0.73	0.68	0.66	0.65	0.65	0.65
$p_1(3,t)$	0.00	0.00	0.00	0.61	0.67	0.70	0.71	0.71	0.72
$q(3,t)$	0.00	0.00	0.61	0.63	0.64	0.65	0.65	0.65	0.65
$r(3,t)$	0.00	0.00	0.00	0.85	0.78	0.76	0.75	0.74	0.74
$p_1(4,t)$	0.00	0.00	0.00	0.00	0.52	0.58	0.61	0.62	0.63
$q(4,t)$	0.00	0.00	0.00	0.52	0.56	0.57	0.58	0.59	0.59
$r(4,t)$	0.00	0.00	0.00	0.00	0.87	0.85	0.85	0.84	0.84

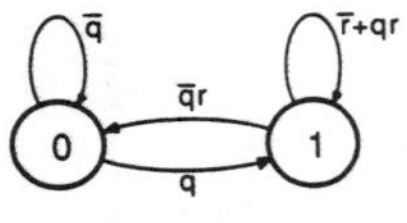

Fig. 3. The state transition diagram of a single buffer. q and r denote $q(k, t)$ and $r(k, t)$, respectively.

Before we generalize (1)–(7) for multibuffered networks in the next section, we note that (4) and (5) can be rewritten as follows, where $\bar{z}(k, t)$ is defined as $1.0 - z(k, t)$ for any variable z.

$$p_0(k, t + 1) = p_0(k, t)\bar{q}(k, t) + p_1(k, t)\bar{q}(k, t)r(k, t) \tag{8}$$

$$p_1(k, t + 1) = p_0(k, t)q(k, t)$$
$$+ p_1(k, t)[q(k, t)r(k, t) + \bar{r}(k, t)]. \tag{9}$$

Note that in single-buffered networks, a buffer has two possible states: the empty state (state 0) and the full state (state 1). In (8) and (9), we can interpret p_i as the probability that a buffer is in state i, and all other terms as transition probabilities of moving from one state to another. Then (8) and (9) can be represented by the state transition diagram of a single buffer in Fig. 3. This interpretation lends itself easily to the generalization of the single-buffered network analysis for multibuffered networks, as can be seen in the next section.

IV. PERFORMANCE ANALYSIS OF MULTIBUFFERED DELTA NETWORKS

In this section, we consider multibuffered delta networks. In a multibuffered network, each input port of every SE has a finite number of buffers so that multiple packets can be placed. (As before, one buffer can hold a single packet.) For some discussions in this section, we need to distinguish between the buffers for an input port of an SE as a whole and an individual buffer slot. We shall refer to the former as a "buffer module," and the latter as a "buffer" in this section. We shall also denote a buffer module of size m as an m-buffer.

In a multibuffered network, a packet is able to move forward one stage if either there is at least one empty buffer at the next stage, or a packet in the full buffer module at the next stage is able to move forward. We note that a packet in an infinite-buffered network is always able to move forward regardless

of the status of the buffer module at the next stage, rendering the analysis much simpler than the one for finite-buffered networks. We have already introduced the new variable m to denote the size of a buffer module. The variables n, $q(k, t)$, S, and d defined in the previous section for single-buffered networks remain the same for multibuffered networks. The earlier definitions of p and r need to be modified to reflect multiple buffers.

m = buffer module size (number of buffers at each input port of an SE).

$p_j(k, t)$ = probability that there are j packets in a buffer module of an SE at stage k at the beginning of the tth stage cycle, $0 \leq j \leq m$ and $\sum_{j=0}^{m} p_j(k, t) = 1.0$ (e.g., p_0, $\bar{p}_0$, p_m, and $\bar{p}_m$ are the probabilities that a buffer module is empty, not empty, full, and not full, respectively).

$r(k, t)$ = probability that a packet in a buffer of an SE at stage k is able to move forward during the tth stage cycle, given that there is at least one packet in the buffer module (i.e., the buffer module is not empty).

With multiple buffers at an input port of each switch, $q(k, t)$, probability that a packet is ready to come to a buffer of an SE at stage k during the tth stage cycle, is related to the number of input ports with nonempty buffer modules at the beginning of the tth stage cycle, of a switch in stage $k - 1$ feeding the buffer module under consideration. By a similar reasoning for (1) and (2), $q(k, t)$ and $r(k, t)$ for an m-buffer can be expressed as follows. As for single-buffered networks, the following three equations for multibuffered networks assume the independence of packets in different buffer modules as well as within a buffer module. As can be seen later in Section V, quite interestingly, this independence assumption for multibuffered networks holds much better than the case of single-buffered networks.

$$q(k, t) = 1 - [1 - \bar{p}_0(k - 1, t)/a]^a, \qquad 2 \leq k \leq n$$

$$r(k, t) = [q(k + 1, t)/\bar{p}_0(k, t)][\bar{p}_m(k + 1, t)$$
$$+ p_m(k + 1, t)r(k + 1, t)], \qquad 1 \leq k \leq n - 1.$$

Likewise, $r(n, t)$ can be expressed as

$$r(n, t) = q(n + 1, t)/p_0(n, t).$$

The major difference between a single-buffer and an m-buffer is in the different possible number of buffer states. While a single-buffer can have only two possible states, full or empty, an m-buffer can be in one of $m + 1$ possible states, containing j packets (state j) with the probability $p_j(k, t)$, for $0 \leq j \leq m$. Since only a single packet may be transmitted between stages per stage cycle, an m-buffer can change its state only among adjacent neighbor states (-1 or $+1$) plus the old state itself as can be seen in Fig. 4. Thus, a nonboundary state j (neither full or empty) can be reached from three previous states, $j - 1$, j, $j + 1$. State 0 does not have -1 neighbor state, whereas state m does not have $+1$ neighbor state. The transition probabilities of moving from one state to another can be obtained by considering the ways in which one could move between the two states and the associated probabilities for movements. For example, a buffer module remains in the

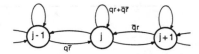

Fig. 4. The state transition diagram of a buffer module of size m in an intermediate state j, $1 < j < m$.

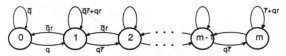

Fig. 5. The complete state transition diagram of a buffer module of size m.

old state j, if there is one departure and one arrival (with the probability $q \cdot r$), or no departure and no arrival (with the probability $\bar{q} \cdot \bar{r}$).

In general, the probability $p_j(k, t+1)$ that a buffer module is in a nonboundary state j at the beginning of stage cycle $t+1$ is the sum of the (mutually exclusive) probabilities that the same buffer module was in state $j-1$, j, or $j+1$ during stage cycle t, each multiplied by the transition probability to state j. We thus have for a nonboundary state j,

$$p_j(k, t+1) = p_{j-1}(k, t)q(k, t)\bar{r}(k, t)$$
$$+ p_j(k, t)[q(k, t)r(k, t) + \bar{q}(k, t)\bar{r}(k, t)]$$
$$+ p_{j+1}(k, t)\bar{q}(k, t)r(k, t).$$

The complete state transition diagram of an m-buffer is given in Fig. 5.

In summary, the following set of state equations holds for m-buffered delta networks, where m is the size of the buffer module at each input port.

$$q(k, t) = 1 - [1 - \bar{p}_0(k-1, t)/a]^a, \quad 2 \le k \le n \tag{10}$$

$$r(k, t) = [q(k+1, t)/\bar{p}_0(k, t)][\bar{p}_m(k+1, t)$$
$$+ p_m(k+1, t)r(k+1, t)], \quad 1 \le k \le n-1 \tag{11}$$

$$r(n, t) = q(n+1, t)/\bar{p}_0(n, t) \tag{12}$$

$$p_j(k, t+1) = q(k, t)[p_{j-1}(k, t)\bar{r}(k, t) + p_j(k, t)r(k, t)]$$
$$+ \bar{q}(k, t)[p_j(k, t)\bar{r}(k, t) + p_{j+1}(k, t)r(k, t)],$$
$$2 \le j \le m-1, 1 \le k \le n \tag{13}$$

$$p_0(k, t+1) = \bar{q}(k, t)[p_0(k, t) + p_1(k, t)r(k, t)],$$
$$1 \le k \le n \tag{14}$$

$$p_1(k, t+1) = q(k, t)[p_0(k, t) + p_1(k, t)r(k, t)]$$
$$+ \bar{q}(k, t)[p_1(k, t)\bar{r}(k, t) + p_2(k, t)r(k, t)],$$
$$1 \le k \le n \tag{15}$$

$$p_m(k, t+1) = q(k, t)[p_{m-1}(k, t)\bar{r}(k, t) + p_m(k, t)r(k, t)]$$
$$+ p_m(k, t)\bar{r}(k, t), \quad 1 \le k \le n. \tag{16}$$

From (14)–(16), it can be verified that the probability of

an m-buffer module being in any one of the possible $m+1$ states, $\sum_{j=0}^{m} p_j(k, t+1)$, is 1.0.

In the steady-state, a packet arrives at an output port of the network with probability $\bar{p}_0(n)r(n)$, which is the normalized throughput S.

$$S = \bar{p}_0(n)r(n). \tag{17}$$

Let $R(k)$ be the average probability that a packet in the buffer of an SE in stage k is able to move forward. Then the normalized delay d can be given as

$$d = \frac{1}{n} \sum_{k=1}^{n} \frac{1}{R(k)},$$

$$\text{where } R(k) = r(k) \sum_{i=1}^{m} [p_i(k)/\bar{p}_0(k)]\frac{1}{i}. \tag{18}$$

Note that (10)–(18) reduce to (1)–(7) when m is 1.

V. PERFORMANCE RESULTS AND COMPARISONS

Analytic Results

Equations (10)–(18) for m-buffered networks of the previous section are very powerful, in that they can be used to determine the normalized throughput and the normalized delay of buffered delta networks with the following four parameters:
1) a = SE size,
2) n = number of stages = $\log_a N$ (N = network size),
3) m = buffer size, i.e., the size of the buffer at each input port of an SE,
4) $q(1)$ = input load applied to the network.

Among the many possible variations of the parameters, the most interesting cases are computed and plotted in Figs. 6–12. Figs. 6 and 7 show normalized throughput versus network size, and normalized delay versus network size, respectively, for various buffer sizes. It is seen that the normalized throughput decreases as the network size increases, and also as the buffer size decreases. The normalized delay decreases as the network size increases, and as the buffer size decreases. However, we observe that the normalized delay of single-buffered delta networks slightly increases as the network size increases. (For the detailed analytic results of single-buffered networks, refer to [1], [11], [3], and [13].)

Figs. 8 and 9 show normalized throughput versus buffer size, and normalized delay versus buffer size, respectively, for various network sizes. The normalized throughput reaches the saturation point very quickly after the buffer size of six. The increase in normalized throughput is very significant up to the buffer size of 3–4. The normalized delay increases almost linearly with the buffer size, especially for not very large networks. These results confirm the well-known observation from simulations [1], [11], "Adding buffers to a packet switching networks can increase throughput. A word of warning—don't make them too large. For most application, the number of buffers should be limited to one, two, or three."

Fig. 10 plots normalized throughput versus switch size for various buffer sizes for multibuffered (4096 × 4096) delta networks. Notice that the throughput does not increase monoton-

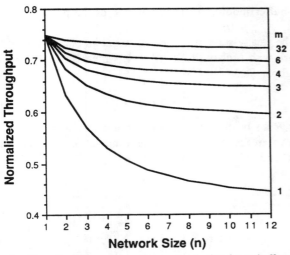

Fig. 6.   Normalized throughput versus network size for m-buffered delta networks ($a = 2$, $q(1) = 1.0$).

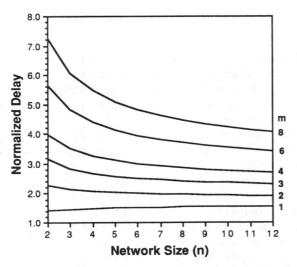

Fig. 7.   Normalized delay versus network size for m-buffered delta networks ($a = 2$, $q(1) = 1.0$).

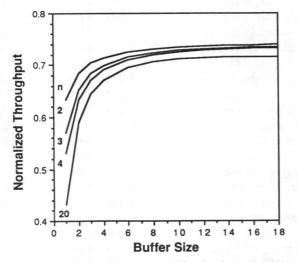

Fig. 8.   Normalized throughput versus buffer size for buffered delta networks ($a = 2$, $q(1) = 1.0$).

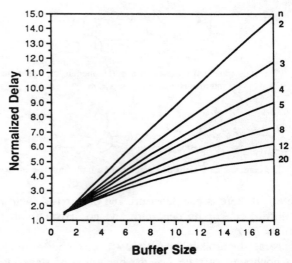

Fig. 9.   Normalized delay versus buffer size for buffered delta networks ($a = 2$, $q(1) = 1.0$).

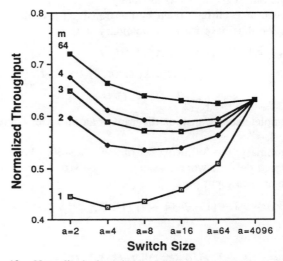

Fig. 10.   Normalized throughput versus switch size for m-buffered (4096×4096) delta networks ($q(1) = 1.0$).

ically with the size of SE as in unbuffered networks. This phenomenon has also been observed in [12] by simulations.

Finally, for (1024×1024) delta networks, normalized throughput versus input load, and normalized delay versus input load are shown in Figs. 11 and 12, respectively. The effects of buffer size on the normalized throughput and also on the normalized delay become apparent only after the input load reaches 0.6–0.7.

Simulation Results

In order to validate the analysis presented in the previous section, we did some simulations of multibuffered delta networks. The basic assumptions for the analysis were implemented in the simulator as follows.

• The N processors generate packets at each stage cycle with probability $q(1)$ (input load).

• The destination of each packet generated at each stage cycle by processors is set randomly by a random number generator (one out of 0 to $N - 1$), to simulate uniform traffic.

• If there is a conflict among packets within an SE, one

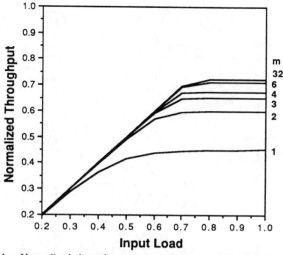

Fig. 11. Normalized throughput versus input load ($q(1)$) for m-buffered (1024×1024) delta networks ($a = 2$).

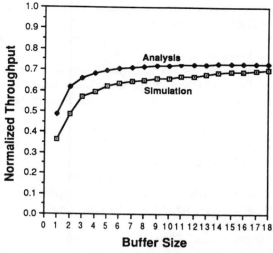

Fig. 13. Normalized throughput versus buffer size for buffered delta networks ($a = 2$, $q(1) = 1.0$, $n = 6$).

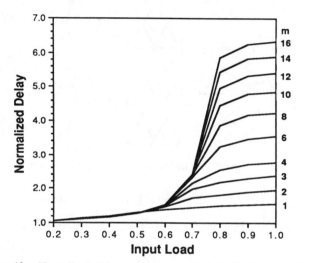

Fig. 12. Normalized delay versus input load ($q(1)$) for m-buffered (1024×1024) delta networks ($a = 2$).

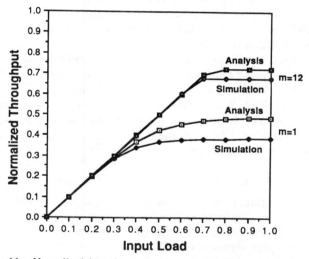

Fig. 14. Normalized throughput versus input load ($q(1)$) for 1-buffered and 12-buffered delta networks ($a = 2$, $n = 6$).

packet is selected randomly again by another random number generator.

• The throughput and the delay are measured at each output port of the network, and averaged over the network size and simulation time span to get the normalized throughput and the normalized delay of the network. In addition, in the simulator, first-in-first-out (FIFO) policy was used for each buffer module.

Sample simulation results are shown in Figs. 13 and 14. Fig. 13 shows normalized throughput versus buffer size for a (64×64) multibuffered delta network based on (2×2) SE's with the input load of 1.0. Fig. 14 plots normalized throughput versus input load for the same network for the buffer size of 1 and of 12.

Analytic results are more optimistic than simulation results. We believe that this is the effect of the highly idealistic uniform traffic model for the analysis. It is well known that a network performs better for uniform traffic than for nonuniform traffic, since with uniform traffic, network load is well distributed resulting in fewer path conflicts.

The analysis is based on several uniformity (or randomness) and independence assumptions. It is those assumptions that make the analysis simple, easy to understand, and easy to compute the result. Some important assumptions are that the packets arriving at each input port of the network are destined uniformly (or randomly) for all output links of the network, the "a" buffer modules in the same ($a \times a$) switch are statistically independent, and packets in a buffer module are independent. However, these assumptions are highly idealistic, and thus the analysis always produces the optimal or upper bound results.

In practice, the uniform traffic assumption is very hard to satisfy in a strict sense for several reasons. First, even though a source node does generate uniform destination requests, some requests may not be able to enter the network (and dropped) when the buffer at the first stage to which the source is connected is full. This disturbs the uniform traffic assumption, and will be more significant as the input load increases. This fact explains some of the observations we can make from Fig. 14. Notice that with single-buffers, there is no difference in normalized throughput between analysis and simulation until the input load reaches 0.3. For the 12-buffered network, no difference is observed until the input load reaches 0.7. Differences increase as the input load increases, and more significantly so for the single-buffered network. Of course, these are also due to the fact that lower input loads inherently create less conflicts in the network compared to higher input loads.

Second, even if the packets are uniform in their destination requests when they enter the network, they may experience somewhat different queueing delays depending upon destined paths and the state of the network. Thus, the head buffers of different buffer modules at each stage may contain requests generated at different stage cycles by source nodes, again disturbing the uniformity.

On a positive note, Figs. 13 and 14 indicate that the independence assumption of multiple buffers for an input port of an SE is indeed reasonable. Although we present results in Fig. 14 on only two different buffer sizes for the clarity of the graph, we have observed that the differences between analytic results and simulation results monotonically decrease as the buffer size increases. This can also be clearly seen in Fig. 13. In addition to the validity of the independence assumption among multiple buffers at each input port of an SE, another reason may well be that the traffic uniformity is less disturbed when more buffers are available, since less packets are dropped in that case. For the simulation, we used the network size of 64. The computing time for larger network sizes was simply overwhelming.

VI. APPLICATION TO NETWORKS OF A DIFFERENT TYPE

In this section, we illustrate the generality of our analytic model of multibuffered delta networks by applying it to multibuffered networks of a different type. For this purpose, multibuffered plus–minus-2^i (PM2I) networks are chosen.

PM2I networks include the augmented data manipulator (ADM), the inverse ADM (IADM), and the gamma network, which are described in detail in [14]–[18]. The $(N \times N)$ ADM consists of $(\log_2 N)+1$ stages, and each stage contains (3×3) SE's except for the first and last stages. The first stage consists of (1×3) SE's, and the last stage (3×1) SE's. The (3×3) SE is not a crossbar but a multiplexer–demultiplexer pair (or a bus with three inputs/outputs), which accepts only one input at a time and directs it to one of three possible outputs. The IADM network is identical to the ADM except that the stages are traversed in reverse order, and the gamma network is identical to the IADM except that each bus switch is replaced by a (3×3) crossbar switch. The (8×8) IADM network (or gamma network) is shown in Fig. 15.

By noting the differences between delta and PM2I networks in the type and size of an SE and the number of switching stages, the following set of state equations can be formulated for multibuffered PM2I networks. Note that for PM2I networks, stages are numbered from 0 to n (unlike from 1 to n for delta networks in previous sections). So the input load applied to a network is $q(0, t)$.

$$q(1, t) = \bar{p}_0(0, t)/3 \qquad (19)$$

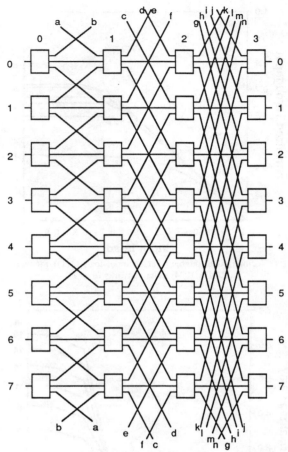

Fig. 15. The (8×8) IADM network (or gamma network).

$$q(k, t) = \begin{cases} 1 - [1 - \bar{p}_0(k-1, t)/3]^3 & : \quad \text{Gamma} \\ [1 - p_0(k-1, t)^3]/3 & : \quad \text{ADM/IADM,} \end{cases}$$
$$2 \le k \le n \quad (20)$$

$$r(0, t) = \bar{p}_m(1, t) + p_m(1, t)r(1, t) \qquad (21)$$

$$r(k, t) = [q(k+1, t)/\bar{p}_0(k, t)][\bar{p}_m(k+1, t)$$
$$+ p_m(k+1, t)r(k+1, t)], \qquad 1 \le k \le n-1$$
$$(22)$$

$$r(n, t) = [1/3 - p_0(n, t)^3/3]/\bar{p}_0(n, t). \qquad (23)$$

Eqs. (14)–(16) for

$$p_j(k, t+1), \qquad 0 \le j \le m, \qquad 0 \le k \le n \quad (24)$$

$$S = \bar{p}_0(0)r(0) \qquad (25)$$

$$d = \frac{1}{n+1} \sum_{k=0}^{n} \frac{1}{R(k)},$$

$$\text{where } R(k) = r(k) \sum_{i=1}^{m} [p_i(k)/\bar{p}_0(k)]\frac{1}{i}. \quad (26)$$

VII. CONCLUSIONS

We introduced a new model for the performance analyses of buffered packet-switching networks in a multiprocessor environment. We used the model to analyze the performance of multibuffered delta networks.

Previously the performance behavior of multibuffered networks had been known for limited cases only through simulations except for the boundary cases (either single or infinite buffers). We compared the analytic results to simulation results, and verified that they agree closely with each other.

The analytic model is general enough to handle networks with arbitrary buffer sizes and switch sizes. It is simple yet powerful enough to be applicable to other types of networks as well. We illustrated the generality of the model by applying it to multibuffered PM2I networks which are multiple-path networks. Previously no attempt had been made to estimate the performance of such networks.

ACKNOWLEDGMENT

We would like to thank the anonymous referees for their constructive comments which helped us to significantly improve this paper, and Y. Kim for lending us his simulation program of buffered banyan networks. Y.-C. Jenq deserves credit for his elegant analytic model of single-buffered banyan networks, which served as the basis of our work reported herein.

REFERENCES

[1] D. M. Dias and J. R. Jump, "Analysis and simulation of buffered delta networks," *IEEE Trans. Comput.*, vol. C-30, no. 4, pp. 273–282, Apr. 1981.

[2] C. Y. Chin and K. Hwang, "Packet switching networks for multiprocessors and data flow computers," *IEEE Trans. Comput.*, vol. C-33, no. 11, pp. 991–1003, Nov. 1984.

[3] Y. C. Jenq, "Performance analysis of a packet switch based on single-buffered banyan network," *IEEE J. Select. Areas Commun.*, vol. SAC-3, no. 6, pp. 1014–1021, Dec. 1983.

[4] J. S. Turner, "Design of an integrated services packet network," *IEEE J. Select. Areas Commun.*, vol. SAC-4, no. 8, pp. 1373–1380, Nov. 1986.

[5] ——, "Design of a broadcast packet switching network," in *Proc. IEEE INFOCOM*, Apr. 1986, pp. 667–675.

[6] J. H. Patel, "Performance of processor–memory interconnections for multiprocessors," *IEEE Trans. Comput.*, vol. C-30, no. 10, pp. 771–780, Oct. 1981.

[7] C. P. Kruskal and M. Snir, "The performance of multistage interconnection networks for multiprocessors," *IEEE Trans. Comput.*, vol. C-32, no. 12, pp. 1091–1098, Dec. 1983.

[8] M. Lee and C. L. Wu, "Performance analysis of circuit switching baseline interconnection networks," in *Proc. 11th Annu. Comput. Architecture Conf.*, 1984, pp. 82–90.

[9] M. Kumar and J. R. Jump, "Performance of unbuffered shuffle-exchange networks," *IEEE Trans. Comput.*, vol. C-35, no. 6, pp. 573–578, June 1986.

[10] A. Varma and C. S. Raghavendra, "Performance analysis of a redundant-path interconnection network," in *Proc. 1985 Int. Conf. Parallel Processing*, 1985, pp. 474–479.

[11] D. M. Dias and J. R. Jump, "Packet switching interconnection networks for module systems," *IEEE Comput. Mag.*, vol. 14, no. 12, pp. 43–53, Dec. 1981.

[12] R. G. Bubenik and J. S. Turner, "Performance of a broadcast packet switch," Tech. Rep. WUCS-86-10, Washington Univ., Comput. Sci. Dep., June 1986.

[13] H. Yoon, K. Y. Lee, and M. T. Liu, "Performance analysis and comparison of packet switching interconnection networks," in *Proc. 1987 Int. Conf. Parallel Processing*, Aug. 1987, pp. 542–545.

[14] T. Feng, "Data manipulating functions in parallel processors and their implementations," *IEEE Trans. Comput.*, vol. C-30, no. 3, pp. 309–318, Mar. 1974.

[15] H. J. Siegel and S. D. Smith, "Study of multistage SIMD interconnection networks," in *Proc. 5th Symp. Comput. Architecture*, Apr. 1978, pp. 223–229.

[16] H. J. Siegel, "Interconnection networks for SIMD machines," *IEEE Comput. Mag.*, no. 12, pp. 57–65, June 1979.

[17] S. D. Smith, H. J. Siegel, R. J. McMillen, and G. B. Adams III, "Use of the augmented data manipulator multistage network for SIMD machines," in *Proc. Int. Conf. Parallel Processing*, Aug. 1980, pp. 75–78.

[18] D. S. Parker and C. S. Raghavendra, "The gamma network," *IEEE Trans. Comput.*, vol. C-33, no. 4, pp. 367–373, Apr. 1984.

Performance of Buffered Banyan Networks Under Nonuniform Traffic Patterns

HYONG SOK KIM AND ALBERTO LEON-GARCIA

Abstract—This paper presents an analytical method to evaluate the performance of the buffered Banyan packet-switching network under nonuniform traffic patterns. It is shown that the nonuniform traffic can have a detrimental effect on the performance of the network. The analytical model is extended to evaluate the performance of multibuffer and parallel Banyan networks. These modified networks are shown to have better throughput capacity and delay performance than the single-buffer Banyan network.

I. INTRODUCTION

VARIOUS switching architectures have been proposed to replace today's time-division switches. Turner and Wyatt [1] propose the use of multistage networks for the packet switch in a network architecture based entirely on packet switching. The multistage network has also been of major interest to computer researchers for its application in parallel processing computers. The Banyan network belongs to the class of multistage networks, and has been considered as a candidate for a fast packet-switching fabric in the communication network.

The Banyan network was first proposed by Goke and Lipovski [2] to interconnect multiple processors and memory modules for parallel processing computers. The basic component of the Banyan network is the two-input, two-output switching element (2×2 crossbar) which operates synchronously for packet switching. A packet at either input port can be passed by the switching element to any of two output ports. However, packets at both input ports cannot be passed to the same output at the same time. Thus, only one packet can be passed, and the other one is either delayed or rejected, depending on whether buffering is available at the input port. A Banyan network is constructed by arranging the switching elements into stages with data paths or links which connect the output port of a switching element in a stage to an input port of a switching element in the next stage, as shown in Fig. 1.

Packets are routed from source port to destination port without the centralized control of switching elements. Switching elements at the nth stage only require the nth bit of the destination address of a packet in order to route it to a proper destination. If the nth bit is "zero" or "one," switching elements route a packet to the upper output port or the lower output port, respectively. The particular interconnection pattern of Banyan networks allows this routing scheme to forward a packet to the proper destination.

A. Review of Related Work

Patel [3] analyzed the performance of unbuffered delta networks in which the interconnection pattern is a permuted form of Banyan network, and in which a uniform traffic load is applied where a

Paper approved by the Editor for Wide Area Networks of the IEEE Communications Society. Manuscript received May 11, 1988; revised February 8, 1989. This work was supported by a grant from the Natural Science and Engineering Research Council (NSERC) of Canada. This paper was presented in part at INFOCOM'88, New Orleans, LA, March 1988.

The authors are with the Department of Electrical Engineering, University of Toronto, Toronto, Ont., Canada.

IEEE Log Number 9035435.

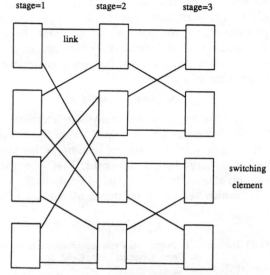

Fig. 1. Structure of Banyan network.

packet at an input port is equally likely to go to any output port. Since he showed that the performance is independent of the choice of interconnection pattern, his analysis could be applied to unbuffered Banyan networks. In Patel's analysis, the throughput of an unbuffered delta network under uniform traffic load is expressed as a quadratic recurrence relation. Kruskal and Snir provided asymptotic solutions for this recurrence relation [4].

Jenq [5] extended Patel's work by analyzing the performance of the single-buffered Banyan network. Similar work was done by Dias and Jump [6]. It was assumed, in these analyses, that the network operates under uniform traffic pattern.

Last, Kruskal *et al.* and Kumar *et al.* [4], [7] introduced the modified delta networks to improve the throughput capacity. These modified networks contain multiple links and multiple buffers. Furthermore, they also showed that one could achieve even higher throughput by adding networks in parallel. They analyzed the performances of the unbuffered modified network with multiple link and parallel networks. The performance evaluation of modified buffered networks was obtained by a means of simulation [7]. Both performance analyses again made the assumption of operating under a uniform traffic pattern.

B. Overview of Paper

Although the performance of Banyan networks under a uniform traffic model is interesting to note, it does not necessarily represent the realistic view of a traffic pattern in a real system. Nonuniform traffic conditions can reflect the traffic patterns of an integrated services network where a wide range of bandwidths needs to be accommodated. Therefore, the performance of the Banyan network under nonuniform traffic becomes an important issue to be studied.

Recently, Wu [8] presented an analysis of single-buffer Banyan networks under nonuniform traffic conditions. In this paper, we present an alternative method for evaluating the performance of buffered Banyan networks. The method presented here increases the number

Reprinted from *IEEE Trans. Commun.*, vol. 38, no. 5, pp. 648–658, May 1990.

of variables that we evaluate by iteration. In doing so, it avoids making assumptions about the values of certain parameters.

The relevant performance figures of merit for the Banyan network are the throughput and the delay. The throughput is defined to be the number of output packets per clock cycle that exit from each output port. The delay is defined to be the time taken for a packet to reach the destination port from the input port. The minimum delay is achieved when a packet proceeds to the next stage without waiting anywhere in the entire network. Thus, the minimum delay is n clock cycles where n is the number of stages in the network.

Under uniform traffic patterns, all throughputs of destination ports and the average delays of all possible source–destination paths are equal. Therefore, the throughput and the delay of the network are obtained from a single destination port and a single source–destination path, respectively. However, the performance figures of the network under a nonuniform traffic pattern require different considerations. Obviously, some source-to-destination paths have better performance than other pairs for nonuniform traffic patterns. We are interested in the maximum allowable throughput at a destination port which has the worst congestion among all the destination ports without driving the network to instability. We define this throughput to be the maximum throughput. We also define the delay of the source-to-destination path which has the longest delay of all paths to be the maximum delay.

In Sections II and III, the structure and the operation of a buffered Banyan networks are described. A mathematical model is developed, and the appropriate steady-state probabilities are obtained to evaluate the performance of the network under nonuniform traffic pattern. The dependence of the performance parameters under a variety of nonuniform traffic patterns is then discussed.

In Section IV, parallel Banyan networks are introduced as a means of increasing throughput performance while maintaining low delays. The analytical method of Section II is modified in order to apply to the parallel Banyan network. Then, the performance of the network under nonuniform traffic patterns is discussed and compared to the performance of the single Banyan network.

II. BUFFERED BANYAN NETWORK

A. The Structure and Operation of the Single-Buffer Banyan Network

The single-buffer Banyan network is a multistage network with stages consisting of an array of switching elements. Each switching element is a 2×2 crossbar switch with a single buffer on each of its input links. The switching element is linked to its adjacent stages in a particular interconnection pattern such that a one-to-one path can be established from any of the input ports to any of the output ports.

We assume that the network operates synchronously. In the first part of the clock cycle, control signals are passed across the network from the last stage toward the first stage, so that every port of the network can determine whether to send or hold its packets. Then, in the second part of the clock cycle, packets move in accordance with the control signals, and this ends the clock cycle. The whole process is repeated in every clock cycle. Since packets can be prevented from advancing, an input buffer is required at every input port at the first stage of the network in order to handle the backup of packets. The Banyan network of size 8×8 is illustrated in Fig. 2.

Packets arrive at an input buffer controller at a rate of λ packets per clock cycle. In the beginning of the clock cycle, if the corresponding buffer of the switching element at the first stage is able to accept a packet, the input buffer controller (IBC) will move a packet into that buffer from its buffer. Arriving packets are placed in the IBC queue according to the order of their arrivals, and they are passed to the first stage of the Banyan network in a first-in, first-out (FIFO) mode. The size of the IBC queue is assumed to be infinite.

The switching element in the first stage sends the packet according to the first bit of the destination address. "Zero" and "one" will route the packet to the upper output port and to the lower output port, respectively. The nth stage switching element will perform the same function according to the nth bit of the destination address. A packet is allowed to advance to the next stage if the buffer in the next

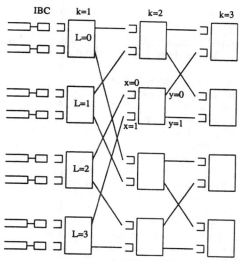

Fig. 2. Buffered Banyan network.

stage is empty or if the buffer is about to become available because its contents advance at the same clock cycle.

A conflict arises when there are packets in both the upper and lower buffers that are destined to the same output port. In this case, one packet is chosen randomly to advance to the next stage, and the other one remains in the buffer. It can be easily seen that the ability for a packet to move forward depends on all stages of the network succeeding the current stage.

B. Modeling and Analysis

Assuming a synchronous operation, as mentioned in the previous section, we can model the whole network as a Markov chain. However, as the size of network increases, the number of states in the Markov chain becomes prohibitively large. The number of states increases exponentially with the number of stages in the network. We use the following queueing network decomposition approach. First, we choose to model each switching element individually; then we describe the relationship among different switching elements through average flow constraints.

Here are the assumptions made for the analysis of the single-buffer Banyan network. Packets are generated at each source node by independent Poisson processes according to arrival rates specified in a load matrix. Each processor generates packet from source i to destination j with rate λ_{ij} at each cycle. The load matrix with its elements λ_{ij} describes the traffic pattern. The network operates synchronously, and a clock cycle is defined to be a time unit required to pass a packet through a stage. All packets are of the same length. There is a single buffer at each port of a switching element. Packets are removed immediately at the output ports. No errors are introduced in the operation of the switching elements and the entire network.

C. Modeling a Single Queue

We first introduce the following notation. Fig. 2 shows how the ports of the network are specified by parameters k, l, x, y.

P_{klxy} Probability that a packet at klx is destined to y.

Q_{klx} Steady-state probability that a packet is ready to come to the port klx.

B_{klx} Steady-state probability that there is a packet in the buffer of the port klx.

r_{klx} Steady-state probability that a packet in buffer klx can advance to the next stage given that there is a packet in the buffer at klx.

r_{klxy} Steady-state probability that a packet at port klx is allowed to advance to the next stage given that there is a packet in the buffer at klx destined to y.

λ_{klxy} Internal throughput defined as the average number of packets per clock cycle passed from a particular port klx to kly.

Now we model a single buffer in the kl switching element shown in Fig. 3 as a Markov chain. Let us define the states of the buffer as 0, 1.

0: There is no packet in the buffer.

1: There is one packet in the buffer.

As shown in Fig. 3, we model the single buffer as being driven by a Bernoulli process with probability Q_{klx} of a packet arrival from the $k-1$th stage and with geometric departures. The probability of departure is determined by the possibility of conflict from the other buffer in the switching element (B_{klx}, P_{klxy}) and the availability of buffer space at the next switching element (r_{klx}).

The two-state Markov chain model of a buffer is shown in Fig. 4 with transition probability from state i to state j, P_{ij} $(i, j = 0, 1)$. When the buffer is in state 0 and there is no packet arrival, the buffer will remain empty at the next clock cycle. However, if a packet arrives at the buffer, it will remain in the buffer for one or more clock cycles until it is allowed to advance. When the buffer is in state 1, it will remain in the same state if there is a packet ready to enter or when there is no packet arrival, but the packet presently in the buffer is not allowed to advance to next stage. Finally, transition from state 1 to state 0 occurs if there is no packet arrival and the packet in the buffer is allowed to advance to the next stage.

These cases are described by the following transition equations:

$$P_{00} = 1 - Q_{klx}$$

$$P_{01} = Q_{klx}$$

$$P_{10} = (1 - Q_{klx})r_{klx}$$

$$P_{11} = Q_{klx} + (1 - Q_{klx})(1 - r_{klx}).$$

From the transition diagram and the equations for transition probabilities, the steady-state probability of the Markov chain is found to be as follows:

$$P^0 = 1 - P^1$$

$$P^1 = \frac{Q_{klx}}{(1 - Q_{klx})r_{klx} + Q_{klx}}. \tag{2.1}$$

These correspond to $1 - B_{klx}$ and B_{klx}, respectively.

The state of buffers at the $(k+1)$th stage depends on the state of buffers at the kth stage at previous clock cycles. However, in order to simplify the analysis, we assume statistical independence between the states of buffers at the $(k+1)$th stage from states of buffers at the kth stage. In other words, the event that buffer space is available at the $(k+1)$th stage and the event that a packet does not encounter a conflict or the conflict is resolved in the packet's favor are assumed to be independent.

Now, we obtain equations for r_{klx}, r_{klxy}, and Q_{klx}. Let $(k+1)\hat{l}\hat{y}$ denote the input port of the $\hat{l}$th switching element at the $k+1$th stage linked to the output port of kly at the kth stage. Also, let $k\bar{l}\bar{y}$ denote the output port of the $\bar{l}$th switching element at the kth stage linked to the input port of the lth switching element at the $k+1$th stage. x represents the complement of x. The notation is shown in Fig. 5.

A packet at the buffer of the klx port destined to the y output port is allowed to advance when the following conditions are satisfied:

1) The buffer at $(k+1)\hat{l}\hat{y}$ can accept a packet and
2) the packet can pass from klx to kly. The buffer at $(k+1)\hat{l}\hat{y}$ can accept a packet when
   a) there is no packet at $(k+1)\hat{l}\hat{y}$ or
   b) there is a packet at $(k+1)\hat{l}\hat{y}$ and it is allowed to advance to the next stage in the same clock cycle. The packet can pass from klx to kly when
   i) there is no packet at klx or
   ii) there is a packet at klx destined to y or
   iii) there is a packet at klx destined to y, and the conflict is resolved in favor of packet at klx. These conditions translate

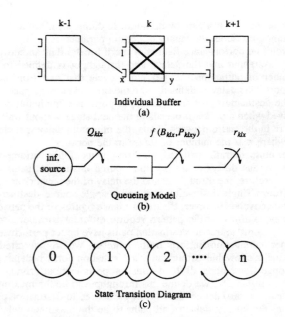

Fig. 3. Queueing model. (a) Individual buffer. (b) Queueing model. (c) State transition diagram.

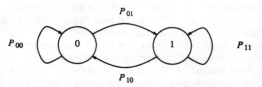

Fig. 4. Markovian model of a node.

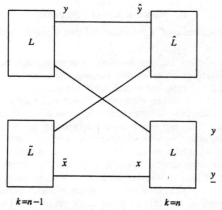

Fig. 5. Notation for linked switching elements at 209, 9.

into the following equation for r_{klxy}:

$$r_{klxy} = \left(1 - B_{klx} + B_{klx}P_{klxy} + \frac{1}{2}B_{klx}P_{klxy}\right)$$

$$\cdot (1 - B_{(k+1)\hat{l}\hat{y}} + B_{(k+1)\hat{l}\hat{y}}r_{(k+1)\hat{l}\hat{y}})$$

$$= \left(1 - \frac{1}{2}B_{klx}P_{klxy}\right)(1 - B_{(k+1)\hat{l}\hat{y}} + B_{(k+1)\hat{l}\hat{y}}r_{(k+1)\hat{l}\hat{y}}). \tag{2.2}$$

r_{klx} can then be obtained from r_{klxy} as follows:

$$r_{klx} = r_{klxy}P_{klxy} + r_{klx\bar{y}}P_{klx\bar{y}}. \tag{2.3}$$

Let EQ_{klx} be the event that a packet is ready to come to the buffe at klx. Also, let

$E^1_{(k-1)\tilde{l}x\tilde{x}}$ = an event that there is a packet at $(k-1)\tilde{l}x$ and it is destined to $\tilde{x}$ linked to klx

$E^0_{(k-1)\tilde{l}x}$ = an event that there is no packet at $(k-1)\tilde{l}x$.

We then have that

$$EQ_{klx} = E^1_{(k-1)\tilde{l}0\tilde{x}} \cap E^0_{(k-1)\tilde{l}1} + E^0_{(k-1)\tilde{l}0} \cap E^1_{(k-1)\tilde{l}1\tilde{x}}$$
$$+ E^1_{(k-1)\tilde{l}0\tilde{x}} \cap E^1_{(k-1)\tilde{l}1\tilde{x}} + E^1_{(k-1)\tilde{l}0\tilde{x}}$$
$$\cap E^1_{(k-1)\tilde{l}1\tilde{x}} + E^1_{(k-1)\tilde{l}0\tilde{x}} \cap E^1_{(k-1)\tilde{l}1\tilde{x}}.$$

This translates into the following equation for Q_{klx}, the probability of the event EQ_{klx}:

$$Q_{klx} = (1 - B_{(k-1)\tilde{l}0})B_{(k-1)\tilde{l}1}P_{(k-1)\tilde{l}1\tilde{x}}$$
$$+ B_{(k-1)\tilde{l}0}(1 - B_{(k-1)\tilde{l}1})P_{(k-1)\tilde{l}0\tilde{x}}$$
$$+ B_{(k-1)\tilde{l}0}B_{(k-1)\tilde{l}1}(P_{(k-1)\tilde{l}0\tilde{x}}P_{(k-1)\tilde{l}0\tilde{x}}$$
$$+ P_{(k-1)\tilde{l}1\tilde{x}}P_{(k-1)\tilde{l}1\tilde{x}} + P_{(k-1)\tilde{l}0\tilde{x}}P_{(k-1)\tilde{l}1\tilde{x}}). \quad (2.4)$$

Now, we are interested in computing Q_{klx} at the first stage, subject to input from the IBC (input buffer controller). Q_{klx} at $k = 1$ is shown to equal

$$Q_{1lx} = \frac{\lambda_{1lxy} + \lambda_{1lxy}}{1 - B_{1lx} + B_{1lx}r_{1lx}} \quad (2.5)$$

where $\lambda_{1lxy} + \lambda_{1lxy}$ is the input rate of packets into the buffer at klx [10].

If we assume that the output buffer is always ready to remove packets immediately at the last stage ($k = n$), then r_{nlxy} is found, from (2.2), to be

$$r_{nlxy} = \left(1 - B_{nlx} + B_{nlx}P_{nlxy} + \frac{1}{2}B_{nlx}P_{nlxy}\right)$$
$$= \left(1 - \frac{1}{2}B_{klx}P_{klxy}\right). \quad (2.6)$$

Now, we derive equations for the throughput, delay, and IBC buffer delay. The throughput is defined to be the average number of departing packets per clock cycle from the last stage of an output link. Since we assume an infinite number of buffers at the IBC, the flow of packets is conserved under steady state for stable system traffic loads. We define the flow rate of packets in the interconnection links as the internal throughput, which is the average number of packets per clock cycle passed from a particular port klx to kly:

$$\lambda_{klxy} = P_{klxy}B_{klx}r_{klx}. \quad (2.7)$$

Equation (2.7) is used to adjust the error resulting from the initial guess in the numerical algorithm described in next section.

The throughput of the network at the output port nly is obtained as follows:

throughput = E(number of packets received at the output port nly)

$$= B_{klx}P_{klxy}r_{klxy} + B_{klx}P_{klxy}r_{klxy}. \quad (2.8)$$

The maximum throughput capacity of the network is computed by increasing the elements of the load matrix until Q_{1lx} becomes 1.0 for some $1lx$. $Q_{1lx} = 1.0$ means that the packets at this input buffer controller are always ready to come to the source port every clock cycle, and indeed, the corresponding IBC buffer contents tend to approach infinity as time progresses.

Packet delay is defined to be the number of clock cycles a packet takes to reach the destination port from the source port. Packet delay is computed for every pair (i, j) of input port i and output port j. Packet delay (i, j) is computed by summing the stage delays along the path taken by packets arriving at i and destined for j. The delay

at a stage is obtained as follows:

$$\text{expected stage delay at } klx = \sum_{k=1}^{\infty} kP(1-P)^{k-1} = \frac{1}{P}$$

where P is the probability that a packet can advance successfully at the given stage in one clock cycle. Here, we assume that the sequence of advance attempts constitutes a sequence of Bernoulli trials. The delay is given in (2.9)

$$\text{delay for a path} = \sum_{\text{all stages}} \frac{1}{r_{klxy}}. \quad (2.9)$$

The minimum possible delay is equal to the number of stages in the network since at least one time unit is spent in each stage, even when there is no waiting.

The IBC buffer delay is given by the following equation [10]:

$$\text{IBC buffer delay} = \frac{2 - (\lambda_{klxy} + \lambda_{klxy})}{2[1 - B_{klx} + B_{klx}r_{klx} - (\lambda_{klxy} + \lambda_{klxy})]}. \quad (2.10)$$

Since an incoming packet has to stay in the IBC buffer at least one clock cycle, even if the buffer at the first stage is available, the minimum IBC buffer delay is one clock cycle. Hence, the total delay is obtained by adding the stage delay and the IBC buffer delay:

$$\text{total delay} = \sum_{\text{all stages}} \frac{1}{r_{klxy}} + \frac{2 - (\lambda_{klxy} + \lambda_{klxy})}{2[1 - B_{klx} + B_{klx}r_{klx} - (\lambda_{klxy} + \lambda_{klxy})]}. \quad (2.11)$$

D. Joining the Single Queues into a Network: Solution by Iterative Method

The single-queue analyses are made consistent by forcing the single-queue variables to yield certain known long-term flows. An iterative procedure for doing this is presented in this section.

The input traffic is described by the load matrix $L = [\lambda(i, j)]$ where $\lambda(i, j)$ is the traffic load originating from input port i and destined for the output port j. From the load matrix, the sum of the matrix elements in a column i and in a row j represents incoming traffic to input port i and outgoing traffic from output port j, respectively. The steady-state flow at the kl switching element along xy, λ_{klxy} can easily be computed from a given load matrix obtained from the interconnection pattern of the network.

The objective of the analysis is to determine the values of B_{klx}, r_{klx}, r_{klxy}, and P_{klxy}. Since the equation describing the dynamics of the network is described by recurrence relations, the solution is obtained by an iterative method. We introduce the following notation for the iteration algorithm.

$r_k^{(j)}$ Vector with all r_{klx} and r_{klxy} of the kth stage as its components at the jth iteration.

$Q_k^{(j)}$ Vector with all Q_{klx} of the kth stage as its components at the jth iteration.

$B_k^{(j)}$ Vector with all B_{klx} of the kth stage as its components at the jth iteration.

Let $Z = f(X, Y)$ denote the fact that the vector Z is a function of the vectors X and Y. Then, the equations derived in the previous sections imply the following functional dependences:

$$r_k = f(B_k, P_k, B_{(k+1)}, r_{(k+1)}) \quad \text{from (2.2)-(2.3)}$$

$$Q_k = f(B_k, r_k) \quad \text{from (2.1)}$$

$$Q_{(k)} = f(B_{(k-1)}, P_{(k-1)}) \quad \text{from (2.4)}$$

$$B_k = f(Q_k, r_k) \quad \text{from (2.1)}$$

183

loop 1 (counter=m)

    loop 2 (counter=i)

        loop 3 (counter=j)

            k=last stage

$$r_k^{(j)} = f(\underline{B}_k^{(j-1)}, \underline{P}^{(m-1)}) \quad \text{from eqn. 2.3, 2.6}$$
$$\underline{Q}_k^{(j)} = f(\underline{B}_k^{(j-1)}, \underline{r}_k^{(j)}) \quad \text{from eqn. 2.1}$$

            from k=last-stage to k=2

$$r_k^{(j)} = f(\underline{B}_k^{(j-1)}, \underline{P}^{(m-1)}, B_{(k+1)}^{(j-1)}, r_{(k+1)}^{(j-1)}) \quad \text{from eqn. 2.2–2.3}$$
$$\underline{Q}_k^{(j)} = f(\underline{B}_k^{(j-1)}, \underline{r}_k^{(j)}) \quad \text{from eqn. 2.1}$$

            from k=1 to k=last-stage-1

$$B_k^{(j)} = f(\underline{Q}_k^{(j-1)}, r_k^{(j)}) \quad \text{from eqn. 2.1}$$
$$\underline{Q}_{(k+1)}^{(j)} = f(\underline{B}_k^{(j)}, \underline{P}_k^{(m-1)}) \quad \text{from eqn. 2.4}$$

            at k=last-stage

$$B_k^{(j)} = f(\underline{Q}_k^{(j-1)}, r_k^{(j)}) \quad \text{from eqn. 2.1}$$

            if $| \underline{Q}^j - \underline{Q}^{(j-1)} | <$ error tolerance
               $| \underline{B}^j - \underline{B}^{(j-1)} | <$ error tolerance
               $| \underline{r}^j - \underline{r}^{(j-1)} | <$ error tolerance
            then exit loop 3
            else continue loop 3

        end loop 3

$$\underline{Q}^{(i+1)} = f(\underline{B}^{(j)}, \underline{r}^{(j)}) \quad \text{from eqn. 2.5}$$

        if $| \underline{Q}^{(i+1)} - \underline{Q}^{(i)} | <$ error tolerance
        then exit loop 2
        else continue loop 2

    end loop 2

$$\underline{P}_k^{(m+1)} = f(\underline{B}_k^{(j)}, \underline{r}^{(j)}) \quad \text{from eqn. 2.7}$$

    if $| \underline{P}_k^{(m+1)} - \underline{P}_k^{(m)} | <$ error tolerance
    then exit loop 3 and compute throughput and delay
    else continue loop 3

end loop 3

(a)

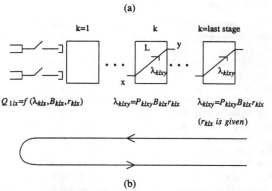

$$Q_{1lx} = f(\lambda_{klx}, B_{klx}, r_{klx}) \qquad \lambda_{klxy} = P_{klxy} B_{klx} r_{klx} \qquad \lambda_{klxy} = P_{klxy} B_{klx} r_{klx}$$

$$(r_{klx} \text{ is given})$$

(b)

Fig. 6. (a) Pseudocode for numerical iteration algorithm. (b) Numerical iteration at $411, 4$.

$$Q_1 = f(B_1, r_1) \quad \text{from (2.5)}$$

$$P_k = f(B_k, r_k) \quad \text{from (2.7)}.$$

The iteration algorithm consists of three layers of looping as shown in Fig. 6(a) and (b). We define m, i, and j as iteration counters for first loop (outmost loop), second loop (middle loop), and third loop (innermost loop), respectively.

We obtain appropriate initial guesses for $P_k^{(0)}$, $B_k^{(0)}$, and $Q_1^{(0)}$ to start the iteration. The third loop starts with $Q_k^{(0)}$ and $P_k^{(0)}$ as given. Using the equations describing the dynamics of the network, we obtain $r_k^{(j)}$ (2.2)–(2.3), $Q_k^{(j)}$ (2.4), and $B_k^{(j)}$ (2.1) from the procedure shown in Fig. 6(a). This loop continues until values of $r_k^{(j)}$, $Q_k^{(j)}$, and $B_k^{(j)}$ converge. When these values converge, we enter the second loop and obtain a new value of $Q_1^{(i)}$ from (2.5). The second loop continues until $Q_1^{(i)}$ converges. When $Q_1^{(i)}$ converges, we obtain new values of P_k from (2.7) and continue until P_k converges.

After all values have converged, the throughput and the packet delay are computed according to (2.8) and (2.11).

E. Results

The analytical model presented in the previous section can be applied to any kind of traffic pattern since the model assumes an arbitrary general traffic pattern. In this section, we define particular traffic patterns for which the network is analyzed. First, simulation results are shown to be in agreement with the analytical model. Then, the performance of the single-buffer Banyan network is analyzed under several traffic patterns of interest.

1) Traffic Patterns: The uniform traffic pattern is defined to be the traffic pattern in which every source port has the same rate of incoming packets and they are destined to every destination port with the same probability $1/n$ where n is the number of nodes. The performance parameters of the single-buffer Banyan network have been obtained by the analytical method in the previous section, and the results agree with Jenq's results [5]. Nonuniform traffic patterns are then defined to be any kind of traffic pattern other than the uniform traffic pattern.

Of the many possible nonuniform traffic patterns, we have selected to study the network under a few nonuniform traffic patterns of interest. One particular nonuniform traffic pattern is the single-source-to-single-destination (SSSD) traffic pattern in which each input source sends all its packets to a single output destination. Therefore, the SSSD traffic pattern drives the Banyan network with a nonuniform traffic pattern. Banyan networks require fewer switching elements because their structure allows links between stages to be shared by different traffic paths. Thus, the SSSD traffic may place extreme loads on common links shared by other traffics. Obviously, conflict due to packet collision occurs more frequently as the traffic intensifies in the shared links. Wide-band traffic such as large data or image can be modeled by SSSD traffic patterns.

The maximum conflict traffic pattern can be modeled by matching the appropriate source port to the destination port such that all the internal links are shared by different pairs of SSSD traffic paths. This traffic pattern has the maximum conflicts among all possible traffic patterns consisting of n distinct pairs of SSSD traffic paths. The bottleneck is created at the shared links since many traffic paths are concentrated at these links. One of the maximum conflict traffic patterns is described by the load matrix given below. Fig. 7 shows the maximum conflict traffic pattern in the network of four stages.

$$
i \text{ source} \quad
\begin{array}{c}
\\ \\ \\ \\ \\
\end{array}
\overset{\textstyle j \text{ destination}}{
\begin{bmatrix}
\lambda_{00} & 0 & 0 & \cdots & 0 \\
0 & \lambda_{11} & 0 & \cdots & 0 \\
0 & 0 & \lambda_{22} & \cdots & 0 \\
\cdots & \cdots & \cdots & \cdots & \cdots \\
0 & 0 & 0 & \cdots & \lambda_{nn}
\end{bmatrix}}.
$$

Another significant traffic pattern involves one SSSD path embedded in a uniform traffic pattern. The load matrix of this traffic pattern has an element e_{ij} where i, j are the source port and destination port of single-source, single-destination traffic, respectively. Also, elements e_{in}, e_{mj} are zero where n, m are not j, i, respectively. The rest of the matrix elements have the same loads defined by uniform traffic. We call this traffic pattern a mixed traffic pattern. The mixed traffic pattern is described by the load matrix as below:

$$
\begin{bmatrix}
\lambda & 0 & \lambda & \lambda & \cdots & \lambda \\
0 & \lambda_{\text{SSSD}} & 0 & 0 & \cdots & 0 \\
\lambda & 0 & \lambda & \lambda & \cdots & \lambda \\
\cdots & \cdots & \cdots & \cdots & \cdots & \cdots \\
\lambda & 0 & \lambda & \lambda & \cdots & \lambda
\end{bmatrix}.
$$

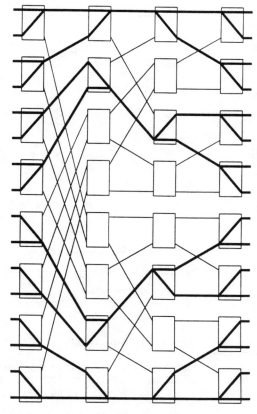

Fig. 7. Maximum conflict traffic pattern at 208, 9.

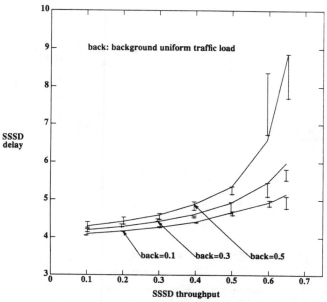

Fig. 8. Simulation results of single-buffer network (SSSD delay versus SSSD traffic).

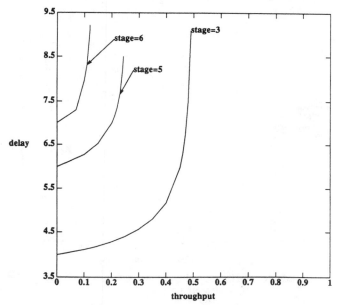

Fig. 9. Single-buffer network under maximum conflict traffic pattern.

This traffic pattern may reflect the network with one dedicated video channel and many voice or data channels. The performance of the network is evaluated under this mixed traffic, and the effect of SSSD traffic on the uniform background traffic and vice versa are studied.

2) Simulation Results: Simulation results are compared to analytical results obtained from numerical analysis for a network of three stages with the mixed traffic pattern. Fig. 8 shows the SSSD traffic delay versus the SSSD throughput load at light, medium, and heavy background uniform traffic. The 95% confidence intervals are computed by the method of Batch mean [9] for each simulation run. The analytical results rest inside the 95% confidence interval for most of the range of parameters tested. The analytical results seem to rest inside the confidence interval when the network is utilized up to 85 ∼ 90% of its maximum throughput. Above this utilization factor, analytical results tend to underestimate the delay found in the simulation results. At a very light traffic load, the analytical results differ from simulation results only by 0.5 ∼ 0.9%. For medium and heavy loads, they differ by 2% and 8 ∼ 9%, respectively. Other traffic patterns have been simulated and compared to analytical results. They also yield accurate results, except at a very heavy load.

3) Analytical Results:

a) Maximum Conflict Traffic Pattern: Fig. 9 shows the throughput and the delay performance of a single-buffer Banyan network of different sizes with the maximum conflict traffic pattern. It shows the effect of the bottleneck created by the conflicts arising in the shared link paths. Under the maximum conflict traffic pattern, the network of size 8×8, 32×32, 64×64 can achieve maximum throughput of 0.49, 0.25, and 0.12 packets/clock cycle, respectively, as shown in Fig. 9. As the size of the network increases, the rate of increase in delay increases. Beyond the capacity of the network, infinite delay results.

The drop in throughput capacity for the maximum conflict traffic pattern compared to the network under uniform traffic ranges from 12 to 75% for a network of size from 8×8 to 64×64, respectively. The maximum throughput under the maximum conflict traffic pattern is almost negligible for larger networks. This is as expected since the links of the larger network are shared by more source-to-destination paths than the smaller network, resulting in greater congestion at the bottlenecks. The number of paths that can share a link increases linearly with the size of the network and exponentially with the number of stages.

b) Mixed Traffic Patterns: Fig. 10 shows the maximum uniform background traffic capacity of the network under the mixed traffic pattern. It can be seen that the maximum uniform background traffic remains constant for an interval, and then decreases linearly as the SSSD throughput increases. The first interval of the curve represents the region where SSSD traffic does not have much degrading effect on the background traffics. The second interval represents the region where the network becomes congested and the frequency of conflicts rises due to the SSSD traffic load on the shared link. Thus, the maximum throughput capacity of the background traffic drops.

Fig. 11 shows the effect of SSSD traffic on the uniform background traffic. As SSSD traffic increases from 0.1 packets/clock cycle to 0.5, there is a significant drop in the maximum throughput capacity of the

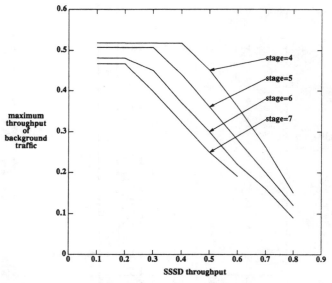

Fig. 10. Maximum throughput of single-buffer network under mixed traffic pattern.

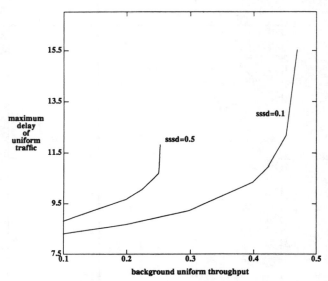

Fig. 11. Effect of SSSD traffic on background uniform traffic in single-buffer network (stages = 7).

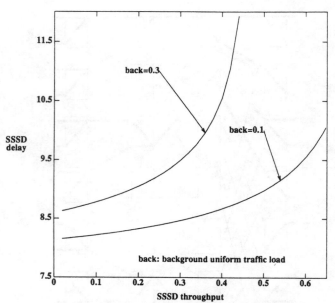

Fig. 12. Effect of background uniform traffic on SSSD traffic in single-buffer network (stages = 7).

background traffic. This degradation of the performance is caused by conflicts in the link shared with the SSSD traffic.

Fig. 12 shows how the background uniform traffic affects the delay and throughput of the SSSD traffic. For the network of seven stages, a small change in background traffic causes a large change in SSSD traffic delay and throughput. The performance of the larger network is more sensitive to the change in background traffic than the smaller network. This is due to the fact that the links of the larger network are shared by more source-to-destination paths.

III. MULTIBUFFER BANYAN NETWORK

A. Introduction

We have seen in Section II that the single-buffer Banyan network suffers from high delay and low throughput when it operates under heavy traffic or highly nonuniform traffic. This degradation in performance is due to the inability of packets to move forward because of conflicts at the switching elements, and to the lack of buffer space at the input ports. One way to reduce the latter effect is to increase the number of buffers in the output port of the switching elements. The throughput is expected to increase by the addition of extra buffers in the switching element.

The analysis of the previous section is easily extended to handle the case where each queue has multiple buffers [10]. We will limit the discussion to the effects on performance of an increased buffer size of two.

B. Results

1) Maximum Conflict Traffic Pattern: Fig. 13 shows the maximum throughput and the delay of the multibuffer Banyan network under the maximum conflict traffic pattern. Each curve represents the delay versus the throughput of the network of different sizes. It is easily seen from Fig. 13 that only minor improvement over the single-buffer Banyan network is obtained in the throughput capacity. The improvement is only noticeable in the network of small sizes. From Section II, we know that the maximum conflict traffic pattern causes the traffic to share the common link in the first stage with two SSSD traffics. Then, the number of SSSD traffics sharing the common link increases by a factor of two for the next stage. This trend continues until the middle stage of the network. From Fig. 7, we notice that the maximum conflict traffic pattern causes no conflict in the switching elements of the second half of the network. Thus, the first few stages are where most of the conflicts occur. The network under the maximum conflict traffic pattern becomes congested at links shared by many SSSD paths. Then, additional buffers are not enough to relieve the congestion. As the size of the network increases, the number of SSSD paths sharing common links increases exponentially. Thus, additional buffer space has even less effect in relieving the congestion in the large network. As we will see in the next section, the multibuffer Banyan network significantly increases the maximum throughput in the other traffic pattern.

2) Mixed Traffic Pattern: Now, we analyze the network under a mixed traffic pattern consisting of a single SSSD path and uniform background traffic.

From Fig. 14, it is easily seen that the multibuffer Banyan network has $10 \sim 15\%$ improvement on the maximum throughput of the background uniform traffic for a given SSSD traffic load, and the throughput capacity increases slightly as the size of the network increases.

Obviously, the low SSSD traffic load does not degrade the performance of the network. As the SSSD traffic intensifies, its neighboring traffics are affected since more conflicts are caused by a higher SSSD traffic load. We are interested in finding at what load level of SSSD traffic the background traffic begins to be affected. We call

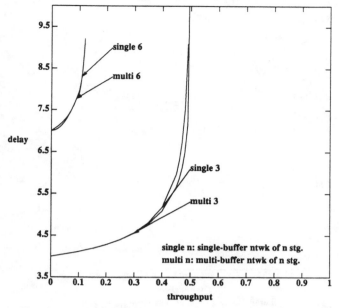

Fig. 13.   Multibuffer network under maximum conflict traffic pattern.

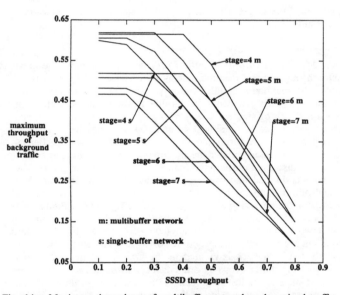

Fig. 14.   Maximum throughput of multibuffer network under mixed traffic pattern.

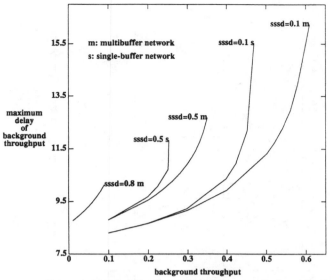

Fig. 15.   Effect of SSSD Traffic on background uniform traffic in multibuffer network (stages = 7).

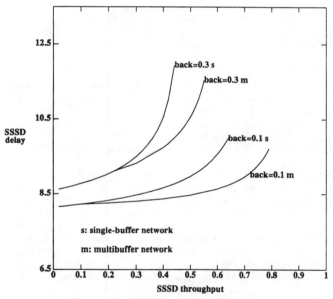

Fig. 16.   Effect of background uniform traffic on SSSD traffic in multibuffer network (stages = 7).

this SSSD traffic load a breakpoint. Fig. 14 shows the effect of the SSSD traffic on the uniform background traffic patterns. The maximum throughput of the background traffic does not degrade until the SSSD traffic reaches 0.5, 0.4, 0.3, and 0.2 for a network of size $8 \times 8, 16 \times 16, 32 \times 32$, and 64×64, respectively. These points correspond to the breakpoints of the SSSD traffic. These breakpoints of the multibuffer Banyan network do not differ much from breakpoints of the single-buffer Banyan network.

A switch in the multibuffer network is more likely to find a space in the buffer of the next stage than in the single-buffer network when it needs to transfer a packet in the multibuffer Banyan network due to its additional buffer space. Hence, the throughput capacity is increased in the multibuffer Banyan network due to the reduction of blocking of packets caused by a lack of buffers. Fig. 15 shows the maximum delay of the background traffic as a function of the throughput of the background traffic for different SSSD traffic loads. Near the maximum throughput, the buffers tend to be full and the packet has to stay in the buffer longer since another packet in the buffer has to transferred first. The rate of increase in delay of the single-buffer

network is higher than that of the multibuffer network. However, the difference in the rate of increase in delay diminishes as the size of the network becomes large. As we discussed earlier, the number of conflicts in the large network is higher than in the smaller network due to increased sharing of the common link in the larger network. Thus, the effect of the additional buffer diminishes as the network gets larger and the traffic intensifies.

Fig. 16 shows the effect of the background uniform traffic on the SSSD traffic. Again, the SSSD traffic in the multibuffer network is less sensitive to the presence of the background traffic than in the single-buffer network.

IV. PARALLEL BUFFERED BANYAN NETWORK

A. Introduction

One way to improve the performance of the buffered Banyan network is to place additional buffered Banyan networks in parallel, as shown in Fig. 17. N networks in parallel offer a packet of N different paths to the destination [4]. Therefore, the introduction redundancy

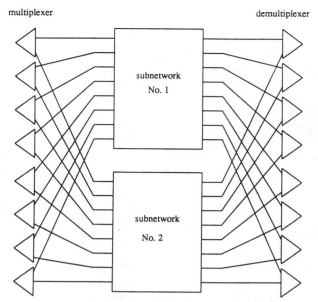

Fig. 17.   Parallel Banyan network.

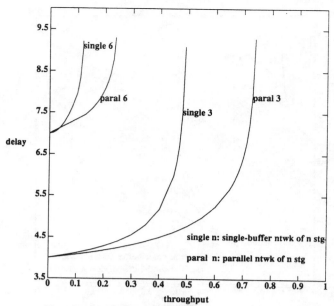

Fig. 18.   Parallel network under maximum traffic pattern.

reduces the load offered to each network. It should be noted, however, that the use of the parallel networks introduces the problem of packets delivered out of sequence.

B. Structure and Operation of Parallel Banyan Network

The parallel network is constructed by placing additional buffered Banyan networks in parallel. We define a subnetwork to be a single Banyan network in the parallel network. The source ports of N subnetworks are joined by the $1 \times N$ multiplexer, and their destination ports are joined by the $N \times 1$ demultiplexer. The structure of a subnetwork is identical to the Banyan network described in Section II.

If one or more packets are ready to come into a multiplexer, only one is forwarded to the randomly selected subnetwork among all subnetworks, and the rest of them wait until the next clock cycle to be delivered. Therefore, if a particular subnetwork receives a packet from the multiplexer, then other subnetworks do not receive any packet from the multiplexer in the same clock cycle. The packet proceeds to the last stage in exactly the same way as in the single-buffer Banyan network. The other difference in the operation arises in the last stage. In the last stage, two kinds of conflicts may arise. The first kind of conflict is caused by packets with the same destination address in different subnetworks. The second kind of conflict is caused by packets destined to the same output port in a switching element of a subnetwork. These conflicts result when more than one packet is destined to the same output port. Therefore, at this stage, the first conflict is resolved by selecting a subnetwork randomly among all subnetworks with packets of the same destination. Then, the conflict in the selected subnetwork is resolved by choosing a packet randomly if there is more than one packet destined to the same destination port. The chosen packet advances to the destination port, and the rest of them stay in the buffer until the next clock cycle to get an opportunity to advance to the destination port.

Since the input traffic is divided equally among all subnetworks, each subnetwork operates with reduced load λ/N where N is the number of subnetworks. Thus, lower delay and higher throughput are possible.

As we will see in the next section, the heavy traffic causes congestion at the last stage due to the conflicts between subnetworks. Therefore, we replace the demultiplexers by the infinite buffers at the last stage in order to relieve congestion at the last stage of the parallel network. Then, the performance of the parallel network is evaluated and the effects of the infinite buffer on different performance parameters are studied.

The analysis of the parallel network is easily obtained from the

analysis of the single network [10]. We will limit the discussion to the results.

C. Results

1) Maximum Conflict Traffic Pattern: Fig. 18 shows the maximum throughput and maximum delay of the parallel Banyan network under the maximum conflict traffic pattern. The improvement due to adding an extra network in parallel is shown to be significant. For a network of size 8×8 and 64×64, the maximum throughput capacity is increased by almost 40 and 100%, respectively. The parallel network outperforms the single-buffer and multibuffer Banyan network, particularly under the maximum conflict traffic pattern.

As we discussed earlier, the load at the input port of the parallel network is divided into two such that the subnetwork operates as if it had only half of the load. The only factor preventing the parallel network from operating as two independent single-buffer Banyan networks is the demultiplexer at the output port of the last stage in each subnetwork. Therefore, the parallel network has a situation where conflicts are possible at the output port of the last stage. However, usually the bottleneck is created in the first few stages. The occupancy of the buffer drops as one advances through each stage from the first stage to the last stage. Thus, the possibility of the conflict at the last stage is reduced due to the lighter load of the subnetworks.

Fig. 19 shows the maximum throughput and maximum delay of the parallel Banyan network with an infinite buffer at the last stage under the maximum conflict traffic pattern. As we expected, the infinite buffer in the small size network reduces the traffic congestion at the last stage, which is due to the conflict between subnetworks. However, the improvement is insignificant for the larger size network.

2) Mixed Traffic Pattern: Fig. 20 shows the maximum throughput capacity of the background traffic for a given SSSD traffic load. The parallel network achieves about 30% improvement in maximum throughput capacity of the background traffic pattern over the single network, and about 8% improvement over the multibuffer network. However, the major improvement comes from shifting the breakpoint locations. The breakpoints of the SSSD traffic ranges from $0.6 \sim 0.7$, $0.6 \sim 0.7$, $0.55 \sim 0.6$, and $0.5 \sim 0.6$ for the network of size 16×16, 32×32, 64×64, and 128×128, respectively. Thus, the background traffic is not influenced by the presence of the SSSD traffic until a higher load compared to that of the single-buffer Banyan network is carried.

Fig. 21 shows the effect of the infinite buffer under the traffic condition of Fig. 20. The infinite buffer improves the throughput performance significantly at a heavy load.

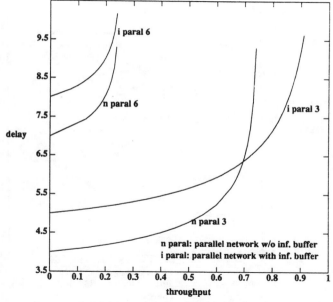

Fig. 19. Parallel network under maximum conflict traffic pattern.

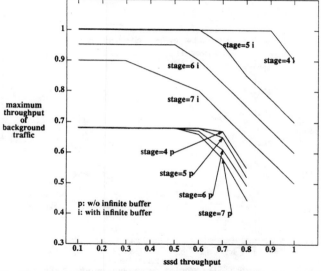

Fig. 21. Maximum throughput of parallel network under mixed traffic pattern.

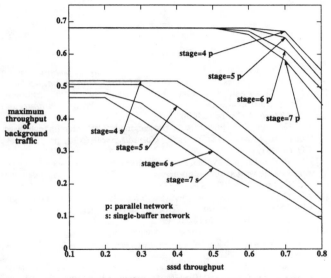

Fig. 20. Maximum throughput of parallel network under mixed traffic pattern.

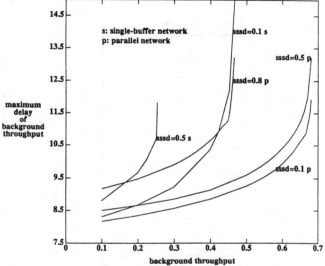

Fig. 22. Effect of SSSD traffic on background uniform traffic in parallel network (stages = 7).

Fig. 22 shows graphs of the uniform background traffic throughput versus maximum delay of background traffic for various SSSD traffic loads. Three curves are plotted for each of the parallel and single-buffer networks. These curves represent different SSSD traffic loads of 0.1, 0.5, and 0.8, respectively. As depicted in Fig. 22, the parallel network has lower delay and higher throughput capacity than the single-buffer network. The improvement due to the infinite buffer is shown in Fig. 23. Since the congestion at the last stage becomes severe with a heavy load, the infinite buffer increases the throughput capacity by relieving the congestion at the last stage.

As the size of the network increases, the gap in the maximum delay of the background traffic between the parallel and the single-buffer networks increases. This behavior suggests that the advantage of the parallel network becomes more apparent for larger size networks. Furthermore, the rate of increase in delay is higher in the single-buffer network than in the parallel network. Even at high SSSD traffic load, the operable range of the parallel network for the background traffic remains up to 0.42 for the network of size 128 × 128, whereas the operable range of the single-buffer network is merely up to 0.22.

The effect of the background traffic on the SSSD traffic is depicted in Fig. 24. As before, the parallel network outperforms the single-buffer network, with an improvement in the throughput capacity and delay. Finally, the effect of the infinite buffer is shown in Fig. 25.

V. CONCLUSION

An analytical method is developed to evaluate the performance of the single-buffer Banyan network under nonuniform traffic patterns. Each port of a switching element is modeled as a Markov chain, and the relationship between switching elements is described by average flow constraints. A similar modeling approach is used to find the analytical method to evaluate the performance of the multibuffer Banyan network and the parallel Banyan network. The results from these analytical methods are compared to the results obtained from the simulation, and they verify the accuracy of the analytical results in the operable range of the network.

The network is analyzed under several nonuniform traffic patterns of interest. This analysis has shown that the single-buffer Banyan network suffers from degradation of performance caused by nonuniformity of the traffic pattern. As the size of the network increases, the detrimental effect of the nonuniform traffic patterns becomes more pronounced.

The single-buffer Banyan network has been modified by using additional buffers and by placing Banyan networks in parallel. The

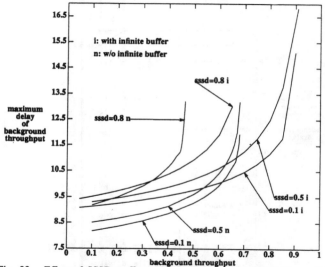

Fig. 23. Effect of SSSD traffic on background uniform traffic in parallel network (infinite buffer).

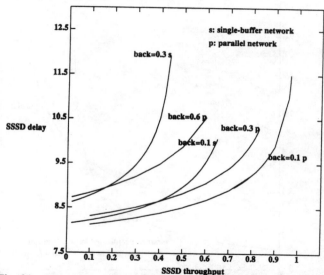

Fig. 24. Effect of background uniform traffic on SSSD traffic in parallel network (stages = 7).

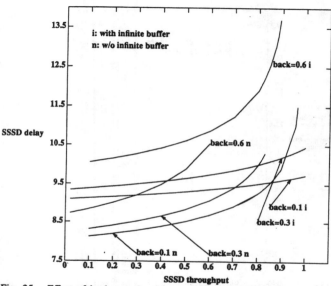

Fig. 25. Effect of background uniform traffic on SSSD traffic in parallel network (infinite buffer).

multibuffer Banyan network shows an improvement in throughput capacity, although this improvement is considered minor when the size of the network becomes large. The parallel Banyan network obtains a significant improvement in both throughput capacity and delay. However, the cost of the implementation is doubled since an additional network is required.

REFERENCES

[1] J. S. Turner and L. F. Wyatt, "A packet network architecture for integrated services," in *Proc. IEEE GLOBECOM'83*, pp. 45–50.
[2] L. R. Goke and G. J. Lipovski, "Banyan network for partitioning multiprocessor systems," in *Proc. 1st Annu. Symp. Comput. Architecture*, 1973, pp. 21–30.
[3] J. H. Patel, "Performance of processor-memory interconnection for multiprocessors," *IEEE Trans. Comput.*, vol. C-30, pp. 771–780, Oct. 1981.
[4] C. P. Kruskal and M. Snir, "The performance of multistage interconnection networks for multiprocessors," *IEEE Trans. Comput.*, vol. C-32, pp. 1091–1098, Dec. 1983.
[5] Y. C. Jenq, "Performance analysis of a packet switch based on single-buffered Banyan network," *IEEE J. Select. Areas Commun.*, vol. SAC-1, pp. 1014–1021, Dec. 1983.
[6] D. M. Dias and J. R. Jump, "Analysis and simulation of buffered delta network," *IEEE Trans. Comput.*, vol. C-30, pp. 273–282, Apr. 1981.
[7] M. Kumar and J. R. Jump, "Generalized delta networks," in *Proc. 1983 IEEE Int. Conf. Parallel Processing*, Aug. 1983, pp. 10–18.
[8] L. T. Wu, "Mixing traffic in a buffered Banyan network," in *Proc. 9th Data Commun. Symp.*, Whistler Mountain, BC, Canada, Sept. 1985.
[9] A. M. Law and W. D. Kelton, *Simulation Modeling and Analysis*. New York: McGraw-Hill, 1982, pp. 295–297.
[10] H. S. Kim, "Performance of buffered Banyan networks under nonuniform traffic patterns," M.A.Sc. thesis, Dep. Elec. Eng., Univ. Toronto, Toronto, Ont., Canada, Dec. 1986.

DESIGN AND ANALYSIS OF BUFFERED CROSSBARS AND BANYANS WITH CUT-THROUGH SWITCHING

Ted Szymanski and Chien Fang

Department of Electrical Engineering and
Center for Telecommunications Research
Columbia University
New York, NY, 10027
e-mail: teds@sirius.ctr.columbia.edu

Abstract: The design and approximate analyses of discrete time buffered crossbar and banyans with cut-through switching are presented. The crossbar switches can contain either (1) input FIFO queueing, (2) input "bypass" queueing where the FIFO discipline is relaxed, (3) a novel scheme called "restricted output queueing" where the number of simultaneous arrivals to an output queue is upper bounded, or (4) a novel combination of input FIFO and restricted output queueing. An analysis for the delay distribution of a packet leaving the network is presented. It is shown that restricted output queueing (or combined input and restricted output queueing) can rival the performance of pure output queueing, while requiring far less hardware; typically, limiting the number of simultaneous arrivals to an output queue to 2 or 3 is sufficient to ensure near optimal performance. It is shown that for light to moderate loads, cut-through switching offers significant improvements since most buffers are empty and cut-through occurs often. A comparison indicates that cut-through switching offers larger performance improvements when newer pin-limited GaAs Integrated circuits are used, rather than conventional CMOS or ECL, due to the similarity to bit-serial transmission in the newer and faster ICs with fewer pins.

1. Introduction

Every analysis of discrete time buffered banyans presented to date assumes that a packet must spend at least one clock cycle in each stage. Using cut-through switching [28], a packet that arrives at an empty queue in one stage can proceed directly to the next stage in the same clock cycle if the link it must traverse is available, thereby minimizing the queueing delays. At light loads cut-through switching behaves like circuit switching, offering low latency. At heavy loads it behaves like packet switching, offering higher bandwidth at the expense of increased queueing delays. The design and approximate analyses for the performance of buffered crossbar and banyans with various queueing schemes and cut-through switching are proposed.

Large discrete time crossbars with input FIFO queueing have a maximum normalized throughput of about 0.586 due to the "Head-of-Line" (HOL) blocking inherent with input FIFO queueing [8,11]. We present an accurate analysis of crossbars and banyans using input FIFO queueing with queues of finite size (by adapting the crossbar analysis in [8]), and an approximate analysis of crossbars and banyans using "input bypass queueing", where the FIFO discipline is relaxed. It is shown analytically that input bypass queueing can approach the throughput of pure output queueing, but has higher delays at heavy loads. The only expense of bypass queueing is that the time slot duration must be increased slightly to allow a queue controller to search the input queue when attempting to find a packet to transmit. (Input bypass queueing was briefly mentioned in [2,8,11] and others, where it was argued that this scheme will be better than input FIFO queueing, but none have presented an analysis for this case. Simulations of banyans using bypass queueing were presented in [2].)

Single stage switches with output queueing and banyans built with nodes containing output queueing can have essentially optimal throughput [3,11,14], since the HOL blocking due to input FIFO queueing is removed by placing the queues at the outputs of the nodes. However, 'pure output queueing' allows an output queue to simultaneously receive as many packets as it has room for, and this property requires either (1) a fully connected network linking the inputs to the output queues, or (2) a regular crossbar which is operated at a much higher speed, or (3) a very fast bus. The required speedup for case (2) is equal to the maximum number of simultaneous packets the output queue can receive.

In this paper, we propose and analyse an obvious but novel scheme which we call 'restricted output' queueing, where an upper limit is placed on the number of simultaneous packet receptions at an output queue. It is shown that single stage switches built with input queues and restricted output queues which can receive between 2 and 4 simultaneous packet arrivals can offer comparable performance to crossbars with pure output queueing, while requiring significantly less hardware. Each output queue can be implemented with standard off-the-shelf multiport RAMs which allow between 2 and 4 simultaneous reads/writes. It is also shown that multistage banyans built with nodes containing restricted output queueing can offer essentially identical performance to banyans built with nodes containing pure output queueing, while also requiring far less hardware. Using our scheme, within each node the crosspoint matrix joining the input ports to the output queues is less expensive than the design in [14], especially as the node size increases. Furthermore, our scheme can simulate pure output queueing using fewer buffers than the design in [14].

It is well known that non-random traffic patterns and hot-spots can reduce the performance of banyans considerably [19]. The degradation due to hot-spots can be removed by (1) combining packets within each node directed to the same hot-spot, (2) "dropping" packets when the performance has degraded, thereby freeing up buffers, or (3) by eliminating hot-spots through the use of distributed software structures [25]. We assume that most hot-spot traffic is removed by using the distributed software structures. We propose a novel scheme for handling remaining hot-spots, using bypass queueing (where the FIFO queueing discipline is relaxed.)

To reduce the degradation due to uniformly distributed but non-random (i.e., deterministic) traffic patterns and to add fault tolerance, we assume throughout this paper that a "Distribution Network" precedes the banyan [2,17]. The use of a distribution network reduces (and can provably eliminate) the performance degradation of a banyan due to non-random uniformly distributed traffic patterns [17].

The number of nodes used in the construction of a network affects the number of integrated circuits needed and the 2 dimensional area and the 3 dimensional volume of printed circuit boards, and hence directly affects the cost of the network. We presents graphs indicating optimal node sizes for various network sizes and various loads. We also propose a reasonable 3 dimensional construction for the banyan based switch with a distribution network.

† This research was supported by NSF Grant CDR-88-11111 through CTR.

Reprinted from *Proc. IEEE Supercomputing*, 1990, pp. 264–273,
November 1990.

To date there exists no simple general analytic model to estimate the delay distribution of a packet leaving a banyan, or to estimate the probability a packet's age exceeds some threshold. We present a simple and general approximate analysis that can be used to compute the delay distribution of packets leaving a series of discrete time FIFO queues with cut-through. For single stage switches with restricted output queueing, the delay distribution is exact. At light-moderate loads, the analytic delay distribution for banyans is reasonably accurate. Our analysis has comparable accuracy to the complicated model derived by Kruskal et. al [14], but ours is simpler and more general. Using these delay distributions, the probability that a packet leaving a banyan is delayed by more than an arbitrary threshold can be estimated.

A final word on the accuracy of our analytic models. Previously, a number of approximate Markov chain analyses for the performance of buffered banyans have been presented (see the references). These analyses all make the simplifying assumption that all buffers/queues in the network are independent. Dias showed that this independence assumption results in relatively large errors for large fully loaded banyans [3], but it remains the only approach (other than possibly state aggregation-disaggregation techniques) to make the model of multistage switches tractable.

Using z-transforms, many researchers have obtained analytic models for single stage switches which are exact as the switch size $N \rightarrow \infty$ [,17,12,8]. The queues must be of infinite size, and the models yield equilibrium solutions only. Such models are not necessarily useful when modelling multistage switches, which have small finite capacity queues within the nodes and which usually have small nodes. Even if one has an exact analytic model for a single node, the primary difficulty in modelling banyans is developing a tandem queue type model where one queue feeds another and accurately modelling the dependencies between these queues. The approximations made here usually limit the accuracy of analytic models for banyans to about 20-40 % at heavy loads. For example, Mitra et. al [17] and Kruskal et. al. [14,15] both developed exact models for single stage switches with output queues of infinite capacity, but the approximations made in the analysis of a banyan resulted in large errors; these models always yield a throughput of unity due to the infinite queue assumption, and in [17] the analysis over-estimated the delay by up to a factor of 2.

Our analysis for pure or restricted output queueing in single stage switches (of finite size and with finite capacity queues) is exact, since the output queues are easily seen to be independent. While we make the independence assumption in our analysis of banyans, we present an extensive set of simulations which show that our analyses are reasonably accurate, usually within 1 % at light loads and typically within 5 % at moderate loads. Another advantage of our Markov chain models is that they yield both transient or equilibrium solutions; the transient solution is useful for estimating the time taken for a set of packets to reach their destinations. We also point out that banyans will not usually be operated at heavy loads, so that model inaccuracies at heavy loads may not be an important issue.

All of these queueing schemes, without cut-through switching, have been analysed and discussed in a related paper [23]. We do not compare the performances with and without cut-through due to space constraints.

This paper is organized as follows. Section 2 includes a review of banyan networks, dilated banyan networks, our cost and delay modelling assumptions, distribution networks and presents our scheme for avoiding degradation due to hot-spot contention. Section 3 gives a brief review of discrete time Markov chains and presents efficient solutions for the equilibrium state of relevant models. Section 4 presents analyses for packet switched banyans with input, output and combined input-output queueing. Section 5 presents an analysis for external queues. Section 6 illustrates the delay distribution calculation. Section 7 includes a comparison of banyans built with equivalent hardware but with nodes of various sizes. Section 8 contains some concluding remarks.

2. Background

Multistage banyans have a number of desirable features; their cost is $O(N\log N)$ nodes / links and they are "self-routing". A square $k^n \times k^n$ SW-banyan network of size N consists of $n = \log_k N$ stages, where each stage consists of N/k nodes, where each node is a crossbar switch of size $k \times k$. (Note that the name "SW-banyan" refers to a specific network topology while the term "banyan" refers to all networks that have a unique path between each input port and each output port.) Our approximate analyses are applicable to any arbitrary $\log N$ stage banyan (i.e., SW-banyan, Omega, etc) provided that the same independence assumption is made for each network.

2.1. Cost and Delay Modelling Assumptions

Assume that each node is built on a single *integrated circuit* (IC) and that all nodes in any one network are of the same size. When comparing the delays of networks built with nodes of different sizes, some normalization is required. Due to pin limitations, it is usually assumed that each IC has a fixed amount of I/O bandwidth (see [5] for a justification). If the bandwidth is evenly distributed over all incident links, then the delay in propagating a fixed sized packet over one link between two $k \times k$ nodes is proportional k. Assuming that the time slot duration is equal to the time it takes to transmit a fixed sized packet between two nodes, then the slot duration for the entire switch will be proportional to the degree of the nodes used in the switch.

Two approaches can be used to increase the link bandwidth of a node. First, a faster technology such as ECL or GaAs can be used. The second approach is to use more ICs, operating in parallel, to implement each node [5]. This approach is basically a "bit-slice" technique, where each IC implements a fraction of the bandwidth of each $k \times k$ node. However, one 'master' IC may have to send control signals to the 'slave' ICs to assure that each IC in one node is in the same state [5].

2.2. Dilated/Replicated Banyan Networks

Two approaches that may improve the performance of banyans are (1) replace each link by multiple parallel links (*i.e., a dilated banyan*), and (2) operate many copies of the network in parallel (*i.e., a replicated banyan*) [11].

A d-dilated banyan can be implemented by space multiplexing each link in a regular banyan so that it can support up to d simultaneous packet transfers. An analysis of packet switched dilated banyans is presented in [22,23], where it was concluded that dilations are not cost effective given the pin limitation assumption. Hence, dilations will no longer be considered.

Another common approach to improve the performance of banyan networks is to *replicate* a banyan r times and to operate all r copies in parallel. A *2-replicated banyan* is shown in fig. 1b.

2.3. Distribution Networks

Wise has presented an efficient 3 dimensional construction of banyans, consisting of two stages of orthogonal printed circuit boards [27]. To completely randomize all traffic in an $\log N$ stage banyan, the distribution network must be another $\log N$ stage banyan less one stage, so that the combination is similar to the $2\log N - 1$ stage Benes network [17]. Since the combination is basically two back-to-back banyans, a 3D construction consists of 4 stages of orthogonal printed circuit boards (two for each banyan).

In some applications it may not be cost effective to provide complete randomization, and simply adding one stage of printed circuit boards may be sufficient. The 3D organization of a 4096×4096 network, built with 4 stages of 8×8 nodes and with 2 extra stages of nodes for randomization (and fault tolerance) is shown in fig. 1c. Each printed circuit board contains a 2 stage 64×64 banyan, with 16 integrated circuits. Assuming the boards are placed about 1/3 of an inch apart, each printed circuit board should be a square with a width of about 24 inches. The entire switch then has dimensions of 2' x 2' x 6', has a total of 3072 integrated circuits, and dissipates at most about 15,000 watts (assuming at most 5 watts per IC).

2.4. A Scheme to Avoid Tree Saturation

When much traffic is directed to a single output, all the buffers in the fan-in tree leading to the output become full, and the networks throughput can drop to 1 or 2 %. Randomization networks cannot solve this problem, and much effort has been directed to find hardware efficient solutions. (One could of course start dropping packets, but this will eventually lead to a flood of re-submissions.) We propose an efficient solution which can guarantee non-hot-spot traffic a fraction of all buffers in the banyan. Our solution is far less expensive than combining in multistage banyans [7,19]. Our solution is based on bypass queueing, and applies to input or output queueing within a node.

We describe our solution for nodes with output bypass queueing (the scheme for input bypass queueing is similar). In this scheme, over each link a queue controller attempts to send the HOL packet into the next stage. If the next stage refuses to accept the packet, the queue controller then attempts to send the second packet in line to the next stage. This process repeats until one packet is accepted or there are no more packets to check. In practice, we should set a limit on the number of attempts.

We do not present an analysis of this queueing scheme in this paper, but its performance must be better than that of output FIFO queueing, which we do analyse. Assume for now there is one hot-spot. Suppose all queues within a banyan have capacity 4, and we wish to guarantee non-hot-spot traffic 2 buffers out of the 4, in every queue. Thus each queue can contain at most 2 packets directed to any destination; the queue controller simply will not accept any more packets from the previous stage if they have the same destination. Therefore, the queues in the fan-in tree can never fill up with hot spot traffic, thus guaranteeing some buffer space for non-hot-spot traffic.

The queue controller need not keep track of destinations. An alternative scheme is the following. Rather than making the decision to accept a packet based on its destination tag, the queue controller should examine the routing digit the packet will use in the next stage. (With input bypass queueing the controller checks the routing digit for the current stage.) If the queue already contains 2 packets with the same routing digit (the assumed limit on hot-spot traffic), it will only accept new packets if they have different routing digits. Our simulations indicate that these schemes are quite effective, when applied to either input bypass or output bypass queueing.

3. Equilibrium Probabilities of Certain Discrete-Time Processes

Our analytic models are based on discrete time Markov chains. Given a state transition diagram and state transition probabilities, there are algorithms for computing the transient response and the equilibrium state for Markov chains. To localize common features of our models (and to avoid interleaving the analysis of banyans with the standard analytic techniques for discrete time Markov chains), the equilibrium analyses of some relevant discrete-time processes will be presented in this section. Readers familiar with such material can proceed to section 4.

In each of the following, let q_j, $(0 \leq j \leq qs)$ denote the probability that a queue has j packets in the steady state, and let q'_j denote temporary (unnormalized) values.

3.1. Single Arrivals, Single Departures

Consider a discrete-time birth-death process which represents a queue capable of holding qs packets, with at most one arrival and one departure in each time interval. A state transition diagram of a birth-death process with state dependent transition probabilities is shown in fig. 2a (the probabilities of remaining in each state are not explicitly shown). Let λ_i denote the transition probability from

state i to state $i+1$, and let μ_i denote that transition probability from state i to $i-1$ The equations for the steady state probabilities are well known and are given by the following (let $q'_0 = 1.0$ initially).

$$q'_i = \prod_{j=0}^{i} \frac{\lambda_j}{\mu_{j+1}} \cdot q'_0 \qquad (1)$$

Eq. (2) normalizes the q' values so that the sum of all probabilities is unity.

$$q_i = q'_i / \sum_{j=0}^{qs} q'_j \qquad (2)$$

3.2. Multiple Arrivals, Single Departures

Consider a discrete-time process which represents a queue capable of holding qs packets, and with at most k arrivals and one departure in each time interval. Let $\lambda_{i,j}$ denote the transition probability from state i to state j for $j > i$ and let $\mu_{i,j}$ denote the transition probability from state i to state j for $j < i$. The equilibrium probabilities of the queue states are given by the following (let $q'_0 = 1.0$). Eq. (2) is used to normalize the q' values.

$$q'_k = \frac{1}{\mu_{k,k-1}} \sum_{j=k-ar}^{k-1} \left[q'_j \sum_{i=k-j}^{ar} \lambda_{i,j} \right] \qquad (3)$$

3.3. Queue States After a Single Departure

Given q_i, $(0 \leq i \leq qs)$, which represent the probabilities that a queue has i packets initially, it will be necessary to find the queue state probabilities after one packet is removed with probability μ. The desired probabilities are given by the following;

$$q'_0 = q_0 + q_1 \cdot \mu \qquad (4)$$

$$q'_i = q_i \cdot (1-\mu) + q_{i+1} \cdot \mu \qquad (5)$$

$$q'_{qs} = q_{qs} \cdot (1-\mu) \qquad (6)$$

3.4. Transient Solutions

The preceding models can also be solved to yield the transient solution, i.e., the states of the systems for each time step, rather than an equilibrium state. In this case, we compute the probabilities for the next time slot based on their values for the current time slot, without using the equilibrium solutions. The probability a queue has i packets at time $t+1$ is simply equal to

$$q_{i,t+1} = \sum_{j=0}^{qs} q_{j,t} \cdot t_{j,i,t} \qquad (7)$$

where $t_{j,i,t}$ is the transition probability from state j to state i at time t.

The transient solution can be used to yield a network's response to periodic offered loads (i.e., when the offered load is periodic in time). In this case, we have observed that the network's response (i.e., throughput) is also periodic. The queue state distributions are also periodic, making the average delay calculation tricky. The average delay can be approximated by the average queue size times the average service time, averaged over the period. This expression gives the same average delay over each clock cycle in the period, but it appears to be the only tractable method.

All of our Markov chain models to be presented can be solved using the transient analysis, to compute system response to periodic offered loads, or bursts of traffic, or to compute the time taken to empty a set of h packets per processor (the "non-regenerative" traffic model considered by Mitra [17]).

4. Markov Chain Analysis for Packet Switched Banyans

Consider the single input buffered banyan examined by Jenq [10]. Each 2×2 node in the banyan now has a single buffer on each of its inputs. Jenq assumed that the network is synchronized with a global clock whose period is referred to as a clock cycle (or time slot). In the first portion of a clock cycle control signals are passed from the last stage towards the first stage, so that every packet knows whether it should move forward one stage or stay in the same buffer. In the second portion of a clock cycle the packets move in accordance with the control signals and the clock cycle ends. The whole process repeats every cycle and packets continue to be shifted into and out of the network. Jenq assumed that there is no cut through switching between nodes; i.e., a packet that arrives at a node must spend at least one clock cycle at each node.

A packet in a buffer selects an output link based on the appropriate bit in its destination tag. If only one packet selects a link, then that packet can be transmitted to the next stage if the receiving buffer in the next stage has sufficient room. If two packets select the same output link, then one of the packets (selected at random) can be transmitted to the next stage if the receiving buffer has sufficient room, and the other packet remains in its buffer. In order for a receiving buffer to have sufficient room, either it is empty at the start of the clock cycle or else it contains a packet that will move forward during the same cycle.

In the preceding example, the clock period will be proportional to $logN$ since the control signals must propagate through all $logN$ stages during each clock cycle before the data transfers can proceed. A better scheme is to operate the network asynchronously (or to stagger the clocks for each stage) to allow each stage to start transmitting a packet it contains as soon as it receives permission from the following stage, so that the clock cycle is independent of $logN$ and equal to the time required to transmit a packet between two nodes. A similar scheme was discussed in [2], where the entire network was synchronized at the bit and packet level. However, an approximate analysis can still assume Jenq's mode of operation, i.e., that a global clock exists to define synchronous clock cycles.

4.1. Input FIFO Queueing

In this section a Markov chain analysis for packet switched $k^n \times k^n$ banyans, with a single FIFO queue of size qs for each incoming link to a node, will be derived. A 4×4 crossbar with input queues is shown in fig. 3a. An $N \times N$ crossbar requires N^2 crosspoints, which limits the maximum size of the crossbar. However, smaller crossbars can be used as the nodes in banyan network of arbitrary size. Assume that there is cut through switching between nodes; a packet that arrives at an empty input queue in stage s can be sent forward to the next stage $s+1$ if it wins contention for the output port and if the next queue is willing to receive it. When a packet enters a non-empty input queue, it moves forward to occupy the first available buffer nearest the head of the queue. In this manner, a packet will not spend any more time than is necessary within the queue.

Let Pr [e] denote the probability that event e occurs, and given a probability x, let $\overline{x}$ denote the probability $1-x$. The following variables are defined.

$q_{i,s}$ = Pr [a stage s input queue contains i packets at the beginning of a clock cycle]

ui_s = Pr [a stage s input queue will offer a packet to a stage s crossbar switch during a clock cycle]

uo_s = Pr [a stage s output link will offer a packet to a stage $s+1$ input queue during a clock cycle]

ar_s = Pr [a stage s queue is able to receive a packet that may be offered to it]

mv_s = Pr [a packet in a stage s input queue will move forward during a clock cycle, given that the buffer/queue is not empty]

We have tried to make the presentation "user friendly". The ui and uo mnemonics denote the *utilization* on the input and output links of a stage s node respectively; ar denotes *able to receive*, and mv denotes *move forward*.

At the beginning of a clock cycle a crossbar input in stage s has a packet with probability $ui_s = \overline{q}_{0,s} + q_{0,s} \cdot uo_{s-1}$, and the probability that output link is selected by at least one HOL packet is given by (8).

$$uo_s = 1 - (1 - ui_s/k)^k \quad (8)$$

Assuming that all packets are uniformly distributed over the output links of a node in every clock cycle, it is straight forward to derive eq. (8) [18].

A packet initially in this stage or one that just arrived to an empty queue at this stage will move forward if it won the contention for an outgoing link (given by uo_s) and if the next stage queue is able to receive a packet (given by ar_{s+1}). Therefore the conditional probability a packet moves forward is given by

$$mv_s = uo_s \cdot ar_{s+1}/ui_s \quad (9)$$

During a clock cycle, an input queue in stage s moves a packet forward with probability mv_s, and it will be able to receive an incoming packet with probability given by the following

$$ar_{s+1} = 1 - q_{qs,s+1} \cdot (1 - mv_{s+1}) \quad (10)$$

The boundary condition is $ar_{n+1} = 1$, which assumes that a packet that reaches its destination is removed instantaneously. (It is straight forward to model the effects of external queues following a banyan network.)

The birth-death process in section 3.1 is used to model a queue in stage s, with the following state transition probabilities.;

$$\lambda_j = uo_{s-1} \cdot (1 - mv_s) \quad \text{for } 0 \le j < qs \quad (11)$$

$$\mu_j = mv_s \cdot (1 - uo_{s-1}) \quad \text{for } 1 \le j \le qs \quad (12)$$

The 'local' equilibrium probabilities for a queue in stage s, assuming that the states of the adjacent stages are static, can be computed directly using the equations (1) and (2) in section 3.1. The 'global' equilibrium probabilities for all variables in every stage (as a function of the load offered to the first stage, uo_0) are found by iteratively computing all variables until they converge (which typically requires a few seconds of computer time). (Note that the global equilibrium forces the traffic flowing through all stages to be equal, since we assume that packets are never dropped once accepted into the network.)

The normalized throughput, defined as the average number of packets that an output link in any stage will move forward during any clock cycle, is given by

$$tp = ui_s \cdot mv_s \quad (13)$$

for any stage s. The normalized delay is defined as the delay through the network divided by the number of stages, yielding the average delay per stage. Using Little's Theorem the normalized delay is given by

$$nd = \frac{1}{n} \cdot \sum_{s=1}^{n} \left[\left(\sum_{i=1}^{qs} q_{i,s} \cdot i \right) \cdot \frac{1}{uo_s} \right] \quad (14)$$

The inner summation in (14) represent the average number of packets in the queue. The latter term represents the average absolute service time that the HOL packet leaving the queue experiences.

4.2. A Refined Analysis for Input FIFO Queueing

The previous analysis makes the common assumption that all queues in the system are independent. We will now adapt the previous analysis to model dependencies between the HOL packets in all queues in any one node. Hui and Aurthurs [8] have presented the following equilibrium analysis for large discrete time crossbar switch with input queueing;

$$(2 - \rho)\lambda^2 - 2(1 + \rho)\lambda + 2\rho = 0 \quad (15)$$

$$\frac{1}{m} = 1 + \frac{\lambda}{2(1 - \lambda)} \quad (16)$$

where λ is the probability that a packet is received in an input queue, ρ is the probability that a queue has a fresh HOL packet given that the queue did not experience HOL blocking in the previous cycle, and where m is the probability that a packet at the head of a queue will move forward.

Hui et. al. used the generating function approach to analyse a queue of infinite capacity. We adapt their analysis to handle a queue of finite capacity, so the model can then be used in the analysis of a banyan.

A queue will receive a fresh HOL packet if it was empty and a packet arrives, or if it had one packet which moves forward and one new packet arrives, or if it has 2 or more packets and the HOL packet moves forward. These events are expressed in the numerator of eq. (17). A queue did not experience HOL blocking if it was empty or if it was not empty and its packet moves forward. These events are expressed in the denominator of eq. (17).

$$\rho = \frac{\left[q_{0,s} \cdot uo_{s-1} + q_{1,s} \cdot mv_s \cdot uo_{s-1} \right] + \sum_{j=2}^{qs} q_{j,s} \cdot mv_s}{q_{0,s} + (1 - q_{0,s}) \cdot mv_s} \quad (17)$$

Given ρ, then λ can be computed using eq. (15), and then m can be computed using eq. (16). By setting

$$uo_s = m \cdot ui_s \quad (18)$$

then the HOL blocking effect within any one node can be modelled with greater accuracy.

Note that the approximate analysis in [8] for HOL blocking is accurate only for large crossbars (i.e., $\geq 16 \times 16$), and hence it will overestimate the HOL blocking in small (i.e., 2×2) crossbars. Our analysis is reasonably accurate when the carried load is less than 0.586 (the maximum throughput of a large crossbar with input FIFO queueing), and when the nodes are of size ≥ 16. Our simulations indicate that it is also quite accurate even for very small switches, provided the load is less than 0.586. The reason for the accuracy for small nodes is probably as follows; the pessimistic throughput estimate due to HOL blocking cancels the optimistic throughput due modelling stages of switches.

4.3. Extension to Input Bypass Queueing

In each time slot the input queue controller examines up to ws packets nearest the head of the queue when attempting to find one to transmit forward. Each time slot has ws phases; in the first phase all HOL packets compete for access to output links. In the ith phase, input queues that have not yet found a packet to transmit to an idle output port attempt to map the ith packet it contains (if any) to an idle output port. Assume that a packet transmission starts after it has found an idle output port, and that the transmission time is much larger than the duration of each phase. Due to the cut-through switching, assume that any packet which arrives to an empty queue can compete for access to the output links in the first phase (i.e., with all other HOL packets).

Let $ui_{p,s}$ denote the probability that an input queue in stage s will attempt to find an idle output port for a packet during phase p. Let $ui_{0,s} = q_{0,s} \cdot uo_{s-1} + \overline{q}_{0,s}$ as before. Let $uo_{p,s}$ denote the probability that an output port in stage s will have a packet to offer the next stage, after phase p. During phase p ($1 \leq p \leq ws$), the probability that an input port will attempt to find an output port is given by

$$ui_{p,s} = \left[\sum_{j=p}^{qs} q_{j,s} \right] \cdot (1 - uo_{p-1,s}) \quad (19)$$

where $uo_{0,s} = 0$. The first term is the probability that the queue has p or more packets, and the second term is the probability that any particular input queue has not yet found a packet to map to an idle output port. The probability that the packet will find an idle output port and win contention for it is given by

$$uo_{p,s} = \left[1 - (1 - ui_{p,s}/k)^k \right] \cdot (1 - uo_{p-1,s}) + uo_{p-1,s} \quad (20)$$

Therefore, the probability a packet moves forward during the time slot is given by eq. (8), with $uo_{ws,s}$ replacing uo_s. The rest of the analysis in section 4.1 remains unchanged. The analysis is approximate, since it still assumes that all packets are randomly distributed over the node's outputs, and our model assumes state-independent transition probabilities. The equilibrium analysis exhibits some numeric instabilities for large ws. In this case, the transient analysis should be used, and iterated until the model converges to its equilibrium point.

4.4. Numerical Results

First, to generate the 95% confidence intervals for all simulations to be presented, the "method of subruns" was used. The network would be first simulated until a steady state was reached, which was arbitrarily defined as the time when at least 100,000 packets where received at the outputs. Then statistics would be gathered over 8 subruns, were each subrun required at least 100,000 packets to be received at the outputs. Hence, each point required at least 900,000 packet receptions.

The analysis and simulations of a 16×16 crossbar with an input queue of size 16 at each input port is shown in fig. 4. Throughout the paper dotted curves are analytic results, solid curves are simulations. The analysis for input FIFO queueing is quite accurate, as expected.

The upper and lower bounds on the performance of input bypass queueing are given by pure output queueing and input FIFO queueing respectively. With bypass queueing, the crossbar throughput increases significantly as the window size ws increases, with diminishing improvements for $ws > 4$. For $ws =$ input queue size, the throughput of input bypass queueing is essentially identical to that of pure output queueing, but the delays are larger at heavy loads, since input bypass queueing no longer services packets within the node in a FIFO order. For light loads, the crossbar and the banyans all exhibit very low delays when cut-through switching is used.

The analytic and simulations results for a $4^3 \times 4^3$ banyan, with various offered loads, is shown in fig. 5. The analysis in section 4.1 yields throughputs which are optimistic by about 0.1, and they are not shown on fig. 5a; Our refined analysis which models HOL blocking has accurate throughputs, but pessimistic delays for large queues.

It is well known that as the input FIFO queue size increases the throughput increases significantly, but the improvements diminish as the queue size increases beyond two [3]. Dias has concluded that the queue size within a node should be small to avoid excessive delays in the network [3], and this conclusion has been repeatedly drawn (i.e., see [26]). However, all of these researchers have ignored the fact that nodes with larger queues carry considerably more traffic. From fig. 5 it is obvious that for a fixed carried load, networks with larger queues always have comparable delays to networks with smaller queues. We therefore conclude that the queues within a node should be as large as possible, given that the node size is fixed.

4.5. Restricted Output Queueing

A 4×4 switch with an output queue for size 4 for each outgoing link is shown in fig. 3b; Each output queue can receive up to 4 packets simultaneously. In general, an $N \times N$ switch with output queueing requires $min(N, qs) \cdot N^2$ crosspoints. A 4×4 switch with input queues and output queues of size 4, but with at most 2 simultaneous arrivals to an output queue, is shown in fig. 3c. In general, such an $N \times N$ switch requires $min(N, qs, L) \cdot N^2$ crosspoints, where L is the upper limit on the number of simultaneous arrivals. For pure output queueing, the node in fig. 3b requires 64 crosspoints, while with restricted output queueing the node requires 32 crosspoints. Some of the crosspoints in fig. 3 are redundant, but we include them for fault tolerance.

The crosspoint savings due to restricted output queueing grows very rapidly as the node size and queue size increase. The parameter L is effectively an upper bound on the amount of hardware required to implement restricted output queueing, or equivalently an upper bound on relative speedup required to implement restricted output queueing.

At the beginning of a clock cycle, each output queue in stage s will attempt to transmit a packet into the next stage. An output queue in the next stage will accept up to L packets destined for it if it has sufficient room. Otherwise, it will accept as many packets as it has room for by selecting these at random from the set of packets destined for it. "Pure" or unrestricted output queueing is modelled by setting L to the output queue size. Packets that are not accepted into the queue remain at the output queue in the preceding stage. Packets that are accepted into an output queue move forward to occupy the first available empty buffer closest to the head of the queue. Due to the cut-through switching, a packet that arrives at an empty output queue in stage s can be propagated through to stage $s+1$ in the same clock cycle if next stage is willing to accept it. Therefore, a packet spends only the minimum amount of time in any queue. Packets that are not accepted into the queue in the next stage remain where they were. This stage then propagates control signals to the previous stage, allowing it to transmit forward.

The following variables are defined.

ui_s = Pr [an incoming link to a node in stage s will offer a packet to its output queue during a clock cycle]

$rx_{i,s}$ = Pr [i packets will be offered to a stage s output queue during a clock cycle]

mv_s = Pr [a packet in a stage $s-1$ output queue will move forward during a clock cycle, given that the queue is not empty]

ar_s = Pr [a stage s output queue is able to receive a packet during a clock cycle]

$q_{i,s}$ = Pr [a stage s output queue contains i packets at the beginning of a clock cycle]

$q'_{i,s}$ = Pr [a stage s output queue contains i packets after possibly transmitting a packet forward during a clock cycle]

Our analysis for a single node is exact. Hluchyj and Karol have presented an analysis for a single node with pure output queueing [11]. For a single node, our answers agree exactly with those given by their analysis, since the models are essentially identical. However, out model has been generalized to include restricted output queueing, and our model naturally handles multistage networks (although the independence assumption has been made for tractability).

We first compute the state of an output queue in stage s after possibly moving one packet forward, (denoted $q'_{i,s}$), using eq. (4)-(6) with departure probability $\mu = mv_{s+1}$. We then compute the probability that a packet from any other the previous stages will move forward into this queue, as follows (let $ui_s = \overline{q}_{0,s-1} + q_0 \cdot (1 - rx_{0,s-1})$).

We first initialize rx as follows.

$$rx_{i,s} = \binom{k}{i}(ui_s/k)^i(1 - ui_s/k)^{k-i} \qquad (21)$$

for $i < L$, and

$$rx_{L,s} = \sum_{i=L}^{k}\binom{k}{i}(ui_s/k)^i(1 - ui_s/k)^{k-i} \qquad (22)$$

and $rx_i = 0$ for $i > L$. An equation which computes the conditional move probability by considering the fraction of packets which move forward successfully is given by the following

$$mv_s = ui_s^{-1}\sum_{i=0}^{qs}\sum_{r=0}^{L}q'_{i,s,t} \cdot rx_r \cdot min(r,qs-i) \qquad (23)$$

The first summation in (23) considers every possible initial queue state and the second summation considers every possible number of arrivals.

The discrete-time process in section 3.2 is used to model an output queue in stage s, with the following state transition probabilities.

$$\mu_i = mv_{s+1}\cdot(1 - ui_s)^k \qquad (24)$$

for $1 \le i \le qs$.

$$\lambda_{i,i+j} = (1 - mv_{s+1})\cdot rx_j + mv_{s+1}\cdot rx_{j+1} \qquad (25)$$

for $i+j \le qs$, and

$$\lambda_{i,i+j} = (1 - mv_{s+1})rx_j + \sum_{x=j+1}^{qs-k} rx_x \qquad (26)$$

for $i+j = qs$, which reflects the event that a queue cannot receive more packets than it has room for, but it can receive up to L packets simultaneously.

The local equilibrium probabilities of an output queue in stage s (based on the states of its adjacent stages) can then be computed using eq. (3) and (2). The global equilibrium probabilities for all variables in every stage are found by iteratively computing all variables until they converge, which typically requires a few seconds of computer time. The normalized throughput is given by

$$tp = ui_s \cdot mv_{s+1} \qquad (27)$$

for any stage s and the normalized delay is given by

$$nd = \frac{1}{n}\cdot\sum_{s=1}^{n}\left[\left(\sum_{i=1}^{qs}q_{i,s}\cdot i\right)\cdot\frac{1}{uo_s}\right] \qquad (28)$$

The derivation of eq. (28) is similar the the derivation of eq. (14) in the preceding analysis.

4.6. Numeric Results

The analytic and simulations results for a 16×16 switch with output queues of size 16 are shown in fig. 6. This graph indicates that $L = 3$ or 4 offers essentially identical performance to $L = 16$, as expected. In other words, it is very rare for more than 3 or 4 packets to arrive at an output queue simultaneously.

For $L = 2$, there is a reasonable probability that a packet will not be accepted into the output queue. To maximize performance when $L = 2$, we have removed some of the buffers from the output side and placed them on the input side, yielding a combined input and restricted output queueing scheme. We assume a packet can cut-through the input queue.

Illias and Denzel have presented an analysis for a single switch with input queueing and pure output queueing, with input queues of infinite capacity and output queues of finite capacity [12]. Their analysis is exact as the switch size approaches infinity. Our model differs from theirs since we use restricted output queueing. Our scheme requires far fewer crosspoints (or "speedup") to achieve the same throughput.

The performance of a switch with combined input FIFO queueing and restricted output queueing, with input queues of size 4 and output queues of size 12 are also shown on fig. 6. (A discussion of how to analytically interface an input queue to an output queue within one crossbar is discussed in section 5, and the analysis of a banyan node is similar.)

Our analysis and simulations indicate that the throughput remains essentially constant as the input queue size increases and the output queue size decreases (so the same number of buffers is used), until the output queue becomes "too small" (i.e., 4 or fewer buffers), at which point the throughput drops. Since in a real application traffic may not always be uniform, a good design would probably have an equal number of buffers on each side. Extensive simulations indicate that once the input and output queue sizes are fixed, $L = 2$ is optimal. The throughput increase when $L = 3$ or $L = 4$ is almost always less than 0.1 percent, when input and restricted output queueing is used!

196

The performance of a $4^3 \times 4^3$ banyan ($N=64$), at various offered loads, is shown in figs. 7. Observe that the analysis is reasonably accurate, typically to within 5 percent, in the underload region. Hence, these models will be useful for the approximate but reasonably accurate analysis of banyans operating under realistic loads.

In a multistage banyan network using restricted output queueing, there is no need for any input queues within each node, since a packet can remain queued in the last stage if it is temporarily not accepted into the next stage. Our analysis and simulations indicate that in a multistage banyan, there are no significant differences between restricted output queueing or combined input and restricted output queueing, provided external queues precede the banyan to reduce packet loss probabilities (see section 5).

5. External Queues

In a real application a network may require external input queues to avoid packet losses when new packets are not accepted into the network in any one clock cycle. Assume that in each clock cycle, each external input FIFO queue is offered a new packet with probability a (the *offered load*) and that each network input port is able to accept a packet from the external queue with probability d. (It is straight forward to model multiple arrivals rather than single arrivals, and similarly for multiple departures). The offered load to the network (which we have denoted uo_0) is set equal to the probability that an external queue is not empty.

The state transition diagram for an input FIFO queue is given in fig. 2a. Let q_i be the probability that the queue contains i packets. Assume a packet that arrives into an empty queue can cut-through in the same clock cycle experiencing zero waiting time. In this case, the state transition probabilities for the birth death model in section 3.1 are given by $\lambda_i = a \cdot (1-d)$ for $0 \leq i \leq qs$ and $\mu_i = d \cdot (1-a)$. (This input queue model is used when analysing combined input-output queueing within a node, when a packet can cut-through an empty input queue, as discussed in section 4.6) The offered load to the network uo_0 is then given by

$$uo_0 = (1 - q_0) + q_0 \cdot a \qquad (29)$$

and the average delay in the queue is then

$$\sum_{i=0}^{qs} q_i \cdot i \cdot \frac{1}{uo} \qquad (30)$$

The model is solved as described in section 3. In the steady state, the probability that a packet is lost due to input queue overflow must also be equal to $a - tp$ for constant a, where tp is the normalized throughput. An external output queue can be handled in the same manner.

6. Delay Distribution

Kruskal, Snir and Weiss have derived a very complicated analysis for the delay distribution of packets leaving a banyan built with nodes containing output queues of infinite size only [15]. Due to its complexity, their analysis is not easily extendible to banyans built with nodes containing other queueing variations. Furthermore, their final equations made use of a number of parameters that were determined from simulations. Our approach, while approximate, is considerably simpler and considerably more general, and yields accuracy comparable to that in [15].

Let mv_{s+1} denote the steady state probability that a packet at a stage s buffer/queue will move forward during any clock cycle. Assuming that the event of moving forward at any stage is independent from cycle to cycle, the probability that a packet spends k clock cycles in a stage s queue which contains l packets, denoted $pdf_s(k,l)$ for $1 \leq k \leq \infty$, is given by the negative binomial distribution

$$pdf_s(k,l) = \binom{k-1}{l-1} mv^l (1-mv)^{k-l} \qquad (31)$$

which reflects the event that $l-1$ packets move forward in the first $k-1$ clock cycles and the packet in question moves forward on the kth clock cycle.

The expected distribution of the delay at stage s, denoted $Pdf_s(k)$, (where k is the delay), is given by

$$Pdf_s(k) = \sum_{l=1}^{qs} pdf_s(k,l) \cdot q_{l,s} \cdot \frac{1}{1 - q_{0,s} + q_{0,s} \cdot a \cdot mv} \qquad (32)$$

if $k \geq 1$ and

$$Pdf_s(0) = \frac{mv \cdot q_{0,s} \cdot a}{1 - q_{0,s} + q_{0,s} \cdot a \cdot mv_s}$$

Eq. (32) is a weighted average of the delay distribution for packets which experience queueing delays in a stage, and packets which cut-through a stage with no delay. In eq. (32) a is the probability one or more packets are offered to a queue in one time slot. The distribution of the total delay through an n-stage banyan is given by

$$Pdf = Pdf_1 * Pdf_2 * \cdots * Pdf_n \qquad (33)$$

where '*' denotes discrete convolution. (For combined input-output queueing within each node, the delay through each queue must be considered.) The results for a 16×16 crossbar and a $16^2 \times 16^2$ banyan with nodes having output queues of size 8, for an offered load of 0.5 appear in fig. 8. In general, we have observed that the analysis for multistage banyans is reasonably accurate for light/moderate loads. The above analysis is exact for single stage switches with restricted output queueing.

The delay in any external queue at the inputs or outputs of the banyan network can be included in a similar manner. The technique can clearly be applied to analyse the delay distributions of queueing in other discrete time systems (such as rings or busses) given the queue state probabilities and the move probability.

7. Comparisons Of Various Networks With Equivalent Cost

An $N \times N$ banyan built with 2×2 nodes requires $N/2$ nodes per stage and $n = \log_2 N$ stages, and an $N \times N$ banyan built with $k \times k$ nodes requires N/k nodes per stage and $\log_k N$ stages. Given $nN/2$ nodes, one can build one $N \times N$ banyan using 2×2 nodes or in general $(\log_2 k \cdot k)/2$ banyans built with $k \times k$ nodes. Franklin concluded that to maximize bandwidth the nodes should be as large as possible [5]. (The largest crossbars would use bit-serial rather than bit-parallel links, and they would also have the largest delays; see section 2.1.) Delays depend on the node size, and Kruskal and Snir presented graphs outlining the domains where various node sizes are best, assuming output queues of infinite capacity Our models are much more accurate when used to compute optimal parameter values, since they model back-pressure due to finite queue sizes.

The total delay through the network is modelled as

$$td = \frac{1}{f_c} \cdot \left[w + \log N - 1 \right] + nd \cdot \log N \cdot \frac{w}{f_c} \qquad (34)$$

where f_c is the clock rate, w is the number of words in a packet, nd is the normalized delay (i.e., per stage), and $\log N$ is the number of stages. The term in brackets is the the minimum transmission time through the network, i.e., assuming no waiting was experienced in any stage. The second term is the average total delay. For low loads the first term in (34) dominates and at high loads the second term dominates. In the intermediate regions the delay may be over estimated somewhat.

We assume 256 pin CMOS IC's running at 50 Mhz and 32 pin GaAs IC's running at 125 Mhz are used. Three packet sizes were considered; 128, 256 and 512 bits with 24 addressing bits and 20 bits of error correction. The number of words in a packet is given by the number of bits in the packet divided by the number of IO pins per link. For example, a 2x2 node built on a CMOS IC with 256 pins would have 64 IO pins per link (ignoring power and ground). Therefore, a 128 bit packet with 44 bits of overhead would require 3 64-bit words (when 2x2 nodes are used), 6 32-bit words (when 4x4 nodes are used), and 11 16-bit words (when 8x8 nodes are used).

For the same cost as 1 banyan built with 2×2 nodes one can built $(k\log k)/2$ banyans built with $k×k$ nodes. We will therefore compare a single network built of 2x2 nodes against 4 parallel banyans built with 4x4 nodes, against 12 banyans built with 8x8 nodes, etc. We also evaluate each option at the same real aggregate carried load (i.e., we adjust the offered loads in each option to reflect the clock speed and the number of IO pins per link). For roughly comparable hardware costs, we assume that queue sizes of 8, 4, 2 and 1 were used for 2x2, 4x4, 8x8 and 16x16 nodes respectively.

Figure 9 shows that smaller switches perform better for smaller network sizes and at light loads. As the load (or network size) increases, parallel networks built with larger nodes become better; For CMOS technology, 2x2 nodes are best at light loads and for small N. For intermediate loads and network sizes, the 4x4 nodes are best, and at very high loads and large network sizes the 8x8 nodes are best. For GaAs technology, the domains of best performance are similar, except that they are shifted to make smaller nodes better more often.

The results are in general agreement with Kruskal and Snir's except that the region where 2x2 nodes are best is greatly expanded and that of 8x8 nodes is vastly diminished. This is partly due to the fact that our queue sizes decrease as the nodes gets larger, whereas in [14] an infinite queue size was assumed. Also when cut-through switching is used the fact that banyans built with smaller nodes have many more stages is no longer a drawback, since packets can cut-through most stages very often at light loads.

8. Conclusions

Approximate analyses for the performance of crossbars and banyans with various queueing schemes and cut through switching were presented. Of theses schemes, restricted output queueing and combined input plus restricted output queueing are novel.

A simple analysis for the delay distribution of a packet leaving the network was presented. The probability a packet's age exceeds a threshold can be computed from the delay distribution. In practice a banyan will be lightly loaded (i.e., 50 % [2]) so that temporary traffic surges can be handled without overload; These analyses are reasonably accurate for light/moderate network loads.

It was shown that input bypass queueing can have comparable throughput to output queueing, without requiring any extra crosspoints. The only "speedup" needed is time to allow a queue controller to search the input queue, which is insignificant for large enough packets. However, it has higher delays since the packets are not serviced in FIFO order. A window size of about 4 appears to best, with diminishing returns for larger window sizes.

For restricted output queueing only, allowing 3 simultaneous arrivals to a queue is sufficient to obtain about 95 % of the performance of pure output queueing, and 4 arrivals obtains > 99 %. Thus, a speedup by a factor of 3 or 4 offers essentially identical performance to pure output queueing. (The Knockout switch [24] is unbuffered, but a Knockout switch with queues at the output would be one way to implement restricted output queueing.) In combined input plus restricted output queueing, allowing two simultaneous arrivals to a queue is optimal, i.e., a "speedup" by a factor of 2 is optimal; any further speedup results in negligible improvement.

Finally, our analytic models can be extended to handle multiple traffic classes, and to "movable boundary" type schemes, where each traffic class is offered a fraction of the total switch bandwidth [21,22].

9. References

[1] H. Ahmadi and W.E. Denzel, "A Survey of Modern High Performance Switching Techniques", IEEE JSAC, VOL SAC-7, No. 7, Sept. 1989, pp. 1091-1103

[2] R.G. Bubenik and J.S. Turner, "Performance of a Broadcast Packet Switch", IEEE Trans. Comm., Vol. 37, No. 1, pp. 60-69

[3] D.M. Dias and J.R. Jump, "Analysis and Simulation of Buffered Delta Networks", IEEE Trans. Comput., Vol. C-30, No. 4, April 1981, pp. 273-282

[4] D.M. Dias and M. Kumar, "Packet Switching in Log N Multistage Networks", Proc. Globecom 1984, pp. 5.2.1-5.2.7

[5] M.A. Franklin, D.F. Wann and W.J. Thomas, "Pin Limitations and Partitioning of VLSI Interconnection Networks", IEEE Trans. Comput., Vol. C-31, No. 11, Nov. 1982, pp. 1109-1116

[6] M.A. Franklin, "A VLSI Performance Comparison of Banyan and Crossbar Communications Networks", IEEE Trans. Comput., Vol. C-30, No. 4, April 1981, pp. 283-290

[7] A. Gottieb et. al, "The NYU Ultracomputer: Designing a MIMD, Shared-Memory Parallel Machine", IEEE Trans. Comput., Vol. C-32, No. 2, Feb. 1983, pp. 175-189

[8] J. Hui and E. Arthurs, "A Broadband Packet Switch for Integrated Transport", IEEE Journal on Sel. Areas of Comm., Vol. SAC-5, No. 8, Oct. 1987, pp. 1264-1273

[9] M. Ilyas and H.T. Mouftah, "Towards Performance Improvement of Cut-Through Switching in Computer Networks", Performance Evaluation, 6, 1986, pp. 125-133

[10] Y-C. Jenq, "Performance Analysis of a Packet Switch Based on Single-Buffered Banyan Network", IEEE Journal Selected Areas of Comm., Vol. SAC-1, No. 6, pp. 1014-1021, Dec. 1983

[11] M.G. Hluchyj and M.J. Karol, "Queueing in High Performance Packet Switching", IEEE JSAC, Vol. 6, No. 9, Dec. 1988, pp. 1587-1597

[12] I. Iliadis and W.E. Denzel, "Performance of Packet Switches with Input and Output Queueing", IEEE Int. Conf. Comm. (Supercom/ICC'90), 1990

[13] H.S. Kim and A. Leon-Garcia, "Performance of Buffered Banyan Networks Under Nonuniform Traffic Patterns", IEEE Infocom '88,

[14] C.P. Kruskal and M. Snir, "The Performance of Multistage Interconnection Networks for Multiprocessors", IEEE Trans. Comput., Vol C-32, pp. 1091-1098, Dec. 1983

[15] C.P. Kruskal, M. Snir and A. Weiss, "The Distribution of Waiting Times in Clocked Multistage Interconnection Networks", IEEE Trans. Comput., Vol. 37, No. 11, Nov. 1988, pp. 1337-1352

[16] M.A. Marsan, G. Balbo, G. Chiola, A. Ciccardi, G. Conte, "Estimating the Average Delay in a Delta Network Operating According to the Cut-Through Packet Switching Technique", in Performance of Distributed and Parallel Systems, Ed. T. Hasegawa et. al., North-Holland, 1989, pp. 491-510

[17] D. Mitra and Cielsak, "Randomized Routing on an Extension of the Omega Network", JACM, Vol. 34, No. 4, Oct. 1987, pp. 802-824

[18] J.H. Patel, "Processor Memory Interconnections for Multiprocessors", IEEE Trans. Comput., Vol. C-30, No. 10, Oct. 1981, pp. 771-780

[19] G.F. Pfister and V.A. Norton, "Hot Spot Contention and Combining in Multistage Interconnection Networks", 1985 IEEE Int. Conf. Parallel Processing, pp. 790-797

[20] S.Z. Shaikh, M. Schwartz and T.H. Szymanski, "Analysis, Control and Design of Crossbar and Banyan Based Broadband Packet Switches for Integrated Traffic", IEEE Supercom/ICC'90, pp. 761-765

[21] S.Z. Shaikh, M. Schwartz and T.H. Szymanski, "Performance Analysis and Design of Banyan Network Based Broadband Packet Switches for Integrated Traffic", IEEE Globecom' 89, pp. 1154-1158

[22] T.H. Szymanski and S. Shaikh, "Markov Chain Analysis of Banyan-Based Fast-Packet Switches with Arbitrary Switch Sizes, Queue Sizes, Link Multiplicities and Speedups", Infocom' 89, pp. 960-971

[23] T.H. Szymanski and S. Shaikh, "Markov Chain Analysis of Banyan-Based Fast-Packet Switches with Input, Output and Combined Input-Output Queueing", submitted for publication, to appear as a CTR Tech. Report

[24] Y.S. Yeh. M.G. Hluchyj, A.S. Acampora, "The Knockout Switch: A Simple, Modular Architecture for High Performance Packet Switching", Int. Switching Symp., 1987, pp. 801-808

[25] P. Yew, N. Tzeng and D.H. Lawrie, "Distributed Hot-Spot Addressing in Large Scale Multiprocessors", 1986 IEEE Int. Conf. on Parallel Processing, pp. 51-58

[26] H. Yoon, K.Y. Lee and M.T. Lui, "Performance Analysis of Multibuffered Packet-Switched Networks in Multiprocessor Systems", IEEE Trans. Comput., Vol. 31, No. 3, March 1990, pp. 319-327

[27] D. Wise, "Layouts for Compact Banyan/FFT networks", Proc. CMU Conf. VLSI Systems and Computations, Computer Science Press, 1981, pp. 186-195

[28] P.Kermani and L.Kleinrock, "Virtual Cut-Through: a New Computer Communication Switching Technique", Computer networks, 3, pp. 267-286, 1979

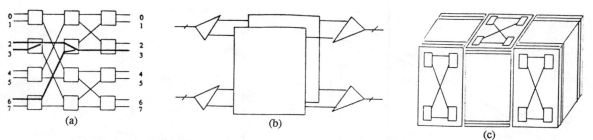

Figure 1: (a) A $2^3 \times 2^3$ SW-banyan network connecting 8 inputs and 8 outputs. Partial paths are shown by the thick lines. Link contention occurs in the first and second stages. (b) A 2 replicated network. (c) Efficient 3D construction of a banyan-based fast-packet switch. A 4-stage banyan network is preceded by multiple 2 stage banyans as the distribution network.

Figure 2: State diagram for a discrete-time birth-death process which represents a queue capable of holding at most 3 packets, with single arrivals and single departures per clock cycle.

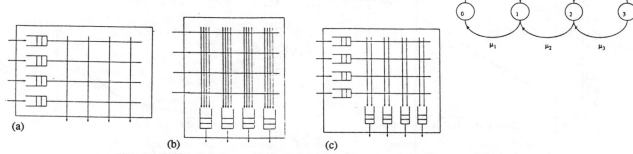

Figure 3: (a) A 4×4 switch with input queueing. (b) A 4×4 switch with output queueing. (c) A 4×4 switch with input queueing and restricted output queueing.

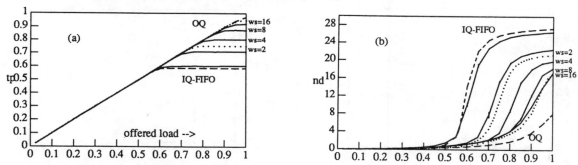

Figure 4: (a) *tp* and (b) *nd* versus offered load for a 16×16 switch with qs = 16 and input bypass & input FIFO queueing; Simulations are shown for ws = 2, 4, 8 and 16; analysis are shown for ws = 1, 2 and 16.

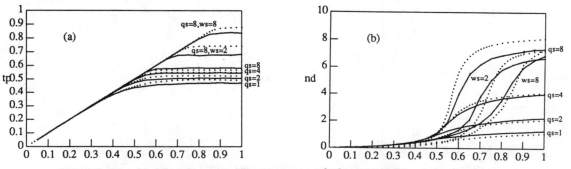

Figure 5: (a) *tp* and (b) *nd* versus offered load for a $4^3 \times 4^3$ banyan with input bypass & input FIFO queueing. In (a) *tp* curves are from section 4.2. In (b), the analytic curves for *nd* are from 4.1. Analysis and simulations are shown for input bypass & input FIFO queueing, qs=8, for ws=2 and ws=8.

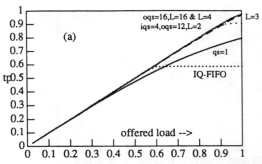
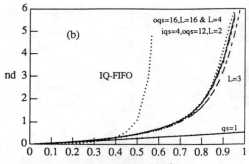

Figure 6: (a) *tp* and (b) *nd* versus offered load for a 16×16 switch with pure and restricted output queueing. Analysis and simulations are shown for qs= 1 & 16; Dashed curves are analyses for restricted output queueing with for qs=16, L=3 and qs=16, L=4.

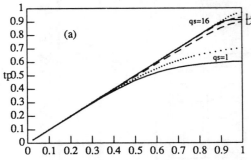
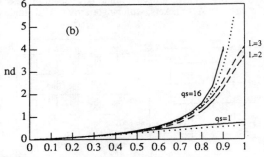

Figure 7: (a) *tp* and (b) *nd* versus offered load for a $4^3×4^3$ banyan with output queueing. Simulations and analysis are shown for qs = 1 & 16. Dashed curves are analytic results for restricted output queueing for qs=8, with L=3 and qs=8 & L=2.

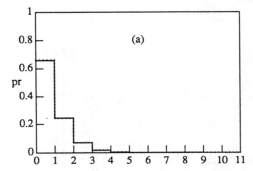
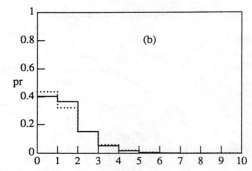

Figure 8: Analysis and simulations showing the probability distribution of the delays for (a) 16×16 switch and (b) $16^2×16^2$ banyan with pure and restricted output queueing, qs = 8, L = 4 & 8, and an external load of 0.5. The graphs for L = 4 & 8 are superimposed on each other.

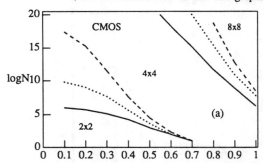
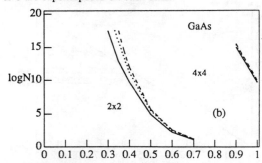

Figure 9: Domain of best performance for (a) CMOS and (b) GaAs based replicated square banyan networks with input bypass queueing and cut-through. Results for 3 different packet sizes, 128, 256 & 512 bits (plus 44 bits header) are shown as solid, dotted & dashed curves respectively;

Performance Analysis of Buffered Banyan Networks

Thomas H. Theimer, Erwin P. Rathgeb, and Manfred N. Huber, *Member, IEEE*

Abstract—Banyan networks are used in multiprocessor computer applications as well as in new, high performance packet switch architectures. In this paper, we will give a classification of the most common Banyan networks and outline an analysis approach for the rather general class of Delta-*b* networks with multiple buffers. Based on this approach, we will discuss the effects of the approximations involved and present a refined analysis algorithm for the special case of a single buffered Delta-2 network. The results of both algorithms will be compared with simulation results to assess their accuracy.

I. INTRODUCTION

IN ADDITION to the existing services offered by today's Integrated Services Digital Network (ISDN), future networks should also include broad-band services like video communication, graphic applications and high speed data communication. For this integrated broad-band communication network (IBCN), the asynchronous transfer mode (ATM) has been proposed as a well suited switching principle.

The ATM technique allows the integration of various services with different bandwidth requirements into a uniform network since the information transfer is based on packets (cells). Future services with requirements which are unknown today can be added easily. The characteristic features of ATM are high speed links, simplified protocols, and high capacity switching nodes. The required throughput can only be achieved by a distributed and highly parallel architecture of the switching nodes.

In several publications [5], [11], [20], [21] Banyan networks have been proposed for the implementation of an ATM switching node. These interconnection networks are well known from multiprocessor computer applications where they are used for the interconnection of multiple processors and memories. Early work in the performance analysis of unbuffered Banyan networks was done by Patel [16], who derived a recursive formula for the evaluation of the network throughput. Since his results were indicating that the maximum throughput is rapidly decreasing with the size of the network, several performance enhancements have been proposed, e.g., the use of parallel links and multiple subnetworks [3], [12].

A significant performance improvement can be achieved if buffers are located in each switching element of the network

Paper approved by the Editor for Communication Switching of the IEEE Communications Society. Manuscript received May 30, 1989; revised October 10, 1989. A version of this paper was presented at the International Seminar on Performance of Distributed and Parallel Systems, Kyoto, Japan, December 7–9, 1988. This work was performed while E. P. Rathgeb and M. N. Huber were with the Institute of Communications Switching and Data Technics at the University of Stuttgart.

T. H. Theimer is with the Institute of Communications Switching and Data Technics, University of Stuttgart, Seidenstr. 36, D-7000 Stuttgart 1, Germany.

E. P. Rathgeb is with the Information Networking Research Laboratory at Bellcore, Morristown, NJ 07960-1910.

M. N. Huber is with the Central Laboratory of the Public Communication Networks Group of Siemens AG, Hofmannstr. 51, D-8000 Munich, Germany.

IEEE Log Number 9040882.

because in this case collisions can only occur if several packets require the use of the same output port of a switching element. Consequently, buffered Banyan networks are attractive for the implementation of interconnection networks with very high throughput requirements.

The performance of buffered Banyan networks has been discussed in various publications. Dias and Jump [2] as well as Jenq [8] have analyzed networks consisting of 2×2 switching elements with single-packet input buffers. The method of Jenq has been extended in [7] to allow the analysis of switching elements with an arbitrary (finite) buffer size and more than two inputs. Kruskal and Snir [9] considered the case of infinite capacity output buffers, and their work has been continued in [10] where an approximation for the waiting time of packets in all stages of the network is derived.

A further aspect for the performance evaluation of Banyan networks is the influence of the external traffic patterns which are reflecting the mapping of connections to the internal traffic flows within the network. A first approach to the analysis of single buffered Banyan networks in a nonuniform traffic environment is described in [23], and it shows that certain nonuniform traffic patterns can have a crucial influence on the performance of the network.

In this paper, we focus on the performance analysis of buffered Banyan networks in a uniform traffic environment. After a general classification of Banyan networks, the concepts for the analysis of Delta-*b* networks, a subclass of the Banyan networks, are briefly described, followed by a discussion of the approximations which are necessary to make the analysis feasible. In the sequel, we will develop a refined approach for the analysis of single buffered Delta-2 networks where we try to eliminate the main approximations having an influence on the accuracy of the results. Finally, we compare the accuracy of several analytical methods to the corresponding simulation results.

II. CLASSIFICATION OF BANYAN NETWORKS

In the literature, several detailed surveys of interconnection networks for various applications can be found in [4], [15], [18], [19]. According to the terminology of Feng [4], Banyan networks have a dynamic topology and belong to the class of multistage interconnection networks. They have been defined by Goke and Lipovsky [6] using graph theoretical methods, but in a simplifying approach we shall consider a Banyan network as a multistage interconnection network which has the property that there is exactly one path between any input port and any output port (single path network). The general class of Banyan networks can be structured into several subclasses.

An *L-level Banyan* is a Banyan where only adjacent stages are connected by links. This means that each path leads exactly through *L* stages. The *L*-level Banyan class can be subdivided into *Regular* and *Irregular Banyans*. Regular Banyans are constructed from a single type of basic elements with *F* inputs

Reprinted from *IEEE Trans. Commun.*, vol. 39, no. 2, pp. 269–277,
February 1991.

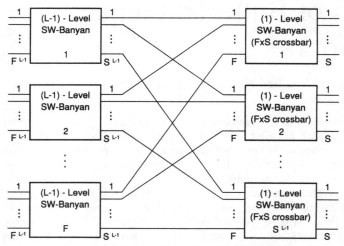

Fig. 1.   Recursive construction of an (L)-level SW-banyan.

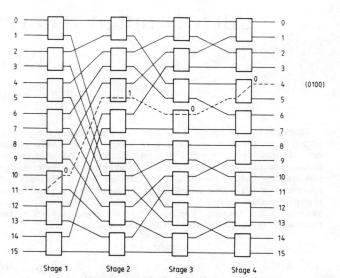

Fig. 2.   Delta-2 network with 4 stages and 16 input terminals.

and S outputs, whereas the type of the elements of an Irregular Banyan is varying throughout the network. The Generalized Delta-networks [3] are an example of Irregular Banyans.

For an economical implementation of switching networks regular Banyans will be preferred because they are composed of identical elements. Two subclasses of regular Banyans are the CC-Banyans and the SW-Banyans (see [6]).

We will restrict our further discussions to the SW-Banyan class, because it includes most of the existing implementations.

As pointed out in [6], large Banyan networks can be constructed in a modular way from smaller ones using appropriate interconnection schemes. In particular, an SW-Banyan is synthesized recursively from a basic crossbar structure with F inputs and S outputs, which is one of the simplest Banyan structures and can be regarded as a (1)-level SW-Banyan. To obtain SW-Banyan structures with L levels, several $(L-1)$-level SW-Banyans are completely interconnected with an extra stage of $(F \times S)$ crossbars as depicted in Fig. 1. The interconnection scheme between the $(L-1)$-level SW-Banyans and the $(F \times S)$ crossbars must be canonical, i.e., output i of all SW-Banyans must be connected to crossbar number i.

Delta networks as defined by Patel [16] have the topological structure of SW-Banyans. They are constructed of n stages corresponding to the L levels of the SW-Banyan. Each stage consists of several switching elements with a input and b output ports, so the network has a^n input terminals and b^n output terminals.

In addition to the topological structure, the definition of Delta networks includes the application of a specific routing scheme called "self routing." Using this scheme, the movement of the packets through a Delta network is controlled by a destination address which must be unique for every output terminal of the network. The destination address is a number of base b with n digits corresponding to the n stages of the network. Each digit directly specifies which output port of a switching element in a specific stage will receive the packet.

This *digit controlled routing* (self routing) mechanism has the advantage that it can be implemented in hardware and that a central control is only needed for the initial setup of a virtual connection. The switching functions are distributed over all switching elements and can be done in parallel.

Rectangular Delta networks are constructed from switching elements which have the same number of inputs and outputs

($a = b$). Consequently, the number of switching elements per stage is constant and the number of network inputs is equal to the number of network outputs. A rectangular Delta network composed of b-input switching elements is called *Delta-b network*.

An example of a Delta-2 network with 4 stages is shown in Fig. 2. The dashed line indicates the path of a packet from input terminal 11 to output terminal 4, which has the binary destination address "0100."

Bidelta networks (Bidirectional Delta networks) have a special topological structure; they remain a Delta network even if the network input links are interpreted as network output links and vice versa. All Bidelta networks are topologically equivalent and can be transformed into each other by relabelling the switching elements and the links [3].

Many of the well known interconnection networks including the Baseline [22], Reverse Baseline [22], Omega [13], Flip [1], Indirect Binary n-Cube [17], and Modified Data Manipulator [22] networks belong to the class of Bidelta networks. The network depicted in Fig. 2 has the topology of a Baseline network.

III. PERFORMANCE ANALYSIS

The performance of an interconnection network is usually characterized in terms of its throughput, the mean transfer time of the packets and the packet loss probability. These characteristics must be related to the offered traffic load of the network, which we define as the number of arriving packets in an interval divided by the maximum possible number of arrivals within this interval. Consequently, the throughput is given by the number of accepted packets divided by the maximum number of arrivals, and the packet loss probability can be derived from the offered traffic load and the corresponding throughput.

We make the following assumptions for the operation and the environment of the interconnection network.

1) The network is operated synchronously, that means the packets are transmitted only at the beginning of a time slot given by the packet clock and thus the time axis is considered to be discrete. This reflects the situation in an ATM environment where all packets have a fixed length and fit exactly into one time slot.

2) A "backpressure mechanism" ensures that no packets are

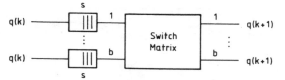

Fig. 3. Basic model of a switching element.

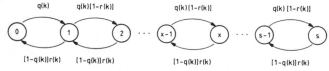

Fig. 4. State diagram of the Markov chain.

lost within the network. Thus, a packet can only leave its buffer if the corresponding destination buffer at the next stage is able to accept it.

3) The arrival process at each input of the network is a simple Bernoulli process, i.e., the probability that a packet arrives within a time slot is constant and the arrivals are independent of each other. This implies that the interarrival time between two packets is geometrically distributed with a minimum distance of one time slot.

4) Each input link is offered the same traffic load, and the destination addresses of the packets are distributed uniformly (at random) over all output links of the network. This is usually referred to as a uniform traffic environment.

5) There is no blocking at the output links of the network. This means that the output links have at least the same speed as the internal links.

It is well known that an exact analysis of buffered Banyan networks is not feasible even in a uniform traffic environment, since it leads to a multidimensional state space with one dimension for every buffer of the network which cannot be handled with reasonable effort. The complexity of the analysis can only be reduced by introducing certain independence assumptions which allow a decomposition of the network into single switching elements. This approach has the advantage that the switching elements can be analyzed separately, but the results are no longer exact due to the approximations involved.

In the following section, we give an outline of the basic analysis method for Delta networks consisting of switching elements with an arbitrary number of inputs and arbitrary number of buffer places at each input of the switching elements. This analysis is an extension of the method of Jenq [8], and a detailed description of this approach is given in [7].

A. Analysis of Delta-b Networks with Multiple Buffers

Fig. 3 shows the basic model of a switching element, which mainly consists of b input buffers and a nonblocking switching matrix to connect the input buffers to the output ports.

As a consequence of the uniform traffic environment, the switching elements in a particular stage as well as the various buffers within these elements are considered as statistically indistinguishable. Therefore, it is sufficient for the analysis to investigate just one buffer in every stage. This buffer can be modeled by the simple Markov chain shown in Fig. 4, if we assume that within each time slot a packet arrives at stage k with probability $q(k)$, and the first packet in the queue leaves the buffer with probability $r(k)$. From this Markov chain the state probabilities $p_i(k)$ of the buffer can be derived as a

function of $q(k)$ and $r(k)$ [7]

$$p_i(k) = \frac{1}{1 - r(k)} [Q(k)]^i p_0(k) \tag{1}$$

with

$$p_0(k) = \left\{ 1 + \frac{1}{1 - r(k)} \sum_{v=1}^{s} [Q(k)]^v \right\}^{-1} \tag{2}$$

and

$$Q(k) = \frac{q(k)[1 - r(k)]}{[1 - q(k)]r(k)}. \tag{3}$$

In the next step, two aspects of the operation of the network have to be considered. The first one is, that we assume a backpressure mechanism which prevents any packet loss within the network. This implies that the steady-state throughput, which is given by the probability that a packet is transmitted on a link in one time slot, is the same on any link within the network. The second one is, that the buffer is assumed to support simultaneous read and write access. Thus, even a full buffer is able to accept a packet if another packet is leaving the buffer in the same time slot. With these considerations, the probability $p^a(k)$ that a buffer at stage k is ready to accept a packet can be expressed as a function of the state probabilities and of $r(k)$.

$$p^a(k) = 1 - p_s(k) + p_s(k)r(k). \tag{4}$$

The influence of the switching matrix is modeled using a formula which has been derived in [2], [16]. Thus, we obtain a simple relationship between the probability $p_0(k)$ that an input buffer at stage k is empty and the probability $q(k + 1)$ that a packet is offered to a buffer at the next stage

$$q(k + 1) = 1 - \left(1 - \frac{1 - p_0(k)}{b} \right)^b. \tag{5}$$

The probability $q(k + 1)$ is less than $1 - p_0(k)$ due to the blocking effects that occur if more than one packet is destined to a specific output of the crossbar in one clock cycle. After analyzing the Markov chain depicted in Fig. 4 this equation can be used to compute the state probabilities of stage $k - 1$ from the state probabilities of stage k.

With the assumption for the boundary of the network (assumption 5) the analysis can be started at the last stage of the network and the complete network can be analyzed successively without any iterations. Various performance measures like the mean total transfer time of a packet can be derived from the state probabilities.

Fig. 5 shows the result of this analysis for a Delta-8 network with 64 inputs and buffersize 10 in comparison with simulation results. The low accuracy at high loads results from several assumptions that have been made in the analysis. First, it was assumed that the packets arrive at the buffers independent of each other. In reality, in case of blocking the packets remain in their buffers in the preceeding stage and contend for the same output in the next clock cycle, i.e., there are correlations between consecutive clock cycles as well as between the states of the buffers in the adjacent stages. Second, the states of the buffers within a switching element were assumed to be independent of each other, which is clearly an approximation. In the next sections, we will present an enhanced approximation for a special case, which includes a refined modeling of the relation-

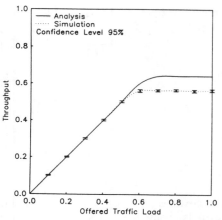 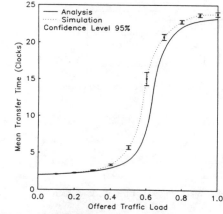

Fig. 5. Analysis of a Delta-8 network with 2 stages and buffersize 10.

ship between the stages as well as among the buffers within the stages.

B. Analysis of Single Buffered Delta-2 Networks

1) Refined Model: In order to identify the essential approximations which are reducing the accuracy of the analysis, we consider a buffer that has received a packet in the previous clock cycle. This packet can be forwarded in the current clock cycle, if the destination buffer is ready to accept it and if there is no conflict with the other buffer of the SE. If the packet is unable to leave the buffer, it has to wait for the next clock cycle to start a new attempt.

Consequently, there is a correlation between the buffered packets in consecutive clock cycles. Furthermore, the probability that a packet is able to leave the buffer significantly decreases if the packet is blocked, because either another packet has entered the destination buffer or the destination buffer is blocked itself. In both cases the destination buffer is occupied by a packet.

To take these effects of blocking into account, we introduce an additional state to characterize a buffer that contains a blocked packet. Consequently, each buffer has three possible states:

- state "0"   buffer is empty;
- state "n"   buffer contains a packet which arrived in the last clock cycle;
- state "b"   buffer contains a packet which has been delayed for at least one clock cycle.

As a further enhancement the dependencies between both buffers of a SE are modeled explicitly by a two-dimensional Markov chain leading to a state space with 9 states per SE. In order to simplify the computation of the state transitions between successive clock cycles, each transition will be subdivided into two logical steps. In the first step (phase 1), the buffers which contain a packet send a request to the next stage indicating that a packet is available for transmission. The packet is forwarded, if there are no collisions with other packets and if the destination buffer at the next stage is ready to accept a packet.

After the completion of phase 1 each buffer has only two possible states. A buffer containing a packet that could not be forwarded during phase 1 is entering a blocked state and the buffer is empty in all other cases.

In the second step (phase 2), the previous stage offers new packets to the buffers at stage k. These packets are accepted only by those buffers which are empty after phase 1. The probability that a packet is offered to a buffer is assumed to be independent of its state. After the packets have entered the buffers, the current clock period is completed. It should be noted that this separation is only logical, because in a real system both phases are performed more or less simultaneously in all SE's.

2) State Space of a Switching Element: After the description of the refined model, the state space of a SE will be derived. Since each SE contains two buffers which are mutually dependent, a two-dimensional state space is required to characterize a SE. In the following, the index (x, y) will be used to identify the state of a SE whose upper buffer is in state x and whose lower buffer is in state y:

$$p_{x,y}(t, K)$$
$$= P\{\text{a SE at stage } k \text{ is in state } x \text{ (upper) and } y$$
$$\text{(lower buffer) at the beginning of clock cycle } t;$$
$$x, y \in 0, n, b\}.$$

The state of a SE after the completion of phase 1 will be characterized by a tilde ($\sim$):

$$\tilde{p}_{x,y}(t, k)$$
$$= P\{\text{the buffers of a SE at stage } k \text{ are in state } x$$
$$\text{(upper) and } y \text{ (lower buffer) at the beginning of}$$
$$\text{phase 2 of clock cycle } t; x, y \in 0, b\}.$$

3) State Transitions of a Switching Element: The transition probabilities of phase 1 depend on the state of the SE. Therefore, the following notation will be introduced to describe the particular transitions:

$$r_{x,0}(t, k) = P\{\text{the packet in the upper buffer of a SE at stage}$$
$$k \text{ can be forwarded during clock cycle } t$$
$$|\text{the lower buffer is empty and the}$$
$$\text{upper buffer is in state } x; x \in n, b\}$$
$$r_{n,n}^{1,0}(t, k) = P\{\text{the packet in the upper buffer of a SE at stage}$$
$$k \text{ can be forwarded during clock cycle } t$$
$$|\text{both buffers are in state } n\}.$$

The suffix $(1, 0)$ indicates that the packet in the upper buffer can be forwarded ("1"), while the packet in the lower buffer has to

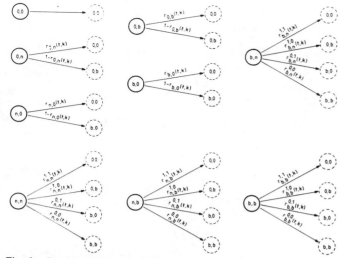

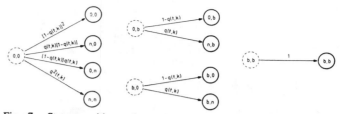

Fig. 7. State transitions of a switching element during phase 2; the intermediate states are denoted by dashed circles and the states at clock event $t + 1$ are indicated by bold circles.

Fig. 6. State transitions of a SE during phase 1; the states at clock event t are denoted by bold circles, and the intermediate states are indicated by dashed circles.

wait ("0"). As there are many possible combinations, this notation should serve as an example.

To simplify the mathematical expressions for the state transitions illustrated in Fig. 6, we introduce the matrix $\vec{P}(t, k)$ containing the state probabilities $p_{x,y}(t, k)$

$$\vec{P}(t, k) = \begin{pmatrix} p_{0,0}(t, k) & p_{n,0}(t, k) & p_{b,0}(t, k) \\ p_{0,n}(t, k) & p_{n,n}(t, k) & p_{b,n}(t, k) \\ p_{0,b}(t, k) & p_{n,b}(t, k) & p_{b,b}(t, k) \end{pmatrix}. \quad (6)$$

In the following, the symbol "$\odot$" denotes a multiplication of two matrices analogous to the scalar product of vectors:

$$\begin{pmatrix} a_{00} & \cdots & a_{x0} \\ \vdots & & \vdots \\ a_{0y} & \cdots & a_{xy} \end{pmatrix} \odot \begin{pmatrix} b_{00} & \cdots & b_{x0} \\ \vdots & & \vdots \\ b_{0y} & \cdots & b_{xy} \end{pmatrix} := \sum_{i,j} a_{ij} \cdot b_{ij}. \quad (7)$$

Using Fig. 6, the state probabilities at the end of phase 1 are calculated from the above defined transition probabilities and the matrix $\vec{P}(t, k)$:

$$\tilde{p}_{0,0}(t, k) = \begin{pmatrix} 1 & r_{n,0}(t, k) & r_{b,0}(t, k) \\ r_{0,n}(t, k) & r_{n,n}^{1,1}(t, k) & r_{b,n}^{1,1}(t, k) \\ r_{0,b}(t, k) & r_{n,b}^{1,1}(t, k) & r_{b,b}^{1,1}(t, k) \end{pmatrix}$$
$$\odot \vec{P}(t, k) \quad (8)$$

$$\tilde{p}_{0,b}(t, k) = \begin{pmatrix} 0 & 0 & 0 \\ 1 - r_{0,n}(t, k) & r_{n,n}^{1,0}(t, k) & r_{b,n}^{1,0}(t, k) \\ 1 - r_{0,b}(t, k) & r_{n,b}^{1,0}(t, k) & r_{b,b}^{1,0}(t, k) \end{pmatrix}$$
$$\odot \vec{P}(t, k) \quad (9)$$

$$\tilde{p}_{b,0}(t, k) = \begin{pmatrix} 0 & 1 - r_{n,0}(t, k) & 1 - r_{b,0}(t, k) \\ 0 & r_{n,n}^{0,1}(t, k) & r_{b,n}^{0,1}(t, k) \\ 0 & r_{n,b}^{0,1}(t, k) & r_{b,b}^{0,1}(t, k) \end{pmatrix}$$
$$\odot \vec{P}(t, k) \quad (10)$$

$$\tilde{p}_{b,b}(t, k) = \begin{pmatrix} 0 & 0 & 0 \\ 0 & r_{n,n}^{0,0}(t, k) & r_{b,n}^{0,0}(t, k) \\ 0 & r_{n,b}^{0,0}(t, k) & r_{b,b}^{0,0}(t, k) \end{pmatrix} \odot \vec{P}(t, k). \quad (11)$$

If phase 1 of clock cycle t is completed, new packets are offered to each buffer of stage k with the probability $q(t, k)$ which is assumed to be independent of the state of the SE. The packets are accepted by those buffers being empty after phase 1, leading to the state transitions illustrated in Fig. 7.

The state probabilities at clock event $t + 1$ are easily derived from Fig. 7:

$$p_{0,0}(t + 1, k) = [1 - q(t, k)]^2 \cdot \tilde{p}_{0,0}(t, k) \quad (12)$$
$$p_{0,n}(t + 1, k) = [1 - q(t, k)] q(t, k) \cdot \tilde{p}_{0,0}(t, k) \quad (13)$$
$$p_{n,0}(t + 1, k) = q(t, k)[1 - q(t, k)] \cdot \tilde{p}_{0,0}(t, k) \quad (14)$$
$$p_{n,n}(t + 1, k) = q^2(t, k) \cdot \tilde{p}_{0,0}(t, k) \quad (15)$$
$$p_{0,b}(t + 1, k) = [1 - q(t, k)] \cdot \tilde{p}_{0,b}(t, k) \quad (16)$$
$$p_{b,0}(t + 1, k) = [1 - q(t, k)] \cdot \tilde{p}_{b,0}(t, k) \quad (17)$$
$$p_{n,b}(t + 1, k) = q(t, k) \cdot \tilde{p}_{0,b}(t, k) \quad (18)$$
$$p_{b,n}(t + 1, k) = q(t, k) \cdot \tilde{p}_{b,0}(t, k) \quad (19)$$
$$p_{b,b}(t + 1, k) = \tilde{p}_{b,b}(t, k). \quad (20)$$

If the transition probabilities are known, the state probabilities of the buffers can be computed iteratively. Therefore, the calculation of the transition probabilities of phase 1 and phase 2 will be discussed briefly in the next section. The explicit mathematical expressions are derived in the Appendix.

4) Transition Probabilities: The probability that a packet is able to leave the buffer depends on the probability of a collision with a packet from the other buffer and on the probability that the switching element in the next stage is ready to accept the packet. Considering the probability that a packet can be accepted by its destination buffer at stage $k + 1$, there are two cases to distinguish.

- If none of the buffers of a SE at stage k contains a blocked packet, there is no relationship between the states of these buffers and the states of the corresponding destination buffers at stage $k + 1$. These buffers can have any of the three possible states (0, n, b).
- If a packet was blocked at stage k, the destination buffer of this packet at stage $k + 1$ always contains a packet. The destination buffer is in state "n," if it received a packet from the other buffer of the SE at stage k in the previous clock cycle, otherwise it must be blocked.

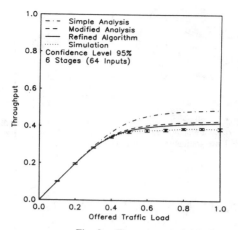 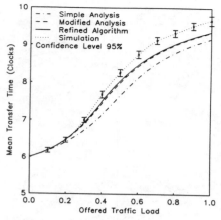

Fig. 8. Throughput and delay of a single buffered Delta-2 network with 6 stages.

It should be noted that a blocked packet also has an influence on a packet which arrives at the other buffer of the same SE. If the new packet is destined for the same output link as the blocked packet (probability 0.5), both packets have the same probability to be forwarded to the next stage.

The probability $q(t, k)$ that a packet is offered to a buffer at stage k can be computed considering the fact that no packets are lost on the internal links of the network due to the backpressure mechanism. Consequently, the probability that a packet is entering a buffer at stage k is equal to the probability that a packet is transmitted on an output link of a SE at stage $k - 1$, which can be calculated from the state probabilities of stage $k - 1$ and from the transition probabilities of phase 1. As a packet can only enter a buffer if the buffer is ready to accept it, the probability $q(t, k)$ is easily derived from the probability that a packet is entering a buffer at stage k.

5) Algorithm: Due to the complexity of the state transitions, the analysis of the network is done iteratively. Each step of the iteration starts with the calculation of the transition probabilities of phase 1 and phase 2 (see Appendix). Afterwards, the state probabilities of the intermediate states [(8)–(11)] and the state probabilities at the next clock cycle [(12)–(20)] can be computed using the transition probabilities. These steps are repeated until the steady state is reached, i.e., until the throughput on the output links of the network is equal to the throughput on the input links.

In the steady state the mean transfer time $T_f(k)$ of a packet at stage k can be calculated using Little's Law:

$$T_f(k) = \lim_{t \to \infty} \frac{1 - [p_{0,0}(t, k) + p_{0,n}(t, k) + p_{0,b}(t, k)]}{\rho(t, k)}$$

(21)

with the throughput $\rho(t, k)$ as defined in (60) [see Appendix].

The total transfer time of a packet through the network is given by

$$T_f = \sum_{k=1}^{n} T_f(k).$$

(22)

IV. RESULTS

In this section we investigate the accuracy of different analysis methods for single buffered Delta-2 networks. Therefore, the results of

- the *simple analysis* outlined in Section III-A which includes the analysis described by Jenq [8],
- a *modified analysis* which uses 3 states per buffer, but preserves the assumption that the buffers of a SE are independent, and
- the *refined algorithm* described in Section III-B

are compared to the results of an event-by-event simulation. The confidence intervals of the simulation results refer to a confidence level of 95%.

Fig. 8 shows the normalized throughput and the mean transfer time of a single buffered Delta-2 network with 6 stages. It is obvious, that the accuracy of all methods is very good as long as the offered traffic load is small. However, if the traffic load approaches the maximum throughput of the network, the accuracy of the *simple analysis* method is insufficient. This is due to the fact, that many packets are blocked mainly at the first stages of the network at high traffic rates. Consequently, the detailed modeling of the blocking effects is resulting in a significant improvement of the accuracy. The considerations of the dependencies between the buffers of a SE leads to a further (though small) improvement of the results.

The influence of the size of the network on the performance is depicted in Fig. 9. Since every additional stage of the network introduces further collisions, the maximum throughput decreases if the size of the network is extended. Although the *simple analysis* method clearly overestimates the performance of large networks, the *refined algorithm* presented in this paper is a good approximation for the performance evaluation of single buffered Delta-2 networks.

V. CONCLUSION

The comparison in Section IV shows clearly that the accuracy of the analysis is mainly determined by the modeling of the dependencies between consecutive clock cycles. The simple analysis assumes that a packet is destined for any output with the same probability, even if it has been blocked. In reality, blocked packets hunt for the same output in consecutive clock cycles. This effect explains the rather poor accuracy of the simple analysis for high loads, i.e., if many packets are blocked somewhere within the network.

The introduction of the "blocked" state, which models the dependencies between consecutive clock cycles in a rather gen-

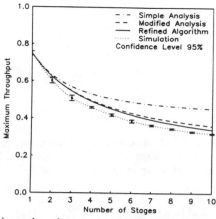

Fig. 9. Maximum throughput versus number of stages of a single buffered Delta-2 network.

eral way while still keeping the evaluation analytically tractable, results in a significant improvement of the accuracy without requiring considerably more computational effort.

$$p_a^n(t, k) = \left[\begin{pmatrix} 0 & r_{n,0}(t, k) & 0 \\ 0 & r_{n,n}^{1,0}(t, k) + r_{n,n}^{1,1}(t, k) & 0 \\ 0 & r_{n,b}^{1,0}(t, k) + r_{n,b}^{1,1}(t, k) & 0 \end{pmatrix} \odot \vec{P}(t, k) \right] / \left[p_{n,0}(t, k) + p_{n,n}(t, k) + p_{n,b}(t, k) \right] \quad (23)$$

$$p_a^b(t, k) = \left[\begin{pmatrix} 0 & 0 & r_{b,0}(t, k) \\ 0 & 0 & r_{b,n}^{1,0}(t, k) + r_{b,n}^{1,1}(t, k) \\ 0 & 0 & r_{b,b}^{1,0}(t, k) + r_{b,b}^{1,1}(t, k) \end{pmatrix} \odot \vec{P}(t, k) \right] / \left[p_{b,0}(t, k) + p_{b,n}(t, k) + p_{b,b}(t, k) \right] \quad (24)$$

$$p_a(t, k) = \begin{pmatrix} 1 & r_{n,0}(t, k) & r_{b,0}(t, k) \\ 1 & r_{n,n}^{1,0}(t, k) + r_{n,n}^{1,1}(t, k) & r_{b,n}^{1,0}(t, k) + r_{b,n}^{1,1}(t, k) \\ 1 & r_{n,b}^{1,0}(t, k) + r_{n,b}^{1,1}(t, k) & r_{b,b}^{1,0}(t, k) + r_{b,b}^{1,1}(t, k) \end{pmatrix} \odot \vec{P}(t, k). \quad (25)$$

Possible extensions of the analysis algorithm include the consideration of switching elements with multiple buffers and perhaps even the consideration of networks built of switching elements with more than two inputs. Furthermore, it would be interesting to evaluate the influence of a connection oriented environment resulting in more complex traffic patterns at the inputs.

APPENDIX

The transition probabilities of phase 1 and phase 2 are derived in this Appendix and the explicit mathematical expressions are presented.

A. Phase I

The probability that a packet is able to leave the buffer depends on the probability of a collision with a packet from the other buffer and on the probability that the switching element in the next stage is ready to accept the packet. If two packets are contending for the same output of a SE, one of the packets is randomly selected for transmission and the other packet has to wait for the next clock cycle.

Considering the probability that a buffer is able to accept a packet, there are three cases to distinguish.

- The buffer contains a "new" packet (state n):

$$p_a^n(t, k) = P\{\text{the upper buffer of a switching element}$$
$$\text{at stage } k \text{ is ready to accept a packet during}$$
$$\text{clock cycle } t \mid \text{the buffer is in state } n\}.$$

- The buffer contains a "blocked" packet (state b):

$$p_a^b(t, k) = P\{\text{the upper buffer of a switching element}$$
$$\text{at stage } k \text{ is ready to accept a packet during}$$
$$\text{clock cycle } t \mid \text{the buffer is in state } b\}.$$

- The buffer is in any of the three possible states $(0, n, b)$:

$$p_a(t, k)$$
$$= P\{\text{the upper buffer of a switching element at stage}$$
$$k \text{ is ready to accept a packet during clock cycle } t\}.$$

Due to the internal symmetry of the traffic, both buffers of a SE are statistically equivalent. Therefore, it is sufficient for the calculation of the above defined probabilities to consider only the upper buffer of a SE:

If none of the buffers of a SE at stage k contains a blocked packet, there is no relationship between the states of these buffers and the states of the corresponding destination buffers at stage $k + 1$. In this case, the probability that the destination buffers at stage $k + 1$ are able to accept a packet is given by $p_a(t, k + 1)$, because these buffers can have any of the three possible states $(0, n, b)$.

With $p_a(t, k + 1)$ resulting from (25), the transition probabilities refering to those states where none of the buffers is blocked can be computed for the stages $1 \cdots n - 1$:

$$r_{0,n}(t, k) = p_a(t, k + 1) \quad (26)$$

$$r_{n,0}(t, k) = p_a(t, k + 1) \quad (27)$$

$$r_{n,n}^{0,0}(t, k) = 0.5[1 - p_a(t, k + 1)]^2$$
$$+ 0.5[1 - p_a(t, k + 1)] \quad (28)$$

$$r_{n,n}^{0,1}(t, k) = 0.5[1 - p_a(t, k + 1)] p_a(t, k + 1)$$
$$+ 0.25 p_a(t, k + 1)] \quad (29)$$

$$r_{n,n}^{1,0}(t, k) = 0.5 p_a(t, k + 1)[1 - p_a(t, k + 1)]$$
$$+ 0.25 p_a(t, k + 1)] \quad (30)$$

$$r_{n,n}^{1,1}(t, k) = 0.5[p_a(t, k + 1)]^2. \quad (31)$$

As there is no blocking at the output links of the last stage, we obtain

$$r_{0,n}(t, n) = 1.0 \tag{32}$$

$$r_{n,0}(t, n) = 1.0 \tag{33}$$

$$r_{n,n}^{0,0}(t, n) = 0.0 \tag{34}$$

$$r_{n,n}^{0,1}(t, n) = 0.25 \tag{35}$$

$$r_{n,n}^{1,0}(t, n) = 0.25 \tag{36}$$

$$r_{n,n}^{1,1}(t, n) = 0.5 \tag{37}$$

If a packet was blocked at stage k, the destination buffer of this packet at stage $k + 1$ always contains a packet. The destination buffer is in state "n," if it received a packet from the other buffer of the SE at stage k in the previous clock cycle, otherwise it must be blocked. The probability that the destination buffer received a packet from a *specific* buffer of the SE at stage k in the previous clock cycle is given by $0.5\rho(t - 1, k + 1)$ where $\rho(t, k)$ is the throughput on an input link of a SE at stage k as defined in (60).

Consequently, the transition probabilities referring to the case that one buffer is blocked are given by

$$r_{0,b}(t, k) = 0.5\rho(t - 1, k + 1) \cdot p_a^n(t, k + 1)$$
$$+ [1 - 0.5\rho(t - 1, k + 1)] \cdot p_a^b(t, k + 1) \tag{38}$$

$$r_{b,0}(t, k) = 0.5\rho(t - 1, k + 1) \cdot p_a^n(t, k + 1)$$
$$+ [1 - 0.5\rho(t - 1, k + 1)] \cdot p_a^b(t, k + 1) \tag{39}$$

$$r_{n,b}^{0,0}(t, k) = 0.5[1 - r_{n,0}(t, k)] \cdot [1 - r_{0,b}(t, k)]$$
$$+ 0.5[1 - r_{0,b}(t, k)] \tag{40}$$

$$r_{n,b}^{0,1}(t, k) = 0.5[1 - r_{n,0}(t, k)] \cdot r_{0,b}(t, k)$$
$$+ 0.25 r_{0,b}(t, k) \tag{41}$$

$$r_{n,b}^{1,0}(t, k) = 0.5 r_{n,0}(t, k) \cdot [1 - r_{0,b}(t, k)]$$
$$+ 0.25 r_{0,b}(t, k) \tag{42}$$

$$r_{n,b}^{1,1}(t, k) = 0.5 r_{n,0}(t, k) r_{0,b}(t, k) \tag{43}$$

$$r_{b,n}^{0,0}(t, k) = 0.5[1 - r_{b,0}(t, k)] \cdot [1 - r_{0,n}(t, k)]$$
$$+ 0.5[1 - r_{b,0}(t, k)] \tag{44}$$

$$r_{b,n}^{0,1}(t, k) = 0.5[1 - r_{b,0}(t, k)] \cdot r_{0,n}(t, k)$$
$$+ 0.25 r_{b,0}(t, k) \tag{45}$$

$$r_{b,n}^{1,0}(t, k) = 0.5 r_{b,0}(t, k) \cdot [1 - r_{0,n}(t, k)]$$
$$+ 0.25 r_{b,0}(t, k) \tag{46}$$

$$r_{b,n}^{1,1}(t, k) = 0.5 r_{b,0}(t, k) r_{0,n}(t, k). \tag{47}$$

Again, the transition probabilities of the SE's at the last stage differ from the other stages:

$$r_{n,b}^{0,0}(t, n) = 0.0 \tag{48}$$

$$r_{n,b}^{0,1}(t, n) = 0.25 \tag{49}$$

$$r_{n,b}^{1,0}(t, n) = 0.25 \tag{50}$$

$$r_{n,b}^{1,1}(t, n) = 0.5 \tag{51}$$

$$r_{b,n}^{0,0}(t, n) = 0.0 \tag{52}$$

$$r_{b,n}^{0,1}(t, n) = 0.25 \tag{53}$$

$$r_{b,n}^{1,0}(t, n) = 0.25 \tag{54}$$

$$r_{b,n}^{1,1}(t, n) = 0.5 \tag{55}$$

If both buffers of a SE at stage k are blocked, the corresponding destination buffers at stage $k + 1$ are blocked, too. In this case, we obtain the following transition probabilities for the stages $1 \cdots n - 1$:

$$r_{b,b}^{0,0}(t, k) = 0.5[1 - p_a^b(t, k + 1)]^2$$
$$+ 0.5[1 - p_a^b(t, k + 1)] \tag{56}$$

$$r_{b,b}^{0,1}(t, k) = 0.5[1 - p_a^b(t, k + 1)] p_a^b(t, k + 1)$$
$$+ 0.25 p_a^b(t, k + 1) \tag{57}$$

$$r_{b,b}^{1,0}(t, k) = 0.5 p_a^b(t, k + 1)[1 - p_a^b(t, k + 1)]$$
$$+ 0.25 p_a^b(t, k + 1) \tag{58}$$

$$r_{b,b}^{1,1}(t, k) = 0.5[p_a^b(t, k + 1)]^2. \tag{59}$$

Due to the assumption that there is no blocking at the output links of the switch fabric, it is impossible that both buffers of a SE at the last stage are in a blocked state.

B. Phase 2

During phase 2, packets from stage $k - 1$ are entering the buffers at stage k. The throughput on an input link of a SE at stage k is defined as the probability that a packet is received on that link and is given by

$$\rho(t, k) = q(t, k) \cdot p_a(t, k). \tag{60}$$

Due to the backpressure mechanism no packets are lost on the internal links of the switching network. Therefore, the probability that a packet is transmitted on an output link of a SE at stage k is equal to the throughput $\rho(t, k + 1)$ on an input link of stage $k + 1$:

$$\rho(t, k + 1) = \begin{pmatrix} 0 & 0.5 r_{n,0}(t, k) & 0.5 r_{b,0}(t, k) \\ 0.5 r_{0,n}(t, k) & r_{n,n}^{1,0}(t, k) + r_{n,n}^{1,1}(t, k) & r_{n,b}^{1,0}(t, k) + r_{b,b}^{1,1}(t, k) \\ 0.5 r_{0,b}(t, k) & r_{n,b}^{1,0}(t, k) + r_{n,b}^{1,1}(t, k) & r_{b,b}^{1,0}(t, k) + r_{b,b}^{1,1}(t, k) \end{pmatrix} \odot \vec{P}(t, k). \tag{61}$$

Note that (61) has been simplified using the symmetry of the internal traffic flow.

$\rho(t, k)$ can be computed from (61) at the end of phase 1. Thus, from (60) we obtain the probability $q(t, k)$ that a packet is offered to an input buffer of stage k:

$$q(t, k) = \frac{\rho(t, k)}{p_a(t, k)}, \quad k = 2 \cdots n \tag{62}$$

with $\rho(t, k)$ from (61) and $p_a(t, k)$ from (25).

$q(t, 1)$ is equal to the offered traffic load at the network inputs and must be specified as an input parameter for the analysis.

REFERENCES

[1] K. E. Batcher, "The flip network in STARAN," *Int. Conf. Parallel Process.*, 1976, pp. 65–71.

[2] D. M. Dias and J. R. Jump, "Analysis and simulation of buffered delta networks," *IEEE Trans. Comput.*, vol. C-30, pp. 273–282, Apr. 1981.

[3] D. M. Dias and M. Kumar, "Packet switching in *nlogn* multistage networks," in *Proc. IEEE GLOBECOM'84*, Atlanta, GA, 1984, pp. 114–120.

[4] T. Feng, "A survey of interconnection networks," *IEEE Comput.*, vol. 14, pp. 12–27, Dec. 1981.

[5] S. Giorcelli, C. Demichelis, G. Giandonato, and R. Melen, "Experimenting with fast packet switching techniques in first generation ISDN environment," in *Proc. ISS'87*, Phoenix, AZ, 1987, pp. 388–394.

[6] L. R. Goke and G. J. Lipovski, "Banyan networks for partitioning multiprocessor systems," *First Ann. Symp. Comput. Archit.*, 1973, pp. 21–28.

[7] M. N. Huber, E. P. Rathgeb, and T. H. Theimer, "Banyan networks in an ATM-environment," in *Proc. ICCC'88*, Tel Aviv, Israel, 1988, pp. 167–174.

[8] Y.-C. Jenq, "Performance analysis of a packet switch based on single-buffered Banyan network," *IEEE J. Select. Areas Commun.*, vol. SAC-1, pp. 1014–1021, Dec. 1983.

[9] C. P. Kruskal and M. Snir, "The performance of multistage interconnection networks for multiprocessors," *IEEE Trans. Comput.*, vol. C-32, pp. 1091–1098, Dec. 1983.

[10] C. P. Kruskal, M. Snir, and A. Weiss, "The distribution of waiting times in clocked multistage interconnection networks," *IEEE Trans. Comput.*, vol. C-37, pp. 1337–1352, Nov. 1988.

[11] J. J. Kulzer and W. A. Montgomery, "Statistical switching architectures for future services," in *Proc. ISS'84*, Florence, Italy, 1984, paper 43.A.1.

[12] M. Kumar and J. R. Jump, "Performance of unbuffered shuffle exchange networks," *IEEE Trans. Comput.*, vol. C-35, pp. 573–577, June 1986.

[13] D. H. Lawrie, "Access and alignment of data in an array processor," *IEEE Trans. Comput.*, vol. C-24, pp. 1145–1155, Dec. 1975.

[14] G. W. R. Luderer, J. J. Mansell, E. J. Messerli, R. E. Staehler, and A. K. Vaidya, "Wideband packet technology for switching systems," in *Proc. ISS'87*, Phoenix, AZ, 1987, pp. 448–454.

[15] R. J. McMillen, "A survey of interconnection networks," in *Proc. GLOBECOM'84*, Atlanta, GA, 1984, pp. 105–113.

[16] J. H. Patel, "Performance of processor-memory interconnections for multiprocessors," *IEEE Trans. Comput.*, vol. C-30, pp. 771–780, Oct. 1981.

[17] M. C. Pease, "The indirect binary *n*-cube microprocessor array," *IEEE Trans. Comput.*, vol. C-26, pp. 458–473, May 1977.

[18] H. J. Siegel, R. J. McMillen, and P. T. Mueller, "A survey of interconnection methods for reconfigurable parallel processing systems," in *AFIPS Conf. Proc.*, vol. 48, 1979 NCC, pp. 529–542.

[19] K. J. Thurber, "Interconnection networks—A survey and assessment," in *AFIPS Conf. Proc.*, vol. 43, 1974 NCC, pp. 909–919.

[20] J. S. Turner, "Design of an integrated services packet network," *Ninth Data Commun. Symp.*, in *ACM SigComm Comput. Rev.*, vol. 15, Sept. 1985, pp. 124–133.

[21] ——, "New directions in communications," 1986 *Int. Zurich Sem. Dig. Commun.*, Zurich, Switzerland, 1986, pp. 25–32.

[22] C.-L. Wu and T.-Y. Feng, "On a class of multistage interconnection networks," *IEEE Trans. Comput.*, vol. C-29, pp. 694–702, Aug. 1980.

[23] L. T. Wu, "Mixing traffic in a buffered banyan network," *Ninth Data Commun. Symp.*, in *ACM SigComm Comput. Commun. Rev.*, vol. 15, Sept. 1985, pp. 134–139.

Priority Performance of Banyan-based Broadband-ISDN Switches.

Srinivasan Tridandapani and James S. Meditch

Department of Electrical Engineering (FT-10)
University of Washington
Seattle, WA-98195

Abstract

A practical fast packet switching module for use in B-ISDN switching architectures is described. The module is built around a self-routing banyan network which employs dilations and replications to enhance throughput. The differing requirements of delay and loss sensitivity for various traffic types are handled by assigning priorities to packets. A new result in the form of the throughput equation of lower priority packets in a banyan network which is improved by dilation, replication and internal speed-up is presented. The delay versus throughput analysis shows that the switch module is capable of handling integrated voice-data-video traffic efficiently.

I. INTRODUCTION

Multistage interconnection networks (MIN's) have played a key role in a number of switching architectures proposed over the last decade. They have found application in both telecommunication switching [1],[2], [3], [4]. and in processor-processor and processor-memory interconnections in massively parallel computers [5],[6],[7]. In this study, we are concerned chiefly with the performance of MIN's applied to Broadband Integrated Services Digital Network (B-ISDN) switching architectures. The analysis is, however, applicable to MIN's used in parallel processing environments.

We believe that priorities need to be assigned to different types of traffic in a B-ISDN switch. Indeed, a recent report on ATM standardization has recommended the use of priority bits in cells [8]. The argument also holds for interprocessor and processor-memory interconnections where critical traffic, such as software interrupts, has been identified as requiring higher priority over non-critical data transactions. It is also the case that fast packet switching offers fairness and bandwidth utilization that circuit switching does not. We, therefore, study a variation of the FP switch in [9] where all classes of traffic are packet switched and a special field in the packets, called the priority field, specifies which packet is successful in the case of contention within the MIN.

The throughput equation for a packet switched banyan network for a single traffic type was first derived by Patel [10]. Szymanski and Hamacher [11] utilized a different method to obtain the same relationships. We follow a method similar to the one in [11] to obtain the throughput of two-priority traffic in a banyan network. This throughput equation is then extended to an arbitrary number of distinct traffic types C where the classification is based solely on priority considerations; the lower the class number of a packet, the higher is its priority in terms of contention resolution within the MIN and in terms of queueing at the input ports of the MIN.

Noting that the throughput of a banyan network with internal blocking-and-loss is very low for most practical purposes, a number of schemes to reduce internal congestion and improve throughput have been proposed. One of the most obvious methods for improving throughput is the use of larger switch nodes, an example being the work of Ta and Meditch [12] who determined the performance improvements obtained through the use of 4×4 switching blocks over 2×2 switching elements. *D-dilated* and *r-replicated* banyan networks were studied in [11] ,[13] and [14]. Both of these reduce internal congestion and lead to significant improvements in throughput. Other options studied in the past include internal speed-up and the use of bit-parallel internal links [15].

We study the throughput of class-II (and in general any arbitrary class-C) packets in banyan networks, which are built with $k \times k$ crossbar nodes, and improved by d-dilation, r-replication and s-internal-speedup. We then use this generalized throughput equation to study the throughput versus delay behavior of a new fast packet switching architecture. In our study of the performance of this switching architecture, we assume that traffic at the input ports of the switch is uniformly distributed. We further assume that there is no correlation in the traffic since each input would carry a traffic pattern which results from multiplexing packetized traffic from several diverse sources. A recent study by Li [16] has shown that if more than 40 calls equally share a link, then the effect of input correlation can be neglected for crossbars. Our analysis, based on these assumptions, shows that the switch can readily support voice, data and video.

The remainder of this paper is organized as follows. In Section II, we provide various definitions and modifications related to banyan networks. We then describe our proposed switching architecture. An analysis of the performance of class-C packets in a banyan network is provided in Section III. The treatment includes extension to dilation, replication and speed-up. We provide

Reprinted from *Proc. IEEE INFOCOM '91*, pp. 711–720,
April 1991.

numerical results on the performance of the switch architecture in Section IV. In Section V, we summarize our work and indicate some areas for further study.

II. SWITCH ARCHITECTURE

In this section, we describe briefly the switch architecture (Fig. 1) which is similar to that of the FP switch in [9]. In Section 2.1, we provide some definitions which are necessary for our presentation. Section 2.2 describes the switch architecture's three major components: the Input Processor (IP), the Multi-stage Interconnection Network (MIN) and the Output Processor (OP). The MIN connects the IPs and the OPs of the switch.

II.1   Preliminaries

We list here some of the definitions related to banyan networks. Most of these definitions and assumptions have appeared in the literature in various forms [13], [11].

A *network* is a directed graph consisting of the following: 1) IP's which have indegree 0; 2) OP's which have outdegree 0; 3) switches which have indegree > 0 and outdegree > 0. Each edge represents one or more lines going from a node to its successor.

A *banyan network* is defined by Goke and Lipovsky [17] to be a network with a unique path from each IP to each OP. This definition implies that the set of paths leading to an OP in the network forms a tree and that the set of paths leading from an IP also forms a tree. An *n-stage multistage network* is a network in which the switches can be arranged in stages with all IP's attached to the inputs of stage 1 switches, all outputs from stage i are connected to inputs to stage $i + 1$, and all OP's are connected to outputs from stage n. Under this definition, the IP's are stage 0 and OP's are stage $n + 1$. A *uniform network* is a multistage network in which all switches in the same stage have the same number of input links and the same number of output links. A *square network of degree k* is a network built with $k \times k$ switches (hereafter referred to as a $k^n \times k^n$ banyan). Figure 1 shows a $2^3 \times 2^3$ banyan network.

A *link contention* is said to occur when two or more packets arrive at one switch and request the same output link. The *contention* is resolved by selecting one packet to be forwarded, and a negative acknowledgement (NAK) is sent back to the other input.

We define *multi-class* traffic as a collection of distinct traffic types which are distinguished on the basis of priority. A traffic type of priority class-C has lower priority than all traffic of class-c ($c = 1, 2 ... C - 1$). Under the condition of multi-class traffic, the link contention resolution scheme behaves differently; packets are selected at random within a class of traffic, but when different classes of traffic contend for the same line, the packets with the highest priority win the contention.

While banyan networks are very attractive in terms of their simplicity, considerations such as performance and fault tolerance often dictate the use of more complicated networks. As mentioned earlier, one easy way of improving the performance of a banyan network is by using large switch nodes. In fact, the best improvement is obtained if the banyan network is replaced by a crossbar.

The obvious drawback is the increased number of crosspoints. We consider three other types of improvement strategies which may be applied to banyan networks in order to reduce internal contention and thereby improve throughput.

The first strategy, referred to as *speed-up*, is implemented by using internal link speeds which are s (where $s = 2^i, i = 1, 2, ...$) times faster than the external line speeds. This can be realized by two methods. One is the use of bit-parallel lines within the MIN [9] and the other is the use of a faster clock. A prototype 16-bit parallel fast packet switch has been reported recently [18]. With internal speed-up of either type, the result is that the internal load on the MIN is reduced by s, the speed-up factor.

The second strategy, referred to as replication, is implemented by using a number of parallel MIN's. The r-*replication* of a network G is defined to be a network consisting of r identical copies of G. In [14], three techniques for using $r = 2^m (m = 1, 2, 3, ...)$ banyan subnetworks of size $N \times N$ each in parallel in order to obtain a network of size $N \times N$ are discussed. These three techniques, i.e., the randomly-loaded, the multiple-loaded and the selectively-loaded techniques, differ primarily in the policy used to distribute the incoming packets among the subnetworks. We choose the first technique, viz., the randomly loaded network for our switch since the other two techniques are not suitable for multi-class traffic.

The third, and perhaps most interesting, strategy for reducing internal contention and improving the throughput of a banyan network, is the use of multiple links. The *d-dilation* of a network G is defined to be the network obtained from G by replacing each edge by d distinct edges. A packet entering a switch may exit using any of the d edges going to the desired successor switch at the next stage. Some form of control is required at the individual nodes in order to decide which of the d identical edges of the logical link are to be used. It has been shown [19] that if a $d = 8$ dilation factor is used, then the throughput of the banyan network with as many as 4096 inputs and 4096 outputs is close to 1.

II.2   Fast Packet Switch

The switch architecture is similar to the FP switch described in [9]. Some minor modifications have been made primarily to accomodate replication and dilation in the MIN. A detailed description of the modified architecture can be found in [20]. We first describe the packet format, following which we briefly describe the three blocks of the switch and finally we describe the ACK/NAK protocol.

A packet contains the following fields: flag (F), network address (NA), packet type (T), supplementary control bits (CON), and a priority field (P). The information (INFO) field carries the payload in the form of data bits. Additional fields such as time stamp (TS) indicating the amount of time a packet has spent in the switch network may be added. This information is required in the design of control strategies which drop packets based on their age.

The IP's primary function is to queue input packets, attach a system switching address to each packet based on the packets network address, attach a system switching return address to each packet so that NAKs may be routed to the correct IP, and then route them to the MIN.

Additional functions such as serial-to-parallel conversion of the packets, or buffering may be performed depending on the method of speed-up, if any, that is used in the MIN.

The MIN provides the interconnection between the IP and OP. It consists of a number of node processors (NP) which are connected as a banyan network. By using self-routing, the NP routes packets to the appropriate output of the switch. If we assume the the dilation $d=2$ and if more than 2 packets request to be forwarded to the same logical output link, then 2 of these packets are chosen at random to be forwarded to the next stage and NAKs are sent back to the previous NP's from which the other packets originated. This procedure is actually more complicated in the case of multi-class traffic. This is because lower priority packets can fail to be forwarded to the next stage in two cases: 1) when there are more than d packets of the priority being considered destined for a tagged logical output and no packets of higher priority destined for the same output and 2) when there are $k, k < d$ higher priority packets destined for the tagged logical output and more than $d - k$ packets of the priority class being considered destined for the same tagged logical output.

The OP performs the reverse functions of the IP. Again the OP has buffering capability since more than one packet may arrive at an OP in a one time slot.

We now describe the use of a contention resolution scheme which is employed in our switch. Such a scheme is needed since more than d (where d is the dilation factor) packets may contend for the same output port in which case d of the packets are selected at random to win the contention and the other packets are blocked.

The switch employs a NAK signal to indicate blocking and a time-out to imply an ACK. When an IP sends a packet into the MIN, it retains a copy of the packet. At the end of the current slot, if the IP does not receive a NAK signal, it will discard the copy of the packet. However, when a packet loses its path contention, a NAK will be sent back to the source. The system switching return address of each packet is used by the NP's to send the packet back to the correct source IP. The IP will then then try to transmit the packet in the next idle slot. For a voice packet, if the packet delay exceeds the delay constraint, the IP will discard the packet.

III. PERFORMANCE ANALYSIS

The performance analysis of the switch is presented in three parts. First, we derive the throughput of class-C packets in a $k^n \times k^n$ banyan network. We then extend this result to the case of a d-dilated, r-replicated, s-internally-sped-up banyan network. Finally, using these throughput equations, we derive queueing results which permit us to establish the delay versus throughput performance characteristics of the switch under priority traffic.

In deriving analytical results for the throughput of the MIN, we make a number of assumptions. Most of these are the standard assumptions commonly used in the literature. The switch is assumed to be a synchronous system in which all packets have the same length, and the total time taken for a packet to be transmitted across the MIN and a NAK, should one be generated, to propagate back, is equal to one time slot of the MIN. Packet arrivals at the inputs to the MIN are independent, identically dis-

tributed Bernoulli random variables, and are equiprobably destined to all network outputs. We assume that this is true even in the case of video and voice traffic, since the actual input traffic to the switch is taken to be the result of merging packetized traffic from a number of diverse sources [16]. We further assume that this is true despite the queueing at the input ports. This effectively means that we will be obtaining optimistic results for the delay and throughput of the switch. There are no internal buffers within the switch nodes or between the switch nodes, which means that if link contention occurs in a d-dilated node, then d of the packets are chosen to be forwarded to the next stage, and NAKs are generated for the rest of the packets. Within any class, the choice of successful packets is purely random, and across classes the choice is on the basis of priority.

III.1 Throughput of $k^n \times k^n$ Unbuffered Banyan Networks

We first consider packet switching networks built of $k \times k$ unbuffered switches with the topology of an n-stage square banyan network. We define $P_s(c, n)$ to be the normalized success probability (or throughput) of class-c packets in an n-stage banyan network. The event that there are exactly r packets of class-c on any particular input at the mth stage of the network is represented as (c, r, m) and we define $P(c, r, m) = \Pr \{there\ are\ r\ class\text{-}c\ packets\ at\ the\ logical\ input\ to\ an\ mth\ stage\ switch\}$.

We first present the throughput rate for class-I packets. The analytical model for this purpose was first developed in [11].

In the undilated case where d=1 and the input arrival rate of class-I packets is p_1, the boundary conditions are

$$P(1, 1, 0) = p_1$$

and

$$P(1, 0, 0) = 1 - p_1$$

and the recurrence equation giving the packet arrival rate to the input of the $(m+1)st$ stage given the packet arrival rate to the input of the mth stage is

$$P(1, 1, m + 1) = \sum_{i=1}^{k} \binom{k}{i} \left(P(1, 1, m)\right)^i \left(P(1, 0, m)\right)^{k-i}$$
$$\cdot \sum_{j=1}^{i} \binom{i}{j} \left(\frac{1}{k}\right)^j \left(1 - \frac{1}{k}\right)^{i-j} \quad (1)$$

Then the throughput of class-I packets in an undilated, unbuffered, n- stage banyan network is given by

$$P_s(1, n) = P(1, 1, n)/P(1, 1, 0) \quad (2)$$

We now extend (1) and (2) to the case of a second priority class. When class-II packets are introduced, the boundary conditions are

$$P(1, 1, 0) = p_1$$
$$P(2, 1, 0) = p_2$$
$$P(1, 1, m) + P(2, 1, m) \leq 1$$
$$P((2, 1, m) \cap (1, 1, m)) = 0$$

where p_1 and p_2 are the input arrival rates of class-I and class-II packets, respectively. The last condition states that the events (1,1,m) and (2,1,m) are mutually exclusive.

The probability that there are x class-I and y class-II packets at the inputs of an mth stage $k \times k$ switch, noting that $x + y \leq k$, is

$$\frac{k!}{x!y!(k-x-y)!} \quad \left(1 - P(1,1,m) - P(2,1,m)\right)^{k-x-y}$$
$$\cdot \left(P(2,1,m)\right)^y \left(P(1,1,m)\right)^x$$

Given that there are x class-I packets and y class-II packets at the inputs of an mth stage $k \times k$ switch, the probability that z class-II packets and no class-I packets are destined for a tagged output is

$$\left(\frac{1}{k}\right)^z \left(1 - \frac{1}{k}\right)^{y-z} \left(1 - \frac{1}{k}\right)^x \quad (3)$$

Summing (3) over all z from 1 through y, i.e., over the number of ways of having at least 1 packet of class-II heading for the tagged output, we obtain

$$\left(1 - \frac{1}{k}\right)^x \sum_{z=1}^{y} \binom{y}{z} \left(\frac{1}{k}\right)^z \left(1 - \frac{1}{k}\right)^{y-z}$$

Finally, summing over all possible combinations of x and y, we find that

$$P(2,1,m+1) = \sum_{x=0}^{k} \sum_{y=1}^{k-x} \frac{k!}{x!y!(k-x-y)!}$$
$$\cdot \left(P(1,1,m)\right)^x \left(P(2,1,m)\right)^y$$
$$\cdot \left(1 - P(1,1,m) - P(2,1,m)\right)^{k-x-y}$$
$$\cdot \left(1 - \frac{1}{k}\right)^x \sum_{z=1}^{y} \binom{y}{z} \left(\frac{1}{k}\right)^z \left(1 - \frac{1}{k}\right)^{y-z}$$

This can be generalized to the class-C case to give

$$P(C,1,m+1) = \sum_{x=0}^{k} \sum_{y=1}^{k-x} \frac{k!}{x!y!(k-x-y)!}$$
$$\cdot \left(Q(C,1,m)\right)^x \left(P(C,1,m)\right)^y$$
$$\cdot \left(1 - Q(C,1,m) - P(C,1,m)\right)^{k-x-y}$$
$$\cdot \left(1 - \frac{1}{k}\right)^x \sum_{z=1}^{y} \binom{y}{z} \left(\frac{1}{k}\right)^z \left(1 - \frac{1}{k}\right)^{y-z}$$

where

$$Q(C,1,m) = \sum_{i=1}^{C-1} P(i,1,m)$$

and

$$0 \leq Q(C,1,m) + P(C,1,m) \leq 1 \quad (4)$$

The resulting throughput of class-C packets in an undilated, unbuffered, n-stage banyan network is then given by

$$P_s(C,n) = P(C,1,n)/P(C,1,0) \quad (5)$$

III.2 Throughput of d-Dilated $k^n \times k^n$ Unbuffered Banyan Networks

We now consider the throughput of a d-dilated network. Again $P(c,r,m)$ is the probability that r class-c packets appear on the d identical links of a logical input to an mth stage $k \times k$ switch. We will assume that when more than one packet is destined for a tagged logical link, the packets are assigned at random to the d lines of the link. If $i > d$ packets head for the same output, then d of these packets are chosen at random to be forward to the next stage and $(i - d)$ of the packets are dropped.

We consider first the case when there are packets of only one class. This result was first developed in [11]. It is obvious that $r \leq d$. In this case, events occurring on each of the k logical input links of a logical $k \times k$ switch must be examined. We define an array of variables $r_p, p \in 1 \ldots k$, where r_p denotes the number of requests that arrive on the pth logical input link of a logical $k \times k$ switch in stage m. For example, if each of k logical input links carry zero requests, then each $r_p = 0$, and this event occurs with probability $P(1,0,m)^k$. The probability that i requests arrive at the input of a stage m, $k \times k$ switch is then

$$\left(\sum_{\sum_x r_x = i} \prod_x P(1,r_x,m)\right) \quad (6)$$

This notation is the same as that used in [11]. The initial conditions for this case are

$$P(1,1,0) = p$$
$$P(1,0,0) = 1 - p(1,1,0)$$

and

$$P(1,j,0) = 0 \qquad \text{for } j \neq 1 \text{ and } j \neq 0$$

The probability that j of these requests from (6) are directed to a particular output link is

$$\binom{i}{j} \left(\frac{1}{k}\right)^j \left(1 - \frac{1}{k}\right)^{i-j}$$

In order to have j packets on a tagged logical output, the condition $j \leq i \leq kd$ must be satisfied. Therefore, summing over all i from j to kd, with $j < d$, we get

$$P(1,j,m+1) = \sum_{i=j}^{kd} \left(\sum_{\sum_x r_x = i} \prod_x P(1,r_x,m)\right)$$
$$\cdot \binom{i}{j} \left(\frac{1}{k}\right)^j \left(1 - \frac{1}{k}\right)^{i-j}$$

and, for $j = d$,

$$P(1,j,m+1) = \sum_{i=j}^{kd} \left(\sum_{\sum_x r_x = i} \prod_x P(1,r_x,m)\right)$$
$$\cdot \sum_{t=d}^{i} \binom{i}{t} \left(\frac{1}{k}\right)^t \left(1 - \frac{1}{k}\right)^{i-t} \quad (7)$$

In (7), the term $\sum_{t=d}^{i} \binom{i}{t}(\frac{1}{k})(1 - \frac{1}{k})^{i-t}$ enumerates the number of ways that at least d packets go to the tagged

213

logical output. The throughput of class-I packets in a d-dilated, unbuffered, m-stage banyan network is, therefore,

$$P_s(1,m) = \frac{\sum_{i=1}^{d} iP(1,i,m)}{P(1,1,0)} \qquad (8)$$

We now extend (7) and (8) to the class-II traffic case. The derivation of the recurrence equation for this case is complicated by the fact that there are four different conditions that we need to account for: 1) when all class-II packets which head for a tagged output are successful, 2) when all class-II packets which head for a tagged output are dropped because of contention with class-I packets, 3) when some of the class-II packets which head for a tagged output are dropped because of contention with class-I packets and 4) when some class-II packets which head for a tagged output are dropped not due to contention with class-I packets, but as a result of there being more than d packets heading for the output port being considered. Hence, we need to modify our state vector to represent the number of class-I packets and the number of class-II packets on a logical link. We, therefore, let the term $P(2,i,j,m)$ denote the probability that there are j class-II packets and i class-I packets at a logical input to a switch in stage m. When this is extended to an arbitrary class-C, we use the term $P(C,i,j,m)$ to represent the probability that there are j class-C packets and j class-c (where $c < C$ as explained below), at a logical input to a switch in stage m. The associated initial conditions are

$$\begin{aligned}
P(1,1,0,0) &= p_1 \\
P(2,0,1,0) &= p_2 \\
P(1,j,i,0) &= 0 \quad \text{for } j \neq 1 \text{ and } j \neq 0, \forall i \\
P(2,i,j,0) &= 0 \quad \text{for } j \neq 1 \text{ and } j \neq 0, \forall j
\end{aligned}$$

and

$$P(2,i,j,m) = 0 \quad \text{for } i+j > d$$

The probability that a class-I packets and b class-II packets arrive at the input of the mth stage, $k \times k$ d-dilated node is given by

$$\left(\sum_{\substack{\sum_x r_x = a \\ \sum_s s_x = b}} \prod_x P(2, r_x, s_x, m) \right) \qquad (9)$$

The probability that j of these b class-II packets and i of these class-I packets are destined for a tagged logical output and the rest are destined for other outputs is

$$\binom{b}{j}\left(\frac{1}{k}\right)^j\left(1-\frac{1}{j}\right)^{b-j}\binom{a}{i}\left(\frac{1}{k}\right)^i\left(1-\frac{1}{i}\right)^{a-i} \qquad (10)$$

If $j < d$, then the probability that fewer than $(d-j)$ class-I packets go to the tagged output, i.e., the case where class-II packets are not dropped, is specified by

$$\sum_{c=0}^{d-j} \binom{a}{c}\left(\frac{1}{k}\right)^c\left(1-\frac{1}{k}\right)^{a-c} \qquad (11)$$

Given that there are $(d-j)$ output lines of a tagged

logical link occupied by class-I packets, the probability that more than j class-II packets go to the tagged logical link is

$$\sum_{z=j}^{b} \binom{b}{z}\left(\frac{1}{k}\right)^z\left(1-\frac{1}{k}\right)^{b-j} \qquad (12)$$

Combining (9), (10), (11) and (12), we get for $j < d$ and $i < d-j$ that

$$P(2,i,j,m+1) = \sum_{b=j}^{kd}\sum_{a=0}^{kd-b} \left(\sum_{\substack{\sum_x r_x=a \\ \sum_x s_x=b}} \prod_x P(2,r_x,s_x,m) \right)$$
$$\cdot \binom{b}{j}\left(\frac{1}{k}\right)^j\left(1-\frac{1}{k}\right)^{b-j}$$
$$\cdot \binom{a}{i}\left(\frac{1}{k}\right)^i\left(1-\frac{1}{k}\right)^{a-i},$$

for $j < d$ and $i = d-j$ that

$$P(2,i,j,m+1) = \sum_{b=j}^{kd}\sum_{a=0}^{kd-b} \left(\sum_{\substack{\sum_x r_x=a \\ \sum_x s_x=b}} \prod_x P(2,r_x,s_x,m) \right)$$
$$\cdot \sum_{z=j}^{b}\binom{b}{z}\left(\frac{1}{k}\right)^z\left(1-\frac{1}{k}\right)^{b-j}$$
$$\cdot \binom{a}{d-j}\left(\frac{1}{k}\right)^{d-j}\left(1-\frac{1}{k}\right)^{a-(d-j)}$$

for $j = d$ that

$$P(2,0,j,m+1) = \sum_{b=j}^{kd}\sum_{a=0}^{kd-b} \left(\sum_{\substack{\sum_x r_x=a \\ \sum_x s_x=b}} \prod_x P(2,r_x,s_x,m) \right)$$
$$\cdot \sum_{c=d}^{b}\binom{b}{c}\left(\frac{1}{k}\right)^c\left(1-\frac{1}{k}\right)^{b-c}\left(1-\frac{1}{k}\right)^{a},$$

and for $i = d$ that

$$P(2,i,0,m+1) = \sum_{b=j}^{kd}\sum_{a=0}^{kd-b} \left(\sum_{\substack{\sum_x r_x=a \\ \sum_x s_x=b}} \prod_x P(2,r_x,s_x,m) \right)$$
$$\sum_{z=d}^{a}\binom{a}{z}\left(\frac{1}{k}\right)^z\left(1-\frac{1}{k}\right)^{a-z} \qquad (13)$$

Therefore, the throughput of class-II packets in a d-dilated, unbuffered, n-stage banyan network is given by

$$P_s(2,n) = \sum_{j=1}^{d}\sum_{i=0}^{d-j} \frac{jP(2,i,j,n)}{P(2,0,1,0)} \qquad (14)$$

This can be extended to the class-C case by simply replacing $P(2,i,j,n)$ in equations (13) and (14) with $P(C,i,j,n)$, where $P(C,i,j,m)$ is defined as the probability that there are j class-C packets and i packets from classes which have lower numbers than C, i.e., i packets with higher priority. The resulting throughput for class-C packets in an n-stage d-dilated unbuffered banyan network is then given by

$$P_s(C,n) = \sum_{j=1}^{d} \sum_{i=0}^{d-j} \frac{jP(C,i,j,n)}{P(C,0,1,0)} \qquad (15)$$

III.3 Throughput of r-Replicated, d-Dilated, s-Sped-up $k^n \times k^n$ Unbuffered Banyan Networks

We extend the results of the previous subsection to characterize the throughput of d-dilated, r-replicated, s-internally-sped-up banyan network. Again, we refer to the assumptions in [11]. We assume that input packets arrive with probability $P'(C,1,0)$ and are randomly sent to one of the r switching planes. Then the probability that a request is issued at the input of one particular plane is $P'(C,1,0)/r$ and the probability that a request is issued at the input of one particular plane in a particular internal time slot is $\frac{P'(C,1,0)}{rs}$. Further, the probability that the final stage of the plane has a request is given by (13) where $\frac{P'(C,1,0)}{rs}$ replaces $P(C,1,0)$. Finally, the throughput of the entire network is then specified by

$$P(C,1,0) = \frac{P'(C,1,0)}{rs}$$
$$P(C,0,0) = 1 - P(C,1,0)$$
$$P(C,j,0) = 0, \qquad \text{for } j \neq 0 \text{ and } j \neq 1$$

and

$$P_s(C,m) = \sum_{j=1}^{d} \sum_{i=0}^{i=d-j} \frac{jP(C,i,j,n)}{P(C,0,1,0)} \qquad (16)$$

III.4 Input Queueing Results

The ACK/NAK based retransmission protocol implies that packets are queued at the IP's of the switch. We will ignore the constant delays, suffered by all packets, such as serial-to-parallel conversion when bit parallel lines are used or, the synchronization delays when internal speed-up is employed, since these are the same for all cases. In analyzing the delay that packets incur in the switch, we consider only input queueing. Although queueing at the output also occurs, we ignore this delay under the assumption that there is no back-pressure, i.e, packets are removed at the output ports immediately upon arrival.

We assume that there is non-preemptive priority queueing at the IP's. Since a blocked packet will be retransmitted in succeeding slots until it is successful, we can model this processor as a single server, discrete-time system which has Bernoulli arrivals and a geometric service time.

Let $\alpha = [\alpha_1, \alpha_2, ..., \alpha_C]$ represent the input arrival rates of the C classes at the IP's of the switch. The geometric service time $\overline{b} = [\overline{b}_1, \overline{b}_2, ..., \overline{b}_C]$ is a function of the throughput rates for each class. The mean and second moment of the service times are given by

$$\overline{b}_c = \sum_{i=1}^{\infty} iP_s(c,n)(1-P_s(c,n))^{i-1}$$

$$\overline{b^2}_c = \sum_{i=1}^{\infty} i^2 P_s(c,n)(1-P_s(c,n))^{i-1} \qquad (17)$$

respectively, where the probability of a packet arrival at the MIN inputs is the product of the input arrival rate at the IPs and the average service time for each class, i.e., $p_c = \alpha_c \overline{b}_c$. Using the above definitions, the waiting time in the IP for each class, $\overline{w}_c$, is given by

$$\overline{w}_c = \frac{\sum_{i=1}^{C} \alpha_i(\overline{b^2}_i + \overline{b}_i)}{2(1 - \sum_{i=1}^{c-1} p_i)(1 - \sum_{i=1}^{c} p_i)} \qquad (18)$$

Equation (18) gives the waiting time for a discrete-time M/Geom/1 queueing system operating with a non-preemptive priority service scheme as derived in [15]. Therefore, the total delay that a class-c packet incurs is

$$\overline{t}_c = \overline{b}_c + \overline{w}_c \qquad (19)$$

IV. NUMERICAL RESULTS

In this section, we study the performance of our proposed switch architecture. The numerical results are presented in two parts.

First, we study the throughput behavior of a packet switching, blocking-and-loss banyan network for two-priority traffic, and examine the effect that the proposed switch improvements have on this throughput behavior. We then study the throughput versus delay characteristics of a 64×64 switch module, which can be used as a reasonable sized building block for larger switches, when the "ACK/NAK with retransmission" protocol is employed to eliminate packet losses.

For the sake of simplicity, we will refer to high priority packets as video packets and low priority packets as voice packets. In Fig. 2, we consider a banyan MIN without dilation and replication. The intensity of input traffic is 0.5 each for video and voice packets. We note that the sum of the throughputs of video and voice packets is the same as the throughput would be if there were only one class of packets with an input arrival rate of 1.0. Figure 3 further emphasizes this *invariance* in total throughput for the case of a 64×64, 2-replicated banyan network built around 2-dilated, 2×2 crossbar nodes. Here, the input arrival rate of video packets is increased from 0.0 to 1.0, while the input arrival rate of voice packets is decreased from 1.0 to 0.0 so that the total input arrival rate is kept at 1.0.

Noticing that the total throughput remains the same regardless of the traffic mixture leads us to two important conclusions. Firstly, we see that if a third class of packets, say data, is introduced, this new class sees the total contention from higher priority packets to be the same, independent of the actual ratio of video and voice packets. In fact, this observation is precisely what developed the rationale behind the generalization of the throughput equations to an arbitrary class C. Secondly, it also indicates that the actual computation of the throughput of lower priority packets is simplified. In fact, by using the throughput equation of the first priority packets alone, one can determine the throughput of various classes for any given traffic mixture at the inputs. We illustrate this argument with an example. Let us suppose that the traffic presented to the inputs of a 2×2 crossbar is 0.3 for video and 0.7 for voice packets. Using the throughput equations derived in Section III, we see that the unnormalized throughput of the crossbar for video packets is 0.2775 and for voice packets the corresponding figure is

0.4725. We also notice that $0.2775 + 0.4725 = 0.75$ which would have been exactly the throughput had there been only one class of packets, say video. Therefore, if we can find the throughput for video packets when the input rate for video packets is 1.0 and and the throughput when the input rate for video packets is 0.3, then the difference of these two results gives us the throughput of voice packets for the mixture of traffic considered above. An intuitive explanation for this phenomenon is the following. If contention occurs at an output port of a d-dilated crossbar between any n packets $(n > d)$, then at most d packets will be successful and $n - d$ packets will be dropped. Priority only affects the choice of packets which will be successful and not the number of successful packets. The above arguments for 2×2 crossbars carries over directly to banyan networks since the latter actually consist of a number of stages of such crossbar nodes.

We now look at the throughput of 64×64 banyan networks, which employ a variety of modifications. In Table 1 and Fig. 4, we show the throughputs for video and voice packets when the network is improved by dilation and replication. Again, we notice the invariance in throughput for all cases. It can also be seen that a dilation factor of 2 is roughly equivalent to a replication factor of 4 in terms of throughput improvement, and this, therefore, suggests that dilation may be preferable to replication. However, dilation increases crosspoint complexity more than replication. It has been shown by Bushnell [21] that a 2-dilated, 2-replicated, 64×64 switch module is a reasonable choice from the point of both performance and ease of fabrication. We note that, for this switch module, the maximum throughput is 0.9150. Figure 5 shows the block-and-loss throughput of voice packets for this switch when video packets occupy a portion of the input bandwidth. The packet loss probability for voice appears to be quite high as the the video packet input rate is increased from 0.3 to 0.7. However, the retransmission protocol discussed below shows that packet loss can be kept at almost zero at the cost of only a nominal delay.

In studying the delay vs. throughput characteristics of the above switch, we make the following assumptions. We introduce data as priority 3 traffic and assume that the external lines will be operated at the SONET rate of 155.52 Mbps. Further, we take the packet size to be 1024 bits, in which case the slot duration is approximately 6.6 μs.

Currently most central offices are being designed to handle 10,000 voice circuits. In this study we assume that the switch module will be required to switch no more 10,000 voice channels. This is actually a worst case assumption since we are implying that all 10,000 channels that the switch may be called upon to support are being switched through one 64×64 module. If we assume that the average bandwidth of a voice channel is 24 Kbps, then 10,000 channels correspond to less than 2.5 percent of the capacity of a 64×64 module operating at SONET rates. This essentially means that the major portion of the bandwidth will be occupied by video and data traffic. Consequently, we study the delays experienced by video and data packets while showing that the delay incurred by the voice packets is reasonable.

Figure 6 depicts the delay vs. throughput of video packets when data packets occupy a portion of the input bandwidth. In the absence of data packets, we note that, with a delay of less than 8 slots, a throughput of 0.80 can be maintained. At SONET rates, this corre-

sponds to a switch capacity of 7.96 Gbps and a delay of approximately 52 μs. If the assumed average bandwidth of a statistically multiplexed video channel is about 30 Mbps, then this means that the switch can support up to 265 video channels. Adding 10,000 voice channels (a normalized bandwidth of less than 0.025) does not raise the delay appreciably. Figure 7 shows the delay vs. throughput of data packets when video occupies a portion of the input bandwidth. Again, we ignore the effects of voice packets since the bandwidth occupied by voice is very small. From Figs. 6 and 7, we see that when the video input bandwidth is 0.6 and the data input bandwidth is 0.20, the delays for video and data packets are less than 3 and 10, respectively. At this operating point, if SONET rates are considered, the switch supports approximately 199 video channels and about 1.99 Gbps of data.

We now consider a different operating point, where the video input bandwidth is 0.3 and the input bandwidth of data is 0.5. In this case, video packets are delayed by less than 2 slots ($13.2 \mu s$), and data packets are delayed by about 10 slots ($66 \mu s$). This operating point corresponds to approximately 99 video channels and 4.97 Gbps of data.

For the two operating points considered above, the effect of voice traffic on video and data traffic is negligible, but the effect of video and data traffic on voice traffic needs to be quantified. We, therefore, study the delay vs. throughput of voice packets for these cases. Figure 8 shows the delay for voice packets when video and data packets occupy a portion of the input bandwidth. For the first case, when the video input bandwidth is 0.6 and the data input bandwidth is 0.2, we see that the delay experienced by voice packets is less than 6 slots when the switch is supporting 10,000 voice channels (0.025 input bandwidth). For the second case, with the same number of voice channels, and video and data input bandwidth of 0.3 and 0.5, respectively, we find that the delay is less than 2 slots. In the worst case then, the average delay for voice is at most $39.6 \mu s$.

V. CONCLUSION

In this paper, we have presented a fast packet switching architecture that is suitable for B-ISDN application. The essential component of this switch is a banyan MIN improved by modifications such as dilation and replication. Packet losses due to contention within the MIN are eliminited by an ACK/NAK retransmission protocol. Packet switching is employed for all classes of traffic.

In order to study the performance of the switch, we have extended the analysis techniques developed in [11] to account for the case of priority. We have obtained a generalized throughput equation for class-II traffic in a d-dilated, r-replicated, s-internally-sped-up internally blocking banyan network. We further extended this equation to accommodate an arbitrary class-C. Invariance in total throughput was observed for any mixture of priority traffic. By using the generalized throughput equation, we were able to study the delay versus throughput behavior of low priority packets in a reasonably sized switch. We considered a 64×64 switch module which is a practical sized building block for large switch architectures. We note that the module is capable of handling large volumes of integrated voice-data-video traffic efficiently and can operate at 80 percent of its capacity with nominal delays.

Our work has uncovered the following areas for further research. Multicasting in banyan type networks has been studied in [22]. Our work has shown that the throughput of banyan networks can be enhanced considerably using modifications such as dilation, replication and internal speed-up. The analysis of multicasting in banyan networks with such improvements is a challenging problem. The problem can be complicated further if priorities are included in multi-point traffic.

In analyzing the performance of the switch, we assumed the absence of back-pressure from the output ports. In general, there will be queueing at the outputs. If dilation or replication is used, then the output queue has to be modelled as a $M^{[x]}/D/1$ queueing system, i.e., an $M/D/1$ queueing system with bulk arrivals. Again the study may be performed with priorities assigned to various classes of traffic.

References

1. A. Huang, and S. Knauer, "Starlite: A wideband digital switch", in *Proc. GLOBECOM '84*, Nov. 1984, pp. 121–125.

2. J. Y. Hui, and E. Arthurs, "A broadband packet switch for integrated transport", *IEEE J. Select. Areas Commun.*, vol. SAC–5, pp. 1264–1273, Oct. 1987.

3. J. S. Turner, and L. F. Wyatt, "A packet network architecture for integrated services", in *Proc. GLOBECOM '83*, Nov. 1983, pp. 45–50.

4. A. K. Vaidya, and M. A. Pashan, "Technology advances in wideband packet switching", in *Proc. IEEE GLOBECOM 1988*, Nov. 1988, pp. 668–671.

5. K. E. Batcher, "The multidimensional access memory in STARAN", *IEEE Trans. Comput.*, vol. C-26, no. 2, pp. 174–177, Feb. 1977.

6. H. F. Jordan, "Performance measurement of HEP – A pipelined MIMD computer", in *Proc. 10th Ann. Symp. Comput. Arch.*, Jun. 1983, pp. 207–212.

7. R. D. Rettberg, et. al., "The monarch parallel processor hardware design", *IEEE Comput. Mag.*, vol. 23, no. 4, pp. 18–30, Apr. 1990.

8. CCITT, New Draft Recom. I. 150, "B-ISDN ATM layer functionality and specifications", Committee XVIII, Jan. 1990.

9. X. Jiang, and J. S. Meditch, "Integrated services fast packet switching", in *Proc. IEEE GLOBECOM '89*, Nov. 1989, pp. 1478–1482.

10. J. H. Patel, "Performance of processor-memory interconnections for multiprocessors", *IEEE Trans. Comput.*, vol. C-30, pp. 771–780, Oct. 1981.

11. T. H. Szymanski, and V. C. Hamacher, "On the permutation capability of multistage interconnection networks", *IEEE Trans. Comput.*, vol. C-36, no. 7, pp. 810–822, Jul. 1987.

12. Q. Ta, and J. S. Meditch, "A high speed integrated services switch based on 4 × 4 switching elements", in *Proc. INFOCOM '90*, Jun. 1990, pp. 1164–1171.

13. C. P. Kruskal, and M. Snir, "The performance of multistage interconnection networks for multiprocessors", *IEEE Trans. Comput.*, vol. C–32, no. 12, pp. 1091 – 1098, Dec. 1983.

14. M. Kumar, and J. R. Jump, "Performance of unbuffered shuffle–exchange networks", *IEEE Trans. Comput.*, vol. C–35, no. 6, pp. 573 – 577, Jun. 1986.

15. X. Jiang, *High Speed Switch Architectures for Integrated Voice-Data-Video Services*. Ph.D. thesis, Dept. of Elec. Engr., Univ. of Wash., Seattle, WA, Feb. 1989.

16. S.-Q. Li, "Performance of a non-blocking space-division switch with correlated input traffic", Technical Report CU/CTR/TR-152-89-31, Center for Telecommun. Res., Columbia Univ., New York, NY, 1988.

17. R. Goke, and G. J. Lipovski, "Banyan networks for partitioning multiprocessor systems", in *Proc. 1st Intl. Symp. Comp. Arch.,*, Dec. 1973, pp. 21–28.

18. Y. Kato, K. T. Shimoe, and K. Murakami, "Development of a high speed ATM switching LSIC", in *Proc. ICC '90*, Jun. 1990, pp. 562-566.

19. K. M. Buhler, *Design of Dependable Broadband Switching Architectures*. M.S. thesis, Dept. of Elec. Engr., Univ. of Wash., Seattle, WA, Aug. 1989.

20. S. Tridandapani, *Analysis of Priority Traffic in Generalized Banyan Networks for Broadband-ISDN Switching*. M.S. thesis, Dept. of Elec. Engr., Univ. of Wash., Seattle, WA, Jul. 1990.

21. E. T. Bushnell, *Dilated Multistage Interconnection Networks for Fast Packet Switching*. M.S. thesis, Dept. of Elec. Engr., Univ. of Wash., Seattle, WA, May 1990.

22. C.-L. Tarng, *High Performance Copy Networks for Multicast Fast Packet Switching Systems*. Ph.D. thesis, Dept. of Elec. Engr., Univ. of Wash., Seattle, WA, June 1990.

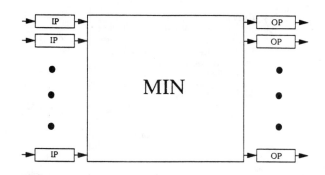

Figure 1. Switch architecture

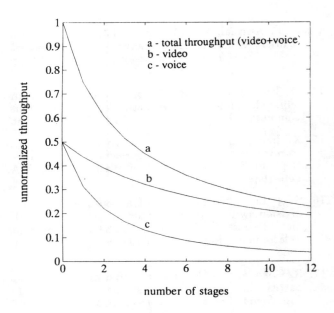

Figure 2. Unnormalized throughput of video and voice packets in a banyan network built around 2 × 2 nodes

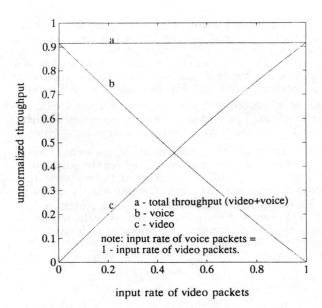

Figure 3. Unnormalized throughput of voice and video packets in a 64 × 64, 2-dilated, 2-replicated banyan network built around 2 × 2 nodes

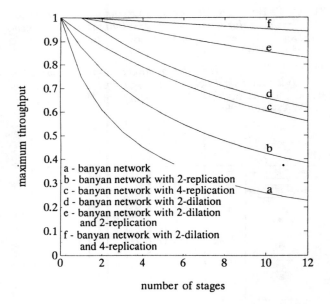

Figure 4. Throughput of class-I packets with blocking and loss

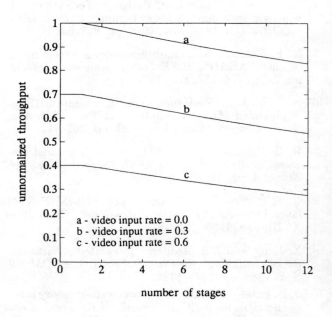

Figure 5. Unnormalized throughput of voice packets when video occupies a portion of the input bandwidth in a 2-dilated, 2-replicated banyan network with blocking and loss

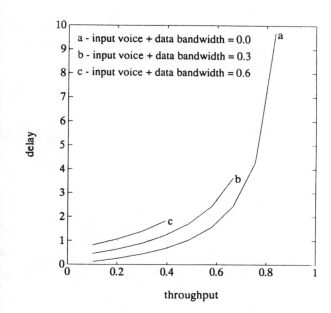

Figure 6. Delay versus throughput for video packets in a 64 × 64, 2-dilated, 2-replicated banyan network when data occupies a portion of the input bandwidth

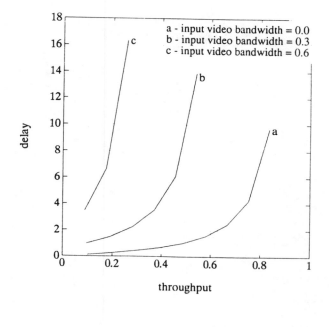

Figure 7. Delay versus throughput for data packets in a 64 × 64, 2-replicated, 2-dilated, banyan network when video occupies a portion of the input bandwidth

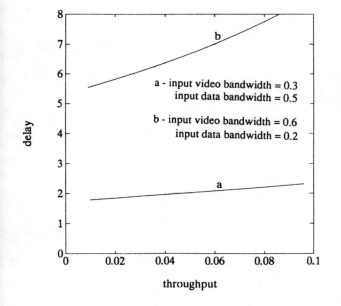

Figure 8. Delay versus throughput for voice packets in a 64 × 64, 2-replicated, 2-dilated, banyan network when video and data occupy a portion of the input bandwidth

| Dilation | Input rate | | Throughput | | | | | |
| | | | replication = 1 | | replication = 2 | | replication = 4 | |
	Video	Voice	Video	Voice	Video	Voice	Video	Voice
D = 1	0.0000	1.0000	0.0000	0.3594	0.0000	0.5466	0.0000	0.7161
	0.2500	0.7500	0.7161	0.2405	0.8382	0.4494	0.9131	0.6504
	0.5000	0.5000	0.5466	0.1722	0.7161	0.3770	0.8382	0.5940
	0.7500	0.2500	0.4361	0.1292	0.6215	0.3217	0.7730	0.5452
	1.0000	0.0000	0.3594	0.0000	0.5466	0.0000	0.7161	0.0000
D = 2	0.0000	1.0000	0.0000	0.7682	0.0000	0.9150	0.0000	0.9741
	0.2500	0.7500	0.9741	0.6996	0.9929	0.8890	0.9981	0.9660
	0.5000	0.5000	0.9150	0.6214	0.9741	0.8559	0.9929	0.9551
	0.7500	0.2500	0.8426	0.5451	0.9472	0.8183	0.9847	0.9419
	1.0000	0.0000	0.7682	0.0000	0.9150	0.0000	0.9741	0.0000

Table 1. Normalized throughput of voice and video packets in a 64 × 64 banyan network with various modifications.

Part 5
Knockout-Based Switching

Introduction

This chapter is concerned with a unique packet-switching architecture known as the knockout switch. The concept is described in the first paper, "The Knockout Switch: A Simple, Modular Architecture for High-Performance Packet Switching," by Y.-S. Yeh, M. G. Hluchyj, and A. S. Acampora.

Each of the N inputs in a knockout switch is connected to a separate bus that feeds each of N "bus interfaces" (one for each output). Packets arriving at an output's bus interface from the various buses first enter packet filters that allow only packets destined for that output to pass through. Each of up to N arriving packets then pass through a knockout concentrator that selects a maximum of L packets out of the arriving N packets ($L < N$). The selected packets then enter a shared buffer for final delivery to the output.

The knockout switch gets its name from the concentrator, which implements a knockout tournament in a parallel fashion using simple switching elements. The authors list this ability to handle multiple arrivals to the same output as one of the two basic functions of a switch architecture (the other being routing). A simple performance analysis shows that even by allowing only up to L of the N arriving packets to pass through the concentrator packet, loss can be kept extremely small. For example, for an infinite-size switch at 90% load with $L = 8$, the packet loss rate is less than 10^{-6}.

Yeh and his coauthors go on to describe advantages of the knockout switch in terms of modularity, maintainability, fault tolerance, and complexity. The knockout switch and principle have been extended in a number of other papers. A knockout switch for variable-length packets is discussed in [1]. The knockout switch is discussed in the context of multicasting in [2], in the context of nonuniform traffic in [5], and in the context of photonic switching in [6].

The knockout switch can be used as a switch in itself, but it can also be used as a building block in more elaborate switch architectures. This can be seen in the next two papers. The first is "Performance Analysis of a Growable Architecture for Broad-Band Packet (ATM) Switching," by M. J. Karol and C. L. I.

The growable architecture analyzed by Karol and I is discussed in a companion paper [3] by Eng, Karol, and Yeh. The switch consists of three stages of switching modules. The first two stages form a self-routing interconnection fabric. The output stage can be implemented via the knockout switch or other architectural choices such as a shared memory switch or a STARLITE switch [4].

Since output queueing is used, Karol and I focus on cell loss probability. There are two components to such loss. One is *knockout loss*. Each output module has m inputs, so if more than m cells arrive in a time slot for the same output module,

they are dropped by their input modules. This limitation on the number of arriving packets is similar to what occurs in a knockout switch. The second source of packet loss is due to the use of a fast but suboptimal routing algorithm in the first two stages. Occasionally a cell (packet) may be dropped at its input module if the routing algorithm cannot make a path assignment, even if fewer than m cells arrived in the time slot for the same output module. This *scheduling loss* is discussed in this paper.

A performance analysis is presented for cell-loss probability for arbitrary patterns of independent cell arrivals. This is basically a worst-case bounding study. The analysis is done both with and without trunking. An analysis for uniform offered traffic, which provides a tighter bound on cell-loss probability, is also included in this paper. The analyses allow the determination of switch module size for a given offered traffic load and desired blocking probability.

The second paper in this collection to use the knockout switch as a building block is "Performance of Output-Buffered Banyan Networks with Arbitrary Buffer Sizes," by H. S. Kim, I. Widjaja, and A. Leon-Garcia. In it they consider a self-routing Banyan network whose 2×2 switching elements are constructed out of 2×2 knockout switches. The advantage of this is that a maximal throughput of 1.0 can be obtained. By way of contrast, in finite-sized Banyan networks with input queueing and single-buffer switching elements, the maximal throughput is 0.45. The 2×2 knockout switch is also simple to implement and lacks the concentrator of larger knockout switches. Moreover, there is no internal switch speedup.

Kim and his coauthors consider two switching strategies for this architecture. In the back-pressure model, a cell may be detained at a switching element if the buffer in the switching element in the next stage that it should enter is full. This form of blocking propagates backward from the outputs to the inputs. In the block-and-lost model, blocked cells are lost.

A simulation study is presented for both strategies. A model that can be solved by iteration and that has its roots in Jenq's early work is developed for the back-pressure model. It is found that at heavy loads the block-and-lost policy has lower mean delay than the back-pressure policy at the cost of packet loss. It is pointed out that if each output link has 16 packet buffers, the loss rate is less than 10^{-6} for moderate loads up to 0.7.

References

[1] K. Y. Eng, M. G. Hluchyj, and Y. S. Yeh, "A knockout switch for variable-length packets," *Proc. IEEE Int. Conf. Commun. '87*, pp. 22.6.1–22.6.6, June 1987.

221

[2] K. Y. Eng, M. G. Hluchyj and Y. S. Yeh, "Multicast and Broadcast Services in a knockout packet switch," *Proc. of IEEE INFOCOM '88*, New Orleans, La., pp. 29–34, April 1988.

[3] K. Y. Eng, M. J. Karol, and Y.-S. Yeh, "A growable packet (ATM) switch architecture: design principles and applications," *IEEE Trans. Commun.*, vol. 40, no. 2, pp. 423–430, Feb. 1992.

[4] A. Huang and S. C. Knauer, "STARLITE: A wideband digital switch," *Proc. IEEE Globecom '84*, pp. 121–125, Nov. 1984.

[5] H. Yoon, M. T. Liu, and K. Y. Lee, "The knockout switch under non-uniform traffic," *Proc. IEEE Globecom '88*, pp. 1628–1634, 1988.

[6] K. Y. Eng, "A photonic knockout switch for high-speed packet networks," *Proc. IEEE Globecom '87*, pp. 1861–1865, 1987.

The Knockout Switch: A Simple, Modular Architecture for High-Performance Packet Switching

YU-SHUAN YEH, FELLOW, IEEE, MICHAEL G. HLUCHYJ, AND ANTHONY S. ACAMPORA

Abstract—A new, high-performance packet-switching architecture, called the Knockout Switch, is proposed. The Knockout Switch uses a fully interconnected switch fabric topology (i.e., each input has a direct path to every output) so that no switch blocking occurs where packets destined for one output interfere with (i.e., block or delay) packets going to different outputs. It is only at each output of the switch that one encounters the unavoidable congestion caused by multiple packets simultaneously arriving on different inputs all destined for the same output. Taking advantage of the inevitability of lost packets in a packet-switching network, the Knockout Switch uses a novel concentrator design at each output to reduce the number of separate buffers needed to receive simultaneously arriving packets. Following the concentrator, a shared buffer architecture provides complete sharing of all buffer memory at each output and ensures that all packets are placed on the output line on a first-in first-out basis.

The Knockout Switch architecture has low latency, and is self-routing and nonblocking. Moreover, its simple interconnection topology allows for easy modular growth along with minimal disruption and easy repair for any fault. Possible applications include interconnects for multiprocessing systems, high-speed local and metropolitan area networks, and local or toll switches for integrated traffic loads.

I. INTRODUCTION

WE are currently witnessing an explosive growth in the deployment of optical fiber within buildings, across cities, states, and countries, and even between continents. In addition, continuing advances are being made in optical communications, with current transmission rates on a single fiber in the Gbit/s realm and still more than a thousand-fold increase in capacity theoretically possible. Lagging this tremendous progress in transmission, however, are the advances made in switching technology necessary to build wideband integrated communication networks. While at some time in the future such switching may be achieved through optical means, at present we are lacking suitable electronic architectures to carry us into the next decade.

To put the communication aspects of switching in proper perspective, we note that every type of switch architecture interconnecting N inputs with N outputs must perform two basic functions. First, it must route the traffic arriving on its inputs to the appropriate outputs. This may be accomplished with a single-stage, crossbarlike inter-connect using N^2 simple switch elements or, at the other extreme, with an $N\log_2 N$ multistage interconnect comprised of 2×2 switch elements. Second, the switch must deal with output contention, where traffic arriving at the same time on two or more inputs may be destined for a common output. With circuit switching, output contention is prevented by using a controller that schedules the arrivals to avoid conflict. The classical time-space-time switch falls into this category [1]. Here, each input to the switch is preceded by a time slot interchange to rearrange the time sequence of the time-multiplexed traffic so that, when presented to the space switch, the data appearing at the N inputs are always destined for distinct outputs. With packet switching, however, packet arrivals to the switch are unscheduled, each containing a header bearing address information used to route the packet through the switch. Without the coordination afforded by a central scheduler, a packet switch must recognize conflict among its inputs and internally store, or buffer, all but at most one of several simultaneously arriving packets destined for a common output; thereby leading to statistical delay, or latency, within the switch. Dealing with output contention in a packet switch can be more complicated than the switch fabric used to route packets to the proper outputs.

The use of high-performance packet switching for building wideband integrated communication networks has received much attention [2]–[7]. Current approaches in the design of high-performance packet switches typically employ binary switch elements appropriately interconnected and arranged to form a multistage switch, and have tended to emphasize a reduction in the number of switch elements needed to a value below that of a fully connected arrangement [8]. Unfortunately, in addition to congestion at the outputs, these element-efficient switches can also congest at each of the binary switch points, thereby requiring that additional measures be taken, such as buffering within each element, flow control between elements, and/or speed-up of the switch fabric itself. The complexity of the buffering and flow control required within the switch element far exceeds that of the basic switching mechanism used to route the inputs to the outputs. Moreover, the delay encountered within the switch fabric is greater than the unavoidable component caused by output congestion alone. Other important areas in which multistage switches have difficulties are ease of modular growth and the ability to easily locate and repair faults with minimal disruption to the operation of the

Manuscript received November 5, 1986; revised April 13, 1987. This paper was presented at the 1987 International Switching Symposium, Phoenix, AZ, March 15–20, 1987.

Y.-S. Yeh and A. S. Acampora are with AT&T Bell Laboratories, Holmdel, NJ 07733.

M. G. Hluchyj was with AT&T Bell Laboratories, Holmdel, NJ. He is now with Codex Corporation, Canton, MA 02021.

IEEE Log Number 8716301.

Reprinted from *IEEE J. Selected Areas Commun.*, vol. SAC-5, no. 8, pp. 1274–1283, October 1987.

switch. The latter is particularly important when one considers that the hardware and software investment associated with switch maintenance often exceeds that of the switch fabric.

In this paper, we propose a new, high-performance packet-switching architecture, which we call the Knockout Switch. The switch has low latency, is self-routing, and is nonblocking. Its architecture provides for simple modular growth, fault tolerance, and easy maintenance procedures. Although based on a fully interconnected topology, the complexity of the Knockout Switch grows only linearly with N within the range of practical interest. Specifically, for $N \leq 1000$, the complexity of the switch is dominated by the buffering requirements and the input/output functions (timing recovery, address look-up and translation, etc.) present in any packet switch. Only when N is much greater than 1000 do we see the N^2 complexity of the interconnection fabric begin to dominate the overall complexity of the switch. Still, with $N = 1000$ and a data rate of 50 Mbits/s on each input line, a total switch capacity of 50 Gbits/s results. This is about 10 times the capacity of the current generation of digital central office switches. To achieve this performance, the Knockout Switch exploits one key observation: in any practical packet switching system, packet loss within the network is unavoidable (e.g., caused by bit errors on the transmission lines).

The Knockout Switch uses a fully interconnected topology to passively broadcast all input packets to all outputs. Preceding each output port is a bus interface that performs several functions. First, by means of a filtering operation, all packets not intended for the output are discarded; this operation effectively achieves the switching function in a fully distributed manner, but requires N simple filtering elements per output (hence, N^2 for the entire switch). Next, each bus interface queues the desired packets in a set of buffers that are shared, for that output, among all input lines. A simple arrangement permits the buffers to be filled in a cyclical manner and served to the output on a first-in first-out basis. Since any delay in the switch arises from output congestion only, the Knockout Switch provides the lowest latency possible in any switching arrangement [9].

The most novel aspect of the Knockout Switch, however, is its use of a knockout contention scheme in the bus interface that funnels the packets accepted from a possibly large number of packet filters (one for each of N inputs) into a far smaller number L of shared, parallel buffers, thereby vastly reducing the overall complexity of the switch. The scheme is analogous to a tournament. All inputs having packets simultaneously arriving to a particular output contend for the right to place data in the first of several parallel buffers. All losers then contend for the second buffer, and so on through buffer L; at this point, any remaining losers, having been given L attempts at victory, are simply discarded. We show that the probability of losing data in this manner is extremely small. For example, with a 90 percent load and N arbitrarily large,

$L = 8$ guarantees a lost packet rate of less than 10^{-6}, comparable with the loss expected from other sources (e.g., channel errors and buffer overflows in a packet switching network); $L = 12$ guarantees a lost packet probability of less than 10^{-10}.

For each bus interface, it is shown that the total size needed for each shared buffer is 40 packets (8 parallel buffers, each 5 packets deep) to maintain an overflow probability of 10^{-6} at an 84 percent load, again comparable to other loss mechanisms. The Knockout Switch is modular, permitting evolutionary growth from small to large configurations; is easily maintained, with faults having minimal disruption to the overall operation of the switch; and can be made fault tolerant with little additional hardware. The characteristics of the Knockout Switch make it well suited for several potential applications: 1) the interconnect for a large multiprocessing system, 2) high-capacity local and metropolitan area networking, and 3) local or toll switching.

In Section II, we describe the architecture of the Knockout Switch in greater detail. In Section III, we describe how it can grow in a modular fashion, is amenable to simple maintenance procedures, and can be made fault tolerant. We also examine in Section III the Knockout Switch implementation complexity. Finally, in Section IV we state our conclusions.

II. KNOCKOUT SWITCH DESCRIPTION

The Knockout Switch is an N-input N-output packet switch with all inputs and outputs operating at the same bit rate. Fixed-length packets arrive on the N inputs in a time-slotted fashion as shown in Fig. 1, with each packet containing the address of the output port on the switch to which it is destined. This addressing information is used by the packet switch to route each incoming packet to its appropriate output. Since the output port address for an arriving packet can be determined via a table look-up prior to the packet entering the switch fabric, the Knockout Switch has applications to both datagram and virtual circuit packet networks.

Aside from having control over the average number of packet arrivals destined for a given output, we assume no control over the specific arrival times of packets on the inputs and their associated output addresses. In other words, there is no time-slot specific scheduling that prevents two or more packets from arriving on different inputs in the same time slot destined for the same output. Hence, to avoid (or at least provide a sufficiently small probability of) lost packets, at a minimum, packet buffering must be provided in the switch to smooth fluctuations in packet arrivals destined for the same output.

A. Interconnection Fabric

The interconnection fabric for the Knockout Switch has two basic characteristics: 1) each input has a separate broadcast bus, and 2) each output has access to the packets arriving on all inputs. Fig. 2 illustrates these two characteristics where each of the N inputs is placed directly

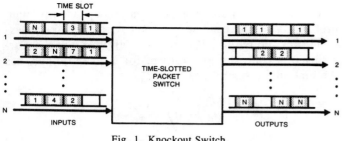

Fig. 1. Knockout Switch.

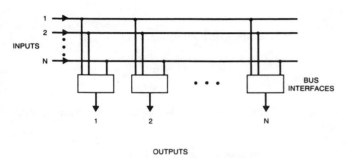

Fig. 2. Interconnection fabric.

on a separate broadcast bus and each output passively interfaces to the complete set of N buses. This simple structure provides us with several important features.

First, with each input having a direct path to every output, no switch blocking occurs where packets destined for one output interfere with (i.e., delay or block) packets going to other outputs. The only congestion in the switch takes place at the interface to each output where, as mentioned, packets can arrive simultaneously on different inputs destined for the same output. Without *a priori* scheduling of packet arrivals, this type of congestion is unavoidable, and dealing with it typically represents the greatest source of complexity within the packet switch. The focus of the Knockout Switch architecture is one of minimizing this complexity.

In addition to the above property, the switch architecture is modular: the N broadcast buses can reside on an equipment backplane with the circuitry for each of the N input/output pairs placed on a single plug-in circuit card. Hence, the switch can grow modularly from 2×2 up to $N \times N$ by adding additional circuit cards. We shall see later that the penalty of having N buses to which all outputs have to interface is not particularly significant. Most of the complexity of the bus interface is associated with the buffering of packets, which for each output does not grow with N. We show in Section III that the architecture for the bus interface also provides us with a simple way to grow the switch from $N \times N$ to $JN \times JN$, $J = 2, 3, 4 \cdots$.

Note from Fig. 2 that the bus structure has the desirable characteristic that each bus is unidirectional and driven by only one input. This allows for a higher transmission rate on the buses, with regeneration used as needed to com-

pensate for losses, and a design more tolerant of faults compared to a shared parallel bus accessed by all inputs. In addition, the packet buffering and bus access control circuitry of the parallel bus is replaced in this architecture by, at most, an elastic buffer at each input used to synchronize the time slots from the individual input lines.

Finally, although it would require a more complex design of the packet filters than that which is described in the next section, the interconnection architecture of the Knockout Switch lends itself to broadcast and multicast features. Since every input is available at the interface to every output, arriving packets can be addressed to and received by multiple outputs.

B. Bus Interface

Fig. 3 illustrates the architecture of the bus interface associated with each output of the switch. The bus interface has three major components. At the top of the figure there are a row of N *packet filters*. Here the address of every packet broadcast on each of the N buses is examined, with packets addressed to the output allowed to pass on to the concentrator and all others blocked. The *concentrator* then achieves an N to L ($L \ll N$) concentration of the inputs lines, wherein up to L packets making it through the packet filters in each time slot will emerge at the outputs of the concentrator. These L concentrator outputs then enter a *shared buffer* composed of a shifter and L separate FIFO buffers. The shared buffer allows complete sharing of the L FIFO buffers and provides the equivalent of a single queue with L inputs and one output, operating under a first-in first-out queueing discipline. In the remainder of this section, we will expand on each of these three parts of the bus interface.

Packet Filters: Fig. 4 shows the format of the packets as they enter the packet filters from the broadcast buses. The beginning of each packet contains the address of the output on the switch for which the packet is destined, followed by a single activity bit. The destination output address contains $\log_2 N$ bits with each output having a unique address. The activity bit indicates the presence (logic 1) or absence (logic 0) of a packet in the arriving time slot and plays an important role in the operation of the concentrator.

At the start of every time slot, the path through each of the N packet filters is open, initially allowing all arriving packets to pass through to the concentrator. As the address bits for each arriving packet enter the row of N packet filters, they are compared bit-by-bit against the output address for the bus interface. If at any time the address for an arriving packet differs from that of the bus interface, the further progress of the packet to the concentrator is blocked. That is, the output of the filter is set at logic 0 for the remainder of the time slot. By the end of the output address, the filter will have either blocked the packet, and hence also set its activity bit to 0, or, if the addresses matched, allowed the packet to continue on to the concentrator. Note that even though a portion of the address bits of a blocked packet may pass through the

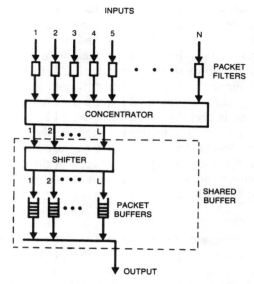

Fig. 3. Bus interface.

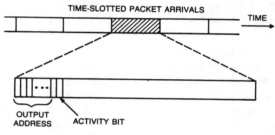

Fig. 4. Packet format.

filter and into the concentrator, these bits no longer serve any useful function and are ignored by the rest of the bus interface.

Concentrator: All packets making it through the packet filters enter the concentrator, which achieves an N to L concentration. Specifically, if there are k packets arriving in a time slot for a given output, these k packets, after passing through the concentrator, will emerge from the concentrator on outputs 1 to k, when $k \leq L$. If $k > L$, then all L outputs of the concentrator will have packets and $k - L$ packets will be dropped (i.e., lost) within the concentrator.

That packets can be dropped within the concentrator should not be of great concern. With any packet switching network, packet loss is inevitable: caused by transmission line errors (particularly in the address portion of the packet), buffer overflows, and network failures. In all cases, recovery is made possible by retransmission protocols, which, as the probability of losing packets in the network decreases, are more efficiently handled end to end rather than link by link [10], [11]. For the Knockout Switch, we implicitly assume all packet retransmissions are initiated at the endpoints of the network, so that the switch is not burdened with this task.

We must ensure, however, that the probability of losing a packet within the concentrator is no greater than that of losing a packet elsewhere in the network. If in each time

slot a packet arrives at each input independently with probability ρ, and each such packet is equally likely destined for each output, P_k, the probability of k packets arriving in a time slot all destined for a given output, has the binomial probabilities

$$P_k = \binom{N}{k}\left(\frac{\rho}{N}\right)^k\left(1 - \frac{\rho}{N}\right)^{N-k}$$

$$k = 0, 1, \cdots, N. \qquad (1)$$

It follows then that the probability of a packet being dropped in a concentrator with N inputs and L outputs is given by

$$\Pr[\text{packet loss}] = \frac{1}{\rho}\sum_{k=L+1}^{N}(k - L)\binom{N}{k}$$

$$\cdot\left(\frac{\rho}{N}\right)^k\left(1 - \frac{\rho}{N}\right)^{N-k}. \qquad (2)$$

Taking the limit as $N \to \infty$, we obtain after some manipulation

$$\Pr[\text{packet loss}] = \left[1 - \frac{L}{\rho}\right]\left[1 - \sum_{k=0}^{L}\frac{\rho^k e^{-\rho}}{k!}\right]$$

$$+ \frac{\rho^L e^{-\rho}}{L!}. \qquad (3)$$

Using (2) and (3), Fig. 5 shows for $\rho = 0.9$ (i.e., a 90 percent load) a plot of the probability of packet loss versus L, the number of outputs on the concentrator, for $N = 16, 32, 64$, and infinity. Note that a concentrator with only eight outputs achieves a probability of lost packet less than 10^{-6} for arbitrarily large N. This is comparable to the lost packet probability resulting from transmission errors for 500-bit packets and a bit error rate of 10^{-9}. Also note from Fig. 5 that each additional output added to the concentrator beyond eight results in an order of magnitude decrease in the lost packet probability. Hence, independent of the number of inputs N, a concentrator with 12 outputs will have a lost packet probability less than 10^{-10}. Fig. 6 illustrates, for $N \to \infty$, that the required number of concentrator outputs is not particularly sensitive to the load on the switch, up to and including a load of 100 percent. It is also important to note that, assuming independent packet arrivals on each input, the simple, homogeneous model used in the analysis corresponds to the worst case, making the lost packet probability performance results shown in Figs. 5 and 6 upper bounds on any set of heterogeneous arrival statistics [12].

The basic building block of the concentrator is a simple 2×2 contention switch shown in Fig. 7(a). The two inputs contend for the "winner" output according to their activity bits. If only one input has an arriving packet (indicated by the activity bit = 1), it is routed to the winner (left) output. If both inputs have arriving packets, one input is routed to the winner output and the other input is routed to the loser output. If both inputs have no arriving packets, we do not care except that the activity bit for both should remain at logic 0 at the switch outputs.

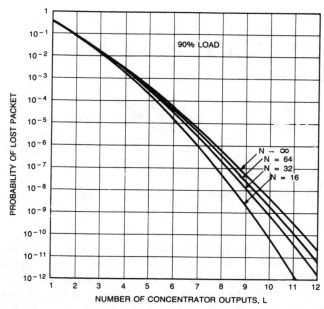

Fig. 5. Lost packet performance of concentrator.

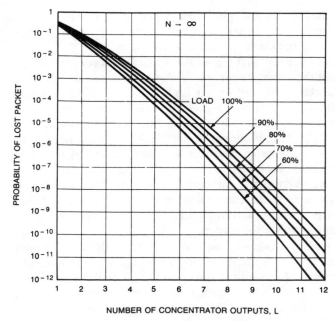

Fig. 6. Lost packet performance of concentrator for various loads.

The above requirements are met by a switch with the two states shown in Fig. 7(b). Here, the switch examines the activity bit for only the left input. If the activity bit is a 1, the left input is routed to the winner output and the right input is routed to the loser output. If the activity bit is a 0, the right input is routed to the winner output, and no path is provided through the switch for the left input. Such a switch can be realized with as few as 16 gates, and having a latency of at most one bit. Note that priority is given to the packet on the left input to the 2 × 2 switch element. To avoid this, the switch element can be designed so that it alternates between selecting the left and right inputs as winners when packets arrive to both in the

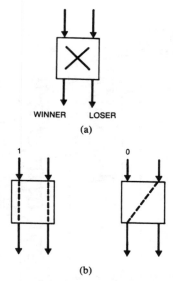

Fig. 7. (a) The 2 × 2 contention switch. (b) States of 2 × 2 contention switch.

same time slot. However, suppose the priority structure of the 2 × 2 switch element were maintained and (as described below) the concentrator were designed so that one input, say the Nth, always received lowest priority for exiting a concentrator output. The packet loss probability for this worst case input, as $N \to \infty$, is given by

$$\text{Pr [packet loss for worst case input]} = 1 - \sum_{k=0}^{L-1} \frac{\rho^k e^{-\rho}}{k!}. \tag{4}$$

Comparing the results of (4) to the lost packet probability averaged over all inputs, as given by (3) and shown in Fig. 6, we find that the worst case lost packet probability is about a factor of 10 greater than the average. This greater packet loss probability, however, can be easily compensated for by adding an additional output to the concentrator.

Fig. 8 shows the design of an 8-input 4-output concentrator composed of these simple 2 × 2 switch elements and single-input/single-output 1-bit delay elements (marked by "D"). At the input to the concentrator (upper left side of Fig. 8), the N outputs from the packet filters are paired and enter a row of $N/2$ switch elements. One may view this first stage of switching as the first round of a tournament with N players, where the winner of each match emerges from the left side of the 2 × 2 switch element and the loser emerges from the right side. The $N/2$ winners from the first round advance to the second round where they compete in pairs as before using a row of $N/4$ switch elements. The winners in the second round advance to the third round, and this continues until two compete for the championship: that is, the right to exit the first output of the concentrator. Note that if there is at least one packet arriving on an input to the concentrator, a packet will exit the first output of the concentrator.

A tournament with only a single tree-structured com-

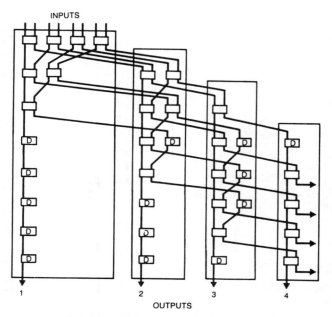

Fig. 8. The 8-input/4-output concentrator.

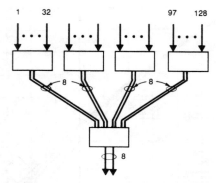

Fig. 9. The 128-to-8 concentrator constructed from 32-to-8 concentrator chips.

petition leading to a single winner is sometimes referred to as a single knockout tournament: lose one match and you are knocked out of the tournament. In a double knockout tournament, the $N - 1$ losers from the first section of competition compete in a second section, which produces a second place finisher (i.e., a second output for the concentrator) and $N - 2$ losers. As Fig. 8 illustrates, the losers from the first section can begin competing in the second section before the competition is finished in the first. Whenever there are an odd number of players in a round, one player must wait and compete in a later round in the section. In the concentrator, a simple delay element serves this function.

For a concentrator with N inputs and L outputs, there are L sections of competition, one for each output. A packet entering the concentrator is given L opportunities to exit through a concentrator output: a packet losing L times is knocked out of the competition and is discarded by the concentrator. In all cases, however, packets are only lost if more than L arrive in any one time slot. As we have seen, for $L \geq 8$, this is a low probability event.

For $N \gg L$, each section of the concentrator contains approximately N switch elements for a total concentrator complexity of $16NL$ gates. For $N = 64$ and $L = 8$, this corresponds to a relatively modest 8000 gates. Once a concentrator microcircuit is fabricated, Fig. 9 illustrates how several identical chips can be interconnected to form a larger concentrator. The approach is to select, based on an acceptable lost packet probability, the number of concentrator outputs L and then fabricate a chip with KL inputs (K an integer ≥ 2) and L outputs. A $K^j L$ input, L output concentrator can be formed by interconnecting $J + 1$ rows of KL-to-L concentrator chips in a treelike structure, with the ith row (counting from the bottom) containing K^{i-1} chips. For the example illustrated in Fig. 9, $L = 8$, $K = 4$, and $J = 2$.

Shared Buffer: The architecture of the bus interface focuses, to the extent possible, on reducing the complexity of packet buffering. This is done first by using a concentrator to reduce the number of inputs that must be buffered simultaneously. Second, through the use of a shared buffer structure, complete sharing of all packet buffer memory within the bus interface is made possible. This is accomplished while still providing a first-in first-out queueing discipline for the arriving packets and keeping the latency through the bus interface to a minimum.

Since in any given time slot up to L packets can emerge from the concentrator, the buffer within the bus interface must be capable of storing up to L packets within a single time slot. To permit high-speed low-latency operation of the Knockout Switch, the bus interface uses L separate FIFO buffers as shown in Fig. 3. A simple technique allows complete sharing of the L buffers, and at the same time provides a first-in first-out queueing discipline for all packets arriving to the output. The latter ensures fairness for access to the output and, more importantly, that successive packets arriving on an input do not get out of sequence within the switch.

As Fig. 3 shows, the L outputs from the concentrator first enter a "shifter" having L inputs and L outputs. The purpose of the shifter is to provide a circular shift of the inputs to the outputs so that the L separate buffers are filled in a cyclic fashion. This is illustrated in Fig. 10 for $L = 8$. Here, in the first time slot, five packets arrive for the output and, after passing through the concentrator, enter the first five inputs to the shifter. For this time slot, the shifter simply routes the packets straight through to the first five outputs, from which they enter buffers 1–5. In the second time slot, four packets arrive and enter the shifter on inputs 1–4. Having in the previous time slot left off by filling buffer 5, the shifter circular shifts the inputs five outputs to the right so that the arriving packets enter buffers 6, 7, 8, and 1. In the third time slot, the inputs are shifted one output to the right so that buffer 2 will receive the next arriving packet from the first output of the concentrator.

The shifter is a switch with L states: circular shift the inputs $0, 1, \cdots,$ or $L - 1$ outputs to the right. Letting S_i denote the state of the shifter and k_i the number of packets

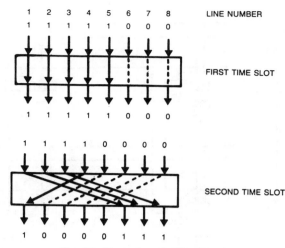

LINE NUMBER

FIRST TIME SLOT

SECOND TIME SLOT

Fig. 10. Illustration of shifter function.

exiting the concentrator in time slot i, we have

$$S_{i+1} = (S_i + k_i) \bmod L \qquad (5)$$

with $S_1 = 0$. A shifter can be constructed with $(L/2) \log_2 L$ 2×2 switch elements in the form of an omega network [13], controlled by a simple state machine obeying (5).

The flow of packets, stored in buffers $1-L$, onto the output line may be viewed as being controlled by a token. The buffer holding the token has the right to transmit one packet in the next time slot, with buffer 1 initially holding the token in the first time slot. If a buffer is empty, it will hold the token indefinitely until a new packet arrives and is transmitted. After this, the token is handed to the next buffer and wraps around in a cyclic fashion.

With the above described means for storing and removing packets for the L buffers, the shared buffer architecture has the following two characteristics.

1) Packets are stored in and removed from the L buffers in a cyclic fashion. At any time, the number of stored packets in each buffer does not differ by more than one among the L buffers. Buffer overflow only occurs when all L buffers are full. We therefore achieve the equivalent of a single buffer shared by all L outputs of the concentrator.

2) The token is held by the buffer with the highest number of stored packets, and the packet to be transmitted next is the one that has waited the longest in the bus interface. Hence, the shared buffer architecture forms the equivalent of a single-server first-in first-out queue with L inputs.

The total amount of buffering required in each bus interface depends on the assumed model for packet arrivals and the restrictions on lost packets resulting from buffer overflows. Using the homogeneous arrival model assumed previously (a packet arrives in a time slot at each input independently with probability ρ and each arriving packet is equally likely destined for each output), the probability of having k packets arrive in a time slot, all destined for the same output, has the binomial probability assignment given by (1). Since the fraction of arriving

packets lost with the concentrator is extremely small for $L \geq 8$, we make the conservative, but accurate, approximation that all arriving packets enter the shared buffer (i.e., none are lost in the concentrator). Also, so that our chosen buffer size is valid for all N, we assume a worst case situation by allowing $N \rightarrow \infty$. Under these conditions, the number of arriving packets to the shared buffer in each time slot has the Poisson probability assignment

$$\Pr[k \text{ arriving packets}] = \frac{\rho^k e^{-\rho}}{k!}. \qquad (6)$$

Since the arriving packets are fixed length, the transmission time for each packet is deterministic, equal to one time slot. Hence, for an infinite buffer, we can model the queueing process at each output as an $M/D/1$ queue [14] having mean queue size

$$\overline{Q} = \frac{\rho^2}{2(1-\rho)} \qquad (7)$$

and steady-state probabilities[1]

$$\Pr(Q = 0) = (1 - \rho) e^\rho \qquad (8)$$

$$\Pr(Q = 1) = (1 - \rho) e^\rho (e^\rho - 1 - \rho) \qquad (9)$$

$$\vdots$$

$$\Pr(Q = n) = (1 - \rho) \sum_{j=1}^{n+1} (-1)^{n+1-j}$$
$$\cdot e^{j\rho} \left[\frac{(j\rho)^{n+1-j}}{(n+1-j)!} + \frac{(j\rho)^{n-j}}{(n-j)!} \right]$$
$$\text{for } n \geq 2 \qquad (10)$$

where the second factor in (10) is ignored for $j = n + 1$.

In Fig. 11 we plot the steady-state probability that the queue size exceeds M for loads varying from 70 to 99 percent. From Figs. 6 and 11, we can conclude that with 8 buffers, each 5 packets deep for a total of 40 packets, the probability of losing a packet within a bus interface is less than 10^{-6} for an 84 percent load on the switch. Models with more bursty arrivals (e.g., those used to model packet voice) will require a larger total buffer to achieve the same lost packet probability [15].

III. MODULARITY, MAINTAINABILITY, FAULT TOLERANCE, AND COMPLEXITY

Often, in the design of a switch, the cost and complexity of the switch fabric play a secondary role to other factors, such as the ability to grow the switch in a modular fashion, to maintain the switch without great difficulty, and to provide the required degree of fault tolerance. In this section we describe how each of these important features can be satisfied with the Knockout Switch architecture. In addition, we show that, like most space-division

[1]The steady-state probabilities in Section 5.1.5 of [14] are for the *total number* of packets in an $M/D/1$ system. We are interested in *queue size*; hence, the modification to (8)-(10).

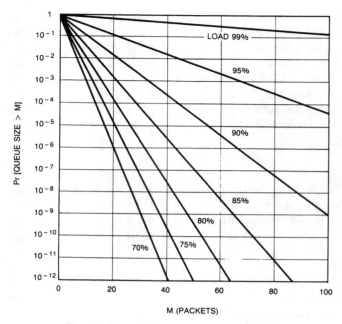

Fig. 11. Shared buffer queueing performance.

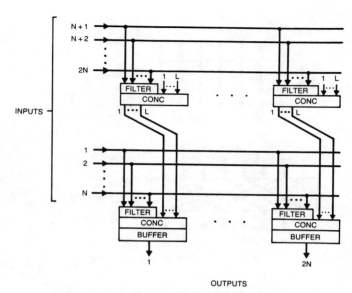

Fig. 12. Modular growth of Knockout Switch.

packet-switching architectures, the complexity of the Knockout Switch results not from the number of gates required in the design of the switch fabric, but rather from pin limitations on the circuit cards and VLSI chips.

A. Modularity

We mentioned earlier that, in addition to growing gracefully from 2×2 to $N \times N$, the Knockout Switch can grow modularly from $N \times N$ to $JN \times JN$, $J = 2, 3, \cdots$. One way to do this is illustrated in Fig. 12 where we have provided each concentrator in the switch with L additional inputs for a total of $N + L$ inputs and L outputs. The interface for each output in a $JN \times JN$ Knockout Switch consists of J separate N-bus interfaces daisy chained together. Specifically, each of the J interfaces for one output contains a row of N packet filters and an $(N + L)$-to-L concentrator, with only the first interface (for buses 1–N) also containing the shared buffer structure with shifter and L FIFO buffers. These J individual components for each output are connected together by attaching the L outputs of the concentrator in the jth interface ($j = 2, 3, \cdots, J$) to the L extra inputs on the concentrator in the $j - 1$st interface. In effect, we have a convenient way of growing the Knockout Switch using a single $(N + L)$-to-L concentrator design and the same shared buffer (one for each output) independent of the switch size.

B. Maintainability

In the layout of the Knockout Switch, the electronics for each of the N input/output pairs can be placed on a separate interface module, with the switch then consisting of N identical interface modules interconnected by N broadcast buses (one for each input). Note that any fault on an interface module will disrupt traffic only on its input

or output and can be repaired by simply replacing the failed interface module. Even while the switch is being repaired, all other inputs and outputs can operate as usual. This is in direct contrast to a multistage switch where a failed intermediate switch point can affect multiple input/output paths and may also be difficult to locate.

C. Fault Tolerance

In many switching applications, stringent operational requirements necessitate a design that is tolerant of faults. This may be achieved by duplicating the entire switch fabric to serve as a standby in the event of a failure. With the Knockout Switch, since all interface modules are identical, a fault tolerant design can be achieved by providing only a single spare interface module attached to the N broadcast buses. This spare module could take over the operation of any one of the N interface modules if a failure occurs. Again, service would only be disrupted on the input or output attached to the failed module, and only as long as it takes to locate the fault and switch over its input and output to the spare module.

D. Complexity

As described in Section II and illustrated in Fig. 3, each bus interface in the Knockout Switch consists of a row of N packet filters, an N-to-L concentrator, and a shared buffer. The shared buffer, containing a simple shifter and L separate FIFO buffers, does not grow with the switch size N. The number of packet filters and the concentrator size, however, both grow in direct proportion to N. A packet filter can be designed with only five gates, giving us a total of $5N$ gates in the bus interface. The concentrator requires about NL simple, 16-gate 2×2 switch elements for a total of $16NL$ gates.

Table I shows, for various N, the combined packet filter and concentrator complexity of each bus interface on the switch in terms of the total number of gates and input/

TABLE I
COMBINED PACKET FILTER AND CONCENTRATOR COMPLEXITY

Switch Size, N	Gates	I/O Connections
128	17 000	136
256	34 000	264
512	68 000	520
1024	136 000	1032

output connections. The number of gates grows as $(5 + 16L)N$ and the required number of input/output connections grows as $N + L$, where $L = 8$ in Table I. At $N = 1024$, the 136 000 gate requirement in Table I is within reach of today's VLSI technology. However, the corresponding 1032-pin requirement for the VLSI chip far exceeds the capabilities that exist today, where the upper limit is in the 128–256 range. Still, the packet filters and concentrator for, say, 128 inputs can easily be integrated onto a single VLSI chip, and then several of these identical chips can be interconnected as shown in Fig. 9 to form a larger combined packet filter and concentrator. Only the chips at the top of the tree have their packet filters enabled, all others function only as concentrators. Also, as Fig. 12 illustrates, circuit card pin limitations on the backplane connector can be overcome by daisy chaining the packet filters and concentrators of multiple bus interface cards to form a larger switch. Hence, both VLSI and circuit card pin restrictions do not limit the growth of the Knockout Switch. In addition, the broadcast buses shown in Figs. 2 and 12 are unidirectional, with the packet framing clock (used to maintain time slot synchrony in the switch) traveling on a separate bus along the same path. Hence, the broadcast buses and frame clock bus can be latched and regenerated as needed to grow the switch.

IV. CONCLUSIONS

We proposed a new, high-performance packet switch architecture: the Knockout Switch. It provides direct interconnection paths from the switch inputs to the outputs, allowing us to greatly simplify the buffer design and thus achieve a more efficient switch. We observed that packet loss in any network is inevitable, whether it is caused by transmission errors or buffer overflow. By allowing the packet switch itself to introduce a small amount of additional packet loss, the concentrator required at each switch output can be reduced from $N \times N$ to $N \times 8$ for arbitrary large N. An $N \times 8$ concentrator is then designed based on knockout matches commonly used in tournaments. A new buffer sharing scheme joins the buffers associated with each of the concentrator output lines to form a simple first-in first-out buffer. The lost packet rate of the Knockout Switch can be made as small as desired and the latency is the smallest achievable by any switch. Moreover, the interconnect allows for easy modular growth, simple maintenance procedures, and a design that can be made fault tolerant. With each of $N = 1000$ inputs operating at 50 Mbits/s, possible applications for a 50 Gbit/s Knockout packet switch include interconnects of multiprocessors, high-speed local and metropolitan area networking, and local or toll switches for integrated traffic loads. With the advent of optical backplanes and integrated optoelectronic devices, the line speed and overall switch capacity could grow much larger.

REFERENCES

[1] H. Inose, *An Introduction to Digital Integrated Communications Systems*. Tokyo, Japan: University of Tokyo Press, 1979.
[2] M. Decina and D. Vlack, Eds., Special Issue on Packet Switched Voice and Data Communication, *IEEE J. Select. Areas Commun.*, vol. SAC-1, Dec. 1983.
[3] J. S. Turner and L. F. Wyatt, "A packet network architecture for integrated services," in *GLOBECOM '84 Conf. Rec.*, Nov. 1983, pp. 45–50.
[4] J. J. Kulzer and W. A. Montgomery, "Statistical switching architectures for future services," presented at the 1984 Int. Switching Symp., Session 43, May 1984.
[5] A. Huang and S. Knauer, "STARLITE: A wideband digital switch," in *GLOBECOM '84 Conf. Rec.*, Nov. 1984, pp. 121–125.
[6] J. S. Turner, "Design of a broadcast packet network," in *Proc. IEEE INFOCOM '86*, Apr. 1986, pp. 667–675.
[7] ——, "Design of an integrated services packet network," *IEEE J. Select. Areas Commun.*, vol. SAC-4, pp. 1373–1380, Nov. 1986.
[8] D. M. Dias and M. Kumar, "Packet switching in N log N multistage networks," in *GLOBECOM '84 Conf. Rec.*, Nov. 1984, pp. 114–120.
[9] M. J. Karol, M. G. Hluchyj, and S. P. Morgan, "Input vs. output queueing in a space-division packet switch," in *GLOBECOM '86 Conf. Rec.*, Dec. 1986, pp. 659–665.
[10] D. F. Kuhl, "Error recovery protocols: Link by link vs. edge by edge," in *Proc. IEEE INFOCOM '83*, Apr. 1983, pp. 319–324.
[11] W. L. Hoberecht, "A layered network protocol for packet voice and data integration," *IEEE J. Select. Areas Commun.*, vol. SAC-1, pp. 1006–1013, Dec. 1983.
[12] W. Hoeffding, "On the distribution of the number of successes in independent trials," *Ann. Math. Statist.*, vol. 27, pp. 713–721, 1956.
[13] D. H. Lawrie, "Access and alignment of data in an array processor," *IEEE Trans. Comput.*, vol. C-24, pp. 1145–1155, Dec. 1975.
[14] D. Gross and C. M. Harris, *Fundamentals of Queueing Theory*. New York: Wiley, 1974.
[15] K. Sriram and W. Whitt, "Characterizing superposition arrival processes and the performance of multiplexers for voice and data," in *GLOBECOM '85 Conf. Rec.*, Dec. 1985.

Performance Analysis of a Growable Architecture for Broad-Band Packet (ATM) Switching

Mark J. Karol, *Senior Member, IEEE,* and Chih-Lin I, *Member, IEEE*

Abstract— A growable architecture for broad-band packet (ATM) switching consisting of a memoryless, self-routing interconnect fabric and modest-size packet switch modules, was recently proposed by Eng, Karol, and Yeh. In this paper, we examine the performance of this architecture. We focus on the cell loss probability, because the architecture attains the best possible delay-throughput performance if the packet switch modules use output queueing. There are two sources of cell loss in the switch. First, cells are dropped if too many simultaneous arrivals are destined to a group of output ports. Second, because a simple, distributed path-assignment controller is used for speed and efficiency, cells are dropped when the controller cannot "schedule" a path through the switch. We compute an upper bound on the cell loss probability for *arbitrary patterns of independent cell arrivals,* possibly including isochronous circuit connections, and show that both sources of cell loss can be made negligibly small. For example, to guarantee less than 10^{-9} cell loss probability, this growable architecture requires packet switch modules of dimension 47×16, 45×16, 42×16, and 39×16 for 100, 90, 80, and 70% traffic loads, respectively. The analytic techniques we use to bound the cell loss probabilities are applicable to other output queueing architectures.

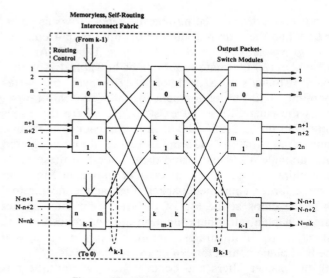

Fig. 1. The proposed switch architecture.

I. INTRODUCTION

THE asynchronous transfer mode (ATM), based on fast packet switching, is currently being studied in standards bodies and is likely to become the switching and multiplexing technique for broad-band ISDN (BISDN). A key element in the migration toward BISDN is the $N \times N$ central switch that provides a high-performance packet interconnect from N input ports (e.g., $N \approx 1000$) to N output ports. Such a broad-band ATM switch may need to support total throughputs ranging from several gigabits per second to hundreds of gigabits per second, with a latency measured in the tens of microseconds and a very small (e.g., less than 10^{-9}) cell[1] loss probability. These goals are attainable with highly parallel switch fabrics that use simple per cell processing distributed among many high-speed VLSI circuits.

As important as the high throughput, low delay, and small cell loss probability, is the modular growth of an ATM switch. The switch architecture should allow the construction of arbitrary-size switches using modest-size modules, so that N can be tailored to grow gracefully as additional users are

brought on-line. Switch growability is essential because a limited number of I/O pins are practical per VLSI chip and board.

Eng, Karol, and Yeh [1] describe a growable switch architecture that meets both the performance and the modularity requirements. The switch is constructed in a three-stage configuration, as in Fig. 1. The first two stages form a memoryless, self-routing interconnect fabric. A simple, distributed path-assignment control is executed at the input modules to avoid internal switch fabric congestion and to minimize the switch fabric complexity [2]. Also, the simple routing of cells avoids the complexity of optimal scheduling [3]. The output stage, composed of modest-size (e.g., 42×16) packet switch modules, can be implemented, among other choices, with either a shared-memory switch, the STARLITE switch [4] or the knockout switch [5]. Cells are queued only in the output stage in order to achieve the best possible delay/throughput performance [6].

In this paper, we examine the performance of this growable ATM packet switch architecture. The results and analytic techniques, however, are more general, and are applicable to other output queueing architectures. The organization of this paper is as follows. A description of the switch architecture is provided in Section II. An upper bound on the cell loss probability for *arbitrary patterns of independent arrivals* is computed in Section III. The bounds are presented for switches with and without internal path trunking, and for both finite and infinite switch dimensions. Using a traditional, uniform traffic

Paper approved by the Editor for Communications Switching of the IEEE Communications Society. Manuscript received September 11, 1989. This paper was presented in part at GLOBECOM '89, Dallas, TX, November 27–30, 1989.

The authors are with AT&T Bell Laboratories, Holmdel, NJ 07733.

IEEE Log Number 9106315.

[1] ATM uses fixed-length cells consisting of a 48 octet (byte) information field and a 5 octet header.

Reprinted from *IEEE Trans. Commun.,* vol. 40, no. 2, pp. 431–439, February 1992.

model, a tighter bound on the cell loss probability is derived in Section IV. Numerical results and conclusions are presented in Sections V and VI, respectively.

II. THE SWITCH ARCHITECTURE

To facilitate the reading of this paper, we start by highlighting the main features of the switch. A complete description of the switch architecture is given in [1].

The switch fabric, illustrated in Fig. 1, operates synchronously with fixed-length cells arriving at the N input ports in a time-slotted fashion. The modules of the first stage provide an expansion by a factor of m/n; that is, each has n input ports and m output ports where $m \geq n$. There are $k = N/n$ such input modules. The second stage is composed of m modules, each of size $k \times k$. Together these two stages form a memoryless interconnect fabric.

At the last stage, the output modules are $m \times n$ packet switches. It is sufficient to route a cell to any one of the m input lines of the output module that contains its destined output port. The output module will send the cell to its appropriate output port using the address in the cell header. Output packet switch modules also buffer cells if there are several waiting for the same output line. Since an output module only has m input lines, if more than m cells destined to an output module arrive in the same time slot, then the excess cells are dropped at their input modules. This is referred to as the generalized knockout principle [5], [1].

A good routing algorithm has to direct the arriving cells through the interconnect fabric to their respective output modules without path conflicts. Since $m \geq n$, the existence of such path assignments is guaranteed by the nonblocking property of the three-stage Clos network [2]. That is, if no more than m cells that arrive in a time slot are destined to the same output module, then an optimal routing algorithm exists which assigns paths through the fabric for all cells to their destined output modules. If more than m cells that arrive in a time slot are destined to the same output module, then the additional cells are dropped at their input modules; this loss will be referred to as the knockout loss since it is the result of the generalized knockout principle. The knockout loss would be the only cell loss if an optimal routing algorithm were used.

Cells are assigned paths on a slot-by-slot basis; at 150 Mb/s, a 53 octet (byte) time slot lasts for only a few microseconds. Therefore, instead of an optimal (but slow) routing algorithm, the switch uses a distributed, fast, suboptimal algorithm [1]. As a result, a cell might not be assigned a path through the switch (and will be dropped at its input module) even though fewer than m other cells are destined to the same output module. Thus, besides the knockout loss of more than m cells arriving destined to an output module, there is additional loss due to the suboptimal scheduling. This additional loss is the main subject of this paper.

The proposed routing scheme divides a time slot into k minislots. In every minislot, each input module is given permission to schedule its cells to a particular output module. For instance, in the first minislot, input module i tries to schedule its cells destined to output module i ($i = 0, \cdots, k -$

1). Then, in the hth minislot ($h = 1, \cdots, k$), input module i attempts to schedule its cells destined to output module $((i+h-1) \bmod k)$. This path-assignment process operates like a daisy chain from one minislot to the next. In the hth minislot, input module i schedules paths for its cells destined to output module $((i + h - 1) \bmod k)$ using only local information: the statuse (i.e., busy or free) of its m outgoing paths (i.e., from the ith input module), and the status of the m incoming paths to the $((i + h - 1) \bmod k)$th output module. The latter information is passed down to the ith input module from the $((i - 1) \bmod k)$th one.

In the hth minislot, cells at the ith input module that are destined to the $((i + h - 1) \bmod k)$th output module select routing paths according to the status, either occupied or available, of the outgoing paths from the input module and the incoming paths to the output module. We use vectors A_i and $B_{(i+h-1) \bmod k}$, m binary elements each, to represent the status of the corresponding input module and output module paths; a binary value 0 indicates an available path, while a 1 means that the path is occupied. The paths leaving the $(k - 1)$th input module and the paths entering the $(k - 1)$th output module are circled in Fig. 1. Note that each outgoing (and incoming) path corresponds to a particular second-stage module. A cell can be routed from the ith input module to the $((i+h-1) \bmod k)$th output module if there is an available path between the two via a common second-stage module. Also, at the beginning of a time slot, paths can be held for isochronous, circuit traffic by setting to 1 the appropriate elements in A_i and B_j.

Within each time slot, the cell loss probability increases as more and more cells are assigned transmission paths through the switch fabric. Consequently, the last input–output module pairs in the scheduling process (i.e., in the kth minislot, from the ith input module to the $((i + k - 1) \bmod k)$th output module) tend to encounter the most occupied paths, and hence lose the most cells. It may be desirable for fairness, at the expense of perhaps complexity, to vary the initial input–output path-assignment pairings. In this paper, we derive an upper bound on the cell loss probability for the worst case, last-scheduled input–output ports. Therefore, a growable switch can be designed with any scheduling sequence in each time slot, and the cell loss probability for arrivals on all lines will be at most the values presented here.

III. CELL LOSS PROBABILITY: ARBITRARY PATTERNS OF INDEPENDENT CELL ARRIVALS

We assume the output packet switch modules use output queueing, so that the overall switch has the best possible delay/throughput performance [6]. Consequently, we only need to compute the cell loss probability. There are two sources of cell loss in the switch: the knockout loss and the loss due to suboptimal scheduling. We neglect any cell loss that occurs in the output packet switch modules due to buffer overflow. This additional cell loss can be independently computed for any desired buffer size, any packet switch module, and any traffic module (e.g., [7]–[12]).

In this section, we compute an upper bound on the cell loss probability for *arbitrary patterns of independent cell arrivals*.

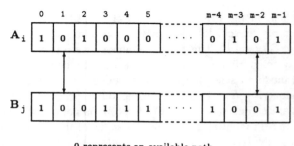

0 represents an available path

1 represents an occupied path

Fig. 2.   Straight matching path selection.

The bound is for the worst case, last-scheduled input–output ports. In Section IV we obtain a tighter bound by assuming a traditional, uniform traffic model. Here, we only assume that the arrivals at different input ports are independent.

Let $P_{I,j}$ be the probability a cell arrives at input port I ($0 \leq I \leq N-1$) destined to output module j ($0 \leq j \leq k-1$). The capital I in $P_{I,j}$ refers to an input *port*, whereas lowercase i is used when we refer to an input *module*. Since the scheduling is done on a slot-by-slot basis, we allow $P_{I,j}$ to change from one time slot to the next.

We first study the case illustrated in Fig. 1, in which internal paths between modules are not trunked. We then analyze the situation with internal path trunking between modules in successive stages, later depicted in Fig. 3.

A. No Internal Path Trunking Between Modules in Successive Stages

A cell can be routed from input module i to output module j, along a path through a second-stage module, only if the corresponding elements in A_i and B_j both have value 0, as illustrated in Fig. 2. We refer to this as a *straight matching*. If there are several available paths (i.e., more than one "matching"), one is selected at random with equal probability. The performance of the scheduling algorithm will be different than indicated here if a nonrandom selection of routing paths is made when several matchings exist. We have not investigated this situation.

Suppose $(m - r)$ elements of A_i and $(m - s)$ elements of B_j have value 0. This corresponds to r occupied outgoing paths from input module i, and s occupied incoming paths to output module j. We assume these "0" elements in A_i and B_j are randomly located with independent, uniform distributions. This assumption is appropriate since there is no global coordination in the routing scheme, and since a path is selected at random when several matchings exist. In attempting to assign a path for a cell from input module i to output module j, the locations of the "0" elements in A_i and B_j will not be independent if other cells have already been scheduled using A_i and B_j. These dependencies, though, increase the probability of finding straight matchings by forcing some 1's in A_i to line up with 1's in B_j. In computing an upper bound on the cell loss probability, we will ignore the advantages of such dependencies.

There are c straight matchings (representing c available paths from A_i to B_j) if the distribution of the 0's in B_j

is such that exactly c of them are aligned with 0's of A_i, and the remaining $(m - s - c)$ 0's of B_j are aligned with 1's of A_i. This scenario of c straight matchings can occur in $\binom{m-r}{c}\binom{r}{m-s-c}$ ways out of the total $\binom{m}{m-s}$ possible distributions of $(m - s)$ 0's in B_j. Thus, given r and s, the probability $P_{C|R,S}(c|r,s)$ of having c straight matchings is hypergeometrically distributed:

$$P_{C|R,S}(c|r,s) = \frac{\binom{m-r}{c}\binom{r}{m-s-c}}{\binom{m}{m-s}}$$

$$\max\{0, (m-r-s)\} \leq c \leq \min\{(m-r), (m-s)\}.$$

(1)

When $(r + s)$ is less than m, there will be at least $(m - r - s)$ straight matchings, and obviously the number of straight matchings cannot exceed either $(m - r)$ or $(m - s)$. The range of c in (1) is determined by these constraints.

From (1), the probability of having no straight matchings for given r and s, such that $r + s \geq m$, is

$$P_{C|R,S}(0|r,s) = \frac{\binom{m-r}{0}\binom{r}{m-s}}{\binom{m}{m-s}} = \frac{r!\,s!}{(r+s-m)!\,m!}$$

$$r \leq m, \qquad s \leq m, \qquad (s+r) \geq m. \tag{2}$$

Suppose a cell destined for the jth output module arrives at the ith input module. The cell will be dropped at the input, during the path-assignment algorithm, if there are no straight matchings of 0's when the vector of available input module paths A_i is compared with B_j, the vector of available output module paths. This cell loss probability is given by (2), assuming there are r occupied input module paths and s occupied output module paths.

Since (2) is a nondecreasing function of r, we obtain an upper bound on the cell loss probability P_{loss}, *for all possible arrival patterns*, by setting $r = (n-1)$ in (2). That is, we assume all other input ports of the ith input module have cell arrivals, and these $(n - 1)$ other cells are already successfully scheduled. If the traffic includes some multicast connections, then additional cells may have been assigned paths and a value greater than $(n - 1)$ need be used for r in (2); the following analysis, otherwise is unchanged. We will not consider multicast connections in this paper.

Let S represent the number of cells destined to the jth output module among the arrivals at the other $(N - 1) = (nk - 1)$ input lines. Then, since (2) is a nondecreasing function of s, and since at most m cells can be scheduled to an output module, we bound the conditional cell loss probability $P_{\text{loss}|S}(s)$ by

$$P_{\text{loss}|S}(s) \leq \begin{cases} \frac{(n-1)!\,s!}{(n-1+s-m)!\,m!} & m-n+1 \leq s \leq m \\ 1 & m+1 \leq s \leq N-1. \end{cases} \tag{3}$$

Multiplying (3) by $P_S(s)$ and summing over s, we obtain

$$P_{\text{loss}} \leq \sum_{s=m-n+1}^{m} P_S(s)\frac{(n-1)!\,s!}{(n-1+s-m)!\,m!} + \sum_{s=m+1}^{N-1} P_S(s).$$

(4)

Since $\frac{(n-1)!\,s!}{(n-1+s-m)!\,m!}$ is a nondecreasing function of s, by defining

$$W_s \triangleq \begin{cases} \dfrac{(n-1)!\,s!}{(n-1+s-m)!\,m!} \\ \qquad s = m-n+1 \\[2ex] \dfrac{(n-1)!\,s!}{(n-1+s-m)!\,m!} - \dfrac{(n-1)!\,(s-1)!}{(n-1+(s-1)-m)!\,m!} \\ \qquad m-n+1 < s \le m, \end{cases} \tag{5}$$

we can write the right-hand side of (4) as

$$\sum_{s=m-n+1}^{m} P_S(s)\frac{(n-1)!\,s!}{(n-1+s-m)!\,m!} + \sum_{s=m+1}^{N-1} P_S(s) =$$

$$W_{m-n+1} \sum_{s=m-n+1}^{N-1} P_S(s) + W_{m-n+2} \sum_{s=m-n+2}^{N-1} P_S(s)$$

$$+ \cdots + W_m \sum_{s=m}^{N-1} P_S(s) \tag{6}$$

where each of the W_s is nonnegative. Alternatively, we can write (6) as

$$\sum_{s=m-n+1}^{m} P_S(s)\frac{(n-1)!\,s!}{(n-1+s-m)!\,m!} + \sum_{s=m+1}^{N-1} P_S(s) =$$

$$W_{m-n+1} + W_{m-n+2} + \cdots + W_m$$

$$- \Pr(0 \le S \le m-n)W_{m-n+1}$$

$$- \Pr(0 \le S \le m-n+1)W_{m-n+2}$$

$$- \cdots - \Pr(0 \le S \le m-1)W_m. \tag{7}$$

To obtain an upper bound on the summation in (7), we will lower bound the terms $\Pr(0 \le S \le z)$—for $z = m-n$, $m-n+1, \cdots$, and $m-1$—by using a result from [13]. This same result was used in [14] to bound the packet loss probability for the knockout switch.

In [13], Hoeffding studies the distribution of the number of successes in independent trials. Since the arrivals at different input ports are assumed independent, [13] shows that the probabilities $\Pr(0 \le S \le z)$ are minimized when $P_{I,j} = E\{S\}/(N-1)$ for $0 \le I \le (N-1)$, provided $z \ge E\{S\}$. Because an output packet switch module only has n output ports, the traffic pattern must be such that $E\{S\} \le n$. So, if $(m-n) \ge n$, or equivalently $m \ge 2n$, then the uniform distribution for $P_{I,j}$ will minimize all terms $\Pr(0 \le S \le z)$ in (7). In other words, if $m \ge 2n$, then the probabilities $\Pr(0 \le S \le z)$ are minimized when S is binomially distributed:

$$P_S(s) = \binom{N-1}{s}\left(\frac{np}{N-1}\right)^s \left(1 - \frac{np}{N-1}\right)^{N-1-s}$$
$$0 \le s \le N-1 \tag{8}$$

where $p = \frac{E\{S\}}{n}$ is the normalized offered load to output module j. Substituting (8) for $P_S(s)$ in (4), we obtain an upper bound on the cell loss probability for all possible traffic arrival patterns:

$$P_{\text{loss}} \le \sum_{s=m-n+1}^{m-1} \binom{N-1}{s}\left(\frac{np}{N-1}\right)^s \left(1 - \frac{np}{N-1}\right)^{N-1-s}$$

$$\cdot \frac{(n-1)!\,s!}{(n-1+s-m)!\,m!} + \sum_{s=m}^{N-1} \binom{N-1}{s}$$

$$\cdot \left(\frac{np}{N-1}\right)^s \left(1 - \frac{np}{N-1}\right)^{N-1-s} \tag{9}$$

In (9), the second summation represents the knockout loss component of P_{loss}. The first summation in (9) is the additional loss due to suboptimal scheduling.

With n and m fixed (so that the $N \times N$ switch is composed of small packet switch modules), $N \to \infty$ as $k \to \infty$. Taking the limit of (9) as $k \to \infty$, we obtain an upper bound on the cell loss probability for arbitrarily large switch dimensions and for all possible traffic arrival patterns:

$$P_{\text{loss}} \le \sum_{s=m-n+1}^{m-1} e^{-np}\frac{(np)^s}{s!}\frac{(n-1)!\,s!}{(n-1+s-m)!\,m!}$$

$$+ \sum_{s=m}^{\infty} e^{-np}\frac{(np)^s}{s!}$$

$$= 1 + \sum_{s=m-n+1}^{m-1} e^{-np}\frac{(np)^s(n-1)!}{(n-1+s-m)!\,m!}$$

$$- \sum_{s=0}^{m-1} e^{-np}\frac{(np)^s}{s!}. \tag{10}$$

Setting $n = 1$ in (10), the knockout loss $1 - \sum_{s=0}^{m-1}(e^{-p})\frac{p^s}{s!}$ agrees with the formula for the worst-case-input knockout loss in [5].

To compare the knockout loss to the "knockout + scheduling" loss, note in (10) that $\frac{(np)^s}{(n-1+s-m)!}$ is maximized when $s = \lceil np+m-n \rceil$ where $\lceil v \rceil$ is the smallest integer greater than or equal to v. Hence, we can bound P_{loss} by

$$P_{\text{loss}} \le 1 + e^{-np}\frac{n!}{m!}\frac{(np)^{\lceil np+m-n \rceil}}{(\lceil np-1 \rceil)!} - \sum_{s=0}^{m} e^{-np}\frac{(np)^s}{s!}. \tag{11}$$

In (11), since $e^{-np}\frac{(np)^{\lceil np+m-n \rceil}}{(\lceil np-1 \rceil)!}$ is a nondecreasing function of p, setting $p = 1$ (i.e., 100% loading) yields

$$P_{\text{loss}} \le 1 + \frac{n^{m+1}}{m!}e^{-n} - \sum_{s=0}^{m} e^{-n}\frac{n^s}{s!}$$

$$= \sum_{s=m}^{\infty} e^{-n}\frac{n^s}{s!} + (n-1)\frac{n^m}{m!}e^{-n}. \tag{12}$$

Finally, using (12) and lower bounding P_{loss} by the knockout loss, we have

$$\Pr(m \le S) \le P_{\text{loss}} \le \Pr(m \le S) + (n-1)\Pr(S = m). \tag{13}$$

In (13), $\Pr(m \le S)$ is the knockout loss and $(n-1)\Pr(S = m)$ is the scheduling loss.

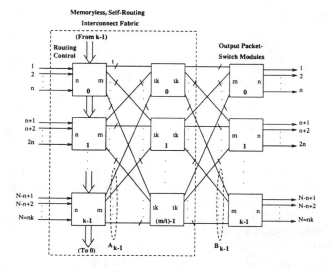

Fig. 3. Modified switch architecture with path trunking.

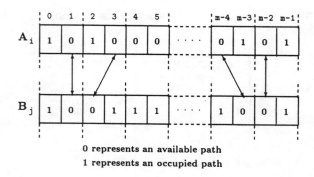

0 represents an available path
1 represents an occupied path

Fig. 4. Diagonal matching path selection.

Before analyzing the growable switch architecture with internal path trunking between stages, note that a space-to-time transformation [1] makes random scheduling of the time-space-time (TST) switch [3] similar to the "matching process" (shown in Fig. 2) used in this suboptimal scheduling algorithm. However, they are not identical. The time slot matching in the TST switch is symmetric in that the maximum number of unavailable slots at the inputs and outputs are the same. The path matching here, though, is not symmetric: the maximum number of unavailable paths from an input module is n and the maximum number of the unavailable paths to an output module is m. Besides the asymmetry, the performance of the TST circuit switch is examined via steady-state analysis, whereas other analytic techniques are needed for this ATM packet switch because it routes independent, random cells on a slot-by-slot basis. Cells not scheduled during their arrival time slot are dropped before the next slot commences, and the states of all the paths are "cleared" for new cells at the beginning of every time slot.

B. Internal Path Trunking Between Stages

So far, we have assumed that a straight matching between A_i and B_j is needed to establish a path through the switch fabric from the ith input module to the jth output module. This requirement can be relaxed by trunking the internal paths. In Fig. 3, there are t lines between each of the first- and second-stage modules (likewise, between the second and third stages). Note that there are $M = m/t$ trunks associated with each input (or output) module, one to each of the M second-stage modules. The analysis in the previous subsection was for $t = 1$ and $M = m$.

The cell loss probability will be lowered by trunking the internal paths between stages. Since a cell can use any of the t paths between modules, *diagonal matchings* (shown in Fig. 4 for $t = 2$), in addition to the straight matchings, can be used to establish a path from an input module to an output module. As t increases, the number of additional paths increases and the cell loss probability approaches the knockout

loss, which occurs exactly when $t = m$. The price paid for this improvement is an increase in the fabric complexity; the size of the second-stage modules increases from $k \times k$ to $tk \times tk$. In the extreme case where $t = m$ and $M = 1$, the second stage becomes a single fully connected $mk \times mk$ module (i.e., a $N\left(\frac{m}{n}\right) \times N\left(\frac{m}{n}\right)$ module), and there is no loss due to suboptimal scheduling. In this subsection, we will upper bound the cell loss probability with various trunk sizes t for arbitrary patterns of independent cell arrivals, so that a judicious decision can be made on whether the increase in complexity is justified by the performance improvement. As before, we assume matchings are made at random when several possibilities exist.

To obtain an upper bound on the cell loss probability for all possible arrival patterns, we follow a procedure similar to that used in the previous subsection. The only difference is that we must compute the probability of having no straight nor diagonal matchings when a cell destined for the jth output module arrives at the ith input module.

Let R_0 represent the number of fully occupied outgoing trunks from input module i. Given that there are R occupied outgoing paths from input module i, the conditional distribution $P_{R_0|R}(r_0|r)$ is

$$P_{R_0|R}(r_0|r) = \frac{\sum_{h=0}^{\left\lfloor \frac{r}{t}-r_0 \right\rfloor} (-1)^h \binom{r_0+h}{h} \binom{M}{r_0+h} \binom{m-t(r_0+h)}{r-t(r_0+h)}}{\binom{m}{r}}$$
$$\cdot \max\{0, r-(m-M)\} \le r_0 \le \lfloor r/t \rfloor. \quad (14)$$

The denominator of (14) is the number of ways to distribute the r occupied paths among the m outgoing paths from input module i. The numerator of (14) is the number of ways to obtain exactly r_0 fully occupied trunks among the M outgoing trunks [15].

Neither straight matchings nor diagonal matchings can be made if and only if the available incoming lines to output module j all come from the second-stage modules that correspond to these r_0 fully occupied input trunks. In other words, no matchings exist if the other $(M - r_0)$ trunks to output module j are all fully occupied. When $s \ge t(M - r_0)$, there are $\binom{tr_0}{s-t(M-r_0)}$ ways for this scenario to occur; hence, in place of (2), the probability of having no straight nor diagonal matchings for given r and s is

$$P_{D|R_S}(0|r,s) = \sum_{r_0} P_{R_0|R}(r_0|r) \frac{\binom{tr_0}{s-t(M-r_0)}}{\binom{m}{s}}. \tag{15}$$

Setting $r = (n-1)$ in (14) and (15) and substituting $P_{D|R,S}(0|n-1,s)$ for $P_{C|R,S}(0|n-1,s) = \frac{(n-1)!\,s!}{(n-1+s-m)!\,m!}$ in (9) and (10) yields the following bounds on the cell loss probability for all possible traffic arrival patterns:

$$P_{\text{loss}}(t) \leq \sum_{s=m-n+1}^{m-1} \binom{N-1}{s} \left(\frac{np}{N-1}\right)^s$$
$$\cdot \left(1 - \frac{np}{N-1}\right)^{N-1-s} P_{D|R,S}(0|n-1,s)$$
$$+ \sum_{s=m}^{N-1} \binom{N-1}{s} \left(\frac{np}{N-1}\right)^s \left(1 - \frac{np}{N-1}\right)^{N-1-s}, \tag{16}$$

and as $N \to \infty$,

$$P_{\text{loss}}(t) \leq 1 + \sum_{s=m-n+1}^{m-1} e^{-np} \frac{(np)^s}{s!} P_{D|R,S}(0|n-1,s)$$
$$- \sum_{s=0}^{m-1} e^{-np} \frac{(np)^s}{s!}. \tag{17}$$

IV. Cell Loss Probability: Tighter Bound for a Uniform Traffic Model

In the previous section, we computed an upper bound on the cell loss probability for arbitrary patterns of independent arrivals by determining the worst possible pattern. In this section, we compute a tighter bound for a traditional, uniform traffic model with independent, identically distributed cell arrivals to the switch. In particular, let p ($0 \leq p \leq 1$) represent the probability of cell arrival at an input port in a time slot. We assume the arrivals at different input ports are independent and the cells arrive with uniformly distributed destination addresses, so that a cell is destined to any one of the k output modules with probability $1/k$. For the cell loss calculations we need only specify the destination output *module*, not the destination output *port*.

Suppose a cell destined for the jth output module arrives at the ith input module. Under this model, the number of cells R that arrive at input module i on the other $(n-1)$ input ports is binomially distributed with parameters $(n-1)$ and p:

$$P_R(r) = \binom{n-1}{r} p^r (1-p)^{n-1-r} \qquad 0 \leq r \leq (n-1). \tag{18}$$

Of these R cells, the number X that are also destined to the jth output module is binomially distributed:

$$P_{X|R}(x|r) = \binom{r}{x} \left(\frac{1}{k}\right)^x \left(1 - \frac{1}{k}\right)^{r-x} \qquad 0 \leq x \leq r. \tag{19}$$

Similarly, the number of cells Y from other input modules (i.e., not the ith) that are destined to the jth output module is binomially distributed:

$$P_Y(y) = \binom{n(k-1)}{y} \left(\frac{p}{k}\right)^y \left(1 - \frac{p}{k}\right)^{n(k-1)-y}$$
$$0 \leq y \leq n(k-1). \tag{20}$$

Consequently, the total number S of cells destined to the jth output module among the arrivals at the other $(N-1) = (nk-1)$ input lines is

$$S = X + Y. \tag{21}$$

A. No Internal Path Trunking Between Modules in Successive Stages

As in the previous section, we bound P_{loss} by assuming that all R cells and all S cells, up to a maximum of m, are already successfully scheduled. Noting that X 1's in A_i line up with X 1's in B_j, and using (2), (18), (19), and (20), we obtain the following upper bound on P_{loss}:

$$P_{\text{loss}} \leq \sum_{r=0}^{n-1} P_R(r) \sum_{x=0}^{r} P_{X|R}(x|r) \left[\sum_{y=m-r}^{m-x-1} \right.$$
$$\left. \cdot P_Y(y) \frac{(r-x)!\,y!}{(r+y-m)!\,(m-x)!} + \sum_{y=m-x}^{n(k-1)} P_Y(y) \right]. \tag{22}$$

Taking the limit of (22) as $k \to \infty$—corresponding to $N \to \infty$ with n and m fixed—we obtain

$$P_{\text{loss}} \leq \sum_{r=0}^{n-1} P_R(r) \left[\sum_{y=m-r}^{m-1} e^{-np} \frac{(np)^y}{y!} \frac{r!\,y!}{(r+y-m)!\,m!} \right.$$
$$\left. + \sum_{y=m}^{\infty} e^{-np} \frac{(np)^y}{y!} \right]$$
$$= 1 + \sum_{r=1}^{n-1} P_R(r) \sum_{y=m-r}^{m-1} e^{-np} \frac{(np)^y\,r!}{(r+y-m)!\,m!}$$
$$- \sum_{y=0}^{m-1} e^{-np} \frac{(np)^y}{y!}. \tag{23}$$

At 100% loading, (23) agrees with (10).

B. Internal Path Trunking Between Stages

Substituting $P_{D|R,S}(0|r,x+y)$ in (22) and (23) yields the following bounds on the cell loss probability:

$$P_{\text{loss}}(t) \leq \sum_{r=0}^{n-1} P_R(r) \sum_{x=0}^{r} P_{X|R}(x|r) \left[\sum_{y=\max\{0,(m-x-r)\}}^{m-x-1} \right.$$
$$\left. \cdot P_Y(y) P_{D|R,S}(0|r,x+y) + \sum_{y=m-x}^{n(k-1)} P_Y(y) \right], \tag{24}$$

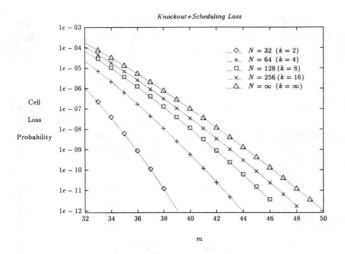

Fig. 5. Cell loss probability as a function of expansion parameter m for various switch sizes N at 90% load. Output group size $n = 16$, and no trunking ($t = 1$).

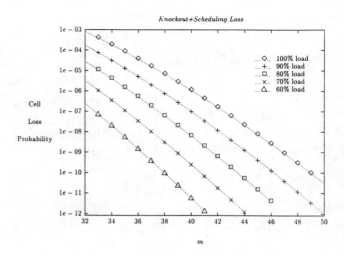

Fig. 6. Cell loss probability as a function of expansion parameter m for various traffic loads. Switch size $N = \infty$, output group size $n = 16$, and no trunking ($t = 1$).

and as $N \to \infty$,

$$
P_{\text{loss}}(t) \leq \sum_{r=0}^{n-1} P_R(r) \left[\sum_{y=m-r}^{m-1} e^{-np} \frac{(np)^y}{y!} P_{D|R,S}(0|r,y) \right.
$$
$$
\left. + \sum_{y=m}^{\infty} e^{-np} \frac{(np)^y}{y!} \right]
$$
$$
= 1 + \sum_{r=1}^{n-1} P_R(r) \sum_{y=m-r}^{m-1} e^{-np} \frac{(np)^y}{y!} P_{D|R,S}(0|r,y)
$$
$$
- \sum_{y=0}^{m-1} e^{-np} \frac{(np)^y}{y!}. \tag{25}
$$

The summation over y in (24) begins at $\max\{0, (m - x - r)\}$ instead of $m - r$, as in (22) because here, for simplicity, we do not exploit the property that X 1's in $\boldsymbol{A}_i$ line up with X 1's in $\boldsymbol{B}_j$. At 100% loading, (25) agrees with (17).

V. NUMERICAL RESULTS

For an output group size $n = 16$, Fig. 5 shows an upper bound on the cell loss probability P_{loss} for *arbitrary patterns of independent cell arrivals*. P_{loss} is shown for 90% load (i.e., $p = 0.9$) as a function of the expansion parameter m for various switch sizes N, and no internal trunking of paths ($t = 1$). The results for $N < \infty$ and $N = \infty$ are from (9) and (10), respectively. Since the $N = \infty$ cell loss probability is greater than the loss probability for $N < \infty$, and since the motivation for this architecture is to grow to large switch dimensions, we focus our attention on the $N = \infty$ results here.

Using the same output group size $n = 16$, Fig. 6 shows an upper bound on the cell loss probability P_{loss} for various loadings of an output module. Taking $m = 41, 38, 36$, and 34 will guarantee that the cell loss probability is below 10^{-6}, for all cell arrival patterns, for 100, 90, 80, and 70% loads, respectively. To guarantee less than 10^{-9} cell loss probability, output packet switch modules of dimension 47×16, 45×16, 42×16, and 39×16 are required for 100, 90, 80, and 70% loads, respectively.

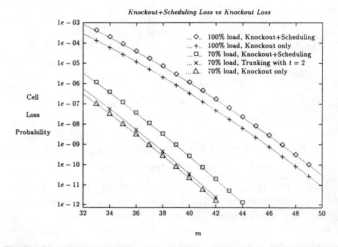

Fig. 7. Upper and lower bounds on cell loss probability as a function of expansion parameter m for 100 and 70% traffic loads. Switch size $N = \infty$, and output group size $n = 16$.

Fig. 7 compares the lower bound (due to knockout loss only) and the upper bound (including both the knockout and the scheduling loss) on the cell loss probability P_{loss}. Note that little improvement would be gained by using an optimal (but slow) scheduling algorithm to achieve the knockout loss lower bound. Besides, the additional loss introduced by suboptimal scheduling is easily compensated for by simply increasing the expansion parameter m by one or two; that is, adding one or two additional input ports to each output packet switch module. Alternatively, the scheduling loss can be greatly reduced by trunking the internal paths between stages, as shown in Fig. 7 for a trunk size of two.

Fig. 8 shows the expansion ratio m/n required to guarantee that the cell loss probability P_{loss} is below various cell loss criteria, and illustrates the advantage gained by grouping outputs.

Finally, Fig. 9 compares the bounds obtained using the arbitrary cell arrival model and the traditional, uniform traffic model. As expected, the uniform model leads to a tighter

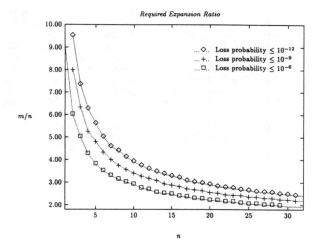

Fig. 8. The required expansion ratio m/n as a function of output group size n for various cell loss probabilities at 90% load. Switch size $N = \infty$, and no trunking $(t = 1)$.

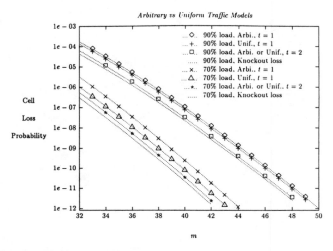

Fig. 9. Cell loss probability as a function of expansion parameter m for the arbitrary cell arrival model and the traditional, uniform model. Switch size $N = \infty$, and output group size $n = 16$.

bound on the cell loss probability, and the two bounds coincide as the loading approaches 100%. With an internal trunk size of two (i.e., $t = 2$), the two bounds agree even at a 70% load.

VI. SUMMARY

We have derived upper and lower bounds on the cell loss probability for the growable ATM switch architecture recently proposed by Eng, Karol, and Yeh. The bounds were obtained for a traditional, uniform traffic model, and more important, for *arbitrary patterns of independent cell arrivals*. The analysis holds for isochronous, circuit traffic, and can be extended to handle multicast connections. We focused on the cell loss probability because the architecture attains the best possible delay-throughput performance if the output packet switch modules use output queueing. The results of our analysis indicate that both the knockout loss (lower bounding the cell loss probability) and the loss due to suboptimal scheduling can be made negligibly small. The closeness of the upper and lower bounds suggest little improvement would be gained by using an optimal (but slow) scheduling algorithm. Rather, one or two

additional input ports could be added to each packet switch module, or internal paths could be trunked, to compensate for the suboptimal scheduling. To guarantee that the cell loss probability is below 10^{-9}, the architecture requires packet switch modules of dimension 47×16, 45×16, 42×16, and 39×16 for 100, 90, 80, and 70% traffic loads, respectively.

REFERENCES

[1] K. Y. Eng, M. J. Karol, and Y. S. Yeh, "A growable packet (ATM) switch architecture: Design principles and applications," *GLOBECOM'89 Conf. Rec.*, Nov. 1989, pp. 1159–1165.
[2] V. E. Benes, "On rearrangeable three-stage connecting networks," *Bell Syst. Tech. J.*, pp. 1481–1492, Sept. 1962.
[3] C. Rose and M. G. Hluchyj, "The performance of random and optimal scheduling in a time-multiplex switch," *IEEE Trans. Commun.*, vol. COM-35, pp. 813–817, Aug. 1987.
[4] A. Huang and S. C. Knauer, "STARLITE: A wideband digital switch," *GOBECOM'84 Conf. Rec.*, Nov. 1984, pp. 121–125.
[5] Y. S. Yeh, M. G. Hluchyj, and A. S. Acapora, "The knockout switch: A simple, modular architecture for high-performance packet switching," *IEEE Select. Areas Commun.*, vol. SAC-5, pp. 1274–1283, Oct. 1987.
[6] M. J. Karol, M. G. Hluchyj, and S. P. Morgan, "Input versus output queueing on a space-division packet switch," *IEEE Trans. Commun.*,

vol. COM-35, pp. 1347–1356, Dec. 1987. Also in *GLOBECOM'86 Conf. Rec.*, Dec. 1986, pp. 659–665.

[7] M. G. Hluchyj and M. J. Karol, "Queueing in high-performance packet switching," *IEEE Select. Areas Commun.*, vol. 6, pp. 1587–1597, Dec. 1988.

[8] T.-C. Hou and D. M. Lucantoni, "Buffer sizing for synchronous self-routing broadband packet switches with bursty traffic," *Int. J. Digital Analog Cabled Syst.*, vol. 2, no. 4, pp. 253–260, Oct.–Dec. 1989.

[9] M. Hirano and N. Watanabe, "Characteristics of a cell multiplexer for bursty ATM traffic," in *ICC'89 Conf. Rec.*, June 1989, pp. 399–403.

[10] L. Dittmann and S. B. Jacobsen, "Statistical multiplexing of identical bursty sources in an ATM network," *GLOBECOM'88 Conf. Rec.*, Nov. 1988, pp. 1293–1297.

[11] A. E. Eckberg, "The single server queue with periodic arrival process and deterministic service times," *IEEE Trans. Commun.*, vol. COM-27, pp. 556–562, March 1979.

[12] M. J. Karol and M. G. Hluchyj, "Using a packet switch for circuit-switched traffic: A queueing system with periodic input traffic," *IEEE Trans. Commun.*, vol. 37, pp. 623–625, June 1989.

[13] W. Hoeffding, "On the distribution of the number of successes in independent trials," *Ann. Math, Statist.*, vol. 27, pp. 713–721, 1956.

[14] M. J. Karol and M. G. Hluchyj, "The knockout switch: Principles and performance," in *Proc. 12th Conf. Local Comput. Networks*, Oct. 1987, pp. 16–22.

[15] W. Feller, *An Introduction to Probability Theory and Its Applications, Volume I.* New York: Wiley, 1968.

PERFORMANCE OF OUTPUT-BUFFERED BANYAN NETWORKS
WITH ARBITRARY BUFFER SIZES

Hyong S. Kim[†], Indra Widjaja and Alberto Leon-Garcia

[†] Department of Electrical and
Computer Engineering,
Carnegie Mellon University,
Pittsburgh, PA 15213

Department of Electrical Engineering,
University of Toronto,
Toronto, Canada M5S 1A4

Abstract

This paper reports a queueing analysis and a simulation study of a switch fabric based on a buffered banyan structure whereby buffers are placed at the output links of each switching element. When buffers are located at the input links, it is well known that the maximum throughput is limited to approximately 0.45 under uniform input traffic pattern. This bottleneck is due to the head of the line (HOL) contention at each switching element and is intrinsic to input queueing. We propose a buffered banyan switch built from smaller knockout switches which are output-buffered switches. With small knockout switches as the basic switching elements, the complexity of the overall switch fabric is manageable and no internal clock speedup is required. Furthermore, it is shown that with the proposed output-buffered banyan switch, a maximum throughput of 1 can be achieved.

1   Introduction

In recent years, design of efficient ATM switching systems for broadband integrated service digital network (BISDN) has received considerable attention. Among the desirable goals of the design are switches that have low delays, distributed routing, high throughput and low hardware complexity. The buffered banyan switch has been considered as a candidate for the switching fabric since it has some desirable features such as the *self-routing* property and low hardware complexity [1], [2]. This paper reports a simulation study of the switch fabric based on a buffered banyan structure whereby buffers are placed at the output ports of each switching element.

A buffered banyan switch of size $N \times N$ consists of $\log_k(N)$ stages, where each switching element is a $k \times k$ switch with a buffer located at each of its input or output links. When buffers are located at the input links, it is well known that the maximum achievable throughput is limited to approximately 0.45 under uniform input traffic pattern for large values of N with a single-buffered binary banyan switch ($k = 2$) [3]. It has also been shown that a multiple-buffered banyan switch does not achieve a major improvement in performance as compared to that of a single-buffered banyan switch [4]. In fact, even with infinite input buffers, the maximum throughput at each 2×2 switching element is limited to approximately 0.75 [5]. The bottleneck arises when two or more packets contend for the same buffer in a switching element. Since only one packet can be admitted to the buffer in one clock cycle, the others are blocked and will have to retry in the next clock cycle. When a packet at the head of a buffer loses a contention, it impedes the rest of the packets in the same buffer from progressing forward if packets are served on a FCFS basis. This bottleneck is called the head of the line (HOL) contention at each switching element and is intrinsic to input queueing [6].

In this paper, we propose a buffered banyan switch built from smaller *knockout* switches which are output-buffered switches. It is shown that with output-buffered banyans, a maximum throughput of 1 can be achieved. Previously, Kruskal and Snir have discussed buffered banyans with output-buffered switches for the case where the buffer capacity is infinite [7]. They model each buffer by an $M/G/1$ queueing system that leads to the Pollaczek-Khinchine mean-value formula to obtain the expected number of packets in the buffer. The performance of the network is then obtained easily and is then verified by simulation. In a real application, it is desirable to have small buffer sizes since the number of buffers is proportional to the number of switches which would obviously affect the complexity of the switching fabric. We consider two implementations: back-pressure and block-and-lost. In the back-pressure model, a packet destined to the next switching element that already has a full buffer cannot progress any further. This packet is said to be back-pressured and has to wait for the next clock cycle to resubmit itself. In the block-and-lost model, a packet destined to the next switching element can always progress in one clock cycle. However, packets will be lost when the buffer is full. Naturally, packet loss probability is another important performance measure for the block-and-lost model.

The outline of this paper is as follows. In Section 2, we discuss the buffered banyan switch and the switching element employing the knockout architecture. An important feature of the design is that it avoids any speed up; that is, the switch fabric runs at the same clock speed as the external clock speed. In Section 3, we present the performance analysis of the back-pressure and block-and-lost models. Various performance measures such as mean delay and blocking probability are presented. Additionally, tradeoffs between buffer size vs. mean delay, and buffer size vs. blocking probability are discussed. In Section 4, we present an approximate Markov chain analysis for the back-pressure model. Concluding remarks are made in section 5.

Reprinted from *Proc. IEEE INFOCOM '91*, pp. 701–710,
April 1991.

2 The Switch Fabric

Throughout the paper, we assume a switch fabric of size $N \times N$. Packets are assumed to be of a constant length; these "packets" are called "cells" in an ATM environment. ATM specifies fixed-length packets comprising 5 octets for the header and 48 octets for the information payload. We also assume that all input links to the network are slotted and synchronized with bit rates equal to 150 Mb/s. The resulting packet slot time is approximately 2.8 μsec. Thus, the switch fabric has to be designed in such a way so that it can handle approximately 350,000 packets per second per input port.

2.1 Buffered Banyan Switches

In the buffered banyan switch, each switching element of size $k \times k$ has a buffer at each of its input or output links. Figure 1 shows an 8×8 buffered banyan switch built from 2×2 switching elements with buffers located at the output links. Routing of packets is decentralized and performed at each switching element. A switching element in the ith stage upon receiving a packet determines which output link to route the packet by examining the ith most significant bit of the destination address which is located in the packet's header. Bit zero selects the upper output link; bit one selects the lower one.

The buffered banyan has input buffer controllers (IBC's) added to the front end of the switch. The IBC's with sufficient buffer space are required in the case of the back-pressure model with finite buffer size to prevent packet loss. When the input loads are moderate, back-pressure occurs very rarely. Packets at the head of queues pass through each stage without much resistance. As a result, the average queue length at each IBC is very small when traffic is moderate. On the other hand, when the input loads are high, the buffers at the switching elements become full. The back-pressure mechanism takes effect and propagates from the output ports towards the input ports. At the input ports, the cells are back pressured to the IBC's resulting in a high average queue length at each IBC. For the case of finite buffer size at the IBC, packet loss is unavoidable at the inputs although packets would never be dropped within the switch.

When the block-and-lost model is implemented, the IBC's are not required since packets arriving at the inputs will always be accepted to the buffers at the first stage regardless of the states of the buffers. If packets are destined to a full buffer, they will be dropped and lost. In this case, packet loss occurs within the switch for the case of finite buffer size.

2.2 The Switching Element

There are two major types of switching architectures to realize the switching elements: input queueing and output queueing switches. Input queueing switches have buffers located at the input links and could be run at the same speed as the external clock speed. However, packets destined to

the same buffer have to win a contention since only one of them could be accepted at one time slot. This would reduce the overall network throughput to approximately 0.45, as already mentioned earlier. Output queueing switches have buffers located at the output links and have to be run k times higher than the external clock speed, assuming a switching matrix of size $k \times k$. With output queueing switches, however, maximum throughput of 1 can be achieved since packets can always progress in each time slot. In order to take advantage of output queueing without increasing the speed of the switch, we employ the knockout switch architecture [8].

The knockout switch has a fully-connected topology in the sense that every input has an exclusive path to every output so that the switch is nonblocking. Each output port has a bus interface which can receive packets from any input port. The bus interface is constructed from three major components: packet filters, a concentrator, and a shared buffer. Each of the bus interfaces is connected to N input ports. The packet filter allows packets with addresses corresponding to the output address of the bus interface to pass into the concentrator. If the address does not correspond to that of the output, the packet is blocked from the concentrator. The purpose of the concentrator is to concentrate N input lines to L output lines. The knockout principle states that if arrivals on different input lines are statistically independent, then the probability that more than L (say, $L = 8$) packets arrive simultaneously to a given output is exceedingly small, even for arbitrarily large N. After the concentrator, the packets enter the shared buffer such that it can accept L packets in one time slot. Further details can be found in [8]. Figure 2 shows the 2×2 knockout switch used as the switching element. The bus interface becomes simple for the case of 2 input and output ports. The packet filters only detect one bit of the address of each packet and no concentrators are required since there are only two possible contenders which can easily be selected simultaneously. The shared buffer can accept two packets in a time slot by providing two pointers shifted by a packet length so that two simultaneous transfers of data can be accommodated.

3 Simulation Study

In this section we present the results of a simulation study of output-buffered banyan networks with arbitrary buffer size. We assume each switching element to be of size 2×2. The switching element is implemented by the knockout switch so that it is equivalent to the crossbar switch with buffers placed at the output ports in terms of performance. We also make the following uniform traffic assumptions:

- At each time slot, a packet arrives with a probability p to an input port independent of previous arrivals and other input ports.

- The destination address of a packet is chosen randomly over the set of all output port addresses according to uniform distribution.

In a real communication environment, the traffic load at each input port may not be the same. We assume that the traffic loads may be balanced by a distributor preceding the banyan network if they are unbalanced originally.

In our simulation study, we gather the results for banyan networks with 64 input ports and 64 output ports. The simulation results are provided with 95% confidence interval by running each simulation point 5 times.

3.1 The Back-pressure Model

In the back-pressure model, packets are allowed to move forward to the next stage if there are buffers available to store them. In particular, suppose at the beginning of time slot t_i, a packet requests to move forward to the next stage, then that packet is able to move forward only if either the buffer in that next stage (say, buffer x) has one more free packet slot to store, or the buffer in that next stage is full and a packet in that buffer is able to move forward in the same time slot. In the second case, buffer x is full at time slot t_i and t_{i+1} since a packet leaving that buffer is immediately replaced by a new packet from the previous stage. When two packets in the ith stage contend for the same buffer in the $i + 1$st stage and there is only one more packet slot ready for use in that buffer, one packet will be accepted at random, the other will be rejected by means of upstream control signals. The rejected packet stays in a buffer of ith stage and would have to try to make request again in the next time slot.

It is easy to see that the buffer size plays an important part in determining the maximum throughput achievable. With buffer size 1, input queueing and output queueing become identical, and the performance degenerates to that of Jenq's result. When the buffer size is larger, more packets are able to move forward in each time slot, increasing the maximum throughput achievable. In the extreme case, when buffer size is infinite, packets are always able to move forward in each time slot. In this case, a maximum throughput of 1 can be attained.

The performance of output-buffered banyan networks with infinite buffer size has been investigated by Kruskal and Snir. They obtained the average transit time of a packet through the $\log_k N$ stages of a banyan network of degree k. The average transit time is given by

$$T = \log_k N \cdot \left(t_c + t_c \frac{(1 - 1/k)p}{2(1 - p)} \right),$$

where t_c is the slot time of the switch.

In our simulation, we assume the IBC's to operate in a cut-through mode; that is, a packet does not have to be buffered completely in an IBC before being advanced to the first stage of the banyan switch whenever possible. The effect of buffer size B is shown in figure 3 whereby the throughput (or carried load) is plotted as a function of the offered load for various buffer sizes. When the buffer size B is 2, the maximum achievable throughput is about 0.55. Increasing the buffer size to 16 increases the maximum throughput to approximately 0.9. The delay results are shown in figure

4 for various buffer sizes B (2, 4, 8, etc.) where delay is defined as the queueing time plus the service time. As can be seen, delay increases appreciably even when the throughput is significantly below 1 when the buffer size is equal to 2. When $B = 16$, delay only starts to increase appreciably when the throughput is about 0.9 and above.

3.2 The Block-and-lost Model

This section addresses the block-and-lost implementation. As already mentioned earlier, packets are always able to move forward in each time slot. However, packets attempting to move to a buffer that is already full will be dropped and lost. Since packets arriving at the input ports are always able to move forward, no IBC's are required in the block-and-lost implementation.

Figure 5 shows the carried load versus the offered load for various buffer sizes. Note that the block-and-lost model offers a higher carried load than that of the back-pressure model for a high offered load for any buffer size. In addition, from figure 6, we note that the block-and-lost model results in lower delays than in the case of the back-pressure model when the throughput is high. This lower delay is obtained at the cost of increasing the packet loss probability. In Fig. 7, the packet loss probability is plotted as a function of the load for various values of buffer sizes. With 16 packet buffers per output link, the loss rate is below 10^{-6} for loads up to 0.7.

4 Queueing Analysis of the Back-pressure Model

Assuming a synchronous operation, as mentioned in the previous section, we can model the whole network as a Markov chain. However, as the size of network increases, the number of states in the Markov chain becomes prohibitively large. The number of states increases exponentially with the number of stages in the network. We use the following queueing network decomposition approach developed by [3]. First, we choose to model each switching element individually; We then describe the relationship among different switching elements through probabilistic means.

Here are the assumptions made for the analysis.

1. Packets are generated at each source node by independent Bernoulli processes. Each processor generates packet from a source to a destination with probability p at each cycle.

2. The network operates synchronously and a clock cycle is defined to be a time unit required to pass a packet through a stage.

3. Packets are removed immediately at the output ports.

4. No errors are introduced in the operation of switching elements and the entire network.

4.1 Modeling a Single Queue

In this section, we model buffers in each stages as a Markov chain with states corresponding to the number of packets in the buffer. Before proceeding to the analysis, we first introduce the following notation.

- Q_k^i = steady state probability that i packets are ready to come to a buffer at the kth stage.

- B_k^i = steady state probability that there are i packets in the buffer of the kth stage.

- r_k = steady state probability that a packet in kth stage buffer can advance to the next stage given that there is a packet in the kth stage buffer.

4.1.1 Intermediate Stage Buffers

We now model the buffer in the kth stage ($0 < k <$ last stage) switching element as a Markov chain. Let m be the number of buffers in an output port of the switching element. We model the buffer as being driven by a packet arrival process with probability Q_k^i of packet arrivals from the $k-1$th stage and with geometric departures. The probability of departure is determined by the possibility of the availability of buffer space at the next switching element (r_k). The $(m+1)$ state Markov chain model of the buffer is shown in figure 8 with transition probability from state i to state j, P_{ij} ($i,j = 0,1,...m$).

The transition probabilities, P_{ij} are described by the following transition equations.

$$P_{00} = Q_k^0, P_{01} = Q_k^1, P_{02} = Q_k^2, P_{03} = 0 \qquad (1)$$

$$\cdots$$

$$P_{10} = Q_k^0 r_k$$
$$P_{11} = Q_k^1 r_k + Q_k^0(1 - r_k)$$
$$P_{12} = Q_k^1(1 - r_k) + Q_k^2 r_k$$
$$P_{13} = Q_k^2(1 - r_k)$$
$$P_{14} = 0$$

$$\cdots$$
$$\cdots$$

The state transition diagram gives the following balance equations:

$$B_k^0 = B_k^0 Q_k^0 + B_k^1 Q_k^0 r_k \qquad (2)$$
$$B_k^1 = B_k^0 Q_k^1 + B_k^1(Q_k^1 r_k + Q_k^0(1 - r_k)) + B_k^2 Q_k^0 r_k$$
$$B_k^2 = B_k^0 Q_k^2 + B_k^1(Q_k^1(1 - r_k) + Q_k^2 r_k) + B_k^2(Q_k^1 r_k + Q_k^0(1 - r_k))$$
$$\qquad + B_k^3 Q_k^0 r_k$$
$$B_k^3 = B_k^1 Q_k^2(1 - r_k) + B_k^2(Q_k^1(1 - r_k) + Q_k^2 r_k)$$
$$\qquad + B_k^3(Q_k^1 r_k + Q_k^0(1 - r_k)) + B_k^4 Q_k^0 r_k$$
$$B_k^4 = B_k^2 Q_k^2(1 - r_k) + B_k^3(Q_k^1(1 - r_k) + Q_k^2 r_k)$$
$$\qquad + B_k^4(Q_k^1 r_k + Q_k^0(1 - r_k)) + B_k^5 Q_k^0 r_k$$

$$\cdots$$

Rather than inverting the transition matrix we obtain the steady-state queue size probabilities directly by simple numerical method from the above balance equations.

$$B_k^1 = \frac{1 - Q_k^0}{Q_k^0 r_k} B_k^0 \qquad (3)$$

$$B_k^2 = \frac{-1}{Q_k^0 r_k}(Q_k^1 B_k^0 + (Q_k^1 r_k + Q_k^0(1 - r_k) - 1)B_k^1)$$

$$B_k^3 = \frac{-1}{Q_k^0 r_k}(Q_k^2 B_k^0 + (Q_k^1(1 - r_k) + Q_k^2 r_k))B_k^1$$
$$\qquad + (Q_k^1 r_k + Q_k^0(1 - r_k) - 1)B_k^2)$$

$$B_k^4 = \frac{-1}{Q_k^0 r_k}(Q_k^2(1 - r_k)B_k^1 + (Q_k^1(1 - r_k) + Q_k^2 r_k))B_k^2$$
$$\qquad + (Q_k^1 r_k + Q_k^0(1 - r_k) - 1)B_k^3)$$

$$\cdots$$

$$B_k^i = \frac{-1}{Q_k^0 r_k}(Q_k^2(1 - r_k)B_k^{i-3} + (Q_k^1(1 - r_k) + Q_k^2 r_k))B_k^{i-2}$$
$$\qquad + (Q_k^1 r_k + Q_k^0(1 - r_k) - 1)B_k^{i-1}))$$

$$\cdots$$

where

$$B_k^0 = \frac{1}{1 + \sum B_k^i / B_k^0} \qquad (4)$$

The state of buffers at the $(k+1)$th stage depends on the state of buffers at the kth stage at previous clock cycles. However, in order to simplify the analysis, we assume statistical independence between the states of buffers at the $(k+1)$th stage from states of buffers at the kth stage. In other words, the event that buffer space is available at the $(k+1)$th stage and the event that a packet does not encounter a conflict or the conflict is resolved in the packet's favor are assumed to be independent.

We now obtain equations for r_k, and Q_k^i. A packet at the buffer of kth stage port is allowed to advance in following situations.

1. There is no packet in the adjacent input port of the switching element and buffer at the $(k+1)$th stage can accept a packet if

   - buffer at the $(k+1)$th stage is not full or
   - buffer at the $(k+1)$th stage have m packets and a packet in the buffer is allowed to advance to the next stage in the same clock cycle.

2. There is a packet in the adjacent input port of the switching element and it is directed to the different output port and the buffer at the $(k+1)$th stage can accept a packet if

   - buffer at the $(k+1)$th stage is not full or
   - buffer at the $(k+1)$th stage have m packets and a packet in the buffer is allowed to advance to the next stage in the same clock cycle.

3. There is a packet in the adjacent input port of the switching element and it is directed to the same output port and the buffer at the $(k+1)$th stage can accept both packets if

- buffer at the $(k+1)$th stage have at least two packet space or
- buffer at the $(k+1)$th stage have $(m-1)$ packets and a packet in the buffer is allowed to advance to the next stage in the same clock cycle.

4. There is a packet in the adjacent input port of the switching element and it is directed to the same output port and the buffer at the $(k+1)$th stage can accept only one packet if

   - the packet in the tagged buffer wins the contention with the packet in the adjacent input port.
   - buffer at the $(k+1)$th stage have $(m-1)$ packets and a packet in the buffer is not allowed to advance to the next stage in the same clock cycle or
   - buffer at the $(k+1)$th stage have (m) packets and a packet in the buffer is allowed to advance to the next stage in the same clock cycle.

These conditions translate into the following equation for r_k.

$$
\begin{aligned}
r_k = & (B_k^0 + 0.5(1 - B_k^0))(1 - B_{k+1}^m + B_{k+1}^m r_{k+1}) \\
& + 0.5(1 - B_k^0)(1 - B_{k+1}^m - B_{k+1}^{m-1} + B_{k+1}^{m-1} r_{k+1}) \\
& + 0.25(1 - B_k^0)(B_{k+1}^{m-1}(1 - r_{k+1}) + B_{k+1}^m r_{k+1}) \quad (5)
\end{aligned}
$$

We obtain following equation for Q_k^i using B_k^i of both input ports of the switching element. Both input ports of the switching element have the same B_k^i since the traffic load is assumed to be balanced.

$$
\begin{aligned}
Q_{k+1}^0 = & 1 - Q_{k+1}^1 - Q_{k+1}^2 \\
Q_{k+1}^1 = & 0.5 B_k^0(1 - B_k^0) + 0.5(1 - B_k^0)B_k^0 \\
& + (1 - B_k^0)(1 - B_k^0)(0.25 + 0.25) \\
= & B_k^0(1 - B_k^0) + 0.5(1 - B_k^0)(1 - B_k^0) \\
Q_{k+1}^2 = & 0.25(1 - B_k^0)(1 - B_k^0) \quad (6)
\end{aligned}
$$

4.1.2 IBC Buffers

We now model the IBC buffers as a Markov chain. We model the buffer as being driven by a Bernoulli process with probability p of packet arrivals from the source and with geometric departures. The probability of departure is determined by the possibility of the availability of buffer space at the switching element (r_0) in the first stage.

The $m+1$ state Markov chain model of the buffer is shown in figure 9 with transition probability from state i to state j, The transition probabilities, P_{ij} are described by the following transition equations.

$$
P_{00} = 1 - p, P_{01} = p, P_{02} = 0 \quad (7)
$$

$$
\cdots
$$

$$
\begin{aligned}
P_{10} &= (1-p)r_0 \\
P_{11} &= (1-p)(1-r_0) + pr_0 \\
P_{12} &= p(1-r_0)
\end{aligned}
$$

$$
P_{13} = 0
$$

$$
\cdots
$$

The state transition diagram gives the above balance equations:

$$
\begin{aligned}
B_0^0 =\ & B_0^0(1-p) + B_0^1(1-p)r_0 \quad (8) \\
B_0^1 =\ & B_0^0 p + B_0^1(pr_0 + (1-p)(1-r_0)) + B_0^2(1-p)r_0 \\
B_0^2 =\ & B_0^1(p(1-r_0) + pr_0) + B_0^2(pr_0 + (1-p)(1-r_0)) \\
& + B_0^3(1-p)r_0 \\
B_0^3 =\ & B_0^2(p(1-r_0) + pr_0) + B_0^3(pr_0 + (1-p)(1-r_0)) \\
& + B_0^4(1-p)r_0 \\
B_0^4 =\ & B_0^3(p(1-r_0) + pr_0) + B_0^4(pr_0 + (1-p)(1-r_0)) \\
& + B_0^5(1-p)r_0
\end{aligned}
$$

$$
\cdots
$$

We again obtain the steady-state queue size probabilities directly by simple numerical method from the following balance equations.

$$
B_0^1 = \frac{1 - (1-p)}{(1-p)r_0} B_0^0 \quad (9)
$$

$$
B_0^2 = \frac{-1}{(1-p)r_0}(pB_0^0 + (pr_0 + (1-p)(1-r_0) - 1)B_0^1)
$$

$$
\begin{aligned}
B_0^3 = & \frac{-1}{(1-p)r_0}((p(1-r_0) + pr_0)B_0^1 \\
& + (pr_0 + (1-p)(1-r_0) - 1)B_0^2)
\end{aligned}
$$

$$
\begin{aligned}
B_0^4 = & \frac{-1}{(1-p)r_0}((p(1-r_0) + pr_0)B_0^2 \\
& + (pr_0 + (1-p)(1-r_0) - 1)B_0^3)
\end{aligned}
$$

$$
\cdots
$$

$$
\begin{aligned}
B_0^i = & \frac{-1}{(1-p)r_0}((p(1-r_0) + pr_0)B_0^{i-2} \\
& + (pr_0 + (1-p)(1-r_0) - 1)B_0^{i-1}))
\end{aligned}
$$

$$
\cdots
$$

where

$$
B_0^0 = \frac{1}{1 + \sum B_0^i / B_0^0} \quad (10)
$$

The equation for r_0 is same as that of equation 5.

4.1.3 Last Stage Buffers

We now look at the buffers at the last stage. The markov chain model remains the same as that of the intermediate stages. However $r_n = 1$ and we obtain following equations for B_n^i.

$$
B_n^1 = \frac{1 - Q_n^0}{Q_n^0} B_n^0 \quad (11)
$$

$$
B_n^2 = \frac{-1}{Q_n^0}(Q_n^1 B_n^0 + (Q_n^1 - 1)B_n^1)
$$

$$
B_n^3 = \frac{-1}{Q_n^0}(Q_n^2 B_n^0 + Q_n^2 B_n^1 + (Q_n^1 - 1)B_n^2)
$$

$$
B_n^4 = \frac{-1}{Q_n^0}(Q_n^2 B_n^2 + (Q_n^1 - 1)B_n^3)
$$

$$B_n^i = \frac{-1}{Q_n^0}(Q_n^2 B_n^{i-2} + (Q_n^1 - 1)B_n^{i-1}))$$

where

$$B_n^0 = \frac{1}{1 + \sum B_n^i / B_n^0} \quad (12)$$

4.1.4 Throughput and Delay

We now derive equations for the throughput and delay. The throughput is defined to be the average number of departing packets per clock cycle from the last stage of an output link.

The throughput of the network at the output port of the last stage is obtained as follows.

$$throughput = 1 - B_n^0 \quad (13)$$

since the packets at the last stage are removed immediately. The packet loss probability is given by the following equation.

$$\text{packet loss probability} = \frac{p - throughput}{p} \quad (14)$$

Packet delay is defined to be the number of clock cycles a packet takes to reach the destination port from the source port. The delay at a stage is obtained using Little's formula as follows.

$$\text{expected stage delay at } k\text{th stage} = \sum_{i=1}^{m} \frac{i B_k^i}{p} \quad (15)$$

The total delay is then

$$\text{total delay} = \sum_{k=0}^{\text{last stage}} \sum_{i=1}^{m} \frac{i B_k^i}{\lambda} \quad (16)$$

The minimum possible delay is equal to $n+1$, where $n+1$ is the number of stages with buffers in the network including the IBC buffer since at least one time unit is spent in each buffers even when there is no waiting.

4.2 Joining the Single Queues into a Network: Solution by Iterative Method

Since the equation describing the dynamics of the network is described by recurrence relations, the solution is obtained by an iterative method. The single-queue analyses are made consistent by forcing the queueing model equations to yield steady state probabilities by iterative method. This iterative procedure is presented in this section.

The input traffic is described by p, load to the input port. The objective of the analysis is to determine the values of B_k^i, steady state probability of buffer occupancy at the kth stage.

Equation 5 describe r_k as a function of B_{k+1}^i and B_k^i as following.

$$r_k = f(B_{k+1}^i, B_k^i) \text{ for } 0 \leq k < n \quad (17)$$

Equations 9, 3, 11, and 6 correspond to following functions.

$$\begin{aligned}
B_0^i &= f(p, r_0) \quad (18) \\
B_k^i &= f(Q_k^i, r_k) \text{ for } 1 \leq k < n \\
B_n^i &= f(Q_n^i) \\
Q_{k+1}^i &= f(B_k^i)
\end{aligned}$$

The iteration algorithm consists of one loop as shown in figure 10. We obtain appropriate initial guesses for B_k^i to start the iteration. The algorithm then updates B_n^i and Q_n^i for the IBC buffers. Using these values, the algorithm recomputes B_k^i and Q_{k+1}^i from the first stage to the last stage. These values are compared to the values obtained from previous computation and the algorithm stops when these parameters converge to a certain value.

4.3 Analytical Results

Simulation results are compared with analytical results obtained from numerical analysis for a network of six stages under uniform traffic pattern. Fig. 11,12,13 show the traffic delay versus the traffic load for various buffer sizes. From the figures we find that the analytical results is accurate when the buffer size is relatively large. When buffer size is small such as 4 as shown in the figure 11, the analytical results presents optimistic performance of the network compared to that of the simulation results.

Fig. 14 shows the packet loss probability versus the IBC buffer size with a load at 0.9. Several curves with different internal buffer sizes are plotted. We can observe that the packet loss probability reaches a minimum value after which no improvement is obtained by increasing the size of IBC buffers. As we increase the internal buffer size, the packet loss probability reduces significantly. With the internal buffer of size 8, we could achieve a very low packet loss probability (i.e. 10^{-10}) with small IBC buffers.

There are several papers [3, 4, 5, 9] on the performance of the buffered banyan network using similar queueing model which is derived by assuming statistical independency of buffers in different stages. The buffer of switching elements are modelled as a markov chain and then these buffers at the different stages are related by a probabilistic means rather than modelling the entire network by a markov chain. Shaik [9] reports that analytical results obtained by Jenq [3] gives an optimistic performance as much as 28% compared to that of the simulation results. It seems that such a model adopted by our analysis and [3, 9] fails to account for the correlation between buffers at the different stages when the buffer size is small and the load is high. Thus, the results from an approximate analysis would be useful in some situations where it is difficult to obtain the results through simulation. We could obtain a low packet loss probability (i.e. 10^{-9}) only through this approximate analysis since simulation may take a long time to generate credible results.

5 CONCLUSION

We introduced two models of output-buffered banyans for ATM switching: back-pressure and block-and-lost models. We varied the buffer size and studied the performance measures of interest for both models.

We also modelled the switching network as a Markov chain and compared the analytical results to the simulation results. The Markov chain analysis for the switching network with small size buffer seems to give optimistic performance compared to the simulation results. Next step in our study is to model the switching network such that the gap between the analytical results and the simulation results could be narrowed by modifying the independence assumptions.

Another work of interest is to utilize a bigger knockout switch (4×4, or 8×8) as the switching element since knockouts with these sizes still have reasonable complexity. We believe total buffer requirements can be reduced with bigger knockout switches since more statistical gains can be obtained with bigger switching elements.

Performance of the output-buffered banyan switch is expected to degrade under nonuniform traffic. This degradation is true for many banyan switches since all the links in banyan switches are shared by many input and output port pairs. The modification of the output-buffered banyan switch to prevent the performance degradation under nonuniform traffic will be the subject of a forthcoming paper.

References

[1] J. Turner, "Design of a broadcast packet switching network," in *Proc. of INFOCOM '86*, IEEE, Mar. 1986, pp. 667–675.

[2] J. Turner, "Design of an integrated services packet network," in *Proc. of 9th ACM Data Communications Symposium*, ACM, Sept. 1985, pp. 124–133.

[3] Y. Jenq, "Performance analysis of a packet switch based on single-buffered Banyan network," *IEEE Journal on Selected Areas in Communications*, vol. SAC-1, pp. 1014–1021, Dec. 1983.

[4] H. Kim and A. Leon-Garcia, "Performance of buffered banyan networks under nonuniform traffic patterns," in *Proc. of INFOCOM '88*, (New Orleans, Louisiana), IEEE, Mar. 1988, pp. 344–352.

[5] D. Dias and J. Jump, "Analysis and simulation of buffered Delta network," *IEEE Trans. on Computer*, vol. C-30, pp. 273–282, April 1981.

[6] M. Karol, M. Hluchyj, and S. Morgan, "Input versus output queueing on a space-division packet switch," *IEEE Trans. on Communications*, vol. COM-35, pp. 1347–1356, Dec. 1987.

[7] C. Kruskal and M. Snir, "The performance of multistage interconnection networks for multiprocessors," *IEEE Trans. on Computer*, vol. C-32, pp. 1091–1098, Dec. 1983.

[8] Y. Yeh, M. Hluchyj, and A. Acampora, "The Knockout switch: A simple, modular architecture for high-performance packet switching," *IEEE Journal on Selected Areas in Communications*, vol. SAC-5, pp. 1274–1283, Oct. 1987.

[9] T. Szymanski and S. Shaikh, "Markov chain analysis of packet-switched banyans with arbitrary switches, queue sizes, link multiplicities and speedups packet switching," in *Proc. of INFOCOM '89*, (Ottawa, Canada), IEEE, April 1989, pp. 960–971.

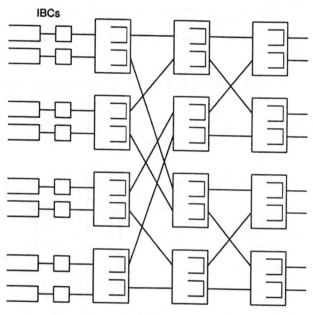

Fig. 1. An 8 x 8 output-buffered banyan switch.

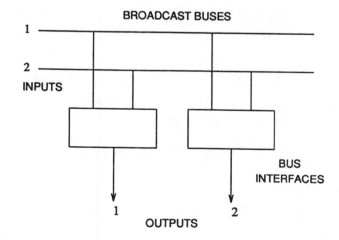

Fig. 2. The 2x2 knockout switch.

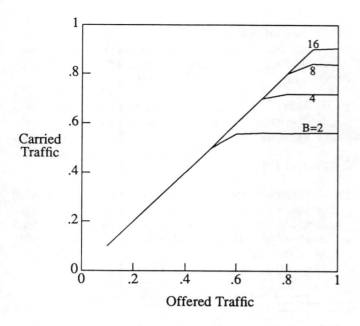

Figure 3: Throughput of the back-pressure banyan

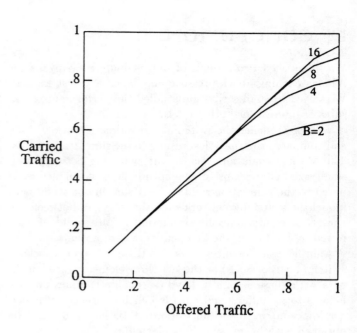

Figure 5: Throughput of the block-and-lost banyan

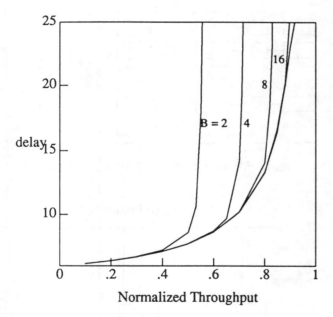

Figure 4: Delay of the back-pressure banyan

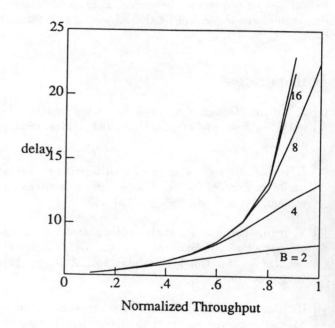

Figure 6: Delay of the block-and-lost banyan

248

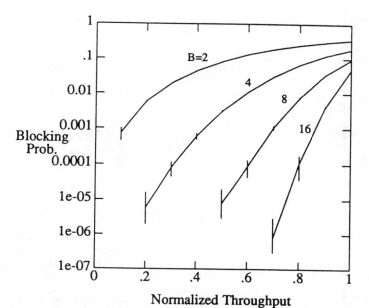

Figure 7:   Packet loss probability of the block-and-loss banyan

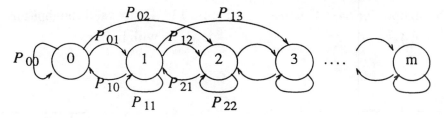

Figure 8: Markov state transition diagram for intermediate stage buffers

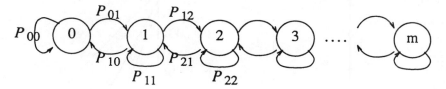

Figure 9: Markov state transition diagram for IBC buffers

loop (counter=m)

    from k=n-1 to k=0

$$r_k(m) = f(B_{k+1}^i (m-1), B_k^i(m-1), p) \quad \textit{from eqn. 5}$$

    from k=1 to k=n (last stage)

$$B_k^i(m) = f(Q_k^i(m-1), r_k(m-1) \quad \textit{from eqn. 3}$$
$$Q_{(k+1)}^i(m) = f(B_k i(m)) \quad \textit{from eqn. 6}$$

    if $| B_k^i(m) - B_k^i(m-1) | <$ **error tolerance**
    then exit loop and compute throughput and delay
    else continue loop

end loop

Figure 10: Pseudo Code for Numerical Iteration Algorithm

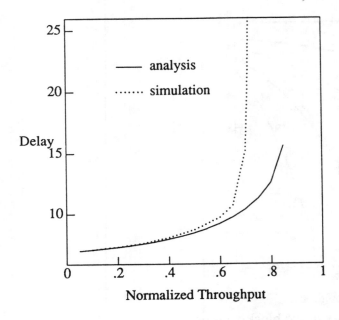

Figure 11: Delay vs. throughput for 64x64 switch with buffer size 4

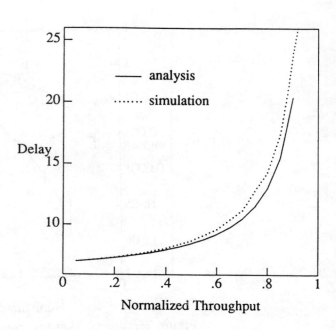

Figure 13: Delay vs. throughput for 64x64 switch with buffer size 16

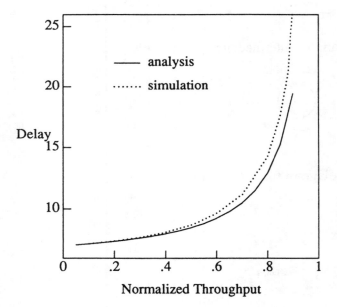

Figure 12: Delay vs. throughput for 64x64 switch with buffer size 8

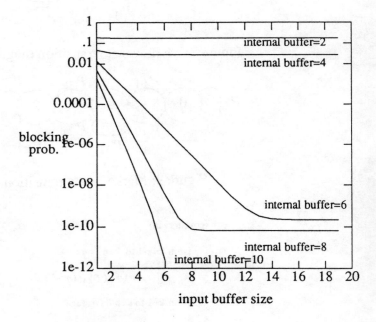

Figure 14: Packet loss probability of the back-pressure model

INTRODUCTION

There has been a growing interest in developing high-speed switches for multicasting [7]. In a multicast transmission one source transmits the same information to multiple destinations. Applications for multicasting include [1]

- Teleconferencing
- Entertainment video
- LAN bridging
- Distributed data processing

This part contains a number of papers that explore switching technology for multicasting.

The first paper is "Nonblocking Copy Networks for Multicast Packet Switching," by T. T. Lee. This paper gives a broad technological overview of copy networks used to produce multiple copies of input packets. The multiple packet copies leaving the copy network will enter a point to point switch for delivery.

Lee cites proposals for multicasting such as the starlite system of A. Huang and S. Knauer [2] and the broadcast packet switch of J. S. Turner [3, 4] (see the following article). He goes on to propose a "nonblocking, self-routing copy network with constant latency."

The heart of Lee's switch is a broadcast Banyan network. In theory a broadcast Banyan network is a Banyan-like network that accepts a single packet with multiple Banyan output destinations and produces copies of the packet at the appropriate outputs. A straightforward implementation leads to a number of problems mentioned in the paper. These problems can be avoided if the input packet destination addresses are constrained to a contiguous interval in the address space. This solution is allowable, since copy network output addresses generated by a single input packet are "ficticious" [1]. One simply wants the desired number of copies to appear at some subset of copy network outputs. Furthermore, for nonblocking, monotonicity and concentration requirements must be satisfied.

Lee's multicast switch replicates packets through an encoding and decoding process. In the encoding process a running adder network and dummy address encoders transform the set of copies to be produced into a set of monotone address intervals, which are inserted into packet headers. These packets enter the broadcast Banyan network. The decoding process involves a Boolean splitting algorithm for the broadcast Banyan network.

Lee's paper gives a detailed account of the implementation of this copy network. Mentioned are expansion networks used for applications needing nonsquare networks. Also mentioned are two schemes for removing idle inputs between active in-

puts for nonblocking. A performance analysis using Chernoff bounds for overflow probabilities appears at the end of this paper. Overflow here is the occurrence of a larger number of copy requests than outputs.

A second architectural approach to multicast switching appears in "Performance of a Broadcast Packet Switch," by R. G. Bubenik and J. S. Turner. The actual switching architecture evaluated in this paper was described in [5]. Major components of the broadcast packet switch include a *copy network* to produce duplicate packets and a *distribution network,* whose randomizing effect allows a *routing network* to operate with less congestion. These three networks are based on appropriately modified Banyan-like networks.

After describing the basic architecture, the results of a simulation study are presented; these results highlight the performance improvements possible with certain architectural features. One is the use of cut-through switching, which allows a packet entering a buffer to begin leaving the buffer after only the header has been received. In traditional store and forward packet networks, a packet must be completely received before it can be retransmitted. Bubenik and Turner demonstrate a performance improvement in throughput using cut-through switching.

The authors go on to show that the use of 4×4 switching nodes rather than 2×2 switching nodes may not lead to a performance improvement because of input queueing blocking. To alleviate this problem, they propose a *bypass queueing* discipline that relaxes the FIFO discipline to boost throughput (see also the input bypass strategy of Szymanski and Shaikh in Part 4). Finally, the performance of the copy network in terms of delay and throughput versus fanout (number of copies produced) is discussed. This discussion includes conditions that prevent overload situations.

The next multicast performance evaluation paper is "Performance Analysis of a Multicast Switch," by J. F. Hayes, R. Breault, and M. K. Mehmet-Ali. In it the authors provide a queueing analysis that emphasizes delay for a "multicast switch" with input port buffering. The switch here is a generic system, which, from a single input packet, produces copies at multiple outputs.

A number of different service disciplines are considered. Under pure work conservation, packet copies from one input packet may reach the output ports over several slots. Alternatively, one may require all of the packets to reach the outputs in one or a fixed number of slots; otherwise one tries again. A queueing analysis for delay based on a model consisting of a set of M/G/1 queues is examined. The service time at one queue (input port) depends on the traffic at the other queues (input ports). A simulation study of the work-conserving discipline is also presented. Hayes and his coauthors note that multicast switches with the same number of inputs and outputs saturate roughly between 0.6 and 0.85 (recall the 0.586 figure for unicasting in the nonblocking switch analysis chapter).

The next paper, "First Stage Multicasting in a Growable Packet (ATM) Switch," discusses modifications to the growable packet switch of Part 5 to allow for multicasting. It is authored by D. J. Marchok and C. E. Rohrs.

Marchok and Rohrs offer two routing policies for multicasting. The first policy implements multicasting in the first stage. It acts as if the first-stage input switching element has additional input lines. Routing is the same as for unicast packets. In the second policy multicasting is performed in the first and third stages. Because each output switching element has a limited number of output lines, additional multicasting takes place at the first stage. This policy performs better than the first.

Marchok and Rohrs present an analysis of loss probability for the second policy (which can be extended to the first policy). Three sources of loss are examined. It is found that reasonable performance can be achieved with only a small change in the first- and third-stage switching element dimensions. This performance is with the caveat that performance degrades if a substantial amount of multicasting takes place at a single input-switching element.

The next paper is somewhat different from the previous ones in this part in that it discusses scheduling algorithms for multicasting rather than hardware implementation. Called "Call Scheduling Algorithms in a Multicast Switch," it is authored by C.-K. Kim and T. T. Lee.

Kim and Lee consider a time-multiplex switch performing multicasting. Time-multiplex switches are extensively used in present-day digital telephone systems [6], although not with multicasting. A time-multiplexed switch performing multicasting copies the information in an input time slot into multiple-output time slots.

The authors first calculate lower and upper bounds on random scheduling. A large gap between the lower and upper bounds for large fanouts (packet replications) motivates the work on sophisticated scheduling algorithms.

Two classes of scheduling algorithms are studied: One involves call packing and the other involves call splitting. First a number of call-packing algorithms similar to ones used for memory management in computer systems are considered. They attempt to choose a slot assignment while placing the fewest restrictions on future assignments. However, a simulation study of call packing shows only a minor performance improvement.

Call-splitting algorithms are then studied. These take the output ports to be addressed and break them into subsets, which are addressed in different slots. In *fixed splitting*, a multicast is uniformly split. In *greedy splitting*, multicasts are split whenever possible down to a minimum fanout of subcalls. Greedy splitting is shown to be more consistent in performance than fixed splitting, which is sensitive to frame size.

Much is not known about multicasting, and a number of topics for future research are presented at the end of Kim and Lee's paper.

REFERENCES

[1] T. T. Lee, "Nonblocking Copy Networks for Multicast Packet Switching," *IEEE J. on Selected Areas in Commun.*, vol. 6, no. 9, pp. 1455–1467, Dec. 1988. Appears in this part.

[2] A. Huang and S. Knauer, "Starlite: A wideband digital switch," *Proc. IEEE Globecom '84*, pp. 121–125, 1984.

[3] J. S. Turner, "Design of a broadcast packet switching network," *Proc. IEEE INFOCOM '86*, Miami, Fla., pp. 667–675, April 1986.

[4] J. S. Turner, "Design of an integrated service packet network," *IEEE J. Selected Areas in Commun.*, Nov. 1986.

[5] J. S. Turner, "Design of a broadcast packet switching network," *IEEE Trans. Commun.*, vol. 36, pp. 734–743, June 1988.

[6] M. Schwartz, *Telecommunication Networks: Protocols, Modeling and Analysis*, Reading, Mass.: Addison-Wesley, 1987.

[7] M. Ahahad, *Multicast Communications in Distributed Systems*, Washington, D.C.: IEEE Computer Society Press, 1990.

Nonblocking Copy Networks for Multicast Packet Switching

TONY T. LEE, SENIOR MEMBER, IEEE

Abstract—In addition to handling point-to-point connections, a broadband packet network should be able to provide multipoint communications that are required by a wide range of applications such as teleconferencing, entertainment video, LAN bridging, and distributed data processing. The essential component to enhance the connection capability of a packet network is a *multicast packet switch*, capable of packet replications and switching, which is usually a serial combination of a copy network and a point-to-point switch. The copy network replicates input packets from various sources simultaneously, and then copies of broadcast packets are routed to their final destination by the switch. A nonblocking, self-routing copy network with constant latency is proposed in this paper. Packet replications are accomplished by two fundamental processes, an encoding process and a decoding process. The encoding process transforms the set of copy numbers, specified in the headers of incoming packets, into a set of monotone address intervals which form new packet headers. This process is carried out by a running adder network and a set of dummy address encoders. The decoding process performs the packet replication according to the Boolean interval splitting algorithm through the broadcast banyan network. Finally, the destinations of copies are determined by the trunk number translators. At each stage of the broadcast banyan network, the decision making is based on a two-bit header information. This yields minimum complexity in the switch nodes.

I. INTRODUCTION

THE goal of a broadband packet network is to provide flexible communication for handling multipoint connections in addition to point-to-point connections. A wide class of applications, such as teleconferencing, entertainment video, LAN bridging, and distributed data processing require multipoint communications. The essential component in the network to achieve this is a *multicast packet switch* which is capable of packet replication and switching.

A multicast packet switch is usually a serial combination of a *copy network* and a point-to-point switch as illustrated in Fig. 1. The copy network replicates input packets from various sources simultaneously, and then copies of broadcast packets are routed to their final destination by the point-to-point switch. Current designs of packet switch systems to support multipoint communications include the *Starlite System* proposed by A. Huang and S. Knauer [4], and the *Broadcast Packet Switch* developed by J. S. Turner [7], [8]. Both systems are parallel

Manuscript received November 1, 1987; revised June 2, 1988. This paper was presented at the 1988 International Zurich Seminar on Digital Communications, Zurich, Switzerland, ETH, March 8–10, 1988.

The author is with Bell Communications Research, Morristown, NJ 07960.

IEEE Log Number 8824385.

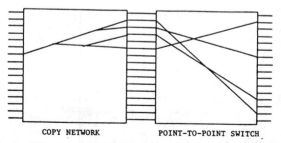

Fig. 1. A multicast packet switch consists of a copy network and a point-to-point switch.

processors with distributed control, and are constructed out of interconnected networks—regularly interconnected identical switch cells [9]. These simple structures lend themselves very well to cost-effective VLSI implementation.

The Starlite copy network is a receiver initiated system. The inputs to the network are original source packets and empty packets generated by receivers. The copy network consists of a *sorting network* [1] and a *broadcast network*. Initially, the sorting network sorts the input packets on their source addresses. At the outputs of the sorting network, the original source packet and the associated empty packets with the same source address will appear contiguously. The broadcast network then replicates the data in each source packet and inserts them into the empty data fields of subsequent empty packets until another source packet is encountered. This process is illustrated in Fig. 2. It is easy to notice that the process of packet replication in this design assumes the synchronization of the source and destinations. An "empty packet setup procedure" is also required. It is not feasible to implement this approach in a broadband packet network where packets may usually experience delay variations due to buffering, multiplexing, and switching in a multiple-hop connection.

The copy network of Turner's broadcast packet switch employs a *banyan network* and a set of *broadcast and group translators* (BGT's). The header of an input broadcast packet contains two fields, a *fan-out* indicating the number of copies requested by the packet, and a *broadcast channel number* (BCN) used by the BGT's to determine the final destination of copy packets. Fig. 3 depicts an example of this copy network, which replicates packets by splitting the fan-out. Using a table lookup method, the trunk number translation is performed by the BGT's when

Reprinted from *IEEE J. Selected Areas Commun.*, vol. 6, no. 9, pp. 1455–1467, December 1988.

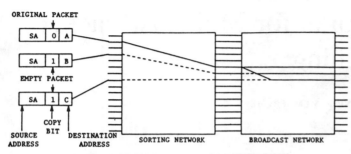

Fig. 2. Copy network of Starlite system.

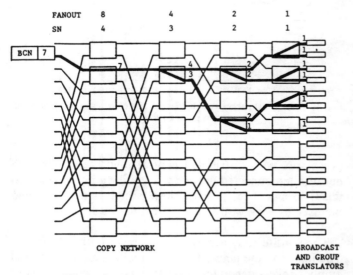

Fig. 3. Copy network of broadcast packet switch.

copies appear at the outputs. Packet collisions cannot be avoided in this scheme, and it is blocking, meaning that two packets may arrive at the same node and attempt to leave on the same output link at the same time. Buffers are required for every internal node to prevent packet loss.

A nonblocking, self-routing copy network with constant latency is proposed in this paper. The basic structure consists of the following components.

• *Running Adder Network (RAN):* It generates running sums of *copy numbers*, specified in the headers of input packets.

• *Dummy Address Encoder (DAE):* It takes adjacent running sums to form a new packet header.

• *Broadcast Banyan Network (BBN):* It is a banyan network with broadcast switch nodes capable of packet replications based on two-bit header information.

• *Trunk Number Translator (TNT):* It determines the outgoing trunk number for each copy packet.

Fig. 4 illustrates the basic structure of our copy network, which can be considered as a cascaded combination of a packet header encoder and a decoder. When broadcast packets are received at the running adder network, the *number of copies* (CN's) specified in the packet headers are added up recursively. The dummy address encoders form new headers consisting of two fields: a *dummy address interval* and an *index reference* (IR). The dummy

address interval, formed by adjacent running sums, is represented by two binary numbers, namely, the *minimum* (MIN) and *maximum* (MAX). The index reference is set equal to the minimum of the address interval, used later by the trunk number translators to determine the *copy index* (CI). The broadcast banyan network replicates packets according to a *Boolean interval splitting algorithm* based on the address interval in the new header. When copies finally appear at the outputs, the TNT's compute the copy index for each copy from the output address and index reference. The *broadcast channel number* (BCN) together with copy index form a unique identifier for each copy. The TNT's then translate this identifier into a *trunk number* (TN), which is added to the packet header and used by the switch to route the packet to its final destination.

References to related articles are cited and background materials are provided in the course of the presentation to make this paper self-contained. Section II begins with a brief introduction to the self-routing and nonblocking properties of banyan networks as a point-to-point switch. Following that, these properties are generalized to broadcast banyan networks along the same principle. The basic approach to synthesize our nonblocking copy network is detailed in Section III. Broadcast networks with larger fan-out than fan-in are desirable in many services applications such as entertainment video. The topology of expansion networks that preserve the self-routing and nonblocking properties of this structure are presented in Section IV. Concentration of active inputs is one of the criteria for the network being nonblocking. Two techniques, one with prioritization and the other without, are discussed in Section V. They were both developed for the Starlite switch and can also be incorporated into our copy network. Section VI is devoted to the estimation of overflow probabilities of the inputs of copy network by means of Chernoff bound. Finally, conclusions of present work and suggestions for future research are summarized in Section VII. W. H. Bechmann had proved a sufficient nonblocking condition of banyan network by introducing an elegant numbering scheme [2], which provides an insight into its topological properties. The proof is paraphrased and appended at the end of this paper.

II. BROADCAST BANYAN NETWORKS

The routing algorithm and nonblocking condition of broadcast banyan networks are discussed in this section. Based on these fundamental properties, the design of a nonblocking copy network is detailed in the next section. A banyan network is an $n \times n$ interconnection network with $N = \log_2 n$ stages. Each stage contains $n/2$ nodes and each node is a 2×2 switch element. The banyan network has been used as a point-to-point switch, based on the so-called *self-routing algorithm* in both Starlite switch (without internal buffering) and Fast Packet Switch (with internal buffering). Fig. 5 illustrates a four-stage banyan network and the self-routing algorithm. The header of each incoming packet contains an n-bit desti-

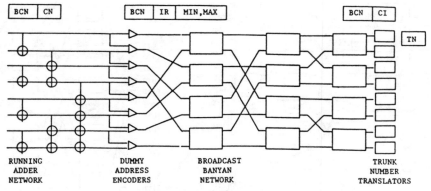

Fig. 4. The basic structure of a nonblocking copy network.

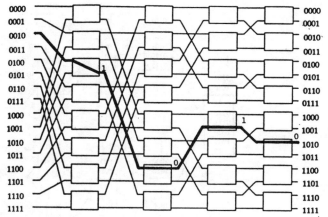

Fig. 5. A four-stage banyan network and the self-routing algorithm for point-to-point switching.

nation address, and the packet routing is done by decoding the header. That is, a node at stage k sends the packet out on link 0 (up) or link 1 (down) according to the kth bit of the header. This property is due to the topology of banyan networks: the path from any input to an output is uniquely determined by the output address.

An interconnection network is nonblocking if the packet routing is not store-and-forward, and internal buffers are not needed within the switch nodes. The banyan network itself is a blocking network. Two packets with different destination addresses may be routed through the same internal link at the same time. However, it is known that if the incoming packets with distinct destination addresses are arranged in an ascending or a descending order, then the banyan network becomes nonblocking. The Starlite switch is a nonblocking network, which is realized by combining a Batcher sorting network and a banyan network. The sorting network sorts the incoming packets on their destination addresses. The nonblocking condition of the banyan network is formally stated as follows [4].

NB1: A banyan network, as a point-to-point switch, is nonblocking if the active inputs $x_1, \cdots, x_k$ and the corresponding outputs $y_1, \cdots, y_k$ satisfy the following.

1) (Monotone): $y_1 < y_2 < \cdots < y_k$, or $y_1 > y_2 > \cdots > y_k$.

2) (Concentration): Any input between two active in-

puts is also active. That is, $x_i \leq w \leq x_j$ implies input w is active.

By induction on the stages, it is easy to see that the above statement is a sufficient nonblocking condition. A rigorous proof is given by W. H. Bechmann [see Appendix].

A. Generalized Self-Routing Algorithm

A broadcast banyan network is a banyan network with switch nodes which are capable of packet replications. A packet arriving at each broadcast switch node can be either routed to one of the output links, or it can be replicated and sent out on both links. The uncertainty of making a decision with three possibilities is $\log_2 3 = 1.585$, meaning that the minimum header information provided by a packet for each broadcast node is 2 bits. Bearing this lower bound in mind, we are going to show how to synthesize an algorithm for packet replications that minimizes the logic complexity of switch nodes.

Fig. 6 illustrates a straightforward generalization of the above self-routing algorithm to multiple addresses. A broadcast packet contains a set of arbitrary n-bit destination addresses. When a packet arrives at a node in stage k, the packet routing and replication are determined by the kth bit of all the destination addresses in the header. If they are all "0" or all "1," then the packet will be sent out on link 0 or link 1, respectively. Otherwise, the packet and its replica are sent out on both links with the following modified header information: the header of the packet sent out on link 0 or link 1, contains these addresses in the original header with their kth bit equal to 0 or 1, respectively. The set of paths from any input to a set of outputs form a (binary) tree embedded in the network, and it will be called an *input–output tree*. The modification of packet headers is performed by the node whenever the packet is replicated.

There are several problems which may arise in implementing this generalized self-routing algorithm. First, the packet header contains a variable number of addresses and the switch nodes have to read all of them. Second, the process of packet header modification depends on the entire set of addresses. This would become a burden on switch nodes. Finally, the input–output trees generated by

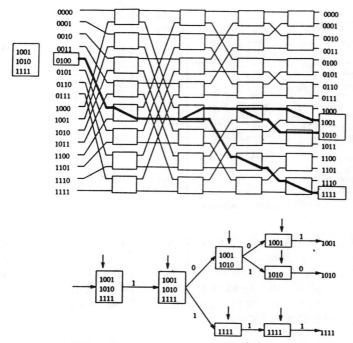

Fig. 6. An input–output tree generated by generalized self-routing algorithm.

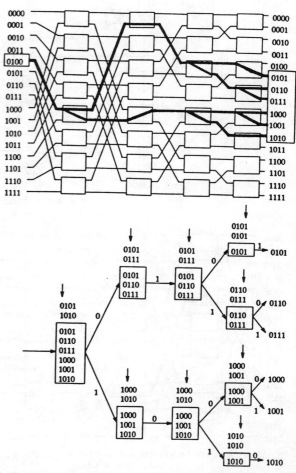

Fig. 7. The Boolean interval splitting algorithm generates the equivalent input–output tree of a packet with interval addresses.

an arbitrary set of input broadcast packets are not link-independent in general and the network is obviously blocking. All these problems are due to the irregularity of the set of destination addresses in the packet header. Fortunately, these problems can be avoided if the set of addresses is restricted to be an interval. This point is elaborated below. This interval constraint on addresses will not affect packet replication because addresses for the copy network are fictitious and are not used for switching. The switching of replicated packets is performed in the succeeding switching network based on actual addresses.

B. Boolean Interval Splitting Algorithm

An *address interval* is a set of contiguous N-bit binary numbers, which can be represented by two numbers, namely, the *minimum* and *maximum*. Suppose that a node at stage k received a packet with the header containing an address interval specified by the two binary numbers: min $(k - 1) = m_1 \cdots m_N$ and max $(k - 1) = M_1 \cdots M_N$ where the argument $(k - 1)$ denotes a node in stage $(k - 1)$ from where the packet came to stage k. Following the generalized self-routing algorithm, the direction for packet routing is described as follows.

1) If $m_k = M_k = 0$ or $m_k = M_k = 1$, then send the packet out on link 0 or 1, respectively.

2) If $m_k = 0$ and $M_k = 1$, then replicate the packet, modify the headers (according to the scheme described below) and send both packets out on both links.

It is easy to realize from these rules that $m_i = M_i$, $i = 1, \cdots, k - 1$ holds for every packet which arrives at stage k. Therefore, the event $m_k = 1$ and $M_k = 0$ is impossible, a don't care condition, due to the min–max representation of the address intervals. Fig. 7 illustrates the

Boolean interval splitting algorithm which generates the equivalent input–output tree of a broadcast packet with interval destination addresses.

The modification of a packet header is simply splitting the original address interval into two subintervals, which can be expressed by the following recursion:

$$\min (k) = \min (k - 1) = m_1 \cdots m_N,$$

$$\max (k) = M_1 \cdots M_{k-1} 01 \cdots 1 \qquad (1a)$$

for the packet sent out on link 0, and

$$\min (k) = m_1 \cdots m_{k-1} 10 \cdots 0,$$

$$\max (k) = \max (k - 1) = M_1 \cdots M_N \qquad (1b)$$

for the packet sent out on link 1 at stage k.

This modification procedure is a routine operation, meaning that it is independent of the header contents and can be considered as a built-in hardware function. Fig. 8 illustrates the switch node logic at stage k to perform this Boolean interval splitting algorithm. The decision at each node is based on a two-bit information provided by the packet header, which is the lower bound that we mentioned at the beginning of Section II-A). Thus, the logic

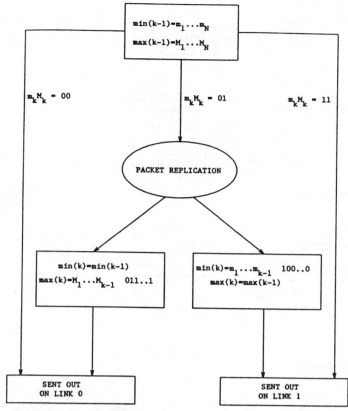

Fig. 8. The switch node logic at stage k of a broadcast banyan network.

$$x_1 = 7 \quad Y_1 = \{1,3\}$$
$$x_2 = 8 \quad Y_2 = \{4,5,6\}$$
$$x_3 = 9 \quad Y_3 = \{7,8,10,13,14\}.$$

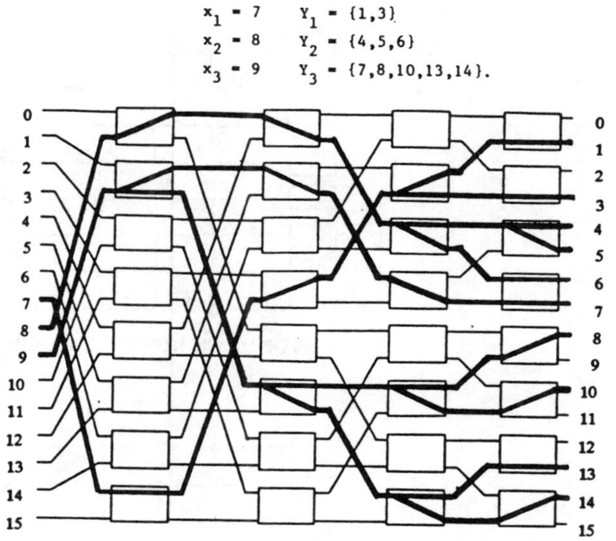

Fig. 9. An example to demonstrate the nonblocking condition of broadcast banyan networks.

complexity of broadcast switch nodes reached its minimum.

C. Nonblocking Condition of Broadcast Banyan Networks

As we described earlier, a set of input broadcast packets generates a set of input–output trees embedded in the broadcast banyan network. The network is nonblocking if this set of trees is always link-independent. The following is a sufficient condition, generalized from *NB1*, which assures that the broadcast banyan network is nonblocking.

NB2: A broadcast banyan network is nonblocking if the active inputs $x_1, \cdots, x_k$ and the corresponding sets of outputs $Y_1, \cdots, Y_k$ satisfy the following:
  1) (Monotone): $Y_1 < Y_2 \cdots < Y_k$, or $Y_1 > Y_2 > \cdots > Y_k$.
  2) (Concentration): Same as *NB1*.

The above inequality $Y_i < Y_j$ indicates that every output address in Y_i is less than every output address in Y_j. An example with active inputs $x_1 = 7$, $x_2 = 8$, $x_3 = 9$, and corresponding outputs $Y_1 = \{1, 3\}$, $Y_2 = \{4, 5, 6\}$, $Y_3 = \{7, 8, 10, 13, 14\}$ is illustrated in Fig. 9. The proof of *NB2* is straightforward. We know from *NB1* that the set of input–output paths $\langle x_1, y_1 \rangle, \cdots, \langle x_k, y_k \rangle$ are link-independent for an arbitrary choice of output addresses $y_1 \in Y_1, \cdots, y_k \in Y_k$. It follows that the set of input–output trees embedded in the broadcast banyan network must be link-independent, and *NB2* is established.

III. THE BASIC STRUCTURE OF COPY NETWORKS

The basic architecture of the copy network has been described in Section I and is shown in Fig. 4. Packet replications are accomplished by two fundamental processes, an encoding process and a decoding process. The encoding process transforms the set of copy numbers, specified in the headers of incoming packets, into a set of monotone address intervals which form new packet headers. This process is carried out by a running adder network and a set of dummy address encoders. The decoding process performs the packet replication according to the Boolean interval splitting algorithm through the broadcast banyan network described in the previous section. Finally, the destinations of copies are determined by the trunk number translators.

A. Encoding Process

The running adder is a $\log_2 n$-stage network constructed from $n/2 \log_2 n$ nodes. It can be arranged as a top-down or a bottom-up recursive structure as illustrated in Fig. 10. Each node is an adder with two inputs and two outputs where the vertical line is simply a pass. The main function of the running adder network is to assign output addresses for each broadcast packet according to the requested number of copies. It first generates the running sums of copy numbers specified by input packets at each port. The dummy address encoders then form the new headers from adjacent running sums. The new header consists of two fields: one designates the dummy address interval, which is represented by two n-bit binary numbers, the minimum and maximum, and the other contains an index reference which is equal to the minimum of the address interval.

If allocation of output addresses begins with address 0, then the following sequence of dummy address intervals will be generated by a top-down running adder:

$$(0, S_0 - 1), (S_0, S_1 - 1), \cdots, (S_{n-2}, S_{n-1} - 1)$$

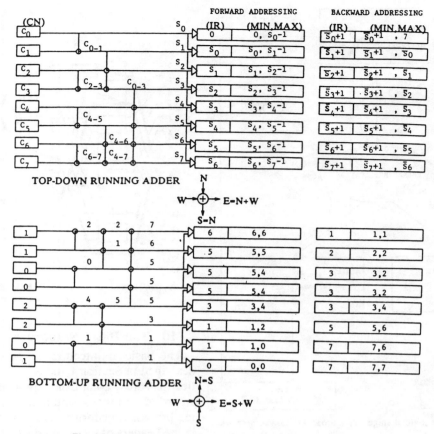

Fig. 10. Running adder networks and dummy address encoders.

where S_i is the ith running sum of the copy numbers. This sequence results in a *forward address assignment*. Suppose $\overline{S}_i$ indicates the 1's complement of S_i, which is formed by changing all 0's in S_i to 1's and 1's to 0's. Then, the sequence

$$(\overline{S}_0 + 1, n - 1), (\overline{S}_1 + 1, \overline{S}_0), \cdots, (\overline{S}_{n-1} + 1, \overline{S}_{n-2})$$

allocates addresses starting from the bottom of the outputs, a *backward address assignment*. The length of each interval is equal to the corresponding copy number in both addressing schemes. It also should be noticed that both sequences are monotone to satisfy the nonblocking condition described previously.

It is evident, as shown in Fig. 10, that the structures of top-down and bottom-up running adders are "complementary" to each other, which suggests that they can be implemented on a single chip. Although only one adder is needed at any time-slot, the other one can be a substitution. If only one running adder is used, say a top-down network, then the high numbered inputs are more likely to overflow than the lower numbered ones. Furthermore, forward or backward address assignment tends to concentrate copy packets on one side, low or high, respectively, at the outputs of the copy network. This of course may increase the potential buffer overflow at the inputs of succeeding point-to-point switch. This asymmetry can be alleviated by using *dual running adder networks*, a top-

down network with forward address assignment and a bottom-up network with backward address assignment. The two networks can be operated alternately from time-slot to time-slot. As a result, fairness for inputs and arbitration for outputs can therefore be achieved simultaneously. The adder also collects useful traffic information that is essential for flow control of broadcast channels, by counting the total number of copies at every time-slot.

B. Decoding Process

The decoding process is carried out by a broadcast banyan network and a set of trunk number translators.

The procedure of packet replication and routing within a switch node at stage k is illustrated in Fig. 11. The minimum and maximum in the packet header are bit-interleaved in the actual implementation of the address interval splitting algorithm. An *activity bit A* is posted in front of the address field to indicate that the packet is active ($A = 1$) or empty ($A = 0$). Fig. 12 illustrates the logic diagram of a switch node at stage k of a broadcast banyan network. The truth table of a *packet controller* is given in Fig. 13. It controls the procedure of packet replication and header modification. The packet routing is controlled by the *route controller* with truth table given in Fig. 14.

When packets emerge from the banyan network, the address interval in their header contains only one address at the outputs of stage N, the last stage. That is, according

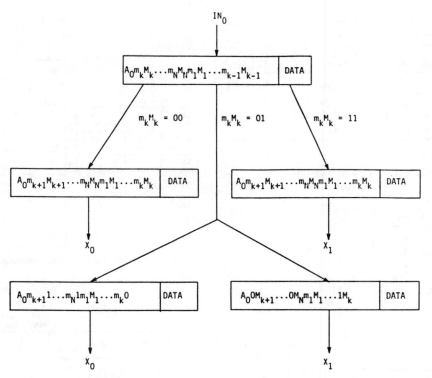

Fig. 11. The procedure of packet replication and routing within a node at stage k.

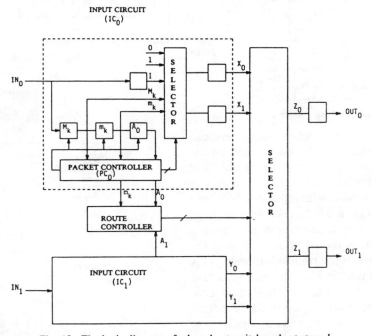

Fig. 12. The logic diagram of a broadcast switch node at stage k.

to Boolean interval splitting algorithm, we have

$$\min (N) = \max (N) = \text{output address}. \quad (2)$$

The packets belonging to the same broadcast channel should be distinguished by copy indexes. Recall that the index reference is initially set equal to the minimum of address interval. Thus, the copy index of each packet is determined by

$$\text{copy index} = \text{output address} - \text{index reference}. \quad (3)$$

This point is illustrated in Fig. 15.

The function of trunk number translators is to assign a new address to each copy of a broadcast packet. Then the copies are routed to their final destination by the succeed-

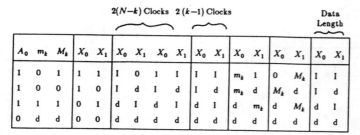

Fig. 13. The truth table of a packet controller.

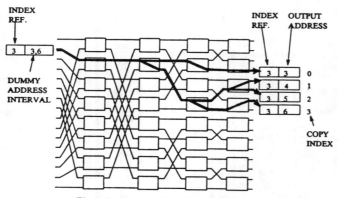

Fig. 14. The truth table of route controller.

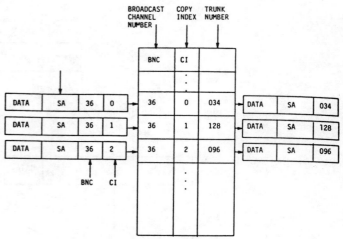

Fig. 16. Trunk number translation by table look-up.

Fig. 15. Computation of copy indexes.

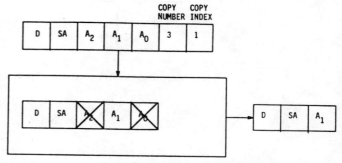

Fig. 17. Trunk number selection.

ing point-to-point switch. Trunk number assignment can be accomplished by a simple table look-up based on the identifier consisting of the broadcast channel number and the copy index associated with each packet. The trunk number translators contain a table with attributes BCN and CI as its primary key. When a TNT receives a copy of broadcast packet, it first converts the output address and IR into CI, and then replaces the BCN and CI with the corresponding trunk number in the translation table. This translation process is illustrated in Fig. 16. An alternative is trunk number selection. Suppose the BCN field contains a list of all trunk numbers belonging to the broadcast channel. The copy index can then be used as a pointer to select one of the trunk numbers in the list and erase the rest. The two parameters CN and CI are inputs to a state machine, which controls the shifting of a shift register and the loading of a buffer. This selection process is illustrated in Fig. 17. The number of copies and destinations of a broadcast channel may change frequently during a connection. Updating this information is fairly easy with both schemes.

Fig. 18 demonstrates the distributed parallel processes of packet replications described in this section.

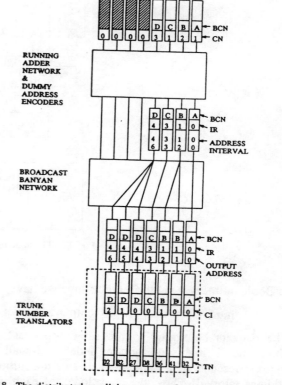

Fig. 18. The distributed parallel processes of packet replications.

IV. EXPANSION NETWORKS

The packet replications are divergent branching processes. Nonsquare networks are more appropriate in many practical applications such as entertainment video distribution and audio broadcast. An N-stage expansion network with $m = 2^M$ inputs and $n = 2^N$ outputs can be arranged as a combination of a set of 2^M $(N - M)$-stage binary trees and a set of 2^{N-M} M-stage square banyan networks. Fig. 19 illustrates an expansion network with 4 inputs and 16 outputs. The binary trees are made up of switch elements with one input and two outputs. The switch node logic is the same as the banyan network. Each input line is expanded to 2^{N-M} lines by an $(N - M)$-stage binary tree to achieve the desired expansion ratio. The remaining M stages contains square banyan networks entirely.

The interconnection of 2^M binary trees and 2^{N-M} banyan networks is similar to the link system of a multistage crossbar switch. The outputs of a binary tree can be uniquely identified by a 2-tuple $(x_1 \cdots x_M, y_1 \cdots y_{N-M})$ where $x_1 \cdots x_M$ is the top-down numbering of the binary trees and $y_1 \cdots y_{N-M}$ is the local address of each output within the binary tree. The input of a banyan network can also be identified by a 2-tuple $(a_1 \cdots a_{N-M}, b_1 \cdots b_M)$ where $a_1 \cdots a_{N-M}$ is the top-down numbering of the banyan network and $b_1 \cdots b_M$ is the local address of the input. The binary trees and banyan networks are transpose interconnected according to this numbering scheme. That is, the two ports $(x_1 \cdots x_M, y_1 \cdots y_{N-M})$ and $(a_1 \cdots a_{N-M}, b_1 \cdots b_M)$ are interconnected by a link if $(x_1 \cdots x_M, y_1 \cdots y_{N-M}) = (b_1 \cdots b_M, a_1 \cdots a_{N-M})$.

The topology of an expansion network is uniquely determined for a given fan-in m and fan-out n. The outputs of an expansion network can be concentrated and multiplexed into trunk number translators as illustrated in Fig. 20. We can use this to reduce the overflow probability of a copy network. It should be noticed that the perfect shuffle of the inputs to each banyan network is replaced by the perfect shuffle of inputs to the binary trees. It is easy to show that the $(N - M)$ stage binary trees with the transpose interconnection preserve the ordering of the inputs. Therefore, the expansion network is nonblocking if the destinations of inputs are monotone and concentrated in the manner of *NB2*.

V. INPUT CONCENTRATION

To satisfy the nonblocking condition, idle inputs between active inputs must be eliminated. Recall that the concentration criterion of *NB2* is the same as its counterpart in *NB1*, which means that the same input concentration technique for point-to-point switch can be applied to the copy network. There are two approaches proposed in the Starlite switch to concentrate the active inputs [4]. Both approaches are based on the activity bit preceding the packet header.

A. Concentration without Prioritization

The first approach uses a concentrator network consisting of a running adder network and a *reverse banyan network*

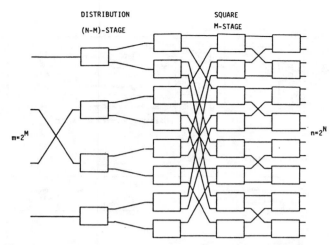

Fig. 19. An expansion network consisting of binary trees and banyan networks.

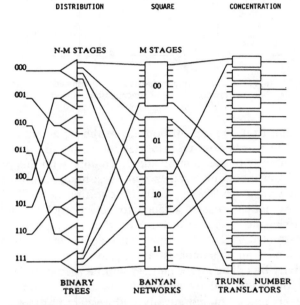

Fig. 20. An expansion network with output concentration.

as illustrated in Fig. 21(a). The running adder computes the *routing address* based on the active bit by counting the total number of active packets above the line. The reverse banyan network is a mirror image banyan network. It ignores inactive packets and routes active packets based on the running sum of activity bits. Thus, all active packets will emerge contiguously at the outputs. For all input sets based on this addressing scheme, it is easy to see from *NB1*, that this concentrator network is also nonblocking. A copy network with this concentrator is illustrated in Fig. 21(b). Notice that fairness for inputs and arbitration for outputs are achieved by dual running adder networks in this design.

B. Concentration with Priority Sorting

The second approach involves priority sorting. The top-down (or bottom-up) running adder of the copy network allocates the output addresses to the incoming packets

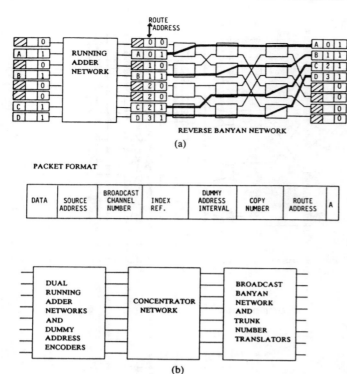

(a)

PACKET FORMAT

| DATA | SOURCE ADDRESS | BROADCAST CHANNEL NUMBER | INDEX REF. | DUMMY ADDRESS INTERVAL | COPY NUMBER | ROUTE ADDRESS | A |

(b)

Fig. 21. (a) An input concentrator consists of a running adder network and a reverse banyan network. (b) Input concentration without prioritization.

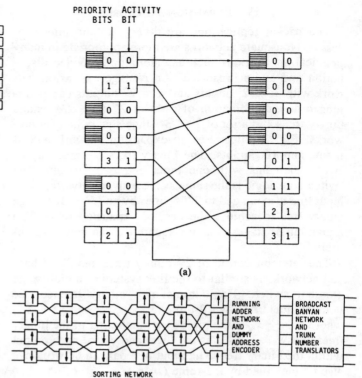

(a)

(b)

Fig. 22. (a) A sorting network sorts the priority bits and activity bit. (b) Input concentration with priority sorting.

starting from the top (or bottom) of the input lines. This top-down or bottom-up prioritization is due to the intrinsic property of the running adder. To cope with this property, the incoming packets can be sorted, according to their *priority indexes* by a Batcher sorting network. The priority index is posted adjacent to the activity bit. A lower index means higher priority. A Batcher sorting network is constructed by 2×2 sorting elements [1] which compare the binary numbers consisting of priority indexes and the activity bits. The larger number will be routed to the output line indicated by an arrow. As a result, the active packets appear monotonically with respect to their priorities at the outputs of sorting network. The sorting process is illustrated in Fig. 22(a). A copy network with priority sorting is shown in Fig. 22(b).

VI. CHERNOFF BOUND OF OVERFLOW PROBABILITIES

Overflow occurs when the total number of copies requested by input packets exceeds the number of output ports. This can be detected by the running adder and overflow packets can be retransmitted. A rudimentary analysis to characterize the behavior of the copy network is given in this section. To simplify the discussion, we assume the copy network is a loss system, meaning that packets are discarded when an overflow occurs.

Recall that inputs to our copy network are served in a sequentially ordered manner that can be either top-down or bottom-up. If no other fairness mechanism, such as dual running adder networks or sorting network, is provided to compensate for this biased priority scheme, then overflow probabilities are monotone increasing with respect to input addresses for homogeneous input traffic. These overflow probabilities can be estimated by Chernoff bound [3], which provides a crude assessment of the copy network's performance. A complete performance analysis for waiting system, that takes retransmission of overflow packets into account, is yet to be developed.

Suppose the network under consideration has m inputs and n outputs. Input processes are assumed to be two-state Markov chains. At any given time-slot, an input line is active with probability ρ and it is idle with probability $1 - \rho$ where ρ is the *offered load* per each input line. Let Y_i be a random variable representing the number of copies requested by an incoming packet at the head of the ith input line. Since inputs are homogeneous, the Y_i are independent identically distributed random variables with density function $q(k) = P_r\{Y_i = k\}, k = 1, 2, \cdots$. The moment-generating function of Y_i is the expectation of e^{sY_i}, which will be denoted by $Q(s)$. Let X_i be the random variable of the number of copies generated by ith input line regardless if it is active or idle. It is obvious that

$$p(k) = P_r\{X_i = k\} = \begin{cases} 1 - \rho, & k = 0, \\ \rho q(k), & k = 1, 2, \cdots. \end{cases} \quad (4)$$

The moment-generating function of X_i is then given by

$$P(s) = E[e^{sX_i}] = \sum_k p(k)e^{sk} = (1 - \rho) + \rho Q(s).$$

$$(5)$$

Thus, the *effective offered load* for each input line is $\rho' = E[X_i] = \dot{P}(0) = \rho\dot{Q}(0) = \rho E[Y_i]$. The overflow probability of ith input line, denoted by $p_b(i)$, can be estimated by Chernoff bound as follows:

$$p_b(i) = P_r\{X_1 + \cdots + X_i > n\} \le e^{-sn}[P(s)]^i \quad (6)$$

where s is a free parameter. We may solve for the value of s that provides the tightest upper bound of $p_b(i)$ by taking the derivative of the right-hand side of (6), which is a convex function. This yields

$$i\frac{dP(s)}{ds} - nP(s) = 0. \quad (7)$$

Two distributions of Y_i are considered, the first one assumes it is a constant and the second one is geometrically distributed. The results are discussed below.

A. Constant Number of Copies

Suppose every broadcast packet requests a constant number of copies, say K. Then the moment-generating function of X_i can be expressed by

$$P(s) = (1 - \rho) + \rho e^{sK}. \quad (8)$$

The stationary value of parameter s, by solving (7), is given explicitly as

$$s = \frac{1}{K} \ln \frac{n(1 - \rho)}{(iK - n)\rho}. \quad (9)$$

Therefore, from (6), the overflow probability of ith input line is bounded by

$$p_b(i) \le \left(\frac{iK\rho}{n}\right)^{n/K}\left(\frac{iK(1 - \rho)}{iK - n}\right)^{i-(n/K)}, \quad \text{for } i > \frac{n}{K} \quad (10)$$

and it is obvious that $p_b(i) = 0$, for $i \le n/K$.

Since the X_i are statistically independent, the generating function of the random variable $X_1 + \cdots + X_i$ is the ith power of (8), yielding

$$P^i(s) = [(1 - \rho) + \rho e^{sK}]^i$$
$$= \sum_{j=0}^{i} \binom{i}{j}(1 - \rho)^{i-j}\rho^j e^{sjK}. \quad (11)$$

In this particular case, the overflow probability can also be determined exactly by

$$p_b(i) = \sum_{j=[n/K]+1}^{i} \binom{i}{j}(1 - \rho)^{i-j}\rho^j \quad (12)$$

where $[n/K]$ is the integer part of n/K. We evaluated the overflow probabilities of a 64×64 copy network as an example. The Chernoff bound is compared to the true value in Fig. 23 for fixed effective offered load $\rho' = 0.6$ with various values of K. The results show that the overflow probabilities, as a measure of the performance, are quite sensitive to the copy number K.

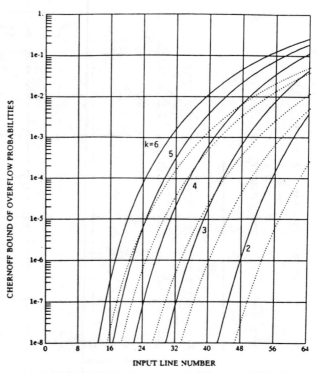

Fig. 23. Overflow probabilities of a 64×64 network with constant copy number assumption and fixed effected offered load $\rho' = 0.6$. The solid curves are Chernoff bound and the dotted curves are exact evaluation.

B. Geometric Distributed Number of Copies

Suppose the density function of the random variables Y_i is given by

$$q(k) = P_r\{Y_i = k\} = \frac{(1 - q)q^{k-1}}{1 - q^K}, \quad 1 \le k \le K \quad (13)$$

where K is the maximum allowable number of copies that can be specified in the header of a packet. The moment-generating function of Y_i is given by

$$Q(s) = \sum_{k=1}^{K} q(k)e^{sk} = \frac{e^s(1 - q)(1 - q^K e^{sK})}{(1 - q^K)(1 - qe^s)}. \quad (14)$$

The average number of copies is $\dot{Q}(0)$. This yields

$$E[Y_i] = \frac{1}{1 - q} - \frac{Kq^K}{1 - q^K}. \quad (15)$$

Similar to the previous procedure, the stationary value of parameter s can be obtained by substituting (14) into (5), and then solving (7) numerically. The Chernoff bound of overflow probabilities for both 64×64 network and 64×128 network are sketched in Fig. 24. The results demonstrate the significant improvements in performance by expanding the fan-out of a copy network.

Fig. 25 illustrates that a multicast packet switch can be arranged as a combination of a number of smaller copy networks and a large point-to-point switch. The distributed running adder in a small copy network can be re-

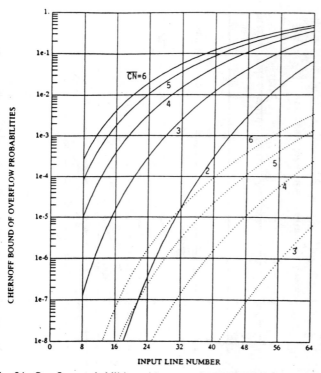

Fig. 24. Overflow probabilities with geometric distributed copy number assumption. The solid curves are results of 64 × 64 network, and the dotted curves are results of 64 × 128 network with fixed effective offered load $\rho' = 0.6$.

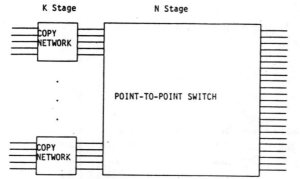

Fig. 25. Partitioning the copy network to reduce the complexity.

placed by serial adders to reduce the complexity. The results of performance analysis can be useful in the determination of the size of copy networks.

VII. CONCLUSIONS

This paper has described a design of nonblocking, self-routing copy networks for multicast packet switches. The main function of a copy network is packet replication, which is realized by cascaded encoding and decoding processes in our design. The encoding process is carried out by a running adder network and a set of dummy address encoders, which generates dummy address intervals as new headers for incoming packets. Then, a broadcast banyan network decodes these address intervals in packet headers according to a Boolean interval splitting algorithm that we have developed. At each node of the broad-

cast banyan network, the decision making is based on a two-bit header information. This yields minimum complexity of switch nodes. The address intervals of a set of inputs to the banyan network are monotone, which assures that the network is nonblocking. The computation of copy index consists of simply subtracting the index reference, initially set equal to the minimum of address interval, from the output address of each packet. The copy index and broadcast channel number combined together form a unique identifier, which is invoked by the trunk number translator to determine the packet's final destination.

The total number of copies, carried by the network, is counted by the running adder on per time-slot basis. This information provides useful traffic statistics for flow control of broadcast channels. To be flexible in many applications, we have studied nonsquare expansion network with larger fan-out than fan-in, which is an extension of the basic banyan topology and possesses the same routing properties as that of the banyan network.

There are two problems inherent in the design of a space-division-based multicast packet switch. One is the occurrence of overflows of the copy network when the total number of copy requests exceeds the number of outputs. The other is output port conflicts in the routing network when multiple packets request the same output port concurrently. The architecture of a multicast broadband packet switch is proposed in [6]. The solution to the first problem is to regulate the input packets, according to the number of copies requested and their priority, so to prevent the copy network from overflowing. The problem of output port conflicts in the routing network can be handled by a contention resolution algorithm reported in [5]. Future research should be focused on overflow control and performance analysis of the overall structure of multicast packet switches with respect to various services.

APPENDIX

Proof of NB1 [2]

In stage k of an N-stage banyan network, for $0 < k < N$, there are 2^{k-1} remaining subnetworks. Each node in stage k can be uniquely represented by two binary numbers $(a_{N-k} \cdots a_1, b_1 \cdots b_{k-1})$. Intuitively, $a_{N-k} \cdots a_1$ is the label of the node numbered from the top within the subnetwork, and $b_1 \cdots b_{k-1}$ is the label of the subnetwork numbered. Fig. 26 illustrates this numbering scheme introduced by Bechmann for a four-stage banyan network.

The node in stage $k + 1$ attached to the node $(a_{N-k} \cdots a_1, b_1 \cdots b_{k-1})$ in stage k by the output link b_k (0 or 1) is represented by $(a_{N-k-1} \cdots a_1, b_1 \cdots b_{k-1}b_k)$. Thus, the path from an input $x = a_N \cdots a_1$ to an output $y = b_1 \cdots b_N$, denoted by $\langle x, y \rangle$, consists the following sequence of nodes: $(a_{N-1} \cdots a_1, \phi)$, $(a_{N-2} \cdots a_1, b_1)$, $\cdots$, $(a_{N-k} \cdots a_1, b_1 \cdots b_{k-1})$, $\cdots$, $(\phi, b_1 \cdots b_{N-1})$. Suppose that two packets, one coming from input $x = a_N \cdots a_1$ to output $y = b_1 \cdots b_N$ and the other

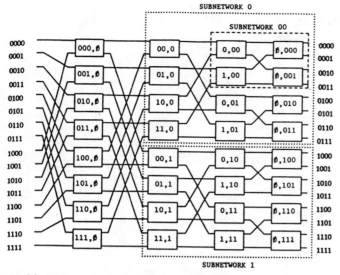

Fig. 26. Beckmann's numbering scheme of a four-stage banyan network.

coming from input $x' = a'_N \cdots a'_1$ to output $y' = b'_1 \cdots b'_N$, collide in stage k. That is, the two paths $\langle x, y \rangle$ and $\langle x', y' \rangle$ are merged at the same node $(a_{N-k} \cdots a_1, b_1 \cdots b_{k-1}) = (a'_{N-k} \cdots a'_1, b'_1 \cdots b'_{k-1})$, and shared the same output link $b_k = b'_k$. Then, we have

$$a_{N-k} \cdots a_1 = a'_{N-k} \cdots a'_1, \qquad (A.1)$$

and

$$b_1 \cdots b_k = b'_1 \cdots b'_k. \qquad (A.2)$$

Since inputs to banyan network are assumed to be monotone (increasing or decreasing) without gaps. It follows that the total number of active inputs between (and including) inputs x and x', $x' - x + 1$, must be less than or equal to the total number of outputs between y and y', $y' - y + 1$. We have

$$x' - x \leq y' - y. \qquad (A.3)$$

According to (A.1) and (A.2), we obtain

$$\begin{aligned} x' - x &= a'_N \cdots a'_1 - a_N \cdots a_1 \\ &= 2^{N-k}(a_N \cdots a'_{N-k+1} - a_N \cdots a_{N-k+1}) \\ &\geq 2^{N-k}, \end{aligned} \qquad (A.4)$$

and

$$\begin{aligned} y' - y &= b'_1 \cdots b'_N - b_1 \cdots b_N \\ &= b'_{k+1} \cdots b'_N - b_{k+1} \cdots b_N \leq 2^{N-k} - 1. \end{aligned}$$

$$(A.5)$$

But (A.3), (A.4) and (A.5) together imply $0 \leq -1$, a contradiction. Therefore, the two paths $\langle x, y \rangle$ and $\langle x', y' \rangle$ must be link-independent, and the assertion in *NB1* is established.

ACKNOWLEDGMENT

I would like to thank E. Arthurs, R. P. Singh, and L. T. Wu for many discussions during the course of this work. I also want to thank W. D. Sincoskie for supporting this work, who also suggested the study of nonsquare networks. I am grateful for the comments I received from J. S. Turner, whose thoughts and work are much appreciated. I particularly want to thank H. J. Chao who helped with the logic design of broadcast switch nodes.

REFERENCES

[1] K. E. Batcher, "Sorting networks and their applications," *AFIPS Proc. Spring Joint Comput. Conf.*, 1968, pp. 307–314.
[2] W. H. Bechmann, private communication.
[3] R. G. Gallager, *Information Theory and Reliable Communication.* New York: Wiley, 1968.
[4] A Huang and S. Knauer, "Starlite: A wideband digital switch," in *Proc. IEEE GLOBECOM '84*, pp. 121–125.
[5] J. Hui and E. Arthurs, "A broadband packet switch for integrated transport," *IEEE J. Select. Areas Commun.*, vol. SAC-5, pp. 1264–1273, Oct. 1987.
[6] T. T. Lee, R. Boorstyn, and E. Arthurs, "The architecture of a multicast broadband packet switch," presented at Proc. INFOCOM '88, New Orleans, LA.
[7] J. S. Turner, "Design of a broadcast packet switching network," in *Proc. INFOCOM, '86*, pp. 667–675.
[8] ——, "Design of an integrated service packet network," *IEEE J. Select. Areas Commun.*, Nov. 1986.
[9] C. L. Wu and T. Y. Feng, "On a class of multistage interconnection networks," *IEEE Trans. Comput.*, vol. C-29, Aug. 1980.

Performance of a Broadcast Packet Switch

RICHARD G. BUBENIK AND JONATHAN S. TURNER, MEMBER, IEEE

Abstract—This paper reports the results of a simulation study undertaken to evaluate a high performance packet switching fabric supporting point-to-point and multipoint communications. This switching fabric contains several components each based on conventional binary routing networks. The most novel element is the *copy network* which performs the packet replication needed for multipoint connections. We present results characterizing the performance of the copy network, in particular, quantifying its dependence on *fanout* and the location of active sources. We also evaluate several architectural alternatives for conventional binary routing networks. For example, we quantify the performance gains obtainable by using *cut-through switching* in the context of binary routing networks with small buffers. One surprising result is that networks constructed from nodes with more than two input and output ports can perform less well than those constructed from binary nodes. We quantify and explain this result, showing that it is a consequence of a subtle effect of the FIFO queueing discipline used in the nodes. We also show that substantially better performance can be obtained by relaxing the strict FIFO discipline.

I. INTRODUCTION

IN [17], Turner describes a packet switched communications system, which supports multipoint connections in addition to conventional point-to-point connections. The principal component of this system is a packet switch called a *switch module* which replicates and distributes multipoint packets and routes point-to-point packets. This paper reports on a simulation study of the switch module undertaken to characterize its performance. Some of these results were reported previously in Bubenik's masters thesis [1].

The switch module described in [17] uses a buffered binary routing network as one of its primary components. This network is one of a class that includes the banyan, delta, shuffle-exchange, and omega networks among others. When these networks are operated in a buffered mode, the distinctions between them become unimportant and we will therefore refer to them all simply as binary routing networks. The performance of such networks has been studied extensively. See, for example, [2]–[4], [6]–[8], [11], [13]. Our work differs from previous studies in several respects. The most important is that we evaluate the performance of a *copy network*; this is a binary routing network modified to perform the packet replication required for multipoint connections, rather than packet routing. This is, to our knowledge, the first study of a network of this sort. Our work also differs from most previous studies in that we consider binary routing networks with a limited amount of buffering (a few packets worth) at each node, an explicit flow control mechanism between nodes and *cut-through switching* [10]. In cut-through switching, a

Paper approved by the Editor for Communication Switching of the IEEE Communications Society. Manuscript received August 25, 1986; revised October 30, 1987 and April 1, 1988. This work was supported under grants from Bell Communications Research, Bell Northern Research, Italtel, SIT, NEC, and the National Science Foundation DCI 8600947. This paper was presented in part at ICC '87, Seattle, WA, June 1987. This work was performed at Washington University, St. Louis, MO.

The authors are with the Department of Computer Science, Washington University, St Louis, MO 63103.

IEEE Log Number 8824904.

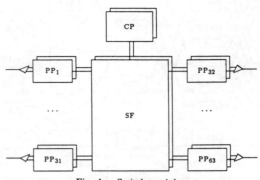

Fig. 1. Switch module.

packet need not be fully buffered at every stage of the network, but may proceed directly to the next stage, once the header has been decoded. This yields significant performance gains. We also consider several architectural tradeoffs in order to assess their effect on system performance.

We note that most analytical studies of systems like the ones we consider here make simplifying assumptions that reduce their usefulness. The work of Jenq [8] is of the most direct relevance, but his analysis applies only to "single-buffered" networks without cut-through. While we have used Jenq's work to validate our simulation results in those cases where the two are comparable, these cases are of only limited practical interest.

Section II briefly describes the architecture and operation of a single switch module. Section III gives the results of simulations of the routing and distribution networks, including results on several architectural variations of the basic design. Section IV gives results for the copy network under a variety of conditions and in Section V we offer our conclusions and outline directions for future research.

II. SYSTEM DESCRIPTION

Reference [17] describes a broadcast packet switching system comprising a number of components called switch modules or SM's. A network made up of these switch modules can support multipoint connections suitable for applications such as entertainment video distribution, LAN interconnection, and voice/video teleconferencing. In this paper, we study the performance of a single SM using a special-purpose simulation program designed for this purpose. The simulator was written in the C + + programming language [14].

We briefly review the operation of the switch module here. The reader is referred to [17] for further details. The overall structure is shown in Fig. 1. The SM terminates up to 63 fiber optic links (FOL). Data are carried on the FOL's in the form of large fixed-length packets of approximately 5000 bits each. The *packet processors* (PP) perform link-level protocol functions, including the determination of how each packet is routed. The switching fabric (SF) is a parallel packet switching interconnection network that provides routing of point-to-point packets plus replication and distribution of multipoint packets. The *connection processor* (CP) is responsible for establishing connections, including both point-to-point and multipoint connections. It will not be of concern to us here.

Reprinted from *IEEE Trans. Commun.*, vol. 37, no. 1, pp. 60–69, January 1989.

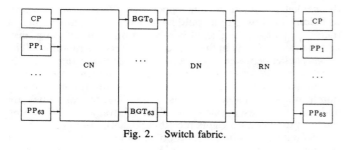

Fig. 2. Switch fabric.

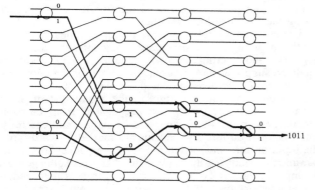

Fig. 3. 16 × 16 binary routing network.

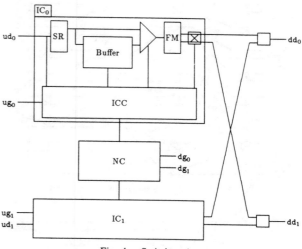

Fig. 4. Switch node.

When a packet enters the SM, it is reformatted by the addition of several new fields, the routing field being the only one that concerns us here. The routing field (RF) contains information needed to process a packet within the switch fabric. In the case of point-to-point packets, it includes an outgoing link number which is used to route the packet through the switch fabric. In the case of multipoint packets, it includes a *fanout* field which specifies the number of outgoing links that must receive copies of the packets.

A block diagram of the switch fabric (SF) is given in Fig. 2. It contains four major components, a *copy network*, a set of *broadcast and group translators*, a *distribution network*, and a *routing network*. The SF runs at a clock rate of 25 Mbits/s and has eight-bit side internal data paths. This gives an effective bit rate of 200 Mbits/s on the internal data paths, or two times the speed of the FOL's. An occupancy of 80 percent on the FOL's translates to a 40 percent occupancy on the internal data paths, which keeps contention and delay low.

When a multipoint packet having k destinations passes through the copy network (CN), it is replicated so that k copies of that packet emerge from the CN. Point-to-point packets pass through the CN without change. The principal function of the broadcast and group translators (BGT) is to assign an outgoing link number to the multiple copies of multipoint packets. As the BGT's effect on performance is completely deterministic, we will ignore them here. The reader is referred to [17] for details.

The distribution and routing networks (DN, RN) move the packets to the proper outgoing PP. The RN is a 64 × 64 binary routing network. Its structure is illustrated in Fig. 3 which shows a 16 × 16 version. The key property of such networks is that they are *bit addressable*—that is, the route a packet takes through the network is determined by successive bits of its destination address. The figure shows paths from two different input ports to output port 1011. Note that at the first stage the packet is routed out the lower port of the node (corresponding to the first one bit of the destination address), at the second stage it is routed out the upper port (corresponding to the zero bit), and in the third and fourth stages it is routed out the lower ports. The self-routing property is shared by a variety of different although similar interconnection

patterns, including the so-called banyan, delta, shuffle-exchange, and omega networks [5].

Buffers are provided at the inputs of each node. Each buffer holds one or more complete packets. The simulation results given below evaluate several possible buffer sizes. A packet may pass through a node without being buffered at all if the desired output port is available when the packet first arrives. Indeed, in a lightly loaded network, a packet can pass through the RN with no buffering at all. In this case it encounters a delay of just a few clock times in each node.

The data paths between nodes are eight bits wide. In addition, there is a single upstream control lead used to implement a simple flow control mechanism. This prevents loss of packets due to buffer overflows within the fabric. The entire network is operated synchronously, both on a bit basis and a packet basis—that is, all packets entering a given stage do so during the same clock cycle.

Fig. 4 is a more detailed picture of a single switch node. Data enters the node on the upstream data lines (ud_0, ud_1) and leaves on the downstream data lines (dd_0, dd_1). The *node control* (NC), at the center of the figure monitors the availability of the downstream neighbors via the downstream grant lines (dg_0, dg_1) and arbitrates access to the outgoing links. The two *input controllers* (IC) receive packets from their upstream neighbors, request the appropriate output link (or links) from the node control, buffer packets if necessary and apply upstream flow control using the upstream grant lines (ug_1, ug_2) as needed to prevent buffer overflow. Nodes can also be constructed with more than two input and output ports. In the next section, we compare the performance of networks constructed from two port nodes and four port nodes.

The entire switch fabric operates synchronously using a fixed length packet cycle. Each packet cycle can be viewed as having two phases. During the first phase of the cycle, grant signals propagate from the outgoing packet processors back through to the front of the RN, DN, and CN. The outgoing packet processors always accept new packets (if they do not have room in the outgoing buffer, packets are lost). Each node in the switch fabric examines its downstream grant lines and the state of its buffers, then generates the appropriate upstream grant signal. An IC in a node can assert its upstream grant line if it has at least one empty buffer slot. If its buffer is occupied, it attempts to reserve one or more output ports for use during the next cycle (the requested port or ports is determined by the first buffered packet). If the node control grants the reservation request, the IC is assured of being able to move one packet forward during the next cycle and will therefore assert its upstream grant line. Hence, the only case in which the

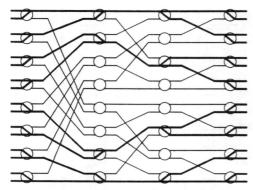

Fig. 5.   Congestion in binary routing networks.

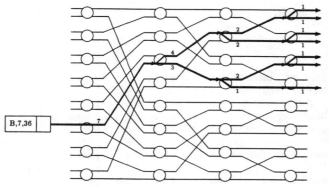

Fig. 6.   16 × 16 copy network.

upstream grant line is not asserted is when an IC has full input buffers and has its reservation request denied. After the grant signals ripple back through to the inputs of the switch fabric, packets flow forward. This is the second phase of the packet cycle. It is possible for the two phases to overlap in time, so the effective length of a packet cycle is just the time required to transmit a packet across one of the internal data paths.

One problem with binary routing networks is that they can become congested in the presence of certain traffic patterns. This is illustrated in Fig. 5, which shows a traffic pattern corresponding to several *communities of interest*. In this pattern, all traffic entering the first four inputs is destined for the first four outputs, all traffic entering the second group of four inputs is destined for the second group of four outputs, and so forth. Note that with this pattern, only one fourth of the links joining the second and third stages are carrying traffic. Thus, if the inputs are heavily loaded the internal links will be hopelessly overloaded and traffic will back up. In a 64 × 64 network, there are six stages and the links between the third and fourth stages can in the worst case by carrying all the traffic on just eight of the 64 links. The purpose of the DN is to eliminate this problem by evenly distributing packets it receives across all its outputs. It has the same internal structure as the RN, but its nodes ignore the destination addresses on packets and route them alternately to each of their output ports. This strategy is modified if one or both ports is unavailable. In this case, the first port to become available is used. This approach is designed to break up any communities of interest and make the combination of the DN and RN robust in the face of pathological traffic patterns. This configuration is evaluated in the next section.

The structure of the copy network (CN) is the same as that of the RN and DN. The CN's function is to make copies of multipoint packets as they pass through, as illustrated in Fig. 6. The packet entering at left has a fanout of 7 meaning that it will be sent to seven different output links. At the first stage,

the packet is routed out the upper port. This is an arbitrary decision—the lower link could have been used at this point. At the second stage, the packet is sent out on both outgoing links and the fanout fields in the outgoing packets are modified. The upper packet generates four copies and the lower one three. In general, a node in the copy network will replicate a packet of its current fanout value exceeds one-half the number of CN output ports reachable from that node. The fanout values are split as evenly as possible, with an arbitrary decision being made as to which port gets the "bigger half" in the case of an odd fanout value. When a multipoint packet is not replicated, it is routed arbitrarily to one of the node's two output ports. Point-to-point packets are routed through the CN arbitrarily, taking the "path of least resistance."

A simulation model was written that includes components for the incoming packet processor, copy network, distribution network, and routing network. The packet processor was modeled simply as an infinite queue. Each of the networks in the switch fabric was modeled explicitly. The simulation mimics the synchronous nature of the actual system. During each packet cycle, grants are propagated from outputs to inputs, new packets are generated by packet sources feeding into the input buffers and packets are moved forward through the input buffers and switch fabric. Each packet source has a parameter specifying the probability that a packet is generated during a particular cycle. Each cycle is treated independently and at most one packet enters a particular input buffer during a packet cycle. The results described in the next section cover several different configurations in addition to the one just described. In particular, in most of the runs not all the components described above are included. Also several variations on the basic design described above are explored. Details of these variations will be given as the results are presented. We note that the simulation results described below are independent of most of the specific choices of system parameters mentioned above. In particular, they are independent of the choice of packet length and relative speed of the switch fabric data paths and fiber optic links.

III. EVALUATION OF THE ROUTING AND DISTRIBUTION NETWORKS

This section describes a basic set of simulations, selected from an extensive simulation study described fully in [1]. We focus here on the routing and distribution networks. Results for the copy network appear in the following section. Unless otherwise stated, the quoted results are for a switch fabric with 64 input and output ports and two port switch nodes.

The first set of results, shown in Fig. 7, is for a configuration consisting of a set of input buffers feeding directly into a routing network. The CN and DN have been omitted to permit a detailed examination of the RN. This delay plot (and all subsequent ones) gives delay in packet times where a packet time is just the time required to transmit a packet across one of the internal data paths. Assuming 5000 bit packets and a 200 Mbits/s data rate on the internal data paths, one packet time is 25 μs. Two sets of delay curves are given. The curves that flatten out at large offered loads give the delay through the RN alone, while the other curves give the total delay through the input buffers and the RN. Delays are measured from the time the front end of a packet enters a component to the time the front end leaves the component. Hence, they neglect the time needed to buffer the packet as it exits the switch fabric (that time is one packet time). Notice also that the simulation ignores details such as the pipeline delay experienced by a packet as it passes through a node, since this is a constant which is completely determined by details of the node design. These simulations were run under the *uniform traffic assumption*—that is, packets generated at the sources were assigned random destination addresses with each destination address equally likely.

In these plots, offered load is expressed as a fraction of the

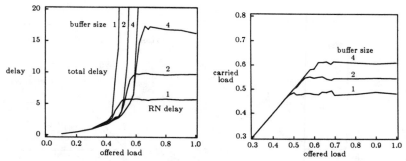

Fig. 7. Delay and throughput curves for routing network.

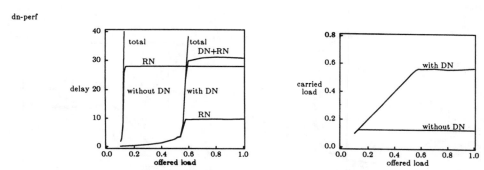

Fig. 8. Performance in the presence of communities of interest.

capacity of the switch fabric under ideal conditions. Under such ideal conditions, the switch fabric is capable of passing a packet out of every output port during every packet cycle. What has been plotted is the offered load measured at the packet sources. Carried load is the fraction of the ideal capacity that the system actually delivers and is measured at the outputs of the switch fabric.

The results in Fig. 7 give the delay for nodes with 1, 2, and 4 buffer slots per input port. As one would expect, increasing the number of buffers increases the throughput of the network and reduces total delay. The carried load curves show clearly the effect of node buffer size on network throughput. As mentioned earlier, the internal data paths operate at twice the rate of the external links meaning that the offered loads within the switch fabric cannot exceed 0.5. If the external links carry a load of 0.8, the internal loading is 0.4. The carried load plot shows that even with a buffer size of one, the RN has sufficient throughput to carry this load. On the other hand, the delay plot shows that the safety margin is unacceptably small with this buffer size. A buffer size of two is probably required to provide an adequate margin against occasional traffic surges. It is worth noting that at low loads, most of the delay is incurred within the RN while at high loads the RN delay becomes constant while the input buffer delay becomes very large. This is a pattern that will be reappear in later results. Note that at an offered load of 0.4, the total delay is approximately two packet times, which translates to about 50 μs, assuming 5000 bit packets and a 200 Mbits/s data rate.

The next set of results demonstrates the effectiveness of the distribution network in breaking up pathological traffic patterns. See Fig. 8. The curves labeled "without DN" show the performance of the RN when subjected to a traffic pattern calculated to cause extensive congestion. In particular, traffic was constrained to eight "communities of interest" similar to those shown in Fig. 5. This pattern forces all the traffic though just eight of the 64 links in the middle stage of the RN. As expected, this leads to a maximum throughput of about 0.125. The curves labeled "with DN" show the performance when the DN is inserted in front of the RN and the same traffic

pattern is applied. Note that the throughput in this case matches that of the RN-only configuration with uniform traffic. Also note that at loads of less than 0.5, the DN adds very little delay. Apparently, at these loads packets encounter little or no "back pressure" from the RN and simply flow through the DN taking the path of least resistance.

A. Effect of Cut-Through

Perhaps the principal difference between the routing network described here and buffered binary routing networks discussed elsewhere is the use of *cut-through operation* in the switch nodes. In most previous work, packets are buffered completely at each node in the switch fabric before being advanced to the next stage. We relax this restriction, allowing a packet to proceed immediately to the outgoing link whenever possible. Cut-through switching has also been employed in the fast packet switch described in [12], [15], [18]. The concept of cut-through switching was first discussed in print in [10], although in a different context. Their analysis of the performance of cut-through switching shows significant advantages, but does not apply to the specifics of our situation. To quantify the advantage of cut-through switching we performed a simulation of the routing network in which cut-through was not used. This simulation is for a 64 port routing network with two slot buffers at the node inputs. The resulting comparison appears in Fig. 9.

Notice that at low loads, the total delay without cut-through is at least six packet times, corresponding to the enforced buffering at each stage of the simulated network. When cut-through is used, the total delay is less than a single packet time. At an offered load of 0.4, the total delay without cut-through is about eight packet times, while with cut-through, it is about two. We also note that the use of cut-through increases the throughput of the switch fabric from 0.5 to 0.55. The delay advantage is significant in the target application as it translates to a delay of 50 μs versus 200 μs. The advantage is even more striking than this comparison reveals, since (as we will see later) most of the delay in the switch fabric comprising the CN, DN, and RN occurs in the RN when cut-through is

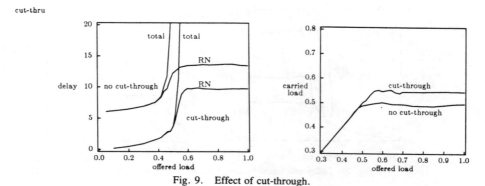

Fig. 9. Effect of cut-through.

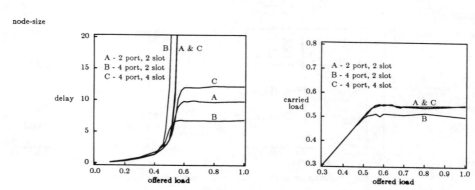

Fig. 10. Effect of node size.

used. Without cut-through the delay would be at least 18 packet times, whereas with cut-through, it is less than three packet times.

Given the advantages obtained with cut-through switching, it is perhaps surprising that it is not universally used. To learn the reason it is not, one must look more closely at specific applications of buffered binary routing networks. In most cases where they have been proposed, the intended application is an interconnection network in a parallel computing system. Typically, these applications involve small packets in which the addressing information may be a large part of the total packet; the advantages of cut-through switching are significantly reduced in such applications and the extra control complexity required of the nodes may not be justified. On the other hand, in communications systems applications where the addressing information is typically a small part of the packet, cut-through is clearly the right choice.

B. Effect of Node Size

We now consider the effect of varying the number of node input and output ports on routing network performance. In particular, we compare the performance of a routing network with 64 input and output ports comprising two port nodes to one comprising four port nodes. Four port nodes appear to offer several advantages. First, they can implement a given size network with half the number of stages required when using two port nodes. This makes them topologically richer and since the links between stages are the places where contention occurs, there is less opportunity for contention with four port nodes. This leads one to expect networks constructed with four port nodes to have larger throughput than networks using two port nodes. One also expects smaller delays, since the number of places a packet can be buffered is smaller. Finally, because there are fewer stages, fewer buffers are needed (assuming the number of buffer slots per node input port is the same in both cases). Hence, larger nodes offer advantages in system complexity as well as performance. The intuitive case in favor of larger nodes is compelling. There is

also analytical support for the contention that larger nodes perform better than smaller ones. In [13], Patel analyzes the performance of unbuffered binary routing networks and shows that larger nodes offer substantial performance advantages.

We examined three different configurations. Network A is a 64 port routing network with two port nodes and two buffer slots per node, network B is a 64 port routing network with four port nodes and two buffer slots per node, network C is a 64 port routing network with four port nodes and four buffer slots per node. In light of the arguments given above, the results shown in Fig. 10 are surprising. Notice that network A outperforms B—it offers higher throughput and smaller total delay. One possible explanation for this behavior is that network B has half the total number of buffer slots that network A has and incurs a performance penalty as a result. Network C however, has the same number of buffer slots as A and while it performs better than B, it is still no better than A.

This apparent anomaly is a subtle consequence of the strict FIFO queueing discipline used in the node buffers. If the output port required by the first packet in a node buffer is unavailable, it and all other packets in that buffer must wait. This is true even if the subsequent packets in the buffer require output ports that are available. This queueing discipline limits the way in which packets can proceed through the network. Moreover, it has a more limiting effect on networks constructed from four port nodes than it does on networks constructed from two port nodes. It is this effect that leads to the apparent anomaly.[1]

C. Effect of Bypass Queueing Discipline

In light of the results just discussed, an obvious next step is to relax the FIFO queueing discipline in the node buffers. We now consider a *bypass queueing discipline* in which a packet

[1] When this effect was first observed, the only likely explanation appeared to be a faulty simulation. It was only after spending several days in a fruitless hunt for simulator bugs that we came to understand the real nature of this effect.

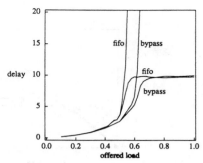
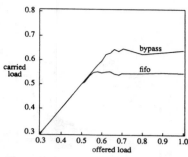

Fig. 11.   Effect of bypass queueing discipline.

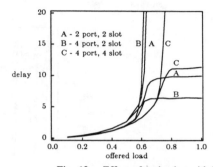
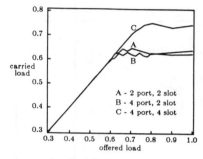

Fig. 12.   Effect of node size with bypass queueing discipline.

other than the first one is allowed to proceed if all its predecessors in the queue are blocked, but it is not. Fig. 11 shows the effect of bypass queueing on a routing network with two port nodes and two buffer slots per node. Note that the throughput of the network increases from about 0.55 to 0.63, a substantial improvement.

We now return to our comparison of two port and four port nodes. Fig. 12 is similar to the earlier comparison, but here all networks use bypass queueing. Here, we observe the expected advantage for four port nodes. In particular, network C achieves a throughput of about 0.75 versus 0.63 for networks A and B. This plot indicates a choice for a network designer. He may use four port nodes to reduce the total number of buffer slots in the network while maintaining the same performance level, or he may use them to increase throughput while holding the number of buffer slots constant. Of course, this tradeoff assumes the bypass queueing discipline, which may be more difficult to implement than the original FIFO discipline. If the original discipline is used, the designer is better off with two port nodes.

We note that the nodes used in the networks described here perform all their buffering on the input side. Another alternative is to buffer at the outputs. In our case, buffering at the input was chosen because it appears simpler to implement. We have not compared the performance of the two alternatives.

IV. PERFORMANCE OF THE COPY NETWORK

We now examine the performance of the copy network (CN). Recall, that the function of the CN is to replicate packets associated with multipoint connections. To our knowledge, ours is the first performance study of networks of this type.

The results in this section are for a 64 port CN with two port nodes, each having two buffer slots per input. The CN is preceded by a set of input buffers. Nothing is connected to the output side, meaning that there is no "back-pressure" at the

last stage of switching. This configuration allows us to focus on properties determined entirely by the CN. In a later section, we look at the CN performance in the context of a fully configured switch fabric.

In our first set of results, packets arrive at all inputs in the CN at the same average rate. The fanout assigned to each packet is selected from a truncated geometric distribution with parameter p. More specifically, the probability that the fanout is k is given by

$$\Pr\,(\text{fanout}=k) = \begin{cases} p(1-p)^{k-1} & 1 \le k < N \\ (1-p)^{N-1} & k=N \end{cases}$$

where $0 \le p \le 1$ and N is the number of ports in the network ($N = 64$ for the results given here). The average fanout obtained with this distribution is given by

$$E\,(\text{fanout}) = \begin{cases} (1/p)[1-(1-p)^N] & 0<p\le 1 \\ N & p=0 \end{cases}$$

Since, each packet entering the CN may give rise to multiple output packets, the offered load for the CN is defined as the product of the input load and the average fanout. Given these definitions, we can now consider the first set of simulation results shown in Fig. 13. The delay plot shows CN delay and total delay for several values of average fanout. The carried load curves are also given for several values of the average fanout. When the average fanout is one, the CN carries the load with no contention and no delay. To understand this, one must realize that to achieve an average fanout of one, the fanout of all packets must be one. Hence, no packet replication occurs and there is no source of contention. When the average fanout becomes two the situation changes dramatically. First, the maximum carried load drops to about 0.865 and there is a corresponding increase in delay. This drop in performance is

271

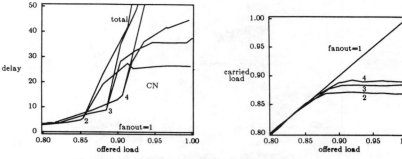

cn-perf

Fig. 13.   CN delay and load with random fanouts.

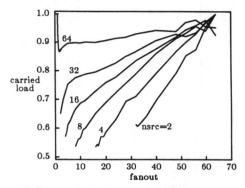

cn-thru1

Fig. 14.   CN throughput with random fanouts.

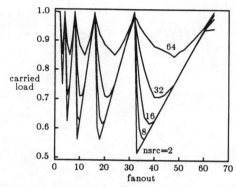

cn-thru2

Fig. 15.   CN throughput for fixed fanouts.

the expected consequence of contention caused by packet replication. The next effect one notices is that as the fanout continues to increase, the performance improves. While this may appear counter-intuitive, it can be explained by observing that for a given value of offered load, a larger fanout means a smaller packet arrival rate (since offered load is the product of input load and fanout). Thus, when the average fanout doubles from two to four, the packet arrival rate at the CN input ports is halved. This effect tends to reduce contention and counteracts the contention-increasing effect of larger fanout.

The last set of results indicates a dependence of performance on the fanout of the sources. We next consider the effect on performance of limiting the number of active CN input ports. In the next set of results, packets were generated only on input ports 0 to *nsrc* − 1 where *nsrc* is a parameter that was varied. This choice of input ports leads to maximum contention in the early stages of the network. The results are shown in Fig. 14. In this plot, fanout is given on the horizontal axis and carried load on the vertical axis. The offered load in all cases is one, so this plot shows the maximum throughput provided by the network under a wide range of different conditions. The best performance is obtained when all inputs are active. In this case, the minimum throughput occurs with a fanout of about 2. When the number of sources is small, there is a lot of contention in the early stages leading to sharply lower throughputs at small fanouts. We note again that as fanouts increase, throughput climbs rapidly due to the lower arrival rates at the input ports which reduce contention. Note that in all cases, the throughput exceeds 0.5.

The results given above are for randomly generated fanouts. Our next set focuses more closely on the effect of different fanouts. In this set, the fanouts were fixed at a constant for each simulation run, rather than being selected at random. The results are shown in Fig. 15. In this plot, offered load is again held at 1, so we see the maximum throughput obtained for various values of fanout and *nsrc*. The difference between this and the previous plot is striking. For each value of *nsrc,* the throughput takes on values close to one whenever the fanout is a power of two. Between powers of two, the throughput drops sharply achieving its minimum value when the number of sources is small. This plot shows more clearly the effect of different fanout values since in each simulation run all packets had the same fanout and consequently were replicated in the same stages of the copy network. This contrasts to the previous set where in each run there was a mixture of different fanout values. The excellent performance at fanouts equal to powers of two is an expected effect as is the sharp deterioration just above powers of two. This deterioration is explained by noting that when the fanout passes a power of two, copying begins one stage earlier in the network than previously, while the packet arrival rate is about the same. This leads to additional contention and the observed performance deterioration. As the fanout continues to increase, the arrival rate decreases (in order to maintain the same offered load), reducing contention in the early stages and improving performance.

We conclude this section with a review of a few analytical results obtained in [1] and discussed further in [20]. We view the traffic imposed on an N port copy network as coming from a collection of sources $S_1, \cdots, S_k$. Each source S_i is described by an ordered triple (P_i, B_i, F_i) where P_i is the input port at which packets from S_i arrive $(0 \leq P_i \leq N)$, B_i is the average arrival rate of packets from $S_i (0 \leq B_i \leq 1)$ and F_i is the required fanout $(1 \leq F_i < N)$. A *traffic configuration* (or simply configuration) is a set of sources. In a *legal configuration* the sum of arrival rates of sources associated with any particular input port must be at most 1. That is, for $i \in [0, N - 1]$

$$\sum_{j,P_j=i} B_j \leq 1.$$

Also, the sum of the products of arrival rate and fanout must be no more than N.

$$\sum_{j=0}^{k} B_j F_j \leq N.$$

Configurations that violate these restrictions exceed the capacities of either the input ports or the output ports of the CN and hence its performance under such conditions is immaterial.

We label the links in the CN with ordered pairs (i, j) where i is the stage number of the link and j is its index within the stage. Stage 0 links are those feeding the nodes in the first stage, stage 1 links connect the nodes in the first and second stages, and so forth. Links are numbered within stages from top to bottom at their left ends, starting from 0. Hence, the first link leaving the top node in the first stage is link $(1, 0)$.

Let $S = (P, B, F)$ be a source. We define the average load induced by S on a link (i, j) in the copy network by

$$\lambda_{ij}(P, B, F)$$

$$= \begin{cases} 0 & \text{if there is no path from link} \\ & (0, P) \text{ to link } (i, j) \\ 2^{-i}B & \text{if there is a path and} \\ & 0 \leq i \leq n - \lceil \log_2 F \rceil \\ 2^{-(n-\lceil \log_2 F \rceil)}B & \text{if there is a path and} \\ & i \geq n - \lceil \log_2 F \rceil \end{cases}$$

This definition simply reflects the behavior of the CN with respect to a particular source. When a packet from source S enters the CN it follows a random path until it reaches stage $f = \lceil \log_2 F \rceil$. Stage f is the first at which replication takes place. For a set of sources S, we define

$$\lambda_{ij}(S) = \sum_{(P,B,F) \in S} \lambda_{ij}(P, B, F).$$

The following theorems are proved in [1].

Theorem 1: Let f be an integer in $[0, \log_2 N]$ and let $S = (S_1, \cdots, S_k)$ be any legal configuration in which $F_i = 2^f$ for $1 \leq i \leq k$. Then, for any link (i, j), $\lambda_{ij}(S) \leq 1$.

Theorem 1 tells us that when all the sources have fanouts equal to the same power of two, there is no legal configuration that can induce an overload on any internal link. If we drop the restriction to powers of two we get the following.

Theorem 2: Let F be an integer in $[1, N]$ and let $S = (S_1, \cdots, S_k)$ be any legal configuration in which $F_i = F$ for $1 \leq i \leq k$. Then, for any link (i, j), $\lambda_{ij}(S) \leq 2$.

So, if the fanouts are all equal, the CN can handle any legal configuration in which the arrival rates on all input ports are ≤ 0.5 without inducing an overload on any internal link. We get a similar result ($\lambda_{ij} \leq 2$) if we insist on fanouts equal to powers of two but drop the restriction that all fanouts be equal. Finally, if we drop all restrictions we obtain the following.

Theorem 3: Let $S = (S_1, \cdots, S_k)$ be any legal configuration. Then, for any link (i, j), $\lambda_{ij}(S) \leq 3$.

This implies that the CN can handle any legal configuration in which the arrival rates on all input ports are $\leq 1/3$ without inducing an overload on any internal link. In each of the above theorems, the bound on λ_{ij} cannot be improved. In Fig. 16 we

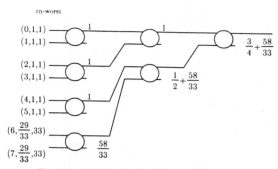

Fig. 16. Pathological traffic pattern for CN.

show a legal traffic configuration for a six-stage CN that induces a load exceeding 2.5 on a link in the center stage. This example generalizes to larger copy networks and induces loads approaching 3 as the network size increases. Fig. 17 shows the performance curves for a simulation of a six-stage CN with this traffic pattern. Note that the maximum throughput for this traffic pattern is about 0.35 and as the offered load continues to increase the throughput drops to about 0.22. At high offered loads multipoint packets must wait a long time in a node where they must replicate because it is rare that both output ports become available simultaneously and if only one is available there is usually a point-to-point packet in the other node buffer that takes it. This suggests that better performance might be obtained at high loads by allowing multipoint packets to be copied serially rather than in parallel. This change would not affect Theorem 3, but could prevent the throughput from dropping at high loads. On the other hand, the implementation of nodes that operate in this way appears substantially more complicated.

V. CLOSING REMARKS

The results presented above allow us to select from among several alternative designs for the system described in [17]. In our recommended configuration, we use two slot buffers at the inputs to each node; we also use cut-through switching and bypass queueing. We use four port nodes in the routing and distribution networks, but two port nodes in the copy network. The performance of this configuration is given in Fig. 18. This is for an input traffic pattern with eight active sources and fanouts selected from a truncated geometric distribution with a mean of eight. This traffic pattern provides a fairly demanding test of the copy network, but is not worst case. Note that at the nominal operating point of 0.4, the delay is just over two packet times (50 μs). Also note that at this load, the delay is split fairly evenly between the CN and the RN and the DN contributes very little delay. Our results give us confidence that this configuration will perform well across a wide range of operating conditions. While there exist pathological traffic patterns that can exceed the capacity of the system, they are sufficiently contrived that it appears unlikely that they would pose an actual threat to a real system.

The simulations results allow us to draw several conclusions concerning the performance implications of alternative node designs. The first is that cut-through switching offers significant performance advantages. In applications where the header is short relative to the packet length, cut-through switching is clearly the method of choice. Without cut-through, the *minimum delay* in the recommended configuration would be twelve packet times (300 μs). Second, two port nodes are the best choice if queueing is performed at the input side of the switch nodes and a strict FIFO queueing discipline is used. The bypass queueing discipline improves performance significantly, especially for larger switch nodes; when bypass queueing is used, large nodes are preferred.

Our results show that the performance of the copy network

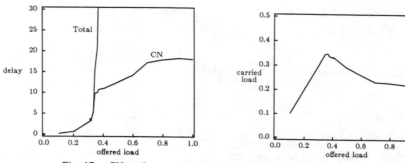

Fig. 17.   CN performance for pathological traffic pattern.

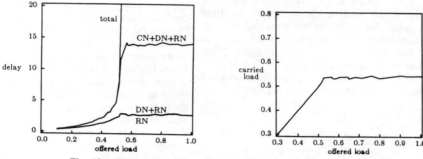

Fig. 18.   Performance of recommended configuration.

is sensitive to the fanout of the incident traffic and also to input ports on which the traffic arrives. There are pathological traffic patterns that can induce a load of 3 on some of the copy network's internal links, but this is as bad as things can get. To make the copy network robust in the fact of arbitrary traffic patterns, there are at least two approaches. The first is to operate the copy network so that the load at the input ports is no more than one third (this may require higher speed operation). The second is to place a second distribution network in front of the copy network. In this configuration, the maximum load obtainable on an internal copy network link is one.

There are many interesting questions that remain to be explored. One area for further research is to seek analytical models of the kinds of networks studied here. An analysis of binary routing networks with cut-through switching and small buffers would be of particular interest. Another direction for additional work is the exploration of other architectural variations. In particular, our use of a separate distribution network is clearly just one of many ways to avoid congestion in binary routing networks. A second approach is described in [19] and many others are possible. A comparison of buffering strategies in the switch nodes would also be worthwhile, we have limited ourselves to buffering at the inputs, but clearly output buffering merits consideration as well. Reference [9] does consider queueing strategies in switching fabrics (in a slightly different context from ours) and shows a clear performance advantage for output buffering over input buffering. Their work does not give any insight as to how output buffering compares to input buffering with bypass queueing. We believe input buffering with bypass queueing to be equivalent to shared buffering and expect it be slightly superior to output buffering, but cannot justify that conjecture at this time.

When it comes to the copy network there remains a wealth of open problems. The most important perhaps is whether or not there exist implementations using binary nodes and $\log_2 N$ stages for which the maximum load that can be induced on an internal link is smaller than three. There are several questions that concern packet replication strategies. 1) Is it best to divide the fanouts evenly as we have done here or to divide them unevenly? 2) Should priority be given to packets that require replication? 3) Should sequential copying be allowed at a node, rather than requiring that both copies proceed in parallel? Finally, we note that the issue of copy networks with nodes having more than two ports remains largely unexplored.

REFERENCES

[1] R. G. Bubenik, "Performance evaluation of a broadcast packet switch," M.S. thesis, Comput. Sci. Dep., Washington Univ., Aug. 1985.

[2] D. M. Dias and J. R. Jump, "Packet switching interconnection networks for modular systems," *Computer,* vol. 14, no. 12, pp. 43–53, Dec. 1981.

[3] D. M. Dias and J. R. Jump, "Analysis and simulation of buffered delta networks," *IEEE Trans. Comput.,* vol. C-30, pp. 273–282, Apr. 1981.

[4] D. M. Dias and Manoj Kumar, "Packet switching in $N \log N$ multistage networks," in *Proc. IEEE GLOBECOM '84,* Dec. 1984, pp. 114–120.

[5] T.-Y. Feng, "A survey of interconnection networks," *Computer,* vol. 14, no. 12, pp. 12–30, Dec. 1983.

[6] M. A. Franklin, "VLSI performance comparison of banyan and crossbar communications networks," *IEEE Trans. Comput.,* vol. C-30, pp. 283–290, Apr. 1981.

[7] L. R. Goke and G. J. Lipovski, "Banyan networks for partitioning multiprocessor systems," in *Proc. 6th Annu. Symp. Comput. Architect.,* Apr. 1979, pp. 182–187.

[8] Y. C. Jenq, "Performance analysis of a packet switch based on a single-buffered banyan network," *IEEE J. Select. Areas Commun.,* vol. SAC-1, no. 6, pp. 1014–1021, Dec. 1983.

[9] M. J. Karol, M. G. Hluchyj, and S. P. Morgan, "Input vs. output queueing in a space division packet switch," in *Proc. IEEE GLOBECOM '86,* Dec. 1986, pp. 659–665.

[10] P. Kermani and L. Kleinrock, "Virtual cut-through: A new computer communication switching technique," *Comput. Networks,* vol. 3, pp. 267–286, 1979.

[11] C. P. Kruskal and M. Snir, "The performance of multistage interconnection networks for multiprocessors," *IEEE Trans. Comput.,* vol. C-32, pp. 1091–1098, Dec. 1983.

[12] J. K. Kulzer and W. A. Montgomery, "Statistical switching architectures for future services," presented at Proc. Int. Switch. Symp., May 1984.

[13] J. H. Patel, "Performance of processor-memory interconnections for multiprocessors," *IEEE Trans. Comput.,* vol. C-30, pp. 771–780, Oct. 1981.

[14] Bjarne Stroustrup, *The C++ Programming Language.* New York: Addison-Wesley, 1986.

[15] J. S. Turner and L. F. Wyatt, "A packet network architecture for integrated services," in *Proc. IEEE GLOBECOM '83,* Nov. 1983, pp. 45–50.

[16] J. S. Turner, "Fast packet switching system," U.S. Pat. 4 491 945, Jan. 15, 1985.[2]

[17] ——, "Design of a broadcast packet switching network," *IEEE Trans. Commun.,* vol. 36, pp. 734–743, June 1988.

[18] ——, "Design of an integrated services packet network," *IEEE J. Select. Areas Commun.,* vol. SAC-4, pp. 1373–1380, Nov. 1986.

[19] J. S. Turner and L. F. Wyatt, "Alternate paths in a self-routing packet switching network," U.S. Pat. 4 550 397, Oct. 29, 1985.

[20] J. S. Turner, "Fluid flow loading analysis of packet switching networks," Washington Univ., Comput. Sci. Dep., Tech. Rep. WUCS-87-16.[3] Also, in Proc. Int. Teletraffic Congr., June 1988.

[21] J. E. Wirsching and T. Kishi, "Conet: A connection network model," *IEEE Trans. Comput.,* vol. C-30, Apr. 1981.

[2] Available upon request from the Commissioner of Patents and Trademarks, Washington, D.C.

[3] Available upon request from the Computer Science Department, Campus Box 1045, Washington University, St. Louis, MO 63130.

Performance Analysis of a Multicast Switch

Jeremiah F. Hayes, *Fellow, IEEE*, Richard Breault, *Member, IEEE*, and Mustafa K. Mehmet-Ali

Abstract—Optical fiber is suited to a wide range of services, several of which would be likely to have a multicast feature whereby information is broadcast to a subset of network nodes. A component of a multicast service would be a multicast switch, a space division switch in which a packet at an input port is routed to a subset of the output ports. In this paper, a study of the performance of a multicast switch is carried out. The primary objective is the determination of the delay under the assumption of input port buffering. The system is modeled as an independent set of $M/G/1$ queues. A key assumption in this analysis is that output port contention is settled by random selection among the contending input ports.

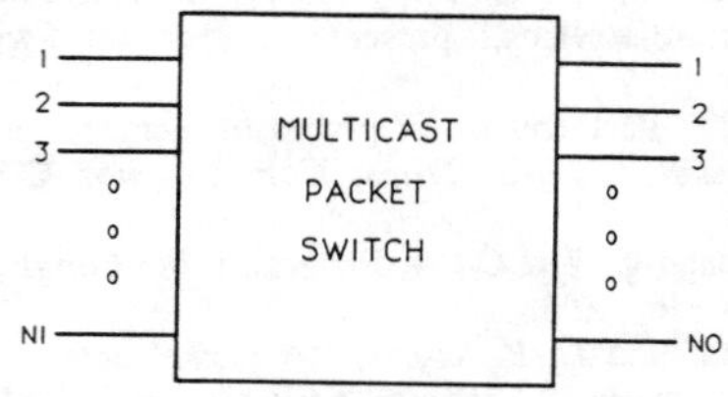

Fig. 1. Multicast packet switch.

I. INTRODUCTION

THE advent of optical fiber as a transmission medium has provided a many-fold increase in the bandwidth that is available. This increased bandwidth makes possible a wide range of services. Indeed, in order to make optical fiber economically viable, a wide range of services must be supported. At this writing, it appears that the range of services will be accommodated in a single transmission and switching format the asynchronous transfer mode (ATM) [1], [2] whereby information is encapsulated in fixed length packets.

As part of the services offered by optical networks, there would be a multicast feature, where the same information is delivered simultaneously to a subset of the network nodes. Multicasting would be appropriate to such applications as entertainment television, teleconferencing, LAN bridging, and distributed data processing. A necessary component of a multicast service is a space division switch which copies packets from a single input port to a selected number of output ports. The purpose of this paper is to present an analysis of the performance of such a multicast packet switch.

Recently, several implementations have been proposed for multicast switches [3]–[7]. The general form is that of a copy network, which replicates the input packet, followed by a non-blocking point-to-point switch which routes the copies to the output ports (see Fig. 1). It may be stretching things a bit, but the same form can be used to describe multicast in the knockout switch [8], in a general way. The model of the multicast switch that we shall analyze is sufficiently abstracted from the details of implementation, that Fig. 1 is sufficient for our purposes.

It is assumed that the switch has *NI* input ports and *NO* output ports. A packet at an input port generates copies for a subset of the output ports, with one copy for each output port. It is assumed that a fixed time interval, called a slot, is required to carry out the steps of copying and switching. Since each input port may be presumed to operate independently, it is possible that copies from different inputs may seek simultaneous access to the same output

port. A key feature of the model that we analyze is the resolution of such conflicts. We assume that, if copies of M different packets are contending for the same output port, one of these is chosen at random with probability $1/M$. This model is in conformity with the architecture proposed in Reference [4] when the, so called, unreserved[1] traffic is being handled. Essentially, the three-phase algorithm presented in [9] is extended to multicast switching.

It is assumed that there is buffering of packets arriving over the input line. Before a packet is processed, all of the copies generated by previous packets must have gained access to output ports. This seems an entirely reasonable way to control congestion within the switch.

Under the assumption of input port buffering, we may consider the residence time of a packet at an input port as its service time in a queueing system. We consider this service time under several different service disciplines. The discipline which we believe to have the greatest interest is that of *pure work conservation*. In this case, copies generated by the same input packet may gain access to output ports over any number of slots. Copies generated in a previous slot continue to contend for output ports until all of the copies generated by a packet have been transmitted. This discipline would give an indication of the best performance that can be realized. Correlations between copies from the same packet are ignored and an output line is used as long as there is one copy in the system destined for it. A contrasting discipline is one where all copies of the same packet must be switched in the same slot. If at least one copy loses the contention for an output port, all copies must be retransmitted. One could consider intermediate discipline for which all copies must be transmitted within a specified number of slots. Because it seems a reasonable mode of operation and because it furnished a bound, we focus upon the work conserving discipline, although, as we shall indicate, the others can be handled by our model as well.

We shall also focus on the so called *infinite storage case* where the load is such that the probability of overflow is negligibly small. As we shall see, our results can also be applied to a situation where only packets which are in service can be buffered. However, this latter case of blocked calls cleared is of lesser interest.

The technique that we use to analyze the performance of the system is standard. It is assumed that the arrival of packets to the input is described by a Poisson process. We model the *NI* inputs as a set of $M/G/1$ queues each of which are independent

Paper approved by the Editor for Optical Switching of the IEEE Communications Society. Manuscript received October 4, 1989; revised February 16, 1990. This work was supported by the Natural Sciences and Engineering Research Council. This work was begun while one of the authors spent a summer at Bellcore.

J. F. Hayes and M. K. Mehmet-Ali are with the Department of Electrical Engineering, Concordia University, Montrèal, PQ H3G 1M8, Canada.

R. Breault is with Bell-Northern Research, Ottawa, Ont., Canada.

IEEE Log Number 9143103.

[1] The case of reserved traffic is considered in a forthcoming paper [10].

Reprinted from *IEEE Trans. Commun.*, vol. 39, no. 4, pp. 581–587, April 1991.

of one another. As is usual in this approach, the basic task is to find the distribution of the service time of a packet. Because of the contention for an output port the service time at a particular input port is a function of the traffic at all of the other ports. In the course of carrying out the analysis, a number of assumptions are necessary, the independence of input queues, for example. In order to justify this as well as the other assumptions that have been made, a Monte–Carlo simulation of the system was devised.

II. CHARACTERIZATION OF INTERFERING TRAFFIC

We make the reasonable assumption that each input port generates copies independently with identical distributions, and that these copies are distributed in a random fashion over the output ports as described in the sequel. Let the number of copies generated by each incoming packet have the probability distribution $\Pr[i \text{ copies}] = P_i$. The probability generating function for the number of copies generated by an incoming packet is given by

$$X(z) = \sum_{i=1}^{\infty} z^i P_i. \tag{1}$$

Consider a packet entering service at a tagged input port. As stated above, the copies generated by this packet contend with copies generated at other input ports for access to output ports. However, when the tagged packet begins service the very same process may have been going on at each of the other input ports for varying periods of time. The number of interfering packets will not be those chosen by random selection from the probability distribution, P_i. There will be a bias toward the higher end of the distribution since these are more likely to be seen by a randomly arriving tagged packet. Accordingly, we may characterize this contending traffic by means of residual distributions. Given that an input port is active the probability distribution of the R residual number of copies encountered by a tagged message is given by [11]

$$Pr[R = j] = r_j = \frac{1}{\overline{X}} \sum_{i=j}^{\infty} P_i \tag{2a}$$

with the corresponding probability generating function

$$\Re(z/\text{Active}) = (z/\overline{X})\{[1 - X(z)]/(1 - z)\} \tag{2b}$$

where $\overline{X}$ is the average number of copies generated by an input port packet, $\overline{X} = X'(1)$. We assume that each of the input ports is active with probability ρ; consequently the probability generating function of the number of interfering copies generated by a single input port as encountered by our tagged packet is given by

$$\Re(z) = \rho(z/\overline{X})\{[1 - X(z)]/(1 - z)\} + 1 - \rho. \tag{3}$$

Equation (3) gives an expression for the pgf of the residual number of conflicting copies coming from one interfering input port. These copies were generated independently at each input port at the beginning of their service time, however in the process of contending for output ports, a certain coupling is created. Nevertheless we assume independence between input ports to keep the complexity of the analysis manageable. The pgf for the residual number of conflicting copies summed over all inputs is then given by

$$\Re_T(z) \triangleq \sum_{l=0}^{NO(NI-1)} Q_l z^l = [\Re(z)]^{NI-1}$$

$$= [\rho(z/\overline{X})\{[1 - X(z)]/(1 - z)\} + 1 - \rho]^{NI-1} \tag{4}$$

where the Q_l is the probability of a total of l residual copies from all sources. $Q_l = \Pr[R_T = l]$

The residual copies are distributed among the output ports in a random fashion. In order to simplify our analysis, it is assumed that this distribution follows the multinomial conditioned on R_T the total number of residual copies. We recognize that there is a nonzero probability that two or more copies from the same input port are generated for the same output port. However, we assume that probability of this event is negligible, particularly for systems of interest, i.e., large numbers of input-output ports. The joint probability distribution of the number of copies on the output ports is given by $\Pr[I_1 = i_1, I_2 = i_2, \cdots, I_{NO} = i_{NO}/R_T = l]$

$$l! \frac{(1/NO)^{i_1}}{i_1!} \frac{(1/NO)^{i_2}}{i_2!} \cdots \frac{(1/NO)^{i_{NO}}}{i_{NO}!}. \tag{5}$$

From this it is straightforward to show that the joint probability generation function conditioned on R_T is given by

$$T(z_1, z_2, \cdots, z_{NO}|R_T) \triangleq E[z_1^{i_1} z_2^{i_2} \cdots z_{NO}^{i_{NO}}/R_T]$$

$$= \left[\sum_{j=1}^{NO} z_j \bigg/ NO\right]^{R_T}. \tag{6}$$

The conditioning on R_T can be removed by averaging over the probability distribution of the residual number of copies. We have

$$T(z_1, z_2, \cdots, z_{NO}) = \sum_{l=0}^{NO[NI-1]} \left[\sum_{j=1}^{NO} z_j \bigg/ NO\right]^l Q_l$$

$$= \Re_T\left(\sum_{j=1}^{NO} z_j \bigg/ NO\right) \tag{7}$$

where $\Re_T(z)$ is given in (4).

III. CALCULATION OF SERVICE DISTRIBUTION

With the joint distribution of the number of interfering copies in hand, we begin our analysis of message delay by focusing on a particular input port. We condition on the event that a packet at this port has generated K copies. As we mentioned above, each of these packets will contend, at the output ports with packets generated by the other $NI - 1$ input ports. We assume that copies are chosen for transmission over the output port in a random fashion and independently from port to port. We begin the analysis by considering the event that a particular set of L copies from the designated input port are chosen. Since the traffic is symmetric, we can choose this set to be the first L, in an arbitrary ordering of the output ports chosen, for example. The probability of this event is given by

$$\Pr(\text{first } L \text{ or more} \,|\, i_1, \cdots, i_L, K) = \prod_{j=1}^{L} 1/(i_j + 1). \tag{8}$$

By averaging over $i_1, i_2, \cdots, i_L$, we find

$$\Pr(\text{first } L \text{ or more} \,|\, K) =$$

$$\sum_{i_1} \sum_{i_2} \cdots \sum_{i_L} \prod_{j=1}^{L} 1/(i_j + 1) \, \text{Prob}(i_1, i_2, \cdots, i_L); \quad L \le K. \tag{9}$$

Notice that the conditioning on K is manifested only through the condition $L \leq K$. Recall that the joint probability generating function of the interfering copies is defined as

$$T(z_1, z_2, \cdots, z_{NO}) \triangleq$$
$$\sum_{i_1} \sum_{i_2} \cdots \sum_{i_{NO}} \prod_{j=1}^{NO} z_j^{i_j} \text{Prob}(i_1, i_2, \cdots, i_{NO}) \quad (10)$$

If we set $z_{L+1} = z_{L+2} = \cdots = z_{NO} = 1$ and we integrate the expression over $z_1, z_2, \cdots, z_L$ we obtain

$$\int_0^1 dz_1 \int_0^1 dz_2 \cdots \int_0^1 dz_L T(z_1, z_2, \cdots, z_L, 1, \cdots, 1) =$$
$$\sum_{i_1} \sum_{i_2} \cdots \sum_{i_L} \prod_{j=1}^{L} z_j^{i_j+1}/(i_j + 1) \text{Prob}(i_1, i_2, \cdots, i_L)|_0^1.$$
$$(11)$$

Thus, from (10) and (11), we find that

Prob(first L or more $| K$) =

$$\int_0^1 dz_1 \int_0^1 dz_2 \cdots \int_0^1 dz_L T(z_1, z_z, \cdots, z_L). \quad (12)$$

At this point we have found the probability that the copies from our particular input port have won contentions at least at the first L output ports of the K chosen by the packet. From the symmetry of the model, this probability is the same for any other set of L or more output ports ($L \leq K$). Thus, (12) gives the probability that any particular set of L *or more* of the K copies have been chosen at output ports in a single trial. The event that is of immediate interest is the probability that *only* the set of L copies gets through. We now proceed to derive the probability of this last event in terms of known quantities.

Let M_L indicate the event that a specific set of L or more of the K generated copies get through to output ports in a slot time. The disposition of the $K-L$ "extra" copies is arbitrary. The event M_L can be expressed as the union of disjoint events, O_{L+j}, which indicates the event that exactly $L + j$ copies get through, including the specified set of L. There are $\binom{K-L}{j}$ such disjoint sets corresponding to the number of ways that the j "extra" copies can be chosen from $K-L$. Further, j may range from 0 to $K-L$. From these considerations, we may write

$$\text{Pr}(M_L | K) = \sum_{j=0}^{K-L} \binom{K-L}{j} \text{Pr}(O_{L+j} | K),$$
$$L = 0, 1, \cdots, K \quad (13)$$

where $\text{Pr}(M_L/K)$ and $\text{Pr}(O_{L+j}/K)$ are the probabilities of the events M_L and O_{L+j}, respectively, conditioned on K. We take advantage of the fact the $\text{Pr}[O_{L+j}/K]$ is the same for all copies. By inverting the set of equations in (13), we obtain the probability of any particular set of L copies getting through.

$$\text{Pr}(O_L | K) = \sum_{j=0}^{K-L} \binom{K-L}{j} (-1)^j \text{Pr}(M_{L+j} | K);$$
$$L = 0, 1, \cdots, K. \quad (14)$$

Up to this point we have been dealing exclusively with a particular set of L output ports chosen from the K output ports chosen by our designated packet. We can find the probability

for any of the possible sets of L of K copies. Let us define this probability to be $P(L | K)$. Since there are $\binom{K}{L}$ possible combinations of L copies on a total of K copies, we get the following expression:

$$P(L | K) = \binom{K}{L} P(O_L | K); \quad L = 0, 1, \cdots, K. \quad (15)$$

There are several possible service disciplines, each of which entail a different distribution of service time. However, each of these service time distributions involve the probability given in (15). As stated above, we are most interested in the pure work conserving discipline, which allows copies to be transmitted in an arbitrary number of intervals. If l copies get through in the first slot, one attempts to get the remaining $K - l$ copies through in the next slot. The process continues until all of the copies have been placed on output ports. The service time is the number of slots that are required to do this. Let $Q(n/K)$ denote the probability that n slots are required to transmit the K copies generated by a packet. The probability that everything can be done in one shot is

$$Q(1/K) = P(K/K). \quad (16a)$$

The event of all of the copies getting through in n slots is the sum of $K + 1$ disjoint events according to the number of copies that get through on the first slot.

$$Q(n/K) = \sum_{l=o}^{K} P(l/K)Q(n - 1/K - l). \quad (16b)$$

In (16b) we assume independence from slot to slot. If there are a large number of input and output ports, we feel that this assumption is reasonable. From equations (16a–b) the probability distribution of the service time in terms of slots can be found.

A service discipline which is the antithesis of work preserving is one in which all of the copies of a packet must get through in the same slot. If not all of the copies can get through in a slot, one tries again for all copies in the next slot. The previously delivered incomplete set is discarded. The probability of completing service in one trial is simply $P(K/K)$. Again, we assume independence from slot to slot and the probability that n slots are required is a geometric distribution given by

$$Q(n/K) = P(K/K)(1 - P(K/K))^{n-1}; \quad n = 1, 2, \cdots. \quad (17)$$

Intermediate between these approaches is one where all of the copies must be transmitted in a specified number of slots otherwise one must start over. For example consider the case where all K copies must be transmitted in two slot times. As in the previous cases

$$Q(1/K) = P(K/K) \quad (18a)$$

and

$$Q(2/K) = \sum_{l=o}^{K} P(l/K)P(K - l/K - l) \quad (18b)$$

If all copies don't get through in these two slots, the process begins again. The probability distribution of the number of slots that are required to get all copies through is given by

$$Q(2n/K) = (1 - Q(2/K))^{n-1}Q(2/K) \quad (19a)$$

and

$$Q(2n + 1/K) = (1 - Q(2/K))^n Q(1/K). \qquad (19b)$$

Once again independence from slot to slot is assumed in performing these calculations.

In all of these service disciplines we have conditioned on the value of K. The complete service distribution is found by averaging over K.

$$Q(n) = \sum_{k=o}^{K} \Pr(K = k) Q(n/K). \qquad (20)$$

The average service time is given by

$$\overline{S} = \sum_{n=o}^{\infty} n Q(n). \qquad (21)$$

The derivation of the service time distribution assumes a particular value of ρ, the probability that a station is occupied [see (2)]. If we assume completely symmetric traffic, this can be related to the average service time. The case of most interest is that of an $M/G/1$ queue with an infinite buffer. Packets that come to a busy server to be stored until the server is available. In this case, from the complete symmetry of the traffic, the probability that a server is occupied is $\rho = \lambda \overline{S}$. Thus, the packet arrival rate for a particular value of ρ is $\lambda = \rho/\overline{S}$. Although it is not noted explicitly, the average service time $\overline{S}$, also depends on ρ [see (2)]. The average delay is given by the Pollacek–Khinchine formula

$$\overline{D} = \overline{S} + \lambda \overline{S^2}/(2(1 - \rho)). \qquad (22)$$

Higher order moments of delay and buffer occupancy can be found by straightforward computation.

In the forgoing we have focused on the case of an infinite buffer. The analysis may also be applied to the case of blocked calls cleared. In this case, the probability of an input port being occupied is

$$\rho = \lambda \overline{S}/(1 - \lambda \overline{S}). \qquad (23)$$

Solving for λ, we find

$$\lambda = \rho/(\overline{S}(1 - \rho)). \qquad (24)$$

Assuming a particular value of ρ, λ can be calculated. The probability of a message being blocked is given by substituting into the Erlang B formula for the value of ρ.

IV. NUMERICAL RESULTS

In order to test the accuracy of the analysis, the results of the analysis were compared to a simulation of the system for the case of most interest—the work conserving discipline and the $M/G/1$ queueing model with infinite buffers. The distribution of the copy generation mechanism is the modified binomial distribution. This distribution can be described as Bernoulli trials on each of the output ports in which the probability of generating a packet on each trial is P. The modification is that the result that no packets are generated is not allowed. The probability distribution is

$$\Pr(j \text{ Copies}) = \binom{NO}{j} P^j (1 - P)^{NO-j} \Big/ \Big[1 - (1 - P)^{NO} \Big]. \qquad (25a)$$

The probability generating function of the modified binomial distribution is given by

$$X(z) = \Big\{ [1 - P + Pz]^{NO} - (1 - P)^{NO} \Big\} \Big/ \Big[1 - (1 - P)^{NO} \Big]. \qquad (25b)$$

This particular distribution was chosen because it provides a way of generating multiple copies which is reasonably mathematically tractable and is easy to implement in a simulation. In any case our focus is upon the accuracy of the analysis. There is no reason to believe that this distribution would lead to results which were more or less accurate than alternative distributions.

The expression in (25b) could be substituted directly into equation (3) for the probability generation function of the residual number of interfering copies. Recall that this expression is used in the calculation of delay [see (4) and (7) and (12)]. Numerical integration can be used to evaluate (12). However, we found an alternative approach to be computationally simpler. If one conditions on the residual number of packets in the system R_T and uses the integrand $T(z_1, z_2, \cdots, z_L / R_T)$ in (12), the integral has a closed form solution. Steps of the analysis are given in the appendix. The solution gives the probability of L or more copies getting through conditioned on K, the number of copies generated by a tagged packet and on R_T. The probability distribution of R_T is computed numerically by an $NI - 1$-fold convolution as indicated in (4) where $X(z)$ is given by (25B). The conditioning on R_T is removed simply by averaging numerically over the probability distribution, given in (A.7).

The analytical expressions were evaluated by a program in C using double precision variables with range 0.56×10^{-308} to 899×10^{308} with precision of 15 decimal digits. Underflow problem occurring in evaluating (A.7) prevented us from doing calculations for large switch sizes. The results of the computation and the simulation are shown in Figs. 2–6 for various switch configurations. The average packet delay is shown as a function of the packet arrival rate for different values of P the probability of a copy being generated at an output port. The labeling on the figures indicates the analytical results as MR and the simulation results as S. The number and lengths of the simulation runs are such that the results are accurate to within 10% with 95% confidence. The numerical calculations required double precision accuracy particularly because of the alternating signs in (14).

The comparison of analysis and simulation shows that the correspondence is quite reasonable. The correspondence improves as the number of input and output ports increases. This is not unexpected since several of the approximations were predicated on a large number of stations. We point out that even for the 4×4 switch the results are reasonably close, the message arrival rate where the knee of the curve is located is predicted. It is noteworthy that, in all of the configurations, the analytical and simulation results are very close in the region below the knees of the curves where the system would be expected to operate. Pressures of time did not allow us to perform a comparison of analysis and simulation for alternative configurations; however one would expect similar accuracy since the assumptions are the same.

We observe also, for the limited number of cases that are involved, that the expansion switches for which $NO > NI$ yield particularly good results. This sort of switch is likely to be used in multicast systems.

It is interesting to compare the total throughput for different cases. In making this comparison one is mindful of the 0.586 bound which has been derived for the unicast case [12]. It is

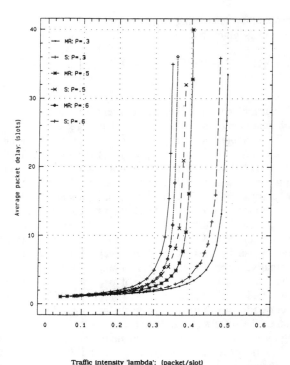

Fig. 2.   Analysis/simulation results delay for a 4 × 4 switch.

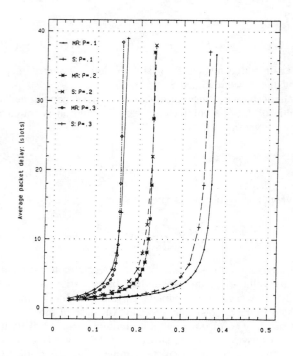

Fig. 4.   Analysis/simulation results delay for a 16 × 16 switch.

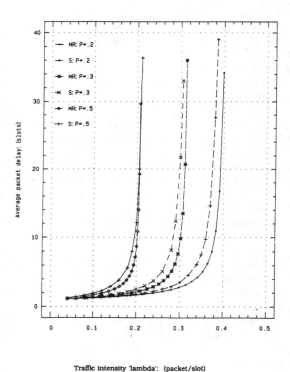

Fig. 3.   Analysis/simulation results delay for an 8 × 8 switch.

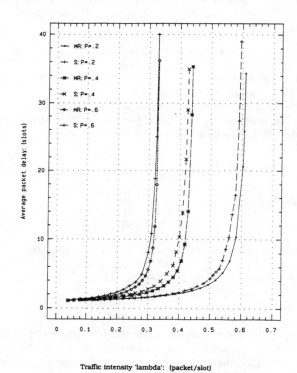

Fig. 5.   Analysis/simulation results delay for a 4 × 8 switch.

difficult to achieve as concrete a result in the present case but our results do exhibit a pattern. The switches with the same number of input and output ports saturation ranges between 0.6 and 0.85, approximately. In general, the higher throughput obtains when more copies are generated per input packet. For example, for eight ports and $P = 0.2$, implying 1.6 copies per packet, the maximum throughput is 0.64; whereas, for $P = 0.5$, or 4 copies per packet, the throughput is 0.85. One can regard the unicast case as a lower bound. For the case where there are more output ports than input ports, the patterns that emerge are higher throughput with the throughput increasing as the number of copies per packet increase.

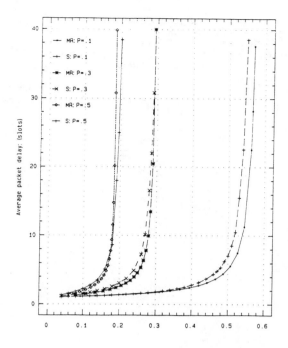

Fig. 6. Analysis/simulation results delay for an 8×16 switch.

Appendix
Integration of the pgf of the Multinomial Distribution

We want to integrate the following expression:

$$\Pr(\text{first } L \text{ or more} \mid R_T, K) =$$

$$\int_0^1 dz_1 \int_0^1 dz_2 \cdots \int_0^1 dz_L \left[\sum_{j=1}^L \frac{1}{NO} z_j + \left(1 - \frac{L}{NO}\right) \right]^{R_T}. \quad \text{(A.1)}$$

By first integrating over z_L from 0 to 1 we get

$$\int_0^1 \left[\sum_{j=1}^L \frac{1}{NO} z_j + \left(1 - \frac{L}{NO}\right) \right]^{R_T} dz_L$$

$$= \frac{NO}{(R_T + 1)} \left[\sum_{j=1}^L \frac{1}{NO} z_j + \left(1 - \frac{L}{NO}\right) \right]^{R_T+1} \Bigg|_0^1 \quad \text{(A.2)}$$

$$= \frac{NO}{(R_T + 1)} \left\{ \left[\sum_{j=1}^{L-1} \frac{1}{NO} z_j + \left(1 - \frac{L-1}{NO}\right) \right]^{R_T+1} - \left[\sum_{j=1}^{L-1} \frac{1}{NO} z_j + \left(1 - \frac{L}{NO}\right) \right]^{R_T+1} \right\}. \quad \text{(A.3)}$$

We can repeat the previous step for z_{L-1}

$$\frac{NO}{(R_T+1)} \int_0^1 \left\{ \left[\sum_{j=1}^{L-1} \frac{1}{NO} z_j + \left(1 - \frac{L-1}{NO}\right) \right]^{R_T+1} - \left[\sum_{j=1}^{L-1} \frac{1}{NO} z_j + \left(1 - \frac{L}{NO}\right) \right]^{R_T+1} \right\} dz_{L-1}$$

$$= \frac{NO^2}{(R_T+1)(R_T+2)} \left\{ \left[\sum_{j=1}^{L-2} \frac{1}{NO} z_j + \left(1 - \frac{L-2}{NO}\right) \right]^{R_T+2} - 2 \left[\sum_{j=1}^{L-2} \frac{1}{NO} z_j + \left(1 - \frac{L-1}{NO}\right) \right]^{R_T+2} + \left[\sum_{j=1}^{L-2} \frac{1}{NO} z_j + \left(1 - \frac{L}{NO}\right) \right]^{R_T+2} \right\}. \quad \text{(A.4)}$$

After integrating over z_{L-2}, it is evident that the resulting term is

$$\frac{NO^3}{(R_T+1)(R_T+2)(R_T+3)}$$

$$\cdot \left\{ \left[\sum_{j=1}^{L-3} \frac{1}{NO} z_j + \left(1 - \frac{L-3}{NO}\right) \right]^{R_T+3} - 3 \left[\sum_{j=1}^{L-3} \frac{1}{NO} z_j + \left(1 - \frac{L-2}{NO}\right) \right]^{R_T+3} + 3 \left[\sum_{j=1}^{L-3} \frac{1}{NO} z_j + \left(1 - \frac{L-1}{NO}\right) \right]^{R_T+3} + \left[\sum_{j=1}^{L-3} \frac{1}{NO} z_j + \left(1 - \frac{L}{NO}\right) \right]^{R_T+3} \right\}. \quad \text{(A.5)}$$

After $n < L$ steps we have

$$\frac{(NO)^n}{\prod_{j=1}^n R_T + j} \sum_{i=0}^n \binom{n}{i} (-1)^i$$

$$\cdot \left[\sum_{j=1}^{L-n} \frac{1}{NO} z_j + \left(1 - \frac{L-n+i}{NO}\right) \right]^{R_T+n} \quad \text{(A.6)}$$

Finally, after all of the integrations have been completed we have

$$\text{Prob(first } L \text{ or more} \mid R_T, K) =$$

$$\frac{(NO)^L}{\prod_{j=1}^L R_T + j} \sum_{i=0}^L \binom{L}{i} (-1)^i [1 - i/NO]^{R_T+L};$$

$$L \leq K. \quad \text{(A.7)}$$

VI. Acknowledgment

The assistance of E. Arthurs and T. T. Lee is gratefully acknowledged.

References

[1] B. Schaffer, "Synchronous and asynchronous transfer modes in the future broadband ISDN," in *Proc. IEEE GLOBECOM '88*, Fort Lauderdale, FL, Nov. 1988, pp. 1552–1558.
[2] B. Eklundh, I. Gard, and G. Leijonhufoud, "A layered architecture for ATM networks," in *Proc. IEEE GLOBECOM '88*, Fort Lauderdale, FL, Nov. 1988, pp. 409–414.
[3] T. T. Lee, "Nonblocking copy networks for multicast packet switching," *IEEE J. Select. Areas Commun.*, vol. 6, pp. 1455–1467, Dec. 1988.
[4] T.T. Lee, R. Boorstyn, and E. Arthurs, "The architecture of a multicast broadband packet switch," in *Proc. IEEE INFOCOM '88*, New Orleans, LA, pp. 1–8.

[5] J. S. Turner, "Design of a broadcast packet switching network," in *Proc. IEEE INFOCOM '86,* pp. 667–675.

[6] ——, "Design of an integrated service network," *IEEE J. Select. Areas Commun.,* vol. SAC-4, Nov. 1986.

[7] A. Huang and S. Knauer, "Starlite: A wideband digital switch," in *Proc. IEEE GLOBECOM '84,* pp. 121–125.

[8] K. Y. Eng, M. G. Hluchyi, and Y. S. Yeh, "Multicast and broadcast services in a knockout packet switch," in *Proc. IEEE INFOCOM '88,* New Orleans, LA, pp. 29–34.

[9] J. Y. Hui and E. Arthurs, "A broadband packet switch for in-tegrated transport." *IEEE J. Select. Areas Commun.,* vol. SAC-5, pp. 1264–1273, Oct. 1987.

[10] C. K. Kim and T. T. Lee, "Call scheduling algorithms in a multicast switch," *IEEE Trans. Commun.,* to be published.

[11] H. M. Taylor and S. Karlin, *An Introduction to Stochastic Modeling.* New York: Academic.

[12] J. Y. Hui and E. Arthurs, "A broadband packet switch for in-tegrated transport," *IEEE J. Select. Areas Commun.,* vol. SAC-5, pp. 1264–1273, Oct. 1987.

FIRST STAGE MULTICASTING IN A GROWABLE PACKET (ATM) SWITCH

Daniel J. Marchok   Charles E. Rohrs

Tellabs Research Center
3702 N. Main Street, Building 2
Mishawaka, IN 46545

ABSTRACT

A three stage packet switch architecture that uses a two stage, self-routing, memoryless, interconnection network to interconnect a third stage of smaller packet switches has been proposed by Eng, Karol, and Yeh. This paper proposes two new routing algorithms that facilitate the multicasting of packets with this architecture. The two routing algorithms, one of which duplicates packets at only the first stage and the other which duplicates packets at the first and third stages, require only minimal alterations to the switch. Having presented the new routing algorithms, the paper analyzes the switch's multicast performance, finding an overbound on the packet loss probability. The analysis, which identifies three sources of packet loss, shows that good multicast performance is maintained with only a small increase in the architecture's expansion, provided that, on the average, the load is not severely concentrated onto any particular input section as a result of multicasting.

1.   Introduction

Many packet switches have been developed for switching ATM cells in the Broadband-ISDN environment [1-9]; however, few of these scale to provide a reasonable solution for switches with a 1000 or more lines. In [10-12], Eng, Karol, I, and Yeh propose a BISDN switch that can be grown from small modules to sizes up to 2048x2048. This growable packet switch (GPS) architecture has three stages, but unlike many of other solutions, the multiple stages do not cause a multiplication of the delay and packet loss. The first two stages form a memoryless, self-routing, interconnection network, while the last stage is comprised of small ATM packet switches of any type.   The interconnection network is used to route packets from an input line to the appropriate output section. The first two stages contribute only one or two cell times to the total delay; thus, the majority of the delay is incurred in the last stage.

Many of the BISDN services listed in the BISDN Baseline Document [13] require multipoint and broadcast communications. These services may be provided with BISDN switches that are capable of multicasting incoming packets to multiple outputs. This paper focuses on adding multicasting capabilities to the GPS architecture and analyzing the resulting performance. The method used provides multicast capability to the GPS with only minimal changes to the GPS architecture and routing algorithm.

The paper is organized as follows: Section 2 reviews the function of the GPS as presented in [10-12]. Section 3 describes two methods for expanding the functionality of the GPS architecture to include multicasting. Section 4 analyzes the cell loss probability of a multicasting GPS (Section 4.1), and investigates the effect of various switch and load parameters on the performance (Section 4.2). The conclusions are in Section 5.

2.   Growable Packet (ATM) Switch Overview

The GPS presented in [10-12] is a Clos like architecture and is shown in Figure 2.1. While the GPS architecture can operate on asynchronous variable length packets, the focus here is on fixed length packets or cells that arrive synchronously. The assumption of synchronous packet arrival is common in ATM switch discussions since synchronization may be treated as a separate problem. The GPS consists of three stages. The last stage is made up of small self-routing packet switches called output modules. A subset of the GPS's out going lines are attached to each of the packet switch modules. The output modules can be any type of ATM packet switch. The function of the first two stages is to route packets to the appropriate output module.   These two stages form a self-routing, memoryless, interconnection network. The network is self-routing in that no central controller or routing table is required to determine a packet's route; although, the network does use a distributed process to determine packet routes for each time-slot. The first two stages are memoryless in that no packets are left in the interconnection network from one time-slot to the next. If a packet cannot be routed in a time-slot, it is discarded. Hence, some loss occurs in the interconnection network. By increasing m, the number of sections in the second stage, the packet loss rate can be made arbitrary small.

Loss in the interconnection network has two sources, knockout loss and routing loss. Knockout loss occurs when, in a given time-slot, more than m packets arrive for a single output packet switch module. Since each output module has only m inputs, only m packets can be successfully routed to a given output module in any time-slot. Routing loss occurs when the routing algorithm fails to find a route through the interconnection network, even though the output module still has a free input. In [11], knockout loss is shown to be the more significant of these two losses.

One implementation of the GPS delays each packet one time-slot at the input. This gives the interconnection network one time-slot during which to determine a route for a packet. While the routings are being determined, the packets that arrived during the preceding time-slot are sent through the interconnection network.

Each input line is attached to a particular first stage or input section. Likewise, each output line is attached to a particular output module. Since a packet's input and output lines are defined at call setup, the packet's input and output sections are also defined. Therefore, the process of routing a packet through the GPS consists of choosing a route from a particular input section to a particular output section. This means that choosing a packet's route through the GPS amounts to choosing the second stage section through which the packet will pass. Any second stage section that meets the following two criteria may be chosen. First, the section must have an unused link to the packet's input section. Second, the second stage section must have a free link to the packet's output section. In the GPS, the first stage performs the routing function.

Reprinted from *Proc. IEEE International Conference on Communications '91*, pp. 1007–1013, June 1991.

In order to facilitate packet routing, Eng, Karol, *et al.* use two sets of link usage vectors, A and B. These vectors are kept by the first stage. The vectors in set A specify which of the links between the first and second stage have been assigned to carry a packet during the next time-slot. Each vector represents the link usage from a single first stage section to the second stage. Each input section stores the A vector that corresponds to its links. The second set of vectors, set B, specifies which links between the second and third stages have been assigned to carry a packet during the next time-slot. Each B vector represents the link usage between all of the center sections and a single output section. These vectors are also stored in the first stage. However, they are shared by all of the stage's sections. Each first stage section can communicate with two other first stage sections. As part of the routing process, each first stage section receives B vectors from one section and pass them on to the next section. The large arrows in Figure 2.1 show the interconnections between first stage sections and the direction the vectors are passed.

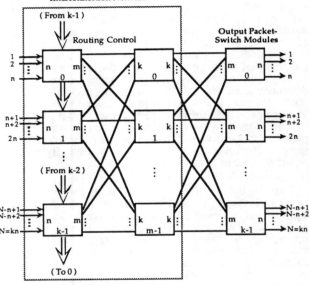

Figure 2.1. The Growable Packet (ATM) Switch as proposed in [10-12].

Figure 2.2 is of an A and a B vector. Vector A_i is kept by input section i. The elements of vector A_i correspond to section i's outgoing links. These links are numbered according to the second stage sections to which they are connected. If a particular outgoing link has already been assigned to carry a packet during the time-slot in question, it is marked with an X. If a link has not been assigned a packet, the corresponding element is left blank. Note in Figure 2.2 vector A_i, links 1, 2, and $m-1$ have been assigned packets. The elements of vector B_j correspond to the links from the second stage to output module j. Again, locations marked with an X correspond to links that have been assigned a packet for the time-slot in question, and blank elements correspond to links which have yet to be assigned.

A packet, which is to go from input section i to output section j, can be assigned a particular second stage section, l, if the lth element of both vector A_i and B_j are blank. Two matches are marked in Figure 2.2. This implies that a packet from input section i to output section j could be assigned a route through either second stage section 3 or m.

The GPS can simultaneously chooses routes for multiple packets; however, to route all the packets, k sequential steps are necessary. Thus, a single time-slot is broken into k mini-slots, where k is the number of output modules. It is assumed that the number of input sections equals the number of output sections. During the routing, each input section always has its own A vector plus one of the B vectors. The B vectors are passed from section to section at the end of each mini-slot. During each mini-slot, each input section determines a route for all of its packets that are destined for the output module corresponding to the B vector it has. The section then marks the A and B vector entries corresponding to the link usages it has assigned. After k mini-slots, all first stage sections have had all the B vectors; thus, the routing is complete.

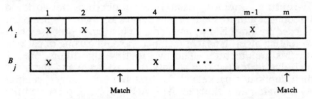

Figure 2.2. Unused routes between input section i and output section j are found by looking for matching blank elements in vectors A_i and B_j. The indices of matching blank elements correspond to second stage sections through which a packet may be assigned a route. (X=already assigned link, Blank=unassigned link).

Not specified in [10-12] is how the second stage sections are informed of the routing decision. One possible method is for the first stage sections to prepend a header to each packet. The header specifies the output module to which the packet is to be routed. With this approach, the second stage sections need to be smart enough to route a packet based on its header.

Another possible approach slightly more complex GPS architecture. In this approach the B vectors are stored in the second stage, with an element of each vector in each section. Each section has the element of each B vector that corresponds to one of the links from it to the last stage. Again during each mini-slot, each first stage section determines its routing to a different output module. However, during these mini-slots, the second stage sections communicate to the first stage. They indicate to the input sections whether or not the link that connects them to the output module of interest has already been reserved by another input section. The input section then lets the second stage section know whether or not it wants to reserve the link. The second stage then knows which input sections are going to send packets and to which output modules they should be routed.

Papers [10-12] show that with a reasonable number of second stage sections, the GPS can achieve a loss probability on the order of 10^{-8} or better. However, the papers do not address the issue of how packets are multicast.

3. Multicasting

The simplest approach for multicasting within the GPS architecture is to perform all of the packet duplication in the first stage. Method (a) shown in Fig. 3.1a performs multicasting via this method. The switch's performance may be improved by reducing the amount of packet duplication for which the first stage is responsible. Method (b) shown in Fig. 3.1b relieves the first stage of some of its responsibility by shifting as much of the packet duplication as possible to the third stage. This is possible because the third stage is made up of packet switches.

3.1. Method (a)

Method (a) performs all the multicasting at the first stage and has a routing process that is very similar to that used in a non-multicast GPS. The only significant difference is that the input sections must look-up the packets' destinations before doing the routing. Once the look-up is complete, the packets are routed just as non-multicast packets are routed. The routing process is equivalent to having more input lines per input section. Thus, each input section simply has more packets to route. This is the simpler of the two multicasting methods.

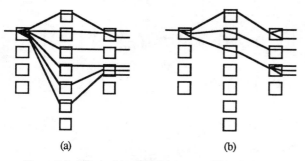

(a)   (b)

Figure 3.1. Figures (a) and (b) show two different methods for multicasting within a GPS .

3.2. Method (b)

Method (b) multicasts at the first and third stages, but does no second stage multicasting. Since each output section has access to only a limited number of output lines, the amount of multicasting it can perform is limited. The remaining multicasting must be performed by the first stage. The link reservation process is similar to that used in a non-multicasting GPS. Two differences are that the third stage must be able to route or multicast the packets it receives and that a first-stage section needs to reserve more than one link for some of its packets.

Two suggested methods for dealing with the third stage multicasting follow: Method one uses, for the third stage sections, packet switches that can determine to which outputs it should multicast the packets it receives. In method two, the first-stage sections determine what multicasting needs to be done and transmit the information to the appropriate output sections. An approach for transmitting the necessary information from the first to the third stage is to store it in a bit vector and to append the bit vector to the packet's header. The bit vector should have one location for each of the n outputs on an output section. If an output is one of the packet's destinations, the corresponding bit in the bit vector is set. These output stage multicast bit vectors are kept in the first stage and appended to the packet routing headers by the first stage. This keeps the complexity in the first stage, which already needs to be fairly intelligent to handle the link reservation and multicast routing.

4. Loss Probability for First Stage Multicasting
4.1. Loss Analysis for First Stage Multicasting

This section analyzes the blocking probability for method (b), the better performing of the two routing methods. The analysis of method (a) can be performed similarly.

Since method (b) routing is similar to the routing used in [10-12], parts of the probability of loss analysis are also similar. As defined in [11], r_i is the number of packets already routed out of input section I_i, and s_k is the number of packets already routed to output section C_k. Both r_i and s_k are overbounded by using for their value the number of packets requesting routing, as oppose to the number successfully routed. Since these bounds are very tight when the packet loss probability is kept low, the bounds will be used for r_i and s_k without further mention. The variable x is defined to be the number of packets that have already requested routing from I_i to C_k. With these variables defined, a general probability of loss equation for the last packet

to be routed may be written as shown in Eq. 4.1. Since the last packet to be routed is the most likely to be blocked, its loss probability may be used as an overbound of the average packet loss probability.

$$P_{loss} = \sum_{s} \sum_{x} P_x \sum_{r} P_{r|x} \, P_{s|x,r} \, P_{loss|s,r,x} \qquad \text{Eq. 4.1}$$

In the analysis done in [11], the blocking probability for the last packet out of an input section was overbounded by setting $r_i = n\text{-}1$. Since an input section has only n inputs, if no packets are multicast, there can be no more then $n\text{-}1$ packets already routed when attempting to route a packet.

On the other hand, when packets are duplicated at the first and third stages, the maximum number of packets that may request routing out of an input section is $n{\cdot}N/n=N$: where n is the number of inputs per input section, N/n is the number of output sections, and N is the total number of output lines attached to the switch. Implied in this statement is that N is divisible by n. With one packet left to be routed, a maximum of $N\text{-}1$ other packets may have needed to be routed from the input section. Hence, $r_{i\ max}$ equals $N\text{-}1$. However, since there are only m middle sections by which packets may be routed, an r_i greater than $m\text{-}1$ guarantees blocking. Since N is typically greater than m, setting r_i to $r_{i\ max}$ results in a probability of loss equal to one. Subsequently, the calculation of the loss probability needs to take into account the distribution of r_i, $p_r(r_i)$. In order to determine $p_r(r_i)$, it is necessary to make some assumptions about the multicast arrival process.

In defining $p_r(r_i)$, the use of third-stage multicasting makes it possible to look only at the output sections that a packet needs to reach, and not at the individual outputs. Let packets arrive with the following characteristics. For a particular input, assume that a packet arrives for a particular output section with probability ρ. Further assume that the same input has the same probability ρ for each output section, and that the probability of an input having an arrival for one output section is independent of whether or not it has an arrival for any other output section. When an input has an arrival for more than one output section, it is assumed that a single multicast packet has arrived. Under these assumptions, the number of output sections, d, that a multicast must reach, i.e. the multicast fan-out, is distributed binomially:

$$p_d(d) = \binom{N/n}{d} \rho^d \left(1 - \rho\right)^{N/n - d} \qquad 0 \le d \le N/n \qquad \text{Eq. 4.2}$$

Notice that the probability of the input being idle is just $p_d(0) = \left(1 - \rho\right)^{N/n}$, while the expected number of packets to be routed out of an input section from a given input line is $\frac{N\rho}{n}$.

If all of the input lines attached to the same input section have the same ρ the density of r_i is also a binomial:

$$p_r(r_i) = \binom{N\text{-}1}{r_i} \rho^{r_i} (1-\rho)^{N\text{-}1-r_i} \qquad 0 \le r_i \le N\text{-}1 \qquad \text{Eq. 4.3}$$

And the probability of r_i given x is the binomial:

$$P_{r|x}(r_i) = \binom{N\text{-}n}{r_i\text{-}x} \rho^{r_i-x} (1-\rho)^{N-n-r_i+x} \qquad x \le r_i \le N\text{-}n \qquad \text{Eq. 4.4}$$

Where the probability density function of x is the following binomial:

$$p_x(x) = \binom{n\text{-}1}{x} \rho^x (1-\rho)^{n\text{-}1\text{-}x} \qquad 0 \le x \le n\text{-}1 \qquad \text{Eq. 4.5}$$

Rather than considering just the uniform case, where all the inputs have the same $\rho's$, we consider a more general case, of which uniform is a special case. Since multicasting is being done in the first stage, the packets arriving on an input section's n input lines may result in an average number of packets to be routed out of the section that is greater than n. The effect that this has on a particular input section's blocking probability may be studied by using two different $\rho's$. The input section from which the packet being routed originates is assumed to have a ρ of ρ_i while the all other inputs have a ρ of ρ_o. The uniform input loading case can be studied by setting ρ_i equal to ρ_o. To study the effect that concentrating part of the load on a particular section has on packet loss, ρ_i can be set greater than ρ_o. Setting ρ_i differently from ρ_o is called input section load concentration. Using these two $\rho's$, the density of x is as shown in Eq. 4.6, while the probability of s_k given x and r_i is as shown in Eq. 4.7.

$$p_x(x) = \binom{n-1}{x} \rho_i^x (1-\rho_i)^{n-1-x} \qquad 0 \le x \le n-1 \qquad \text{Eq. 4.6}$$

$$p_{s|x,r_i}(s_k) = \binom{N-n}{s_k-x} \rho_o^{s_k} (1-\rho_o)^{N-n-s_k} \qquad x \le s_k \le N-n \qquad \text{Eq. 4.7}$$

The overbound for $P_{loss/s,r}$ from reference [11] also holds for the multicast case and is shown in Eq. 4.8. Equation 4.8 is obtained by using combinatorial analysis. A packet cannot be routed if the are no matches between empty locations in the vectors A_i and B_j as shown in Fig. 2.2. Using combinatorial analysis, when $r+s \ge m$, $\binom{r}{m-s}$ of the $\binom{m}{m-s}$ ways to choose the empty locations in B_j cause the empty locations in B_j to match up with marked locations in A_i. The probability of loss given r and s is then simply $\binom{r}{m-s}$ divided by $\binom{m}{m-s}$, which gives Eq. 4.8. This is an overbound because it does not take into account that some of the r_i packets on input section I_i may be part of the s_k packets using output section C_k. If x of the r_i packets go to C_k, the exact equation for $P_{loss/s,r,x}$ is as shown in Eq. 4.9.

$$P_{loss|s,r} \le \begin{cases} 0 & for \quad s < m-r \\ \dfrac{r! \, s!}{(r+s-m)! \, m!} & for \quad m-r \le s \le m \\ 1 & for \quad m+1 \le s_k \le N-1 \; or \; m<r \end{cases} \qquad \text{Eq. 4.8}$$

$$P_{loss|s,r,x} = \begin{cases} 0 & for \quad s < m+x-r \\ \dfrac{(r-x)! \, (s-x)!}{(r+s-m-x)! \, (m-x)!} & for \quad m+x-r \le s \le m \\ 1 & for \quad m+1 \le s_k \le N-1 \; or \; m<r \end{cases} \qquad \text{Eq.4.9}$$

By substituting Equations 4.4, 4.5, 4.7, and 4.9 into Equation 4.1, an overbound on the packet loss probability for the last packet to be routed from input section i is found for the case where $m \ge n$. Simplifying the result of this substitution gives Eq. 4.10.

In order to relate ρ_o and ρ_i to the output line utilization, μ, another assumption must be made. Previously it was assumed that a particular input line has a packet arrive for a particular output section with probability ρ. It was further assumed that the same input has the same probability ρ for each output section, independently of whether or not a packet arrives for any

other output section. Here that assumption is extended to the probability of packet arrivals for output lines. The probability of a packet arriving on a particular input line for a particular output line is ρ_l, where the subscript l stands for the line probability.

$$P_{loss} = \sum_{s=0}^{m-1} \sum_{x=0}^{n-1} \binom{n-1}{x} \left\{ \sum_{r=m-s+x}^{m-1} \binom{N-n}{r_i-x} \rho_i^{r_i} (1-\rho_i)^{N-1-r_i} \cdot \right.$$
$$\left[\binom{N-n}{s-x} \rho_o^{s-x} \left(1-\rho_o\right)^{N-n-s+x} \cdot \left(\frac{(r-x)! \, (s-x)!}{(r+s-m-x)! \, (m-x)!} \right) \right]$$
$$+ \sum_{r=m}^{N-1} \binom{N-n}{r_i-x} \rho_i^{r_i} (1-\rho_i)^{N-1-r_i} \cdot \binom{N-n}{s-x} \rho_o^{s-x} \left(1-\rho_o\right)^{N-n-s+x} \right\}$$
$$+ \sum_{s=m}^{N-1} \sum_{x=0}^{n-1} \binom{n-1}{x} \rho_i^x (1-\rho_i)^{n-1-x} \cdot \binom{N-n}{s-x} \rho_o^{s-x} \left(1-\rho_o\right)^{N-n-s+x}$$
$$\text{Eq.4.10}$$

The calculation of ρ_l from the ρ is as follows. Since the only way for an input line to not have a packet for a particular output section is for it to not have a packet for any of the section's outputs, $(1-\rho)$ equals $(1-\rho_l)^n$. This gives Eq. 4.11.

$$\rho_l = 1 - (1-\rho)^{1/n} \qquad \text{Eq. 4.11}$$

The two section based $\rho's$, ρ_o and ρ_i, have the corresponding line based $\rho's$ of ρ_{lo} and ρ_{li}. These two line based $\rho's$ can be used to calculate the average output line utilization, μ. There are n input lines with a ρ of ρ_{li} and $N-n$ input lines with a ρ of ρ_{lo}. The equation for μ is shown in Eq. 4.12.

$$\mu = (N-n)\rho_{lo} + n\rho_{li} \qquad \text{Eq. 4.12}$$

In the preceding analysis, two equal probability assumptions were made with regard to the probability distribution of arrivals. The assumptions are that the inputs on any particular input section have the same arrival probability distribution and that all but one input section have the same probability distribution. The following shows that, given the independence assumption made earlier, which was also made, these two equal probability assumptions are the worst case distributions with regard to overload and knockout loss.

Overload loss occurs when, during a given time-slot, a single input section has more than m packets to route where m is the number of center sections and is also equal to the number of links leaving an input section. When performing first-stage multicasting, each of the n input lines on a section may need to route a packet to each of the N/n output sections. Hence, the maximum number of packets a single input section may need to route in a given time-slot is N. The actual number of packets that an input section needs to route is the result of N independent trials. Each of the trials has a success probability of p_j; where in the non-equal probability case, p_j may be different for each value of j. For $m \ge \sum_{j=1}^{N} p_j$, the probability of the number of successes being greater than m is maximized when $p_1 = p_2 = \cdots = p_n = \rho_{li}$. This is proved through a simple application of Theorem 5 from [14]. Hence, given the independence assumption, the equal probability assumption causes the worst case overload loss.

Knockout loss occurs when, during a given time-slot, more than m packets arrive to be routed to a single output section. Where m is the number of center sections and links connecting the second stage to a single output section. For the knockout loss case, an argument that is similar to that of the overload case, holds. Each input line may or may not need to route a packet to a particular output section. Hence, the number of packets that need to be routed to a particular output section, during a time-slot, is determined by the outcome of N trials each with a probability p_j. If these trials are independent and if the average number of packets that need to be routed to the output section is less than m, then the worst case set of p_j's is

$p_1=p_2=\cdots=p_N=\rho_{lo}$. This is the equal probability case that occurs when $\rho_{li}=\rho_{lo}$.

4.2. First Stage Multicast Loss Results

The performance of method (b) multicasting, which multicasts at the first and third stage, depends on several variables. These variables include the following: N - the number of lines attached to the switch, n - the number of inputs per first-stage section, m - the number of middle stage sections, μ - the output line utilization, and $\rho_{li}*N$ - the average number of packets leaving the switch as a result of each input line attached to the input section of interest.

A GPS without multicasts has two sources of packet loss: routing loss, which occurs because a non-optimal routing algorithm is being used, and knockout loss, which results from more than m packets arriving simultaneously for a single output section. Since there are only m middle sections, each with one link to any given output section, only m packets can be simultaneously routed to any output section. With the introduction of first-stage multicasting, a third type of blocking, input section overload blocking, is also introduced. Input section overload loss occurs when, because of multicasting, an input section has more than m packets to route during a single time-slot. As with knockout loss, the limit is again m since an input section has only m outgoing links, one to each middle section. These three types of loss comprise Eq. 4.10. Eq. 4.10 is broken into its component parts in Eq. 4.13a-c, which are as follows: Eq. 4.13a - routing loss, Eq. 4.13b - overload loss not including packets that would also be lost due to knockout loss, and Eq. 4.13c - knockout loss. Hence, packets that are lost both because of knockout and overload are included only in 4.13c and not in 4.13b. If these packets are added to the probability of loss in Eq. 4.13b, it simplifies to the probability of overload equation, Eq. 4.13d.

$$P_{loss(routing)} = \sum_{s=0}^{m-1} \sum_{x=0}^{n-1} \binom{n-1}{x} \left\{ \sum_{r=m-s+x}^{m-1} \binom{N-n}{r_i-x} \rho_i^{r_i} (1-\rho_i)^{N-1-r_i} \cdot \right.$$
$$\left. \left[\binom{N-n}{s-x} \rho_o^{s-x} \left(1-\rho_o\right)^{N-n-s+x} \cdot \left(\frac{(r-x)!\,(s-x)!}{(r+s-m-x)!\,(m-x)!}\right) \right] \right\} \text{ Eq.4.13a}$$

$$P_{loss(overload)} - P_{loss(overload\ and\ knockout)} =$$
$$\sum_{s=0}^{m-1} \sum_{r=m}^{N-1} \sum_{x=0}^{n-1} \binom{N-n}{r_i-x} \binom{n-1}{x} \rho_i^{r_i}(1-\rho_i)^{N-1-r_i} \cdot \binom{N-n}{s-x} \rho_o^{s-x}(1-\rho_o)^{N-n-s+x}$$
$$\text{Eq. 4.13b}$$

$$P_{loss(knockout)} =$$
$$\sum_{s=m}^{N-1} \sum_{x=0}^{n-1} \binom{n-1}{x} \rho_i^{x}(1-\rho_i)^{n-1-x} \cdot \binom{N-n}{s-x} \rho_o^{s-x}(1-\rho_o)^{N-n-s+x} \text{ Eq.4.13c}$$

$$P_{loss(overload)} = \sum_{r=m}^{N-1} \binom{N-1}{r_i} \rho_i^{r_i}(1-\rho_i)^{N-1-r_i} \qquad \text{Eq. 4.13d}$$

In [11], the knockout loss is shown to the most significant problem. However, with the addition of multicasting, overload and routing loss are also significant. This can be seen in

Fig. 4.1. The figure, which plots the probability of packet loss versus the load on the input section of interest, shows each of the three types of loss and the total loss probability (Equations 4.10, 4.13a, 4.13c, and 4.13d). The graph of Fig. 4.1 is for a GPS with $N=512$, $n=16$, $m=64$, and $\mu=0.8$.

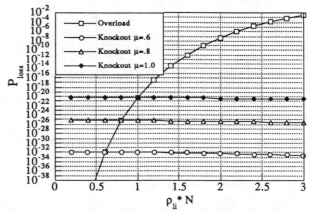

Figure 4.1 This figure, which plots the probability of packet loss versus the load on the input section of interest, shows each of the three types of loss and the total loss probability (Equations 4.10, 4.13a, 4.13c, and 4.13d). The figure is for a GPS with $N=512$, $n=16$, $m=64$, and $\mu=0.8$.

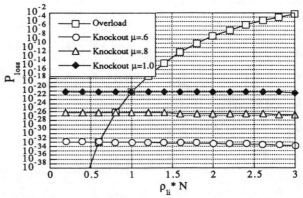

Figure 4.2 Here the knockout and overload loss probabilities are plotted for a variety of output utilizations. The knockout probability is dependent only on $\rho_{li}*N$ and is not effected by changes in the output utilization. Notice that when $\mu = \rho_{li}*N$ the knockout loss equals the overload loss. For this graph, the GPS parameters are $N=512$, $n=16$, and $m=64$.

As would be expected, at a given output utilization, knockout blocking dominates for the $\rho_{li}*N < \mu$ case, and overload loss dominates for the $\rho_{li}*N > \mu$ case. While the knockout and overload losses are the predominate causes of packet loss when $\rho_{li}*N$ is significantly different from the output utilization, near the uniform load point, packet loss is predominantly due to the non-optimal routing. This can be seen in Fig. 4.1. Routing performance could be improved by using a packing rather than a random route selection algorithm. A simple approach to route packing is to choose the lower numbered center section when there exists multiple center sections from which to choose. This would tend to route more packets through the lower numbered center sections. This type of packing would save the routes through the higher numbered center sections for the more difficult to route packets. When the choice between multiple center sections is done randomly, the effect is to spread the load evenly between center sections.

In Fig. 4.1 the average output line utilization is held constant while the portion of that average contributed by the input section of interest is varied. For the uniform case of ρ_o equal to ρ_i, $\rho_{li}*N$ equals μ. When $\rho_{li}*N$ equals μ, the knockout loss equals the overload loss. This is further illustrated by Fig. 4.2, which plots the knockout and overload loss probabilities for a variety of output utilizations. Since the overload probability is dependent only on $\rho_{li}*N$ and not on the output utilization, it is the same for each of the three utilizations. Notice that when $\mu = \rho_{li}*N$, the knockout loss equals the overload loss. The GPS parameters for Fig. 4.2 are $N=512$, $n=16$, and $m=64$.

Because the overload probability is unaffected by changes in output line utilization and because overload loss dominates when $\rho_{li}*N > \mu$, the total loss probability is relatively unchanged by changes in output line usage when $\rho_{li}*N > \mu$. Figure 4.3 shows how the packet loss probability is affected by changing output line utilizations on a GPS with the parameters: $N=512$, $n=16$, and $m=64$. While utilization has little effect on packet loss for $\rho_{li}*N > \mu$, it can be seen in Fig. 4.3 that the packet loss for $\rho_{li}*N < \mu$ is significantly effected by the switch utilization. This is because packet loss at these lower input section load concentrations is mainly due to the knockout loss.

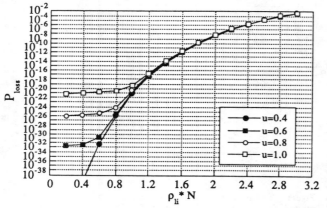

Figure 4.3 Plot of packet loss probability for various output line utilizations on a GPS with the parameters: $N=512$, $n=16$, and $m=64$.

Figure 4.4 looks at another variable that affects the packet loss probability, which is the size of the switch. The figure plots the probability of loss for various size switches with the parameters: $n=16$, $m=64$, and $\mu=0.8$. The switch is grown by adding input and output sections of 16 lines each. As the switch size increases, so does the likelihood of loss due to overload and knockout. Likewise, routing loss increases. Routing loss becomes more prevalent as the number of packets emanating from the same input section or destined for the same output section is increased. Another phenomena is that with respect to the total loss, routing loss becomes less of a factor as N increases. For example, for $N=128$, $n=16$, $m=64$, and $\mu=\rho_{li}*N=\rho_{lo}*N=0.8$ the routing loss probability is $2.5x10^{-27}$ while the knockout and overload loss probabilities are each $1.22x10^{-31}$; this is about a four order of magnitude difference. For $N=512$, $n=16$, $m=64$, and $\mu=\rho_{li}*N=\rho_{lo}*N=0.8$ the routing loss probability is $7.3x10^{-24}$ while the knockout and

overload loss probabilities are each $6.53x10^{-26}$, which is about a two order of magnitude difference. The phenomena can also be seen in Fig 4.4. For larger switch sizes, the total loss curves follow the knockout and overload curves more closely. Notice the more abrupt inflection point near $\rho_{li}*N = 0.8$ in the larger switches' loss curves.

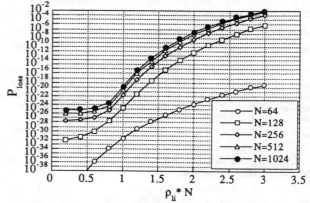

Figure 4.4 Another variable that affects the packet loss probability is the size of the switch. The figure is for various size switches all with the parameters: $n=16$, $m=64$, and $\mu=0.8$.

As has been shown in [11], the probability of loss is very dependent on the ratio $m{:}n$. Figure 4.5 plots the blocking probability versus input load for various $m{:}n$ ratios. The fixed switch parameters are $N=512$, $n=16$, and $\mu=0.8$. The maximum input section load that needs to be supported is perhaps the greatest single factor in determining the amount of expansion, the ratio of $m{:}n$, needed.

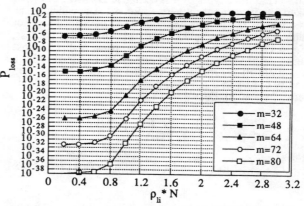

Figure 4.5 The blocking probability is plotted versus input load for various $m{:}n$ ratios. The fixed switch parameters are $N=512$ and $n=16$.

Method (a) and method(b) multicasting allow packets to be multicast by a GPS switch with only minor modifications in the routing algorithm in [1]. These methods provide good performance when the amount of multicasting on any particular input section is small. However, if the amount of multicasting on a particular input section is large, $\rho_{li}*N > 2.0$, a significant increase in the expansion factor, the ratio of $m{:}n$, is needed. This is shown in the following example, which uses numbers from Fig. 4.5. Figure 4.5 is for the $\mu=0.8$, $n=16$, and $N=512$ case. At $\rho_{li}*N=\mu=0.8$, an expansion factor of three provides a $P_{loss} < 10^{-13}$ while at $\rho_{li}*N=2.0$ an expansion factor of nearly five is required achieve the same P_{loss}. Figure 4.6 compares the packet loss overbound of [11] for the non-multicasting GPS with the multicasting GPS's packet loss overbound of Eq. 4.10. For this comparison, the load concentration is set to one, *i.e.* ρ_i is set equal to ρ_o. As can be seen in the figure, these two packet loss probabilities are very close and intersect near μ equal *0.6*. While the probability of blocking in the non-multicasting GPS is always lower, the bounds cross because at low utilizations the non-multicast bound becomes looser. The bound becomes looser because it assumes r equal to *n-1*, which is less likely at lower utilizations. The plots are for a GPS with the parameters *N=512* and *n=16*.

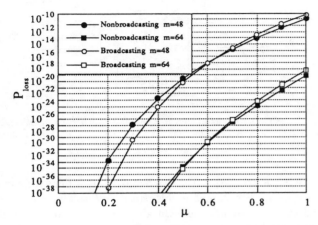

Figure 4.6 This figure compares the non-multicasting GPS packet loss overbound of [11] with the packet loss overbound for a multicasting GPS as calculated in Section 4.1. An input section load concentration of one, *i.e.* ρ_i equal to ρ_o, is used for the multicasting GPS. Both the expansion ratio of 48:16 and 64:16 are plotted. The plots are for a GPS with the parameters *N=512* and *n=16*.

5. Conclusions

This paper has presented two methods of multicasting with the growable packet switch (GPS) architecture presented in [10-12]. Both methods require only minimal modifications to the GPS architecture. The analysis shows that if multicasts are limited as to avoid large amounts of multicasting from a single input section, performance equivalent to that of the non-multicasting GPS can be achieved with a minimal increase in the ratio *m:n*. The key factor limiting the GPS's performance is the amount of multicasting or load concentration on a given input section. As the load concentration increases, overload blocking increases rapidly.

The analysis also showed that near a load concentration of one, routing loss is a factor. Routing loss can be improved through the use of a packing algorithm during the link reservation process.

The methods presented in this paper add multicasting capability to the GPS architecture. The additional expansion

needed is minimal when the input section load concentration caused by multicasting is kept low. In a BISDN network, this could be done by call refusal during call set-up. Thus, the GPS architecture with a changed routing algorithm provides a viable, growable architecture not only for non-multicast packets, but also for multicast packets.

6. References

[1] Hamid Ahmadi and Wolfgang E. Denzel, "A survey of Modern High Performance Switching Techniques," *IEEE J. Select. Areas in Commun.*, Vol. 7, No. 7, pp. 1091-1103, Sept. 1989.

[2] A. Huang and S. Knauer, "Starlight: A wideband digital switch." in *Proc. IEEE Globecom '84*, pp. 121-125.

[3] K. Y. Eng, M. G. Hluchyj, and Y. S. Yeh, "Multicast and Broadcast Services in a Knockout Packet Switch," in *Proc. IEEE Infocom '88*, pp. 29-34.

[4] Tony T. Lee, "Nonblocking Copy Networks for Multicast Packet Switching," *IEEE J. Select. Areas in Commun.*, Vol. SAC-6, No. 9, pp. 1455-1467, Dec. 1988.

[5] Tony T. Lee, "A Modular Architecture for Very Large Packet Switches," in *Proc. IEEE Globecom '89*, pp. 1801-1809.

[6] Chin-Tau Lea, "A New Broadcast Switching Network," *IEEE Trans. Comm..*, Vol. 36, No. 10, pp. 1128-1137, Oct. 1988.

[7] I. Gopal, I. Cidon, and H. Meleis, "Paris: An approach to integrated private networks." in *Proc. ICC '87* pp. 764-773.

[8] Joseph Y. Hui, *Switching and Traffic Theory for Integrated Broadband Networks*, Kluwer Academic Publishers, Boston, Massachusetts, 1990.

[9] J. Giacopelli, W. Sincoskie, and M. Littlewood, "Sunshine: A High Performance Self-Routing Broadband Packet Switch Architecture," in *Proc. International Switching Symposium '90*.

[10] Kai Y. Eng, Mark J. Karol, and Y. S. Yeh, "A Growable Packet (ATM) Switch Architecture: Design Principles and Applications," in *Proc. IEEE Globecom '89*, pp. 1159-1165.

[11] M. J. Karol and Chih-Lin I, "Performance Analysis of a Growable Architecture for Broadband Packet (ATM) Switching," in *Proc. IEEE Globecom '89*, pp. 1173-1180.

[12] Kai Y. Eng, Mark J. Karol, and Chih-Lin I, "A Modular Broadband (ATM) Switch Architecture with Optimum Performance," in *Proc. International Switching Symposium '90*, Vol. 4, pp. 1-6.

[13] American National Standards Institute, *Broadband Aspects of ISDN Baseline*, T1S1.5 Technical Sub-Committee, February 1990.

[14] W. Hoeffding, "On the Distribution of the Number of Successes in Independent Trials," *Ann. Math. Statist.*, Vol. 27, pp. 713-721, 1956.

Call Scheduling Algorithms in a Multicast Switch

Chong-Kwon Kim and Tony T. Lee, *Senior Member, IEEE*

Abstract—**Multicast switching is emerging as a new switching technology that can provide efficient transport in a broad-band network for video and other multipoint communication services. In this paper, we develop and analyze call scheduling algorithms for a multicast switch. In particular, we examine two general classes of scheduling algorithms: call packing algorithms and call splitting algorithms. Performance improvement by call packing algorithms examined in this paper is shown to be negligible. In contrast, call splitting algorithms can provide significantly lower blocking by reducing the level of output port contention. However, excessive call splitting could degrade performance because of the additional load introduced to the input ports. We present a simple call splitting algorithm called greedy splitting which achieves the performance approaching the best performance.**

I. INTRODUCTION

A SWITCH with a flexible point-to-multipoint connection capability, called a *multicast switch*, can connect any input port to any subset of its output ports [1]–[3]. A multicast switch is an important component for supporting the growing demands of flexible multipoint communication services such as video distribution and teleconferencing. However, previous studies on multicast switching have been focused primarily on the design of switch hardware configurations, and the performance of scheduling algorithms for a multicast switch has not been extensively studied. In this paper, we investigate the performance of circuit switched traffic in a multicast switch using the call blocking probability as a performance measure.

The particular switch architecture investigated in this paper is a time-multiplexed switch (TMS). In a TMS, an input/output line is divided into repetitive frames and each frame is further partitioned into a fixed number of time slots (or slots for short). A point-to-point switch establishes a circuit switched call by associating two empty slots (an empty slot in the input frame and another in the output frame) in the same position. In a multicast switch with time-multiplexed input signals, the establishment of a connection for a multicast call requires empty slots available in the same position *both* in the input frame and in *every* output frame (see Fig. 1). Therefore, a multicast call generally encounters a higher blocking probability than a point-to-point call because of the increased output port contention. This paper presents efficient call scheduling algorithms for multicast switches to enhance multicast switch performance.

Paper approved by the Editor for Communication Switching of the IEEE Communications Society. Manuscript received September 22, 1988; revised February 17, 1989. This paper was presented in part at INFOCOM'90, San Francisco, CA, June 1990.

C.-K. Kim is with the Department of Computer Science and Statistics, Seoul National University, Seoul, Korea.

T. T. Lee is with the Department of Electrical Engineering, Polytechnic University, Brooklyn, NY 11201.

IEEE Log Number 9106360.

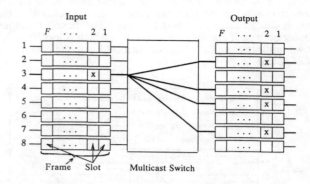

Fig. 1. A block diagram of an 8 × 8 multicast switch. Each frame consists of F time slots. A call from input 3 to outputs $\{2, 4, 5, 7\}$ is connected through the second slot.

Two general classes of call scheduling algorithms for a multicast switch are studied in this paper: *call packing algorithms* and *call splitting algorithms.* Call packing algorithms are the rules to select a proper slot position for a new call such that the new connection imposes the fewest restrictions on future call connections [4], [5]. Contrary to call packing algorithms, which have been studied to improve the performance of a point-to-point switch, call splitting algorithms are unique to a multicast switch. To schedule a multicast call, its input must be connected to all the requested outputs in one identical slot position. However, a multicast call, which cannot be switched at one slot position, can be connected by using more than one slot positions each of which connects the input and a subset of the addressed output ports. Note that call splitting increases input port loading; a call uses σ free input slots instead of a single slot when the call is split into σ subcalls. Based on this observation, we develop call splitting algorithms which partition the original outputs of a multicast call and generate a number of subcalls each of which connects the same input to a subset of outputs.

The performance of scheduling algorithms for point-to-point communication systems has been addressed in [6], [7]. These studies showed that the performance improvement from the optimal scheduling is not significant. Specifically, in a point-to-point TMS, optimal scheduling achieves only a 10 to 15% increase in the permissible offered load over a purely random scheduling algorithm [7]. However, in the case of multicast calls, our results show that the gap between optimal and random scheduling algorithms increases rapidly as the *fan-out*, the number of destinations of a call, increases. Furthermore, call splitting algorithms can significantly improve the performance of a multicast switch, but call packing algorithms cannot. In particular, we show that a simple call splitting algorithm, which we call *greedy splitting*, achieves the

Reprinted from *IEEE Trans. Commun.*, vol. 40, no. 3, pp. 625–635, March 1992.

performance approaching the optimum for a multicast switch.

The rest of the paper is organized as follows. In Section II, we derive the lower and upper bounds for call blocking probability in a multicast switch in order to assess the potential of efficient scheduling algorithms. Section III introduces several call packing algorithms and compares their performance. The advantage and overhead of call splitting algorithms are discussed in Section IV, and conclusions are given in Section V.

II. POTENTIAL PERFORMANCE OF EFFICIENT SCHEDULING ALGORITHMS

To justify the need for studies on multicast call scheduling techniques, we first assess the potential performance of efficient scheduling algorithms by comparing the performance upper bound of a multicast switch and the performance of the simplest scheduling algorithm. The difference between the upper bound and the performance of the simplest algorithm provides the margin for the improvement that can be achieved by more sophisticated call scheduling algorithms. A large margin suggests the need for further research on efficient call scheduling algorithms for multicast switches.

Call scheduling is constrained by the availability of input and output time slots. Suppose a multicast switch has N input ports and N output ports, and each input and output frame consists of F slots. Let $T = [t_{ij}]$ be the traffic matrix where t_{ij} represents the reserved traffic load from input port i to output port j in terms of a number of time slots per frame. A call request of fan-out D from an input port i to D output ports, $j_1, \cdots, j_D$ is denoted as $i \to (j_1, j_2, \cdots, j_D)$. The *capacity constraints* for this particular call request are

$$\sum_{k=1}^{N} t_{ik} < F, \tag{1}$$

and

$$\sum_{k=1}^{N} t_{kj_m} < F, \quad m = 1, \cdots, D. \tag{2}$$

The above conditions guarantee that at least one idle slot exists in the input frame and in the every requested output frame. *Overflow blocking* is call blocking due to the lack of idle slots in one or more of involved input/output ports.

A time-multiplexed point-to-point switch where each N input and N output frame has F slots, can schedule a new call request if the above capacity constraints are satisfied *and* rearrangement of existing calls is allowed. Such a switch is called a rearrangeably nonblocking switch [4]. However, the capacity constraint is not sufficient for scheduling multicast calls because an input must be connected to multiple output ports at identical time slots. Even though idle slots are available in the input frame and in the every requested output frame, a multicast call still can be blocked due to *slot contention*, when the idle slots do not match properly. *Slot contention blocking* is the blocking due to mismatched idle slots given that idle slots are available in the input and the requested output

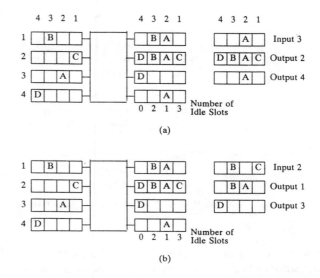

Fig. 2. Examples of overflow blocking and slot contention blocking. In this configuration, a call $A: 3 \to (1, 2, 4)$ is already assigned to the second slot. The connection patterns of calls B, C, and D are also displayed. (a) A new call $F: 3 \to (2, 4)$ is blocked due to the overflow at output 2. (b) A new call $F: 2 \to (1, 3)$ is blocked due to slot contention.

frames. Examples of overflow blocking and slot contention blocking are shown in Fig. 2.

Since a new call violating the capacity constraint cannot be scheduled by any algorithm, overflow blocking provides a lower bound B_L of call blocking probability. However, overflow blocking is not a tight lower bound because all scheduling algorithms, in reality, are not free from slot contention blocking. A random scheduling algorithm, which places a multicast call in a randomly selected slot, is the simplest scheduling algorithm. We use the performance of the random scheduling algorithm as an upper bound B_U of blocking probability because the random scheduling algorithm is the simplest scheduling algorithm and more sophisticated algorithms should perform better than the random scheduling algorithm.

To simplify the analysis, we assume that traffic is uniformly distributed over input and output ports, and the call arrival process at each input is Poisson with arrival rate λ and the holding time of each call is exponentially distributed with mean $1/\mu$. This paper deals with homogeneous traffic only where all calls have the same fixed bandwidth and fan-out. We also ignore the correlation between input and output frames in our analysis and assume that the number and position of busy slots in one frame are independent of other frames. Under these assumptions, each frame can then be modeled by a continuous-time Markov chain, as shown in Fig. 3. The number of assigned (busy) slots in a time frame determines the state of each input or output port. The symmetry in a point-to-point switch allows an input link and an output link to be modeled by the same Markov chain. However, in a multicast switch output frames are loaded with D times as many busy slots as input frames. Note that we cannot use the arrival rate λ directly in the analysis of the Markov chain because some calls fail to enter the system due to call blocking. The *effective arrival rate* (or throughput) to an input port $(\lambda_I(\cdot))$

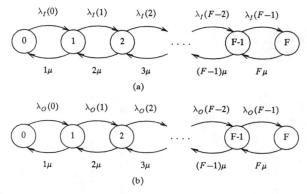

(a)

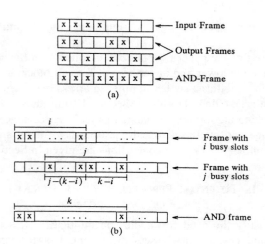

(b)

Fig. 3. The continuous-time Markov chains of input and output. (a) Input. (b) Output.

and to an output port $(\lambda_O(\cdot))$ are the arrival rates of successful calls. Since calls are blocked by either frame overflow or slot contention, the effective arrival rates are determined by the blocking probability, which in turn depends on the effective arrival rates. Hence, we derive the steady state probability distribution of the Markov chain using an iterative numerical method.

A. Lower Bound

For the lower bound, we only need to compute the frame overflow probability to derive the effective arrival rates. Let $p(k)$ and $q(k)$ be the steady-state probabilities that k slots are busy in an input frame and an output frame, respectively. A call which arrives at an input port of state k $(0 \leq k < F)$ is accepted if all the addressed D destinations have at least one idle slot. The effective arrival rate at the input of state k, $\lambda_I(k)$, is

$$\lambda_I(k) = \begin{cases} \lambda \cdot (1 - q(F))^D & 0 \leq k < F \\ 0 & k = F. \end{cases} \quad (3)$$

Suppose that an output in state k $(0 \leq k < F)$ is one of the D outputs requested by a call. This call is accepted if there are idle slots in both the input and all the other $D - 1$ output frames. Therefore, the effective arrival rate at the output in state k is

$$\lambda_O(k) = \begin{cases} D \cdot \lambda \cdot (1 - p(F)) \cdot (1 - q(F))^{D-1} & 0 \leq k < F \\ 0 & k = F. \end{cases} \quad (4)$$

Note that the average arrival rate of an output port is D times larger than that of an input port because a call is multicast from one source to D destinations.

The lower bound of blocking probability is defined to be the probability that an arrived call finds one or more requested frames are saturated. The lower bound B_L is

$$B_L = 1 - (1 - p(F)) \cdot (1 - q(F))^D. \quad (5)$$

B. Upper Bound

Contrary to the lower bound calculation, we must consider the positions of idle slots in the input and output frames to

Fig. 4. Construction of an AND-Frame. (a) Reserved slots are denoted by x (false). (b) Superposition of two frames of size F with i and j busy slots. The AND frame has k busy slots if there are $j - (k - i)$ overlapped busy slots.

compute the upper bound. Suppose a call request from input i to a set of outputs $\{j_1, j_2, \cdots, j_D\}$. A slot group for the call is a set of slots from the input frame i and every output frame $j_1, \cdots, j_D$ that occur at the same temporal position (See Fig. 4). A slot group can accommodate a call if all member slots are free. Thus, we say that a slot group is busy if any member slot is busy. Let the idle/busy state of a slot be denoted by true/false boolean logic. Then, the idle/busy state of a slot group is obtained by performing AND operations with its member slots. An AND-frame is then defined to be a logical frame generated by performing AND operations to each slot group. Fig. 4(a) shows an example of an AND-frame constructed from three involved frames.

Given two frames of size F with i and j randomly distributed busy slots $(0 \leq i, j \leq F)$, the conditional probability that there are k busy slots in their corresponding AND-frame [see Fig. 4(b)] is given by

$$\pi(k|i,j) =$$
$$\begin{cases} \dfrac{\binom{i}{k-j} \cdot \binom{F-i}{k-i}}{\binom{F}{j}} & \text{if } \max(i,j) \leq k \leq \min(i+j,F) \\ 0 & \text{otherwise.} \end{cases} \quad (6)$$

The probability that an AND frame generated from m output frames has exactly k busy slots, denoted by $q_m(k)$, is recursively determined by

$$q_m(k) = \sum_{i=0}^{k} \sum_{j=k-i}^{k} \pi(k|i,j) \cdot q_{m-1}(j) \cdot q(i), \quad (7)$$

with the initial condition $q_1(k) = q(k)$. Another equation of interest is the conditional probability that there are k busy slots in an AND-frame given that one of the participating frames has l $(\leq k)$ busy slots. This conditional probability, $q_m(k|l)$, can be similarly derived by the following recursion:

$$q_m(k|l) = \sum_{i=0}^{k} \sum_{j=\max(k-i,l)}^{k} \pi(k|i,j) \cdot q_{m-1}(j|l) \cdot q(i), \quad (8)$$

with the initial condition $q_1(k|l) = \delta(k - l)$.

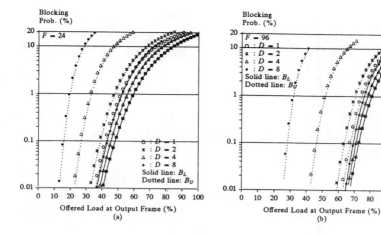

Fig. 5. The lower and upper bounds versus the offered load at the output frames.

To derive the effective arrival rate of the random scheduling algorithm, we first derive the probability of idle slot groups from (6)–(8). A call arriving at an input with k busy slots is blocked if there is no idle slot in the AND-frame generated from the input and the D requested outputs. The probability of this event is $\pi(F|k,j)$ if the AND-frame of the D outputs has j busy slots. It follows that the conditional blocking probability given that the input is in state k (i.e., the input frame has k busy slots), $P_{B|I=k}$, is

$$P_{B|I=k} = \sum_{j=F-k}^{F} \pi(F|k,j) \cdot q_D(j). \tag{9}$$

Therefore, the effective arrival rate is

$$\lambda_I(k) = \begin{cases} \lambda \cdot \left(1 - P_{B|I=k}\right) & 0 \le k < F \\ 0 & k = F. \end{cases} \tag{10}$$

A call destined to an output port of state k is blocked if there is no idle slot in the AND-frame generated from the target output, the other $D-1$ outputs, and the input. We first derive the conditional probability distribution of the number of busy slots in the AND-frame generated from the D output frames. This conditional probability given that one participating frame has k busy slots is determined by (8). Hence, the effective call arrival rate is as shown in (11) below

Equation (9) is the blocking probability of a call arriving at an input of state k. Removing the condition on the input port state, we obtain the upper bound B_U.

$$B_U = \sum_{k=0}^{F-1} P_{B|I=k} \cdot p(k) + p(F). \tag{12}$$

We compare the lower and upper bounds of call blocking probability by varying several parameters including the fan-out (D), frame size (F) and offered load. Note that the offered load is the load measured at the output ports. In Fig. 5, the lower and upper bounds are plotted by solid lines and dotted lines, respectively. Comparing the two graphs of $F = 24$, and $F = 96$, we can observe that a switch with a larger frame achieves better performance. Note that B_U as well as B_L would reduce to 0 as $F \to \infty$ if the offered load is less than 100%. However, B_U approaches to 0 more slowly than B_L, and for large fan-out traffic there is a significant difference between the two bounds in the range of frame sizes we investigated.

In Fig. 5, B_L for $D = 1$, and 2 are almost identical, and B_L increases only slightly with regard to D when $D > 2$. However, the upper bound B_U increases rapidly as the fan-out increases, indicating that idle slot groups deplete quickly at large fan-outs. As pointed out in [7], the gap between the lower bound and the upper bound is not large at $D = 1$, i.e., in a point-to-point switch. In a multicast switch, because the upper bound increases more rapidly than the lower bound as the fan-out increases, the difference between the two bounds is quite significant at larger fan-outs. The large gap suggests the need for further research on efficient call scheduling algorithms for multicast switches.

III. PERFORMANCE OF CALL PACKING ALGORITHMS

In the previous section, we found that the upper bound B_U, the blocking probability of random scheduling, is much larger than the lower bound B_L. This result indicates that slot contention is a major factor in multicast call blocking when calls are placed randomly into frames. We may reduce slot contention by assigning calls to time slots more effectively.

$$\lambda_O(k) = \begin{cases} D \cdot \lambda \cdot \left(1 - \sum_{i=0}^{F} \left(\sum_{j=\max(F-i,k)}^{F} \pi(F|i,j) \cdot q_D(j|k) \right) \cdot p(i) \right) & 0 \le k < F \\ 0 & k = F. \end{cases} \tag{11}$$

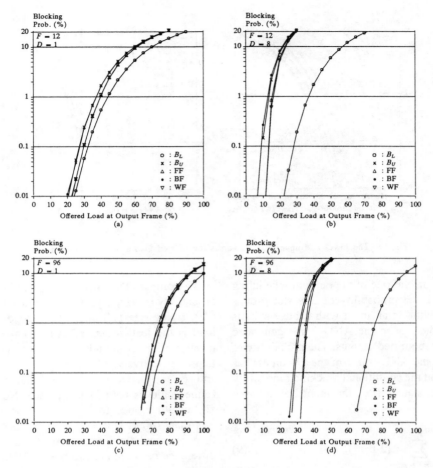

Fig. 6. Comparison of call packing algorithms. $N = 256$.

In this section, we introduce a number of more sophisticated call packing algorithms and compare their performance while using the two performance bounds B_U and B_L as references.

We have developed *First Fit* (FF), *Best Fit* (BF), and *Worst Fit* (WF) call packing algorithms, which are analogous to the First Fit, Best Fit, and Worst Fit storage placement strategies of computer operating systems [8]. These packing algorithms search for idle slot groups based on different criteria. The FF algorithm is deterministic; the idle slot search sequence is fixed regardless of the current slot assignment configuration. The BF and WF decide the search sequences adaptively based on the number of idle slots in every column of the output frames.

The FF, BF, and WF policies can be best explained by an example. Suppose that the existing calls in a 4×4 multicast switch are arranged as in Fig. 2. The number of idle slots in each column of output frames is shown at the bottom.

- *FF:* The FF is a simple algorithm. It starts to search sequentially from the first column and assigns a call to the first idle slot group encountered.

- *BF:* The BF is the most intuitively appealing policy. A column which fits a call most tightly (a column with the fewest idle slots) without contention is selected (Ties are broken arbitrarily). In the above example illustrated in Fig. 2, the search sequence is $4 \rightarrow 2 \rightarrow 3 \rightarrow 1$. The BF tries to fill already highly utilized columns as much as

possible, in order to reduce potential slot contention in other lightly utilized columns, for the later call requests.

- *WF:* Contrary to the BF, the WF selects the column with the largest idle capacity first. The search sequence is $1 \rightarrow 3 \rightarrow 2 \rightarrow 4$ in the example.

The performance of these algorithms is studied by computer simulation. Two parameters, the number of input or output ports (N) and the frame size (F), determine the size and speed of a switch. The number of input or output ports is fixed at $N = 256$ and the frame size is varied from $F = 6$ to 96 in our simulation.

Fig. 6(a) and (b) compare the performance of the three call packing algorithms for point-to-point traffic ($D = 1$) and for multicast traffic ($D = 8$) in a switch with the frame size $F = 12$. In both cases, the performance of the BF policy is almost the same as the upper bound, and the difference between the FF and the WF is negligible. The poor performance of the BF is somewhat surprising, but this result is consistent with the result reported in the literature of computer memory management systems [8]. The BF policy in memory management suffers from having many small fragments which cannot satisfy most memory requests. We further investigated the performance of call packing algorithms with different frame sizes of $F = 6$, 12, 24, 48, 72, and 96. [The results of the $F = 96$ case are shown in Fig. 6(c) and (d).] Our results show similar performance patterns at all investigated frame sizes.

In Fig. 6, the FF and WF perform better than the BF at small offered loads. However, the improvement becomes smaller as the offered load increases. In addition, for $D = 8$ traffic, the performance improvement by the FF or WF policy is negligible relative to the large gap between B_U and B_L. In summary, our simulation shows that the call packing algorithms we examined do *not* effectively improve the performance of a multicast switch.

IV. PERFORMANCE OF CALL SPLITTING ALGORITHMS

As shown in Fig. 5, the gap between B_U and B_L is not great when $D = 1$. However, it increases rapidly as the fan-out increases, which suggests the potential advantage of call splitting. Because slot contention is less severe for traffic of smaller fan-out, it is possible to reduce the call blocking probability by splitting a large fan-out call into subcalls with smaller fan-out.

It should be noted that call splitting increases the input port loading since each subcall is considered as one individual call and it occupies one input time slot. This additional input loading may offset the advantage of reduced output contention if the maximum number of subcalls is not restricted. Because there is a tradeoff between the benefit of reducing slot contention probability and the overhead of increasing input frame utilization, the splitting parameter, which controls the maximum number of subcalls that can be split from an original call, must be carefully chosen.

A. Fixed Call Splitting Algorithm

To clearly identify the effect of call splitting and the tradeoff relation between the benefit and the overhead due to call splitting, we first study the performance of a simple call splitting algorithm called the *fixed splitting* (FS) algorithm. With FS, the D destinations of an original call are randomly partitioned into σ subsets of the same size such that the fan-out of each subcall is $d = D/\sigma$. We obtain the simplest call splitting algorithm using this restrictive assumption of uniform and deterministic splitting (i.e., splitting is done independent of the condition of a switch). An example of the FS algorithm is shown in Fig. 7. Suppose slots are assigned as in Fig. 7(a) before a new call $F: 1 \rightarrow (1, 2, 3, 4)$, arrives at the system. The call is blocked because there is no idle slot group. Suppose that $\sigma = 2$ and the call is split into two subcalls $F_1: 1 \rightarrow (1, 2)$ and $F_2: 1 \rightarrow (3, 4)$. Then the two subcalls can be assigned at the first and third columns, respectively, as shown in Fig. 7(b).

The Markov chain of an input port under FS is shown in Fig. 8. Since every call is partitioned into σ subcalls, $0, \sigma, 2\sigma, \cdots$ slots are allocated in the input frame as the $0, 1, 2, \cdots$ calls are connected. The Markov chain of an output port is still the same as the one shown in Fig. 3(b). The effective arrival rates, $\lambda_I(\cdot)$ and $\lambda_O(\cdot)$, are determined as follows.

Suppose that a call arrives at an input frame at state k. The call is split into σ subcalls which are randomly ordered and labeled by $C_1, C_2, \cdots$, and C_σ.[1] The call is blocked if

[1] Even though the sequence of subcalls is arbitrarily determined, the sequence is fixed and the connections of subcalls must be done in the fixed sequence. The strict ordering eliminates the possibility of backtracking in the call connection process.

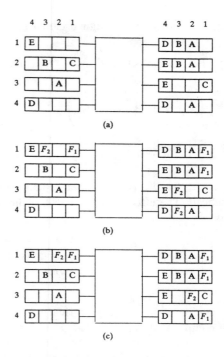

Fig. 7. Examples of fixed splitting and greedy splitting. (a) Before the assignment of a new call $F: 1 \rightarrow (1, 2, 3, 4)$, (b) FS, (c) GS.

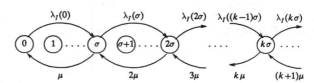

Fig. 8. The continuous-time Markov chain of an input frame when a call is split into σ subcalls.

$k \geq F - \sigma + 1$ because the connection of all subcalls requires σ idle input slots. The probability that the first subcall C_1 is blocked, $P_{B_1|I=k}$, is

$$P_{B_1|I=k} = \sum_{j=F-k}^{F} \pi(F|k, j) \cdot q_d(j). \tag{13}$$

If C_1 is blocked, then the original call is blocked and the assignment process is finished. (It may be possible to connect the call if the call is split in different ways; we do not allow resplitting of calls in the interest of simplicity). If C_1 is connected, it occupies one idle slot of the input frame, which now has $k + 1$ busy slots. Therefore, the blocking probability of the second subcall (C_2) after the connection of C_1, $P_{B_2|I=k}$, is

$$P_{B_2|I=k} = \sum_{j=F-k-1}^{F} \pi(F|k+1, j) \cdot q_d(j). \tag{14}$$

Similarly, the blocking probability of C_σ, $P_{B_\sigma|I=k}$, is

$$P_{B_\sigma|I=k} = \sum_{j=F-k-\sigma+1}^{F} \pi(F|k+\sigma-1, j) \cdot q_d(j). \tag{15}$$

The blocking probability of the original call, $P_{B|I=k}$, is

$$P_{B|I=k} = \sum_{s=1}^{\sigma} \Pr(C_s \text{ is blocked}; C_1, \cdots, C_{s-1} \text{ are allocated})$$

$$= \sum_{s=1}^{\sigma} \left(P_{B_s|I=k} \cdot \prod_{r=1}^{s-1} P_{A_r|I=k} \right)$$

$$= 1 - \prod_{s=1}^{\sigma} P_{A_s|I=k} \qquad (16)$$

where $P_{A_s|I=k} = 1 - P_{B_s|I=k}$ is the acceptance probability of the sth subcall. The effective arrival rate at the input port of state k is

$$\lambda_I(k) = \begin{cases} \lambda \cdot \left(1 - P_{B|I=k}\right) & 0 \le k < F - \sigma + 1, \\ & k \text{ is divisible by } \sigma \\ 0 & \text{otherwise.} \end{cases} \qquad (17)$$

The computation of the effective arrival rate at an output port is more complicated. Suppose that an output of state k is one of the D destinations of a call. The target output port can belong to any one of σ subcalls with the equal probability of $1/\sigma$. Suppose that the call arrives at an input of state i and that the target output port belongs to the mth subcall. The call blocking probability under these conditions is

$$\Pr(\text{the call is blocked}|I - i, O = k, S = m) =$$

$$1 - \prod_{s=1}^{\sigma} P_{A_s|I=i,O=k,S=m}$$

where

$$P_{A_s|I=i,O=k,S=m} =$$

$$\begin{cases} 1 - \displaystyle\sum_{j=\max(F-i-s+1,k)}^{F} \pi(F|i+s-1,j) \cdot q_d(j) & s \ne m \\ 1 - \displaystyle\sum_{j=\max(F-i-s+1,k)}^{F} \pi(F|i+s-1,j) \cdot q_d(j|k) & s = m \end{cases} \qquad (18)$$

Removing the conditions on the subcall association of the target output port and the state of input port, we obtain the blocking probability at the output port of state k

$$P_{B|O=k} = \sum_{i=0}^{F} \sum_{m=1}^{\sigma} \Pr(\text{blocked}|I = i, O = k, S = m)$$
$$\cdot \Pr(S = m) \cdot p(i)$$

$$= \sum_{i=0}^{F-\sigma} \left(\sum_{m=1}^{\sigma} \left(\frac{1}{\sigma}\right) \cdot \left(1 - \prod_{s=1}^{\sigma} P_{A_s|I=i,O=k,S=m}\right) \right)$$

$$\cdot p(i) + \sum_{i=F-\sigma+1}^{F} p(i). \qquad (19)$$

The effective arrival rate is

$$\lambda_O(k) = \begin{cases} D \cdot \lambda \cdot \left(1 - P_{B|O=k}\right) & 0 \le k < F \\ 0 & k = F. \end{cases} \qquad (20)$$

Given the conditional call blocking probability at an input port (16), we obtain the blocking probability BP as

$$\text{BP} = \sum_{k=0}^{F-\sigma} P_{B|I=k} \cdot p(k) + \sum_{k=F-\sigma+1}^{F} p(k). \qquad (21)$$

The numerical results of FS at $D = 4$, illustrated in Fig. 9(a) and (b), show that the benefit of FS is more pronounced at larger frame sizes. For example, the performance improvement by two-splitting ($\sigma = 2$) is more significant at $F = 48$ than at $F = 12$. Four-splitting ($\sigma = 4$) is an overkill for $D = 4$ traffic; the blocking probability of four-splitting is even higher than B_U. Apparently, the additional input load overwhelms the benefit of reduced slot contention in this case. However, at $F = 48$, four-splitting performs better than nonsplitting when the offered load is large. In summary, the results show the advantage of FS with an appropriate splitting parameter (here, $\sigma = 2$). This advantage becomes even more evident at larger frames and greater offered loads.

Fig. 9(c) and (d), which show the performance of FS at $D = 8$, indicate that the benefit of call splitting is more pronounced when the fan-out is large. It is important to note that the optimal value of the splitting parameter varies with respect to the frame size and switch utilization. For example, four-splitting, which performs worse than non-splitting at $F = 12$ when the offered load is small, performs best at $F = 48$. Suppose the required blocking probability is in the range of $0.1\% \le \text{BP} \le 1.0\%$. At $F = 12$, two-splitting performs best for the given BP. However, at $F = 48$, four-splitting performs better than two-splitting. Hence, the optimal splitting parameter should be adaptively determined based on the offered load, frame size, and the required blocking probability.

B. Greedy Call Splitting Algorithm

As discussed above, FS splits a multicast uniformly into a fixed number of subcalls. Also, FS is a deterministic procedure which ignores the existing slot assignment configuration of a switch. To investigate a more flexible and practical splitting strategy, we develop a simple adaptive splitting algorithm which we call *greedy splitting* (GS). GS splits calls whenever possible regardless of other splitting possibilities or the consequence of the current splitting. However, GS controls the granularity of subcalls by a parameter d_m, which specifies the minimum size (fan-out) of subcalls. Thus, it also controls the maximum number of subcalls (σ_M) that can be generated from a single call; $\sigma_M = \lceil D/d_m \rceil$, i.e., the smallest integer greater than or equal to D/d_m.

Suppose that a call of fan-out D arrives at a switch. The switch searches slot columns which can connect at least d_m destinations. If a slot column that can accept δ ($\ge d_m$) destinations is found, then the switch allocates all δ slots in that column and continues to search for the next slot columns to connect the remaining $(D - \delta)$ destinations. The minimum subcall size criteria is applied in every step except the last step. At the completion step, a column connecting all remaining destinations is accepted regardless of the subcall size.

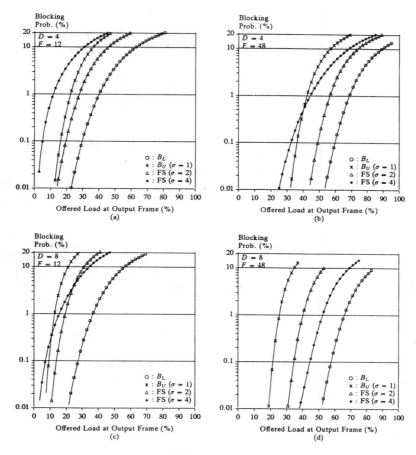

Fig. 9. Performance of FS.

To illustrate the above procedure, suppose that a new call request $F : 1 \rightarrow (1, 2, 3, 4)$ arrive at a switch whose time slots are assigned as Fig. 7(a). Suppose that $d_m = 2$ and we sequentially search all columns starting from the first column. The first column accommodates three destinations, $\{1, 2, 4\}$ and a subcall $F_1 : 1 \rightarrow (1, 2, 4)$ is generated. The second column can connect the remaining destination 3, and the second subcall $F_2 : 1 \rightarrow (3)$ is created. Slot assignments after the connections of the two subcalls are shown in Fig. 7(c).

We used computer simulation to study the performance of GS and compared the performance of GS and FS. The simulation was conducted by varying the fan-out ($D = 4, 8$) and the frame size ($F = 6, 12, 24, 48, 72, 96$). We also used different values of the splitting parameters, σ, and d_m (σ_M), to study the effect of subcall granularity.

Fig. 10 compares the performance of FS with the performance of GS, which is shown by the dotted lines in the figure. In both Fig. 10(a) and (b), GS outperforms FS and further improves the switch performance significantly. The performance improvement by GS becomes more evident when the fan-out is larger. In our simulation, the performance of GS is close to the lower bound, the best performance limit of a multicast switch. Thus, we may conclude that the call splitting algorithms have a greater impact on the performance of a multicast switch than the call packing algorithms we considered.

Note that the input port overhead of FS increases linearly

in proportion to the splitting parameter σ. At $D = 4$, four-splitting performs worse than two-splitting, indicating the negative return of excessive splittings in FS. However, in GS, the input loading does not increase linearly in proportion to the parameter σ_M, which limits the maximum number of splitting. Contrary to FS, the performance of GS at $\sigma_M = 4$ is always better than

$\sigma_M = 2$. This indicates that GS can avoid unnecessary call splitting and reduces the splitting overhead.

Fig. 11(a) and (b), which plot the maximum offered loads at 1% call blocking probability as functions of the frame size, compare the performance of splitting algorithms at $D = 4$ and $D = 8$, respectively. Both graphs indicate the advantage of large frames; a switch with a large frame achieves the better switch utilization than a switch with a small frame for the same call blocking probability. Even though the performance of all algorithms improves as the frame size increases, the rates of improvement are not the same. For example, FS with $\sigma = 4$ performs most poorly at a small frame of $F = 12$ in both Fig. 11(a) and (b). Splitting of a call into four subcalls is too excessive at $F = 12$ because each scheduled call would occupy $1/3$ of the total time slots. However, the negative effect of call splitting decreases and the performance of FS with $\sigma = 4$ improves rapidly as the frame size increases. For $D = 8$ traffic, FS with $\sigma = 4$ performs second best when $F \geq 24$. Contrary to FS, whose performance is sensitive to the frame size, GS shows the consistent performance. GS with $\sigma_M = 4$

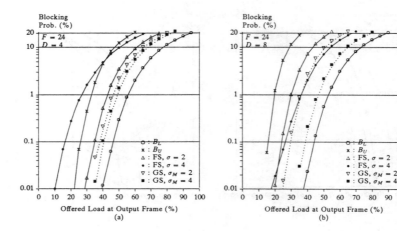

Fig. 10.   Performance of GS.

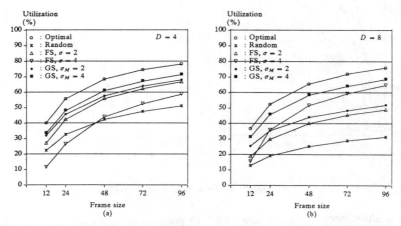

Fig. 11.   Effect of frame size on the performance of call splitting algorithms. $N = 256$.

achieves the best performance at all frame sizes investigated in this paper.

V. CONCLUSION

A multicast switch can connect any input port to any subset of its output ports. A multicast call generally encounters a higher blocking probability than a point-to-point call because of higher output port contention. In this paper, we investigated efficient call scheduling algorithms to reduce the blocking probability of a multicast switch. We considered two classes of call scheduling algorithms: call packing algorithms and call splitting algorithms. While call packing algorithms have been studied for point-to-point switches, the call splitting algorithms are unique to a multicast switch.

Our performance study shows that the call packing algorithms examined in this paper are *not* effective in significantly improving the performance of a multicast switch. We found that slot contention blocking is the predominant factor of blocking a multicast call, which suggest that call splitting may be an efficient strategy to handle multicast calls. We investigated the performance of call splitting algorithms and developed an adaptive call splitting algorithm called greedy splitting. Our results show that greedy splitting can improve the performance by avoiding unnecessary splitting. We found

in our simulation that the performance of greedy splitting is close to the lower bound, the best performance limit of a multicast switch.

Even though splitting algorithms can improve the performance of a time multiplexed switch, the implication is the additional complexity in the control structure of a switch. Especially, our result suggests that the optimal splitting parameter for the fixed splitting algorithm should be adaptively determined. Contrary to FS, the performance of greedy splitting is less sensitive to traffic and switch characteristics. However, it certainly requires further investigations to fully understand the dependence of the optimal splitting parameter on the traffic and switch characteristics. Another future research area is the investigation of other switch architectures for multicast traffic. For example, the knockout switch [9], which consists of multiple buses, can easily support broadcast traffic. However the control structure for the flexible multicast switching capability has not yet been proposed for the knockout switch.

APPENDIX A

We assumed that input and output frames are independent in our analysis. However, there is dependency between input and output frames since each call is scheduled into the slots of the same position in the input and all the requested output frames.

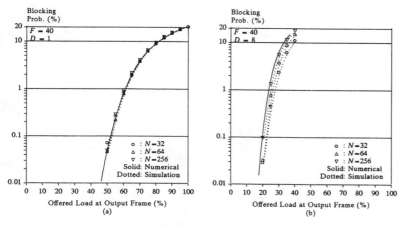

Fig. 12. B_U versus the offered load. Comparison of numerical and simulation results.

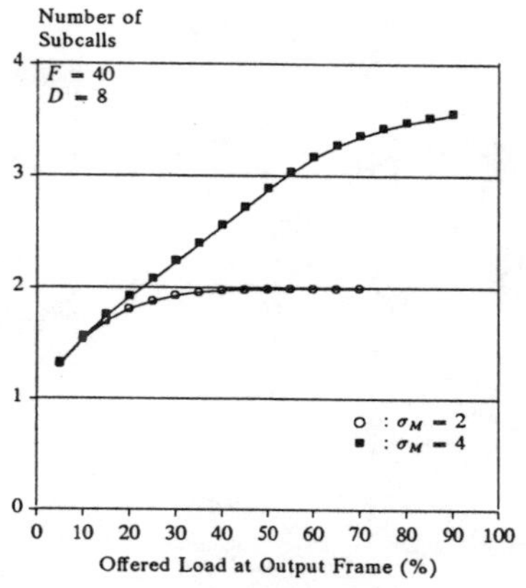

Fig. 13. Average number of subcalls in GS.

The degree of dependency varies depending on the number of input/output ports and the fan-out of traffic. The dependency would be more distinctive when a switch is smaller (fewer input/output ports) and the fan-out is larger.

We studied the effect of the switch size and the fan-out on the frame dependency with the random scheduling algorithm (the upper found, B_U). The dotted lines in Fig. 12 are the simulation results of switch sizes $N = 32$, 64, and 256, respectively. At $D = 1$, all simulation results are very close to the numerical result, which indicates that frame dependency is not strong even in a small switch of $N = 32$. However, when $D = 8$, there is significant difference between the numerical result and the simulation results of small switch sizes. The difference is decreasing as the switch size increases and is not significant when $N = 256$. We performed a similar study with FS and observed that the degree of dependency is determined by the switch size and the fan-out of *subcalls*.

APPENDIX B

Fig. 13, which displays the average number of subcalls generated from an original call, clearly shows the adaptiveness of greedy splitting. In both graphs, the number of splitting increases as the offered load increases. However, the average number of subcalls does not increase linearly. While GS reduces the input port overhead by accepting most calls without splitting at low offered loads, GS performs more call splitting and reduces slot contention blocking at higher offered loads when it is more difficult to accept a new call without splitting.

ACKNOWLEDGMENT

Many discussions with E. Arthurs during the initial stage of the work were very fruitful. We would like to thank G. Herman, L. T. Wu, F. S. Dworak, and G. Gopal for their careful reviews which improve the presentation of the paper. We also thank anonymous referees for invaluable and thoughtful comments.

REFERENCES

[1] A. Huang and S. Knauer, "Starlite: A wideband digital switch," in *Proc. GLOBECOM'84*, 1984, pp. 121–125.
[2] T. T. Lee, R. Boorstyn, and E. Arthurs, "The architecture of a multicast packet switch," in *Proc. INFOCOM 88*, New Orleans, LA, 1988.
[3] J. S. Turner, "Design of a broadcast switching network," in *Proc. INFOCOM'86*, 1986, pp. 667–675.
[4] V. E. Benes, *Mathematical Theory of Connecting Networks and Telephone Traffic*. New York: Academic, 1965.
[5] J. C. Bellamy, *Digital Telephony*. New York: Wiley, 1982.
[6] S. M. Barta and M. I. Honig, "Analysis of a demand assignment TDMA blocking system," *AT&T Bell Lab. Tech. J.*, vol. 63, no. 1, pp. 89–114 1984.
[7] C. Rose and M. G. Hluchyj, "The performance of random and optimal scheduling in a time-multiplex switch," *IEEE Trans. Commun.*, vol. COM-35, pp. 813–817, Aug. 1987.
[8] P. Calingaert, *Operating System Elements*. Englewood Cliffs: Prentice-Hall, 1982.
[9] Y. Yeh, M. Hluchyj, and A. Acampora, "The knockout switch: A simple modular architecture for high-performance packet switching," *IEEE J. Select. Areas Commun.*, vol. SAC-5, pp. 1426–1435, 1987.

INTRODUCTION

This part introduces a type of switching network not seen so far in this collection: toroidal networks. Toroidal networks get their name from the fact that they can be "embedded" (placed) in a natural and planar manner on the surface of a torus (doughnut-type shape). Torii have a number of intriguing mathematical properties [1]. Toroidal networks can also be naturally placed in a rectangular area. In this case most links follow a rectangular (grid) layout, and some links are used to directly connect nodes on the left and right boundaries and nodes on the top and bottom boundaries. This has led to the idea of using toroidal networks as distributed switching networks in a metropolitan area network environment.

The first paper in this collection, by N. F. Maxemchuk in 1985, initiated much of the current interest in toroidal networks for communications. It is entitled "The Manhattan Street Network."

Maxemchuk makes a number of fundamental points in this paper. One is that a mesh-type network, like the toroidal network, should have a higher throughput than linear network topologies such as the bus or ring [2]. But most importantly is the idea that *deflection routing* can be used in conjunction with a regular topology such as a torus.

What is deflection routing? Maxemchuk's Manhattan Street network is a toroidal network with unidirectional links. The links in adjacent rows and columns in the network alternate in direction. Each node thus has two input lines (from two neighboring nodes) and two output lines (to two different neighboring nodes). Time is slotted, and in one slot up to two packets may arrive from neighboring nodes (one per input line). Moreover, each node may transmit up to two packets over its two output lines in a slot (one per line.) If fewer than two packets enter a node from its neighbors in a slot or if two enter and at least one is destined for that node, the local traffic source may insert a packet or packets into the node for transmission.

In Maxemchuk's 1985 paper each node thus may hold at most two packets. In terms of routing for each packet, one or both output lines may correspond to a shortest path to the destination. The basic problem in routing that can occur is that both packets may prefer the same output link. In this case one packet is randomly chosen for the preferred output link and the other packet is "deflected" to the nonpreferred link. This means that packets will occasionally take paths that are not shortest paths. However, the regular topology of the torus and the use of deflection routing leads to a much simpler implementation than in traditional store and forward packet networks for networks with irregular topologies. In particular, relative addressing [7, 8] can be used to simplify routing in a regular toroidal network. The simple implementation suggests that such a network can operate at a very high speed.

Maxemchuk's 1985 paper discusses the topology of the Manhattan Street network, adding nodes to existing Manhattan Street networks and the idea of hierarchical Manhattan Street networks. He also discusses file transfer and implementation.

The next paper in this chapter proposes a toroidal network with bidirectional links. It is entitled "HR^4 Net: A Hierarchical Random-Routing Reliable and Reconfigurable Network for Metropolitan Area" and is authored by F. Borgonovo and E. Cadorin. The use of bidirectional links has certain advantages, such as a higher throughput, shorter shortest-path distances, and increased reliability. The reliability advantage occurs because if a link fails in a unidirectional toroidal network, one node will have two inputs and one output. Thus, without additional measures, packets will arrive at this node at a rate that is higher than the rate at which they can depart. When a bidirectional link in a bidirectional toroidal network fails, every node still has the same number of inputs and outputs.

In this paper Borgonovo and Cadorin describe a deflection routing policy that sends the maximum number of packets on preferred output links. Switching is more complex in HR^4 net than in the Manhattan Street network; as with the potential for four packets in a node, there are more potential switching configurations (Borgonovo and Cadorin list 30).

The paper contains performance information concerning average path length and throughput. The throughput curves are parametrized by output buffer size. Borgonovo and Cadorin also make an observation that toroidal networks need input buffers at the nodes. This improves the blocking probability seen by arriving packets [3].

The first paper devoted exclusively to the performance evaluation of deflection routing networks in this part is "Signal Flow Graphs for Path Enumeration and Deflection Routing Analysis in Multihop Networks." It is authored by E. Ayanoğlu. The paper discusses a highly original approach using signal-flow graphs that is similar to one used for studying convolutional codes.

In this paper the original network graph is transformed into a signal-flow graph. A signal-flow graph is a directed graph with weights that have one to one association with a set of linear equations. The transfer function of the signal-flow graph can be found. The number of paths of a fixed length from a source to a destination can be determined from a Taylor series expansion of the transfer function. Moreover, the mean number of hops in the network as a (rational) function of the deflection probability can be found from the derivative of the transfer function. The mean number of hops as a function of packet-arrival rate can then be computed using a fixed-point algorithm.

Ayanoğlu examines a 64-node Manhattan Street network and 64-node Shuffle net (see the next part) in detail. A commercial computer algebra package is used to calculate the transfer functions. He notes that although both networks have switching elements with two inputs and two outputs, this condition can be relaxed.

The next paper in this part is "Sharp Approximate Models of Deflection Routing in Mesh Networks," by A. G. Greenberg and J. Goodman. In this paper the authors present two performance models for regular (patterned) networks using deflection routing. Although this paper deals with the Manhattan Street network, similar analyses are available for hypercubes [4] and Shuffle nets [5] (see the next part).

The first model is a *one-node* model, which consists of a set of difference equations based on certain independence and relabeling invariance assumptions. This model can be used to compute mean throughput and mean delay. The *one-packet* model is based on the motion of a single packet through the network. The effect of other packets appears as a deflection probability parameter. Accuracy for both models is about 1%. The one-packet model can be used for large networks with up to 10,000 nodes.

It is possible to add output buffers to the nodes in a Manhattan Street network to improve performance [6]. This is studied in "Performance Analysis of Deflection Routing in the Manhattan Street Network," by A. K. Choudhury and V. O. K. Li. An output buffer capable of holding k packets is placed before each of the two output lines in each node. Packets are buffered in the buffer corresponding to their preferred link as long as that buffer is not full.

Choudhury and Li develop an analytic model based on one in [4]. Certain independence assumptions are made in this model. They find that adding even a single output buffer to each output link leads to a significant improvement in mean throughput and delay. A very close match is shown between the analytic results and simulation results.

Because packets traveling between a given source and destination may follow different paths in a toroidal network using deflection routing, a reassembly buffer will be needed at the destination. The reassembly buffer places arriving packets into their correct order before forwarding them. Reassembly in the context of Manhattan Street networks is examined in "Effect of a Finite Reassembly Buffer on the Performance of Deflection Routing," by A. K. Choudhury and N. F. Maxemchuk.

In this paper upper and lower bounds to the probability of overflow from the reassembly buffer are computed. A specific bottleneck model operating under nonuniform traffic is then considered. The distribution of the random component of delay for this model is evaluated. A series of performance curves for overflow probability and mean throughput as functions of arrival rate and reassembly and nodal output buffer size are also presented. A numerical technique that determines link utilization by iteration and the delay distribution through a random walk method is used.

Choudhury and Maxemchuk present mean throughput curves for local area network (LAN), metropolitan area net-

work (MAN), and wide area network (WAN) scenarios. Interestingly, they find a drop in mean throughput at high load that becomes significant as network size increases. They conclude by suggesting the need for further research on end-to-end high-level protocols to ". . . ensure that the oversubscription to links is tightly monitored and kept in the range where the performance is satisfactory."

The paper by Choudhury and Maxemchuk examined nonuniform traffic in a Manhattan Street network. Another look at this issue appears in "Nonuniform Traffic in the Manhattan Street Network," by J. Brassil and R. Cruz. In it they extend a solution approach similar to that of Greenberg and Goodman in this part and Krishna and Hajek in the next part for studying uniform traffic to nonuniform traffic.

Brassil and Cruz's iterative solution method is based on an independence assumption concerning incoming traffic at a node from its two links and local source. Three contention rules are considered. The best of these bases the choice of which packet to deflect on packet age; that is, if two packets in a node prefer the same output link, the older in terms of the number of hops experienced is sent to the output link and the younger is deflected. This policy serves to minimize the maximal delay experienced by packets [3]. Brassil and Cruz's technique requires the solution of $O(N^3)$ equations for N nodes. This is better than the exponential growth for a direct Markov chain solution.

Brassil and Cruz discuss routing around a congested "hot spot" in the network. They find the spreading of packets over a large number of network links to be efficient.

REFERENCES

[1] T. G. Robertazzi, "Toroidal networks," *IEEE Commun. Magazine,* vol. 26, no. 6, pp. 45–50, June 1988.

[2] A. S. Tanenbaum, *Computer Networks,* 2nd ed., Englewood Cliffs, N.J.: Prentice Hall, 1988.

[3] H.-Y. Huang and T. G. Robertazzi, "Performance evaluation of the Manhattan Street network with input buffers," *Proc. IEEE Int. Conf. Commun. '92,* Chicago, Ill., pp. 202–206, 1992.

[4] A. G. Greenberg and B. Hajek, "Deflection routing in hypercube networks," *IEEE Trans. Commun.,* vol. 40, no. 6, pp. 1070–1081, June 1992.

[5] B. Hajek and A. Krishna, "Performance of shuffle-like switching networks with deflection, *Proc. IEEE INFOCOM '90,* San Francisco, Calif., pp. 473–480, June 1990.

[6] N. F. Maxemchuk, "Comparison of deflection and store-and-forward techniques in the Manhattan Street and shuffle-exchange networks," *Proc. IEEE INFOCOM '89,* Ottawa, Canada, pp. 800–809, 1989.

[7] N. F. Maxemchuk, "Regular mesh topologies in local and metropolitan area networks," *AT&T Tech. J.,* vol. 64, no. 7, pp. 1659–1685, Sept. 1985.

[8] N. F. Maxemchuk, "Routing in the Manhattan Street network," *IEEE Trans. Commun.,* vol. COM-35, no. 5, pp. 503–512, May 1987.

THE MANHATTAN STREET NETWORK

N. F. Maxemchuk

AT&T Bell Laboratories
Murray Hill, New Jersey 07974

ABSTRACT

The throughput per user in loop and bus configured local area networks decreases at least linearly with the number of users. These networks cannot be extended to a metropolitan area with many users. Mesh networks increase the throughput of conventional local area networks by decreasing the fraction of the network capacity needed to transmit information between a source and a destination. These networks have multiple paths that increase the reliability of the networks, and have point to point links that can cover a metropolitan area.

The Manhattan Street Network is a two-connected regular mesh network. This network significantly increases the throughput and reliability of local networks. In addition, simple local routing decisions can be made, single nodes can be added with only two wire changes, and hierarchical structures can be created with a small number of connections between remote locations.

1. INTRODUCTION

Loop topologies[1] and random access strategies[2] were first applied to local data networks in the late sixties. These networks trade reliability, total throughput, and, the distance the network can span[3], for simple access and transmission strategies. The increased deployment of optical fibers and CATV systems makes it possible to obtain high bit rate communications over wider areas. Therefore, access protocols that limit the distance a network can span are less desirable. Increases in the number of computer users and the bit rates their devices require, make networks with greater throughputs necessary. In addition, because of advances in VLSI technology and the increasing complexity of devices being connected to local networks, more complex access strategies are reasonable.

Regular mesh networks[4] increase the throughput of local area networks, without increasing the transmission rate, by using a smaller fraction of the network capacity to communicate between nodes and

preventing nodes that communicate frequently from interfering with other nodes. They are also more reliable than conventional local networks and do not have the distance constraints associated with CSMA/CD protocols. The access strategies for these networks are more complex than the current generation of local networks, but are considerably less complex than general store-and-forward networks, such as the ARPA net[5]. In this paper, the access strategy for a slotted loop system is extended to regular meshes. This strategy does not store-and-forward the data at intermediate nodes, but requires packets to be immediately forwarded.

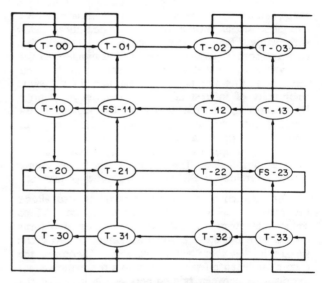

Figure 1   A 16 node MSN with two communities of 7 personal computers (T) and a file server (FS).

The Manhattan Street Network (MSN) is a two-connected regular mesh network. It has the same number of connections per node as a bidirectional loop. This network significantly increases the throughput and reliability of local networks. In addition, simple local routing decisions can be made, single nodes can be

Reprinted from *Proc. IEEE Globecom '85*, pp. 255–261, December 1985.

added with only two wire changes, and hierarchical structures can be created with a small number of connections between distant locations. This network is amenable to storageless access strategies because many of the packets passing through a node do not care which outgoing link they use, and, if a packet is forced to take the wrong link at an intermediate node, the penalty is small. These characteristics of the MSN are explained in this paper. In addition, a file transfer protocol which does not require packet resequencing is shown.

2. TOPOLOGY

The MSN is based on a grid of alternatingly directed streets and avenues, as shown in figure 1. The nodes exist on the corner of a street and an avenue. The rational for this type of a network is that routing from a particular street and avenue to a destination should be straightforward. As in a city with this layout, any destination street and avenue can be found without asking directions, even when some roads are blocked. In addition, it should be possible to layout the network to make sense geographically.

The principle difference between a grid connecting corners with streets and a grid connecting nodes with wires, is that the physical constraints associated with a two dimensional surface can be violated more easily with wires. For instance, consider a network that connects several groups of personal computers to file servers. Assume that the file servers are in the same room and that the personal computers in the same community of interest are in the same physical area. By connecting the file server to the region of the network with the terminals, rather than basing the connections strictly on the physical location of devices, the file server appears to be in the same neighborhood as the terminals. This reduces the interference between terminals in different communities of interest.

The difference in physical constraints also allows the extremes of the grid to be connected. These connections form the grid on the surface of a torus instead of a flat surface. The advantage of this cyclic surface is that there are no corners. Therefore, the maximum distance from a source to a destination is not the distance between two corners of the grid, but the distance between the center and one of the corners. In addition, the graph can be rotated so that the same local routing function, that is only dependent upon the relative location of the present node and the destination, can be used at every node.

For instance, consider a network with r rows and c columns. The current node has coordinates (i_s, j_s), and the destination node has coordinates (i_d, j_d). The current node is considered to be at location $(0,0)$, and the relative location of the destination (i,j), is expressed as:

$$i = [(1-2 (j_s \bmod 2)) (i_d - i_s) + \frac{r}{2} - 1] \bmod r - (\frac{r}{2} - 1)$$
$$j = [(1-2*(i_s \bmod 2)) (j_d - j_s) + \frac{c}{2} - 1] \bmod c - (\frac{c}{2} - 1).$$

The current node is now in the center of the network. The value of i is between $-(\frac{r}{2} - 1)$ and $\frac{r}{2}$, and j is between $-(\frac{c}{2} - 1)$ and $\frac{c}{2}$. The factors $1-2*(j_s \bmod 2)$ and $1-2*(i_s \bmod 2)$ guarantee that the links leaving the current node point toward increasing i and j. The routing decision now depends only on the relative location of the destination and not on the current node.

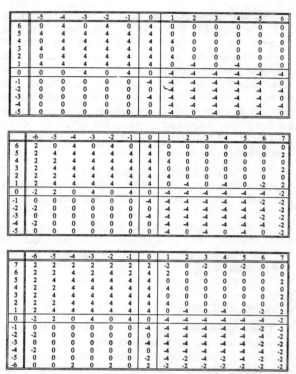

	-5	-4	-3	-2	-1	0	1	2	3	4	5	6
6	0	4	0	4	0	4	0	0	0	0	0	0
5	4	4	4	4	4	4	0	0	0	0	0	0
4	0	4	4	4	4	4	4	0	0	0	0	0
3	4	4	4	4	4	4	4	0	0	0	0	0
2	0	4	4	4	4	4	4	0	0	0	0	0
1	4	4	4	4	4	4	0	-4	0	-4	0	0
0	0	0	4	0	4	0	-4	-4	-4	-4	-4	-4
-1	0	0	0	0	0	-4	-4	-4	-4	-4	-4	0
-2	0	0	0	0	0	-4	-4	-4	-4	-4	-4	-4
-3	0	0	0	0	0	-4	-4	-4	-4	-4	-4	0
-4	0	0	0	0	0	0	-4	-4	-4	-4	-4	0
-5	0	0	0	0	0	0	-4	0	-4	0	-4	0

	-6	-5	-4	-3	-2	-1	0	1	2	3	4	5	6	7
6	2	0	4	0	4	0	4	0	0	0	0	0	0	0
5	2	4	4	4	4	4	4	0	0	0	0	0	0	2
4	2	2	4	4	4	4	4	4	0	0	0	0	0	0
3	2	4	4	4	4	4	4	0	0	0	0	0	0	2
2	2	2	4	4	4	4	4	0	0	0	0	0	0	0
1	2	4	4	4	4	4	4	0	-4	0	-4	0	-2	2
0	-2	2	0	4	0	4	0	-4	-4	-4	-4	-4	-4	-2
-1	0	0	0	0	0	0	-4	-4	-4	-4	-4	-4	-2	-2
-2	-2	0	0	0	0	0	-4	-4	-4	-4	-4	-4	-2	-2
-3	0	0	0	0	0	0	-4	-4	-4	-4	-4	-4	-2	-2
-4	-2	0	0	0	0	0	-4	-4	-4	-4	-4	-4	-2	-2
-5	0	0	0	0	0	0	0	-4	0	-4	0	-4	0	-2

	-6	-5	-4	-3	-2	-1	0	1	2	3	4	5	6	7
7	2	2	2	2	2	2	2	-2	0	-2	0	-2	0	0
6	2	2	4	2	4	2	4	2	0	0	0	0	0	0
5	2	4	4	4	4	4	4	0	0	0	0	0	0	2
4	2	2	4	4	4	4	4	4	0	0	0	0	0	0
3	2	4	4	4	4	4	4	0	0	0	0	0	0	2
2	2	2	4	4	4	4	4	0	0	0	0	0	0	0
1	2	4	4	4	4	4	4	0	-4	0	-4	0	-2	2
0	-2	2	0	4	0	4	0	-4	-4	-4	-4	-4	-4	-2
-1	0	0	0	0	0	0	-4	-4	-4	-4	-4	-4	-2	-2
-2	-2	0	0	0	0	0	-4	-4	-4	-4	-4	-4	-2	-2
-3	0	0	0	0	0	0	-4	-4	-4	-4	-4	-4	-2	-2
-4	-2	0	0	0	0	0	-4	-4	-4	-4	-4	-4	-2	-2
-5	0	0	0	0	0	0	0	-4	-2	-4	-2	-4	0	-2
-6	0	0	2	0	2	0	2	-2	-2	-2	-2	-2	-2	2

Figure 2   Routing preference in a 12x12, 12x14, and 14x14 MSN.

The routing preference from the central node to outlying nodes for a 12x12, 12x14, and 14x14 MSN is shown in figure 2. In this network, the two links emanating from the central node are directed upwards and to the right. The routing preference is the shortest distance from the central node to the destination when the link to the right is taken, minus the shortest path to the destination when the upwards directed link is taken. Therefore, a negative number implies that the right link leads to the shortest path to the destination, and a positive number implies that the upwards link yields the shortest path to the destination. The magnitude of the number shows how much longer the distance to the destination would be if a packet were forced to take a less desirable path. A zero implies

304

that the distance to the destination is the same along either path. The figures show that to get to half of the nodes either path can be taken, to get to a quarter of the nodes the upward path should be taken, and to get to the other quarter of the nodes the right path should be taken. The figures also show that if a packet is forced to take the wrong path, the increase in path length to the destination is never more than four.

Figures 3 and 4 show how nodes may be added to the MSN. Figure 3 shows how two columns are added to the basic square structure within the MSN. The dotted lines show the links that will be broken when the next node is added. Figure 4 shows how the procedure is continued to add nodes to partially full columns. Each time a new node is added, two links are broken and connected to the new node. This is no greater than the number of links that must be broken in the bidirectional loop. Eventually this procedure leads to a network with two additional rows or columns, and the pattern of alternatingly directed rows and columns is preserved.

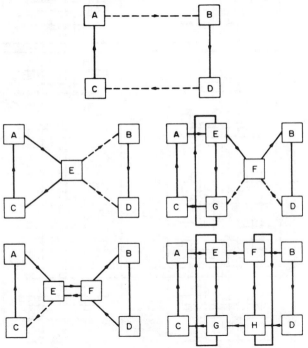

Figure 3   Adding nodes E, F, G, and H one at a time to the basic rectangular structure consisting of Nodes A, B, C, and D in an MSN.

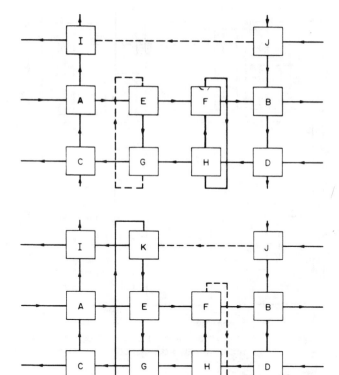

Figure 4   Adding a node K to a partially full column in an MSN.

Hierarchical structures are useful in an MSN to:

— decrease the number of paths between physically distant sections of the network,

— eliminate long paths between communities of interest, and

— prevent traffic between communities of interest form affecting communications in other communities of interest.

Hierarchical structures can be constructed while maintaining the two-connected strategy, as in figure 5. However, this will make routing more complex. An alternative is to connect one or more of the nodes in a local area to a higher level network, as in figure 6. Using this approach, a hierarchical addressing and routing structure, similar to that used in the telephone system, can be used. For example, the address within the local area corresponds to a phone number and the address of the local network on the higher level network corresponds to the area code. When sending a packet within the local area an area code is not required.

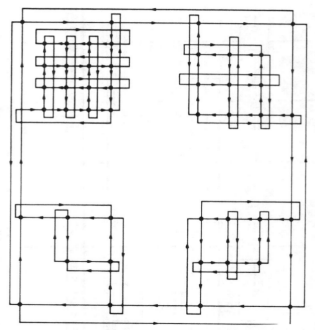

Figure 5   An hierarchical MSN in which all of the nodes are two-connected.

3. EXAMPLE

In this example, an MSN, with as few as 64 nodes, increases the throughput of bus configured networks by a factor of 20 to 25. This comparison assumes that the same rate communication lines are used in all of the networks. A factor of two increase in throughput is obtained because there is twice as much capacity emanating from each node in the MSN. However, the major portion of the increase occurs because messages use only a fraction of the total network capacity. Greater increases are obtained in larger networks.

Two traffic distributions are considered, a uniform distribution and a skewed distribution. In the uniform case, each node sends an equal amount of data to each of the other nodes. The skewed distribution corresponds to what might occur in a network of personal computers and file servers. The network is divided into communities of interest, each consisting of a file server and 7 personal computers. A personal computer directs 80% of its traffic to its own file server and 20% to the other file servers. The computer receives an equal amount of traffic from the file servers.

For each traffic distribution, the throughput for four network topologies is investigated. The first two networks are the conventional broadcast bus and loop configured networks. The throughput of the bus network is calculated assuming that the link utilization can approach one, and is an upper bound on the

achievable throughput. In the loop network, the packets only use the links between the source and destination. The remaining two networks are two-connected networks, the bidirectional loop and the MSN. For the bidirectional loop, the file server is in the middle of the seven personal computers it is servicing. The MSN with 16, 32, 48, and 64 nodes are 4x4, 8x4, 8x6, and 8x8 arrays respectively. The seven personal computers and the file server in a community of interest are arranged in a 4x2 array on the network, as in figure 1.

Traffic on the two-connected networks is placed on a shortest path between the source and destination. If there are several shortest paths between a source and a destination, the path with the smallest flow is selected.

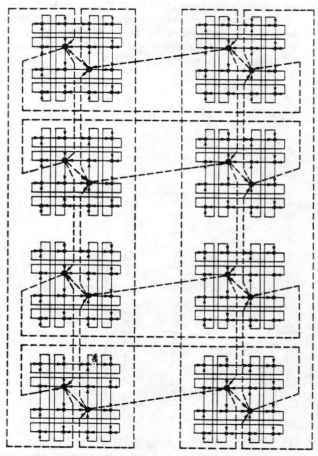

Figure 6   An hierarchical MSN in which the nodes connected to the hierarchical network are four-connected.

306

Traffic with the shortest distance between a source and destination is assigned to the network first. Once a source destination requirement is assigned a path, the path is not changed if a link on the path becomes saturated, and the requirements are not split if two equally good paths exist. The throughput is determined by increasing the traffic levels from the sources until the utilization on any link equals one. This procedure does not lead to the optimum throughput, but gives a reasonably good idea what can be achieved.

The average bit rate a user obtains in a network with ten megabit per second transmission links, and the improvement this represents over a broadcast network, is presented in table 1. For the conventional broadcast and loop network, the fraction of the capacity a user obtains decreases linearly with the number of users, as expected. The loop system provides about twice as much throughput per user as the broadcast network because, on the average, a packet transmitted on this network uses only half of the network capacity. The two-connected networks obtain a factor of two increase in throughput because there is twice as much capacity emanating from each node, and an additional increase because the networks use a smaller fraction of the network capacity to transfer a packet between the sources and destinations.

Both the bidirectional loop and the MSN respond well to the skewed requirements. These networks are capable of allowing complete connectivity while preventing users in different communities of interest from interfering with one another. This characteristic is extremely important in designing large networks. The MSN also respond well to a large group of users with uniform transmission requirements. This occurs because the average distance between users does not increase as rapidly as in the bidirectional loop.

TABLE 1. The Average Megabits per User in Networks with 10 Mbps Channels

Network	16 Nodes Mb/Usr	Imprv.	32 Nodes Mb/Usr	Imprv.	48 Nodes Mb/Usr	Imprv.	64 Nodes Mb/Usr	Imprv.
Uniform Requirements								
Bdcst.	.63		.31		.21		.16	
Loop	1.25	2.00	.63	2.00	.42	2.00	.31	2.00
BiDi Loop	4.69	7.50	2.42	7.75	1.63	7.83	1.23	7.87
MSN	6.52	10.43	4.56	14.59	4.31	20.70	3.94	25.20
Skewed Requirements								
Bdcst	.36		.18		.12		.09	
Loop	.71	2.00	.36	2.00	.24	2.00	.18	2.00
BiDi Loop	2.63	7.37	2.08	11.67	1.80	15.11	1.59	17.82
MSN	2.63	7.37	2.17	12.17	2.12	17.80	2.12	23.76

4. IMPLEMENTATION

There are two links and a local source inputting data to a node, and two links and a local sink removing data from the node. Occasionally, multiple inputs try to transmit data to the same outgoing link.

One way to resolve this problem is to queue packets waiting for a link. The network now assumes the complexity of a store-and-forward network. Not only must potentially large packet queues be maintained, but adaptive routing, flow control, deadlock avoidance and packet resequencing issues must be addressed.

In this section, the slotted system technique, developed for loop communication systems, will be extended to mesh networks with equal in and out degrees. The general strategy guarantees that every packet arriving on an incoming link, and not destined for the node, will be transmitted on one of the outgoing links. Therefore, it is not necessary to maintain a packet queue for the links emanating from the node. The requirement that packets passing through the node take one of the outgoing paths results in longer paths when the shortest path to the destination is busy. However, in the MSN the path to only half of the destinations is increased if a packet is forced to take one path rather than the other. In addition, if a packet is forced to take a less desirable path, the distance to the destination is increased by at most four.

The storage between the local source and the network is also limited. Packets from the local source are only transmitted when one of the outgoing links is not being used by an incoming link. It is assumed that either the local source can be throttled when the network is busy, or that the source provides data at a low rate relative to the network transmission rate. In the latter case, when a packet is lost, it must be recovered by a higher level protocol. If the network delivers packets faster than the local sink can accept them, packets are either transmitted on one of the outgoing links or discarded. In the former case, the network is used for storage. Since new packets cannot enter the network when it is recirculating old packets, this transmission strategy acts as a flow control mechanism. The assumptions on the local source and sink are implicit in all loop configured systems without infinite storage.

The packets of data in a slotted system are fixed size. A node continuously transmits bits on each of the links emanating from the node, and periodically transmits a start of slot indication. The start of slot indication is followed by a packet of data or an empty slot. In the interval between the start of slot transmissions, at most one packet of data is received on each of the incoming links. The packets that are received between start of slot transmissions are forwarded after the start of slot is transmitted. These packets are switched to one of the outgoing links or the local sink before data from the local source is given access to the slot. Since there are the same number of links arriving and leaving from each node, and the local source can be throttled, a queue of packets will not accumulate. The operation of a slotted system without a packet queue is shown in figure 7.

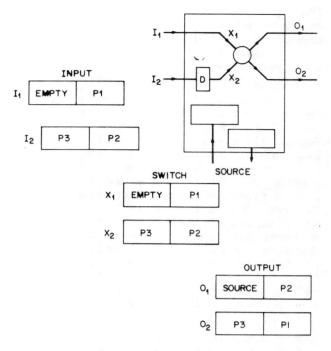

INPUT

I_1 | EMPTY | P1

I_2 | P3 | P2

SWITCH

X_1 | EMPTY | P1

X_2 | P3 | P2

OUTPUT

O_1 | SOURCE | P2

O_2 | P3 | P1

Figure 7   Extension of slotted loop systems to a mesh network.

In the slotted system, a small number of fixed size packet buffers can be inserted at the output channels to reduce the probability of a packet taking a longer path. The probability that a packet must take a longer path is the probability that two arriving packets must take the same path and the buffer for that path is full. Without buffering, one packet must take the longer path whenever two arriving packets want to take the same path. Decreasing the probability that a packet takes a longer path decreases the fraction of the network resources that a packet uses, and increases the throughput of the system. The tradeoff between buffering and system throughput remains to be investigated.

5. FILE TRANSFER

A file transfer consists of several packets from a source to the same destination. In a system in which packets do not take the same path, it is possible that packets are not received in the same order that they are transmitted. Packets may be resequenced at the receiver, however, it is preferable to avoid this task.

One possible solution to this problem is to transmit one packet at a time and wait for an acknowledgement. This reduces the file transfer rate. However, because of the small delays at each node, this solution is not as bad in mesh networks as it is in store-and-forward network. For instance, in a slotted system the average

delay per node is half a slot time, and, the average round trip delay equals the average number of hops between nodes, L, times the slot time. Therefore, there is an average of L slots between each packet in the file, and the file transfer rate equals the channel rate divided by $L+1$. Higher file transfer rates can be achieved by end-to-end protocols that take advantage of the delay characteristics of the system.

Because of the small amount of delay at each node, it is unlikely that packets that take routes that have approximately the same length will arrive out of sequence. This probability can be reduced by allowing a small number of slots between packets in the same file transfer. A simple file transfer protocol, that takes advantage of this characteristic, operates like the window protocols of store-and-forward networks and the go back N protocols of satellite systems. This protocol labels packets in a file transfer with a sequence number and a retry number. At the beginning of a file transfer, the transmitter and receiver start with the same sequenc eand retry number. The sequence number is the order of the packets. The receiver:

— increments its retry number when a packet with the expected retry number and a larger sequence number is received,

— sends a positive acknowledgement if a packet has a sequence less than or equal to the expected number,

— sends a negative acknowledgement with its retry number and expected sequence number if the packet has a larger sequence number than it expects, and

— commits a packet if it has the expected sequence number.

The transmitter:

— stops saving a packet for retransmission when it receives a positive acknowledgement for a packet with a sequence number greater than or equal to the expected number.

— adopts a new retry number and starts retransmitting from a negatively acknowledged sequence number if the a negative acknowledgement with a larger retry number is received.

— periodically retransmits the last packet in a file transfer until it receives an acknowledgement.

The transmitter initially transmits packets in the file transfer in every available slot. However, when it receives negative acknowledgements it increases the number of slots between subsequent packets.

Since this protocol only accepts packets in the correct order, packets do not have to be resequenced at the receiver. Several packets can be in transit between the source and destination. If the receiver misses a packet it must send a negative acknowledgment for every packet with a larger sequence number than expected to be certain that the transmitter receives the negative acknowledgement. The retry number is included so that the transmitter only backs up, and starts retransmitting, when it receives the first negative acknowledgement to an outstanding packet.

The transmitter adaptively changes the number of slots between packets according to network load and the rate of the receiver. When the network is lightly loaded, all packets follow the best path to the receiver, and arrive in sequence. If the receiver can accept packets at this rate, there are no negative acknowledgements, except for infrequent transmission errors, and the file transfer rate equals the channel rate. When the network is heavily loaded, the packets follow different paths, are received out of sequence, and the file transfer rate decreases. If the receiver cannot accept packets as fast as the transmitter can deliver them, the buffer in the interface unit is full when the packets arrive. In the systems described, these packets are directed to one of the output links at the node, and recirculate in the network until the buffer is available. These packets arrive out of sequence, negative acknowledgements are transmitted, and the transmitter slows down.

6. CONCLUSION

The MSN increases the throughput of conventional local area networks by decreasing the fraction of the network capacity needed to transmit information between a source and a destination. This network has multiple paths between each source and destination, thus increasing the reliability of local networks. The network consists of point-to-point links, and can be extended to cover a metropolitan, rather than a local, area.

In general, mesh networks require complex store-and-forward nodes that also route messages, control the flow of data entering the network, resequence packets at the destination, and recover packets with errors. There are characteristics of the local or metropolitan area that allow these functions to be simplified. In the local and metropolitan area environment, regular network topologies make routing straightforward. Lower error rates make it reasonable to recover errors on an end-to-end basis. This allows loop access protocols to be extended to mesh networks, eliminating the need for buffering and additional flow control protocols. The small node delays in these systems also make it reasonable to implement file transfer protocols that do not require packet resequencing.

REFERENCES

[1] E. H. Steward, "A Loop Transmission System," Conference Record of International Conference on Communications, San Francisco, June 1970, pp. 36-1, 36-9.

[2] N. Abramson, "The ALOHA-system - Another Alternative for Computer Communications," University of Hawaii Tech. Rep. B70-1, April 1970, AD707853.

[3] R. M. Metcalf, D. R. Boggs, "Ethernet: Distributed packet switching for local computer networks," Commun. ACM 19, July 1976, pp 395-404.

[4] N. F. Maxemchuk, "Regular Mesh Topologies in Local and Metropolitan Area Networks," to appear in AT&T Bell Laboratories Technical Journal, September 1985.

[5] D. C. Walden, "Experiences in Building, Operating and Using the ARPA Network," Second USA-Japan Computer Conference, Tokyo, Aug. 1975.

HR4-NET: A HIERARCHICAL RANDOM-ROUTING RELIABLE AND RECONFIGURABLE NETWORK FOR METROPOLITAN AREA

F. Borgonovo, E. Cadorin

Dipartimento di Elettronica, Politecnico di Milano and
Centro di Studio per le Telecomunicazioni Spaziali, CNR,
Piazza Leonardo da Vinci 32, 20133 MILANO, Italia.

ABSTRACT

Metropolitan Area Networks constitute a class of networks which is expected to extend to a metropolitan area most of the appealing features of Local Area Networks, expecially those referring to very high speed data transfer capability, flexibility, reconfigurability and relatively low cost. In addition they should be able to take advantage of the traffic locality and to provide different paths between sources and destinations. HR4-NET is a grid-topology network which appears to match the above requirements. It uses a two-level routing scheme based on a very simple routing information and does not store data. The paper describes the net architecture and provides some preliminary results on the network throughput obtained by simulation.

1. INTRODUCTION

HR4-NET is a packet switching communication network which, similarly to the recently proposed Manhattan Street Network[1,2,3], presents features that make it particularly attractive as a proposal in the field of Metropolitan Area Networks.

Metropolitan Area Networks (MANs) constitute a class of networks which is expected to extend to a metropolitan area most of the appealing features of Local Area Networks (LANs), expecially those referring to very high speed data transfer capability, flexibility, reconfigurability and relatively low cost. Unfortunately there are several reasons which prevent from conceiving a MAN just as a big (in the geographical sense) LAN. These reasons will be easily understood by following the design guidelines of LANs.

When technology advances led to the availability of very high speed transmission systems at relatively low cost over short distances and it became economically feasible the connection of several data processing devices, the architectural choices for Lans naturally fell on packet switching systems whit broadcast capability. Packet switching was in fact a well established tecnique for data transfers and the broadcast choice was the simplest way to avoid switching centers as those used in Wide Area Networks (WANs) which are based on expensive machines and have a small throughput if compared to the desired transmission speed. Moreover the broadcast capability allowed to distribute the switching tasks directly to user stations thus providing a key factor as long as the reconfigurability of the network is concerned.

Among different proposals and implementations[4], two ways of achieving the broadcast capability in cable LANs have emerged. They are based either on passive bus transmission technology or on point to point digital transmission systems connected at each station by digital repeaters. While the systems based on passive busses are appealing in affording simple access protocols and granting reliable transmission all over the network, their extension is somewhat limited and active repeaters must be added to obtain a greater coverage. Point to point systems do not suffer the extension limitations of busses but special provisions must be adopted to save the communicating capability of the net when some stations fail or are unpowered.

In a Metropolitan area the connection reliability of a network must be assured at a very high degree and the only mean to attain this goal seems to be to provide multiple paths between each source and destination. Multiple paths may be added to a ring[5] to bypass failed nodes (fig.1) and to assure the broadcast feature among the survivor ones. However this solution may present managing difficulties and results in a poor utilization of the links. Mesh topology networks on the contrary may take advantage from possible multiple paths by equalizing the load of the links expecially in a metropolitan area where a high degree of traffic

Reprinted from *Proc. IEEE INFOCOM '87*, pp. 320–326,
March/April 1987.

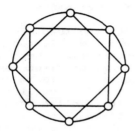

Fig. 1 - Ring with multiple paths.

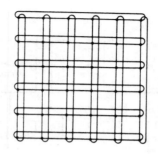

Fig. 2 - Topology of the HR⁴-NET.

locality can be expected. In general mesh networks require complex switching centers able to route packets throughout the net and to temporarily store them in queues awaiting for processing or transmission resources. Handling buffers queues and determining the routing are often time consuming tasks so that, even if a dedicated processor is used ,these operations may constitute a bottleneck in the network throughput expecially when, as is the case for MANs, transmission lines have a speed of tens or hundreds of Mb/s.

A solution for the above outlined difficulties can be obtained by reducing both the need for complex buffer handling and the complexity of routing tasks so that they can be performed by dedicated hardware running at line speed.

The choice of the routing procedure in mesh networks has a big impact on the performance of the whole net. In particular the shortest path routing technique minimizes the transmission capacity used by each packet and, if the load is balanced among the nodes, it may be effective in maximizing the throughput of the net. However transmission queues are needed to manage concurrent requests of the same transmission link. To avoid the overflows of queues the routing algorithm must provide alternative paths which must be cleverly selected in order to avoid a dramatic reduction in the throughput at high load. This routing technique is quite complex and may be too much time consuming to be usefully applied in MANs. In addition it requires some knowledge on nodes locations and this knowledge must be updated if nodes are added or dropped from the net.

The above drawbacks are completely eliminated by a totally random routing procedure[6], i.e. the procedure which selects at random the output link of every packet. It can be performed fast enough to cope with very fast line speeds. It does not require any knowledge about the location of nodes and can avoid the need for input and output queues. In fact it is always possible to switch without conflicts the input packets to the output links provided that the links have the same speed and that those in input do not exceed in number those in output. Unfortunately the network throughput allowed by the random routing is very low if compared with the shortest-path throughput; moreover their ratio decreases as the size of the network increases as it will be shown in Sec. 3.

In this work we propose a mesh network topology and a routing procedure able to maintain the advantages of the random routing, i.e. simplicity, no need for queues, no need for topology knowledge, still achieving a network throughput which favorably compares with the shortest-path upper limit. The network is presented in Sec. 2. In Sec. 3 the results obtained by extensive simulations are reported and compared. The effect of adding output buffers queues is also investigated.

2. NETWORK DESCRIPTION

HR⁴-NET is a regular grid-topology network like the Manhattan Street Network[1,2,3], but differently from that each node is linked to each of its four neighbors through full-duplex connection lines (fig.2). The simple two-level hierarchical routing scheme adopted does not require to store packets, is throughput effective and works properly if any links fail or nodes are added to the network.

Full-duplex communication links, all at the same speed, guarantee that at every node the capacity of the outgoing links does not exceed the capacity of the incoming ones should any nodes fail, and prevent from packet losses as no storage is provided. In addition, full-duplex links provide greater connectivity than unidirectional links do and allow an almost continous check of the neighbor nodes, thus granting the network the possibility for a fast dinamic reconfiguration upon node or link failures.

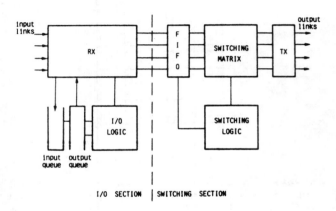

Fig. 3 - Functional block scheme of the node.

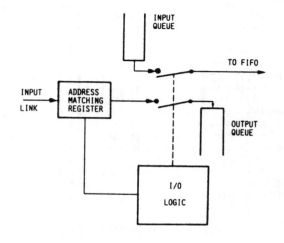

Fig. 4 - I/O section.

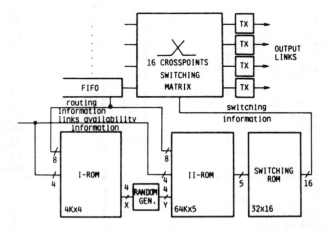

Fig. 5 - Switching section.

A node, whose functional block scheme is depicted in fig. 3, works as follows.

The switching and transmission processes take place on a time slotted basis, i. e. each node synchronizes the transmission of packets in frames of fixed length equal to T. The I/O Section performs on every input link the I/O operations usually encountered in slotted rings (fig. 4). Packets addressed to the node are extracted. Packets originating at the node are accepted from the input queue as long as an input slot is recognized empty or a packet has been extracted. Thus at most four packet per slot may enter the network at each node. No routing is performed at this stage.

The four frames (empty or not) entering the switching section are stored in the FIFO buffers to compensate for different frame alignments. After that, the routing information (see below) is obtained and packets are contemporarily switched to the outputs according to the routing rules. The switching process is performed by a 4x4 switching matrix (fig. 5). The switching configuration is obtained on the basis of the routing information by addressing a 16 bit word ROM, where each bit specifies the logical value of the corresponding matrix crosspoint.

The routing mechanism organizes the network into different rings structures at two levels. Low level(L) rings (the streets) are connected by high level (H) rings (the avenues) at each node (corners). Each L-ring is identified by its own address number (ring address) known by all the nodes belonging to that ring and used as routing information. Thus it must be specified in the address information fields of each packet.

At each node, a packet is said to be an high-level packet (H-packet) if its destination ring address does not match the ring address of the node, meaning that the packet must be relaid over the H-ring. Otherwise it is said a low-level packet (L-packet) meaning that the packet must be relaid over the L-ring to reach its destination. The set of packet types (H,L or O if the slot is empty) at the input links constitutes the Routing Information (RI).

A packet is said to be satisfied by a routing procedure if it is routed to the output links according to the above stated desiderata (i.e. an H-packet is forwarded to an H-link and an L-packet to an L-link), provided that the output link is not the packet incoming link. It may happen that the packets arriving at a node can be satisfied by different switching configurations. It is also possible that,

Table 1 - Sample of switching configurations.

RI — L_1	L_2	H_1	H_2	X	Y 1	2	3	4	5	6	7	8	9	M
0	L	H	0	1	╪									0
H	L	H	L	1	╫									0
0	0	0	L	2	⌣	⌣								0
L	L	L	H	3	╫	╪	╫							1
H	H	L	L	4	⌣	⌣⌣	⌣⌣	⌣						0
H	0	L	0	4	⌣	⌣⌣	⌣	⌣						0
H	H	L	H	6	⌣	⌣⌣	╪	⌣⌣	╪	⌣				1
H	H	H	H	9	⌣	⌣⌣	╪	╪	╫	╪	⌣⌣	╪	⌣	2

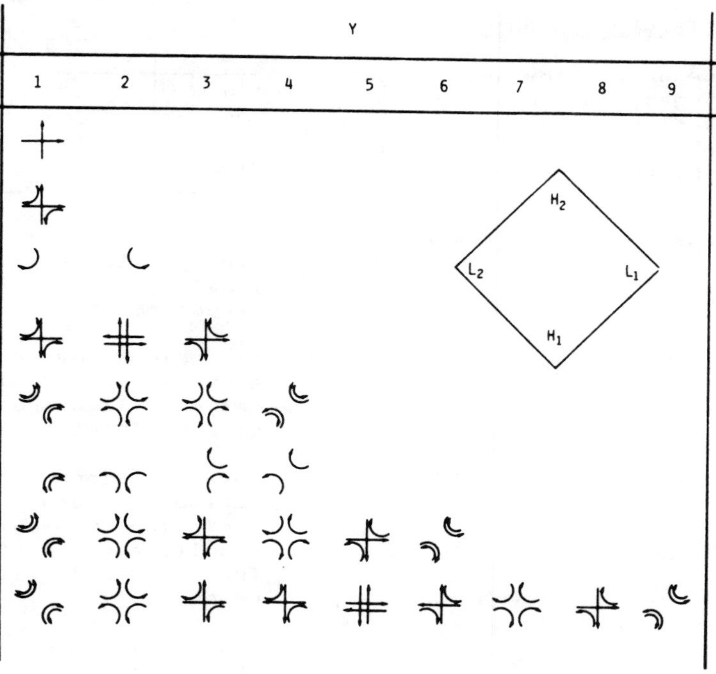

(A diamond diagram is shown to the right of the table with vertices labelled H_2 (top), L_2 (left), L_1 (right), H_1 (bottom).)

due to conflicting desiderata (e.g. more than two packets are H-packets) and the lack of queueing room, some of them cannot be satisfied and must be misdirected. The HR$^+$routing strategy chooses randomly among these possible alternatives, trying to satisfy locally the maximum number of packets.

In detail, among the switching configurations allowed by a given Routing Information pattern, the algorithm considers only those, say 1, 2, ... X, which satisfy the maximum number of packets. The final choice is given by the outcome of a random variable Y, $1 \leq Y \leq X$. As an example, Table 1 shows the X switching configurations corresponding to a sample of given Routing Information. In any case the number M of misdirected packets is not greater than two, while its uniform average when four packets are always present is .75.

The described strategy tends to minimize the traveling delay at low load. In fact if a packet is always satisfied, it follows two rings at most. At high load the random component attempts to balance the resources by using differents paths. The packets suffer a longer traveling delay than if they were stored waiting for the availability of the best route and consequently the throughput is less than the limit achieved with infinite storage. This is the price traded against the difficulty of handling queues and will be

investigated in the next section.

From the practical point of view the switching configurations can be obtained by addressing the II-ROM shown in fig. 5 with the following components:

- the routing information at each input link (3^4 different cases are possible);

- the links availability information (0 = normal, 1 = failed) of each output link (2^4 possibilities);

- the random component Y.

In HR$^+$-NET a dummy ring address can also be implemented to allow nodes to distinguish another type of packet, namely the random packet (R-packet) which can be treated, as far as the switching is concerned, as an empty packet. R-packets travel more or less randomly toward the destination depending on the load balance of the net and are particularly useful in the set-up or reconfiguration phases to inquiry the nodes about their ring addresses.

A node may be added to the net by breaking at first a low level ring. The complexity of the inserted node may be greatly reduced as no routing is required. When several such nodes have been added to different low-level rings, they may be connected through high level rings in order to increase the network connectivity. This procedure assures a modular growth of the net according to traffic needs.

3. RESULTS AND COMPARISONS

We start by comparing the average path length L of a NxN grid topology under HR[4], purely random and shortest-path routing techniques in low load conditions, i.e. when just one packet at a time is using the net.

HR[4] and shortest-path are easily shown to be:

$$L_{HR} = \frac{N^2}{(N+1)}$$

$$L_{SP} = \frac{N^3}{2(N^2-1)} \qquad \text{for N even}$$

$$L_{SP} = \frac{N}{2} \qquad \text{for N odd}$$

Results for the random routing technique are available only for K-connected topologies[7]. In our case the average path length $l_{i,j}$ between a source in (0,0) and a sink in (i,j) is easily shown to satisfy the following system of linear equations:

$$l_{i,j} = 1 + (l_{i+1,j}+l_{i,j+1}+l_{i-1,j}+l_{i,j-1})/4$$

for $\qquad 0 \leq i,j \leq N-1$

with the boundary conditions:

$$l_{0,0} = 0$$
$$l_{N,j} = l_{0,j}$$
$$l_{i,N} = l_{i,0}$$

Averaging over i and j we get:

$$L_{RR} = \sum l_{i,j}/N^2$$

Average path lengths can be used to obtain an upper limit U to the throughput achievable under balanced load when unlimited storage is available at the nodes. U is given by the ratio between the number of links (4N²) and L. Thus

$$U_{HR} = 4(N+1)$$

$$U_{SP} \approx 8N$$

Note that a ring topology with the same overall link capacity gives:

$$U_R = 8$$

The above results show the capability of

TABLE 2
Upper limits to the throughput achievable with unlimited storage at nodes.

	5x5	10x10	20x20
S-P	40	80	160
HR[4]	24	44	84
Random	3.1	2.4	1.9
Ring	8	8	8

the grid topology to take advantage of the locality of traffic better than a ring structure does.

In Table 2 the values of U for the four aforementioned cases and for different network sizes are compared and show the good performance of HR[4]-NET.

Investigations under balanced traffic and variable load conditions has been performed by simulation and the throughput results are reported in figures 6, 7 and 8 for the 5x5, 10x10 and 20x20 topologies respectively.

A sort of congestion effect was suspected so that in order to better investigate the network capacity, we chose to simulate a purely random offered traffic with loss, i.e.: at each slot and at each node four packets are generated each with probability p. Those not accepted by the net in the same slot are discarded (in operating conditions they are necessarily queued). Simulation results show in fact that the throughput reaches the maximum for values of p < 1 depending on the size of the net. After that point the throughput decreases until the saturation condition p=1 is reached. The difference between the maximum throughput and the saturation throughput increases as the size of the net increases and it became sensible for a 20x20 net. Thus in this case and in operating conditions it becomes convenient to throttle the output of the input queue so that the optimum value of p cannot be exceeded.

In Table 3 the values of the maximum throughput S_M, the saturation throughput S_S and the upper limit U are summarized for the reader's convenience. It appears that the ratio S_M/U is almost .5 in the three cases and slightly decreases as the size N increases.

In order to get an idea of the effect that output buffers have on the performance, we made some runs adding a buffer queue of constant length at each output link. The routing strategy has been modified according to the following rules.

As long as room is available in a queue,

314

TABLE 3
Upper limit U, maximum throughput S_M, and
saturation throughput S_S in HR⁴.

	5x5	10x10	20x20
U	24	44	84
S_M	14.2	22.8	39.5
S_S	14.2	21.4	26.9

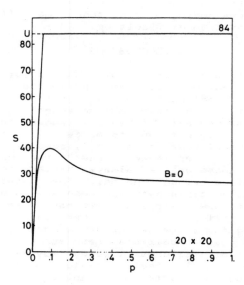

Fig. 8 – Throughput S of a 20x20 topology versus the packet generation probability p

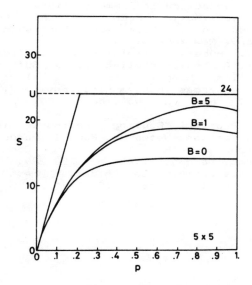

Fig. 6 – Throughput S of a 5x5 topology versus the packet generation probability p for different seizes B of the output buffers.

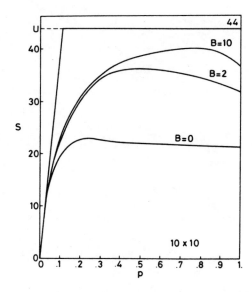

Fig. 7 – Throughput S of a 10x10 topology versus the packet generation probability p for different seizes B of the output buffers.

input packets are directly stored in that output queue provided that it is the correct one according to the basic routing strategy. The remaining packets are switched according to the old rules. Note that the switching process is never restricted to a reduced set of queues as the room corresponding to the no-buffer case is always available.

The results are reported in figures 6, and 7 together with the straigth lines giving the performance when infinite buffering is provided for both the output queues and the newly generated packets. A neat increment and a right shift in the throughput peak were foreseen because the corresponding point must approach the point (1,U) as the buffer length increases without limit. However a dramatic decrease toward the no-buffers case was expected near the point p=1 due to the filling of buffers and the subsequent increase of the random component in routing.

As a matter of fact, the whole buffer availability cannot be filled over 75 percent as the condition for an increase in buffer occupancy at a node is that at least one buffer queue is empty. Thus the last increase must lead to the condition in which three out of the four buffer queues are filled and the other one is empty. Being a certain amount of buffer resources always available even at p=1, routing contentions are less than in the no-buffer case so that the saturation throughput is always better.

Adding output buffers can increase to some

extent the throughput of the net; however the corresponding routing scheme may offer implementation difficulties.

4. CONCLUSIONS

A grid topology network with hierarchical random routing has been presented. It extends to metropolitan areas most of the features which make LANs so appealing in local environment. It has been shown that the throughput performance of the hierarchical routing favorably compares to the limit achieved under the best circumstances and is always better than that of a ring topology. The network provides multiple paths between any couple of nodes. New nodes and links may be added without affecting its operation and allowing for a gradual growth according to traffic needs. Additional results on the effects of failures, topology asymmetry, and changes in routing rules will be presented in a forthcoming paper.

5. REFERENCES

1. N. F. Maxemchuk, "Regular Mesh Topologies in Local and Metropolitan Area Networks", AT&T Technical Journal, Vol. 64, No. 7, September 1985, pp. 1659-1685.

2. N. F. Maxemchuk, "The Manhattan Street Network", Globecom '85, Dec. 2-5, 1985, New Orleans, Louisiana, pp. 252-261.

3. N. F. Maxemchuk, "Routing in The Manhattan Street Network", submitted to IEEE Transaction on Communications.

4. .W. Stallings, "LOCAL NETWORKS An Introduction", Macmillan Publishing Company, New York, 1984.

5. D. E. Huber, W. Steinlin, P. J. Wild, "SILK: An implementation of a Buffer Insertion Ring", IEEE Journal on selected areas in communications, Vol. SAC-1, No. 5, November 1983, pp. 766-774.

6. R. T. Prosser, "Routing Procedures in Communication Networks Part I: Random Procedures", IRE Trans. on Comm. Systems, December 1962, pp. 322-329.

7. L. Kleinrock, "Communication Nets", McGraw-Hill, New York, 1964.

SIGNAL FLOW GRAPHS FOR PATH ENUMERATION AND DEFLECTION ROUTING ANALYSIS IN MULTIHOP NETWORKS

Ender Ayanoğlu

AT&T Bell Laboratories
Holmdel, NJ 07733

ABSTRACT

In this paper, we apply the method of signal flow graphs to path enumeration and deflection routing performance analysis in multihop networks. The method, which is similar to the one used to calculate the distance properties and error performance of convolutional codes, uses a labeling of the edges of the network graph, and the subsequent solution for the transfer function of the resulting signal flow graph. A Taylor series expansion of the transfer function yields the number of paths of a given length between the source and the destination. It is desirable that this number be large so that routing strategies can provide alternate paths to reach a node in case of congestion or link failure, with increased reliability, recoverability, and reduced delay. On the other hand, the derivative of the transfer function yields the average number of hops in the network as a rational function of the probability of deflection of a packet. We then use a fixed point algorithm to solve for the probability of deflection as a function of the arrival rate of packets. Combining the two results, we obtain the average number of hops as a function of the packet arrival rate. We give analysis examples chosen from proposed multihop network architectures.

1   Introduction

A signal flow graph (a linear signal flow graph) is a weighted directed graph which is a one-to-one correspondence with a set of linear equations [1]. These graphs are used in several engineering fields, in particular in linear system theory to calculate transfer functions of linear systems in the Laplace or Z transform domain. In that case, one topologically associates a linear system consisting of multipliers, adders, and rational functions of the transform variable with a signal flow graph, and uses matrix elimination methods or Mason's rule [1] to determine the transfer function between the input and the output of the system. The same method is used in convolutional channel coding theory to calculate distance properties of convolutional codes and their performance bounds [2].

The method of deflection routing (also known as hot-potato routing) is used in high-speed multihop networks due to its simplicity of implementation. Its basic principle is to eliminate the need for storage at intermediate nodes in a high-speed network since this requires high-speed memory, which is either not available or expensive. For example, for lightwave networks, optical storage is not available and, using electronic storage requires optical-to-electrical and electrical-to-optical conversions in addition to high-speed electronic storage, which increases the cost. A practical method to implement lightwave networks while simpli-

fying receiver circuitry is multihop networks. In multihop networks data travel from their source to their destination via several hops through intermediate nodes. A node in such a network has typically the same number of incoming and outgoing links, in addition to the input and output connections to the local node, connected by means of a network interface unit (NIU). The NIU administers the transmission of incoming and locally generated packets to other nodes, as well as the absorption of incoming packets destined for the local node. A multihop network that employs deflection routing uses fixed-size packets, each transmitted within a time slot. At one NIU of such a network, if all the incoming links carry data and none of the incoming packets is destined for the local node, all of them are transmitted on the outgoing links and the locally generated packets are not injected into the network. If any two of the outgoing packets have the same preferred outgoing link on their destination path, the conflict is resolved by transmitting only one of them on its preferred link. The other packet, instead of being stored temporarily at the local node, is then transmitted on one of the other outgoing links, i.e., it is "deflected." This packet then continues its journey toward its destination from the node it now reaches. This increases the number of hops a packet incurs while it travels in the network. However, since deflection routing simplifies implementation, and since the network has high-speed transmission links, the increase in delay and the associated reduction in network throughput may be tolerable. There exist several network topologies for which deflection routing has been proposed for local and metropolitan area networks and implemented as experimental prototypes [3], [4].

Deflection routing was invented in the early days of computers and computer communication, when computation was more expensive relative to the transmission of data. Then, simple routing strategies were developed that were easy to implement but were wasteful of channel capacity. In the simplest such scheme, random routing, when a packet arrives at a node and if it is not destined for that node, it is transmitted on an outgoing link chosen at random. As described above, in deflection routing there is a preferred link for a packet arriving at a node; but if that link is not available, one of the remaining links is chosen for the transmission of that packet. The criterion with which this link is chosen may be closeness to destination or smallest delay, or it may be random. Note that both of these strategies allow loops to be existent on transmission paths. As computation became less expensive, routing strategies that are computationally more intensive but that are more efficient on channel capacity were developed. With the advent of lightwave communication, however, transmission of data became relatively more inexpensive again. Currently deflection routing is being considered for lightwave net-

Reprinted from *Proc. IEEE Globecom '89*, pp. 1022–1029,
November 1989.

works since it enables the elimination of storage at intermediate nodes, which becomes difficult as transmission speeds get higher; and to exploit high bandwidth transmission media, as opposed to storing the packet temporarily at the intermediate node.

In its first part, this paper describes the use of signal flow graphs for determining path length properties of networks for the analysis of routing strategies that allow loops on transmission paths such as random routing or deflection routing. For the analysis and design of such routing strategies, it is important to know the number of paths of a given length from any one node to another. It is desirable that this number be large so that routing strategies can provide alternate paths to reach a node, in case of congestion or link failure, with increased reliability, recoverability, and reduced delay. In the second part of this paper, we extend the method of signal flow graphs to calculate the delay of a packet accepted into the network in terms of the average number of hops as a rational function of the probability of deflection p at a node, in a similar way to calculating the distance properties of convolutional codes under bit errors. We then use this rational function, together with another rational function of p, the average number of don't care decisions for outgoing links on a path, that is calculated with the same technique to solve a set of nonlinear equations that yields the probability of deflection in terms of the packet arrival rate at a node. This enables us to characterize the delay as a function of the packet arrival rate.

For simplicity of description, we restrict ourselves to networks with in- and out-degree equal to 2, i.e., there are two incoming and outgoing links to each NIU in the network. However, this condition is not necessary and can easily be relaxed. In our analysis, we assume a uniform traffic pattern, and that the link utilizations in the incoming links to an NIU is independent as in [5]. Among several possible criteria for choosing which packet to deflect, we analyze the random rule, i.e., the packet to be deflected is chosen randomly among competing ones for an outgoing link. We also assume that a new packet is generated at a local node independently of the utilizations of the incoming links. The calculation results presented here corroborate with simulation and calculation results in [5]. Our method uses a symbolic method to solve a set of linear equations and the subsequent differentiation of a rational function. In this paper, we used a commercially available computer algebra package to carry out the calculations. Computer algebra routines tailored to this application can be written to increase efficiency. Further improvements in computational efficiency can be obtained by using Mason's rule to evaluate the transfer function.

The value of an analytical solution to a problem cannot be overstated. Our technique provides analytical expressions as opposed to a numerical procedure to calculate delay as a function of the probability of deflection, and hence it provides an analytical solution to the deflection routing performance analysis problem.

2 Path Enumeration

We will illustrate the method by means of an example. Consider the simple network in Figure 1. We are interested in finding the number of paths from node a to node b at a distance l, where one unit of distance is one pass over one link.

To do this, we remove the paths outgoing from node b, making it a true sink, and add an auxiliary node a' which will serve as the source, providing a directed edge from this node to a. Associate weight 1 to this edge, and weights D to all the other edges to obtain the directed graph in Figure 2. D is a dummy variable representing one pass through a link. In a signal flow graph, contributions of the traversed edges are multiplicative, therefore this formulation enables one to add the number of passes in terms

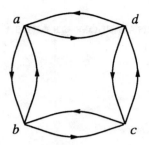

Figure 1: Example network.

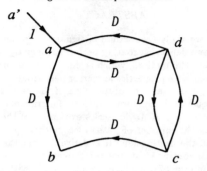

Figure 2: Modified example network for calculating $T_{ba}(D)$, the transfer function from a to b.

of powers of D.

Now, write the node equations (i.e., the weighted paths that terminate at the node)

$$
\begin{aligned}
a &= a' + Dd \\
b &= Da + Dc \\
c &= Dd \\
d &= Da + Dc.
\end{aligned}
\tag{1}
$$

Solving this set of linear equations, one obtains b in terms of a' as

$$
b = \frac{Da'}{1 - 2D^2}
\tag{2}
$$

from which the transfer function $T_{ba}(D) = b/a'$ can be obtained. By expanding the rational function into a geometric series, one obtains

$$
T_{ba}(D) = \frac{D}{1 - 2D^2} = D + 2D^3 + 4D^5 + \cdots = \sum_{l=0}^{\infty} 2^l D^{2l+1}.
\tag{3}
$$

As in convolutional channel coding theory, this series yields significant information on path length properties of the graph in Figure 2. It states that there is one path from node a to node b of length 1, two paths of length 3, four paths of length 5, and so on. In general, from (3) we see that there are 2^l paths of length $2l + 1$ for $l \geq 0$. This graph is simple enough so that the result can be calculated almost by inspection. However, for more complicated graphs, the method of inspection does not work and one has to resort to a systematic procedure. The signal flow graph method described above can be automated by using a symbolic evaluation package like *Macsyma* (trademark of Symbolics, Inc.), *Mathematica* (trademark of Wolfram Research, Inc.), or *Maple* (registered trademark of the University of Waterloo), or by using Mason's rule.

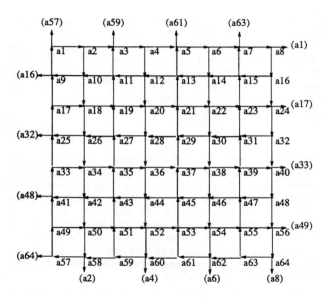

Figure 3: 64-node Manhattan Street Network.

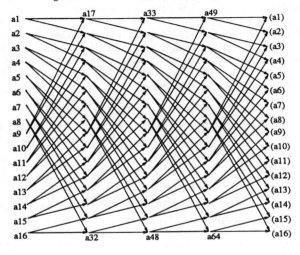

Figure 4: 64-node ShuffleNet.

3 MS and SN: Path Number Comparison

Manhattan Street Network is a toroidal mesh network with in- and out-degree 2, shown in Figure 3 for 64 nodes [3]. It is highly modular and easily growable. The nodes in this network are placed on a grid, and are connected vertically and horizontally in a mesh structure. As in the streets of Manhattan, vertical and horizontal connections alternate in direction. This enables one to easily get back to the original path in the case of a deflection by "walking around the block," with a penalty of only 4 hops. The network is wrapped around a sphere in a toroidal fashion by connecting nodes on the vertical or horizontal edges of the mesh to the nodes on the other vertical and horizontal edge to provide full connectivity. Thus, there are four different basic ways to get back to a node from itself: two by going around the block with a distance of 4 each, and two by going around the torus with a distance of $\sqrt{N}$ each. Although it is desirable for a node to have several short paths to get back to the original path in the case

of a deflection, the existence of a large number of such nodes prevents one from going to other nodes in a smaller number of hops, increasing the average number of hops between nodes.

ShuffleNet is a cylindrical multi-stage shuffle exchange network, shown in Figure 4 for 64 nodes [4]. It can be implemented with any in- and out-degree, but in this paper we will only look at the case where the in- and the out-degrees are equal to 2. Then, SN has 2^k nodes at each side of a column, connected in a perfect shuffle pattern, where k is the number of columns in the network, i.e., $N = k2^k$. The network is cylindrically wrapped around, i.e., nodes to the right of the last column are the same as those to the left of the first one. There is only one basic path that one can get back to oneself, which is by going around the cylinder, the distance of which is k. ShuffleNet is not as modular and as easily growable as the Manhattan Street Network.

In this section, we will compare the 64-node Manhattan Street network with the 64-node ShuffleNet. In the sequel, we will denote the Manhattan Street network as MS, ShuffleNet as SN, 64-node MS as MS-64, and 64-node SN as SN-64.

Using *Macsyma*, we have calculated the transfer functions from node $a1$ to all the other nodes in both networks. (Due to perfect symmetry, this gives information about any node in the network.) For SN-64, there are seven distinct transfer functions which we will denote by $T_1(D)$–$T_7(D)$,

$$
\begin{aligned}
T_1(D) &= D + D^5 + 15D^9 + 225D^{13} + 3375D^{17} + \cdots \\
T_2(D) &= D^2 + 3D^6 + 45D^{10} + 675D^{14} + 10125D^{18} + \cdots \\
T_3(D) &= D^3 + 7D^7 + 105D^{11} + 1575D^{15} + 23625D^{19} + \cdots \\
T_4(D) &= D^4 + 15D^8 + 225D^{12} + 3375D^{16} + 50625D^{20} + \cdots \\
T_5(D) &= 2D^5 + 30D^9 + 450D^{13} + 6750D^{17} + \cdots \quad (4) \\
T_6(D) &= 4D^6 + 60D^{10} + 900D^{14} + 13500D^{18} + \cdots \\
T_7(D) &= 8D^7 + 120D^{11} + 1800D^{15} + 27000D^{19} + \cdots .
\end{aligned}
$$

Due to the symmetry in the network, it can easily be seen that nodes $a17$ and $a18$ have the same transfer function. This is what we denote as $T_1(D)$ above. Similarly, nodes $a33$–$a36$ share the same transfer function, which we have denoted as $T_2(D)$ above. The other node-transfer function relations are as shown below.

$T_1(D)$	$T_2(D)$	$T_3(D)$	$T_4(D)$
$a17, a18$	$a33, \cdots, a36$	$a49, \cdots, a56$	$a2, \cdots, a16$

$T_5(D)$	$T_6(D)$	$T_7(D)$
$a19, \cdots, a32$	$a37, \cdots, a48$	$a57, \cdots, a64$

We show these results in Table 1. The lth entry in the upper row of Table 1 represents the superscript of D in the series in (4), the numbers inside the table represent the nonzero coefficient for that term, and the numbers on the leftmost column represent the number of nodes with that transfer function. Table 2 lists the transfer functions and the number of times they arise for MS-64. A direct interpretation of Tables 1 and 2 is this: the entry in the kth row and lth column gives the number of paths that a node is reachable from $a1$ (and due to symmetry, from any node) in l hops, and there are as many as the kth row entry of the leftmost column of such nodes. On this path, all the nodes (including $a1$ itself), but the destination node can appear more than once. For example, from Table 1, in SN-64 there are sixteen nodes reachable from node $a1$ at a path length 9.

First we calculate the "total" series for both networks by multiplying each series in Tables 1 and 2 by the number of times they appear, and adding the resulting series. This gives, for MS-64

	1	2	3	4	5	6	7	8	9	10
2	1				1				15	
4		1				3				45
8			1				7			
15				1				15		
14					2				30	
12						4				60
8							8			

	11	12	13	14	15	16	17	18	19	20
2			225				3375			
4				675				10125		
8	105				1575				23625	
15		225				3375				50625
14			450				6750			
12				900				13500		
8	120				1800				27000	

Table 1: First 20 terms of the transfer functions for the 64-node ShuffleNet network. Upper row: power of D (the distance), table entries: coefficient for that term (number of paths), leftmost column: number of nodes with that transfer function.

	1	2	3	4	5	6	7	8	9	10
2	1				2				16	
4		1				4				44
8			1				8			
8				1				14		
3				2				16		
12					2				28	
4						2				52
8						4				56
8							6			
4								12		
2									24	

	11	12	13	14	15	16	17	18	19	20
2			200				2896			
4				616				9152		
8	100				1448				21664	
8		208				3128				47152
3		200				2896				43328
12			416				6256			
4				848				12976		
8				832				12512		
8	108				1680				25488	
4		216				3360				50976
2			432				6720			

Table 2: First 20 terms of the transfer functions for the 64-node Manhattan Street network. Upper row: power of D (the distance), table entries: coefficient for that term (number of paths), leftmost column: number of nodes with that transfer function.

$$T_{TMS}(D) = 2D + 4D^2 + 8D^3 + 14D^4 + 28D^5 + 56D^6 + 112D^7 + 208D^8 + 416D^9 + 832D^{10} + 1664D^{11} + 3128D^{12} + 6256D^{13} + 12512D^{14} + 25024D^{15} + 47152D^{16} + 94304D^{17} + 188608D^{18} + 377216D^{19} + 711104D^{20} + \cdots \quad (5)$$

and for SN-64

$$T_{TSN}(D) = 2D + 4D^2 + 8D^3 + 15D^4 + 30D^5 + 60D^6 + 120D^7 + 225D^8 + 450D^9 + 900D^{10} + 1800D^{11} + 3375D^{12} + 6750D^{13} + 13500D^{14} + 27000D^{15} + 50625D^{16} + 101250D^{17} + 202500D^{18} + 405000D^{19} + 759375D^{20} + \cdots. \quad (6)$$

The interpretation of these series is that the coefficient of the lth term equals the number of paths that some node in the network is reachable from $a1$ in l hops. Note that all the coefficients of $T_{TSN}(D)$ are greater than or equal to those of $T_{TMS}(D)$. This result shows that there are more paths from any node in SN-64 to reach the average node than there are in MS-64. A close inspection shows that the coefficients in both of the above series obey the rule

$$c_l = 2 \cdot c_{l-1} - n_l \qquad \text{for } l > 1 \qquad (7)$$

where $c_1 = 2$, and where c_l is the coefficient of the lth term in the series, n_l is the number of paths of length l that originate and close back on $a1$, with $a1$ only appearing at the beginning and at the end of the path. For example, in MS-64, there are two paths that close back on $a1$ of length four, whereas in SN-64 there is one. The coefficients n_l can be obtained by using the signal flow graph method and by transforming $a1$ into two nodes: one with only incoming links and the other with only outgoing links, and treating the latter as the source and the former as the destination in the signal flow graph method.

The series $T_{TMS}(D)$ and $T_{TSN}(D)$ give the number of paths of length l from node $a1$ to any other node regardless of whether the nodes that are reached are distinct or not. One network topology may be considered superior to one another if it has a larger number of distinct nodes that can be reached at l hops. This information can also be deduced from the transfer functions in Tables 1 and 2 by adding the leftmost column entries for the nonzero entries in the table for a particular power of D. We will call the series generated this way as $T_{DMS}(D)$ and $T_{DSN}(D)$. For MS-64, $T_{DMS}(D)$ is given by

$$T_{DMS}(D) = 2D + 4D^2 + 8D^3 + 11D^4 + 14D^5 + 16D^6 + 16D^7 + 15D^8 + 16D^9 + 16D^{10} + 16D^{11} + 15D^{12} + 16D^{13} + 16D^{14} + 16D^{15} + 15D^{16} + 16D^{17} + 16D^{18} + \cdots, \qquad (8)$$

and for SN-64,

$$T_{DSN}(D) = 2D + 4D^2 + 8D^3 + 15D^4 + 16D^5 + 16D^6 + 16D^7 + 15D^8 + 16D^9 + 16D^{10} + 16D^{11} + 15D^{12} + 16D^{13} + 16D^{14} + 16D^{15} + 15D^{16} + 16D^{17} + 16D^{18} + \cdots. \qquad (9)$$

As can be seen from the equations for $T_{TMS}(D)$ and $T_{TSN}(D)$, the number of different nodes that can be reached in l hops in the two networks are equal, except for $l = 4$ and $l = 5$, in which cases there are more nodes in SN-64.

Another way to write $T_{TMS}(D)$ and $T_{TSN}(D)$ is in a sum of products form for the coefficients. This captures how many distinct nodes are reachable in l hops via how many different paths. In this form, $T_{TMS}(D)$ is

$$\begin{aligned}
T_{TMS}(D) = &(2 \cdot 1)D + (4 \cdot 1)D^2 + (8 \cdot 1)D^3 + (8 \cdot 1 + 3 \cdot 2)D^4 + \\
&(14 \cdot 2)D^5 + (12 \cdot 4 + 4 \cdot 2)D^6 + (8 \cdot 8 + 8 \cdot 6)D^7 + \\
&(3 \cdot 16 + 8 \cdot 14 + 4 \cdot 12)D^8 + (12 \cdot 28 + 2 \cdot 24 + 2 \cdot 16)D^9 + \\
&(8 \cdot 56 + 4 \cdot 52 + 4 \cdot 44)D^{10} + (8 \cdot 108 + 8 \cdot 100)D^{11} + \\
&(4 \cdot 216 + 8 \cdot 208 + 3 \cdot 200)D^{12} + (12 \cdot 416 + 2 \cdot 200)D^{13} + \\
&(4 \cdot 848 + 8 \cdot 832 + 4 \cdot 616)D^{14} + (8 \cdot 1680 + 8 \cdot 1448)D^{15} + \\
&(4 \cdot 3360 + 8 \cdot 3128 + 3 \cdot 2896)D^{16} + \\
&(8 \cdot 6720 + 12 \cdot 6256 + 2 \cdot 2896)D^{17} + \\
&(12 \cdot 13500 + 4 \cdot 10125)D^{18} + (8 \cdot 25488 + 8 \cdot 21664)D^{19} + \\
&(4 \cdot 50976 + 3 \cdot 43328 + 8 \cdot 47152)D^{20} + \cdots, \qquad (10)
\end{aligned}$$

and $T_{TSN}(D)$ is

$$\begin{aligned}
T_{TSN}(D) = {} & (2\cdot 1)D + (4\cdot 1)D^2 + (8\cdot 1)D^3 + (15\cdot 2)D^4 + \\
& (14\cdot 2 + 2\cdot 1)D^5 + (12\cdot 4 + 4\cdot 3)D^6 + (8\cdot 8 + 8\cdot 7)D^7 + \\
& (15\cdot 15)D^8 + (14\cdot 30 + 2\cdot 15)D^9 + (12\cdot 60 + 4\cdot 45)D^{10} + \\
& (8\cdot 120 + 8\cdot 105)D^{11} + (15\cdot 225)D^{12} + \\
& (14\cdot 450 + 2\cdot 15)D^{13} + (12\cdot 900 + 4\cdot 675)D^{14} + \\
& (8\cdot 1800 + 8\cdot 1575)D^{15} + (15\cdot 3375)D^{16} + \\
& (14\cdot 6750 + 2\cdot 3375)D^{17} + \\
& (4\cdot 12976 + 8\cdot 12512 + 4\cdot 9152)D^{18} + \\
& (8\cdot 27000 + 8\cdot 23625)D^{19} + (15\cdot 50625)D^{20} + \cdots \quad (11)
\end{aligned}$$

Now, every coefficient has a sum of products form, where in each term the first multiplier in a product is the number of distinct nodes, and the second is the number of paths those distinct nodes are reachable from $a1$ in l hops.

Obviously, a network is superior (in the sense of providing a richer set of connections from a node to *distinct* others) if for every such coefficient in its total transfer function, all the terms have multipliers equal to or greater than the corresponding ones in the other network's total transfer function (where the greater condition holds at least once). By comparing the two series above, we observe that for the first 20 terms, all coefficients of SN-64 possess the above property with the exception of those for $l = 4, 8, 9,$ and 20. We list those below, underlining those contributions for which SN-64 is not superior in the above sense.

Term	MS-64
D^4	$8\cdot 1 + \underline{3\cdot 2}$
D^8	$\underline{3\cdot 16} + 8\cdot 14 + 4\cdot 12$
D^9	$12\cdot 28 + 2\cdot 24 + \underline{2\cdot 16}$
D^{20}	$\underline{4\cdot 50976} + 3\cdot 43328 + 8\cdot 47152$

Term	SN-64
D^4	$8\cdot 1 + \underline{7\cdot 1}$
D^8	$\underline{3\cdot 15} + 8\cdot 15 + 4\cdot 15$
D^9	$12\cdot 30 + 2\cdot 30 + \underline{2\cdot 15}$
D^{20}	$\underline{4\cdot 50625} + 3\cdot 50625 + 8\cdot 50625$

SN-64 is not uniformly superior to MS-64 in the above sense. However, there are very few cases in which MS-64 has more paths to reach the same number of nodes in l hops than SN-64. Therefore, we can say that in most cases SN-64 has more paths to reach the same number of distinct nodes in l hops than MS-64.

Tables 1 and 2 also reveal distance information about the two networks. The distance between two nodes is the power of the first nonzero entry in the transfer function. The maximum distance is the column index for the row whose first nonzero index is the largest in Tables 1 and 2. The maximum distance for SN-64 is 7, whereas the maximum distance for the MS-64 is 9. On the other hand, the weighted distance (sum of distances of all nodes) of SN-64 is 292 and that of MS-64 is 316. This makes the average distance of SN-64 4.63 (as should be expected [4]) whereas that of MS-64 5.02 (again, as should be expected [3]).

The number of distinct transfer functions in a network is a measure of its symmetry. In the two cases we considered, this measure is 7 and 11 for SN-64 and MS-64, respectively.

4 Calculating the Average Number of Hops

We will illustrate the method by means of the same example. Consider the simple network in Figure 1 again. We are interested in finding a transfer function that will enable us to calculate the average number of hops as a function of the probability of deflection at a node.

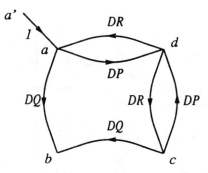

Figure 5: Modified example network for calculating the transfer function $T_{ba}(D, P, Q, R)$.

For that purpose, we again remove the paths leaving node b, making it a true sink, and add an auxiliary node a' which will serve as the source, providing a directed edge from this node to a. Associate weight 1 with this edge. Now, consider the edges going from node a, ab and ad. Note that, taking the edge ab one reaches the node b in a single hop. On the other hand, taking the edge ad one arrives at b at a minimum of three hops. Hence, the edge ab is the preferred one in going from node a to b. We associate the weight DQ with the preferred edge ab, and the weight DP with the other edge, ad. Similarly, we assign the weight DQ to cb, and DP to cd. On the other hand, taking either da or dc one arrives at b in two hops. In other words, a packet arriving at node d destined to go to node b has a "don't care" decision in terms of choosing a link to go out on. In this case, we associate the weight DR to both edges da and dc. The result is in Figure 5.

Now, write the node equations

$$\begin{aligned}
a &= a' + DRd \\
b &= DQa + DQc \\
c &= DRd \\
d &= DPa + DPc.
\end{aligned} \quad (12)$$

By solving these equations, we can calculate the transfer function $T_{ba}(D, P, Q, R)$ as

$$T_{ba}(D, P, Q, R) = \frac{b}{a'} = \frac{DQ}{1 - 2D^2 PR}. \quad (13)$$

By expanding T_{ba} into a geometric series, we obtain

$$T_{ba}(D, P, Q, R) = DQ + 2D^3 PQR + 4D^5 P^2 QR^2 + \cdots. \quad (14)$$

In general, the transfer function $T_{ba}(D, P, Q, R)$ is of the form

$$T_{ba}(D, P, Q, R) = \sum_{i,j,k} a_{ijk} D^i P^j Q^k R^{i-j-k} \quad (15)$$

where a_{ijk} is defined as

a_{ijk} : Number of paths from node a to node b whose length is i which have j deflected and k preferred edges. $\quad (16)$

The transfer function introduced in Section 2 can be derived from (15) as

$$T_{ba}(D) = \sum_i a_i D^i = T_{ba}(D, P, Q, R)\Big|_{\substack{P=1 \\ Q=1 \\ R=1}} \quad (17)$$

321

where

$$a_i = \sum_{j,k} a_{ijk}. \tag{18}$$

Recall that we would like to calculate the average number of hops from node a to node b as a function of the deflection probability p. We define this as $\bar{h}_{ba}(p)$. Using the definition (17) of a_{ijk}, $\bar{h}_{ba}(p)$ can be calculated as

$$\bar{h}_{ba}(p) = \sum_{i,j,k} i \, a_{ijk} \, p^j (1-p)^k \left(\frac{1}{2}\right)^{i-j-k}. \tag{19}$$

From a comparison of (15) and (19), it can be seen that $\bar{h}_{ba}$ can be obtained from T_{ba} as

$$\bar{h}_{ba}(p) = \left. \frac{\partial T_{ba}(D,P,Q,R)}{\partial D} \right|_{\substack{D=1 \\ P=p \\ Q=1-p \\ R=0.5}}. \tag{20}$$

From (13), $\bar{h}_{ba}(p)$ for the example network can be obtained as

$$\bar{h}_{ba}(p) = \frac{1+p}{1-p}. \tag{21}$$

To determine the average number of hops versus the arrival rate, we need the probability of a don't care decision that a packet encounters when it arrives at a node. In order to calculate this quantity, first we define the average number of don't care nodes that a path encounters on its path from a to b as $\bar{d}_{ba}(p)$. Similarly to $\bar{h}_{ba}(p)$, using the definition (17) of a_{ijk}, $\bar{d}_{ba}(p)$ can be calculated as

$$\bar{d}_{ba}(p) = \sum_{i,j,k} (i-j-k) \, a_{ijk} \, p^j (1-p)^k \left(\frac{1}{2}\right)^{i-j-k}. \tag{22}$$

Again, by using (17), we can calculate $\bar{d}_{ba}$ from T_{ba} as

$$\bar{d}_{ba}(p) = \left. R \frac{\partial T_{ba}(D,P,Q,R)}{\partial R} \right|_{\substack{D=1 \\ P=p \\ Q=1-p \\ R=0.5}}. \tag{23}$$

For the example network, we have

$$\bar{d}_{ba}(p) = \frac{p}{1-p}. \tag{24}$$

Then, the probability of a packet having a don't care decision on the path from a to b, $P_x^{ba}(p)$, can be calculated as

$$P_x^{ba}(p) = \frac{\bar{d}_{ba}(p)}{\bar{h}_{ba}(p)}, \tag{25}$$

which, for the example network, is equal to

$$P_x^{ba}(p) = \frac{p}{1+p}. \tag{26}$$

Due to symmetry, we have $\bar{h}_{da}(p) = \bar{h}_{ba}(p)$, $\bar{d}_{da}(p) = \bar{d}_{ba}(p)$, and $P_x^{da}(p) = P_x^{ba}(p)$ in the example network.

Similarly, for node c, we calculate

$$T_{ca}(D,P,Q,R) = \frac{2D^2 QR}{1 - 2D^3 PR}, \tag{27}$$

$$\bar{h}_{ca}(p) = \bar{d}_{ca}(p) = \frac{2}{1-p}, \tag{28}$$

```
Set all edge entries to ∞;
Set edge entries for the destination node to 0;
Set distance = 0;
While (there are edge entries = ∞) do
        For all nodes node with an edge entry = distance do
            newdistance = 1 + min (edge entries of node);
            For all nodes previousnode with an edge to node do
                Edge entry from previousnode to node = newdistance;
            End for
        End for
        distance = distance + 1;
End while
For all nodes node do
        For all nodes nextnode with an edge coming from node do
            If edge entry for nextnode < other next node of node
                label edge between node and nextnode DQ;
            Else if edge entry for nextnode = other next node of node
                label edge between node and nextnode DR;
            Else
                label edge between node and nextnode DP;
            End if
        End for
End for
```

Table 3: Algorithm to label the edges of the network graph for graphs with out-degree 2, given a destination node. "Edge entries of a node" refers to the labels of the edges leaving that node.

and

$$P_x^{ca}(p) = 1. \tag{29}$$

For a network with N nodes, the average number of hops is given as

$$\bar{h}(p) = \frac{1}{N(N-1)} \sum_{n=1}^{N} \sum_{\substack{m=1 \\ m \neq n}}^{N} \bar{h}_{mn}(p) \tag{30}$$

and the average probability of a don't care decision is given as

$$P_x(p) = \frac{1}{N(N-1)} \sum_{n=1}^{N} \sum_{\substack{m=1 \\ m \neq n}}^{N} P_x^{mn}(p). \tag{31}$$

For the example network, we can now calculate the average number of hops as

$$\bar{h}(p) = \frac{1}{3} \left(\bar{h}_{ba}(p) + \bar{h}_{ca}(p) + \bar{h}_{da}(p) \right) = \frac{2(2+p)}{3(1-p)}, \tag{32}$$

and the average probability of a don't care decision as

$$P_x(p) = \frac{1}{3} \left(P_x^{ba}(p) + P_x^{ca}(p) + P_x^{da}(p) \right) = \frac{1+3p}{3(1+p)}. \tag{33}$$

5   Labeling the Edges of the Network Graph

In this method, edges of the network graph need to be labeled with one of the symbols DP, DQ, or DR for each source and destination pair. This can be easily accomplished by backtracking the distance from the destination associated with each edge using breadth-first labeling, in a dynamic fashion. We describe the algorithm in pseudo-code in Table 3.

The algorithm first associates a weight to every edge in the graph, equal to the minimum distance from the originating node of this edge to the destination, provided that that edge is chosen. Then, it examines all the edge entries originating from a node. If they are all equal, it labels all those edges with DR, otherwise it labels the edge with the smaller distance DQ, and the one with the larger distance DP.

6 Probability of Deflection and Packet Arrival Rate

In this section we find the relation between the probability of deflection p and the probability of packet arrival P_g using an analysis of the events of link utilization and packet deflection. This analysis follows the one-packet model in [5]. In this section we will drop the dependence of p on the pertinent variables for simplicity of notation.

First, we observe that a packet takes, on the average, $\bar{h}$ hops to reach its destination. Therefore, the probability that the packet is at its destination, or its probability of absorption, is given as

$$P_a = \frac{1}{\bar{h}}. \tag{34}$$

We denote the link utilization, that is, the probability of a link carrying a packet in a time slot, by ρ. We denote the probability of packet generation in a slot, or the arrival rate, by P_g. By assuming that the two links entering into a node are statistically independent, we observe the probability that both links carry a packet is given by ρ^2, exactly one carries a packet is given by $\rho(1-\rho)$, and neither carries a packet is given by $(1-\rho)^2$. Also note that an arriving packet is injected into the network unless both links relay the traffic destined for other nodes. Under these observations, we can deduce the following equation for the steady state link utilization

$$\begin{aligned} \rho &= (1-\rho)^2\frac{P_g}{2} \\ &+ \rho(1-\rho)\left((1-P_a)(1+P_g) + P_aP_g\right) \\ &+ \rho^2\left((1-P_a)^2 + P_a(1-P_a)(1+P_g) + P_a^2\frac{P_g}{2}\right). \end{aligned} \tag{35}$$

Now, assume that we have a packet with a preferred link going into an intermediate node in the network. Observe that a packet gets deflected only when another packet is present to leave the node. For a packet arriving at a node to meet a packet, one of the three conditions need to be met: *(i)* another packet is carried on the other incoming link but not absorbed, or *(ii)* it is absorbed but a new packet is generated locally, or *(iii)* a packet is not carried on the other incoming link but a packet is generated locally. Also observe that our incoming packet gets deflected if the competing packet does not have a don't care decision for outgoing link selection (probability $1 - P_x$), it has the same preferred link as our packet (probability $1/2$), and the competing packet wins the coin toss (probability $1/2$). Note from (19) that p is defined as the probability of deflection given that a packet has a preferred link. Summarizing these conditions, we get the following equation for deflection probability

$$p = \frac{1}{4}(1-P_x)\left(\rho(1-P_a) + (\rho P_a + 1 - \rho)P_g\right). \tag{36}$$

Since we have expressions for P_a and P_x as rational functions of p, we can solve the nonlinear equations (35) and (36) using a fixed point algorithm. We present this algorithm in Table 4.

Equation (36) is an approximation for p, since it only considers the deflection of a packet already in the network, arriving at an NIU, and ignores the deflection of a packet generated locally.

Set ϵ_ρ to accuracy with which to calculate ρ, e.g., 10^{-4};
Set ϵ_p to accuracy with which to calculate p, e.g., 10^{-4};
Set ΔP_g to increments of P_g desired, e.g., 10^{-3};
Set P_g to 0;
Set ρ^* to 0;
Set p^* to 0;
While $(P_g \leq 1.0)$ do
 Set $\rho = \rho^* - 2\epsilon$;
 Set $p = p^* - 2\epsilon$;
 While $((|\rho^* - \rho| > \epsilon)$ or $(|p^* - p| > \epsilon))$ do
  Set $\rho = \rho^*$;
  Set $p = p^*$;
  Set $\rho^* = $ RHS(35);
  Set $p^* = $ RHS(36);
 End while
 Printout P_g, p;
 Set $P_g = P_g + \Delta P_g$;
End while;

Table 4: Fixed point algorithm to calculate probability of deflection in terms of the packet arrival rate.

However, since in most cases a node relays traffic due to other source-destination pairs, it is a reasonable approximation and, as will be seen later, it gives close results to those obtained by simulation.

7 MS and SN: Average Number of Hops Comparison

In Section 3, we compared MS-64 and SN-64 for their path properties between nodes when the two networks use random or deflection routing. In this section, we compare these two networks for their delay versus packet arrival rate performance under deflection routing.

We have used the technique in Section 4 to calculate the transfer functions $T(D, P, Q, R)$ for MS-64 and SN-64. We then used (32) and (33) to calculate the average probability of hops and the average number of don't cares as functions of p. Finally, we have used the latter results, and (35) and (36) to calculate $p(P_g)$ and $\bar{h}(p)$, and obtain $\bar{h}(P_g)$ from these. Those results are plotted in Figure 6. Also shown in Figure 6 are the simulation results from [5]. Please note the different scales used in plotting different quantities in Figure 6, and the notation $\bar{h}(P_g) = \bar{h}(p(P_g))$.

First, we observe that the number of hops in deflection routing varies within a factor of two for P_g in the interval $[0, 1]$. This is a tolerable variation, and considering the simplicity of implementation of deflection routing, implies that it is a feasible alternative for the solution of high-speed buffer problem in high-speed multihop networks such as lightwave networks. Second, the behavior of $\bar{h}(p)$, the average number of hops versus the probability of deflection, is similar in both networks, except MS-64 has consistently higher number of average hops than SN-64 for a given probability of deflection. On the other hand, MS-64 has slightly lower probability of deflection p for a given probability of packet arrival P_g than SN-64. The basic reason for this can be observed by noting from (36) that p is directly proportional to the probability of not having a don't care decision at a node, $(1 - P_x)$. This number is higher for SN-64 as compared to MS-64 throughout the range of P_g. Since $p(P_g)$ varies between 0 and 0.17 for both networks, the operation is on the flat portion of the $\bar{h}(p)$ curve. Since $p_{SN}(P_g)$ is slightly higher than $p_{MS}(P_g)$ over the interval $P_g \in [0, 1]$, the difference $\bar{h}_{MS}(p) - \bar{h}_{SN}(p)$ remains about the same for this range of P_g. We have also plotted simu-

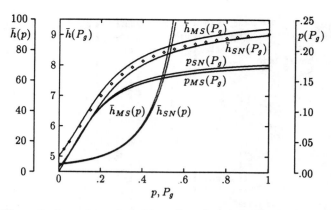

Figure 6: Average number of hops $\bar{h}$, and probability of deflection p, as functions of the probability of packet arrival P_g and the probability of deflection p in 64-node ShuffleNet (SN) and 64-node Manhattan Street Network (MS), and simulation results of Greenberg and Goodman for 64-node Manhattan Street Network, indicated by small rectangles. Note the different scales and the notation that $\bar{h}(P_g) = \bar{h}(p(P_g))$.

lation results for MS-64 from [5] in Figure 6. The agreement of simulation results and our calculations are generally good over the range $P_g \in [0, 1]$, the error increasing with P_g. We believe this can be further improved by a more detailed equation for p than (36), incorporating the deflection of a packet entering into the network.

8 Summary and Conclusions

In this paper we have described the application of the method of signal flow graphs to path enumeration and deflection routing in multihop networks. First, we have applied the method of signal flow graphs to the calculation of path lengths in a multihop network employing random or deflection routing. This calculation provides insight into the number of alternate paths to reach a node in case of congestion or link failure, which provide increased reliability, recoverability, and reduced delay. The method consists of labeling the edges of the network graph with a dummy variable, solving for the transfer function of the resulting signal flow graph, and expanding the transfer function into a Taylor series. The coefficients in this series correspond to the number of paths between the source and the destination with a given length.

Using this method, we have compared 64-node Manhattan Street Network (MS-64) and 64-node ShuffleNet (SN-64), and derived the following conclusions: in SN-64 *(i)* there exist more paths to reach a node from any other in l hops, *(ii)* there exist a larger number of distinct nodes reachable from a node in l hops, *(iii)* in most cases, there are more paths to reach those distinct nodes in l hops than the MS-64 network. In addition, SN-64 has smaller maximum distance and average distance, and is more symmetric as measured by the number of distinct transfer functions than the MS-64.

We then described the application of the method of signal flow graphs to the analysis of deflection routing in multihop net-

works. The method uses a labeling of the edges of the network graph, and the subsequent solution for the transfer function of the resulting signal flow graph, similarly to the technique for the enumeration of paths summarized above. This time, however, the derivatives of this transfer function with respect to the pertinent variables yield the average number of hops in the network, as well as the average probability of don't care decisions for a packet in the network, as rational functions of the probability of deflection of a packet. We then use a fixed point algorithm to solve for the probability of deflection as a function of the arrival rate of packets, as described in [5], which yields the average number of hops as a function of the packet arrival rate.

Using this method, we have analyzed 64-node and 24-node Manhattan Street Network (MS) and ShuffleNet (SN) this time for their deflection routing performance. We have observed that in both of these networks, the increase in average number of hops remains within a factor of two, making deflection routing a viable alternative to solve the high-speed buffer problem in high-speed networks, particularly lightwave networks. We have also observed that the deflection routing performance in terms of the average number of hops is better for SN than MS in the case of both 64 nodes and 24 nodes, although the probability of deflection for MS is lower than the one for SN in both cases. Our calculation results are in good corroboration with simulation and calculation results published in [5] for MS-64.

We have used a general, commercially available computer algebra package to solve for the transfer function. The efficiency of the technique can be improved by using specialized computer algebra routines, or Mason's rule to solve for the transfer function.

9 Acknowledgment

The author would like to thank Richard J. Caballero for his help with *Macsyma* programs, and to Mark Karol for several productive discussions.

References

[1] L. A. Zadeh and C. A. Desoer, *Linear System Theory: The State Space Approach*, McGraw Hill, New York, 1963.

[2] A. J. Viterbi and J. K. Omura, *Principles of Digital Communication and Coding*, McGraw-Hill, New York, 1978.

[3] N. F. Maxemchuk, "Regular Mesh Topologies in Local and Metropolitan Area Networks," *AT&T Technical Journal*, Vol. 64, pp. 1659–1685, September 1985.

[4] A. S. Acampora, M. J. Karol, and M. G. Hluchyj, "Terabit Lightwave Networks: The Multihop Approach," *AT&T Technical Journal*, Vol. 66, pp. 21–34, November 1987.

[5] A. G. Greenberg and J. Goodman, "Sharp Approximate Models of Deflection Routing in Mesh Networks," To be published in the *IEEE Transactions on Communications*.

Sharp Approximate Models of Deflection Routing in Mesh Networks

Albert G. Greenberg and Jonathan Goodman

Abstract— Deflection routing is a simple, decentralized, and adaptive method for routing data packets in communication networks. In this paper we focus on deflection routing in the *Manhattan street network* (a two-dimensional directed mesh), though our analytic approach should apply to any regular network. We present two approximate performance models that give sharp estimates of the steady state throughput and the average packet delay for packets admitted to the network. The results of extensive simulation experiments are reported, which corroborate the models' predictions. The results show that deflection routing is very effective. Two measures of the merit of a network for deflection routing are its diameter and its deflection index. Networks are presented whose diameter and deflection index are near the optimal values.

I. INTRODUCTION

IN a large high-speed communication network, the switching elements ought to be simple so that little time is consumed in computing routing decisions. It is of course desirable that the network route data packets with small delay and achieve high throughput when the demand for it arises. Aiming to design practical, large, high-speed networks, several researchers [1], [2], [4], [5], [15], [20], [22] have proposed a novel decentralized routing scheme that we term *deflection routing*. In particular, deflection routing has been proposed and implemented in geographically distributed computer networks [15] and in communication networks embedded in massively parallel computers [20], [5], [6]. A striking property of deflection routing is that packets, once admitted to the network, are not blocked or queued because of congestion. Instead, congestion causes nodes to misroute packets temporarily.

In this paper, we focus on the performance of deflection routing in the Manhattan street network (Fig. 1), though our analytic approach should apply to any regular network. See [7] for a similar analysis of deflection routing in hypercube networks, [10] for a similar analysis of networks like the shuffle networks discussed below, and [17] for performance comparisons between deflection routing and conventional store and forward routing. We give two ways to estimate the performance of deflection routing, using either a *one node* or a *one packet* model. Both models provide estimates of the

Paper approved by the Editor for Communication Networks of the IEEE Communications Society. Manuscript received March 28, 1990; revised February 10, 1991. This paper was presented in part at the 1986 International Seminar on Teletraffic Analysis and Computer Performance Evaluation, Amsterdam, The Netherlands, June 1986.

A. G. Greenberg is with AT&T Bell Laboratories, Murray Hill, NJ 07974.

J. B. Goodman is with Courant Institute of Mathematical Sciences, New York University, New York, NY 10012.

IEEE Log Number 9206077.

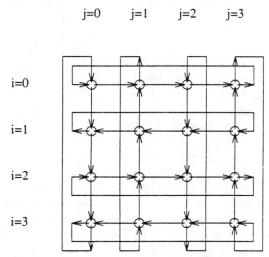

Fig. 1. 4 × 4 Manhattan street network. Nodes are indexed (i, j). Viewed as a communication network, nodes are sources and sinks for packets of data and edges communication links.

throughput in a Manhattan street network with relative error that appears to be ≈1% or less under the full range from very light to very heavy loads; see Tables I and II. The *one packet* model leads to criteria for performance of deflection routing in general networks. We tell how to construct networks that are nearly optimal with respect to two of these criteria.

The Manhattan street network is a directed two-dimensional mesh with toroidal boundaries (Fig. 1). Horizontal and vertical paths alternate in direction. A node operates in a discrete, time slotted fashion. Each link can transmit at most one packet per time slot. A node always attempts to route packets on shortest paths to their destinations.

At each time $t = 0, 1, 2, \cdots$, a given node may receive up to two packets from its neighbors, one for each incoming link. In addition, it may generate one new packet locally. In deflection routing, if at time t the node receives two packets from its neighbors, neither destined for this node, and generates a new packet as well, then the new packet is blocked—either queued or rejected and cleared. A conflict may arise at time t if the node obtains two packets, neither destined to this node, for example, one from one of the two neighbors and one generated locally. If the shortest path calculations performed for the two packets indicate that they should exit the node on different links, then the two exit accordingly at time $t + 1$. Otherwise the node invokes a local conflict resolution rule, which selects one packet to exit on its preferred link, and the other packet

Reprinted from *IEEE Trans. Commun.*, 22, January 1993.

on the other link. We say the misdirected packet is *deflected*. Two conflict resolution rules are introduced and studied here.

To model arrivals of new packets, we assume that at each time slot each node generates a single new packet independently with fixed probability g, and generates no packet with probability $1-g$. If the packet is blocked as just described, then we assume it is rejected and cleared. If $g = 1$ it is as if at each node there is an infinite queue of new packets awaiting transmission in the network. Thus, the model allows us to study saturation conditions, which are critical in design. We assume that the choice of a new packet's destination satisfies a *relabeling invariance* condition, made precise in Section III. That condition is satisfied if, for example, new packets' destinations are chosen independently and uniformly at random from the $n^2 - 1$ nodes not coinciding with their sources (here n is the number of rows in the network, which is equal to the number of columns).

The resulting network model is a Markov chain. However, for the $n \times n$ network, the chain has at least $n^{O(n^2)}$ states so it is intractable to compute performance statistics (for example, steady-state throughput or mean packet delay) via the steady-state probability distribution. We introduce an *independence approximation*, described in Section IV, which reduces the complexity dramatically. This approximation and the assumption of relabeling invariance lead to a system of difference equations, which we refer to as the *one node* model. These equations express the probability distribution describing the packets that a given node emits at time $t + 1$ in terms of the probability distribution describing the packets the node emits at time t. From these distributions, we can easily calculate the instantaneous network input and output rates. The fixed point of the system provides estimates of the steady state values of these distributions, and estimates of the steady state throughput and average packet delay.

Extensive experiments show the latter estimates are quite accurate. We solved the *one node* model, over a wide range of arrival intensities, for the 4×4, 8×8, and 12×12 networks, and the two conflict resolution rules considered. As described in Section IV, we ran a corresponding series of Monte Carlo simulations. We found the relative errors between the model's predictions and the values measured in simulation were typically 1% or less, and were never as large as 2%. About 10 to 20 times the computer time was needed for running the simulations as opposed to solving the model. Since the model is not stochastic it does not suffer form initialization bias and other difficulties associated with Monte Carlo simulation.

In Section V we present a simpler model, the *one packet* model, providing estimates of the network's steady state behavior that turn out to be about as accurate as those obtained from the one node model. The one packet model describes the motion of a single packet through the network; the contribution of other packets appears through a single parameter: the *probability of deflection, p*. Another fixed point formulation gives p as a function of the offered load. With this model we can calculate higher moments of delay. We used the model to investigate the performance of networks with as many as 10^4 nodes. Such large networks are beyond the feasible reach of

the *one node* model, and beyond that of simulation. Overall, the results show that deflection routing in the Manhattan street network works efficiently.

The *one packet* model exposes a few rough characteristics of the network that influence performance. Two such characteristics are the diameter and the *deflection index* of the network. Informally, the deflection index of a network is the maximal number of hops that a single deflection adds to a packet's delay. (Thus, the delay of a packet is at most the length of a shortest path between its source and destination plus the product of the deflection index and the number of deflections experienced in route.) A graph theoretic definition is given in Section VI.

An advantage of the Manhattan street network is that its deflection index is just four. Any graph that is not completely connected must have deflection index at least two. A disadvantage of the Manhattan street network is that its diameter is relatively large: $\sqrt{N}$ where $N = n^2$ is the total number of nodes. In contrast, the binary shuffle graph [3] has optimal diameter $\log_2 N$ and deflection index also $\log_2 N$ where $N = 2^m$ is the number of nodes, $m \geq 1$. In Section VI we show how to reduce a graph's deflection index to four while only doubling its diameter. Specifically, given a regular, degree 2 graph G of N nodes and diameter d, we produce a regular, degree 2 graph G' of $4N$ nodes, diameter at most $2d + 2$, and deflection index at most 4. Applying the construction to binary shuffle graphs, for example, gives a family of graphs, which we call *double shuffle* graphs, with deflection index 4 and diameter $2 \log_2 N$.

A virtue of deflection routing is simplicity. In the Manhattan street network, the routing calculations could easily be implemented in hardware [15]. The number of buffers involved is minimal. As Maxemchuk [15] points out, several problems that require complex solutions in conventional store and forward routing algorithms, such as buffer management and flow control, are handled simply and efficiently in deflection routing. In this regard, the networks controlled by deflection routing are similar to many bus and ring local area networks. Moreover, simulations [17] show that the performance of deflection routing compares favorably with that of store and forward methods.

II. ROUTING

In this section the details of routing with deflection in the Manhattan street network are presented. Simple and effective conflict resolution rules are proposed.

We number the row and column indexes of the $n \times n$ Manhattan street network form 0 to $n - 1$. Even numbered rows point east (towards increasing column indexes), and odd numbered rows west. Even numbered columns point south (towards increasing row indexes), and odd numbered columns north.

At a given time t, a node (i, j), $0 \leq i, j \leq n - 1$, receives up to two packets, one from each incoming link. A packet destined to (i, j) is absorbed. A packet destined to another node (i_d, i_d) has *routing preference*:

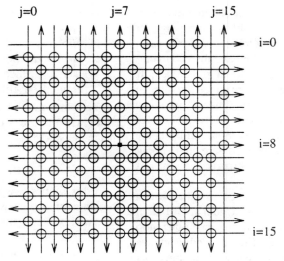

Fig. 2. Circles mark nodes where the routing preference of a packet destined to $(8,7)$ is not "don't care"; $(8,7)$ is marked with a small square.

— *don't care* if both outgoing links lie on the shortest paths from (i, j) to (i_d, i_d),
— *row* if just the outgoing row link lies on a shortest path from (i, j) to (i_d, i_d),
— *column* if just the outgoing column link lies on a shortest path from (i, j) to (i_d, i_d),

Fig. 2 depicts the pattern *don't care* preferences for a packet in route to destination $i_d, i_d = (8, 7)$ in a 16×16 mesh. A node (i, j) is marked with a circle if the packet's preference is not *don't care* at node (i, j), and is left unmarked otherwise. We may regard the four lines of consecutive circles converging at the center of the mesh as delimiting four quadrants. Within a quadrant, the circles occur in a checkerboard pattern. Note that a packet's routing preference is *don't care* at every other hop while moving inside a quadrant.

We say a packet attempts a *critical bend* when its routing preference demands a change in direction. Specifically, a packet at a given node attempts a critical bend at time $t + 1$ if either

• the packets has routing preference *row* at this node and the packet entered the node on the column input at time t, or
• the packet has routing preference *column* at this node and the packet entered the node on the row input at time t.

Fact 1: There is a shortest path from any incoming edge of any node (i, j) to any other node (i_d, j_d) that contains at most three critical bends.

To see this, refer again to Fig. 2 and suppose for definiteness that $(i_d, j_d) = (8, 7)$. The lines $i = 8$ and $j = 7$ divide the network into four quadrants. Label these quadrants, I, II, III, and IV, starting from the upper right and proceeding clockwise. Packets on the III IV boundary moving east and on the III II boundary moving north have no critical bends on their shortest paths. Call these positions zero bend positions. Now, any packet in III that is moving north and east has one critical bend to get onto a zero bend position, so they are one bend positions. A packet in IV moving south or in II moving west is also in a one bend position. Continuing, any packet

in III will be able to bend into a one bend position after one hop so the remaining positions in III are two bend positions. Likewise, any horizontally moving packet in IV or vertically moving packet in II is in a two bend position. Also, any west or south moving packet in I is in a two bend position. Finally, any packet not in a zero, one, or two bend position will be able to bend onto one after one hop.

Routing works as follows. Suppose node (i, j) receives exactly one packet at a time t, and the packet is not destined to (i, j). If the packet's routing preference is *don't care*, then at $t + 1$ it is emitted on either (i, j)'s outgoing row or outgoing column link—the choice is decided by a fair coin toss. Otherwise (i, j) emits the packet on the outgoing link that matches its routing preference. Suppose (i, j) receives two packets at time t, neither destined to (i, j). If both routing preferences are *don't care*, then a fair coin toss pairs the two packets to the two links. If one preference is *don't care*, and one is either *row* or *column*, then the latter is emitted on the matching link, and the former on the other link. If the preferences are different and neither is *don't care*, then each packet is emitted on the outgoing link matching its routing preference. Last, if the two preferences are either both *row* or both *column*, then node (i, j) invokes a *conflict resolution* rule to pair the packets to outgoing links. Three such rules are as follows:

— **random:** A fair coin toss pairs the two packets to the two links.
— **straight-through:** The packet that arrived on the incoming row link is emitted on the outgoing row link and the packet that arrived on the incoming column link is emitted on the outgoing column link.
— **closest to finish:** The packet closer to its destination is matched to its preferred link and the other packet to the other link. If the two are equally close then a fair coin toss pairs the two packets to the two links.

It is worth stressing that the routing rules all take advantage of the high density of *don't care* preferences in the Manhattan street network (Fig. 2). When a packet that does care about which link to take out of a node meets with one that does not, the later defers to the former. One might consider simpler strategies where a packet whose preference is *don't care*, chooses one of the outgoing links at random and then competes for that link as if that link were its original preference. If a conflict over the link arises it is then resolved by one of the three rules just mentioned. Our experiments [9] show such strategies perform *significantly* worse. For example, for a 20×20 network, operating with links 80% utilized, the *random* and *straight-through* rules as defined here attain about 25% greater throughput than their counterparts that convert *don't care* preferences to random ones.

In this paper we investigate just the first two rules, which seem simpler to implement than the third. Under the *random* rule, a packet is susceptible to deflection at every other hop while moving inside a quadrant (cf. Fig. 2), and becomes susceptible at every hop upon reaching the row or column of its destination. Under the *straight-through* rule, a packet already moving in a given direction will continue to do so. Thus, it is susceptible to deflection only when attempting a critical bend. Hence, by Fact 1, at all times (including its first hop) in the network a packet need only escape three possible deflections since it is only susceptible to deflection at a critical bend and there are at most three critical bends on a shortest path. This distinguishes the *straight-through* rule from the *random* rule, where about half the packet's hops can involve a deflection so the number of escapes needed to reach its destination grows with the size of the network. The third *closest to finish* rule has the advantage of provably avoiding deadlock, and it ought to lead to good performance. However, it is more complex than the other two rules, and is not investigated here.

Unfortunately, we do not have a proof that routing under the *straight-through* rule does not deadlock, that is, does not reach a situation where, with probability 1, packets circulate forever without reaching their destinations. One might worry that deadlocks might appear in scenarios that are hard to imagine. In extensive experiments with $n \times n$ networks, with $n \leq 30$, we never detected deadlock. The random component of the *straight-through* rule, which is used when two packets whose routing preferences are both *don't care* meet, appears to be crucial to the rule's freedom from deadlock. We note that the Connection Machine [5], [6] uses a type of deflection routing, which also appears to work well in practice but lacks a formal proof of deadlock freedom.[1]

As described in the following section, we suppose that a node may have one new (locally generated) packet present at each time slot t. Suppose that one is present at a time t and that k ($0 \leq k \leq 2$) outgoing links are needed for the packets that just arrived on the two incoming links:

— If $k = 0$, then the new packet is emitted on the link matching its routing preference or, if that preference is *don't care*, on either the row or the column outgoing link, depending on the outcome of a fair coin toss.

---

[1] One way to rule out deadlock is to take special action with packets that have accumulated a very large number of hops in the network: either dropping such packets, or using new (failsafe but possibly inefficient) routine rules to ensure their eventual transit to their destinations. The HEP [20] used the latter method.

— If $k = 1$, then the conflict resolution rule pairs the two packets to the two outgoing links, *acting as if the new packet arrived on the link not holding a packet requiring further routing at t.*

— If $k = 2$, then the new packet is blocked, that is, not emitted on either link.

The rule for $k = 1$ implies that on its first hop a packet can either win or lose a conflict with a packet already traveling in the network.

III. STOCHASTIC MODEL

In this section we introduce a stochastic model describing the generation of new packets. A key assumption is that the stream of packets generated at one node is statistically identical to that generated at another. To make this precise, we introduce some notation, which also streamlines the discussion of the following section.

For all i and j ($0 \leq i, j \leq n - 1$), define the automorphism $A_{i,j}$ as a relabeling of the mesh, putting (i, j) at the origin:

$$A_{i,j}(u, v) = (w, x)$$

where

$$w = \begin{cases} (u - i) \bmod n & \text{if } j \text{ is even} \\ (i - u) \bmod n & \text{if } j \text{ is odd} \end{cases}$$

$$x = \begin{cases} (v - j) \bmod n & \text{if } i \text{ is even} \\ (j - v) \bmod n & \text{if } i \text{ is odd.} \end{cases}$$

Thus, row i (column j) is renumbered 0, and scanning from row i (column j) in the direction of node (i, j)'s outgoing row (column) link, the other rows (columns) are renumbered 1 through $n - 1$.

Assumption A1 (relabeling invariance): At all times t and for all i, j, u, v ($0 \leq i, j, u, v \leq n - 1$), the probability that node (i, j) generates a new packet destined to node (u, v) at time t is the same as the probability that node $(0, 0)$ generates a new packet destined to node $A_{i,j}(u, v)$ at time t.

Assumption A1 will be satisfied provided the choices of new packet destinations are made symmetrically. A concrete way to satisfy A1, which we shall adopt, is to assume new packets' destinations are chosen uniformly at random among the $n^2 - 1$ nodes not coinciding with their sources. Alternatively, first the distance separating source and destination could be chosen (say, according to a geometric distribution), and then the node could be chosen uniformly at random from among the nodes at that distance. Alternatively, the choices could be skewed (say, favoring nodes whose row indexes are close to that of the source). In brief the choices may depend in any way on the relative position assigned to a destination when the source is viewed as the origin $(0, 0)$.

The stochastic model of the network is simple. We assume that at each time t, each node (i, j) generates a new packet independently with a fixed probability:

$$g = \text{Pr(generating a new packet at a given slot).}$$

The packet's destination is chosen independently with respect to a fixed probability distribution satisfying the relabeling invariance assumption above. We assume that if a new packet generated at a given node at a given time slot cannot be admitted to the network at that slot (because both outgoing links are needed for packets already traveling in the network) then that packet is blocked and cleared (never retried, i.e., forgotten).

A. Throughput and Delay

In this section we obtain useful formulae for the network throughput and mean packet delay, and call attention to an upper bound on throughput.

Focus on node $(0,0)$. Refer to a packet destined to node (i,j) as packet (i,j). Let

$$r_t(i,j) = \Pr(\text{node } (0,0) \text{ emits packet } (i,j)$$
$$\text{on its row output at time } t),$$

and

$$c_t(i,j) = \Pr(\text{node } (0,0) \text{ emits packet } (i,j)$$
$$\text{on its column output at time } t).$$

Here $(i,j) = (0,0)$ means node $(0,0)$ emits no packet on the output in question. Let

$$b_t = \Pr(\text{node } (0,0)$$
$$\text{blocks the new packet generated at time } t$$
$$\mid \text{a new packet is generated a time } t).$$

The row output of node $(0,0)$ is idle with probability $r_t(0,0)$, and the column output is idle with probability $c_t(0,0)$. Applying the relabeling invariance assumption, the instantaneous utilization of any row link is $1 - r_t(0,0)$ and that of any given column link is $1 - c_t(0,0)$. Thus, the instantaneous mean number of packets in the network is given by

$$n^2(1 - r_t(0,0) + 1 - c_t(0,0)). \tag{1}$$

A packet is absorbed over a row link with probability $r_t(0,1)$, namely the probability that node $(0,0)$ emits a packet to node $(0,1)$ at time t. Similarly, a packet is absorbed over a column link with probability $c_t(1,0)$. Hence, the instantaneous output rate is

$$n^2(r_t(0,1) + c_t(1,0)), \tag{2}$$

whereas the instantaneous input rate is

$$n^2(g(1 - b_t)). \tag{3}$$

Let $r(i,j)$, $c(i,j)$, and b denote the corresponding steady state values, i.e., the limits as $t \to \infty$. By (2), at steady state, new packets depart the network at rate $\lambda = n^2(r(0,1) + c(1,0))$, so by (3) the limiting blocking probability $b = 1 - \lambda/(n^2g)$. It follows from Little's law [11], [12], (1), and (2) that the steady-state mean packet delay is given by

$$D = \frac{2 - r(0,0) - c(0,0)}{r(0,1) + c(1,0)}.$$

A simple counting argument provides a basic upper bound on steady-state throughput, which holds for *any* routing scheme, including schemes with infinite capacity buffers, elaborate flow control, etc. A packet must be transmitted an average of at least m times where m is the average distance separating the packet's source and destination. At most E packets can be transmitted per unit time where E is the number of edges in the underlying communication graph ($E = 2n^2$ for the Manhattan street network). Hence the steady-state throughput cannot exceed E/m. In particular, assuming a new packet's destination is chosen uniformly at random from the nodes not coinciding with its source, the steady-state throughput

$$\lambda \lesssim \frac{2n^2}{n/2} = 4n.$$

(Since the graph is connected like a torus, the expected distance separating the source row from the destination row is $\sim n/4$, as is the expected distance separating the two columns.)

IV. ONE NODE MODEL

The key quantities are $r_t(i,j)$ and $c_t(i,j)$, the probability distributions for the packets that node $(0,0)$ emits at time t. Unfortunately, calculating these quantities exactly is intractable even for small networks since the actions of node $(0,0)$ are coupled to the actions of all other nodes. It turns out that the following simple approximation uncouples node $(0,0)$ and gives the matrices r_{t+1} and c_{t+1} as simple functions of just r_t and c_t.

Approximation A2 (independence): For all u, v, w, $x(0 \leq u, v, w, x \leq n-1)$ and all time $t \geq 0$, the probability that at time t node $(0,0)$ receives packet (u,v) on its row input link and packet (w,x) on its column input link is the product of the probabilities of these two events.

Combining this approximation with the assumption (not an approximation) of relabeling invariance gives a *one node* model, and an effective method for calculating approximations for $r_t(i,j)$ and $c_t(i,j)$. Consider the event that at time t node $(0,0)$ receives packets (u,v) and (w,x) on its row and column input links, respectively. In other words, at time t node $(0, n-1)$ transmits packet (u,v) on its row output and node $(n-1,0)$ transmits packet (w,x) on its column output. By Approximation A2, the probability of the two transmissions is the product of the probabilities of each. By relabeling invariance, this product is

$$r_t(A_{0,n-1}(u,v)) \cdot c_t(A_{n-1,0}(w,x)),$$

the product of the probabilities of corresponding transmissions at node $(0,0)$.

Now, suppose that at time t node $(0,0)$ receives packet (u,v) on its row input and packet (w,x) on its column input. Define the conditional probabilities.

$R_{i,j}(u,v,w,x) = $ probability that at time $t+1$ node $(0,0)$ emits packet (i,j) on its *row* output;

$C_{i,j}(u,v,w,x) = $ probability that at time $t+1$ node $(0,0)$ emits packet (i,j) on its *column* output.

TABLE I
ONE NODE MODEL VERSUS MONTE CARLO SIMULATION MEAN AND THROUGHPUT DATA, FOR THE 8 × 8 NETWORK

| | *random* conflict resolution | | | | *straight-through* conflict resolution | | | |
| | simulation | | *one node* model | | simulation | | *one node* model | |
g	λ	D	λ	D	λ	D	λ	D
0.025	1.591	5.235	1.595	5.233	1.594	5.233	1.596	5.233
0.050	3.154	5.479	3.161	5.468	3.158	5.476	3.161	5.469
0.100	6.037	5.991	6.046	5.977	6.045	5.996	6.046	5.980
0.150	8.342	6.526	8.363	6.493	8.349	6.514	8.363	6.496
0.200	9.985	6.998	10.019	6.954	9.996	6.987	10.019	6.955
0.250	11.089	7.387	11.139	7.334	11.114	7.364	11.142	7.330
0.300	11.833	7.700	11.897	7.635	11.864	7.669	11.906	7.627
0.350	12.350	7.983	12.426	7.873	12.392	7.907	12.440	7.861
0.400	12.710	8.144	12.808	8.064	12.771	8.096	12.828	8.048
0.450	12.990	8.301	13.093	8.219	13.051	8.252	13.120	8.199
0.500	13.205	8.429	13.314	8.347	13.280	8.372	13.345	8.324
0.550	13.368	8.540	13.487	8.454	13.452	8.480	13.523	8.428
0.600	13.501	8.635	13.628	8.544	13.589	8.570	13.668	8.516
0.650	13.613	8.710	13.743	8.621	13.707	8.645	13.787	8.591
0.700	13.711	8.775	13.840	8.688	13.804	8.713	13.887	8.657
0.750	13.784	8.838	13.922	8.747	13.886	8.770	13.972	8.714
0.800	13.852	8.891	13.993	8.798	13.956	8.821	14.045	8.764
0.850	13.918	8.932	14.053	8.844	14.016	8.868	14.108	8.808
0.900	13.969	8.973	14.106	8.885	14.080	8.901	14.163	8.848
1.000	14.056	9.043	14.195	8.955	14.171	8.969	14.256	8.916

The quantities R and C depend on the new packet generation rate g. In particular, if packet $(u, v) = (0, 0)$ then the packet is absorbed, and a new one is generated with probability g. A straightforward argument in conditional probability gives, for all $i, j (0 \leq i, j \leq n - 1)$ and all $t \geq 0$,

$$\begin{pmatrix} r_{t+1}(i,j) \\ c_{t+1}(i,j) \end{pmatrix} = \sum_{0 \leq u,v,w,x \leq n-1} \begin{pmatrix} R_{i,j}(u, v, w, x) \\ C_{i,j}(u, v, w, x) \end{pmatrix} r_t(A_{0,n-1}(u, v)) \cdot c_t(A_{n-1,0}(w, x)). \quad (4)$$

This is the central equation of the *one node* model.

Most of the n^4 terms in the sum on the right-hand side of (4) are zero. Ignoring the zeros, for each t, (4) represents a system of $2n^2$ equations, each having about $2n^2$ terms. Given any initial condition determining r_0 and c_0, (4) gives a method for calculating r_t and c_t for $t \geq 1$, which leads easily to a method for calculating the instantaneous values of the blocking probability, the network input rate, and the network output rate (described in the previous section).

In this paper, we focus on the solution, r_* and c_*, as $t \to \infty$. The natural numerical technique to find the solution is iteration: use (4) to produce r_{t+1} and c_{t+1} from r_t and c_t; stop when the change in the system from t to $t + 1$ is small. We adopt the initial condition:

$$\begin{pmatrix} r_0(i,j) \\ c_0(i,j) \end{pmatrix} \equiv \begin{pmatrix} 0 \\ 0 \end{pmatrix}$$

meaning the system starts empty. Of course, it is well known that, for many problems, simple iteration is slow to converge. There are many potentially faster iteration strategies such as overrelaxation [8].

We computed r_* and c_* in this way for a wide range of packet generation rates g, for the 4 × 4, 8 × 8, and 12 × 12 networks, and for both the *random* and the *straight-through* conflict resolution rules. It was assumed that new packet destinations are chosen uniformly at random from the $n^2 - 1$ nodes not coinciding with their sources; this is reflected in terms R and C of (4). Under this assumption, $r_t(i, j) = c_t(j, i)$, so just the r_t terms were computed. To control round off error, the term $r_t(0, 0)$ was computed via

$$r_t(0, 0) = 1 - \sum_{(i,j) \neq (0,0)} r_t(i, j),$$

rather than via (4). The iteration was stopped after the first $t + 1$ such that

$$\max_{0 \leq i,j \leq n-1} \left\{ \frac{r_{t+1}(i,j)}{r_t(i,j)} \right\} - \min_{0 \leq i,j \leq n-1} \left\{ \frac{r_{t+1}(i,j)}{r_t(i,j)} \right\} < 10^{-6}.$$

Table I gives some of the results for the 8 × 8 network. The table also includes corresponding data from Monte Carlo simulations. Each row of simulation data was obtained from a run of $t = 210\,000$ steps, using the batched mean method [15] with 21 batches of 10 000 steps each. The results of the first batch were discarded. Each data value lies within a 90% confidence interval of width less than 1% of the value.

Over this series of experiments, as the packet generation rate g varies from 0.025 to 1.0, the link utilization u (a good measure of the load) varies from 0.06 to 0.99. (Little's law tells us that at steady state the link utilization u satisfies $2n^2u = \lambda D$, so u can be recovered from the table.) Over the whole series, we found the following.

- There is remarkably little to choose between the performance of the *random* and *straight-through* rules. In the next section we see the story appears to be the same for large networks.

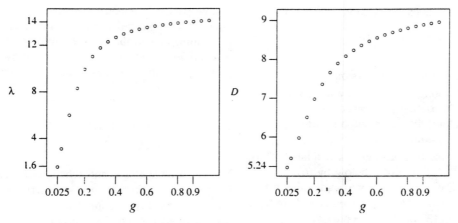

Fig. 3. Graphs of the data of Table I for the random conflict resolution rule. At this scale, the corresponding data from the one node model is indistinguishable from the simulation data.

- The relative difference between data from the one node model and data from simulations is less than 1%. In further experiments on networks up to size 12×12 we found the relative error never exceeded 2%.

In Fig. 3 the throughput and mean delay data are plotted for the *one node* model with *random* conflict resolution. Throughput and delay increase monotonically with the packet generation rate g. In the 8×8 network, the mean distance between source and destination turns out to be 5.03. Applying the reasoning of Section III-C, it follows that the throughput of any routing scheme for the 8×8 mesh is at most

$$\frac{2 \cdot 8^2}{5.03} \cong 25.5$$

and the mean delay is of course at least 5.03. At saturation, where the new packet generation rate g is near 1 and the link utilization is near 1 also, it is as if at each node there is an infinite queue of new packets waiting to be injected into the network. Even in this case the network serves the queues efficiently: with throughput ~ 14 and mean delay ~ 9— values within a factor of two of the ideal values.

Applying Brouwer's fixed point theorem [14] to the system of equations corresponding to the model shows that each system has a fixed point. However, we do not have a proof that this fixed point is unique and that our numerical method for finding it always converges. In our experiments, we observed smooth and relatively rapid convergence to a fixed point, within acceptable error bounds.

V. ONE PACKET MODEL

In this section we present a *one packet* model for estimating the steady-state performance of the network. In this model a single packet performs a random walk in the network. At each node on route to its destination, if the packet is susceptible to deflection then it is either deflected (probability p) or routed correctly (probability $1 - p$). An analysis of this random walk combined with a simple approximate analysis of the real network gives

- the link utilization u as a function of u and p, and
- p itself as a function of u and p.

As in the *one node* model, our approach is to solve for the fixed points u_* and p_* of the two equations. We obtain estimates of the steady state moments of delay in the real network by solving the random walk with $p = p_*$. An estimate of the steady state throughput is derived from the first moment of delay and u_*. For concreteness, the analysis is tailored for the case where new packet destinations are chosen uniformly at random from the $n^2 - 1$ nodes not coinciding with their sources. It is straightforward to generalize the analysis to any choices of destinations satisfying the relabeling invariance assumption.

Consider the following absorbing random walk. At time $t = 0$, a packet is generated at node (i_0, j_0) with destination $(0, 0)$. Assume $(i_0, j_0) \neq (0, 0)$ and the $n^2 - 1$ such choices are equally likely. At each time $t \geq 1$ until the packet reaches $(0, 0)$ and is absorbed:

1) The packet's routing preference is calculated at its current location (i, j).
2) If the preference is *don't care*, then the packet is emitted either on (i, j)'s row output link or on its column output link—the choice is decided by a fair coin toss. If the packet is not susceptible to deflection at (i, j) then it is emitted on its preferred link. Otherwise, the choice of link is decided by a biased coin toss: The packet is emitted on its preferred link with probability $1 - p$, and is emitted on the other link with probability p. All coin tosses are independent of each other.

Under the *random* rule, the packet is susceptible to deflection whenever its routing preference is *don't care*. Under the *straight-through* rule the packet is susceptible whenever attempting a critical bend, and is possibly susceptible on its first hop. (In the analysis of random walk described above, we assume, for simplicity, that the packet's susceptibility to deflection on its first hop is computed as if entered the node on the row input.)

Let D denote the packet's average delay, or expected time until absorption, and $D(i, j)$ the average delay if started at

331

$(i_0, j_0) = (i, j)$. Averaging over the possible starting nodes,

$$D = \frac{1}{n^2 - 1} \sum_{(i,j) \neq (0,0)} D(i, j). \qquad (5)$$

The $D(i, j)$ satisfy simple one step recursions. Let

- (i_1, j_1) be the node along node (i, j)'s outgoing link that matches the packet's routing preference, and
- (i_2, j_2), the node along the other outgoing link.

It the packet's routing preference at (i, j) is *don't care*, it does not matter which of the two links gets which label. Thus, the recursion defining $D(i, j)$ is given below by (6).

To obtain a self-consistent equation for p some other quantities are needed, which also have simple recursive definitions (given in (11), (12), and (13) of the Appendix):

- $\Delta(i, j)$, the average number of nodes the packet visits where its routing preference is *don't care* if started at (i, j), and
- $\Psi(i, j)$, the average number of steps where the packet's routing preference is straight-through (it enters on the row input and has routing preference row, or it enters on the column input and has routing preference column) if started at (i, j).

The most efficient way to compute the $D(i, j)$, $\Delta(i, j)$, and $\Psi(i, j)$ is successive substitution using the defining recursions. Here, in theory and in practice the iterations always converge [8]. An interation is a sweep of the mesh, with cost $O(n^2)$. For p in the range of interest, it turns out that $O(n)$ iterations suffice. For larger networks, we could replace the exact calculations with Brownian motion approximations. An individual packet moving through the network executes a certain random walk. When the network size is large, the random walk is well approximated by Brownian motion (as in [18]). This extra approximation would allow us to give explicit formulas rather than solve the discrete system of equations of the one packet model.

We now use this random walk model to produce an approximate analysis of the real network. The analysis starts with Approximation A2 of Section IV: in steady state a node's two input links are independent. Hence, an input link carries a packet independently with probability u, the steady state link utilization. To this add:

Approximation A3 (random walk delay): The delay of a packet is given as a function of the deflection probability p (5) and (6).

A given link carries a packet at a given slot with probability u. By Little's law, a fraction ua of such packets are immediately absorbed where

$$a = \frac{1}{D}.$$

A new packet is injected into the network with probability g unless both input links carry packets and neither is absorbed. By symmetry and the approximation that a node's two input links are independent, we obtain the following statement of the property that the link utilization is invariant at steady state:

$$u = u^2 \left[(1-a)^2 + a(1-a)(1+g) + a^2 \frac{g}{2} \right]$$
$$+ u(1-u)[(1-a)(1+g) + ag]$$
$$+ (1-u)^2 \frac{g}{2}. \qquad (7)$$

The u^2 term weights events where packets arrive on both input links at a given time, the $u(1-u)$ term events where a packet arrives on exactly one input link, and $(1-u)^2$ events where neither input link carries a packet.

Now, let us consider the characteristics of a typical packet, which we call the "tagged" packet, in the real network at equilibrium. From the quantities $D(i, j)$, $\Psi(i, j)$, and $\Delta(i, j)$, we form estimates of the following:

- the fraction of time the packet's routing preference is don't care,

$$\delta = \frac{\sum_{(i,j)} \Delta(i, j)}{\sum_{(i,j)} D(i, j)}, \qquad (8)$$

- the fraction of time the packet's routing preference is straight-through (row input to row output or column input to column output),

$$\Psi = \frac{\sum_{(i,j)} \Psi(i, j)}{\sum_{(i,j)} D(i, j)}, \qquad (9)$$

- and the fraction of time the packet's routing preference is bend (row input to column output or column input to row output), $1 - \delta - \Psi$.

Suppose the tagged packet is susceptible to deflection at time t in steady state. If it is to be deflected a "competing" packet must be present. A competing packet is present if a packet is carried on the other input link and is not absorbed; a packet is carried on the other input link, it is absorbed, and a new packet is generated locally; or a packet is not carried on the other input link, and one is generated locally. Thus, the tagged packet meets with a competing packet with probability

$$u(1-a) + uag + (1-u)g = u(1-a) + (ua + 1 - u)g.$$

Under the *straight-through* rule, the tagged packet is susceptible if its routing preference is bend. It is deflected if the competing packet's preference is straight-through (probability Ψ). Under the *random* rule, the packet is susceptible if its routing preference is not *don't care*. If susceptible, the tagged

---

$$D(0, 0) = 0$$

$$D(i, j) = 1 + \begin{cases} \frac{1}{2} D(i_1, j_1) + \frac{1}{2} D(i_2, j_2) & \text{if routing preference is } don't\ care \text{ at } (i, j) \\ D(i_1, j_1) & \text{if not susceptible to deflection at } (i, j) \\ (1-p)D(i_1, j_1) + pD(i_2, j_2) & \text{otherwise} \end{cases} \qquad (6)$$

TABLE II
ONE PACKET MODEL VERSUS MONTE CARLO SIMULATION DATA, FOR THE 8 × 8 NETWORK AND THE RANDOM CONFLICT RESOLUTION RULE

| | *random* conflict resolution | | | | | | |
| | simulation | | | | *one packet* model | | |
g	λ	D	σ		λ	D	σ
0.025	1.591	5.235	2.139		1.596	5.231	2.146
0.050	3.154	5.479	2.419		3.161	5.464	2.422
0.100	6.037	5.991	3.022		6.047	5.971	3.001
0.150	8.342	6.526	3.615		8.365	6.488	3.572
0.200	9.985	6.998	4.142		10.020	6.953	4.077
0.250	11.089	7.387	4.576		11.136	7.337	4.490
0.300	11.833	7.700	4.927		11.890	7.643	4.818
0.350	12.350	7.938	5.196		12.413	7.885	5.077
0.400	12.710	8.144	5.418		12.791	8.079	5.283
0.450	12.990	8.301	5.598		13.072	8.236	5.451
0.500	13.205	8.429	5.745		13.288	8.366	5.589
0.550	13.368	8.540	5.867		13.459	8.474	5.704
0.650	13.501	8.635	5.982		13.597	8.566	5.801
0.700	13.711	8.775	6.136		13.805	8.712	5.957
0.750	13.784	8.838	6.205		13.884	8.772	6.020
0.800	13.852	8.891	6.273		13.953	8.824	6.076
0.850	13.918	8.932	6.315		14.013	8.871	6.125
0.900	13.969	8.973	6.363		14.065	8.912	6.619
1.000	14.056	9.043	6.445		14.151	8.983	6.245

packet is involved in a conflict if either its preference is to be routed straight-through (probability $\Psi/(1-\delta)$) and the competing packet's is bend (probability $(1-\Psi-\delta)$), or the tagged packet's preference is bend (probability $(1-\Psi-\delta)/(1-\delta)$) and the competing packet's is straight-through (probability Ψ). If susceptible and involved in a conflict, the tagged packet is deflected with probability $1/2$. Thus, for the *random* rule we estimate the probability of deflection as

$$(\Psi/(1-\delta)(1-\Psi-\delta) + (1-\Psi-\delta)/(1-\delta)\Psi) \cdot 1/2$$
$$= \Psi(1-\Psi-\delta)/(1-\delta).$$

This is summarized in

Approximation A4: A packet's routing preference is don't care with probability δ given in (8) and straight-through with probability Ψ given in (9). Hence, under the random rule, the probability of deflection is

$$p = (u(1-a) + (ua+1-u)g) \cdot \Psi(1-\Psi-\delta)/(1-\delta),$$

and under the *straight-through* rule is

$$p = (u(1-a) + (ua+1-u)g) \cdot \Psi.$$

This completes the analysis. To apply it we compute the fixed points u_* and p_* of (7) and Approximation 4. As in the *one node* model, this is done by iteration. The cost of an iteration is dominated by the calculation of D, Δ, and Ψ, and is $O(n^3)$. In our experiments a small number of iterations sufficed. The results are about as sharp as those obtained from the *one node* model; for throughput and mean delay, the relative difference between the analytical results and corresponding simulation data is typically less than 1%. In Table II data is presented contrasting the analytical results

with results from Monte Carlo simulations, for the 8 × 8 network operating under the *random* rule. (The simulation data in Table II is the same as that in Table I.) We collected corresponding data for the *straight-through* rule, and found the data to be similar in the actual numerical values (as we know the data should be, by Table I) and similar in the discrepancy between analytical predictions and simulation results.

Having obtained p_*, any moment of delay can be calculated using one step recursions similar to (6). In particular, we can calculate the standard deviation of delay σ from the first two moments. Table II also gives the results of that calculation. Note that the standard deviation of delay is relatively small. At light loads, the standard deviation just accounts for the dispersion arising from the destination being initially at a random distance from the source. At heavy loads, the standard deviation is still a fraction of the average delay.

Networks of size as large as 100 × 100 can easily be solved using the *one packet* model. Table III gives the results of solving the *one packet* model for a series of large networks, and contrasts the *random* and *straight-through* rules. It turns out that the two rules are remarkably close in performance.

At saturation ($g, u \approx 1$), under the *straight-through* rule, the probability of deflection p_* is nearly equal to Ψ, the probability of having a straight-through preference. Under the *random* rule, at saturation the probability of deflection is nearly equal to $\Psi(1-\Psi-\delta)/(1-\delta)$. In computing the data of Table III, we found that: under both rules, Ψ is increasing: under the *straight-through* rule, the probability of deflection is increasing (from $p = 0.41$ at $n = 10$ to $p = 0.54$ at $n = 100$), and under the *random* rule this probability is decreasing (from $p = 0.19$ at $n = 10$ to $p = 0.10$ at $n = 100$). Moreover, under the *random* rule, the probability of a bend routing preference $(1-\delta-\Psi)$ is decreasing (from 0.35 at $n = 10$ to 0.12 at $n = 100$).

TABLE III
THROUGHPUT AND DELAY DATA FOR SATURATED $n \times n$ NETWORKS OBTAINED BY SOLVING THE ONE PACKET MODEL

	straight-through conflict resolution		random conflict resolution	
n	λ	D	λ	D
10	17.157	11.610	17.559	11.342
20	41.401	19.296	41.567	19.218
30	67.912	26.485	68.102	26.411
40	97.605	32.769	97.414	32.834
50	127.654	39.155	127.520	39.197
60	159.516	45.125	159.148	45.230
70	191.299	51.219	191.178	51.251
80	224.371	57.039	224.221	57.078
90	257.213	62.975	257.512	62.902
100	291.075	68.703	291.539	68.594

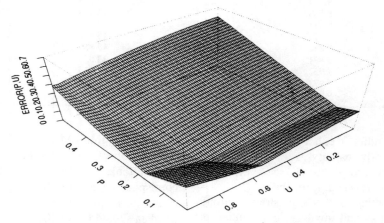

Fig. 4. A view of the mapping associated with the fixed point calculation.

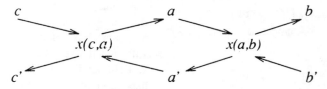

Fig. 5. Transformation of the edges (c, a) and (a, b) in G to several edges in H.

Formally, the equations above determine a mapping F from the unit square to the unit square whose fixed point (u_*, p_*) solves $(u, p) = F(u, p)$. Letting $(u', p') = F(u, p)$, define error (u, p) as $|u' - u| + |p' - p|$. In Fig. 5, we have plotted error (u, p) for the 8×8 network and the *random* rule. The surface looks like a creased sheet, touching 0 only at the point $(u_*, p_*) \approx (.71, .13)$. This is meant as an illustration, not as a proof that F has a unique fixed point.

Working with the one packet model, we can quickly get some insight into the performance of large networks. Suppose the network is large and is saturated: $g, u \approx 1$. Let random variable T denote the packet delay, and $\gamma(s)$ the number of deflections a packet experiences if started at a distance s from its destination. Ignoring rare boundary conditions, a single deflection adds 4 hops to the delay, so

$$T \cong s + 4\gamma(s).$$

The average distance between the packet's source node, (i_0, j_0), and destination node, $(0, 0)$ is $E(s) \cong n/2$, since the average distance from 0 of both i_0 and j_0 is $\cong n/4$. Thus $E(s) \cong n/2$, and the average delay is

$$E(T) = D \cong \frac{n}{2} + 4E\left(\gamma\left(\frac{n}{2}\right)\right).$$

Under the *straight-through* rule, the tagged packet is susceptible to deflection only when attempting a critical bend (i.e, only when its routing preference is bend). Since every shortest path to the destination has at most three critical bends (irrespective of the length of the path), the tagged packet need only win at most three conflicts in a row (arising at the random times when the critical bends are attempted) to ensure that a shortest path is followed. Thus, the one packet model tells us that the number of deflections is stochastically dominated by a geometric random variable with parameter $(1 - p)^3$, which leads to

$$E(\gamma(s)) \leq \frac{1}{(1-p)^3} - 1 \leq \frac{1}{(1-p)^3},$$

for all $s \geq 1$ and all p, $0 \leq p \leq 1$. This in turn leads to

$$D \leq \frac{n}{2} + \frac{4}{(1-p)^3} \quad \text{and} \quad \lambda \gtrsim \frac{2n^2}{n/2 + 4/(1-p)^3}. \quad (10)$$

The bound for λ presumes that the arrival rate parameter g is large enough (say, $g > 4/n$) that the link utilization $u \sim 1$ as n tends to infinity.

It is reasonable to conjecture that as the network size increases δ the estimate of the probability assigned to the don't care routing preference, remains bounded away from 0. Under the *straight-through* rule, when susceptible, a packet is deflected with probability $\Psi \leq 1 - \delta$. This leads us to the conjecture that Ψ is bounded away from 1, and in turn that as the network dimension n tends to infinity, the probability of deflection remains bounded away than 1. It then follows from (10) that the mean delay and throughput tend to $n/2$ and $4n$ as n tends to ∞, i.e., the ideal values (cf. Section III-A). Another consequence of p being bounded away from 1 under the *straight-through* rule is that, irrespective of the initial distance s between a packet's source and destination, all moments of $\gamma(s)$ are bounded.

Let us do a similar calculation for large saturated networks controlled by the *random* rule. Roughly under the *random* rule, the tagged packet is susceptible to deflection no less frequently than every other hop. To obtain a lower bound on the number of deflections γ we can always alter the random walk so that at, say, the even hops the packet is never deflected. If we then observe the packets at odd hops we would see that the distance to the destination decreases by two if the packet is not deflected and increases by two if it is deflected. Applying these observations, the one packet model implies that under the *random* rule

$$E(\gamma(s)) \geq \frac{1}{2} \cdot \frac{sp}{1-2p},$$

if $p < 1/2$, and $E(\gamma(s))$ diverges if $p \geq 1/2$. This leads to

$$D \geq \frac{n}{2} \frac{1}{1-2p} \quad \text{and} \quad \lambda \leq 4n(1-2p).$$

Thus, the one packet model predicts that the *random* rule can attain asymptotically ideal performance if p tends to 0 as the network size increases. This would occur if the probability of a bend routing preference $(1 - \Psi - \delta)$ tends to 0, while the probability of a don't care preference (δ) remains bounded away from 1. This is reasonable, and our numerical experiments indeed suggest $1 - \Psi - \delta$ is decreasing.

VI. OPTIMIZING DIAMETER AND DEFLECTION INDEX

Routing with deflection may be summarized as follows.

1) A node always attempts to emit a packet on an outgoing link that lies on a shortest path to the packet's destination.

2) A conflict arises if the node simultaneously receives packets from two neighbors, where the shortest paths to both destinations use the same outgoing link. A conflict resolution rule is then applied, which selects one packet for transmission on the desired link, and the other for transmission on the other link.

Clearly, the algorithm generalizes to any network whose communication graph is strongly connected and regular (all nodes have identical indegree and outdegree). Degree 2 graphs are of particular interest because as the degree of the graph increases the local conflict resolution problems become increasingly complicated, and the nodes increasingly difficult to build.

One measure of the merit of a graph as host to the routing algorithm is its diameter (the maximum, over each pair of nodes, of the length of a shortest path between the pair). Small diameter contributes to small delay and to large throughput. Another, less standard measure is *deflection index*, defined as follows. Given any two nodes a and b in a directed, strongly connected graph, let $sp(a, b)$ denote the length of a shortest path from a to b. We define the *deflection index* of node a to be the least positive integer r such that

$$sp(b, c) + 1 \leq sp(a, c) + r,$$

for all nodes b and c where a and b are connected by an edge $(sp(a, b) = 1)$, and a, b, and c are mutually distinct. With respect to the routing algorithm, the deflection index is the maximal number of hops a single deflection adds to a packet's delay. Small deflection index, like small diameter, contributes to small delay and to large throughput. In addition, small deflection index reduces the variance of packet delay, with the result that packets arrive more nearly in order. Any graph that is not fully connected must have deflection index ≥ 2; this is attained by the ring connected in the clockwise and the counterclockwise directions (each node has outgoing links to each of its two neighbors).

A binary shuffle graph [3] is one of the many graphs similar to the hypercube. To define the binary shuffle graph, let the number of nodes $N = 2^n$, and let $x_1 x_2 \cdots x_{n-1} x_n$ be the binary representation of the node numbered x, $0 \leq x \leq N - 1$. This node is connected to nodes $x_n x_1 x_2 \cdots x_{n-1}$ and $\overline{x_n} x_1 x_2 \cdots x_{n-1}$ where $\overline{x_n} = 1 - x_n$ denotes the complement of bit x_n. The diameter of the graph and its deflection index are $n = \log_2 N$. (A packet at node $a = (11 \cdots 10)$ destined to node $c = (11 \cdots 11)$ either reaches that destination in one hop, or is deflected to node $b = (01 \cdots 11)$, whose distance from c equals the graph's diameter, $\log_2 N$. This and the fact that the deflection index cannot exceed the diameter shows that the graph's deflection index is $\log_2 N$.) This diameter is optimal among all graphs with indegree and outdegree 2, whereas for large n the deflection index is far from the optimal value 2.

In contrast, the Manhattan graph has diameter $\sqrt{N}$ (far from optimal) where N is the number of nodes, and deflection index 4 (nearly optimal). Applying the construction of the following theorem to the family of binary shuffle graphs gives a family of *double shuffle* graphs that do well on both counts; the deflection index is just 4 and the diameter is about $2 \log N$. A *one packet* model for these graphs predicts that, as the input rate g increases, throughput rises to a value proportional to $N/\log N$ with delay remaining proportional to $\log N$. A simple Little's law argument like that given in Section III-A shows that no graph with indegree and outdegree 2 using any routing strategy whatsoever can have better asymptotic throughput/delay performance.

Theorem: Let G be any directed, strongly connected, regular graph of degree 2 on N nodes, with arbitrary diameter d. It is possible to construct a corresponding directed, strongly connected, regular graph H of degree 2, with total number of nodes $4N$, diameter $\leq 2d + 2$, and deflection index ≤ 4.

Proof: Corresponding to each node a in G, there are two nodes in H, a and a'. For each edge (a, b) in G, we add a new node $x(a, b)$ to H, and edges as described in Fig. 6. A copy of G is embedded in H. Collapsing all edge pairs $(a, x(a, b))$, $(x(a, b), b)$ to (a, b) produces the copy. A copy of G with all edge directions reversed, a graph we call G', is also embedded in H. Collapse all edge pairs $(b', x(a, b))$, $(x(a, b), a')$ to produce this copy. To see that H has deflection index 4, refer to Fig. 6 to see that each link in H belongs to cycle of length 4. To obtain the diameter bound consider a node u in the embedded copy of G and a node v' in the embedded graph G'. To go from u to v', we may take a shortest path from u to v in the embedded copy of G (at most $2d$ hops), and then go from v to v' in two additional hops. It follows that the distance between u and v' is at most $2d + 2$. All other cases yield to a similar argument, showing that the diameter of H is at most $2d + 2$. ∎

Last, we mention a third measure of merit for a graph as host to deflection routing: *density of don't care nodes*. Consider any strongly connected, regular graph of degree two. For any two nodes u and v we say v is a *don't care* node with respect to u if both outgoing edges from v lie on shortest paths to u. Define the density of *don't care* nodes to be the minimum, over all nodes u of the graph, of the fraction of *don't care nodes* with respect to u. The density of *don't care* nodes tends to $1/2$ for the Manhattan street network, whereas it is negligible (small and tending to 0 with the graph size) for the binary shuffle and double shuffle networks. It is reasonable to conjecture that for a given diameter and deflection index, the greater the density of *don't care* nodes the smaller the probability of deflection and the better the graph supports deflection routing.

VII. Conclusion

In this paper we have presented modeling techniques that accurately and cheaply predict the performance of a large routing network. We have used this technique to study deflection routing in the Manhattan street network. Our simpler *one packet* model also predicts that the performance of deflection routing in a general network depends mainly on a few of its gross characteristics: its diameter, its deflection index (number of extra hops imposed on a packet by a single deflection), and its frequency of *cares* (times when a packet is being routed to a node where deflection is possible). We give a network, the "double shuffle", that is optimal with respect to diameter and deflection index.

Maxemchuk [3] pointed out that the routing algorithm can be easily adapted to take advantage of additional buffers associated with a node's two outgoing links. When a conflict arises over a link, and the corresponding buffer is not full, a packet can be enqueued in the buffer, rather than deflected onto another link. We believe this would lead to some improvement in throughput and delay. However, we believe that the buffers need be no larger than the graph's deflection index (4 for the Manhattan graph). A recent simulation study bears this out [17]. It seems unlikely that buffers of larger size would improve performance. Indeed, buffers may hurt: As mentioned in the Introduction, it has been suggested that deflection defuses, "hot spots" that sometimes arise under fixed routing schemes under some types of balanced loads [2]. The greater the buffers, the fewer the deflections, so the less adaptive the algorithm may be to hot spots. Quantifying the advantages and disadvantages of additional buffering is a fascinating topic for future research. However, the numbers presented here show that performance is remarkably good without additional buffers.

The independence assumption underlying our analytic models may be violated in the real network. Apparently, however, such violations contribute little to steady state behavior. Other instances where independence approximations were used with success to produce tractable models of complex queuing systems include studies of the following.

— queues in data networks ("Kleinrock's independence assumption": at each queue, a packet's length is redrawn from a fixed probability distribution)
— breakdowns in production lines with finite storage bin capacity [19], and
— packet routing in omega networks [13].

Though our models give useful and sharp estimates at small cost (compared with simulation), there are gaps in the underlying theory. In particular, we have no proof that the models have unique solutions. Our models are what physicists call self consistent mean field approximations. We know of systems where the mean field approximations have multiple solutions although ours do not seem to. The existence of multiple solutions rather than invalidating the approximation, would give important insight about the system—that it has multiple operating points.

In our model, if a new packet is blocked then it is dropped. It might be of interest to model the situation where blocked packets are instead queued in local buffers. At each slot, a node receives a random number of new packets, which are inserted into the node's local buffer, and the node attempts to inject one packet from the buffer (if there is one), following the rules described above giving priority to packets already moving in the network. The equilibrium behavior of this model could be analyzed approximately via another fixed point approximation: in which each local buffer is modeled as a discrete time $G/M/1$ queue, where at each slot a packet is removed with fixed probability μ.

Appendix

In this short section, we present the recursive definitions of the quantities $\Delta(i, j)$ and $\Psi(i, j)$ that arose in the one packet model, where we follow the progress of a single packet that when susceptible to deflection is deflected with

$$\Delta(0,0) = 0$$

$$\Delta(i,j) = \begin{cases} 1 + \frac{1}{2}\Delta(i^+,j) + \frac{1}{2}\Delta(i,j^+) & \text{if routing preference is } don't\ care \text{ at } (i,j) \\ (1-p)\Delta(i,j^+) + p\Delta(i^+,j) & \text{if routing preference is } row \text{ at } (i,j) \\ (1-p)\Delta(i^+,j) + p\Delta(i,j^+) & \text{if routing preference is } column \text{ at } (i,j) \end{cases} \qquad (11)$$

$$\Psi_{\text{row}}(0,0) = 0$$

$$\Psi_{\text{row}}(i,j) = \begin{cases} \frac{1}{2}\Psi_{\text{row}}(i^+,j) + \frac{1}{2}\Psi_{\text{row}}(i,j^+) & \text{if routing preference is } don't\ care \text{ at } (i,j) \\ 1 + \Psi_{\text{row}}(i,j^+) & \text{if routing preference is } row \text{ at } (i,j) \\ (1-p)\Psi_{\text{row}}(j,i^+) + p\Psi_{\text{row}}(i,j^+) & \text{if routing preference is } column \text{ at } (i,j) \end{cases} \qquad (12)$$

$$\Psi_{\text{row}}(0,0) = 0$$

$$\Psi_{\text{row}}(i,j) = \begin{cases} \frac{1}{2}\Psi_{\text{row}}(i^+,j) + \frac{1}{2}\Psi_{\text{row}}(i,j^+) & \text{if routing preference is } don't\ care \text{ at } (i,j) \\ 1 + (1-p)\Psi_{\text{row}}(i,j^+) + p\Psi_{\text{row}}(j,i^+) & \text{if routing preference is } row \text{ at } (i,j) \\ (1-p)\Psi_{\text{row}}(j,i^+) + p\Psi_{\text{row}}(i,j^+) & \text{if routing preference is } column \text{ at } (i,j) \end{cases} \qquad (13)$$

fixed probability p. Recall that $\Delta(i,j)$ represents the expected number of steps where the packet's routing preference is *don't care*, if started at node (i,j). Let (i,j^+) and (i^+,j) denote the row and column neighbors of node (i,j). Equation (A1) gives the one-step recursion describing $\Delta(i,j)$. Recall that $\Psi(i,j)$ is the expected number of moves where the packet's routing preference is straight-through (row to row or column to column), given that the packet's origin is node (i,j). Let $\Psi_{\text{row}}(i,j)(\Psi_{\text{col}}(i,j))$ denote the expected number of moves where the packet's routing preference is straight-through, given that the packet arrives to (i,j) along node (i,j)'s row (column) input. By convention, at its origin the packet is treated as if it arrived on the node's incoming row link, so $\Psi(i,j) = \Psi_{\text{row}}(i,j)$. By the symmetry of the network and the conflict resolution rules, $\Psi_{\text{col}}(i,j) = \Psi_{\text{row}}(j,i)$, so it suffices to compute Ψ_{row}. Equation (A2) gives the one-step recursion describing $\Psi_{\text{row}}(i,j)$, for the *straight-through* routing rule. We used $\Psi_{\text{col}}(i^+,j) = \Psi_{\text{row}}(j,i^+)$ in the last line of the last equation. Equation (A3) describes $\Psi_{\text{row}}(i,j)$, under the *random* routing rule.

ACKNOWLEDGMENT

We thank P. Doyle, N. Maxemchuk, and M. Reiman for several enlightening discussions. Our original formulae for the one packet model, were slightly different and gave less accurate approximations; one of the referees suggested some alternative formulae, which are now in the paper. Special thanks to D. Mitra for suggesting that an iterative method of performance analysis might work.

REFERENCES

[1] P. Baran, "On distributed computing networks," *IEEE Trans. Commun. Syst.*, Mar. 1964.
[2] A. Borodin, and J. Hopcroft, "Routing, merging, and sorting on parallel models of computation," in *Proc. 14th Ann. ACM Symp. Theory Comput.*, San Francisco, May 1982, pp. 338–344.
[3] N. de Bruijn, "A combinatorial problem," *Nederl. Akad. Wetensh. Proc.*, vol. 29, pp. 758–764, 1946.
[4] P. Y. Chen, D. H. Lawrie, P. -C. Yew, D. A. Padua, "Interconnection networks using shuffles," *IEEE Comput.*, pp. 55–62, Dec. 1981.
[5] W. D. Hillis, *The Connection Machine*. Cambridge, MA: MIT Press, 1985.
[6] ——, "The connection machine," Scientific American, vol. 256, no. 6, June 1987.
[7] B. Hajek, and A. G. Greenberg, "On deflection routing in hypercube networks," *IEEE Trans. Commun.*, vol. 40, pp. 1070–1081, June 1992.
[8] G. Golub, and C. van Loan, *Matrix Computations*. Johns Hopkins Univ. Press, 1983.
[9] J. Goodman, and A. G. Greenberg, "Sharp approximate models of adaptive routing in mesh networks," *Teletraffic. Analy. Comput. Perform. Eval.*, The Cen. for Mathemat. Comput. Sci., June 2–6, 1986, Amsterdam, The Netherlands; O. J. Boxma, J. W. Cohen, and H. C. Tijms, Eds. pp. 255–270, North Holland, 1986.
[10] A. Krishna and B. Hajek, "Performance of shuffle-like switching networks with deflection," in *Proc. INFOCOM' 90*, IEEE Computer Society Press, San Francisco, CA, June 1990.
[11] J. C. P. Little, "A proof of the queueing formula $L = \lambda W$," *Oper. Res.*, vol. 9, pp. 383–387, 1961.
[12] L. Kleinrock, *Queueing Systems*, vol. 1. New York: Wiley, 1975.
[13] C. Kruskal, M. Snir, and A. Weiss, "On the distribution of delays in buffered multistage interconnection networks for uniform and nonuniform traffic," in *Proc. 1984 Int. Conf. Parallel Processing*, Columbus, OH, Aug. 1984, pp. 215–219.
[14] S. Lefschetz, *Introduction to Topology*. Princeton, NJ: Princeton University Press, 1949.
[15] N. F. Maxemchuk, "Regular mesh topologies in local and metropolitan area networks" *AT&T Tech. J.*, vol. 65, no. 7, 1659–1685, Sept. 1985.
[16] ——, "Routing in the Manhattan street network," *IEEE Trans. Commun.*, vol. COM-35, pp. 503–512, May 1987.
[17] ——, "Comparison of deflection and store-and-forward techniques in the Manhattan street and shuffle-exchange networks," in *Proc. IEEE INFOCOM' 89*, Ottawa, Ont., Canada, Apr. 1989, pp. 800–809.

[18] M. I. Reiman, "Open queueing networks in heavy traffic," *Math. Oper. Res.*, vol. 9, no. 3, pp. 451–458, 1984.

[19] B. A. Sevastýanov, "Influence of storage bin capacity on the average standstill time of a production line," in *Theory of Probability and Applications*, 1962.

[20] B. Smith, "Architecture and applications of the HEP multiprocessor computer system," in *Real Time Signal Processing IV, Proc. SPIE*, 1981, pp. 241–248.

[21] H. S. Stone, "Parallel processing with the perfect shuffle," *IEEE Trans. Comput.*, vol. 12, pp. 274–278, 1984.

[22] X.-N. Tan, K. C. Sevcik, and J.-W. Hong, "Optimal routing in the shuffle-exchange networks for multiprocessor systems," in *CompEuro 88—System Design: Concepts, Methods and Tools*, Brussels, Belgium, IEEE Comput. Society Press, Apr. 1988, pp. 255–264.

Performance Analysis of Deflection Routing in the Manhattan Street Network

Abhijit K. Choudhury and Victor O. K. Li*
Communications Sciences Institute
University of Southern California

Abstract

The deflection routing strategy is similar to the "hot-potato" technique and has been proposed for high-speed networks. An analytic model has been developed to study the performance of unbuffered and buffered deflection routing in regular networks, like the Manhattan Street Network. This model is extremely accurate as can be seen from the comparison with simulation results. The results show that the Manhattan Street Network performs well under heavy load; the introduction of a few buffers improves the throughput and delay in the network to a great extent .

1. Introduction

The current thrust towards large high-speed networks has made it desirable to design switching elements that are simple and fast, without sacrificing much in terms of throughput and delay performance. A novel, distributed scheme that has attracted a lot of attention is *deflection routing*. This scheme has been suggested for use in large geographically distributed computer networks, like the Manhattan Street Network [Max87] and SIGNET [BT89] and also for interconnection networks that connect processors and memory modules [TSH89].

Deflection routing is similar to "hot-potato routing" [Bar64]. It is used in networks that have nodes with equal in-degree and out-degree and may be implemented with or without buffers at the node. This scheme allows one to design nodes with bounded buffers and also to ensure that there is no packet loss due to buffer overflows. At any node, the packets in transit are served with absolute priority over packets from the local source. If there is no storage available at the node and there is contention for the same outgoing channel, the packets are temporarily *misrouted*. So, deflection routing wastes some bandwidth in order to gain fast switching capabilities and minimal storage requirements at the nodes.

This scheme has been proposed for routing in the Manhattan Street Network (MS_Net)[Max87] and the effect of buffering packets at the nodes on the per-

formance has been studied [Max89]. An approximate analysis of deflection routing in the MS_Net, under the uniform traffic model has been presented [GG86, BC90]. Deflection routing in hypercube and shuffle-like networks has also been analyzed [GH89, Szy90, TSH89, KH90].

Although the model we develop can be used to analyze deflection routing in any regular network, in this paper we shall focus on the Manhattan Street Network (MS_Net) which is a regular, two-connected network with unidirectional links; the nodes are connected as a grid on a toroidal surface, with the adjacent rows and columns traveling in opposite directions. The MS_Net has certain structural characteristics that make it well suited for deflection routing. A large fraction of the total number of nodes are so located that either outgoing link is on a shortest path to the destination. Since, under deflection routing, a packet that can take either path does not deflect a packet that has a preferred direction, this results in a lower probability of deflection of the packets. In the MS_Net, the penalty incurred due to deflection is four hops. The destination node is never more than three turns away from the source node; hence, as the network becomes larger, a packet travels straight more often than to turn. Since a deflection can occur only when one packet wishes to turn and one wishes to go straight, this implies that the probability of deflection decreases as the network size increases.

In this paper, we develop an analytic model, based on the model developed in [GH89], to study the performance of deflection routing in the MS_Net, under a uniform traffic model with the random conflict resolution rule. This model has been used to study the performance of deflection routing in any regular network (like the Shuffle-Net [CL90]) and under any conflict resolution rule [CL91].

2. Model

In this section, we develop an analytic model to study the performance of deflection routing in a two-connected network, like the MS_Net, under the random conflict resolution rule.

---

*This research was supported in part by the Department of Defense Joint Services Electronics Program under contract number F49620-88-C-0067.

Reprinted from *Proc. IEEE International Conference on Communications*, pp. 1659–1665, June 1991.

2.1 Network Operation

The model assumes a slotted network in which the nodes operate synchronously. The distance between the nodes is considered to be one slot although an extension to networks with longer links is simple. Using the notation in [GH89], at any node, at the beginning of any slot, there are

- U transit packets (received from other nodes during the previous slot and destined for other nodes), and

- V new packets, provided to the node by the local source.

At most one packet arrives per incoming link, which implies that $U \leq 2$. Of the $(U + V)$ packets present, $(U + V - 2)^{+}$ [†] packets are blocked (the transit packets have priority here) and $\min\{V, (2 - U)\}$ of the new packets are accepted. So, $\min\{(U + V), 2\}$ packets are to be transmitted by the node during the slot. The packets are first lined up and an attempt is made to satisfy their preferences. If two packets want the same output link, under the *random conflict resolution* rule, the conflict is decided by tossing a fair coin; in the case of unbuffered deflection routing, the packet that loses the toss does not get the preferred link whereas in buffered deflection routing, as long as the buffer associated with the preferred output link has space, the losing packet also gets its preferred output link, only it is placed in the buffer behind the winning packet. If a packet is not assigned to its preferred link, the packet is said to have been *deflected*. In buffered deflection routing, deflection occurs only when the buffer associated with the preferred link is full.

2.2 Offered Traffic

The traffic model assumed is memoryless, i.e., if the new packet generated at a node in a slot is not accepted by the network, it is rejected and cleared. We assume that the number of new packets, V, offered to a node by the local source, during a slot, has a Bernoulli distribution $Ber(v)$[‡]. Other distributions, like the binomial distribution, could also be used[CL90]. The number of new packets offered is independent from node to node and slot to slot. The destination of the new packet at a node is assumed to be independently and uniformly distributed over the remaining nodes in the network.

2.3 Network Topology

The model assumes a regular network topology with nodes that have equal in-degree and out-degree. Let $Nq(i)$ be the number of nodes that are placed at a distance of i hops from the source node, N being the total number of nodes in the network; so, under the

---

[†]The notation x^{+} means $\max\{x, 0\}$.

[‡]$Ber(p)$ represents a Bernoulli distribution with parameter p, $0 \leq p \leq 1$ and $B(n, p)$ represents a binomial distribution with parameters $n \geq 1$ and $0 \leq p \leq 1$

uniform traffic assumption, $q(i)$ denotes the probability that a new packet has its destination i hops away; the topology affects the analysis through the distribution of $q(i)$ (which varies with the network topology), through the deflection index of the network (the increase in path length due to a single deflection) and through the values of the probabilities of deflection.

3. Analysis of Deflection Routing

We now wish to determine both the transient and the steady state behavior of the MS_Net with the uniform traffic assumption. We wish to derive statistics like throughput per node and the network delay, i.e., the time taken for a packet to travel from the source to the destination once it has entered the network and investigate the improvement in throughput and delay performance when a finite number of output buffers is introduced at each node.

Although the arrivals of packets on different links are not necessarily independent, we conjecture that, under the uniform traffic assumption, they should be nearly so. Also, the choice of output links for the outgoing packets need not be independent; again, we conjecture that, with the uniform traffic model, the choice of outgoing link for a given packet would be independent of the preferences of the other packets. We propose to derive an approximate performance model by pretending that the arrivals on different links are independent and that the selection of output links are made independently; later, we hope that the simulation results will validate our model. So, our model is based on the following two approximations :

Approximation 1. The arrival of a packet on an incoming link is independent of arrivals on the other incoming link.

Approximation 2. The choice of an output link made by a packet is independent of the choice of the other packets.

When there is no storage at the nodes, a conflict between two packets for the same output link results in one of the packets being deflected and the deflected packets have to traverse longer paths to their destination. If there is infinite storage at the nodes, the probability of a packet being deflected becomes zero and the system behaves like a store-and-forward network with shortest path routing. However, when there is a finite amount of storage at each node (Fig.1), contention for the same outgoing link causes one of the contending packets to be stored in the buffer and delayed by one slot instead of being deflected to the other outgoing link, unless the buffer associated with the desired output link is full.

In two-connected networks, the probability of deflection is the same when there are K buffers shared by two output links as when there are K buffers assigned to each output link [Max89]. The state transition diagram of the buffers shown below (Fig. 2)

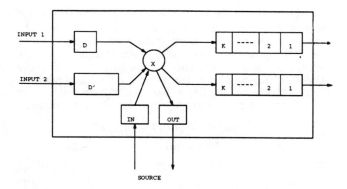

Figure 1: The structure of a node.

demonstrates the fact that the total number of packets stored at a node never exceeds K. Using the notation in [Max89], the states are labeled with the number of messages waiting at each output link and the transitions are labeled by the number of messages directed to each output link by the switch. Let $\bar{Y}$ imply possibilities other than Y and X imply all possibilities; so, $(0,\bar{2})$ is $(0,0)$ or $(0,1)$ and $(X,1)$ is $(0,1)$ or $(1,1)$. Since the total number of messages at the node never exceeds K, the system can be implemented with a total of K buffers, shared by both output links, instead of K buffers at each output link; for the purpose of our analysis, however, we shall assume that there are K buffers at each output link (Fig. 1).

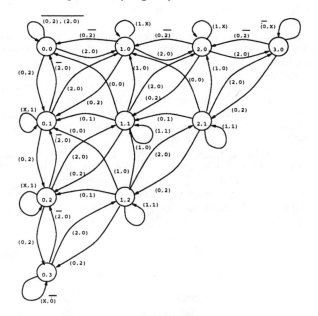

Figure 2: State transition diagram for buffer utilization at a node with 3 buffers assigned to each output link.

When there are buffers at a node, packets are not deflected to the other link unless the buffer for their preferred output link is full. If two packets prefer the same output link, as long as the buffer for that output link is not full, both packets are stored in that buffer; a fair coin toss decides the order in which they are stored

- the packet winning the toss is stored ahead of the packet that loses. The number of packets at a node increases only if two packets are directed to the same output link, there are less than K packets waiting for that link and there are no packets waiting for the other link [Max89]; this is because one packet can be sent out on each output link in every slot.

For a given network, we first derive, the acceptance and deflection probabilities at a given node during a time slot. The buffer occupancy probabilities are computed next. We then use these probabilities to derive an approximate analysis of the whole network.

3.1 *Calculation of Acceptance and Deflection Probabilities*

We focus on one node since, under the uniform traffic model, the behavior of every node in the network is identical. Consider any incoming link at a node during a slot t and let

$$m_t = Pr\{\text{a transit packet is received} \\ \text{on the link in slot } t\}$$

$$m_t(i) = Pr\{\text{a transit packet, } i \text{ hops from its} \\ \text{destination, is received on} \\ \text{the link in slot } t\}$$

Define $\mathbf{m_t} = [m_t(0), m_t(1), \ldots, m_t(d)]$, where d is the diameter of the network being considered. The number of transit packets, U, arriving at a node in a given slot t, has the binomial distribution $B(2, m_t)$, by Approx. 1. The number of new packets, V, generated at a node in a given slot has the Bernoulli distribution $Ber(v)$ and is independent of U. The probability of a typical new packet being accepted is

$$a(m_t, v) = 1 - m_t^2 \qquad (3.1.1)$$

Let $p(i, m_t, v)$ (or $p_o(i, m_t, v)$) be the probability that a typical *transit packet* (or typical *accepted new packet*) loses the conflict resolution process, given that the node is i hops away from the packet's destination. The values of these probabilities depend on the specific network being analyzed and the conflict resolution rule being used. We now derive expressions for $p(i, m_t, v)$ and $p_o(i, m_t, v)$ under the random conflict resolution rule.

The Manhattan Street Network has a large number of nodes where the routing preference is *don't care*, i.e., a packet has no preferred direction. Let $ps(i)$ be the probability that a transit packet, i hops away from its destination, prefers to go straight (e.g. comes into the node along the row link and prefers to go out on the row link) and $px(i)$ be the probability that it prefers to turn. The probability that a transit packet, that is i hops from its destination, has no preferences at a given node is $(1 - ps(i) - px(i))$. Two transit packets will contend for the same output link only if one prefers to go straight and one wishes to turn ; one packet gets its preferred link and the other is deflected, the decision being made by a fair coin toss. Let $pn(i)$ be the probability that a new packet, i hops from its destination has preference for either the row or the column direction: due to the symmetric nature of the topology, the values of the preferences in the two directions will be identical. The probability that new packet, i hops away from its destination, has no preferred direction is $(1 - 2pn(i))$.

Let ms_t (or mx_t) be the probability of there being a packet that wishes to go straight (or turn). Let qn be the probability of there being a new packet with a preferred direction.

$$ms_t = \sum_{r=1}^{d} m_t(r)ps(r)$$

$$mx_t = \sum_{r=1}^{d} m_t(r)px(r)$$

$$qn = \sum_{r=1}^{d} q(r)pn(r)$$

The deflection probabilities now become

$$p(i, m_t, v) = \frac{1}{2}[ms_t px(i) + mx_t ps(i)$$
$$+ 0.5v(1 - m_t)qn[ps(i) + px(i)]]$$
$$(3.1.2)$$

$$p_o(i, m_t, v) = \frac{1}{2}\left[pn(i)\{ms_t + mx_t\}\frac{2(1 - m_t)}{a(m_t, v)}\right]$$
$$(3.1.3)$$

Here $ps(i), px(i)$ and $pn(i)$ represent properties of the specific network being studied and can be obtained by simulation. The distribution of nodes, $q(i)$, from a typical source, can be obtained from the network topology. The penalty incurred for deflecting a packet in an MS_Net is exactly four hops.

3.2 Calculation of Buffer Occupancy Probabilities

Let $\Pi_t = [\pi_t(0, 0), \pi_t(0, 1), \ldots, \pi_t(0, K), \pi_t(1, 0), \ldots, \pi_t(K-1, K), \pi_t(K, K)]$, be the vector of buffer occupancy probabilities, such that $\pi_t(i, j)$ is the probability for the system to be in state (i, j) at the start of slot t.

We apply a transformation on the state space such that the state (x, y) is mapped onto a state w, where

$$w = (K + 1) \cdot x + y$$

It can be shown that this transformation is one-to-one. As a result,

$$
\begin{aligned}
\Pi_t &= [\pi_t(0, 0), \pi_t(0, 1), \ldots, \pi_t(0, K), \\
&\quad \pi_t(1, 0), \ldots, \pi_t(K - 1, K), \pi_t(K, K)] \\
&\equiv [\psi_t(0), \ldots, \psi_t(K), \psi_t(K + 1), \\
&\quad \ldots, \psi_t((K + 1)^2 - 2), \psi_t((K + 1)^2 - 1)] \\
&= \Psi_t \\
\Pi_0 &= \Psi_0 = [1, 0, 0, \ldots, 0]
\end{aligned}
$$

Let $\mathbf{Q}_t$ be the state transition matrix consisting of elements q_t^{ij}, $0 \le i, j \le K$, such that q_t^{ij} denotes the transition probability from state i to state j, during slot t. Suppose G_t^{cd} is the probability that, during time slot t, c packets are switched to one output and d to the other. If, during slot t, the number of transit packets, U, and the number of new packets, V, are distributed as $B(2, m_t)$ and $Ber(v)$ respectively, then, given m_t,

$$
\begin{aligned}
G_t^{20} &= G_t^{02} = m_t^2 P^{tt} + 2m_t(1 - m_t)vP^{tn} \\
G_t^{11} &= m_t^2(1 - 2P^{tt}) + 2m_t(1 - m_t)v(1 - 2P^{tn}) \\
G_t^{10} &= G_t^{01} = \{2m_t(1 - m_t)(1 - v) + (1 - m_t)^2v\}0.5 \\
G_t^{00} &= (1 - m_t)^2(1 - v) \qquad (3.2.1)
\end{aligned}
$$

where, in slot t,

$$P^{tt} = \frac{ms_t \cdot mx_t}{m_t^2} \qquad P^{tn} = \frac{qn \cdot [ms_t + mx_t]}{2m_t}$$

Using these G_t^{cd} and the state transition diagram, the elements q_t^{ij} of the transition matrix $\mathbf{Q}_t$ can be computed. The next step is to evaluate the buffer occupancy probability vector Π_{t+1}, that gives the state probabilities at the start of the slot $(t + 1)$. From the theory of discrete Markov chains,

$$\Psi_{t+1} = \Psi_t Q_t \qquad (3.2.2)$$

Since $\Pi_{t+1} \equiv \Psi_{t+1}$, we have obtained the required buffer occupancy probability vector.

3.3 Approximate Analysis

The transient analysis of buffered deflection routing has two steps at each iteration. At any time slot t, given the buffer occupancy probability vector Π_t and the link utilization vector m_t, the transition probabilities between the various states in the buffer are computed and hence, the buffer occupancy probability vector, Π_{t+1}, is obtained from Π_t. The next step is to use the Π_{t+1} vector to compute the link utilization vector, m_{t+1}, from the set of *update equations*.

Consider a fixed output link and for $0 \le i \le d$, define $q(i), m_t(i)$ and m_t as before. Assume $m_0 \equiv 0$ and $\Pi_0 \equiv [1, 0, 0, \ldots, 0]$, since we assume that the network is initially empty. Given m_t, we can compute m_{t+1}, $t \ge 0$, using the following *update equations*, for $0 \le i \le d$:

$$
\begin{aligned}
m_t &= \sum_{i=1}^{d} m_t(i) \\
X_t(i) &= m_t(i)[1 - p(i, m_t, v)] \\
&\quad + 0.5va(m_t, v)q(i)[1 - p_o(i, m_t, v)] \\
Y_t(i) &= m_t(i)p(i, m_t, v) \\
&\quad + 0.5va(m_t, v)q(i)p_o(i, m_t, v) \\
m_{t+1}(i) &= I_{(k>1)} \sum_{l=0}^{K-1} \sum_{j=0}^{K} \{\pi_{t-l+1}(l, j)X_{t-l}(i + 1) \\
&\quad + \pi_{t-l}(l, j)Y_{t-l-1}(i + 1)\} \\
&\quad + \pi_{t+1}(0, K)Y_t(i - 3) \\
&\quad + \pi_{t-K+1}(K, 0)X_{t-K}(i + 1) \qquad (3.3.1)
\end{aligned}
$$

where $I_{(N)}$ is the indicator function and equals 1 if N is true and 0 otherwise.

$$p(i, m, v) = p_o(i, m, v) = 0, \text{ for } i < 1 \text{ or } i > d$$

$$q(i) = m_t(i) = 0, \text{ for } i < 0 \text{ or } i > d$$

$$\pi_k(l, j) = 0 \text{ for } k < 0$$

To understand the throughput equation, we focus on one output link and its associated buffer; the packets transmitted are either

- transit and new packets present at a node at the beginning of slot $(t - l + 1)$, (when there were l packets in the preferred output buffer and j packets in the other) with destinations at a distance $(i + 1)$ hops from the node, which were not delayed by a competing packet during slot $(t - l + 1)$, for $0 \le l \le K$,

- transit and new packets present at a node at the beginning of slot $(t - l)$ (when there were l packets in the preferred output buffer and j packets in the other) with destinations at a distance $(i + 1)$ hops from the node, which are delayed one slot by a competing packet during slot $(t - l)$, for $0 \le l \le K$,

- transit and new packets present at the node at the beginning of slot $(t + 1)$ (when there are 0 packets waiting in the output buffer of interest and K packets in the other buffer) with destinations at distance $(i - 3)$ from the node, which was deflected away from the other outgoing link to the one under consideration during slot $(t + 1)$,

- transit and new packets present at the node at the beginning of slot $(t - K + 1)$ (when there are K packets waiting in the output buffer of interest and 0 packets in the other buffer) with destinations at distance $(i + 1)$ from the node, which are not deflected during slot $(t - K + 1)$.

Note that if $K = 0$, the above equations reduce to the ones for a system with unbuffered deflection routing. As time t tends to infinity, the vector of buffer occupancy probabilities $\mathbf{\Pi_t}$ reaches a limit $\bar{\Pi}$ and the link utilization vector $\mathbf{m_t}$ reaches a limit, $\bar{m}$. These steady state limits satisfy equations (3.3.1) without the subscripts, for $0 \le i \le d$:

$$\bar{m} = \sum_{i=1}^{d} \bar{m}(i)$$

$$\bar{X}(i) = \bar{m}(i)[1 - p(i, \bar{m}, v)] \\ + 0.5 va(\bar{m}, v)q(i)[1 - p_o(i, \bar{m}, v)]$$

$$\bar{Y}(i) = \bar{m}(i)p(i, \bar{m}, v) + 0.5 va(\bar{m}, v)q(i)p_o(i, \bar{m}, v)$$

$$\bar{m}(i) = I_{(k>1)} \sum_{l=0}^{K-1} \sum_{j=0}^{K} \bar{\pi}(l, j)\{\bar{X}(i+1) + \bar{Y}(i+1)\} \\ + \bar{\pi}(0, K)\bar{Y}(i - 3) + \bar{\pi}(K, 0)\bar{X}(i + 1)$$

$$(3.3.2)$$

The steady-state link utilization, ρ, is given by

$$\rho = \sum_{i=0}^{d} \bar{m}(i) \qquad (3.3.3)$$

At steady-state, the rate at which packets are offered and accepted at the source should equal the rate at which the packets are absorbed from the network at the destination. The steady-state absolute throughput per node per slot, $\Lambda_T(\bar{m}, v)$, is defined to be the average number of slots a source at a node acquires and is given by

$$\Lambda_T(\bar{m}, v) = va(\bar{m}, v) = 2\bar{m}(0) \qquad (3.3.4)$$

The maximum steady-state absolute throughput achievable per node per slot is $2/d_{avg}$, where d_{avg} is the average number of hops a packet has to travel in the given network (or the average number of slots it is transmitted in). We define the normalized throughput per node per slot, $\Lambda_N(\bar{m}, v)$, as the absolute throughput normalized by the maximum that a node can achieve. This normalization compensates for the reduction in absolute throughput when the number of nodes in the network increases and provides an indication of how much of the maximum possible throughput is being achieved [Max89].

$$\Lambda_N(\bar{m}, v) = \frac{va(\bar{m}, v)}{2/d_{avg}} = \frac{2\bar{m}(0)}{2/d_{avg}} \qquad (3.3.5)$$

Using this model, the average delay encountered by the packet in the network (or the average network delay) can be found; however, due to the assumption that the offered traffic is memoryless, the queueing delay encountered by a typical packet, before it can access the network, cannot be found. The average network delay is the sum of the average propagation delay between nodes, the average delay, at the input of a node, in aligning the packets, and the average queueing delay in the output buffer (in the case of unbuffered deflection routing, this last term is zero). We assume that the networks being considered have relatively short links so that the sum of the average propagation delay and the average alignment delay is on the average equal to one slot transmission time. For networks with longer links, this component of the overall delay has to be increased [Max89]. So, the total average network delay encountered by the packet, $D_T(\bar{m}, v)$, is

$$D_T(\bar{m}, v) = \bar{N}(\bar{m}, v)(1 + \bar{N}_Q(\bar{m}, v)) \quad (3.3.6)$$

$$\bar{N}_Q(\bar{m}, v) = \sum_{l=0}^{K} \sum_{j=0}^{K} l \, \bar{\pi}(l, j) \qquad (3.3.7)$$

where $\bar{N}(\bar{m}, v)$ is the average number of nodes visited by a tagged packet and $\bar{N}_Q(\bar{m}, v)$ is the average number of packets waiting in the output buffer when the tagged packet arrives. For unbuffered networks, $\bar{N}_Q(\bar{m}, v) = 0$. To compare networks with different number of nodes, the absolute delay measure, $D_T(\bar{m}, v)$ is normalized with respect to the delay in an unloaded network, d_{avg} (this is equal to the length of the average shortest path).

$$D_N(\bar{m}, v) = \frac{\bar{N}(\bar{m}, v)(1 + \bar{N}_Q(\bar{m}, v))}{d_{avg}} \qquad (3.3.8)$$

Using Little's law, the delay encountered by the packet in the unbuffered network can be obtained; packets are accepted by the network at the rate $m(0)$ per link per slot, the link utilization is $\rho = \sum_{i=0}^{d} m(i)$ and each packet spends, on the average, $D_T(\bar{m}, v)$ slots in the network.

$$\bar{N}(\bar{m}, v) = \frac{\sum_{i=0}^{d} m(i)}{\bar{m}(0)} = (1 + \frac{\bar{m}}{\bar{m}(0)}) \quad (3.3.9)$$

$$D_T(\bar{m}, v) = (1 + \frac{\bar{m}}{\bar{m}(0)})(1 + \bar{N}_Q(\bar{m}, v)) \quad (3.3.10)$$

3.4 Results

There is very little decrease in the throughput of the MS_Net with increasing v. As expected, the addition of buffers at the nodes increases the throughput and reduces the delay. In the MS_Net, the normalized throughput for unbuffered deflection routing is about 55-60% ; it increases to about 75-80% with the addition of just one buffer and to about 90% with the addition of just 4 buffers (Figs. 3, 4). There is a significant reduction in the average network delay as well. As seen in the figures, the analytic results are within 2% of those obtained by simulation.

4. Conclusions

In this paper, we have presented an analytic model to study the performance of deflection routing in regular networks, under the uniform traffic assumption. The model was used to analyze the performance of this routing scheme on the Manhattan Street Network. The results showed that the MS_Net performs well under heavy load; the introduction of a few buffers improves the throughput and delay in the network to a great extent. The results obtained by our model agree with the results obtained by simulation.

This paper studies the performance of deflection routing under uniform load, which implies that the nodes are statistically identical in packet generation and the destinations of the packets are uniformly distributed. However, it remains to be seen how the network performs under non-uniform load. It is expected that the technique, being adaptive in nature, should be able to diffuse some of the "hot spots" that may develop due to unbalanced loading. Future work will be directed to studying the performance of deflection routing under non-uniform loading and to quantifying the effects of buffering on congestion.

References

[Bar64] P. Baran. On Distributed Communication Networks. *IEEE Trans. Commun. Systems*, 12:1–9, March 1964.

[BC90]   J. Brassil and R. Cruz. An Approximate Analysis of Packet Transit Delay for Deflection Routing in the Manhattan Street Network. Submitted to IEEE Trans. Commun, 1990.

[BT89]   A. M. Bignell and T. D. Todd. SIGNET: A New Ultra-High-Speed Lightwave Network Architecture. In *IEEE Pacific Rim Conference on Communications, Computers and Signal Processing*, pages 40–43, June 1989.

[CL90]   A. K. Choudhury and V. O. K. Li. Performance Analysis of Deflection Routing in the Manhattan Street and Minimum-Distance Networks. Submitted to IEEE Transactions on Communications, 1990.

[CL91]   A. K. Choudhury and V. O. K. Li. Effect of Conflict Resolution Rules on the Performance of Deflection Routing. To be submitted for publication, 1991.

[GG86]   A. G. Greenberg and J. Goodman. Sharp Approximate Models of Adaptive Routing in Mesh Networks. In J. W. Cohen O. J. Boxma and H. C. Tijms, editors, *Teletraffic Analysis and Computer Performance Evaluation*, pages 255–270, 1986. revised,1989.

[GH89]   A. G. Greenberg and B. Hajek. Deflection Routing in Hypercube Networks. Submitted to IEEE Transactions in Communications, July 1989.

[KH90]   A. Krishna and B. Hajek. Performance of Shuffle-Like Networks with Deflection. In *INFOCOM '90*, volume 2, pages 473–480, June 1990.

[Max87] N. F. Maxemchuk. Routing in the Manhattan Street Network. *IEEE Trans. Commun.*, COM-35(5):503–512, May 1987.

[Max89] N. F. Maxemchuk. Comparison of Deflection and Store-and-Forward Techniques in the Manhattan Street and Shuffle-Exchange Networks. In *INFOCOM '89*, volume 3, pages 800–809, April 1989.

[Szy90] T. Szymanski. An Analysis of Hot-Potato Routing in a Fiber Optic Packet Switched Hypercube. In *INFOCOM '90*, volume 3, pages 918–925, June 1990.

[TSH89] X. N. Tan, K. C. Sevcik, and J. W. Hong. Optimal Routing in the Shuffle-Exchange Networks for Multiprocessor Systems. In *Comp-Euro 88-System Design: Concepts, Methods and Tools*, pages 255–264. IEEE, Euromicro, IEEE Comput. Soc. Press, Washington, D.C., April 1989.

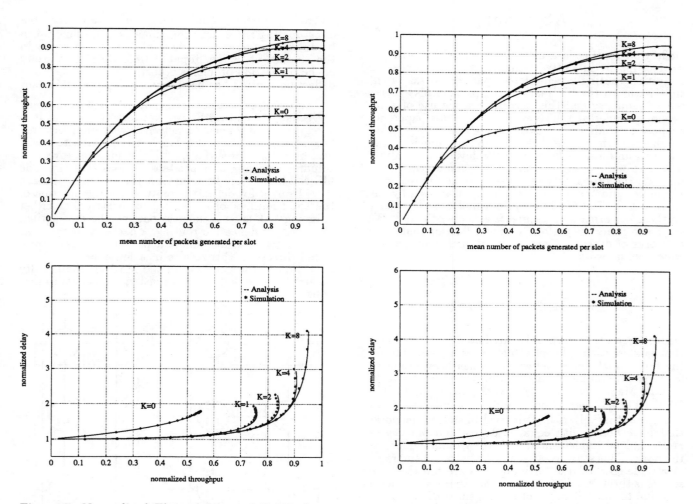

Figure 3: Normalized Throughputs and Delays in an 8x8 MS_Net with K buffers on each output link.

Figure 4: Normalized Throughputs and Delays in a 16x16 MS_Net with K buffers on each output link.

Effect of a Finite Reassembly Buffer on the Performance of Deflection Routing

Abhijit K. Choudhury
University of Southern California
Los Angeles, CA 90089-0272

Nicholas F. Maxemchuk
AT&T Bell Laboratories
Murray Hill, NJ 07974

Abstract

Deflection routing is a distributed switching strategy used primarily in slotted regular networks. Since the deflection routing scheme forces packets to take multiple paths to the destination, the packets may arrive out of sequence and this necessitates the use of a reassembly buffer at the receiver end. In this paper, we propose to study the effect of a finite reassembly buffer on the performance of deflection routing under a specific non-uniform traffic model. We develop an analytic model to understand the effect of oversubscription to links on the probability of packet loss at the reassembly buffer and the effective throughput as seen by the source. The effect of adding output buffers at the nodes on these performance measures is also investigated. We shall also study the performance of deflection routing in networks of different sizes like local, metropolitan and wide area networks.

1. Introduction

Deflection routing is a novel, distributed routing scheme that is attracting a lot of attention as a candidate for the future high-speed networks. It has been proposed for large geographically distributed computer networks like the Manhattan Street Network [Max87] and SIG-NET [BT89]. Deflection routing is similar to "hot-potato routing" [Bar64]. It is used in slotted networks that have nodes with equal in-degree and out-degree and could be implemented with or without buffers at the node. This scheme assigns priority to transit traffic at a node so that if all arriving slots are full the local source is throttled. This ensures that there is no packet loss due to unavailability of buffers since all the packets reaching a node are sent out in the next slot. So, in deflection routing, some bandwidth is traded off for fast switching capabilities and storage requirements at the nodes.

A number of researchers have attempted to characterize the performance of deflection routing in various regular networks, using an uniform traffic model. Some have studied the Manhattan Street Network [Max89, GG86, CL90a, BC90] while others have studied the hypercube [GH89] and shuffle networks [TSH89, KH90,

CL90b]. Some work has also been done in the area of estimating the evacuation time (i.e., the time taken to drain a number of packets placed at the nodes in the absence of further input)[BH85, Haj88]. A study of the congestion problems occurring in networks using deflection routing has been recently reported [Max90].

An example of a network that uses deflection routing is the Manhattan Street Network (MS_Net). It is a regular, two-connected network with unidirectional links. The nodes in the MS_Net are connected as a grid on a toroidal surface, with the adjacent rows and columns traveling in opposite directions (Fig. 1). The MS_Net has certain structural characteristics that make it well suited for deflection routing. A large fraction of the total number of nodes are so located that either outgoing link is on a shortest path to the destination. Since, under deflection routing, a packet that can take either path does not deflect a packet that has a preferred direction, this results in a lower probability of deflection of the packets. In the MS_Net, the penalty incurred due to deflection is constant and equal to four hops.

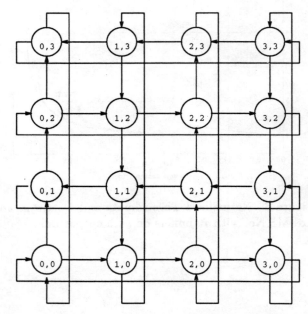

Figure 1: The Manhattan Street Network.

Reprinted from *Proc. IEEE International Conference on Communications*, pp. 1637–1646, June 1991.

Since the deflection routing scheme forces packets to take multiple paths to the destination, the packets may arrive out of sequence and this necessitates the use of a reassembly buffer at the receiver end. In this paper, we propose to study the effect of a finite reassembly buffer on the performance of deflection routing under a specific non-uniform traffic model. We shall also study the effect of using deflection routing in local, metropolitan and wide area networks. The deflection routing scheme is described in Section 2 and the reassembly problem is stated. In Section 3, a simple mathematical model is presented to compute an estimate of the probability of overflow at the reassembly buffer. The distribution of the random component of the delay through the network is evaluated analytically in Section 4. The accuracy of the analytic estimate of the distribution is discussed in Section 5. In Section 6, we present a discussion of the results obtained and finally we conclude with some comments based on the results of this study.

2. The Reassembly Problem in Deflection Routing

Deflection routing is a simple, distributed switching strategy used primarily in slotted regular networks. The scheme requires the nodes of the network to have equal in-degree and out-degree and in each slot, the incoming packets are forwarded on the outgoing links. As a result, if there is contention for an output channel, some of the packets are misrouted or *deflected* and have to travel an additional distance to the destination. The

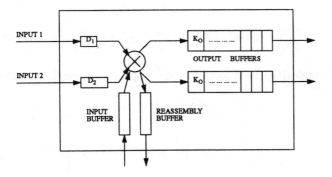

Figure 2: Structure of a node.

performance of this routing strategy could be improved by adding buffers at each output link, in which case deflection occurs only if the output buffer associated with the preferred link is full. The structure of a node is shown in Fig. 2. Buffered deflection routing causes the packets to suffer an additional delay at the nodes due to buffering but the overall delay is reduced, since by avoiding deflection and staying behind in the buffer, the packet is able to avoid the penalty of having to travel four extra hops; this is particularly beneficial if the propagation delay along each hop is large. This scheme ensures that when an output buffer is full, as many packets are sent out in each slot as the number of incoming packets; hence, the nodal buffer loss in this

scheme is zero. This is an extremely desirable feature for the future high speed wide-area networks since an enormous number of packets will be 'on the fly' in each link and nodal buffer loss is expected to be the chief source of packet loss.

However, this attractive feature of deflection routing comes at a certain price. This strategy forces packets to take multiple routes from the source to the destination and since different paths may have different delays, this implies that the packets may arrive at the destination out of sequence. The severity of this problem increases when congestion sets in due to oversubscription to the links. Since every deflection makes a packet travel an extra number of hops, if the propagation delay along each hop is large, the deflected packet could be delayed by a large number of slots. A reassembly buffer is used at the receiver end to hold the arriving packets until the missing packet arrives. All prior work on deflection routing has been based on the assumption of that the receiver has an infinite reassembly buffer and is able to put all the received packets back in sequence. In practice, however, since the reassembly buffer is going to be finite, this may prove to be a source of packet loss, resulting in throughput degradation.

In this paper, we propose to study the effect of a finite reassembly buffer on the packet loss probability in a network using deflection routing. We assume that the receiver drops a packet when it arrives too far out of sequence and initiates a retransmission from the source by sending a control message back. All subsequent packets from the source are dropped by the receiver until the first lost packet is received since we assume that the delay before retransmission is initiated is large and the reassembly buffers are not large enough to buffer up all the packets.

3. Model

In a network using deflection routing, the successive packets in a message could arrive out of sequence. At the receiver, a finite reassembly buffer is maintained in order to put these packets back in proper sequence before delivering them to the host connected to the node. We now develop a simple mathematical model in order to estimate the probability of overflow of the reassembly buffer.

Consider a particular source-destination pair (s, d), where the source s transmits at an average rate of R_s (the source rate is assumed to be less than the average rate at which vacant slots arrive at the source). Suppose that the source transmits the ith packet in slot s_i, where $s_i = s_{i-1} + t_i$. Let t_i, the number of slots between successive transmissions be independent, identically distributed (*i.i.d.*) random variables, with mean $E[t_i] = 1/R_s$ and variance $Var[t_i]$. At the receiver, the ith packet arrives in slot r_i, where $r_i = s_i + P_{sd} + \Delta_i$; the delay through the network consists of a constant component P_{sd}, the length of shortest path between the source s and the destination d and a random component, Δ_i, which is the additional delay (in slots) experienced by the ith packet in travelling from the source to the destination (in general, the distribution of Δ depends on the source-destination pair). Δ_i is assumed to be an *i.i.d.* random variable with mean $E[\Delta]$ and variance $Var[\Delta]$. Assume that the receiver has a finite reassembly buffer of size K_r.

For ease of implementation, we shall assume that the reassembly buffer accepts only packets bearing a sequence number that is within K_r of the last in-sequence packet. A buffer overflow occurs when a packet arrives too far out of sequence for the finite buffer to put it back in sequence. This assumption leads to a pessimistic estimate since, with the same number of buffers and a different buffering strategy, the arrival of a packet that is not within K_r of the last in-sequence packet may not cause an overflow. According to the buffering policy we adopt, the ith packet will be dropped if it arrives before any of the packets with a sequence number $(i - K_r)$ or less, i.e. if $r_i < r_j$, for all $0 \le j \le i - K_r$. Therefore, the probability that the reassembly buffer overflows when the ith packet is received is

$$P_{ov}(i) = Pr\left\{ \bigcup_{j=K_r}^{i} (z_j < 0) \right\} \qquad (3.1)$$

where

$$z_j = r_i - r_{i-j} = \Delta_i - \Delta_{i-j} + \sum_{t=i-j+1}^{i} t_l$$

The random variable z_j has a mean $E[z_j] = \lceil j/R_s \rceil$ and a variance $Var[z_j] = 2Var[\Delta] + jVar[t_l]$ and is independent of i. We conjecture that the overflow probability should be a function of the reassembly buffer size, K_r, the random component of the delay through the network, Δ and the source rate, R_s.

Instead of trying to evaluate the overflow probability exactly, we will attempt to bound the value. A very simple lower bound is obtained by assuming that the union is dominated by the first term, $z_{K_r} < 0$, since z_{K_r} has the lowest mean. So,

$$P_{ov}(i) \ge Pr\{z_{K_r} < 0\} \qquad (3.2)$$

We shall focus on the case where t_i, the time between successive transmissions, is a constant and equal to $1/R_s$. This is the case when packets are inserted periodically by the source and leads to a smaller value of P_{ov} since this decreases the variance of the random variables, z_i.

$$P_{ov}(i) \ge Pr\{z_{K_r} < 0\}$$
$$\ge Pr\left\{ \Delta_{i-K_r} > \Delta_i + \left\lceil \frac{K_r}{R_s} \right\rceil \right\}$$
$$\ge \sum_{n \ge 0} Pr\left\{ \Delta_{i-K_r} > n + \left\lceil \frac{K_r}{R_s} \right\rceil / \Delta_i = n \right\} Pr\{\Delta_i = n\}$$
$$\ge \sum_{n \ge 0} Pr\left\{ \Delta_{i-K_r} > n + \left\lceil \frac{K_r}{R_s} \right\rceil \right\} Pr\{\Delta_i = n\}$$

since delays are independent $\qquad (3.3)$

To carry the analysis further, we need to assume a distribution for the random delay through the network, Δ. In particular, we assume a memoryless distribution since we conjecture that after a deflection, a packet does not retain the memory of how many deflections it suffered before and traverses the network like any other transit packet. Simulations will be used to check the accuracy of our model in a later section. So,

$$Pr[\Delta = n \text{ slots}] = (1 - a)a^n \ , \quad 0 \le a \le 1 \qquad (3.4)$$

$$P_{ov}(i) \ge \sum_{n \ge 0} (1 - a)a^{2n+1+\lceil \frac{K_r}{R_s} \rceil}$$
$$\ge \frac{a^{\lceil \frac{K_r}{R_s} \rceil + 1}}{1 + a} \qquad (3.5)$$

To find an upper bound, we rewrite equation (3.1) in the following way.

$$P_{ov}(i) = 1 - Pr\left\{ \bigcap_{j=K_r}^{i} (z_j \ge 0) \right\} \qquad (3.6)$$

Proceeding as before,

$$Pr\left\{ \bigcap_{j=K_r}^{i} (z_j \ge 0) \right\} = Pr\left\{ \bigcap_{j=K_r}^{i} (\Delta_{i-j} \le \Delta_i + \left\lceil \frac{j}{R_s} \right\rceil) \right\}$$
$$= \sum_{n \ge 0} Pr\left\{ \bigcap_{j=K_r}^{i} (\Delta_{i-j} \le n + \left\lceil \frac{j}{R_s} \right\rceil)/\Delta_i = n \right\} \cdot$$
$$Pr\{\Delta_i = n\}$$
$$= \sum_{n \ge 0} Pr\{\Delta_i = n\} \prod_{j=K_r}^{i} Pr\left\{ \Delta_{i-j} \le n + \left\lceil \frac{j}{R_s} \right\rceil \right\}$$

by independence

$$= \sum_{n \ge 0} (1 - a)a^n \prod_{j=K_r}^{i} (1 - a^{n+1+\lceil \frac{j}{R_s} \rceil})$$

Considering the first order terms of the product only,

$$Pr\left\{ \bigcap_{j=K_r}^{i} (z_j \ge 0) \right\}$$
$$\ge \sum_{n \ge 0} (1 - a)a^n \left\{ 1 - \sum_{j=K_r}^{i} a^{n+1+\lceil \frac{j}{R_s} \rceil} \right\}$$
$$\ge 1 - \frac{a^{\frac{K_r}{R_s}+1}}{1 + a} \left[\frac{1 - a^{\frac{i-K_r+1}{R_s}}}{1 - a^{\frac{1}{R_s}+1}} \right]$$

Thus, we get the following upper bound,

$$P_{ov}(i) \le \frac{a^{\frac{K_r}{R_s}+1}}{1 + a} \left[\frac{1 - a^{\frac{i-K_r+1}{R_s}}}{1 - a^{\frac{1}{R_s}}} \right] \qquad (3.7)$$

Thus, we have obtained upper and lower bounds on the probability that the buffer overflows when the ith packet arrives. To find the steady state values of these bounds, we take the limit as i approaches infinity.

$$\frac{a^{\lceil \frac{K_r}{R_s} \rceil + 1}}{1 + a} \le P_{ov} \le \frac{a^{\frac{K_r}{R_s}+1}}{1 + a} \left[\frac{1}{1 - a^{\frac{1}{R_s}}} \right] \qquad (3.8)$$

Properties of P_{ov}

A closer look at the bounds on P_{ov} indicate that the following interesting facts [Max90].

Fact 1. *The overflow probability, P_{ov}, is invariant to the mean of the delay through the network and is an increasing function of the variance.*

This statement can be proved by observing that P_{ov} is a function of the random variables, z_i, whose mean contains no contribution from the random component of the network delay, Δ. The variance of z_i has a significant contribution from the variance of Δ_i; thus, the probability that z_i is negative depends strongly on the variance of the Δ_i. (Note that $Var[\Delta] = a/(1-a)^2$ is an increasing function of a and so is P_{ov}.)

Fact 2. *For a fixed source rate, the overflow probability, P_{ov}, is invariant to the network rate.*

If the network rate increases by a factor f and the source keeps its mean packet insertion time constant, the new source rate is $R'_s = R_s/f$ packets per slot. As a result of the increase in network rate, the delay through the network (in slots) scales up; let the new delay be denoted by $\Delta' = f\Delta$. Using the distribution defined in (3.4), the distribution of Δ' is given by

$$Pr[\Delta' = n \text{ slots}] = \begin{cases} Pr[\Delta = n/f \text{ slots}] \\ \qquad \text{if } n/f \text{ is an integer} \\ 0 \text{ otherwise} \end{cases}$$
$$= \begin{cases} (1-a)a^{n/f} \\ \qquad \text{if } n/f \text{ is an integer} \\ 0 \text{ otherwise} \end{cases}$$

$$(3.9)$$

The resulting lower bound on the overflow probability is

$$P_{ov} \geq \lim_{i \to \infty} Pr\left\{\Delta'_{i-K_r} > \Delta'_i + \left\lceil \frac{K_r}{R'_s} \right\rceil\right\}$$
$$\geq \lim_{i \to \infty} \sum_{n \geq 0} Pr\left\{\Delta'_{i-K_r} > n + \left\lceil \frac{K_r}{R'_s} \right\rceil / \Delta'_i = n\right\}(1-a)a^{\frac{n}{f}}$$
$$\geq \sum_{n \geq 0}(1-a)a^{\frac{2n+1}{f} + \frac{1}{f}\left\lceil \frac{K_r}{R'_s} \right\rceil}$$
$$\geq \frac{a^{\frac{1}{f}\left\lceil \frac{K_r}{R'_s} \right\rceil + 1}}{1+a}$$
$$\geq \frac{a^{\left\lceil \frac{K_r}{fR'_s} \right\rceil + 1}}{1+a} = \frac{a^{\left\lceil \frac{K_r}{R_s} \right\rceil + 1}}{1+a} \text{ , since } R_s = fR'_s$$

Similarly, the upper bound can be shown to be invariant to the change in the network rate.

Fact 3. *For a given network rate, the overflow probability, P_{ov}, is an increasing function of the source rate.*

Using the same technique as before, we note that if we increase the source rate R_s by a factor while holding the network rate constant, the number of reassembly buffers at the node, K_r, has to increase by the same factor if the probability of overflow has to be kept constant.

Fact 4. *For a given network rate, source rate and reassembly buffer size, the overflow probability, P_{ov}, increases with an increase in the propagation delay (which implies an increase in the penalty due to each deflection).*

Suppose the penalty due to each deflection is constant and equal to δ slots. The corresponding distribution of Δ is given by

$$Pr[\Delta = n\delta \text{ slots}] = (1-a)a^n \text{ , } n = 0, 1, 2, \ldots$$

Note that (3.4) gives the distribution of Δ for $\delta = 1$; in practical networks, δ cannot be less than 2. The resulting lower bound on the overflow probability is

$$P_{ov} \geq \lim_{i \to \infty} Pr\left\{\Delta_{i-K_r} > \Delta_i + \left\lceil \frac{K_r}{R_s} \right\rceil\right\}$$
$$\geq \lim_{i \to \infty} \sum_{n \geq 0} Pr\left\{\Delta_{i-K_r} > n\delta + \left\lceil \frac{K_r}{R_s} \right\rceil / \Delta_i = n\delta\right\} \cdot$$
$$(1-a)a^n$$
$$\geq \sum_{n \geq 0}(1-a)a^{2n+1+\frac{1}{\delta}\left\lceil \frac{K_r}{R_s} \right\rceil}$$
$$\geq \frac{a^{\left\lceil \frac{K_r}{\delta R_s} \right\rceil + 1}}{1+a}$$

Hence, we see that if the penalty due to each deflection increases to δ slots, the size of the reassembly buffer has to be scaled up by a factor of δ to keep the P_{ov} the same.

Since we observe that the difference between the upper and lower bounds of P_{ov} is very small, we shall use the lower bound as an estimate of the probability of overflow, $\hat{P}_{ov}$.

4. Evaluating the Distribution of Network Delay

We now proceed to evaluate the distribution of Δ, the random component of the network delay, for a specific non-uniform traffic scenario, using the Manhattan Street Network (MS_Net) as a sample network that uses deflection routing. With non-uniform loading, it is possible to heavily oversubscribe to links and thereby create severe congestion in the network. In this section, we shall develop a model to understand the effect of oversubscription to links on packet loss at the reassembly buffer and effective throughput as seen by the source. We shall also find out the effect of adding output buffers at the nodes on these performance measures. The simple analytic model yields the distribution of the number of deflections suffered by a tagged packet; this directly translates into the distribution of delay given the increase in path-length due to a deflection. It will be seen that the distribution can be approximated, after the first few terms, by a dominant component, yielding an almost memoryless distribution. The accuracy of the model will be studied in Section 5.

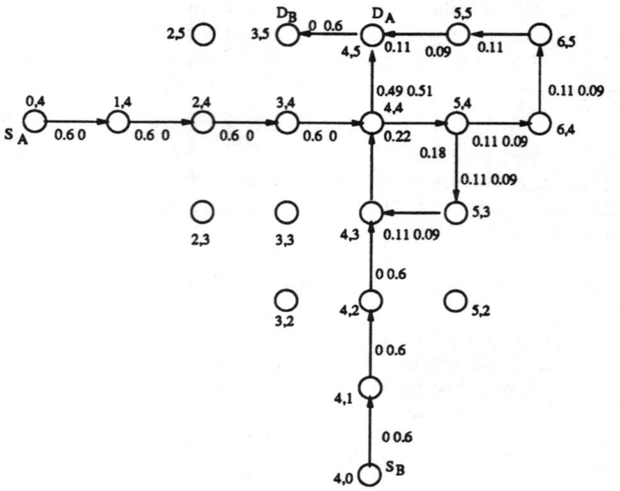

Figure 3: Congestion in a 10×10 MS_Net with non-uniform load, 8 output buffers per node and $r_A = r_b = 0.6$

Consider a 10×10 MS_Net with just two sources - source S_A, at node (0,4), transmitting to destination D_A, at node (4,5), and source S_B, at node (4,0), transmitting to destination D_B, at node (3,5). It can be seen (Fig. 3) that both the sources have unique shortest paths to their destinations and share a link $(4,4) \rightarrow (4,5)$ which turns out to be a bottleneck. Fig. 3 shows the utilization pattern that is generated when both the sources are transmitting at rate 0.6 and there are eight output buffers at each node; the two numbers associated with each link are the utilization due to S_A and S_B respectively.

Once a packet is deflected, it need not come back to the bottleneck; in fact, it could follow a different path to the destination. This feature of deflection routing can be exploited to increase the throughput by over-subscribing to the bottleneck link. However, if both sources transmit in every slot, the bottleneck link is heavily oversubscribed and half the packets arriving at node $(4,4)$ will be deflected. This leads to wide-spread congestion in the network which spreads outward from the congested node and blocks other potential users from accessing the network [Max90]. It is obvious that, in this case, even adding output buffers at the nodes will not alleviate the situation, since due to the heavy oversubscription at the bottleneck, the buffers will soon fill up and we will be faced with the same situation as before. It is, therefore, clear that the oversubscription to links in the network has to be kept at a much lower level so that the deflection probability is low and yet the bottleneck link is fully utilized. Our study shows that under moderate oversubscription, the network per-

formance is acceptable especially if output buffers are used at the nodes.

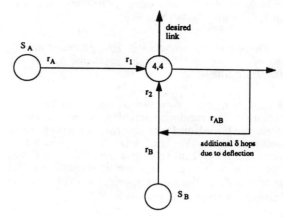

Figure 4: Model for two sources with a single bottleneck.

Consider the simple model of two sources that go through a single bottleneck as shown in Fig. 4. Suppose source S_A transmits at a rate r_A and source S_B transmits at rate r_B; under this simplified model, the percent oversubscription of the bottle-neck link is $(r_1 + r_2 - 1) \times 100\%$, where r_1 and r_2 are the effective utilizations seen by the bottleneck node at its input links. This model assumes that packets from the two sources reach the bottleneck node (4,4) and some of them are deflected; the deflected packets are equally likely to find a different path from the next node to their respective destinations or to travel four additional hops to come back to node (4,4) and compete for their desired link. We also assume that r_{AB}, the traffic due to deflected packets which find their way back to the bottleneck, adds to the traffic from source S_B so that $r_2 = r_B + r_{AB}$. This is obviously not accurate in the case of unbuffered deflection routing or when the number of output buffers, K_o, is low (say 1 or 2) since it does not take into account the additional deflections that a packet may suffer at the nodes surrounding the bottleneck. However, as the number of output buffers increases the model gets closer and closer to reality. The total number of buffers at each node is $K_{node} = 2K_o + K_r$. In general, the size of the reassembly buffer is a multiple of the penalty suffered due to deflection since that would determine the number of deflections that the buffer can handle, in the worst case, before having to drop a packet.

Consider a tagged packet from source S_A: upon reaching node (4,4) for the first time, i.e., along link $(3,4) \rightarrow (4,4)$, it could get deflected by a packet from node (4,3) with probability p_{A_1} whereas after the first deflection, when it approaches node (4,4) from node (4,3), it could get deflected with probability p_{A_2}. Assuming random conflict resolution,

$$p_{A_1} = \frac{r_2}{2} \cdot Pr[\text{ preferred output buffer is full}]$$

$$p_{A_2} = \frac{r_1}{2} \cdot Pr[\text{ preferred output buffer is full}]$$

(4.10)

350

It can be shown that when two links, with utilizations of r_1 and r_2 compete for the same output buffer of size K_o, the probability of the buffer being empty, π_0, and the probability of the buffer being full, π_{K_o}, are given by the following expressions.

$$\pi_0 = \left[\frac{1 - p_2/p_0}{1 - (p_2/p_0)^{K_o+1}}\right]$$
$$\pi_{K_o} = (p_2/p_0)^{K_o}\pi_0 \qquad (4.11)$$

where p_2 and p_0 are the probabilities that, in a given slot, two and zero packets compete for the preferred output buffer respectively.

Let $P_A(n)$ be the probability that a tagged packet from source S_A undergoes n deflections. In a 10×10 MS_Net, a packet, after getting deflected at $(4,4)$, is equally likely to be sent back to $(4,3)$ or to take a different route to the destination. Therefore,

$$P_A(0) = 1 - p_{A_1}$$
$$P_A(1) = p_{A_1}\left[\frac{1}{2} + \frac{1}{2}\cdot(1 - p_{A_2})\right] = p_{A_1}(1 - \frac{1}{2}p_{A_2})$$
$$P_A(n) = p_{A_1}(1 - \frac{1}{2}p_{A_2})(\frac{1}{2}p_{A_2})^{n-1} , \; n \geq 2 \qquad (4.12)$$

Given the distribution of $P_A(n)$, the distribution of the random component in delay through the network, Δ, can be easily computed. In a network, where each deflection increases the distance to the destination by δ slots, the distribution of Δ is given by

$$Pr[\Delta = n\delta \text{ slots}] = P_A(n) , \; \forall n \qquad (4.13)$$

In the MS_Net, for every deflection, the packet has to travel an additional four hops; the number of additional slots, δ, by which the delay increases depends on the inter-node distance (in slots per hop). Once the distribution of Δ is computed, the bounds on P_{ov} can be found using the method in Section 2. Using the lower bound on P_{ov} as an estimate,

$$\hat{P}_{ov} = p_{A_1}(\frac{1}{2}p_{A_2})^{\lceil\frac{K_r}{\delta R_s}\rceil}\left[(1 - p_{A_1}) + \frac{p_{A_1}p_{A_2}}{1 + \frac{1}{2}p_{A_2}}\right] \qquad (4.14)$$

We now proceed to obtain an expression for average throughput, given the estimate of the overflow probability. Let $\bar{B}$ be the average number slots between the start of a transmission and the first time the reassembly buffer overflows , i.e., a packet is dropped. Since source S_A transmits once every $1/r_A$ slots and one packet out of every $1/\hat{P}_{ov}$ packets is dropped on the average,

$$\bar{B} = 1/\left(r_A \cdot \hat{P}_{ov}\right) \qquad (4.15)$$

Let $\bar{I}$ be the amount of time it takes for the information about the packet loss to reach the source and for the first of the retransmitted packets to reach the destination. We shall assume that $\bar{I} = 2P_{sd}$ slots, which implies that the deflections are ignored when sending the information to the source and when the first of the retransmitted packets makes its way to the destination So,

$$\text{Avg. throughput for} S_A = \frac{\bar{B}}{\bar{I} + \bar{B}} \times \text{Rate of} S_A$$
$$= r_A \cdot \frac{1/r_A \cdot \hat{P}_{ov}}{2P_{sd} + 1/r_A \cdot \hat{P}_{ov}}$$
$$= \frac{r_A}{1 + 2P_{sd} \cdot r_A \cdot \hat{P}_{ov}} \qquad (4.16)$$

5. Accuracy of the Model

The model used in Section 4 to evaluate the distribution of Δ, the random component of network delay, is a very simple one and is accurate only under certain special conditions. It assumes that : first, the packets from sources S_A and S_B suffer deflections only at the bottleneck node and that the traffic does not spread too much and secondly, that the deflected packets merge into the stream of packets from source S_B heading towards the bottleneck node. The first assumption is clearly invalid when the number of output buffers at the nodes is low (say 0 or 1) or when the source rate is moderately high [Max90]. In these cases, the traffic tends to spread out around the bottleneck node, thereby creating a large number of possible locations where deflections may take place. Also, unless the source rate of S_B is relatively low and there are a few buffers at the nodes (say, at least 2) the deflected packets are likely to be further deflected at the point where they meet the packets coming from S_B. So, the model is accurate only when the source rates are moderate and there are a few output buffers at the nodes (Fig. 3). However, we would not like to operate at source rates where the number of deflections are significantly high and also we would try to use a few buffers so that the traffic does not spread too much causing other sources to be blocked out. Hence, this simple model proves to be quite adequate in characterizing the behavior of the packets in the region of interest. The reason for studying the reassembly problem is to identify the region where the deflection routing scheme performs satisfactorily and to construct a flow control scheme that will tend to ensure that the system operates in that region. So, even in the region where the model is not accurate, it correctly predicts the direction in which adjustments have to be made to reach satisfactory performance.

Fig. 4 compares the distribution of Δ obtained from the model with those obtained from simulation[*]; the figures show the distributions for values of R_s equal to 0.4, 0.5 and 0.6 and also for values of K_o equal to 0, 2, 4 and 8. Note that the linearity of all the plots bears out our conjecture that the distribution of Δ would be close to geometric. We observe that when $R_s = 0.4$, the analytic model is accurate when we have 2, 4 or 8 buffers. As expected, the model underestimates the

---
[*]By simulation, we do not mean Monte Carlo simulation; we have used a numerical technique that iteratively computes the steady-state utilizations of the links in the network and then uses a random walk approach to compute the distribution of delay.

number of deflections in the case of unbuffered deflection routing; this behavior is seen in all three figures. If $R_s = 0.5$, we observe that the analytic model is still accurate for buffered deflection routing. As we increase the value of R_s, the oversubscription at the bottleneck increases and when one of the links saturate, the nodes around the bottleneck become potential sites for deflection. The analytic model, which just considers one bottleneck node, can no longer predict the system behavior accurately. Once the oversubscription becomes heavy, increasing the number of output buffers ceases to affect the performance, since the heavy loading keeps the buffers full and once that happens the performance is identical irrespective of the number of output buffers.

6. Results

Once the distribution of the random component of network delay, Δ, has been found, we proceed to estimate the probability of buffer overflow, $\hat{P}_{ov}$, using equation (4.14). Fig. 6 shows a plot of the $\hat{P}_{ov}$ against user rates for various sizes of output and reassembly buffers (the distribution used was obtained from the simulation in order to plot the metric over a larger range accurately; in the moderate loading range, say below rate 0.6, the analytic model gives the same plots except for the case of unbuffered routing). The size of the reassembly buffer depends on the number of deflections the system is designed to handle. In the MS_Net, each deflection increases the path length of a packet by four hops. The penalty suffered by the packet depends on the inter-node distance in slots per hop and is four times the number of slots on each hop. The reassembly buffer size should be designed to hold a number of slots equal to a multiple of the penalty due to deflection (in slots). For example, if the penalty due to deflection is 20 slots then a buffer size of 80 slots will be able to handle four deflections. Fig. 6 shows two sets of plots - one for a buffer-size that can handle four deflections and one that can handle three. We notice that when we can handle three deflections, as long as the users do not exceed an average rate of 0.55, with two or more output buffers, the probability of buffer overflow is acceptable ($\approx 10^{-8}$) ; the same is true for a buffer that can handle four deflections if the sources control their transmission rate to below 0.6.

We now consider the effect of a finite reassembly buffer on the effective throughput of the source. Networks of three different sizes are considered - specifically, we look at 16×16 MS_Nets with different internode distances so that the diameter and the average distance between nodes in the network correspond to those found in local area networks (LANs), metropolitan area networks (MANs) and wide area networks (WANs). We define diameter as the maximum length of the shortest path between any two nodes and average distance is defined as the average shortest path between nodes. Suppose the network operates at 150 Mbps and that each slot is 53 bytes long. Assuming that the speed of light in the medium is 0.69 times that of the speed of light in a vacuum, there are approximately 1.7 slots in transit on each kilometer of the network. If we assume

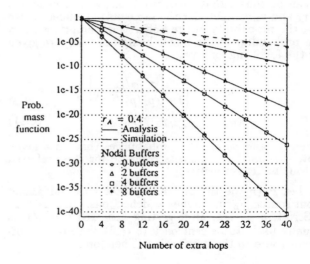

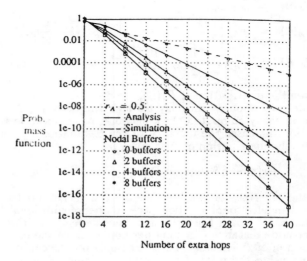

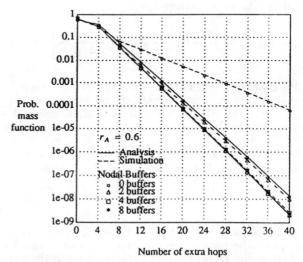

Figure 5: Probability mass function of the random component of network delay for source S_A as a function of the extra number of hops for a given source rate r_A.

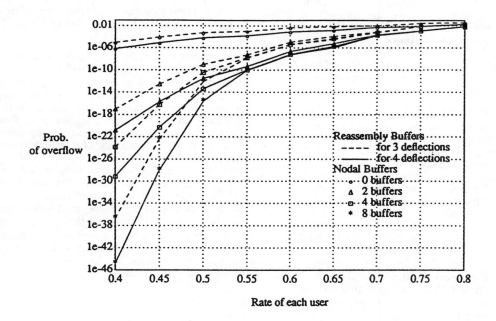

Figure 6: Probabilty of overflow as a function of user rate.

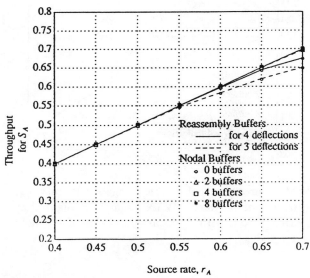

Figure 7: Throughput of source S_A against its rate, r_A, for a local area network (LAN) that is configured as a 16×16 MS_Net, with a propagation delay of one slot between nodes.

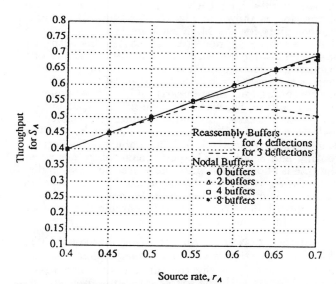

Figure 8: Throughput of source S_A against its rate, r_A, for a metropolitan area network (MAN) that is configured as a 16×16 MS_Net, with a propagation delay of 5 slots between nodes.

that the propagation delay between nodes is 1 slot, we have a MS_Net with a diameter of approximately 10 km. and an average distance of 5.28km - this network would qualify as a LAN. If we increase the inter-node propagation delay to 5 slots, keeping the number of nodes constant, we have a network with a diameter of about 49.7 km. and an average distance of 26.40 km. - this network has the dimensions of a MAN. Similarly, an inter-node propagation delay of 400 slots yields a coast-to -coast WAN with a diameter of about 3979 km. and an average distance of 2111 km.

As the number of slots per hop (i.e., the propagation delay between nodes) increases, the penalty suffered by a deflected packet, in terms of the number of slots by which it is delayed, increases. In the MS_Net, a single deflection delays a packet by four times the number of slots on each hop. This effect is reflected in the size of the reassembly buffer since the receiver needs a buffer size at least equal to the penalty if it hopes to recover from the effect of a single deflection. For a large WAN, like the one we consider, the penalty for a single deflection amounts to 1600 slots or 84.8 Kbytes. However, this estimate assumes just 256 nodes placed on a coast-to-coast MS_Net; in reality, we would be more likely to have at least 4000 nodes. The same penalty for a 4096 node MS_Net is around 460 slots or 24.3 Kbytes. The other effect that a large propagation delay has on this system is that it increases the time between the instant the receiver drops a packet and the instant the same packet, retransmitted by the source, arrives at the receiver. The larger the propagation delay the larger is the recovery time and this reduces the effective throughput of the source.

A plot of the effective throughput of source S_A against the source rate, for a LAN with an inter-node distance of one slot, is shown in Fig. 7. For this and the subsequent plots for the MAN and the WAN, we assume that the source S_A, node (0,4), is contending with just one other source S_B, in a scenario similar to the one discussed in Section 4. We observe that in the LAN, as long as the source rate is in the range 0.4 to 0.7, the effective throughput rises linearly and is almost equal to the source rate, r_A, when we have some output buffers and 12 or 16 reassembly buffers (since the penalty for deflection here is 4 slots, this buffer size can handle 3 or 4 deflections respectively). With unbuffered deflection routing and 3 buffers, the performance degrades when the rate is above 0.65. Since the source-destination distance for S_A is just 5 slots, the amount of time it takes to recover is minimal. Also, we note that the number of reassembly buffers required for acceptable performance is small, just equal to 12 or 16 slots.

For the MAN (Fig. 8), with a penalty equal to 20 slots, we notice that with 80 reassembly buffers (enough for 4 deflections), the effective throughput still increases almost linearly with the source rate. However, with 60 buffers, we note that the effective throughput reaches a peak and then starts dropping; this is because the large recovery time starts to make its presence felt. This effect causes a dramatic degradation of the effective throughput in the plot for the WAN. Also, the WAN, with its penalty equal to 1600 slots, requires enormous amount of buffering to come anywhere close to acceptable performance when the source rate is moderately

high. The other effect that can be seen in Figs. 8,9 but is not so clear in Fig. 7 is the effect of adding output buffers at the nodes - by reducing the probability of deflection, they help to improve the overflow probability and therefore the effective throughput. Their contribution is particularly significant when the load is moderate.

From these results, we conclude that in larger networks the oversubscription to links has to be tightly controlled. In the case of non-uniform traffic, the probability of overflow at the reassembly buffer depends on the source rate. The large penalty for deflection makes it necessary to ensure, using higher level flow-control measures, that the source rates stay in the desired range so as to keep the probability of overflow restricted. In this way, in a large WAN, it may be possible to operate with sufficient buffering to handle just one or two deflections. Since the size of the reassembly buffer required depends upon the source rate, a low rate user would require a smaller reassembly buffer (hence, less expensive equipment) to reach the desired level of performance than a user with a higher rate. Also, we note that as the networks increase in size, it becomes necessary to use a few output buffers to ensure that the probability of deflection is low.

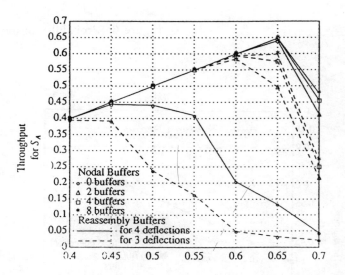

Figure 9: Throughput of source S_A against its rate, r_A, for a wide area network (WAN) that is configured as a 16×16 MS_Net, with a propagation delay of 400 slots between nodes.

7. Conclusion

In this paper, we have studied the effect of a finite reassembly buffer on the performance of deflection routing under a specific non-uniform traffic model. An estimate of the probability of overflow at the reassembly buffer was presented and its properties were studied.

We developed a model to understand the effect of oversubscription to links on the probability of packet loss at the reassembly buffer and the effective throughput as seen by the source. The effect of adding output buffers at the nodes on these performance measures was also investigated. We also saw the effect of increasing the network size on the performance of deflection routing.

When the oversubscription to the links is moderate, the deflection routing scheme was found to perform admirably well with a small number of output buffers and enough reassembly buffers to handle three or four deflections. Since a large penalty due to deflection seriously degrades the effective throughput and greatly increases the reassembly buffer size, in larger networks, it is essential to tightly control oversubscription to the links, using higher level flow control schemes, in order to reduce the probability of overflow and ensure satisfactory performance.

Having studied the effect of oversubscription to the links, the next step of our research will focus on the design of higher level mechanisms that will operate on an end-to-end basis and ensure that the oversubscription to links is tightly monitored and kept in the range where the performance is satisfactory.

References

[Bar64]   P. Baran. On Distributed Communication Networks. *IEEE Trans. Commun. Systems*, 12:1–9, March 1964.

[BC90]   J. Brassil and R. Cruz. An Approximate Analysis of Packet Transit Delay for Deflection Routing in the Manhattan Street Network. Submitted to IEEE Trans. Commun, 1990.

[BH85]   A. Borodin and J. E. Hopcroft. Routing, Merging, and Sorting on Parallel Models of Computation. *Journ. of Computer and System Sciences*, 30:130–145, 1985.

[BT89]   A. M. Bignell and T. D. Todd. SIGNET: A New Ultra-High-Speed Lightwave Network Architecture. In *IEEE Pacific Rim Conference on Communications, Computers and Signal Processing*, pages 40–43, June 1989.

[CL90a]   A. K. Choudhury and V. O. K. Li. Performance Analysis of Deflection Routing in the Manhattan Street Network. Accepted for publication in ICC '91, 1990.

[CL90b]   A. K. Choudhury and V. O. K. Li. Performance Analysis of Deflection Routing in the Manhattan Street and Minimum-Distance Networks. Submitted to IEEE Transactions on Communications, 1990.

[GG86]   A. G. Greenberg and J. Goodman. Sharp Approximate Models of Adaptive Routing in Mesh Networks. In J. W. Cohen O. J. Boxma and H. C. Tijms, editors, *Teletraffic Analysis and Computer Performance Evaluation*, pages 255–270, 1986. revised,1989.

[GH89]   A. G. Greenberg and B. Hajek. Deflection Routing in Hypercube Networks. Submitted to IEEE Transactions in Communications, July 1989.

[Haj88]   B. Hajek. Bounds on Evacuation Time for Deflection Routing. Preprint, November 1988.

[KH90]   A. Krishna and B. Hajek. Performance of Shuffle-Like Networks with Deflection. In *INFOCOM '90*, volume 2, pages 473–480, June 1990.

[Max87]   N. F. Maxemchuk. Routing in the Manhattan Street Network. *IEEE Trans. Commun.*, COM-35(5):503–512, May 1987.

[Max89]   N. F. Maxemchuk. Comparison of Deflection and Store-and-Forward Techniques in the Manhattan Street and Shuffle-Exchange Networks. In *INFOCOM '89*, volume 3, pages 800–809, April 1989.

[Max90]   N. F. Maxemchuk. Problems Arising from Deflection Routing: Live-lock, Lockout, Congestion and Message Reassembly. In *NATO Workshop on Architecture and Performance Issues of High Capacity on Local and Metropolitan Area Networks*, June 1990.

[TSH89]   X. N. Tan, K. C. Sevcik, and J. W. Hong. Optimal Routing in the Shuffle-Exchange Networks for Multiprocessor Systems. In *Comp-Euro 88-System Design: Concepts, Methods and Tools*, pages 255–264. IEEE, Euromicro, IEEE Comput. Soc. Press, Washington, D.C., April 1989.

NONUNIFORM TRAFFIC IN THE MANHATTAN STREET NETWORK

Jack Brassil† and Rene Cruz‡

Dept. of Electrical & Computer Engineering
University of California, San Diego
La Jolla, CA 92093

ABSTRACT

The *Manhattan Street Network* is a toroidal network proposed for application as a Metropolitan Area Network. The network is well suited to use deflection routing, an adaptive routing scheme where nodes attempt to send packets along shortest paths to their destinations. Contention for communication link access is resolved by forcing packets to travel longer paths. We consider a network operating with deflection routing and *nonuniform* traffic. In addition, each communication link may be characterized by an arbitrary but known propagation delay. We study 3 distinct contention resolution rules, focusing on one which gives priority to packets in the network longest. Independence approximations are used to describe the distribution of packet arrivals at each node. A recursive equation is written for each node, from which we derive approximate steady state packet delay distributions. Examples are presented and the independence assumption is discussed.

1. Introduction

A Manhattan Street Network (MSN) is a regular network of degree 2 with N rows and M columns of nodes (Figure 1) [1]. The network is packet switched and time-slotted, with fixed length slots of duration τ_s. We assume that communication links have identical transmission rates, and packets are transmitted error-free but may suffer propagation delay. Maxemchuk [1] proposed a locally implementable, adaptive deflection routing algorithm for an unbuffered MSN (i.e., nodes without store-and-forward buffers). A packet arriving to a node other than its destination is "switched" to an outgoing communication link. Nodes attempt to route arriving packets along shortest paths to their destinations; a packet is *deflected* if link contention causes the packet to be routed along a longer path. Deflection routing on an MSN operating with a uniform traffic demand and zero propagation delay has been studied in [2, 3]. In this paper, we examine packet transit delay in an MSN operating under a nonuniform traffic demand. Our approach is similar to that introduced by Greenberg and Goodman [2] and by Krishna and Hajek [4] to study uniform traffic on the MSN and on networks with shuffle interconnections, respectively. Other authors are also studying nonuniform traffic in networks with deflection routing [5] and the related problems of packet resequencing [6] and congestion [7].

† On leave from AT&T Bell Laboratories, Middletown, NJ.

‡ This work was partially supported by the National Science Foundation under grant NSF NCR-8904029.

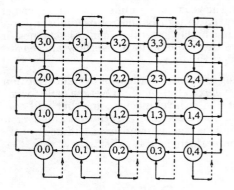

Figure 1 - A 4x5 Manhattan Street Network

2. Node Operation and Routing

We begin by discussing node operation in the absence of propagation delay. Each node receives 0, 1 or 2 packets in each time slot from directly connected nodes. Each node has a separate packet buffer for each input and output link. The link buffer capacities are each exactly 1 packet. At the start of a time slot, each node removes any packet in its input link buffers that has reached its destination. Each node independently generates a new packet to be admitted to the network in each time slot with probability

$$t_{f,g}(x,y) \triangleq P[\text{node } (f,g) \text{ generates a new packet} \qquad (2.1)$$

$$\text{to admit destined for } (x,y)].$$

A node generates at most 1 packet in each time slot. We assume that a node does not generate packets destined for itself (i.e., $t_{f,g}(f,g) = 0$). The new packet is *blocked* (and lost) if both input link buffers contain packets. If only one input link buffer contains a packet and a new packet is generated, the new packet is placed in the unoccupied link buffer. If neither input link buffer is occupied and a new packet is generated, a fair coin toss assigns the new packet to either buffer. Hence, top priority is assigned to routing packets in transit rather than admitting new packets. The routing algorithm is then executed.

Each node attempts to route packets along shortest paths to their destinations. In a network with propagation delay, the "shortest path" distance between 2 nodes is taken to be the least time path. If only 1 of a node's 2 outgoing links is along a shortest path, that link is chosen as the packet's *routing preference*. If both outgoing links are along shortest paths neither link is preferred. If only one packet arrives to a node (i.e., is in an input link buffer) and has a routing preference, it exits via its preferred link. If only

Reprinted from *Proc. IEEE International Conference on Communications*, pp. 1647–1651, June 1991.

one packet arrives and has no routing preference, a fair coin toss assigns the packet to an outgoing link. If two packets arrive and have different routing preferences, they each exit via their preferred links. Otherwise, a contention resolution rule determines link assignments.

We consider 3 contention resolution rules. For brevity, we focus our attention on the best performing rule, the *slotcount* rule [3]. Detailed descriptions of the other 2 rules we studied, known as the *random* and *straight* rules, are found in [2]. Let a packet's *age* be the number of time slots it has spent in the network. We assume that each packet carries a time slot counter, initialized to zero and incremented in each time slot. In the *slotcount* contention rule, packets are assigned to a node's outgoing links as follows. If 2 packets arrive and one has a routing preference and the other does not, the latter *defers* to the former by permitting it to exit on its preferred outgoing link. If 2 packets arrive and have identical routing preference, the packet longest in the network is assigned the preferred outgoing link, and the "younger" packet is deflected. If neither arriving packet has a preferred outgoing link, or both packets prefer the same link and have the same age, a fair coin toss assigns the two arrivals to outgoing links.

We now consider modifications to node operation required in the presence of propagation delay. If propagation delay causes a packet to arrive after the start of a time slot, the packet is held in the input link buffer until the beginning of the next slot. Hence, the required input link buffer capacities must be increased to at most 2 packets. Otherwise, node operation is as described above, with the packets to be "switched" in the current time slot being those packets at the head of the input link buffers.

3. Delay Analysis

Define the following probabilities for packets arriving on the incoming links and departing on the outgoing links of node (f,g) in a time slot:

$$r_{f,g}(x,y,s) \triangleq P[packet\ with\ age\ s\ destined\ for\ (x,y) \quad (3.1a)$$

$$arrives\ on\ the\ input\ row\ link\ of\ (f,g)]$$

$$c_{f,g}(u,v,l) \triangleq P[packet\ with\ age\ l\ destined\ for\ (u,v) \quad (3.1b)$$

$$arrives\ on\ the\ input\ column\ link\ of\ (f,g)]$$

$$rf_{,g}(x,y,s) \triangleq P[packet\ with\ age\ s\ destined\ for\ (x,y) \quad (3.1c)$$

$$departs\ on\ the\ output\ row\ link\ of\ (f,g)]$$

$$cf_{,g}(u,v,l) \triangleq P[packet\ with\ age\ l\ destined\ for\ (u,v) \quad (3.1d)$$

$$departs\ on\ the\ output\ column\ link\ of\ (f,g)].$$

We now present the key simplifying independence approximation: *at each node in each time slot the states of arrivals on the incoming row and column links are independent, and are independent of the generation of any new packets, i.e.*

$$P[\ packet\ with\ age\ s\ destined\ for\ (x,y)\ arrives\ on\ input$$

$$row\ link\ of\ (f,g),\ packet\ with\ age\ l\ destined\ for\ (u,v)$$

$$arrives\ on\ input\ column\ link\ of\ (f,g),\ packet\ destined$$

$$for\ (a,b)\ is\ generated\ at\ (f,g)]$$

$$= r_{f,g}(x,y,s)\ c_{f,g}(u,v,l)\ l_{f,g}(a,b).$$

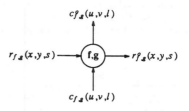

Figure 2 - Packet arrival and departure probabilities at node (f,g)

We may then easily write the transit delay distribution for packets destined to node (x,y):

$$\Delta_{x,y}(s) \triangleq P[packet\ has\ age\ s\ |\ packet\ arrives\ to\ destination\ (x,y)]$$

$$= \frac{r_{x,y}(x,y,s) + c_{x,y}(x,y,s)}{\sum_s r_{x,y}(x,y,s) + c_{x,y}(x,y,s)}. \quad (3.2)$$

Other performance measures are also easily written. We next develop a recursion to solve for the link arrival probabilities. By considering each event that can occur at a node in a time slot and using the above independence approximation, we may write equations for the output link probabilities of node (f,g) in terms of its input link and new arrival probabilities:

$$rf_{,g} = F(r_{f,g}, c_{f,g}, l_{f,g}) \quad (3.3a)$$

$$cf_{,g} = G(c_{f,g}, r_{f,g}, l_{f,g}). \quad (3.3b)$$

The specific expressions for Eq. 3.3 for nodes operating under each of the 3 contention rules are found in the Appendix.

Next we relate the output link probabilities of a node and the input link probabilities of directly connected nodes. Let (f,g) and $(\acute{f},g)$ be the nodes terminating the outgoing row and column links of node (f,g), respectively. Let $\{\chi\}_K$ denote the integer χ modulo K. The Manhattan graph adjacency matrix can be determined from the following table:

(f,g)	$(\acute{f},g)$	(f,g)
even, even	$(\{f+1\}_N,g)$	$(f,\{g+1\}_M)$
even, odd	$(\{f-1\}_N,g)$	$(f,\{g+1\}_M)$
odd, even	$(\{f+1\}_N,g)$	$(f,\{g-1\}_M)$
odd, odd	$(\{f-1\}_N,g)$	$(f,\{g-1\}_M)$

Let the propagation delay on the output row and column links of (f,g) be designated $\delta f_{,g}$ and $\delta f_{,g}$. Non-zero propagation delay causes the input link probabilities of downstream nodes to be a delayed version of the output link probabilities of upstream nodes:

$$r_{f,g}(x,y,s) = rf_{,g}(x,y,s - \left\lceil \delta f_{,g} / \tau_s \right\rceil) \quad (3.4a)$$

$$c_{f,g}(x,y,s) = cf_{,g}(x,y,s - \left\lceil \delta f_{,g} / \tau_s \right\rceil). \quad (3.4b)$$

Network behavior is approximately described by Eqs. 3.3-4, which may be solved iteratively. In the following sections we solve these equations for simple but interesting traffic demands.

4. Example: Updating a Shared Resource

Consider a 4×4 MSN with zero propagation delay. Let each node admit packets destined for node (2,2), except for the destination node itself. Each of the other 15 nodes independently generate packets at rate 1/15. Solving Eqs. 3.3-4 iteratively for a network operating with the *slotcount* rule produces the approximate transit delay distribution and link utilizations found in Figures 3 & 4.

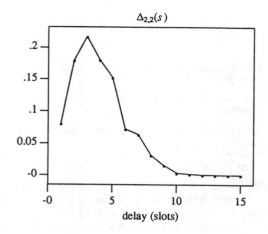

Figure 3 - Transit delay distribution (*mean* = 3.91, *standard deviation* = 1.95) at the destination node (2,2).

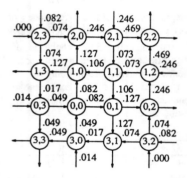

Figure 4 - Row and column link utilizations yield an average blocking probability of $8.3 x 10^{-4}$.

5. Example: Routing around a "hot spot"

Deflection routing naturally forces packets to circumvent congested regions of the network. We now demonstrate how packets are redirected around a congested link (i.e., "hot spot"). A similar example is considered in [7]. Consider the simple demand of Figure 5a. Exactly 2 sources are active, nodes (0,0) and (7,2). Each generates 1 packet in each time slot to send to a distinct destination in an 8x8 MSN with zero propagation delay. Under this traffic demand, solving Eqs. 3.3-4 for nodes operating with the *slotcount* rule yields the approximate link utilizations of Figure 5b. The utilizations presented are those obtained after 32 iterations, starting with an initial condition for Eq. 3.3 corresponding to an empty network (i.e., $r_{f,g} \equiv 0$, $c_{f,g} \equiv 0$ for all (f,g)).

A number of observations are noteworthy. First, in steady state, all packets destined for node (0,4) are deflected at least once. This occurs since each newly admitted packet from source node

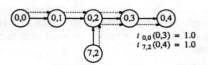

Figure 5a - Two continuously transmitting sources produce localized congestion along the unique shortest paths (dotted lines) to the packets' destinations.

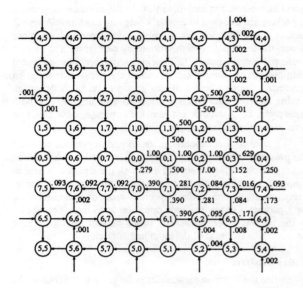

Figure 5b - The resulting link utilizations larger than $5.0 x 10^{-4}$. The mean delay of packets reaching destinations (0,4) and (0,3) are 8.86 and 6.46 slots, respectively.

(7,2) is deflected at node (0,2) by an "older" packet arriving on the node's incoming row link. Second, deflections cause some fraction of admitted packets to eventually traverse the incoming links to the source nodes. This "feedback" causes some offered packets to be blocked. One may view this mechanism as a limited form of backpressure which can automatically throttle sources as congestion develops.

In our model, we have chosen to have nodes place top priority on routing packets in transit rather than admitting new packets. Several authors have noted that such a policy can lead to severe unfairness for certain traffic demands. In our example, node (0,2) is entirely occupied routing transit packets, and would be unable to admit any new packets without discarding arriving packets.

At first glance, the "spreading" of packets over a large number of network links may seem to be an inefficient use of bandwidth. However, we now show that this is not so. For the demand of Figure 5a, an "optimal" routing scheme would route 100% of all offered traffic, fully utilizing 10 links. Yet, the "spreading" of packets shown in Figure 5b combines to consume the bandwidth equivalent of 12.58 links, with only 2.4% of offered packets blocked.

6. Discussion of Independence Assumption

The validity of our model is crucially dependent on the validity of the independence approximation. For example, in Figure 6 we present a comparison of the transit delay distribution taken from our analysis here and a simulation [3] of a fully utilized 8x8 MSN operating with the *slotcount* rule and a uniform traffic demand. Indeed, the approximation is quite accurate.

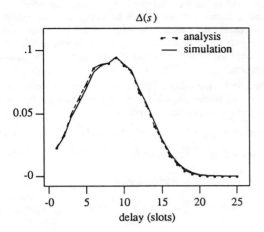

Figure 6 - A comparison of the delay distribution found by solving Eqs. 3.3-4 and a simulation transit delay histogram for a fully utilized network with a uniform traffic demand and no propagation delay.

We now offer an example of a demand where the independence assumption is "strongly" violated. Consider the "streaming source" of Figure 7, where 1 source (a,b) generates packets destined for (x,y) with rate 1 packet/slot. No other sources are active.

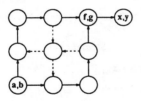

Figure 7 - A "streaming source" violates the assumed independence of arrivals at node (f,g).

Clearly no deflections are possible since, by definition, (a,b) admits at most one packet per time slot. Consider the node (f,g) found on a shortest path from (a,b) to (x,y), with disjoint shortest subpaths from (a,b) incident at (f,g). Arrivals at (f,g) in a time slot are strongly dependent since an arrival on one incoming link precludes an arrival on the other. In an 8x8 MSN using the *slotcount* rule, our model predicts that generated packets are blocked with probability $1.3x10^{-4}$, and 44.0% of admitted packets are deflected at least once. Indeed, this dependency exists to some degree in a large class of traffic demands.

7. Conclusion

An independence assumption has allowed us to derive approximate delay distributions for an MSN operating under a rather general traffic demand. The approach may be generalized to study deflection routing on arbitrary topologies. For an N node MSN supporting the maximum number of conversations, it is necessary to solve $O(N^3)$ equations. While this number grows rapidly, it is smaller than the number needed in a direct solution via a Markov chain, where the number of states increases exponentially [2]. We note that Eqs. 3.3-4 may not have a unique solution, however, no time-homogeneous demand we tested produced multiple solutions. An open question is if a more complex yet still tractable model exists which will produce a better approximation to the actual network behavior.

Acknowledgement - The authors thank Joseph Bannister for comments that were helpful in revising this paper.

References

1. N. F. Maxemchuk, ''The Manhattan Street Network,'' *IEEE Globecom 1985*, pp. 255-261, New Orleans, La., September, 1985.

2. A. G. Greenberg and J. W. Goodman, ''Sharp Approximate Models of Adaptive Routing in Mesh Networks (Preliminary report),'' *Teletraffic Analysis and Computer Performance Evaluation*, Elsevier Science Publishers, B. V., 1986.

3. J. T. Brassil and R. L. Cruz, ''An Approximate Analysis of Packet Transit Delay in the Manhattan Street Network,'' submitted to *IEEE Trans. on Communications*, February, 1990.

4. A. Krishna and B. Hajek, ''Performance of Shuffle-Like Switching Networks with Deflection,'' *Proceedings of IEEE INFOCOM'90*, pp. 473-480, June,1990.

5. J. Bannister, F. Borgonovo, L. Fratta, and M. Gerla, ''A Performance Model of Deflection Routing and its Application to the Topological Design of Multichannel Networks (1/91 Draft),'' in preparation.

6. A. K. Choudhury and N. F. Maxemchuk, ''Effect of a Finite Reassembly Buffer on the Performance of Deflection Routing (1990 Draft),'' in preparation.

7. N. F. Maxemchuk, ''Problems Arising from Deflection Routing: Live-lock, Lockout, Congestion and Message Reassembly,'' *Proceedings of NATO Workshop on Architecture and High Performance Issues of High Capacity Local and Metropolitan Area Networks*, France, June, 1990.

APPENDIX

We now provide the exact form of Eq. 3.3 for nodes operating under each of the 3 contention rules we studied. Each routing algorithm is symmetric w.r.t. input links, so we present only the equation for the ouput row link probabilities; the equations for the output column link probabilities can be obtained by interchanging all row and column labels in the corresponding equations below.

We begin by sketching the development of the equations for a network operating with the *slotcount* rule. Consider any node (f,g) in a time slot. Define the routing preference indicator function:

$$I_{x,y}^{r(c,e)} \triangleq \begin{cases} 1 & \text{packet destined for } (x,y) \text{ prefers the} \\ & \text{output row (column, either) link} \\ 0 & \text{otherwise.} \end{cases} \quad (A.1)$$

Let $\mathbf{E}$ be the event that both input link buffers are occupied when the routing algorithm is executed; one input link buffer holds a packet of age $s-1$ destined for node (x,y), and the other input link buffer holds a packet of age l destined for node (u,v). Note that $(x,y) \neq (f,g)$ and $(u,v) \neq (f,g)$, since packets destined for node (f,g) have been removed prior to executing the routing algorithm. By convention, a newly generated packet placed in a input link buffer has age 0. Given event $\mathbf{E}$, let $I_{s,l}(u,v,x,y)$ be the probability that the packet destined for (x,y) exits on the output row link in the next time slot. By the definition of the *slotcount* rule, we have

$$I_{s,l}(u,v,x,y) \triangleq \begin{cases} I_{s<l}(u,v,x,y) & \text{if } s-1 < l \\ I(u,v,x,y) & \text{if } s-1 = l \\ I_{s>l}(u,v,x,y) & \text{if } s-1 > l \end{cases} \quad \text{(A.2)}$$

where we define

$$I(u,v,x,y) \triangleq \tfrac{1}{2} I^r_{x,y} I^c_{u,v} + \tfrac{1}{2} I^r_{x,y} I^r_{u,v} + I^r_{x,y} I^c_{u,v} \quad \text{(A.3)}$$
$$+ I^r_{x,y} I^t_{u,v} + I^t_{x,y} I^c_{u,v} + \tfrac{1}{2} I^t_{x,y} I^t_{u,v}$$

$$I_{s<l}(u,v,x,y) \triangleq I^c_{u,v} + I^r_{x,y} I^t_{u,v} + \tfrac{1}{2} I^t_{x,y} I^t_{u,v} \quad \text{(A.4)}$$

$$I_{s>l}(u,v,x,y) \triangleq I^r_{x,y} + I^t_{x,y} I^c_{u,v} + \tfrac{1}{2} I^t_{x,y} I^t_{u,v}. \quad \text{(A.5)}$$

Let $\alpha_{f,g}$ ($\beta_{f,g}$) be the probability that no continuing packet arrives on the input row (column) link of (f,g) in a time slot, i.e.

$$\alpha_{f,g} \triangleq 1 - \sum_{\substack{u,v,l \\ (u,v)\neq(f,g)}} r_{f,g}(u,v,l) \quad \text{(A.6)}$$

$$\beta_{f,g} \triangleq 1 - \sum_{\substack{u,v,l \\ (u,v)\neq(f,g)}} c_{f,g}(u,v,l). \quad \text{(A.7)}$$

Let $\gamma_{f,g}$ be the probability that no new packet is generated at (f,g) in a time slot, i.e.

$$\gamma_{f,g} \triangleq 1 - \sum_{u,v} t_{f,g}(u,v). \quad \text{(A.8)}$$

The specific expressions for Eq. 3.3 are as follows:

Slotcount rule

$$r^g_{,g}(x,y,1) = t_{f,g}(x,y)\, \alpha_{f,g}\, \beta_{f,g}\, \left[I^r_{x,y} + \tfrac{1}{2} I^c_{x,y} \right]$$
$$+ t_{f,g}(x,y) \sum_{\substack{u,v,l \\ l>0 \\ (u,v)\neq(f,g)}} \left[\alpha_{f,g}\, c_{f,g}(u,v,l) \right.$$
$$\left. + \beta_{f,g}\, r_{f,g}(u,v,l) \right] I_{1,l}(u,v,x,y)$$

$$r^g_{,g}(x,y,s) = \gamma_{f,g}\, \left[\alpha_{f,g}\, c_{f,g}(x,y,s-1) \quad (x,y) \neq (f,g)\,, s>1 \right.$$
$$\left. + \beta_{f,g}\, r_{f,g}(x,y,s-1) \right] \left[I^r_{x,y} + \tfrac{1}{2} I^c_{x,y} \right]$$
$$+ \sum_{\substack{u,v \\ (u,v)\neq(f,g)}} \sum_{l=1} \left[r_{f,g}(x,y,s-1)\, c_{f,g}(u,v,l) \right.$$
$$\left. + c_{f,g}(x,y,s-1)\, r_{f,g}(u,v,l) \right] I_{s,l}(u,v,x,y)$$
$$+ \sum_{u,v} \left[\beta_{f,g}\, r_{f,g}(x,y,s-1) \right.$$
$$\left. + \alpha_{f,g}\, c_{f,g}(x,y,s-1) \right] t_{f,g}(u,v)\, I_{s,0}(u,v,x,y)$$

$$r^g_{,g}(f,g,s) = 0.$$

The equation for $r^g_{,g}(x,y,1)$ consists of 2 terms corresponding to the events where a new packet is generated and not blocked. The first term corresponds to the event where no continuing packet arrives, and the second term corresponds to the event where exactly 1 continuing packet arrives.

The equation for $r^g_{,g}(x,y,s)$, $s>1$ consists of 3 terms. The first term corresponds to the event that exactly 1 continuing packet arrives and no new packet generated. The second and third terms correspond to both input link buffers being occupied immediately prior to executing the routing algorithm, with 2 link arrivals (i.e. no new packet) and 1 link arrival and 1 new packet, respectively. The final equation states that nodes do not emit packets destined for themselves.

The equations for the other rules we studied are developed similarly and stated below. Again it is helpful to define the following collections of routing preference indicator functions:

$$I_r(u,v,x,y) \triangleq I^r_{x,y} + I^t_{x,y} \left(I^c_{u,v} + \tfrac{1}{2} I^t_{u,v} \right) + I^c_{x,y} I^c_{u,v} \quad \text{(A.9)}$$

$$I_c(u,v,x,y) \triangleq I^c_{x,y} \left(1 - I^r_{u,v} \right) + I^t_{x,y} \left(I^c_{u,v} + \tfrac{1}{2} I^t_{u,v} \right). \quad \text{(A.10)}$$

Random rule

$$r^g_{,g}(x,y,1) = t_{f,g}(x,y)\, \alpha_{f,g}\, \beta_{f,g}\, \left[I^r_{x,y} + \tfrac{1}{2} I^c_{x,y} \right]$$
$$+ t_{f,g}(x,y) \sum_{\substack{u,v,l \\ (u,v)\neq(f,g)}} \left[\alpha_{f,g}\, c_{f,g}(u,v,l) \right.$$
$$\left. + \beta_{f,g}\, r_{f,g}(u,v,l) \right] I(u,v,x,y)$$

$$r^g_{,g}(x,y,s) = \gamma_{f,g}\, \left[\alpha_{f,g}\, c_{f,g}(x,y,s-1) \quad (x,y) \neq (f,g)\,, s>1 \right.$$
$$\left. + \beta_{f,g}\, r_{f,g}(x,y,s-1) \right] \left[I^r_{x,y} + \tfrac{1}{2} I^c_{x,y} \right]$$
$$+ \sum_{\substack{u,v,l \\ (u,v)\neq(f,g)}} \left[r_{f,g}(x,y,s-1)\, c_{f,g}(u,v,l) \right.$$
$$\left. + c_{f,g}(x,y,s-1)\, r_{f,g}(u,v,l) \right] I(u,v,x,y)$$
$$+ \sum_{u,v} \left[\beta_{f,g}\, r_{f,g}(x,y,s-1) \right.$$
$$\left. + \alpha_{f,g}\, c_{f,g}(x,y,s-1) \right] t_{f,g}(u,v)\, I(u,v,x,y)$$

Straight rule

$$r^g_{,g}(x,y,1) = t_{f,g}(x,y)\, \alpha_{f,g}\, \beta_{f,g}\, \left[I^r_{x,y} + \tfrac{1}{2} I^c_{x,y} \right]$$
$$+ t_{f,g}(x,y)\, \alpha_{f,g} \sum_{\substack{u,v,l \\ (u,v)\neq(f,g)}} c_{f,g}(u,v,l)\, I_r(u,v,x,y)$$
$$+ t_{f,g}(x,y)\, \beta_{f,g} \sum_{\substack{u,v,l \\ (u,v)\neq(f,g)}} r_{f,g}(u,v,l)\, I_c(u,v,x,y)$$

$$r^g_{,g}(x,y,s) = \gamma_{f,g}\, \left[\alpha_{f,g}\, c_{f,g}(x,y,s-1) \quad (x,y) \neq (f,g)\,, s>1 \right.$$
$$\left. + \beta_{f,g}\, r_{f,g}(x,y,s-1) \right] \left[I^r_{x,y} + \tfrac{1}{2} I^c_{x,y} \right]$$
$$+ r_{f,g}(x,y,s-1) \sum_{\substack{u,v \\ (u,v)\neq(f,g)}} \beta_{f,g}\, t_{f,g}(u,v)\, I_r(u,v,x,y)$$
$$+ r_{f,g}(x,y,s-1) \sum_{\substack{u,v,l \\ l>0 \\ (u,v)\neq(f,g) \\ (u,v,l)\neq(x,y,s-1)}} c_{f,g}(u,v,l)\, I_r(u,v,x,y)$$
$$+ c_{f,g}(x,y,s-1) \sum_{\substack{u,v \\ (u,v)\neq(f,g)}} \alpha_{f,g}\, t_{f,g}(u,v)\, I_c(u,v,x,y)$$
$$+ c_{f,g}(x,y,s-1) \sum_{\substack{u,v,l \\ (u,v)\neq(f,g) \\ l>0 \\ (u,v,l)\neq(x,y,s-1)}} r_{f,g}(u,v,l)\, I_c(u,v,x,y)$$
$$+ c_{f,g}(x,y,s-1)\, r_{f,g}(x,y,s-1)$$

Part 8
Shuffle Network Switching Networks

INTRODUCTION

Another interconnection network topology that has received attention in terms of its suitability for deflection routing is the shuffle network. Like a toroidal network, it can be implemented with 2 × 2 switching elements (see [1–2] for a study of a 4 × 4 implementation). Moreover, shuffle networks with 2 × 2 switching elements have a diameter that grows logarithmically with the number of switching elements (nodes). This is the smallest rate for networks with 2 × 2 switching elements (see the paper by Krishna and Hajek in this part). The diameter of the Manhattan Street network, by way of comparison, grows as the square root of the number of nodes.

The first paper in this chapter is "Performance of Shuffle-Like Switching Networks with Deflection," by A. Krishna and B. Hajek. In it the authors take a look at a number of shuffle network protocols and implementations. They also emphasize "evacuation" problems. Evacuation refers to the time taken for an initial distribution of packets in the network to exit if there are no further inputs. They describe a class of shuffle ring networks, where (n, k) represents k columns of switching elements, with 2^n switching elements in each column (stage).

In particular, Krishna and Hajek look at "closer to destination" routing, where packets closer to their destinations are given priority in deflection situations. With some assumptions, the time taken to evacuate a $(n, 1)$ shuffle ring is discussed in terms of optimality and bounds.

Krishna and Hajek discuss a (n, n) shuffle ring network ("ShuffleNet" [3]) in terms of a deflection routing policy that gives priority to older packets (in terms of the number of deflections) and, for two packets of the same age, gives priority to the packet closest to the destination. An iterative numerical procedure leading to the calculation of state transition probabilities and delay distributions is described.

Also examined in this paper are some shuffle network variations. One is a crossback network with a link added in the reverse direction of every link in a (n, k) shuffle ring, resulting in 4 × 4 switching elements. An upper bound in evacuation time is found. At the price of added switch complexity, the evacuation time is reduced. Another variant is a stay or shuffle version of the $(n, 1)$ shuffle ring. Here each node has a link to itself. This acts as a buffer for deflected packets, allowing them to be deflected without increasing their distance to the destination.

The other paper in this part is "Shuffle Interconnection Networks with Deflection Routing for ATM Switching: The Closed Loop Shuffleout," by M. Décina, P. Giacomazzi, and A. Pattavina. It proposes a class of switching networks incorporating shuffle networks. The closed-loop version of this class is discussed in this paper, and the open-loop version appears in [4].

Two variants of the closed-loop switch appear in this paper: buffered closed-loop shuffleout (BCSS) and expanded closed-loop shuffleout switch (ECSS). Both switches consist of input and output queues, the shuffle fabric, and feedback paths. The switching elements are 2 × 4, with two outputs connected in the normal shuffle fashion and two leading to output queues. The switching elements implement shortest-path routing and, when contention for outputs occurs, deflection routing. Packets that are not routed to the output queues by the time they reach the end of the shuffle network are recirculated.

A performance analysis for delay and cell-loss probability is presented for the closed-loop architecture. Among the sources of cell loss examined are input and output cell loss, resequencing cell loss (two resequencing schemes are proposed), and cell loss in the ECSS concentrator. This paper also includes a large number of performance curves.

REFERENCES

[1] M. Décina, V. Tecordi, and G. Zanolini, "Throughput and packet loss in deflection routing multichannel-metropolitan area networks," *Proc. IEEE Globecom '91*, pp. 1200–1208, Nov. 1991.

[2] M. Décina, V. Tecordi, and G. Zanolini, "Performance analysis of deflection-routing multichannel metropolitan area networks," *Proc. IEEE INFOCOM '92*, Florence, Italy, pp. 2435–2443, May 1992.

[3] A. S. Acampora and M. J. Karol, "Terabit lightwave networks: The multihop approach," *AT&T Tech. J.* 1987, pp. 21–34.

[4] M. Décina, P. Giacomazzi, and A. Pattavina, "Shuffle interconnection networks with deflection routing for ATM switching: The open-loop shuffleout," *Proc. 13th Int. Teletraffic Congress*, Copenhagen, June 1991.

Peformance of Shuffle-Like Switching Networks with Deflection

Arvind Krishna   Bruce Hajek

Coordinated Science Laboratory and the
Department of Electrical and Computer Engineering
University Of Illinois at Urbana-Champaign
Urbana, IL 61801

Abstract

Four packet-switched networks using shuffle-exchange interconnections and deflection routing are analyzed. The first two are well-known networks based solely on shuffle interconnections, and the other two are variations in which the negative effects of deflection is reduced. Approximate state equations are given under a uniform traffic assumption. The equations predict the distribution of packet delay, and can be used in situations where packets are assigned priorities. The four networks are briefly compared to each other and to Batcher-Banyan sorting networks and hypercube deflection networks.

1 Introduction

Deflection routing, originally called *hot-potato routing* [1], is a technique for maintaining bounded buffers in a packet-switched communication network. If, due to congestion at a switch, not all packets can be sent out along shortest paths to their destinations, some packets are sent out on other links. The penalty is an increase in the distance traveled by packets, and the reward is the simplicity of switch design resulting from the absence of large buffers and routing tables. Traditional store-and-forward networks use extensive computation at the nodes to determine packet routes in order to use transmission bandwidth sparingly. In contrast, deflection leads to simple switches by making liberal use of transmission bandwidth. Since the penalty for longer routes increases as propagation delay becomes more of a factor, we consider deflection routing primarily for networks with a small physical diameter, such as those in a multiprocessor computer system or a packet switch for telecommunications.

We consider deflection routing in this paper for several well-known networks based on shuffle exchanges. In order to unify the description of the networks, in Subsection 2.1 we define the class of *shuffle-ring* networks. Informally, a shuffle-ring network with parameters n and k consists of k columns of switching nodes, with 2^n 2×2 nodes in each stage. The columns are arranged in a ring, with each column connected to the next by a perfect shuffle interconnection. We do not consider the network to be a "multi-stage" network in the strict sense of the word, because packets can be injected by, or destined to, any of the $k2^n$ nodes in the network. The important special cases of a shuffle-exchange network ($k=1$) and ShuffleNet ($k=n$) [2] are examined in Subsections 2.2 and 2.3, respectively.

˙This research was supported by the National Science Foundation under contract NSF ECS 83 52030

One reason for our interest in shuffle-ring networks is that they use 2×2 switching nodes, an important consideration for implementation using optical components. Another reason is that these networks have a diameter that grows typically logarithmically with the number of nodes, which is the smallest rate possible for 2×2 nodes. The networks admit simple routing rules based on packet destinations.

An apparent drawback of deflection routing in shuffle-ring networks is that a single deflection can cause the distance of a packet from its destination to increase significantly. The detrimental effect can be largely reduced by giving priority to packets closer to finishing [3]. We show for shuffle-exchange networks ($k=1$) that (for approximate evolution equations) giving priority to packets closest to their destinations is the optimal priority rule, in a certain sense. We also consider ShuffleNet when priority is based primarily on the number of deflections a packet has undergone, and in the case of ties also on the distance to the respective destinations. This priority rule is proposed for the network under construction at UC Boulder [4]. Our results indicate that this priority rule has the desired effect of reducing the tail of the delay distribution, at the expense of slightly increasing the mean delay.

Another approach for reducing the negative effects of deflection is to consider variations of shuffle networks for which a deflection causes the distance of a packet to its destination to increase only by a small amount. Two such variations are considered in Section 3. One of these (the *cross-back switch*) uses shuffles augmented by inverse shuffles, while the other uses a shuffle augmented by the identity interconnection. The cross-back switch is a slight generalization of the shuffle-exchange and exchange-unshuffle network of Tan et al [5].

We assume in this paper that the destinations of the packets generated at each node are distributed uniformly among the other nodes in the network. Our analysis begins with certain evolution equations, which are derived, roughly speaking, by ignoring the dependence between arrivals on different links of a switch. Similar evolution equations were given by [3,6,7]. We consistently found the solutions of the evolution equations to closely match simulations.

The bulk of the end results that we report in this paper concern the time it takes for the network to *evacuate* when operating synchronously with no packets injected other than those of the initial load. However, the evolution equation method works well in predicting delay and throughput for the networks in steady state operation as well. We illustrate this fact in our consideration of the UC Boulder ShuffleNet design, and we intend to report other delay/throughput results in the near future. We feel that the evacuation time, emphasized here, is an important performance parameter for several reasons. First, some networks support synchronous computation, and the normal operation of such networks consists of repeated evacuations. Secondly, as noted before, all

Reprinted from *Proc. IEEE INFOCOM '90*, pp. 473–480, June 1990.

the networks we consider here are recirculating. However, the networks can be expanded in space in such a way that the evolution that we describe from-slot-to-slot in time applies as well to packets moving from-copy-to-copy in space, through copies of the original network. The evacuation time of the original network dictates how many copies the space expanded network must have. A final reason for our interest in evacuation time is that it requires *transient* analysis of the network, so that the agreement with simulation better confirms the accuracy of our method. The ability to predict transient response is important even if average delay is the main item of interest, for if the offered traffic is bursty the network may continually be in a transient mode.

A comparison of the complexity of several networks using deflection and a network based on the Batcher-Banyan sort [8] is given in Section 4. Complexity is measured in units of switches per packet per slot, and is adjusted by being multiplied by the product of the number of inputs and number of outputs of the component switches. The complexities reported are based on the analysis of Sections 2 and 3.

We close this section by briefly commenting on some other work. Maxemchuk [9,10,11] has extensively studied the performance of deflection routing in a Manhattan street network. Like the shuffle-ring networks, these networks are constructed from 2×2 switching nodes. They have the advantage that a single deflection increases the distance to the destination by only a small amount, and the disadvantage that the network diameter grows as the square root, rather than as the logarithm, of the number of nodes. Finished products using deflection routing include the Hep multiprocessor computer system [12] and the Connection Machine [13]. An interesting treatement of flow control and cut-through for deflection routing is given in the paper of Ngai and Seitz [14].

2 Shuffle Ring Networks

2.1 Network model and operation

An (n, k) shuffle ring network[1] has $k2^n$ nodes and can be conceptually visualized as having k columns of 2^n nodes each. The columns are connected in a unidirectional ring with shuffle interconnections between consecutive columns. Each node is labeled by a pair (c, r), where $0 \le c < k$ and $0 \le r < 2^n$. We often express the row identifier in base 2 notation, so $(c, r) = (c, r_{n-1}, r_{n-2}, \ldots, r_0)$ where $r_i \in \{0, 1\}, 0 \le i < n$. Each node has two outgoing links and two incoming links. The two links going out of node $(c, r_{n-1}, r_{n-2}, \ldots, r_0)$ lead to nodes $(c \oplus 1, r_{n-2}, \ldots, r_0, 0)$ and $(c \oplus 1, r_{n-2}, \ldots, r_0, 1)$, where $c \oplus 1 = (c + 1 \mod k)$.

We will consider packet switching, with routes chosen as follows. Consider a packet starting at node (c, r) and destined for for node (c', r'). Let us first consider the special case when $c' = c \oplus n$. In this case, the packet wishes to pass through the sequence of n nodes (c^i, r^i), $1 \le i \le n$, where $c^i = c \oplus i$ and r^i is given by placing a window of length n over the sequence $r_{n-1}, r_{n-2}, \ldots, r_0, r'_{n-1}, \ldots, r'_1, r'_0$, with the window positioned i clicks from the left-most position. The construction guarantees that there is a link from (c^i, r^i) to (c^{i+1}, r^{i+1}) for $0 \le i < n$, where $(c, r) = (c^0, r^0)$.

In general, let j be the column number such that $c' = c \oplus n \oplus j$. For the first j hops, the packet does not care which

output links it takes. After proceeding through j links, the packet wishes to follow the unique route of n more links described in the special case above. The diameter of this network is $n + k - 1$.

We have described the route a packet nominally takes. However, due to congestion, a packet may be forced to take an undesired link at some node. This packet is said to be *deflected*. After a deflection, the packet begins to follow a new desired route as if it is just starting. If the packet is at distance i, $(i \le n)$ from its destination and is deflected, the distance to destination increases by one less than a multiple of k – typically[2] the smallest multiple of k such that the new distance is at least n.

The network operates synchronously: the time axis is divided into slots corresponding to packet transmission times and each link can relay one packet per time slot. Consider a fixed node at the beginning of a time slot. Since the node has two input links, as many as two packets were received during the previous time slot. Any such packets destined for the node are removed from the network. New packets may be injected into the network at the node, bringing the total to at most two continuing packets. Under deflection routing, the continuing packets have to be assigned to the two outgoing links. Since two packets might desire the same output link, the deflection routing scheme requires a rule for resolving this conflict.

The rule for resolving routing conflicts is based on the states of the packets. The state of a packet is comprised of information that it carries in the form of control bits. As an example, the state of a packet may consist of the destination address, the source address and the number of times the packet has been deflected. The network designer has to decide what information is relevant for resolving conflicts, and include it in the state of a packet. One rule that suggests itself is based on priorities. A priority is computed for each packet based on its state. Packets are then ordered with respect to their priorities. All packets with the same priority can be ordered randomly amongst themselves, all orders being equally likely. The packets are considered one at a time, in order of decreasing priorities. When a given packet is considered, the node examines whether the packet has a preferred link. If that link is free (no packet has been assigned to that link), the packet is assigned to the link. Otherwise, the packet is assigned to a free link, all choices being equally likely.

Note that our conflict resolution rule is a *two-pass* strategy. On the first pass, the node computes the priority of each packet. The node uses the priority to order the packets and in the second pass routes the packets on the output links.

The arrivals of packets on different links of a node are not necessarily independent for the models described in this paper. Also, the choice of output links made by different packets at the same switch are not necessarily independent. However, under the uniform traffic assumption, on the average packets choose each output link of a switch equally often. We will derive an approximate performance analysis by pretending that arrivals on different links are independent, and packets choose the output links of a switch equiprobably and independently of each other. Specifically, we make the following approximations for each node and time slot:

Approximation 2.1 *A switch receives packets on an incoming link independently of whether a packet is received on other links. Also, the state of these packets is drawn independently from a single distribution.*

---

[1] All of the networks described in this paper can be easily generalized to have p-ary shuffle interconnections, with $p \times p$ switches at each node.

[2] It is possible to do a little better in some cases, but on the average the routes we have described are only one link longer than the shortest routes between two nodes in the network.

Approximation 2.2 *The choice of output link of a packet which is yet to reach its destination is randomly, uniformly distributed between the two output links, and this choice is made independently of other packets.*

2.2 Shuffle Exchange Network

We study deflection on the $(n,1)$ shuffle ring network in this section. Suppose each node has two packets for delivery to other nodes in the network. The destination of each packet is chosen uniformly from amongst the other $N-1$ nodes, where $N = 2^n$, and each packet chooses its destination independently of other packets. We determine, under some simplifying approximations, how long the network takes to evacuate.

We use *closest-to-destination* priority routing, which implies that a packet which has fewer links to travel towards its destination gets preference in case there are conflicts. The intuition behind this rule is to try and reduce the number of packets in the network as soon as possible, hopefully decreasing the number of conflicts later.

The performance of the system described above can be determined by a Markov chain analysis. However, the number of states is N^N, making an exact analysis prohibitively expensive (in terms of computational resources). Similarly, simulation also requires large computational resources. Under approximations 2.1 and 2.2, we will derive evolution equations which determine the behaviour of a typical packet in the system. The number of states reduces to $n+1$, where $n = \log_2 N$. These evolution equations will then be used to derive bounds on the evacuation time, suitably defined, for the approximate network. These bounds are tight in the sense that the upper and lower bounds are both $O(n^2)$.

Consider a fixed packet, which has yet to reach its destination, at the beginning of a time slot. The node makes a decision about which output link the packet will traverse during this time slot. This implies that the distance to the packet's destination may decrease by one, or if the packet is deflected the distance to its destination will increase to n. No packet has a distance greater than n from its destination, because there is a route of n links from a node to any other. Thus, we view the packets as performing a random walk on the integers $0, 1, 2, \ldots, n$, where state i corresponds to the distance of a packet from its destination. The packets start at state n at time 0, and their destination is state 0. If a packet is not deflected in a time slot its state decreases by one, and if deflected, the packet goes to state n.

Consider a fixed link and define $p_t(i)$, $0 \le i \le n$ to be the probability that a packet i links from its destination (at the end of the slot) traverses the link during time slot t, and set $\mathbf{p}_t = (p_t(0), p_t(1), \ldots, p_t(n))$. Also, let $p_t(0)$ include the probability that there is no packet on the link, so that $\mathbf{p}_t$ is a probability distribution. Let us determine the deflection probability for a packet. Suppose a packet at distance i ($i > 0$) from its destination arrives on a link. This packet can be deflected by a packet from the other link only if the second packet is in a state $1, \ldots, i$, and the second packet prefers the same output link as the first packet. The second packet prefers the same output link with probability $1/2$. If the second packet is at state i, it will win the conflict with probability $1/2$, and if its state is less than i, it will always win the routing conflict. Thus, the deflection probability for a packet at state i is $\frac{1}{2}\sum_{j=1}^{j=i-1} p_t(j) + \frac{1}{4}p_t(i)$. All deflected packets go to state n. The evolution of $\mathbf{p}_t$ is given by the following *update equation*:

$$
\begin{aligned}
p_{t+1}(0) &= p_t(0) + p_t(1)\left(1 - \tfrac{1}{4}p_t(1)\right) \\
p_{t+1}(i) &= p_t(i+1)\left(1 - \tfrac{1}{2}\sum_{j=1}^{j=i} p_t(j) - \tfrac{1}{4}p_t(i+1)\right) \\
p_{t+1}(n) &= \tfrac{1}{4}\left(\sum_{j=1}^{j=n} p_t(j)\right)^2
\end{aligned}
$$

$$(2.1)$$

where $1 \le i < n$. These equations are similar to the evolution equations numerically evaluated in [3]. The probability mass $p_{t+1}(n)$ is the mass that is deflected at time $t+1$. The expression for $p_{t+1}(n)$ in equation (2.1) follows from the equation below.

$$
p_{t+1}(n) = \frac{1}{4}p_t^2(1) + \sum_{j=2}^{j=n} p_t(j)\left(\frac{1}{2}\sum_{k=1}^{k=j-1} p_t(k) + \frac{1}{4}p_t(j)\right)
$$

We simulated a shuffle exchange graph using deflection routing to compare the behavior of the network with predictions derived from the update equations. Figures 1 and 2 show the averaged results of twenty simulation runs for a 2^9 node shuffle exchange graph, together with the corresponding predictions. The nodes of the network were all filled with two packets at the beginning of the first slot, and then no new packets were added to the network. The destination of each packet in a node was chosen equiprobably from amongst the other $2^9 - 1$ nodes, and independently of all other packets. The close agreement between simulation and predictions apparent in the data presented in figures 1 and 2 is representative of all the data we have observed.

We define the evacuation time, T_{evac}, as the first time, t, at which the expected link utilization in the system, $\sum_{i=1}^{i=n} p_t(i)$, is less than $2^{-(n+1)}$. The intuition behind this definition is that it corresponds to the first time at which there is less than one packet in all the nodes put together. A numerical value for T_{evac} can be computed by iterating the update equation and observing the time, t, at which the required condition on $\mathbf{p}_t$ is met. On the other hand, obtaining an expression for T_{evac} in terms of n requires an explicit evaluation of $p_t(0)$ in terms of $\mathbf{p}_0$.

We introduce a partial order $\prec$ on the link occupation probabilities $\mathbf{p}_t$. This partial order serves two main functions in this paper. Firstly, it is used for a comparison of priority rules, which is summarized in Theorem 2.1. Secondly, it is used to provide bounds on the evacuation time, and these bounds are summarized in Theorem 2.2. The partial order is defined by $\mathbf{p}_t \prec \mathbf{q}_t$ if and only if

$$
\sum_{j=i}^{j=n} p_t(j) \le \sum_{j=i}^{j=n} q_t(j), \quad 1 \le i \le n. \tag{2.2}
$$

Roughly speaking, $\mathbf{p} \prec \mathbf{q}$ corresponds to there being more packets further away from their destinations under distribution $\mathbf{q}$ than under distribution $\mathbf{p}$. This order relation, $\prec$, has the usual properties of a partial order, i.e. it is a transitive and reflexive relation. The next lemma states that the relation $\prec$ is preserved under equation (2.1).

Lemma[3] 2.1 *If $\mathbf{p}_{t_0} \prec \mathbf{q}_{t_0}$ and $\mathbf{p}$ and $\mathbf{q}$ evolve according to equation (2.1), then $\mathbf{p}_t \prec \mathbf{q}_t$, $t \ge t_0$.*

---

[3]Due to space limitations, the proofs of all Lemmas and Theorems is omitted

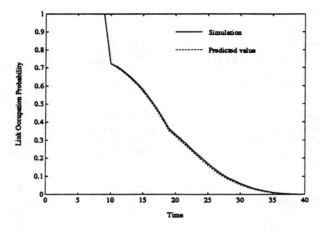

Figure 1: Link Occupation Probabilities in a 2^9 node shuffle exchange graph. The link occupation probability denotes the fraction of the 1024 links used in each time slot. Predicted values are obtained from the update equation.

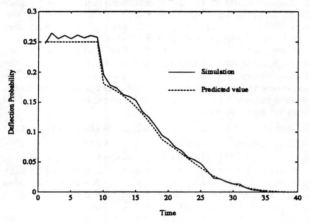

Figure 2: Fraction of packets deflected in a 2^9 node shuffle exchange graph. The deflection probability denotes the fraction of the packets deflected in each time slot. Predicted values are obtained from the update equation.

An important issue that we now address is whether the *closest-to-destination* priority rule is optimal for the evacuation time. Consider a priority rule, δ, which is used to resolve routing conflicts. Suppose two continuing packets are at a node at the beginning of a time slot. One packet is in state i, the other is in state j, and both packets desire the same output link. Then, let δ_{ij} be the probability that the packet in state i wins the routing conflict. It follows that $\delta_{ji} = 1 - \delta_{ij}$. Set p_t^δ to be the probability distribution of packet states under routing rule δ. In general, δ may also be a function of the time slot t. The evolution equations are

$$
\begin{aligned}
p_{t+1}^\delta(0) &= p_t^\delta(0) + p_t^\delta(1)\left(1 - \tfrac{1}{2}\sum_{j=1}^{j=n} p_t^\delta(j)\delta_{j1}\right) \\
p_{t+1}^\delta(i-1) &= p_t^\delta(i)\left(1 - \tfrac{1}{2}\sum_{j=1}^{j=n} p_t^\delta(j)\delta_{ji}\right) \\
p_{t+1}^\delta(n) &= \tfrac{1}{4}\left(\sum_{j=1}^{j=n} p_t^\delta(i)\right)^2
\end{aligned}
$$

$$(2.3)$$

where $1 < i \leq n$. The next lemma states that the closest-to-destination priority rule is the best possible rule amongst the class of rules described above.

Lemma 2.2 *Let* $\mathbf{p}_0 = \mathbf{p}_0^\delta$, *where* $\mathbf{p}_0$ *is a probability distribution on* $0, 1, \ldots, n$. *If* $\mathbf{p}_0$ *evolves according to equation (2.1) and* $\mathbf{p}_0^\delta$ *evolves according to equation (2.3), then* $\mathbf{p}_1 \prec \mathbf{p}_1^\delta$.

Let δ be any local rule which is used to resolve routing conflicts. We define a symmetric rule as any local rule which ignores the link on which packets enter the node and also that packets desire each output link equally often under this rule. These are the intuitive conditions that correspond to approximations 2.1 & 2.2. We define the evacuation time for strategy δ, T_{evac}^δ, in a similar manner to T_{evac}, as the first time t at which the link utilization probability is less than $2^{-(n+1)}$. The following theorem summarizes the reason for choosing closest-to-destination priority routing. The proof of this theorem is an easy consequence of Lemmas 2.1 and 2.2.

Theorem[4] 2.1 *Let* T_{evac} *be the evacuation time of the network under closest-to-destination priority routing, and let* T_{evac}^δ *be the evacuation time for any other symmetric rule,* δ, *used to resolve routing conflicts. Then, under Approximations 2.1 & 2.2,* $T_{evac}^\delta \geq T_{evac}$.

We have found explicit bounds on the evacuation time, T_{evac}, which are summarized in the following theorem.

Theorem 2.2 *Let* $\mathbf{p}_0$ *be given by* $p_0(n) = 1$; $p_0(i) = 0$, $i < n$, *and* $\mathbf{p}_t$ *evolve according to equation (2.1). Let* T_{evac} *be the evacuation time for this system. Then,*

$$
\frac{n^2}{16} \leq T_{evac} \leq \frac{4n^2}{9} + n + n\log_2(n)
$$

Equation (2.1) is hard to solve explicitly in a closed form. However, observe that this equation has the property that if $p_t(j) = 0$, $1 \leq j \leq k$, then for $0 \leq i \leq (k-1)$,

$$
p_{t+i+1}(k-i) = p_{t+i}(k-i+1)\left(1 - \frac{1}{4}p_{t+i}(k-i+1)\right). \quad (2.4)
$$

In order to take advantage of this simplification, consider a system in which the mass from states $1, 2 \ldots, n$ is swept into state n every n time slots. Suppose a sweep is done at time t, setting $p_t(i) = 0$, $1 \leq i \leq n-1$. Then, this system will follow equation (2.4) with $k = n - 1$. Also, the link occupation probabilities of this system dominate those of the original system, in the sense of $\prec$. The evacuation time of this system provides the upper bound in Theorem 2.2. In a similar fashion, we construct another simple system to provide the lower bound in Theorem 2.2.

---

[4]Lemmas 2.1, 2.2, and Theorem 2.1 do not extend to (n, k) shuffle ring networks with $k \geq 2$

2.3   (n,n) Shuffle Ring Networks

We now consider a (n, n) shuffle ring network, as defined in Subsection 2.1. This network has been proposed in [2] as the ShuffleNet and is being implemented as an optical interconnection network at the University of Colorado [4]. The state of a packet includes:

Age – the number of times a packet has been deflected so far.

Distance – the number of links a packet has to traverse before reaching its destination.

We include the *age* in the state to try and ensure that packets do not remain in the network for an extremely long time. If a packet has been deflected a number of times, its age is large. Hence, if we give priority to older packets, the hope is that they will not be deflected and will soon leave the network. The state is written as a two-tuple (age, distance). Packets which do not have a preferred output link (distance $> n$) get lower priorities than the remaining packets. Priority of packets which have a preferred output link is based on a lexicographic ordering of (Age, Distance) as follows:

- Older packets take precedence over younger packets (lower age packets).

- Within the same age, packets with lower distance take precedence over packets with higher distance.

In order to use only a few bits for the *age* information we set a maximum age, m. The minimum age is one. If a packet with age m gets deflected its age remains at m. Otherwise, if a packet gets deflected, its age increases by one. Every packet in the system makes a state transition in every time slot. Possible state transitions for a packet with state (i, j) are:

$j > n \Longrightarrow$ the only transition is to $(i, j - 1)$.

$j \leq n$ and $i < m \Longrightarrow$ the transition could be to $(i, j - 1)$ or $(i + 1, j + n - 1)$.

$j \leq n$ and $i = m \Longrightarrow$ the transition could be to $(m, j - 1)$ or $(m, j + n - 1)$.

Let $p_t(i, j)$ be the probability that a packet j hops from its destination (at the end of the slot), with age i, traverses a given link during time slot t. Set $\mathbf{p}_t$ to be the collection $\{p_t(i, j), 1 \leq i \leq m, 0 \leq j < 2n\}$. The transitions from distance j to distance $(j + n - 1)$ correspond to deflections. A packet with higher priority chooses the same output link as a given packet with probability $\frac{1}{2}$, by approximation 2.2. Then, using approximation 2.1, it is clear that the deflection probability, $d_t(k, l)$, for a packet in state (k, l) at the end of time slot t is nonzero only when $l \leq n$, and is given by

$$d_t(k, l) = \frac{1}{4} p_t(k, l) + \frac{1}{2} \left(\sum_{j=1}^{j=l-1} p_t(k, j) + \sum_{i=k+1}^{i=m} \sum_{j=1}^{j=n} p_t(i, j) \right)$$

(2.5)

Let $\nu_t(i, j)$ represent the external (from the node, not an input link) arrivals to state (i, j) during time slot t. $I_{[.]}$ is the usual indicator function. Now, given $\mathbf{p}_t$ and $\nu_t(., .)$, we can compute $\mathbf{p}_{t+1}$ by the following algorithm.

Procedure Update

1. $\hat{p}_t(i, j) = p_t(i, j) + \nu_t(i, j)$

2. Compute $d_t(i, j)$ according to equation (2.5), using $\hat{p}_t(i, j)$ instead of $p_t(i, j)$.

3. For $0 < j < n$, $p_{t+1}(i, j) = \hat{p}_t(i, j + 1)(1 - d_t(i, j + 1))$.

4. $p_{t+1}(i, j + n - 1) = \hat{p}_t(i, j + n) + I_{[i>1]} \hat{p}_t(i - 1, j) d_t(i - 1, j) + I_{[i=m]} \hat{p}_t(i, j) d_t(i, j)$, $1 \leq j \leq n$.

5. $p_{t+1}(i, 0) = \hat{p}_t(i, 0) + \hat{p}_t(i, 1)(1 - d_t(i, 1))$.

Suppose the desired throughput of the network is τ, and traffic is balanced (all destinations are equally likely for newly generated packets). One way of achieving this is to set $\nu_t(1, j) = \tau/n$, $n \leq j < 2n$, and all other $\nu_t(i, j) = 0$, for $t > 0$. Then, a fixed point, $\mathbf{p}^*$, can be computed for procedure update, with $\mathbf{p}_{t+1} = \mathbf{p}_t = \mathbf{p}^*$. Once $\mathbf{p}^*$ has been determined, all the state transition probabilities are known and the delay distribution can be found for the approximated network. This can be of great help to the network designer to answer questions such as

- How many ages are desirable?

- What is the average delay versus age curve for a fixed throughput?

- What is the delay distribution?

We show some typical results in figures 3 and 4, computed using procedure update. All the results are for a (6,6) shuffle ring network, which has 384 nodes. The *number of ages* represents the maximum number of ages allowed, m. Figure 3 represents an average load case, where each link is utilized approximately in 1 of every 3 time slots. Figure 4 is for a heavily loaded case, where links are utilized in almost every time slot. Similar results can be computed for different sizes of networks. We verified some of these results against simulations, and found the approximations to be accurate within 1%.

These results indicate that more ages result in a higher average delay, but reduce the spread of the delay distribution (since the ninety-ninth percentile is consistently lower for $m = 4$, compared to $m = 1$).

3   Variations on Shuffle Networks

3.1   Crossback Switching Network

The network we describe here is a variant of the (n, k) shuffle ring network obtained by adding a link in the opposite direction alongside each of the original links. Thus, every node has four outgoing and four incoming links. Recall the notation of Subsection 2.1. The four links going out of node $(c, r_{n-1}, r_{n-2}, \ldots, r_0)$ lead to nodes $(c \oplus 1, r_{n-2}, \ldots, r_0, 0)$, $(c \oplus 1, r_{n-2}, \ldots, r_0, 1)$, $(c \ominus 1, 0, r_{n-1}, \ldots, r_2, r_1)$, and $(c \ominus 1, 1, r_{n-1}, \ldots, r_2, r_1)$, where $c \ominus 1 = (c - 1 \mod k)$.

We will consider packet switching, with routes chosen as follows. We classify packets as either *left-packets* or *right-packets*. A right-packet desires exactly the same route as it would in a shuffle ring network. The construction of desired routes for left-packets is similar to the construction of routes in Subsection 2.1, except that these routes use only edges that go from columns c to $c \ominus 1$. We describe the route of left-packets in detail. Consider a left-packet starting at node (c, r)

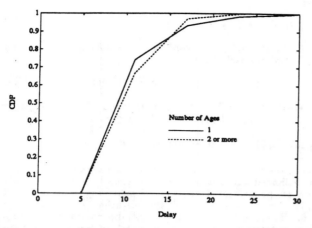

Figure 3: Distribution of delay in a (6,6) shuffle ring network for a throughput, $\tau = 0.03$. This corresponds to 23.04 packets for the network per time slot.

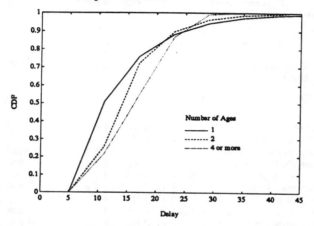

Figure 4: Distribution of delay in a (6,6) shuffle ring network for a throughput, $\tau = 0.05949$. This corresponds to 45.69 packets for the network per time slot.

and destined for node (c', r'). Let us first consider the special case when $c' = c \ominus n$. In this case, the packet wishes to pass through the sequence of n nodes (c^i, r^i), $1 \leq i \leq n$, where $c^i = c \ominus i$ and r^i is given by placing a window of length n over the sequence $r'_{n-1}, r'_{n-2}, \ldots, r'_0, r_{n-1}, \ldots, r_1, r_0$, with the window positioned i clicks from the right-most position. The construction guarantees that there is a link from (c^i, r^i) to (c^{i+1}, r^{i+1}) for $0 \leq i < n$, where $(c, r) = (c^0, r^0)$.

In general, for left-packets, let j be the column number such that $c' = c \ominus n \ominus j$. For the first j hops, the packet does not care which of the two output links it takes in the direction of decreasing column number. After these j links are taken, the packet wishes to follow the unique route of n more links described in the special case above. The longest route of this network is $n + k - 1$.

We have described the route a packet nominally takes. However, due to congestion, a packet may be forced to take an undesired link at some node. This packet is said to be *deflected*. As an example, consider a packet at node (c, r) which wishes to visit node node $(c \oplus 1, r^1)$ next. If the packet

is deflected to node $(c \ominus 1, r^2)$, in the next time slot, this packet will attempt to take any link to column c. If the packet at node (c, r) is deflected to node $(c \oplus 1, r^3)$, in the next time slot, this packet will attempt to take any link to column c. In both the above cases, there is a link to node $(c \oplus 1, r^1)$ from the node reached in column c. In general, if a packet is at distance i from its destination and gets deflected, the distance to its destination increases by one to $i + 1$.

The network operation is largely the same as described in Subsection 2.1. Consider a fixed node at the beginning of a time slot. Since the node has four input links, as many as four packets were received during the previous time slot. Any such packets destined for the node are removed from the network. New packets may be injected into the network at the node, bringing the total to at most four continuing packets. A packet, upon entering the network, is classified as either a left-packet or a right-packet (equiprobably and independently of other packets).

Deflection routing requires a rule for resolving conflicts. We propose a very simple rule for resolving these, based on a *two-pass* strategy by the node at the beginning of each time slot. On the first pass, the node sequentially considers the continuing packets in random order. When a given packet is considered, the node determines if there are any links that are both unassigned and desired by the packet. If so, the packet is assigned to one such link. Otherwise, the packet is not allocated to any link. On the second pass, the controller assigns the remaining packets to the unoccupied links, all choices being equally likely. All of the packets allocated links in the second pass have been deflected.

An analysis of this network is made in the same spirit as the analysis of Section 2.2. and under very similar approximations. We compute an upper bound an upper bound on the deflection probability and this is used to derive an upper bound on the evacuation time. These results are summarized in the following theorem.

Theorem 3.1 *Consider an (n, k) crossback network operating as described above. The evacuation time, T_{evac}, is defined as the time at which all, except a fraction 2^{-n}, of the packets have reached their destinations. Suppose the network began with 4 packets at each node, with the destinations of each packet chosen independently and at random from amongst the other $2^n - 1$ nodes. Then, under the approximations described above, $T_{evac} \leq 6(n + k)$.*

This is in contrast to the shuffle network, where T_{evac} was proportional to n^2. However, the price for this reduction in evacuation time is paid for by increased switch complexity. We will further compare these networks in Section 4 and Table 2.

The shuffle-exchange and exchange-unshuffle networks considered by Tan et al [5] are precisely the crossback networks with $k = 1$. They showed how to route packets along shortest length paths. Such paths are, on average, two to four hops shorter than the nominal paths we use, though the savings come at the expense of increased complexity needed for string matching. Their strategy applies to crossback switches in general.

3.2 Stay-or-Shuffle network

The network we describe here is a variation on the shuffle exchange network $((n, 1)$ shuffle ring). The stay-or-shuffle network is constructed from the $(n, 1)$ shuffle ring by adding links from every node to itself. Every node has three outgoing and three incoming edges. Since there is only one column, we drop the column index from the node labels. Two of the outgoing links from node $(r_{n-1}, \ldots, r_1, r_0)$ lead to nodes $(r_{n-2}, \ldots, r_1, r_0, 0)$, and $(r_{n-2}, \ldots, r_1, r_0, 1)$. We call these links the shuffle links. The third link out of $(r_{n-1}, \ldots, r_1, r_0)$ leads to node $(r_{n-1}, \ldots, r_1, r_0)$, and all such links are referred to as non-shuffle links. The non-shuffle links are equivalent to having one buffer in each node.

Routes are chosen as described in Subsection 2.1 for shuffle ring networks. The difference between these networks lies in what happens to a packet upon being deflected. Deflection occurs if a packet is routed on a link distinct from its preference. A packet which is routed onto a non-shuffle link does not change its distance to destination during the time slot. As in the case of the shuffle exchange graph, a continuing packet always prefers a particular outgoing edge. Let the distance to destination of a packet be k ($k \leq n$). Then, if a packet is routed according to its preference, this distance decreases to $k - 1$. If a packet is deflected onto a non-shuffle link, the packet remains at distance k from its destination. If deflected onto a shuffle link, the distance of the packet from its destination increases to n. Hence, if we decide that the state space of the packets is the distance to destination, then as before, the packets are performing a random walk on the integers $0, 1, \ldots, n$.

A rule is required to resolve conflicts. These conflicts arise when two or three packets at a node prefer the same output link. We use a priority rule to break conflicts. The packets are routed in order of their priority, which is the distance to destination. Packets closer to their destinations get higher priorities, and ties are broken randomly. When it is a packet's turn to be routed, the node examines the preferred link. If that link is free, the packet is assigned that link. Otherwise, the node examines the non-shuffle link from the node to itself. If this link is free, the packet is assigned that link. If neither of these two links are free, the packet is assigned the third link. This rule is a *two-pass* strategy by the node.

An analysis of this network is made in the same spirit as before. We derive approximate evolution equations based on approximations similar to those in Section 2. Set T_{evac} to be the first time that the link occupation probability falls below 2^{-n}. Then, the evolution equations above can be used to evaluate T_{evac}. Some typical values are shown in Table 1. Some of these values were compared to evacuation times for a simulation of the actual network, and they were found to be accurate within 5%.

n	6	7	8	9	10	12	15	20
T_{evac}	15	18	21	24	28	35	47	67

Table 1: Evacuation time for stay-or-shuffle networks: the first row represents n, where the network has 2^n nodes, and the second row contains the evacuation time.

4 Comparison

In this section we compare the complexity of $N \times N$ switching networks, built using the various networks discussed earlier. There are two approaches to building these switching networks from a given network.

1. Let the switching network be the given network. Then, if the network is filled up with packets, it delivers them to their destinations in T_{evac} time units.

2. As described in the introduction, the networks can be expanded in space in such a way that the evolution that we described from-slot-to-slot in time applies as well to packets moving from-copy-to-copy in space, through copies of the original network. The evacuation time of the original network dictates how many copies the space expanded network must have. This approach gives a pipelined version of the original network. The cost of this network is T_{evac} times the original cost, but the throughput is T_{evac} times as much as the original throughput.

A comparison of various networks is shown in Table 2. An example of a Batcher-Banyan[5] network is Starlite [8]. The second column contains estimates on T_{evac} for the various networks. These estimates are first order estimates that fall between the bounds we have proven and agree well with numerical computations for n upto 100. The third column gives the individual node complexity, where a node with n_1 input edges and n_2 output links is considered to have $n_1 n_2$ complexity. Also, if a node in the graph has k input(output) links, it actually has $k + 1$ input(output) links, since one link is needed for the external host to inject (remove) traffic. The fourth column represents the number of packets each node injects into the network. The fifth column shows a measure of the overall complexity, which is given by the product of T_{evac} and node complexity divided by the packets per node per time slot. The comment on *multiple packets* refers to whether multiple packets for the same destination can be delivered in T_{evac} time steps in the network.

5 Conclusion

The similarity of the complexities of the various networks listed in Table 2 is striking. Both the crossback switch and the stay-or-shuffle switches have relatively small values of evacuation time, as we had hoped. However, it appears that the improved performance there may barely, if at all, compensate for the increased complexity of the component switches. This suggests that the search for improved networks should be pursued within the family of binary switching networks. In particular, a network having the approximate evacuation time of the cross-back switch, but made from 2×2 switches, would be quite attractive. The network should admit simple deflection routing, or some other simple form of dynamic routing based on local information.

Our analysis has been based on the approximate evolution equations and simulation. Even in situations with priority classes, the evolution equations matched simulations quite well. The problem of analytically validating the evolution equations without any independence approximations has to date appeared intractable. Nevertheless, we have found the approximate evolution equations to be useful tools for quickly exploring a large class of networks, and networks of fairly large size (over 10^6 nodes) can be readily handled.

Acknowledgements
The authors wish to thank Albert Greenberg and Jon Sauer

[5]The $\frac{1}{2}$ in the column for node complexity refers to the fact that only $\frac{N}{2}$ switches are needed in each stage of Batcher's sorting network. The T_{evac} value refers to the number of stages needed for Batcher's sorting network.

Network type	T_{evac}	Node complexity	Packets per node per time slot	Overall complexity	Comments
Shuffle network	$\frac{1}{4}n^2$	$9 = 3 \times 3$	2	$\frac{9}{8}n^2$	Multiple packets Average performance
Crossback network	$4n$	$25 = 5 \times 5$	4	$25n$	Same as above
Stay-or-shuffle network	$\frac{1}{10}n^2$	$16 = 4 \times 4$	2	$\frac{4}{5}n^2$	Same as above
Batcher-Banyan Network	$\frac{1}{2}n^2$	$2 = 2 \times 2 \times \frac{1}{2}$	1	n^2	No multiple packets All permutations are allowed
Hypercube network	n	$n^2 = n \times n$	n	n^2	Multiple Packets Average performance

Table 2: Comparison of the complexity of $N \times N$ switching networks, $n = \log_2 N$

for many useful discussions.

References

[1] P. Baran, "On distributed communication networks," *IEEE Trans. Communication Systems*, vol. 12, pp. 1–9, 1964.

[2] A. S. Acampora, M. J. Karol, and M. G. Hluchyj, "Terabit lightwave networks: the multihop approach," *AT&T Technical Journal*, pp. 21–34, 1987.

[3] D. H. Lawrie and D. A. Padua, "Analysis of message switching with shuffle-exchanges in multiprocessors," in *The Proceedings of the Workshop on Interconnection Networks for Parallel and Distributed Processing*, pp. 116–123, 1980. reprinted in IEEE Press book, Interconnection Networks, Wu and Feng Eds., 1984, IEEE Computer Society Press.

[4] J. R. Sauer, "An optoelectronic multi-Gb/s packet switching network," February 1989. Preprint, Optoelectronic Systems Center, Univ. of Colorado.

[5] X. N. Tan, K. C. Sevcik, and J. W. Hong, "Optimal routing in the shuffle-exchange networks for multiprocessor systems," in *CompEuro 88 – System Design: Concepts, Methods and Tools*, pp. 255–264, IEEE, Euromicro, April 1988. published by IEEE Comput. Soc. Press, Washington, D.C.

[6] A. G. Greenberg and J. Goodman, "Sharp approximate models of adaptive routing in mesh networks," in *Teletraffic Analysis and Computer Performance Evaluation*, (O. Boxma, J. W. Cohen, and H. C. Tijms, eds.), pp. 255–270, Elsevier, Amsterdam, 1986. revised, 1988.

[7] A. Greenberg and B. Hajek, "Approximate analysis of deflection routing in hypercube networks," August 1989. Submitted to IEEE Trans. on Communications. Also, presented at the TIMS meeting, Osaka, July 1989.

[8] A. Huang and S. Knauer, "Starlite: a wideband digital switch," in *Proceedings of Globecom Conference*, pp. 121–125, IEEE Press, 1984.

[9] N. F. Maxemchuk, "Regular mesh topologies in local and metropolitan area networks," *AT&T Technical Journal*, vol. 65, pp. 1659–1685, September 1985.

[10] N. F. Maxemchuk, "Routing in the manhattan street network," *IEEE Transactions on Communications*, vol. COM-35, pp. 503–512, May 1987.

[11] N. F. Maxemchuk, "Comparison of deflection and store-and-forward techniques in the manhattan street and shuffle-exchange networks," in *Proceedings of IEEE Infocom'89*, pp. 800–809, 1989.

[12] B. Smith, "Architecture and applications of the HEP multiprocessor computer system," in *Real-time signal processing IV–Proc SPIE 298*, (T. F. Tao, ed.), pp. 241–248, Society Photo-Optical Eng, 1981.

[13] W. D. Hillis, *The Connection Machine*. Cambridge, Mass.: MIT Press, 1985.

[14] J. Y. Ngai and C. L. Seitz, "A framework for adaptive routing in multicomputer networks," in *Proc. ACM Symp. Parallel Alg. and Architech.*, June 1989.

Shuffle Interconnection Networks with Deflection Routing for ATM Switching: the Closed-Loop Shuffleout [+]

M. Dècina[*]*, P. Giacomazzi*[*]*, A. Pattavina*[**]

[*] Dept. of Electronics, Polytechnic of Milan/CEFRIEL, Italy
[**] INFOCOM Dept., University "La Sapienza", Rome, Italy

Abstract

A new class of switching architectures for broadband packet networks, called Shuffleout, is introduced in this paper. Its interconnection network is a multistage structure built out of unbuffered 2x4 switching elements. Shuffleout is basically an output-queued architecture in which the number of cells that can be concurrently switched from the inlets to each output queue equals the number of stages in the interconnection network. The switching element operates the cell self-routing adopting a shortest path algorithm which, in case of conflict for interstage links, is coupled with deflection routing. The architecture presented here is called Closed-Loop Shuffleout, since the cells that cross the whole interconnection network without entering the addressed output queues are lost. A different version of this architecture, called Open-Loop Shuffleout is described in a companion paper. The key target of the proposed architecture is coupling the implementation feasibility of a self-routing switch with the desirable traffic performance typical of output queueing.

1. INTRODUCTION

Interconnection networks for asynchronous transfer mode (ATM) applications have been widely investigated in the eighties. Most of the proposals are based on the use of multistage strauctures of switching elements (SEs) with very small size, say 2x2. The basic feature of these SEs is their autonomous capability to route the received packets (or *cells*) towards their respective outlets based on the status of one bit of the outlet address carried by the cell (*bit-based self-routing*). Different classes of interconnection networks can be identified based on the mode of allocating the buffers storing the cells to be switched. Architectures with purely input queueing [1] or purely output queueing [2] or shared internal queueing [3] were proposed based on the adoption of a non-blocking interconnection network built out of unbuffered SEs.. Adopting a blocking interconnection network required to provide each SE with an adequate dedicated storage capability [4,5] in order to provide satisfactory a traffic performance.

Different types of drawbacks characterize each of these proposals, e.g. SE complexity in fabric-buffered solutions,

[+] Work carried out at CEFRIEL, Milan (Italy), and supported by the Italian National Research Council in the frame of the Telecommunication Project.

throughput limitations in input-queued switching fabrics, significant increase in the total number of switching stages in the interconnection network in the shared-queued solution. In particular the output-queued solution proposed in Ref. [2], which provides an optimal traffic performance, raises implementation problems related to the "internal speed-up" N, since up to N cells can concurrently access the output interface.

Solutions that take full advantage of the hardware simplicity and self-routing capability of unbuffered SEs were also proposed that adopt a small internal speed-up, say less than 10. A sorting-banyan structure is generally used in these cases that adopts output queueing together with input queueing [6,7] or with shared internal queueing [8]. Traffic performance is improved in these solutions compared to architectures without speed-up at the expense of an increased hardware complexity of the overall interconnection network.

In this paper we propose a new class of ATM switching architectures called *Shuffleout* [9], which is a multistage self-routing structure based on output queueing. It accomplishes an internal speed-up greater than that characterizing the mentioned mixed queueing strategies in order to obtain quasi-optimal performance, but still less than N so as to limit the hardware complexity. The SE operates an *address-based cell self-routing*, that is in general more complex to be implemented than the bit-based self-routing of banyan networks. Nevertheless, by fully exploiting the properties of the shuffle pattern adopted in each interstage connection, the SE complexity is not increased significantly. Shortest-path routing is applied to cells received in each SE, which is internally unbuffered. Deflection routing is also applied in case of path conflicts within the interconnection network in order to minimize the cell loss.

Two alternative architectures of Shuffleout switches have been conceived, the Open-Loop Shuffleout switch (OSS) and the Closed-Loop Shuffleout switch (CSS). Cells that have not reached the desired output queue after crossing the whole interconnection network are lost in the former switch and recirculated within the network in the latter switch. This paper describes the CSS architecture, whereas the OSS architecture is studied in a companion paper [10].

Section 2 describes two different implementations of the CSS architecture, the Buffered Closed-Loop Shuffleout Switch (BCSS) and the Expanded Closed-Loop Shuffleout Switch (ECSS). In the BCSS each inlet of the first stage in the interconnection network is shared between the link

Reprinted from *Proc. IEEE INFOCOM '91*, pp. 1254–1263, April 1991.

coming from the upstream switch and a link connected to an outlet of the SEs in the last stage. Thus buffers are needed at switch inlets to minimize cell loss in case of concurrent arrivals at the same inlet. In the ECSS the links used to recirculate cells from the outlets of the last stage are fed into dedicated switch inlets. Thus an expansion of the interconnection network size is needed in order to feed concurrently into the network both new cells and recirculated cells. An analytical model is developed in Section 3 to evaluate the performance of BCSS and ECSS, whereas the obtained results in terms of cell loss probability and cell delay are reported in Section 4.

2. THE SWITCH ARCHITECTURE

In this section the BCSS and ECSS architectures are presented, by describing the interconnection pattern among SEs in different stages and the routing procedure carried out by each SE. A detailed hardware scheme for the SE has been designed and its structure is described in a companion paper [10]. Suffice to say that the gate count of s SE is particularly small, on the order of 200 gates per SE for a switch with siza 1024x1024.

2.1 The BCSS architecture

The basic structure of the NxN BCSS interconnection network is represented in Fig. 1 for $N=8$. It includes a number S of stages of 2x4 switching elements (SEs) with adjacent stages interconnected by a shuffle connection pattern. Each stage includes N/2 rows of SEs, ranging from row 0 to row N/2 - 1. Each of the N switch inlets is terminated by an input queue, whereas each of the N switch outlets is terminated by an output queue. Each of these queues is fed by cells received on S links, one per stage. Thus, each SE is connected to two output queues: the SE in row i is connected to output queues $2i$ and $2i+1$. Each SE in stages 1 to $S-1$ is also connected to two SEs in the following stage. SEs in stage S are connected to the input queues, implementing a recirculated structure. The packets entering the SEs of stage S that are not routed to the output queues are transmitted to the input queues, for further trial. Input queues are needed to prevent packet collisions, since two packets (a new arrival and a recirculated one) could be offered at the same time to a SE inlet of the first stage. Priority is given to the recirculated cells, therefore cells in the input queues cannot be transmitted into the network until the recirculated cells keep busy the first stage inlet. Recirculated cells are to be delayed, in order to synchronize their arrival at the input queue with the next arrival of new cells. Therefore, a cell undergoes a delay of one time slot every time it is recirculated.

In order to perform the routing function, each SE needs some information about the switch outlet required by the cells. Therefore, cells are provided with an additional header field, called *tag*, containing the address of the required switch outlet. If the network size is N, the tag is made of a $m =\log_2 N$ bit string. The cell routing within the interconnection network adopts jointly two distinct criteria, i.e. *shortest path* and *deflection routing*. A SE always attempts to route the packets received on its inlets along the shortest path to the required switch outlet. The length of the shortest path (that is the *distance* of a cell to the required switch outlet) is defined as the minimum number of interstage hops needed to reach the required switch outlet. The cell distance ranges from 0 to $m-1$ hops. If the

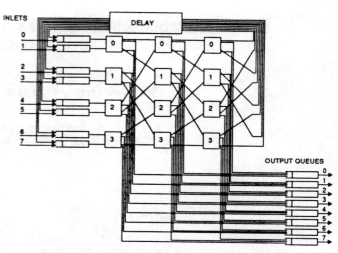

Fig. 1 - The BCSS structure for N=8 and S=3

packet address matches (does not match) with one of the switch outlets reachable through the SE, it is said to require a *local* (*remote*) switch outlet. A packet requiring a local switch outlet in the SE is routed to the appropiate output queue. The same rule applies if the packets entering the SE are addressed to two distinct local switch outlets. If both packets require the same local switch outlet, or if the shortest path route to the required remote switch outlet implies choosing the same interstage link, deflection routing is applied. One cell is routed along the required link (belonging to the shortest path), while the other cell is routed along a vacant interstage link.

In order to apply the deflection routing algorithm, the SE examines the cell distance and *priority*. The priority of a cell is defined as the number of recirculations it has undergone (it is obviously updated at each recirculation). The higher priority packet is routed along the link belonging to the shortest path, while the other one is deflected from it. If both packets have the same priority, the SE deflects the packet further from the required switch outlet. If the two packets have the same distance to their destinations, a random choice is carried out.

In Fig. 2 an example of cell routing within the BCSS is shown. Two cells simultaneously enter the switch from inlets 0 and 1, both requiring outlet 4. Both cells find empty input queues, therefore they concurrently enter the SE in row 0 in the first stage. Since both shortest paths follow the lower interstage link, a conflict occurs. The cells have the same distance to outlet 4, therefore a random choice is carried out, assigning the shortest path to the cell received on inlet 1. The cell winning the conflict reaches at the third stage a SE connected to outlet 4 and exits the network. The deflected cell cannot reach the required switch outlet and it is recirculated. This cell is given an additional delay in order to synchronize its arrival at the input queue with an eventual new cell arrival at the input queue of inlet 4. The recirculated cell is then transmitted to the required switch outlet from the first stage, no care about the new arrivals configuration, since it is given a priority.

Let τ denote the SE latency, that is the time it takes for a packet to propagate through a SE. A packet reaching the output queue from the sth stage ($s=1 \dots S$) undergoes a network delay of $s\tau$. Therefore, packets simultaneously entering the switch may reach the output queues at differ-

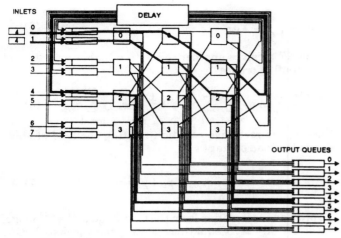

Fig. 2 - An example of cell routing in the BCSS

ent times. Synchronization of arrivals at the output queues is guaranteed by imposing an additional delay of $(S-s)\tau$ to the cells exiting the network from stage s. In such a way cells entering the network at time t_0 and recirculated r times reach the output queues at time $t_0+S\tau + rT$.

This routing procedure implies that cells entering the switch from the same inlet and requiring the same outlet can follow different paths within the network and consequently undergo a different number of recirculations. If a cell enters the switch at time t_0 and exits the network after three recirculations, it reaches the output queue connected to the required switch outlet at time $t_0 + S\tau + 3T$. It may happen that a cell entering the switch at time $t_0 + T$ (from the same inlet and requiring the same outlet) reaches the proper output queue after only one recirculation, at time $t_0 + S\tau + 2T$. In this case an out-of-sequence has occured, since the first cell entering the switch reaches the output queue one time slot later than the second cell.

Two resequencing schemes have been considered in order to keep packet sequence integrity. In the *connection oriented* scheme the output queues perform simultaneously N resequencing processes, one per each input queue from which the cells enter the switch. The drawback of this scheme is the great complexity of the output queues, due to the need for performing N simultaneous resequencing processes. In the *global* resequencing scheme the output queues perform the resequencing algorithm based on the *age* of the cell (that is the number of time slots elapsed between its arrival at the switch and its entrance in the output queue), no care about the input queue from which it entered the switch. A window with width W time slots is defined so that cells reaching the output queues $w \leq W$ time slots after their arrival at the switch are given an additional delay of $W-w$ time slots. Cells *older* than W time slots are discarded. In order to minimize the cell loss due to the resequencing algorithm, a careful design is needed to properly select the resequencing window width.

The *connection oriented* resequencing scheme minimizes W, but increases substantially the output queue complexity. The *global* resequencing algorithm implies larger resequencing windows, but smaller output queue complexity, since only one resequencing process per output queue has to be performed. We will show that employing the *global* resequencing algorithm the window widths needed to reach very low cell loss probability (less than 10^{-9}) are on the

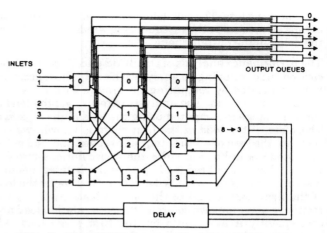

Fig. 3 - The ECSS structure for N=8, P=3 and S=3

order of a few time slots, even for high offered loads. If the network includes a large number of stages, the network cell delay $S\tau$ may exceed one time slot. In general, if the network cell delay consists of t time slots and a fraction of time slot, the delay network is to be designed in order to make the global loop cell delay (that is the time it takes for a cell to cross the network from stage 1 to stage S and to be recirculated) equal to $t+1$ time slots.

2.2 The ECSS architecture

The ECSS structure is shown in Fig. 3 for $N=8$. A subset of the network inlets is used exclusively for recirculations. In the case of Fig. 3 the recirculated cells are transmitted to the *3 auxiliary* switch inlets. In general the network size is N and the auxiliary inlets are P, therefore the number of external inlets is $N-P$. The N feedback lines are to be connected to the P auxiliary switch inlets. Therefore, a concentrator from N to P lines is needed. In general the ECSS requires a lower number of network stages in order to reach the same performance as the BCSS. The drawback is a network expansion, measured as $E=P/(N-P)$ (the number of recirculating links per external input port). The P lower SE outlets, that in the BCSS are connected to the local switch outlets, are vacant. Cells require output addresses ranging from 0 to $N-P-1$, while addresses from $N-P$ to $N-1$ are never required, since they do not correspond to switch outlets.

The cell routing scheme within the ECSS coincides with the one of the BCSS. Also for the ECSS a resequencing algorithm has to be performed in the output queues. Since the ECSS does not include input queues, the cell age coincides with its priority (the number of suffered recirculations). Since the cell priority is in general quite low, the *global* resequencing scheme has to consider very short resequencing windows.

3. PERFORMANCE ANALYSIS

The parameters we use to evaluate the performances of the BCSS and ECSS are the cell delay, that is the time it takes for a cell to cross the switch, and the cell loss probability. In the BCSS the cell loss probability includes various components: the loss in the input queues, the loss in the output queues and finally the loss due to the resequencing algorithm. In the ECSS the cell loss probability includes the loss in the concentrator, the loss in the output queues and the resequencing loss. The cell delay consists of a fixed number of time slots equal to the resequencing window width W and the delay in the output queue.

The performance analysis of the BCSS and ECSS has been carried out developing an analytical model, whose accuracy has been tested by comparing the results with those provided by simulation. The analysis is recursive: the load of the $s+1^{th}$ stage is computed using the load of the sth stage. Therefore, if the load of the first stage is known, we can compute the load of all the stages. However, the load of the first stage is unknown, since the recirculations make it greater than the offered load by an unpredictable amount. A second level of iteration is needed: at first we fix the load of the first stage equal to the offered load. Applying the recursive equations we compute the recirculation load and consequently a new value for the first stage load. We continue the iteration until the first stage load reaches a steady value. In the analysis we also compute the probability distribution of the number of time slots needed by the cells to reach the output queues. In such a way we can compute the cell loss probability due to the resequencing algorithm and the output queues distribution. The analysis has been carried out under the hypothesis of a purely random traffic offered to the switch. That is cell arrivals on SE inlets are independent and each cell is likely to require any remote SE outlet (or local SE outlet if the cell distance is 0) with equal probability.

3.1. The analytical model of the network

Let p_s be the load of the sth stage of the network; p_0 is the mean traffic offered to the network and p_{s+1} is the recirculation load. Moreover, let $q_{s,d,r}$ be the load of the sth stage, due to the cells d hops far from the required switch outlet and recirculated r times. The load p_s and the distribution $q_{s,d,r}$ are related by the following equation :

$$p_s = \sum_{d=0}^{m-1} \sum_{r=0}^{\infty} q_{s,d,r} \qquad (1)$$

(throughout this paper the result of a sum is set to 0 if the lower index is greater than the upper index in the sum). The recursive equation relating p_{s+1} with p_s is drawn as follows. A *tagged* cell can exit the network from stage s only if it has distance $d=0$. Furthermore one of the following conditions must be true:

- the cell is the only one to be routed by the SE, therefore no conflicts can occur, this event has probability:

$$p_a = 1 - p_s \qquad (2)$$

- the SE receives another cell with lower priority,

$$p_b = \sum_{r=0}^{\infty} q_{s,0,r} \sum_{\rho=0}^{r-1} \sum_{d=0}^{m-1} q_{s,d,\rho} \qquad (3)$$

- the SE receives another cell with equal priority and distance $d>0$,

$$p_c = \sum_{r=0}^{\infty} q_{s,0,r} \sum_{d=1}^{m-1} q_{s,d,r} \qquad (4)$$

- the SE receives another cell with equal priority and distance $d=0$ and no conflict occurs, or the considered cell wins the eventual conflict,

$$p_d = \frac{3}{4} \sum_{r=0}^{\infty} q^2_{s,0,r} \qquad (5)$$

- the SE receives another cell with a higher priority and no conflict occurs

$$p_e = \sum_{r=0}^{\infty} q_{s,0,r} \sum_{\rho=r+1}^{\infty} \sum_{d=1}^{m-1} q_{s,d,\rho} + \frac{1}{2} \sum_{r=0}^{\infty} q_{s,0,r} \sum_{\rho=r+1}^{\infty} q_{s,0,\rho} \qquad (6)$$

Finally, the recursive equation relating the load of stage $s+1$ to the load of stage s is:

$$p_{s+1} = p_s - \left(p_a \sum_{r=0}^{\infty} q_{s,0,r} + p_b + p_c + p_d + p_e \right) \qquad (7)$$

To proceed the iteration from the first stage to the Sth stage, we also need a set of equations relating the q_{s+1} distribution with the q_s distribution. It should be noted that performing an interstage hop the cell distance can vary, while the priority cannot change, since it is updated only when the cell is recirculated. The $q_{s+1,d,r}$ distribution is computed as:

$$q_{s+1,d,r} = \sum_{\rho=0}^{\infty} \sum_{\delta=0}^{m-1} P\{q_{s+1,d,r}/q_{s,\delta,\rho}\} q_{s,\delta,\rho} \qquad (8)$$

Computation of $P\{q_{s+1,d,r}/q_{s,d+1,r}\}$

A cell passing from stage s to stage $s+1$ reduces (by one hop) its distance d if it is routed along the shortest path. This happens if one of these conditions is true:
- the cell is the only one to be routed by the SE, therefore no conflicts can occur or
- the SE receives another cell with lesser priority or
- the SE receives another cell with higher priority and no conflict occurs, or
- the SE receives another cell with equal priority and
  - greater distance or
  - lesser distance $d>0$ and no conflict occurs or
  - distance $d=0$ or
  - equal distance, and no conflict occurs. If a conflict occurs the considered cell wins it (it is not deflected).

In short:

$$P\{q_{s+1,d,r}/q_{s,d+1,r}\} = (1-p_s) + \sum_{\rho=0}^{r-1} \sum_{\delta=0}^{m-1} q_{s,\delta,\rho} + \frac{1}{2} \sum_{\rho=r+1}^{\infty} \sum_{\delta=1}^{m-1} q_{s,\delta,\rho} +$$

$$\sum_{\rho=r+1}^{\infty} q_{s,0,\rho} + \sum_{\delta=d+2}^{m-1} q_{s,\delta,r} + \frac{1}{2} \sum_{\delta=1}^{d} q_{s,\delta,r} + \frac{3}{4} q_{s,d+1,r} + q_{s,0,r} \qquad (9)$$

Computation of $P\{q_{s+1,d,r}/q_{s,d,r}\}$

When $d \neq m-1$ the computation is carried out as follows. A cell passing from stage s to stage $s+1$ can retain its original distance only if it is deflected. Cells following the shortest path reduce their distance, while deflected cells generally increase it. Due to the topological features of the shuffle connection pattern, a deflected cell retains its distance only if it is routed along interstage links 0 or $N-1$ (the *extreme* interstage links). Therefore, a cell is deflected and retains its distance if:
- the SE receives another cell with a higher priority and a conflict occurs or
- the SE receives another cell with an equal priority and lesser distance $d>0$ and a conflict occurs or

374

– the SE receives another cell with an equal priority and equal distance, a conflict occurs and the considered cell is deflected,

together with the condition that the cell is to be deflected along an extreme interstage link.

In short, we can state that:

$$P\{q_{s+1,d,r}/q_{s,d,r}\} = \frac{2}{N}\,Q \qquad (10)$$

$$Q = \begin{cases} \left\{ \dfrac{1}{2}\displaystyle\sum_{\rho=r+1}^{\infty}\sum_{\delta=1}^{m-1} q_{s,\delta,\rho} + \left(\dfrac{1}{2}\displaystyle\sum_{\delta=1}^{d-1} q_{s,\delta,r} + \dfrac{1}{4}\,q_{s,d,r} \right) \right\} & d \neq 0 \\[4mm] \dfrac{1}{2}\displaystyle\sum_{\rho=r+1}^{\infty} q_{s,0,\rho} + \dfrac{1}{4}\,q_{s,0,r} & d = 0 \end{cases} \qquad (11)$$

When a cell has maximum distance ($d=m\text{-}1$) to the required switch outlet, every deflection keeps it at distance $m\text{-}1$, no care about the interstage link it follows (extreme or not extreme). Therefore, if $d = m\text{-}1$, Eq.10 applies in which the factor 2/N is removed.

Computation of $P\{q_{s+1,D,r}/q_{s,d,r}\}$, where $D > d$

A cell passing from stage s to stage $s+1$ increases its distance only if it is deflected. A cell is deflected if:
– the SE receives another cell with higher priority and a conflict occurs or
– the SE receives another cell with equal priority and lower distance $d>0$ and a conflict occurs or
– the SE receives another cell with equal priority and equal distance, a conflict occurs and the considered cell is deflected.

Therefore we have:

$$P\{q_{s+1,D,r}/q_{s,d,r}\} = \Gamma_{D,d}\,Q \qquad (12)$$

where Q is given by Eq. 11. The $\Gamma(D,d)$ factor is the probability that a cell gets a new distance $D>d$ after being deflected when it was d hops distant from the required switch outlet. Some approximate expressions for $\Gamma(D,d)$ were obtained, although better results were achieved by numerically evaluating it. The cell on link i requiring outlet j is deflected, and the new distance (D) measured. Examining all the possible couples (i,j), a very good approximation of $\Gamma(D,d)$ is obtained.

3.2 The iterative process for the BCSS

Once the first stage load p_1 and distribution $q_{1,d,r}$ are known, we can compute the recirculation load p_{S+1} and the corresponding $q_{S+1,d,r}$ distribution. As just stated, at the first step of the iteration the first stage load is set equal to the load offered to the switch. At the first step of the iteration we also have to initialize the $q_{1,d,r}$ distribution. For $r\neq0$ $q_{1,d,r}$ is set to 0. For $r=0$, $q_{1,d,0}$ is numerically evaluated: it is computed measuring the distance from the switch inlet i to the switch outlet j, for all the couples (i,j). In such a way we measure the distance distribution for a cell entering the network from a specific inlet chosen with probability 1/N and destined for a specific outlet chosen with the same probability. From now on this distribution will be called $q_{0,d}$. Once the load of stage $S+1$ (the recirculation load) is computed using this method, new values for the first stage load p_1 and $q_{1,d,r}$ distribution are to be found, in order to proceed in the iterative process. Two different types of iterations are considered, one for the BCSS, and one for the ECSS. For the BCSS the first stage load is

computed using the recirculation load as following. A first stage link is occupied by a cell if:
– a cell is recirculated on that link or
– a recirculation does not occur and
   – the input queue is idle and a new cell arrives or
   – the input queue is not idle.

These conditions lead to the equation:

$$p_1 = p_{S+1} + (\,1 - p_{S+1}\,)\,(\,\pi_0\,p_0 + 1 - \pi_0\,) \qquad (13)$$

where π_i ($i=0,...,Q_i$) is the probability that the input queue contains i cells. In order to compute π_0 we have to solve the Markov chain of the input queue, that is shown in Fig. 4. The transition probabilities are:
– a_0: the probability that the number of cells in the input queue decreases. This event occurs if there is not a new arrival at the input queue and the first stage link connected to it is not kept busy by a recirculated cell.
– a_1: the probability that the number of cells in the input queue does not vary. This event occurs if:
   – a new cell arrives at the input queue and the first stage link connected to it is free, or
   – there is not a new arrival, and the first stage link connected to the input queue is kept busy by a recirculated cell.
– a_2: the probability that the number of cells in the input queue increases. This event occurs if a new cell arrives at the input queue and the first stage link connected to the queue is kept busy by a recirculated cell.

Formally we have:

$$a_0 = (\,1 - p_0\,)\,(\,1 - p_{S+1}\,) \qquad (14)$$

$$a_1 = p_0\,(\,1 - p_{S+1}\,) + p_{S+1}\,(\,1 - p_0\,) \qquad (15)$$

$$a_2 = p_0\,p_{S+1} \qquad (16)$$

The probability distribution of the number of cells in the input queues is readily computed:

$$\pi_i = \frac{1 - \dfrac{a_2}{a_0}}{1 - \left(\dfrac{a_2}{a_0}\right)^{Q_i+1}} \left(\frac{a_2}{a_0}\right)^i \qquad (17)$$

where Q_i is defined as the input queue capacity, measured in cells. Replacing Eq. 17 (for $i=0$) in Eq. 13, we can compute a new value for the load of the first stage. The input queue cell loss probability l_{iq} is easily computed as:

$$l_{iq} = \pi_{Q_i} \cdot a_2 \qquad (18)$$

In order to compute a new value for the $q_{1,d,r}$ distribution we have to include a component due to the new cells $q_{0,d}$, and a component due to the recirculated cells $q_{S+1,d,r}$:

$$q_{1,d,r} = \begin{cases} q_{0,d} & \text{if } r = 0 \\ q_{S+1,d,r-1} & \text{if } r \neq 0 \end{cases} \qquad (19)$$

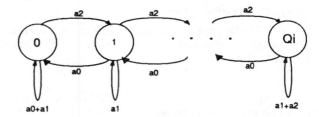

Fig. 4 - Markov chain for the input queue

3.3 The iterative process for the ECSS

At the first step of the iteration the first stage load is set to:

$$p_1 = \frac{p_0}{(1+E)} \qquad (20)$$

since only a subset of the inlets receives new cells. Once the recirculation load p_{S+1} is computed, a new value for the first stage load p_1 is found using Eq. 20. In this equation the first addendum includes the load due to the new arrivals, while the second includes the load due to the recirculated cells.

$$p_1 = \frac{p_0}{(1+E)} + p_{S+1}(1 - l_{co}) \qquad (21)$$

l_∞ is the concentrator loss probability, defined as:

$$l_{co} = \sum_{i=P}^{N-1} \frac{1}{i+1} \binom{N-1}{i} p_{S+1}^i (1 - p_{S+1})^{N-i-1} \qquad (22)$$

In order to compute the $q_{1,d,r}$ distribution we have to consider a component due to the new cells, and a component due to the recirculated cells:

$$q_{1,d,r} = \begin{cases} q_{0,d} \dfrac{N-P}{N} & \text{if } r = 0 \\ q_{0,d} \displaystyle\sum_{\delta=0}^{m-1} q_{S+1,\delta,r-1}(1 - l_\infty) & \text{if } r \neq 0 \end{cases} \qquad (23)$$

It should be noted that in the concentration the cells lose the distance gained in crossing the network and a new distance is assigned, using the $q_{0,d}$ distribution.

3.4 The delay distribution

In order to compute the output queue distribution and the cell loss probability due to the finite resequencing window, it is necessary to find the distribution of the cell delay from the time of its arrival at the switch, until it reaches the output queue. This delay consists of the input queue delay (only for the BCSS) and the network delay.

The input queue delay distribution in the BCSS

Let v_i be the probability that a cell waits i slots in the input queue. The v_i distribution can be computed as:

$$v_i = \sum_{n=0}^{Q_i} v_{i/n} \, \pi_n \qquad (24)$$

where $v_{i/n}$ is the probability that a cell waits i slots in the input queue, given that it finds n cells in the queue at its arrival. The probability that a cell finds n cells in the input queue at its arrival is π_n, just defined in Eq. 17. If a cell finds n cells in the input queue, it waits at least n slots. Furthermore, the distribution of the service time of the cells in the input queue is given by:

$$s_i = P\{\text{service time of } i \text{ slots}\} = (1 - p_{S+1}) \, p_{S+1}^i \quad (25)$$

In fact, a cell in the head of the queue has to wait until the first stage inlet is free from a recirculated cell. Therefore, we can state that:

$$v_{i/n} = s_i^{(* \, n)} * \delta(i - n) \qquad (26)$$

By replacing Eq. 25 in Eq. 26 and Eq. 26 in Eq. 24 the v_i distribution is easily computed.

The network delay

For both BCSS and ECSS, the network delay is the number of recirculations a cell suffers, that coincides with its priority. Once the $q_{s,d,r}$ distribution is known, the distribution of the network delay z_i (measured in time-slots) is defined as:

$$z_i = \frac{1}{\psi} \left[\sum_{d=0}^{m-1} q_{1,d,i} - \sum_{d=0}^{m-1} q_{S+1,d,i} \right] \qquad (27)$$

Where ψ is a normalization factor: it is needed in order for z_i to sum to 1. In the BCSS $z_i \psi$ adds up to $p_0(1 - l_{iq})$, while in the ECSS it adds up to $p_0(1 - l_{co})$, therefore we have:

$$\psi = \begin{cases} p_0(1 - l_{iq}) & \text{BCSS} \\ p_0(1 - l_{co}) & \text{ECSS} \end{cases} \qquad (28)$$

The network delay distribution, including the delay in the input queue for the BCSS is the convolution of v_i and z_i. For the ECSS it coincides with z_i:

$$r_i = \begin{cases} \displaystyle\sum_{k=0}^{\infty} v_k \, z_{i-k} & \text{BCSS} \\ z_i & \text{ECSS} \end{cases} \qquad (29)$$

Once the r_i distribution is known, the cell loss probability due to the cells arriving at the output queue with a delay greater than W time slots is computed as:

$$l_d = \sum_{k=W+1}^{\infty} r_k \qquad (30)$$

3.5 The output queue model

Due to the out of sequence phenomenon, the model of the output queues is complex, since we have to consider the additional amount of queueing necessary to perform the *global* resequencing algorithm. For both BCSS and ECSS we have shown how to compute the distribution r_i of the network (and eventually input queue) delay. The cell resequencing is obtained imposing an additional delay of W-w time slots to the cells entering the output queue with a w time slots delay. A number k of cells, each with an i time slot delay arrive at an output queue with probability:

$$b_k = \binom{S}{k} \left(\frac{p_q \, r_i}{S}\right)^k \left(1 - \frac{p_q \, r_i}{S}\right)^{S-k} \qquad (31)$$

where p_q is the load offered to the output queue, which is given by:

$$p_q = \begin{cases} p_0(1 - l_{iq}) & \text{BCSS} \\ p_0(1 - l_{co}) & \text{ECSS} \end{cases} \qquad (32)$$

We impose to these cells a delay of W-i time slots, therefore the distribution of the number k of cells arrived at the output queues with a delay of i time slots and receiving the resequencing delay is the W-i fold convolution of b_k, which is given by:

$$R^{(i)}(k) = \binom{S(W-i)}{k} \left(\frac{p_q \, r_i}{S}\right)^k \left(1 - \frac{p_q \, r_i}{S}\right)^{S(W-i)-k} \qquad (33)$$

The distribution of the total number of cells being delayed is readily computed by convolving the W distributions $R^{(i)}{}_k$:

$$R^{TOT} = R^{(0)} * R^{(1)} * R^{(2)} * \ldots\ldots * R^{(W-1)} \qquad (34)$$

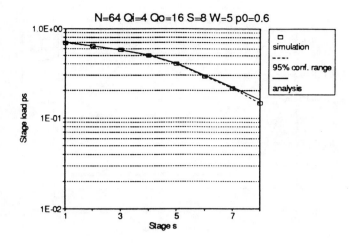

Fig. 5 - *Analysis validation: the stage load*

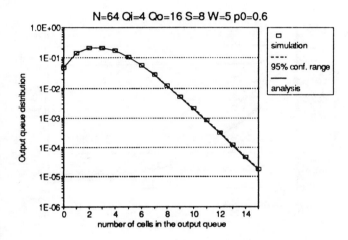

Fig. 6 - *Analysis validation: the output queue distribution*

Once the cells are properly delayed, it is necessary to cope with multiple requests for the same switch outlet: some queueing is needed. Under the hypothesis of random traffic offered to the switch, the queue can be defined as a GEOM/D/1, since the distribution of the arrivals is

$$P\{k \ arrivals\} = \binom{L}{k}\left(\frac{pq}{L}\right)^k\left(1-\frac{pq}{L}\right)^{L-k} \qquad L \leq S \qquad (35)$$

and the server transmits one packet per slot. Each output queue receives cells arriving from S lines, one per stage. If S is high (i.e. greater than 16) it is convenient to concentrate these lines from S to L (with $L<S$), in order to reduce the output queue complexity. The probability distribution g_k of the number of packets in this queue has been computed by solving its Markov chain. Once the g_k distribution is computed, the total number of packets in the output queue (either being delayed or queued) is found by convolving it with the R^{TOT} distribution:

$$b_k = \sum_{j=0}^{\infty} g_j \, r^{TOT}_{k-j} \qquad (36)$$

The output queue cell loss probability (called l_{oq}) is approximately computed as:

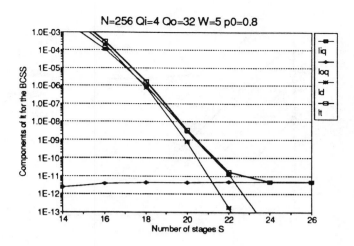

Fig. 7 - *Cell loss components in BCSS*

$$l_{oq} = \sum_{j=Q_{o+1}}^{\infty} b_j \qquad (37)$$

where Q_o is defined as the output queue capacity, measured in cells.

In Fig. 5 it is shown how the analytical results match those obtained by simulation. The stage load is represented for a BCSS with N=64 inlets, input queue capacity Q_i=4 cells, output queue capacity Q_o=16 cells, S=8 stages, resequencing window width W=5 slots and offered load p_0=0.6. In Fig. 6 the output queue distribution is shown for the same network. It should be noted that the distribution is not monotone; this is due to the resequencing function performed in the output queues.

4. PERFORMANCE RESULTS

4.1 The BCSS architecture

We describe the performance of the switch in terms of cell loss probability and cell delay.

The cell loss probability of the switch includes various components:
- the cell loss probability in the input queue l_{iq},
- the *resequencing* cell loss probability l_d, due to the cells that arrive at the output queues with a delay greater than W slots,
- the cell loss probability in the output queues l_{oq}.

Another component of the cell loss probability should be mentioned: the loss in the output queue concentrator. A careful study of the concentrator cell loss probability is carried out in Ref. [2]: it is shown that, under the hypothesis of a purely random traffic the cell loss probability in the concentrator is always lower than 10^{-10}, for L in the order of 11 - 12. From now on we will assume that L is large enough to make the concentrator loss much lower than the other loss components and we will neglect it.

The components of the cell loss probability versus the number of stages are plotted in Fig. 7, for a network with N=256 inlets, Q_i=4 cells, Q_o=32 cells, W=5 time slots and offered load p_0=0.8. The total cell loss probability is defined as:

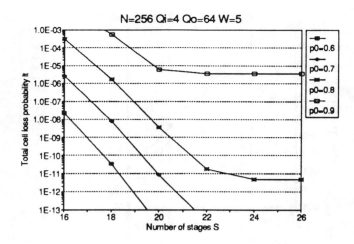

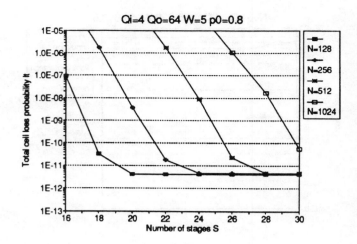

Fig. 8 - Cell loss for different offered loads in BCSS

Fig. 10 - Cell loss for different switch sizes in BCSS

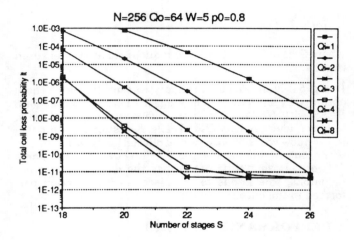

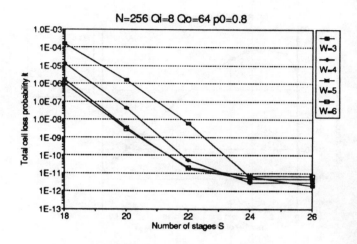

Fig. 9 - Cell loss for different input queue sizes in BCSS

Fig. 11 - Cell loss for different window widths in BCSS

$$l_t = 1 - (1 - l_{iq})(1 - l_d)(1 - l_{oq}) \qquad (38)$$

The l_{iq} and l_d loss components decrease, as the number of network stages increases, owing to the decrease of the recirculation load. In fact, decreasing the recirculation load, the input queues are offered a lighter total load, and both l_{iq} and cell delay decrease. Therefore, for a low numer of stages (in Fig. 7 for $S \le 23$), l_{iq} and l_d dominate over the total cell loss probability. For a high number of stages ($S \ge 24$) the output queue cell loss probability (l_{oq}) dominates. The l_{oq} component is nearly independent from the number of stages, thus leading to the *saturation* of the total cell loss probability.

The total cell loss probability versus the number of stages is shown in Fig. 8 for a network with N=256 inlets, Q_i=4 cells, Q_o=64 cells, W=5 time slots and offered load p_0 ranging from 0.6 to 0.9. For an offered load of 0.9 and 0.8 the saturation due to the output queue loss is recognizable, while for lower loads this phenomenon occurs for very low values of the total cell loss probability.

The effect of the input queue capacity Q_i is shown in Fig. 9, plotting the total cell loss probability versus the number of stages for a network with N=256 inlets, offered load p_0=0.8, Q_o=64 cells and W=5 time slots. A substantial decrease of

l_t is obtained as the input queue capacity increases from 1 cell to 4 cells. A further increase of the input queue capacity does not affect appreciably l_t.

The total cell loss probability versus the number of stages is plotted in Fig. 10 for various network sizes N, ranging from 128 to 1024 inlets. The offered load is p_0=0.8, the input queue capacity is Q_i=4, the output queue capacity is Q_o=64 and W=5 slots. Bigger networks reach the steady value of the total cell loss probability for a higher numer of stages.

Fig. 11 shows how the resequencing window W affects l_t for N=256, p_0=0.8, Q_i=8, Q_o=64. Expanding the resequencing window width W is convenient only when the l_d component is greater than l_{oq}, which is generally true for a small number of stages. In the case of Fig. 11, it is convenient to expand the resequencing window only for $S<24$. Furthermore, the figure shows that there is no advantage in increasing W beyond a certain threshold (W=5), since the dominant cell loss component becomes l_{iq} (see Fig.7 showing the different cell loss components). The cell delay includes various components:
- a $S\tau$ network delay, including the additional delay for synchronizing the arrivals at the output queue,
- a W time slots delay, due to the resequencing process,

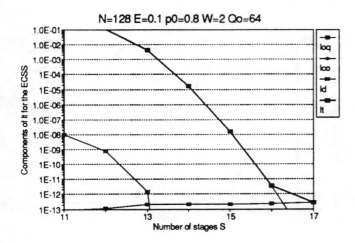

Fig. 12 - Cell loss components in ECSS

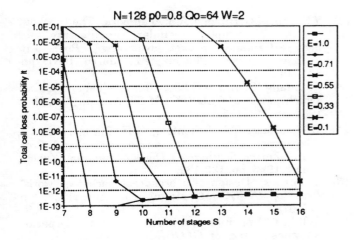

Fig. 13 - Cell loss for different expansion ratios in ECSS

— a random component, consisting of the time elapsed in the output queue due to the multiple arrivals scheduling. In the hypothesis of infinite queue capacity the average output queue delay is given by Ref. [11]):

$$T_q = \frac{p_q}{2 (1 - p_q)} \qquad (39)$$

Eq. 39 can be used to evaluate the average delay in the output queue even for finite queue capacity with good approximation, if l_{oq} is low.

For a network with N=1024 inlets, S=32 stages, offered load p_0=0.9, W=5 time slots, input queue capacity Q_i=4 cells, output queue capacity Q_o=64 cells the average output queue delay is 4.5 time slots. The tag reading time consists of 10 bit delays, while the routing time is pessimistically evaluated 20 bit delays. Therefore, the SE cross delay consists of 30 bit times, for a total of 960 bit times (= 2.26 time slots) in the case of 32 stages. Therefore, the total cell delay is 11.7 cells; assuming a line speed of 150 Mbit/s the delay is approximately 33 μs.

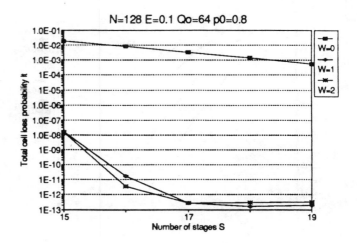

Fig. 14 - Cell loss for different window widths in ECSS

4.2 The ECSS architecture

The components of the cell loss probability for the ECSS are:
— l_{co}: the loss in the concentrator, due to the fact that cells coming from N lines are to be recirculated, while only P recirculations lines are available.
— l_d: the resequencing loss, due to cells entering the output queue with a delay larger than W slots.
— l_{oq}: the loss in the output queue.

The total cell loss probability is given in this case by:

$$l_t = 1 - (1 - l_{co})(1 - l_d)(1 - l_{oq}) \qquad (40)$$

The components of the cell loss probability versus the number of stages are plotted in Fig. 12 for a network with N=128 inlets, a number E=0.1 of recirculating lines per input port, offered load p_0=0.8, W=2 slots, Q_o=64 cells. For a low number of stages (S<17) the l_{co} component dominates, since a small S means a high recirculation load. For a high number of stages ($S{\geq}17$) the l_{oq} component dominates.

The effects of E (the number of recirculating lines per input port) are shown in Fig. 13 for a network with N=128 inlets, offered load p_0=0.8, output queue capacity Q_o=64 cells and resequencing window W=2 slots. A quite low degree of recirculation is needed to obtain very good values for l_t. In the case of Fig. 13, for E=0.71, the total cell loss probability is on the order of 10^{-3} with 8 stages, but it drops below 10^{-9} with only one additional stage.

Very small resequencing windows are needed for the ECSS. In Fig. 14 the effects of W are shown for a network with N=128 inlets, offered load p_0=0.8, E=0.1, Q_o=64 cells. Adopting a 1 slot resequencing window the total cell loss probability is nearly the minimum we can obtain.

The cell loss probability versus the number of stages is plotted in Fig. 15, for a network with 128 inlets, E=0.1, W=2, output queue capacity of 64 cells and offered load ranging from 0.6 to 0.9. For high loads a saturation effect is recognizable: providing a high number of stages the l_c and l_d components become much smaller than l_{oq} (see also Fig. 12).

The total cell loss probability versus the number of stages is plotted in Fig. 16 for a network with offered load 0.8, E=0.1, W=2, output queue capacity of 64 cells and a number

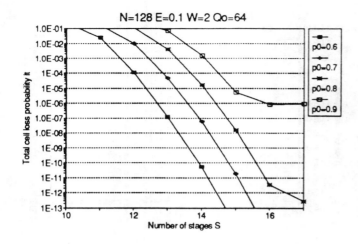

Fig. 15 - *Cell loss for different offered loads in ECSS*

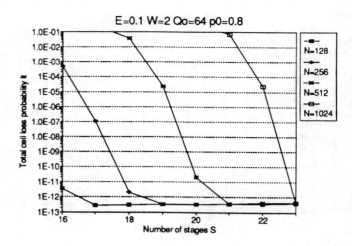

Fig. 16 - *Cell loss for different switch sizes in ECSS*

of inlets ranging from 128 to 1024. The figure shows that bigger networks require a higher number of stages in order to reach a comparable value for l_t.

The components of the cell delay in the ECSS are again the network delay, the resequencing delay and the output queue delay. The number of stages and the resequencing window for the ECSS are smaller than the ones for the BCSS, therefore the global cell switch cross delay is lower for the ECSS.

5. CONCLUSIONS

A new ATM switching architecture has been described in this paper that is based on a multistage shuffle interconnection network of switching elements each operate an address-based self-routing of the cells. The interconnection network accomplishes an internal speed-up equal to the number of stages, which is the key feature to provide very small cell loss probabilities for random traffic even for high offered loads. Compared to the Open-Loop Shuffleout architecture (see Ref. [10]), the Closed-Loop structure is characterized by a much smaller number of stages, thus simplifying the "fan-in" problem in the interconnection between the SE local outlets and the associated output queue. The price to pay is developing additoinal hardware to implement either input queues (BCSS) or adopting an enlarged interconnection network equipped with a concentrator on its outlets (ECSS).

REFERENCES

[1] J. Hui, E. Arthurs, "A broadband packet switch for integrated transport", *J. on Selected Areas in Comm.*, Vol. SAC-5, No. 8, Oct. 1987, pp. 1264-1273.

[2] Y.S. Yeh, M.G. Hluchyj, A.S. Acampora, "The knockout switch: a simple, modular architecture for high-performance packet switching", *Proc. Int. Switching Symp.*, Phoenix, AZ, March 1987, pp. 801-808.

[3] A. Huang, S. Knauer, "Starlite: a wideband digital switch", *Proc. of GLOBECOM 84*, Atlanta, GA, Nov. 1984, pp. 121-125.

[4] R.W. Muise, T.J. Schonfeld, G.H. Zimmerman III, "Experiments in wideband packet technology", *Proc. of Int. Zurich Seminar on Digital Commun.*, Zurich, March 1986, pp. 135-139.

[5] J. Turner, "Design of a broadcast packet network", *Proc. of INFOCOM*, Miami, FL, Apr. 1986, pp. 667-675.

[6] T.T. Lee, "A modular architecture for very large packet switches", *Proc. of GLOBECOM*, Dallas, Tx, Nov. 1989, pp. 1801-1809.

[7] A. Pattavina, "A broadband packet switch with input and output queueing", *Proc. of Int. Switching Symp.*, Stockholm, Sweden, May-June 1990, Vol. 6, pp. 11-16.

[8] J.N. Giacopelli, M. Littlewood, W.D. Sincoskie, "Sunshine: a high performance self-routing broadband packet switch architecture", *Proc. of Int. Switching Symp.*, Stockholm, May-June 1990, Vol. 3, pp. 123-129.

[9] CEFRIEL's patent pending.

[10] M. Decina, P. Giacomazzi, A. Pattavina, "Shuffle interconnection networks with deflection routing for ATM switching: the Open-Loop Shuffleout", *Proc. of 13th International Teletraffic Congress*, Copenhagen, June 1991.

[11] M.J. Karol, M.G. Hluchyj, S.P. Morgan, "Input versus output queueing on a space-division packet switch", *IEEE Trans. on Commun.*, Vol. COM-35, No. 12, Dec. 1987, pp. 1347-1356.

Hot Spots, Packet Trains, and Other Extensions

INTRODUCTION

This concluding part deals with a variety of issues and solutions in high-speed switching and its performance evaluation. The breadth of the ideas described in this part show that there are a great many questions regarding the design and performance evaluation of high-speed switching networks that need further exploration.

The first two papers discuss improving the performance of standard Banyan networks (see also the paper by Yoon et al. in Part 4). The first paper deals with a shortcoming of Banyan routing networks: the fact that there is only a single path between an input and output, so that contention for links is likely to arise. The paper, by G. J. Anido and A. W. Seeto, is entitled "Multipath Interconnection: A Technique for Reducing Congestion within Fast Packet Switching Fabrics." The authors' solution is to cascade a routing network and switching network so that there are N paths from any input to any output.

Anido and Seeto propose four algorithms to choose a path through the multipath interconnection network. Each in its own way attempts to minimize the path cost (where "cost" is measured in terms of loading). Although the algorithms are "static," since path allocations are not changed, the authors suggest that (rearranging) dynamic algorithms could be developed.

Anido and Seeto evaluate the four algorithms using a mixture of analysis and simulation. Specifically, traffic intensity moments on links are determined by simulation. These are then used in a tandem queueing model to calculate mean queue length and blocking probability (for finite buffers).

The authors note that queue performance is heavily influenced by link variance at heavy loads. They also find a limiting throughput of about 0.65. A point of diminishing returns in terms of mean throughput is reached when a stage's buffer size is about 5. Finally, and importantly, Anido and Seeto find that mean queue length and mean delay are bounded for finite buffer capacity and high link loads.

A different approach to improving the performance of Banyan networks appears in the next paper "A Self-Routing Multistage Switching Network for Broadband ISDN," by H. S. Kim and A. Leon-Garcia. They propose a multistage switching network, where the normal Banyan's stages of 2×2 switching elements alternate with stages of "distributor" elements. The distributors assign packets at their inputs to outputs in a sequential manner. This makes efficient use of distributor buffer space and presents a "balanced" load to the switching elements at the next stage. The maximum mean throughput of this switch is close to 100%, which is higher than the 45% maximum throughput of the single buffered Banyan network of Jenq in Part 2, for instance. Broadcasting can also be simply implemented in this network.

Kim and Leon-Garcia also present a queueing analysis of their network. It is subdivided into a different discrete-time Markov chain for each of the first-, the ith-, and the last-stage distributor and network output buffers. The first two chains can be solved through their global balance equations [1-4] and the last chain can be solved recursively. Kim and Leon-Garcia find that beyond a certain network output buffer size, packet-loss probability is not improved. The network has a small delay for loads below 90%.

The next paper investigates switching in multiclass environment (see also the paper by Tridandapani and Meditch in Part 4). There are two schools of thought regarding differentiating among traffic by traffic class. One is that it is simplest to design a network to treat all traffic equally, that is, as a single class. Quality of service is then set to meet the requirements of the most demanding service(s) to use the network. The other school of thought holds that if each class of traffic is treated so that it is guaranteed only to meet it's own quality of service requirements, then efficiencies in network design and operation will result.

This paper is entitled "Analysis, Control, and Design of Crossbar and Banyan-Based Broadband Packet Switches for Integrated Traffic," and it is authored by S. Z. Shaikh, M. Schwartz, and T. H. Szymanski. In it the authors develop a set of equations that can be solved iteratively to model a Banyan network with multiclass traffic. Switching elements are output buffered. A preemptive-like priority scheme is used, in which a higher-priority packet can preempt the servicing of a lower-class packet (see [5] for a treatment of nonpreemptive priority).

Shaikh and his coauthors discuss a simple modification of the network that allows a fair division of throughput between classes. They also discuss means to control the time delay experienced by each class. Finally, a procedure for the computation of the delay distribution for each class is explained.

One interconnection topology that has received a great deal of attention for multiprocessors is the hypercube. An N-dimensional hypercube network is based on an N-dimensional cubic structure. The next paper examines hypercube networks with deflection routing. It is entitled "An Analysis of 'Hot Potato' Routing in a Fiber Optic Packet Switched Hypercube" and is authored by T. Szymanski.

Two architectural implementations, one with fixed wavelength lasers and one with tunable lasers, are suggested in this paper. Also described are several deflection algorithms. An analysis is presented that results in an iterative numerical procedure to calculate the performance. Under it and for uniform traffic, the state of the network is modeled through the state of a single node.

Szymanski's paper goes on to describe routing algorithms where packets closer to their destination are given priority (see the paper by Krishna and Hajek in the previous part). Finally, the evaluation of prioritized multiclass traffic is discussed.

Szymanski finds that the use of deflection routing in a hypercube leads to higher throughputs than the use of shortest-path routing, buffering, and no flow control. Note that a different treatment of deflection routing in hypercube networks appears in [10].

A very practical problem in high-speed packet-switch design is how to scale current designs into the large-capacity versions needed for future broadband networks (see the paper by Karol and I in Part 5). One estimate [6] is that a central office would require switch fabrics with 10,000 high-speed ports and 100,000 terminal connections. One answer to this question appears in "A Modular Architecture for Very Large Packet Switches," by T. T. Lee. In it he describes a switch architecture that is a synthesis of knockout and Banyan-Batcher architectures.

Lee proposes a switch architecture where a switch comprises a number of switch modules and output multiplexers. The switch modules are nonblocking and self-routing. They comprise Batcher sorting networks, binary routing trees, and Banyan routing networks. Among advantages of this approach are independence of switch modules, preservation of nonblocking and self-routing properties, incremental port expansion, and the potential for Terabit switch capacity.

Lee's paper discusses a unified approach for handling contention resolution and output space extension. It concludes with a performance analysis for mean throughput, mean delay, and overflow probability. Close to 100% mean throughput can be achieved.

One of the challenges of designing future broadband networks is evaluating their performance under new and realistic traffic models. One such type of traffic is bursty traffic. This issue is examined in "Stochastic Models for ATM Switching Networks," by A. Descloux. In it the author provides two statistical models of bursty traffic in a high-speed switching environment.

The first model is for a single Banyan network, where cells in a burst occupy consecutive cells (packet train). It leads to a set of difference equations that are solved recursively. The second model involves multiple Banyan networks where cells in a burst are randomly spaced. An iterative solution is developed.

Descloux's paper also discusses cell loss in the Bellcore Sunshine design [7, 8]. Specifically, recirculating trap loss and output queue loss are considered. The analysis of this paper has been used to verify a simulation study of the Sunshine switch. The results indicate that minimal cell spacing and small load variations have a great impact on performance. Buffer requirements also are sensitive to changes in burst size.

Although many initial studies of high-speed switching networks have dealt with uniform traffic loadings, realistic traffic scenarios can be markedly nonuniform. For instance, if an output address (or addresses) is selected more often than other output addresses, congestion can develop in the switching network under this "hot spot" phenomena. This is studied in "On Nonuniform Packet Switched Delta Networks and the Hot Spot Effect," by P. G. Harrison. In it the author examines delta networks (a subclass of Banyan networks) with 2×2 switching elements.

Harrison obtains expressions for the mean delay and delay density for packets traveling through the network. This is based on a closed queueing network model of the interconnection network. The expressions are functions of network normalization constants such as the ones used in the convolution algorithm [4, 9].

These results are used in the latter part of the paper to examine quantitatively the presence of a hot spot in the network. Harrison finds that with the presence of a hot spot ". . . even paths to the hot spot do not saturate until the very last stage, the hot spot itself." He suggests that this implies that large buffers are needed only at the last stage—buffers adjacent to hot spots.

The last paper in this part discusses the analysis of packet-train inputs to a statistical multiplexer. It is entitled "Performance of Statistical Multiplexers with Finite Number of Inputs and Train Arrivals" and is written by Y. Xiong and H. Bruneel.

Xiong and Bruneel review the existing literature on correlated arrivals to statistical multiplexers. In this paper they examine a discrete-time model where the number of consecutive multipacket "messages" on a link and the number of slots in an idle period are geometrically distributed. Individual messages comprise a constant number of packets.

Xiong and Bruneel use generating function techniques [1] to develop a functional equation that can lead to the calculation of mean buffer occupancy and higher moments. However, due to it's computational complexity, this technique is limited to small message sizes. Therefore, an upper bound to the tail distribution of the buffer occupancy is derived. This bound agrees very well with simulation results. The authors find that as message size (and thus correlation) increases, the probability of the queue length exceeding a certain size also increases.

REFERENCES

[1] L. Kleinrock, *Queueing Systems, Vol. I: Theory*, New York: John Wiley, 1975.

[2] R. B. Cooper, *Introduction to Queueing Theory*, New York: North-Holland Publishing Co., 1981.

[3] A. Leon Garcia, *Probability and Random Processes for Electrical Engineering*, Reading, Mass.: Addison-Wesley Publishing Co., 1989.

[4] T. G. Robertazzi, *Computer Networks and Systems: Queueing Theory and Performance Evaluation*, New York: Springer-Verlag, 1990.

[5] S. Z. Shaikh, M. Schwartz and T. H. Szymanski, "Performance, analysis and design of Banyan network–based broadband packet switches for integrated traffic," *Proc. of IEEE Globecom '89*, pp. 1154–1158.

[6] T. T. Lee, "A modular architecture for very large packet switches,"

IEEE Trans. on Commun., vol. 38, no. 7, pp. 1097–1106, July 1990. Appears in this part.

[7] C. Day, J. Giacopelli, and J. Hickey, "Applications of self-routing switches to LATA fiber optic networks," *Proc. Int. Switching Symposium*, Phoenix, Arizona, March 1987.

[8] J. N. Giacopelli, M. Littlewood, and W. D. Sincoskie, "The sunshine switch," *Proc. Int. Switching Symposium '90*, May–June, 1990.

[9] S. C. Bruell and G. Balbo, *Computational Algorithms for Closed Queueing Networks*, New York: North Holland, 1980.

[10] A. G. Greenberg and B. Hajek, "Deflection routing in hypercube networks," *IEEE Trans. Commun.*, vol. 40, no. 6, pp. 1070–1081, June 1992.

Multipath Interconnection: A Technique for Reducing Congestion Within Fast Packet Switching Fabrics

GARY J. ANIDO, MEMBER, IEEE, AND ANTHONY W. SEETO, MEMBER, IEEE

Abstract—Multistage interconnection networks such as the banyan are well suited to multiprocessor and fast packet communication systems. However, the banyan interconnection is prone to internal link congestion resulting in a blocking switch architecture. Several solutions have already been proposed, and implemented, to reduce the severity of link congestion. In general, these solutions offer packets of multiplicity of paths which tend to increase packet delay variability and allow delivery of out-of-sequence packets. This, in turn, may lead to an increase in end-to-end protocol complexity, particularly in the case of real-time services. A solution, called multipath interconnection, is proposed to overcome this difficulty. Multiple (i.e., alternate) paths are provided and one is selected at call setup time. Subsequent packets belonging to the call are constrained to follow the selected path. A number of path selection strategies are presented.

Implementation costs, in terms of relative processing and memory requirements, of each algorithm are compared. The ability of the algorithms to evenly distribute traffic across the internal links is evaluated by means of a simulation study. Queueing behavior is estimated using an approximate model of a switch path. The results indicate that the more able a path selection algorithm is to distribute switch traffic, the greater the limiting throughput. The model is also used to investigate the effect of parameters, such as switch size and buffering capacity, on switch performance.

I. Introduction

MULTISTAGE interconnection networks have found extensive use in multiprocessor applications [1]–[3] and communication systems [4], [5]. Banyan networks, and other members of a class of topologically equivalent multistage networks [6], have a number of desirable characteristics such as self-routing, suitability for VLSI implementation, and path uniqueness [7]. The latter means that connecting each inlet and outlet pair of a banyan network is a unique path. When used for fast packet switching [8], [9] this characteristic ensures that individual packets belonging to the same logical connection (or call) will follow the same route. As a result, out-of-sequence packets are prevented and intracall packet delay variability is expected to be less than would be the case if packets were routed independently through the network. This, in turn, leads to the design of simpler protocols for the syn-chronization of real-time services such as voice and video [10].

The path uniqueness of the banyan interconnection has the disadvantage of resulting in a blocking switch. Since a number of paths may share common links within the network, simultaneous connections of more than one inlet and outlet pair may result in conflicts in the use of these internal links [7]. In general, and for fairly uniform traffic patterns, these conflicts can be resolved by providing a small amount of buffering (typically, a single packet buffer) per switching element [11]. Such a single-buffered banyan network was analyzed in [12] under the assumptions of fixed-length packets and uniform traffic pattern. Uniform traffic strictly implies that the inlet and outlet used by each packet are independently chosen from a uniform distribution. It was found that the limiting throughput available is around 0.45. In a different study [13], a single-buffered banyan network was subjected to a somewhat nonuniform traffic pattern. In addition to a background of uniform traffic, a single point-to-point connection was established (that is, all the packets arriving on one particular inlet were directed to a given outlet). It was found that once the point-to-point traffic exceeded a throughput of about 0.45 the throughput of traffic sharing links with the point-to-point traffic fell towards zero. A simulation study of a packet switch based on the banyan network has shown that the limiting throughput can be increased from 0.45 to about 0.65 as the number of buffers per switching element is increased from 1 to 4 [14].

The use of greater packet buffering will increase not only the maximum throughput of the network, but also the delay incurred [11]. However, if the path followed by a packet through the network could be chosen so as to avoid links which are becoming congested, it may be possible to maintain high throughput and low delays. One means of doing so is to provide multiple paths within the interconnection network itself [15]. Another approach is to precede the interconnection network with a further network, the purpose of which is to establish conditions under which intercall packet conflicts are, in some sense, minimized. The use of a *sorting network* placed before an Omega (or banyan) network can be used to eliminate internal conflicts provided there are no repeated destinations amongst the input packets [4], [5]. The number of stages required by an N-input sorter, such as the Batcher network [16], requires $(\log_2 N + 1)/2$ times as many switching

Manuscript received September 1, 1987; revised June 8, 1988. This paper was presented in part at the Second Australian Teletraffic Research Seminar, University of Adelaide, Australia, November 30–December 1, 1987; and the International Zurich Seminar on Digital Communications, ETH, Zurich, Switzerland, March 8–10, 1988.

The authors are with Systems Development, Overseas Telecommunications Commission, Sydney, 2001 Australia.

IEEE Log Number 8824390.

Reprinted from *IEEE J. Selected Areas Commun.*, vol. 6, no. 9, pp. 1480–1488, December 1988.

stages than the corresponding banyan network. Instead of a presorter, the use of a randomizing, or *distribution*, network has been proposed in [8]. The purpose of this network is to distribute packets evenly throughout the succeeding banyan network. The simulation study cited earlier [14] has shown that such a distribution network is effective in reducing congestion resulting from nonuniform traffic patterns arising from *communities of interest*.

Each of the methods discussed in the preceding paragraph present individual packets with a multiplicity of paths. As mentioned earlier, this may lead to problems with the provision of real-time services owing to the increase in intracall packet delay variance due to the interaction with intercall traffic. A preferred solution may be to make the multiplicity of paths available for random selection at call setup time, yet once a path has been selected, all packets belonging to the call are constrained to follow the same path. Such an approach has been proposed in [17] as the basis for a fast packet switching exchange.

In this paper, we present a generalization of the distribution network approach. In order to avoid the problem of intracall delay variance, packets belonging to a given call are constrained to follow the same path through the switch. Path selection is at call setup and based on an algorithm designed to minimize, in some sense, the cost of the path. A number of path selection algorithms have been identified and are evaluated in this paper. In Section II, the operation of the switching network is described. The path selection algorithms are discussed in Section III together with a comparison of relative processing and memory requirements. An approximation technique using hybrid-simulation is presented in Section IV. Using this technique each path through the switch is modeled as a tandem connection of single server queues with finite buffer space. The utilization of each server, which depends on the traffic pattern and the method used to allocate paths, is determined by simulation. The results of this technique, such as the mean queue length and blocking probability as functions of mean applied load and buffering capacity, are presented in Section V.

II. THE MULTIPATH INTERCONNECTION NETWORK

A fast packet switch based on a multipath interconnection network is shown in Fig. 1. Overall switch operation is controlled by means of a *connection control processor* (CCP) which performs connection control and switch maintenance functions. The CCP, however, performs no per packet processing. *High speed trunk* (HST) lines provide either user access (via suitable network terminations) or access to other switches. Each HST is terminated by a *trunk controller* (TC) which performs link level protocol functions. Call setup packets are forwarded to the CCP which determines the routing information for that call.

The *multipath interconnection network* (MIN) in an ($N \times N$) switching fabric based on the use of self-routing baseline networks [6] with S stages of switching elements. Fig. 2 shows an example of an (8×8) MIN. The

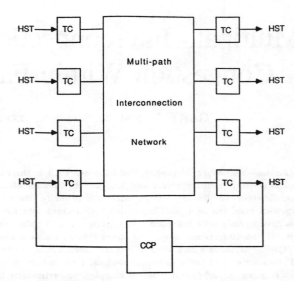

Fig. 1. Fast packet switch configuration. High speed trunk (HST) links terminate on trunk controllers (TC) which perform link level protocol functions. Multipath interconnection network (MIN) performs the switching function. Paths through the MIN are selected by the connection control processor (CCP).

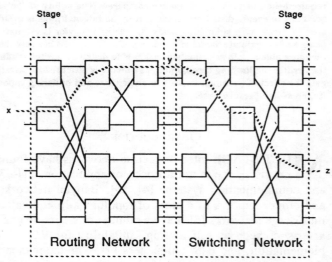

Fig. 2. An (8×8) multipath interconnection network. The dashed line illustrates a path between inlet x and outlet z via intermediate port y.

MIN consists of two baseline networks, the *routing network* (RN) and the *switching network* (SN). The addition of the RN effectively increases the number of paths available to a call from 1 to N. Outputs of the switching elements in one stage are connected to the inputs of switching elements in the next by means of *links*. Links at the input to stage 1 are called *inlets*; the links between the routing and switching networks are called *intermediate ports*, or simply *ports*; and links at the output of stage S are called *outlets*. So that we may consistently refer to links at the input of a stage, we regard outlets as being inputs to a null stage $S + 1$. Both networks consist of $\log_2 N$ stages of $N/2$ switching elements. Switching elements are (2×2) crossbars with buffer space for one or more packets at each input. An element at stage i will switch a packet up or down depending on the ith bit of the header. A free buffer must be available at the next stage to allow the packet to move forward. Flow between stages is controlled by means of hardware handshake lines between corresponding switching elements. In Fig. 2 the networks are shown as mirror images for conceptual simplicity; in practice, both RN and SN could be identical. This of course would simplify implementation.

The purpose of the RN is to route a packet to one of the N inlets of the SN. In its turn, the SN switches the packet to the required outlet. Since every MIN inlet has access to any of the RN outlets, each of which have access to any of the SN outlets, there exists N paths between each MIN inlet and outlet pair. In fact, the topology of the MIN is closely related to that of the Benes network which is a rearrangeable nonblocking network [7]. The CCP therefore has a choice of N possible paths when determining the route to be followed by a particular call.

III. PATH SELECTION ALGORITHM

In this section, we describe four possible path selection algorithms. References to processing and memory requirements assume a single processor CCP with random access memory. Minimum storage is taken to be $O(\log_2 N)$ per word and time requirements are estimated in terms of arithmetic and logic operations on such words.

1) Least-Cost Method: Consider the triplet (x, y, z) where x is an inlet, y is a port, and z is an outlet. Such a triplet uniquely defines a path through the MIN. Associated with the path is a set of links $R = \{(i, j)\}$ (indicated by the dashed line in Fig. 2) where each (i, j) is the jth link of stage i. In particular, $(1, j) = x$, $(S/2 + 1, j) = y$, and $(S + 1, j) = z$. Define the load matrix $\tilde{L}$ by

$$\tilde{L} = [L_{i,j}] \quad i = 1, 2, \cdots S + 1; \quad j = 1, 2, \cdots, N \tag{1}$$

where each element $L_{i,j}$ is the number of established calls on the jth link of stage i. In this method, also called the *optimal method*, the cost of a path is defined as the sum of the load on each link in the associated link set R. The CCP selects the path with the lowest cost. The method entails the following operations.

1) The set of links R associated with each of the N possible paths connecting the inlet and outlet pair is determined.

2) The cost is evaluated for each of the N possible paths.

3) The lowest cost path (with link set R_{opt}) is selected.

4) Each of the $L_{i,j}$ in R_{opt} are incremented.

5) When a call terminates each of the $L_{i,j}$, associated with that call, are decremented.

As a minimum, the least-cost method has a memory requirement of $O(N \log_2 N)$ due to the need to store the load matrix $\tilde{L}$. The processing required to determine path cost for each of the N paths is also $O(N \log_2 N)$. To this must be added the processing and/or memory needed to obtain the link set at each path. Hence, for large N this method could be expensive to implement. On the other hand, since the complete state of the MIN is taken into account, it is expected that the method makes an optimal path selection.

2) Binary Method: Given an inlet and outlet pair (x, z), we can identify each of the N possible paths with a set of ports $P = \{y\}$. Since the intermediate port y uniquely identifies each of the path choices, it is tempting to select a path based solely on the state of these ports. Define a load array $[L_y]$ where each L_y is the number of calls applied to the yth port. Hence, for this method, we define the cost of a path through y as L_y. As in the previous method, the path with the lowest cost is selected. The method also attempts to take into account the structure of the MIN, if not its state, by using a binary search for the port with the lowest load. Implied in the method are the following steps.

1) Partition the set of intermediate ports P into two halves, P_1 containing the first $N/2$ ports and P_2 the remaining ports.

2) Calculate the sum of the link loads in each partition and select the partition with the lowest total cost.

3) Repeat the procedure until each partition contains a single port. The partition with the lesser load then identifies the selected path.

Memory requirements for the binary method are $O(N)$. Processing requirements depend primarily on the partitioning process. The set P is partitioned through at total of $\log_2 N$ steps. At the mth step, there are two partitions each containing $N/2^m$ ports whose loads must be summed and compared. Hence, the number of arithmetic and logic operations required is

$$A = 2 \sum_{m=1}^{\log_2 N} \frac{N}{2^m} = 2N \left(1 - \frac{1}{N}\right) = O(N). \tag{2}$$

Both memory and processing requirements are therefore $O(N)$ in contrast to $O(N \log_2 N)$ for the optimal method.

3) Linear Method: This method is essentially the same as the binary method with the exception that a simple linear search of the set of intermediate ports is used. Unlike either of the previous two methods, the structure of the MIN is ignored, and the sole basis for path selection is the load on the intermediate ports. As for the binary method, the memory and processing requirements are $O(N)$, although actual processing burden would be somewhat less for the linear method due to the simpler search algorithm.

4) Random Method: In this case, all state information is ignored and the path is selected by a uniformly random choice of one of the N intermediate ports. With memory and processing requirements being $O(1)$ this method is clearly the least expensive to implement and forms the basis of a number of earlier works. The distribution network proposed in [8] essentially makes a random path selection for each packet, while the randomization network of [17] establishes paths randomly for each call.

IV. Model and Analytic Approach

This section outlines the model used, and approach taken, to evaluate the relative efficacies of the path selection strategies considered in the previous section. The performance measures of interest are the expected queue length and the maximum throughput through the MIN. In the case of infinite buffering, the maximum throughput is equal to the applied load leading to an unbounded expected queue length. In the case of finite buffering, the blocking probability is also of interest. In addition to considering the different route selection strategies, we wish to study the effect of a number of switch parameters such as the number of stages in the switch and the amount of buffering space available at each stage.

A Markov chain could be used to model the system. However, the number of states involved in a Markov chain representation of even a single banyan matrix is $k^{(N \log_2 N)/2}$ where k is the number of states of a switching element [12]. Since we wish to consider, amongst other parameters, the effect of the buffering capacity per element on queueing behavior, k itself can become large. The computational effort required to handle the resulting very large state-space is clearly unfeasible. Therefore, a hybrid analytic-simulation method has been employed.

In the approach adopted to analyze the performance of the MIN, we decouple the per call path selection process from the per packet switching process. A simulation is first used to determine how, at steady state, traffic is distributed through the switch as a result of the path selection algorithms. Next, a simple analytic model of a switch path, based on a chain of finite waiting place single-server queues, is used to estimate performance metrics such as expected queue length and blocking probability.

A. The Switch Simulator

The first step of the method is to obtain moments of the distribution of the traffic intensities on the links of the MIN as a result of each path selection algorithm. This is produced by a switch simulator. Inputs to the simulator are the size of the user population, the traffic pattern, and the route selection method. Outputs from the simulator are the mean and variance of the traffic intensity on the internal links.

Traffic is modeled using a similar approach to that employed in [18]. The call arrival process is assumed to be Poisson with rate ξ and exponentially distributed call holding times with mean η^{-1}. Let n_k, $k = 1, 2, \cdots$, be the number of calls in progress at transition k, and assume that $n_0 = L$. Then the number of calls in progress at the next transition will increase to $L + 1$ with probability

$$p_{L, L+1} = \begin{cases} \dfrac{\xi}{\xi + L\eta}; & L = 0, 1, \cdots, M - 1 \\ 0, & \text{otherwise} \end{cases} \tag{3}$$

where M is the size of the user population. This is just the probability that, at the next transition of n_k, a new call arrives, given L calls currently in progress. Similarly, the probability that, at the next transition, an existing call terminates is $p_{L, L-1} = 1 - p_{L, L+1}$. The average packet arrival rate while $n_k = L$ is taken to be $\lambda = rL$, r^{-1} being the average packet interarrival time per call.

A newly arriving call has its input chosen, with equal probability, from one of the N inputs to the MIN. The corresponding output from the MIN depends on the nature of the traffic pattern selected. Three traffic patterns are considered as follows.

1) Uniform Pattern: The output is chosen independently, and equiprobably, from one of the N outputs.

2) Horizontal Pattern: A call arriving at input i, $i = 2, 3, \cdots N - 1$ has a corresponding output chosen, with equal probability, from $(i - 1, i, i + 1)$. Calls arriving on either the first $i = 1$ or last $i = N$ inputs have outputs chosen, equiprobably, from $(1, 2, N)$ or $(1, N - 1, N)$, respectively. This pattern is expected to result in a traffic distribution within the switch which is somewhat rougher than would be obtained with the uniform pattern.

3) Diagonal Pattern: In this case, a call arriving on input i, $i = 2, 3, \cdots N - 1$ has an output chosen, equiprobably, from $(N - i, N - i + 1, N - i + 2)$. Calls arriving on inputs $i = 1$ or $i = N$ have outputs chosen, equiprobably, from $(1, N - 1, N)$ or $(1, 2, N)$, respectively. The traffic distribution within the switch obtained by this traffic pattern would be expected to be rougher than that resulting from either of the above traffic patterns.

When a new call arrives (with inlet and outlet determined by the traffic pattern) a path selection algorithm is used to determine the appropriate path. Once the path for a call has been decided, each element of the load matrix $\tilde{L}$ associated with a link used by the path is incremented. In addition, the call is allocated a circuit number to which the path link set is bound. When a call terminates, one of the currently allocated circuit numbers is selected at random. The associated link set is then used to decrement the corresponding elements of the traffic matrix $\tilde{L}$.

The output from the switch simulator are moments of the distribution of the traffic intensities. In particular, the mean link intensity μ_λ can be written as

$$\mu_\lambda = \frac{r}{N(S + 1)} \sum_{i=1}^{S+1} \sum_{j=1}^{N} L_{i,j} \tag{4}$$

and the variance σ_λ^2 is given by

$$\sigma_\lambda^2 = \frac{r^2}{N(S + 1)} \sum_{i=1}^{S+1} \sum_{j=1}^{N} \left(L_{i,j} - \frac{\mu_\lambda}{r} \right)^2 \tag{5}$$

given that each call generates packets at an average rate r packets per second.

In Figs. 3–5, we plot normalized link intensity variance $v = \sigma_\lambda^2 / C$ as a function of normalized mean link intensity $p = \mu_\lambda / C$. Normalization is with respect to the maximum link capacity C. Fig. 3 assumes uniform traffic, Fig. 4 horizontal traffic, and Fig. 5 uses a diagonal pattern. An 8×8 MIN is assumed with link capacity equal to $20r$. Traffic is generated by a population of users each produc-

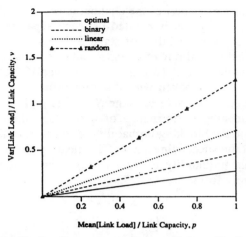

Fig. 3. Normalized variance v versus mean p of the traffic distribution through an (8×8) MIN showing the effect of each path selection algorithm with uniform traffic pattern.

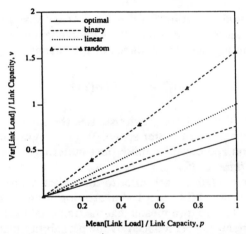

Fig. 4. Normalized variance v versus mean p of the traffic distribution through an (8×8) MIN showing the effect of each path selection algorithm with horizontal traffic pattern.

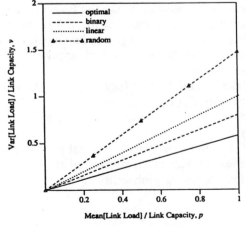

Fig. 5. Normalized variance v versus mean p of the traffic distribution through an (8×8) MIN showing the effect of each path selection algorithm with diagonal traffic pattern.

ing 0.5 Erlangs of traffic. The size of the user population is variable from 0 to 160. In each case, the four path selection algorithms are considered. Three observations can be made in the light of these results. First, for a given selection algorithm and traffic pattern σ_λ^2 is linearly proportional to μ_λ. Second, the constant of proportionality increases as the traffic pattern becomes less uniform. Third, optimal path selection results in consistently the lowest σ_λ^2. This is followed by binary and linear path selection, with random selection yielding the highest σ_λ^2. It is worth noting that, in all cases, binary is not substantially worse than optimal. In view of the $O(\log_2 N)$ reduction in memory and processing costs, the binary selection algorithm may be a useful compromise between implementation cost and variance reduction.

B. Switch Channel Model

The switch simulation discussed in Section IV-A allows us to determine how traffic is distributed across the links of the MIN as a result of the path selection algorithms described in Section III. We now wish to estimate the per packet queueing performance. Two key factors determining the queueing behavior of the MIN are the interaction between packets of different calls sharing common links, and the "back-pressure" exerted on the flow of packets by lack of free buffer space in successive stages. We wish to formulate a simple queueing model of a path through the MIN which accounts for these factors. The set of links R associated with the triplet (x, y, z) will be referred to as a *switch channel*, or simply *channel*. Traffic generated by the users of a switch channel, called *channel traffic*, enters the system at the first stage and proceeds, stage by stage, to the final stage of the channel. This procession depends on the availability of free buffer space at each stage. In addition to channel traffic, individual links of each channel will be offered traffic arising from other channels which share links with the switch channel. This traffic is said to be composed of *transit packets*.

The average arrival rate of channel traffic is λ_0, while the transit traffic sharing the link between the $(i - 1)$th stage and the ith stage has an arrival rate λ_i. Hence, the total traffic applied to this ith link has an arrival rate $\lambda_0 + \lambda_i$. The following assumptions are made to enable the formulation of a simple model of a switch channel.

1) Each packet is assumed to have a length drawn independently from a negative exponential distribution. There are essentially three reasons for this assumption: 1) the resulting Markovian service time distribution greatly simplifies the model; 2) almost without exception, previous work has dealt with the fixed length packet case (for example, [11]–[14]) with variable length packets not being investigated; 3) the assumption of exponentially distributed packets lengths is conservative in comparison to fixed length packets, and hence safe to assume.

2) The arrival process of the transit packets at each stage is independent of the arrival process of the channel packets. This assumption is based on the observation that the set of inlets with a path to one of the links of a $(2 \times$

2) element is disjoint from the corresponding set with a path to the other link of the same element [11]. The independence assumption will be good in those cases where transit packets share only one link in common with the switch channel, and will become less accurate as the number of shared links increase. However, since the path selection algorithms attempt to distribute traffic throughout the MIN, it is reasonable to suppose that sharing of links between switch channels is in some sense minimized.

3) The arrival process at stage i is assumed to be Poisson with rate $\lambda_0 + \lambda_i$ and the average service rate at each stage is μ. This assumption is motivated by the Palm–Khintchine theorem which guarantees that the sum of n independent renewal processes obeys Poisson statistics provided n is sufficiently large [19].

4) Each link within the MIN is offered traffic whose mean arrival rate is independently drawn from a normal distribution with mean μ_λ and variance σ_λ^2.

Under these assumptions each switch channel is statistically identical and can be modeled by a tandem network of $M/M/1$ queues with either finite or infinite buffering at each stage.

The buffer space at stage i is K_i. Three cases will be considered. In the first, an infinite buffer is assumed to exist at each stage. Second, finite buffers are assumed at every stage except the first, which incorporates the input FIFO of the trunk controller and is assumed infinite. Due to hardware flow control exerted between stages, no packets are lost when the buffer in any stage is full. Hence, in both of these cases, the MIN forms a pure delay system. In the third case, a finite buffer exists at each stage including the first, although it is not necessarily of the same size as the other buffers. In this case, the MIN acts as a loss system due to the nonzero blocking probability of the first stage buffer.

1) *Infinite Buffers at Every Stage:* With unlimited buffer capacity at each stage, the service of a packet at stage i is independent of the state of the queue at stage $(i + 1)$. The number of packets that a channel packet can expect to find queued at stage i depends only on the traffic arrival rate $\lambda_0 + \lambda_i$. Hence, the total mean queue length of the switch channel can be written as

$$Q = \sum_{i=1}^{s} \frac{\rho_i}{1 - \rho_i} \qquad (6)$$

where $\rho_i = (\lambda_0 + \lambda_i)/\mu$ is the intensity at stage i. The mean queue length for the entire MIN can be obtained by averaging over all active switch channels.

2) *Infinite Input Buffer, Finite Stage Buffers:* In this case, a buffer of unlimited capacity is still assumed at the input to the channel (at stage 1) but each subsequent stage i is assumed to have finite capacity for K_i, $i = 2, 3 \cdots$ S packets, including the packet being served. Hence, each stage i of the channel model is modeled as a finite $M/M/1/K_i$ queue [20]. With the hardware flow control described earlier, a packet at the head of the queue in stage i cannot be served if the queue in stage $i + 1$ is blocked.

Given that the outlet of the channel is never blocked (that is, no flow control is exerted between separate MIN's), the blocking probability of stage S depends only on the combined arrival rate of channel and transit packets $\lambda_0 + \lambda_i$ and is denoted $P_b(S)$. Since the server of the previous stage will be blocked with this probability, the effective service rate available to stage $S - 1$ is $\mu(1 - P_b(S))$. The queueing performance of each stage therefore depends on the blocking probability, and hence traffic intensity, of the succeeding stage. The mean queue length and blocking probability of the ith stage can be determined iteratively from

$$\rho_i = \frac{\lambda_0 + \lambda_i}{\mu(1 - P_b(i + 1))} \qquad (7a)$$

$$E[q(i)] = \frac{\rho_i}{1 - \rho_i}\left(1 - (K_i + 1) P_b(i)\right) \qquad (7b)$$

$$P_b(i) = \frac{\rho_i^{K_i}(1 - \rho_i)}{1 - \rho_i^{K_i+1}} \qquad (7c)$$

with initial condition $P_b(S + 1) = 0$ and given $K_1 = \infty$. The mean queue length of the switch channel is obtained by summing over the S stages of the channel

$$Q = \sum_{i=1}^{S} E[q(i)]. \qquad (8)$$

In both, this and the previous cases the inlet is assumed to have unlimited buffer space. We now remove this assumption and consider finite inlet buffering.

3) *Finite Buffers at Every Stage:* With finite buffer space provided at each inlet to the switch there is now a nonzero probability that traffic arriving for a channel will be lost. This in turn means that the traffic intensity at each subsequent stage is reduced by the probability that the first stage is full. Hence, by replacing (7a) by

$$\rho_i = \frac{\lambda_0 + \lambda_i}{\mu} \frac{1 - P_b(1)}{1 - P_b(i + 1)}; \quad i = 2, 3, \cdots S \qquad (9)$$

the queueing behavior of the channel can be determined. (Note that we are assuming that the Poisson nature of the arrival process has not been destroyed.) However, (9) requires $P_b(1)$ to be known prior to its calculation by (7c). This problem can be overcome by applying an iterative procedure as follows.

1) Make an initial guess (say, zero) for the inlet blocking probability $P_b(1)$.

2) Using (9) in (7), determine $P_b(i)$ and $E[q(i)]$ for each stage i.

3) With new value of $P_b(1)$ repeat previous step.

4) Repeat previous two steps until the value obtained for the inlet blocking probability converges.

V. RESULTS

In Fig. 6 we show the mean queue length as a function of normalized mean link load $p = \mu_\lambda/C$ with normalized

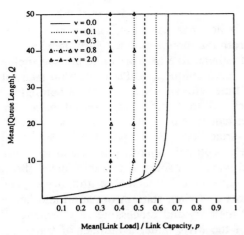

Fig. 6. Mean queue length Q of a switch channel versus mean load p with variance v as a parameter; number of stages $S = 6$, stage 1 with infinite buffer, remaining stages are singly buffered.

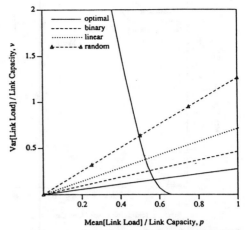

Fig. 7. Variance v versus mean link load p for uniform traffic (as in Fig. 3) superimposed on maximum throughput versus variance (derived from Fig. 6).

link load variance $v = \sigma_\lambda^2/C$ as a parameter. The MIN is assumed to have six stages. Each stage, with the exception of the first stage, has buffer space for one packet. The first stage has unlimited buffer capacity. Under light loads the queue length is independent of link variance. However, as the load increases the queue behavior becomes heavily dependent on the link variance with queue performance improving as the link variance is reduced. The lower bound on mean queue length occurs, as might be expected, when $v = 0$. We wish to relate this behavior back to the path selection algorithms. The maximum throughput as a function of link variance is plotted in Fig. 7. Superimposed on this is the variance versus mean load information earlier shown in Fig. 6 for the uniform traffic pattern case. Similar figures could be drawn for the other traffic patterns. From the results in Fig. 7, we see that maximum throughput occurs at a mean load approaching 0.5 in the case of random path selection, with throughput increasing to about 0.6 for optimal selection.

The limiting throughput in Fig. 7 occurs at a link load of about 0.65, and requires vanishingly small link variance. The path selection algorithm needed to approach this limit would doubtless make use of the rearrangeable nature of the Benes network. In other words, following each call setup or termination, paths could be reselected to jointly minimize the cost, hence load variance, of existing calls. The remaining results will be developed assuming limiting throughput conditions.

The MIN has twice as many stages as a switch using a single baseline network. We would therefore like to see if there is a penalty associated with the increased number of stages. In Fig. 8, the mean queue length is shown with the number of stages as a parameter. Again, we assume each stage is singly buffered, except stage one which has an infinite buffer. As might be expected, the maximum throughput decreases as the number of stages is increased. However, throughput rapidly reaches its limiting value with about 10 stages. This corresponds to only a switch size of (32×32). Hence, for any reasonable switch size,

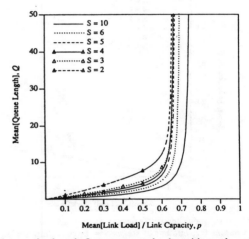

Fig. 8. Mean queue length Q versus mean load p with number of stages S as a parameter; infinite input buffer, remaining stages singly buffered.

the increase in the number of stages due to the use of a MIN has no effect on maximum throughput.

So far the MIN has been singly buffered. We now consider the effect of increasing the buffer capacity of each stage. As before, the first stage has infinite capacity, and the MIN has six stages. Mean queue length is shown in Fig. 9 with buffer size as the parameter. While increasing the buffer capacity results in increased limiting throughput, there is little incentive to using a buffer space greater than about 5. Beyond this throughput improvements become small, and to take advantage of the increased throughput significantly greater queues, and delays, are incurred. A similar conclusion was reached in [21].

The effect of finite input buffering at stage 1 is shown in Figs. 10 and 11. Fig. 10 shows the expected queue length versus mean link load with the input stage buffer size as a parameter. The MIN is assumed to have, as before, six stages, with one buffer per stage. For finite input buffer capacity the expected queue length, and hence delay, remains bounded even for high link loads. The cor-

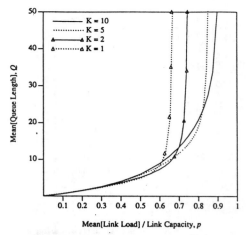

Fig. 9. Mean queue length Q versus mean load p with buffer size per stage K as a parameter; number of stages $S = 6$, infinite input buffer.

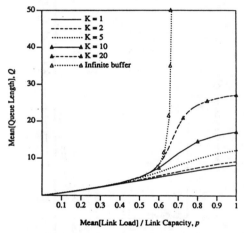

Fig. 10. Mean queue length Q versus mean load p with input buffer size K as a parameter; number of stages $S = 6$, remaining stages singly buffered.

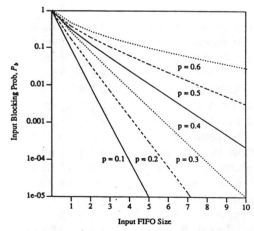

Fig. 11. Blocking probability P_b at input versus input buffer size with mean load p as a parameter.

responding blocking probability is displayed in Fig. 11. An input buffer capacity of at least 10 packets would be required for blocking probability less than, say, 10^{-3}.

VI. CONCLUSIONS

A method has been presented which, by means of a multipath interconnection network, increases the throughput of banyan-like networks without sacrificing the property of path uniqueness. Path selection takes place at call setup time with subsequent packets belonging to the call constrained to follow the selected path. This ensures packet sequence integrity and minimal intracall packet delay variability. Four possible selection algorithms (namely, optimal, binary, linear, and random) were presented which, in some sense, minimized the cost associated with each call. Each algorithm made use of different amounts of switch state information and hence have different memory and processing requirements. The operation of the MIN under a number of traffic patterns was then simulated to determine the effect of the path selection algorithms. It was found that the ability of the path selection algorithm to distribute traffic throughout the switch is inversely related to the cost of implementing the algorithm. The binary algorithm appeared to be a useful compromise.

Next, a queueing model of a path through the MIN was formulated in order to estimate queueing behavior in terms of expected queue lengths and blocking probability. It was found that maximum throughput increases as the variance of the link load distribution decreases. The increase in throughput due to the optimal algorithm, compared to the random algorithm, was about 10 percent. This improvement increases to about 15 percent in the limiting case (corresponding to zero variance). The queueing model was also used to determine the effect of the number of stages in the MIN and the amount of buffering per stage.

The path selection algorithms considered in the current work can be described as static in the sense that, once a path is allocated to a call, the call retains that path until termination. However, the MIN can be operated as a rearrangeable nonblocking switch. Dynamic path selection algorithms, which reallocate paths to existing calls following each call setup or termination, could doubtless be formulated to take advantage of the rearrangeable nature of the MIN. The present work does not, however, address this issue.

ACKNOWLEDGMENT

The authors would like to thank the anonymous reviewers for their helpful suggestions. The authors would also like to thank the Manager Director of OTC for permission to publish this work. The views expressed are those of the authors and not necessarily the official opinion or policy of OTC.

REFERENCES

[1] L. R. Goke and G. J. Lipovski, "Banyan networks for partitioning multiprocessor systems," in *Proc. 1st Ann. Symp. Comp. Arch.*, Apr. 1973, pp. 21–28.
[2] C. Mead and L. Conway, *Introduction to VLSI Systems.* Reading, MA: Addison-Wesley, 1979.
[3] M. A. Franklin, S. A. Kahn, and M. J. Stucki, "Design issues in the

development of a modular multiprocessor communications network,'' in *Proc. 6th Ann. Symp. Comp. Arch.*, Apr. 1979, pp. 182–187.

[4] A. Huang and S. Knauer, ''Starlite: A wideband digital switch,'' in *Proc. IEEE GLOBECOM '84*, Nov. 1984, pp. 5.3.1–5.3.5.

[5] C. Day, J. Giacopelli, and J. Hickey, ''Applications of self-routing switches to data fiber optic networks,'' in *Proc. ISS'87*, vol. 3, Mar. 1987, pp. A7.3.1–A7.3.5.

[6] C.-L. Wu and T.-Y. Feng, ''On a class of multistage interconnection networks,'' *IEEE Trans. Comput.*, vol. C-29, pp. 694–702, Aug. 1980.

[7] T.-Y. Feng, ''A survey of interconnection networks,'' *Computer*, vol. 14, pp. 12–17, Dec. 1981.

[8] J. S. Turner, ''Design of an integrated services packet network,'' in *Proc. 9th ACM Data Commun. Symp.*, Sept. 1985, pp. 124–133.

[9] ——, ''Design of a broadcast packet switching network,'' presented at IEEE INFOCOM '86, Miami, FL, Apr. 1986.

[10] W. A. Montgomery, ''Techniques for packet voice synchronization,'' *IEEE J. Select. Areas Commun.*, vol. SAC-1, pp. 1022–1028, Dec. 1983.

[11] D. M. Dias, and J. R. Jump, ''Analysis and simulation of buffered delta networks,'' *IEEE Trans. Comput.*, vol. C-30, pp. 273–282, Apr. 1981.

[12] Y. C. Jenq, ''Performance analysis of a packet switch based on single-buffered banyan network,'' *IEEE J. Select. Areas Commun.*, vol. SAC-1, pp. 1014–1021, Dec. 1983.

[13] L. J. Wu, ''Mixing traffic in a buffered banyan network,'' in *Proc. 9th ACM Data Commun. Symp.*, Sept. 1985, pp. 134–139.

[14] R. G. Bubenik and J. S. Turner, ''Performance of a broadcast packet switch,'' in *Proc. IEEE ICC'87*, June 1987, pp. 31.6.1–31.6.5.

[15] C. A. Lea, ''The load-sharing banyan network,'' *IEEE Trans. Comput.*, vol. C-35, pp. 1025–1034, Dec. 1986.

[16] K. E. Batcher, ''Sorting networks and their applications,'' in *Proc. AFIPS Conf., 1968 Spring Joint Comp. Conf.*, pp. 307–313.

[17] M. De Prycker and J. Bauwens, ''A switching exchange for an asynchronous time division based network,'' in *Proc. IEEE ICC'87*, June 1987, pp. 22.3.1–22.3.8.

[18] C. J. Weinstein, M. L. Malpass, and M. J. Fischer, ''Data traffic performance of an integrated, circuit- and packet-switched multiplex structure,'' in *Proc. IEEE ICC'79*, June 10–14, 1979, pp. 24.3.1–24.3.5.

[19] T. Suda, H. Miyahara, and T. Hasegawa, ''Performance evaluation of a packetized voice system—Simulation study,'' *IEEE Trans. Commun.*, vol. COM-32, pp. 97–102, Jan. 1984.

[20] L. Kleinrock, *Queueing Systems, Vol. 1: Theory*. Wiley, 1975.

[21] D. M. Dias and J. R. Jump, ''Packet switching interconnection networks for modular systems,'' *Computer*, pp. 43–53, Dec. 1981.

A Self-Routing Multistage Switching Network for Broadband ISDN

HYONG S. KIM AND ALBERTO LEON-GARCIA

Abstract—Many switching networks have been proposed for broadband ISDN [1]. Switching networks which are capable of achieving maximum throughput of 100% either have large internal speedup [2], [3] or require large amounts of hardware [4], [5]. For example, the input–output difference self-routing network [3] and ATOM switch [2] require large internal speed-up factors. In order to maintain the nonblocking property, the internal speed of an $N \times N$ switching network has to be N times higher than the maximum rate of the incoming packets. The internal speed can be reduced by introducing parallel switching planes thereby increasing the hardware complexity. In this paper, we propose a new switching network that approaches a maximum throughput of 100% as buffering is increased. This self-routing switching network consists of simple 2×2 switching elements, distributors, and buffers located between stages and in the output ports. The proposed switching network requires speed-up of two. The structure and the operation of the switching network are described and the performance of the switching network is analyzed.

I. INTRODUCTION

SEVERAL switching systems have been proposed for broadband ISDN but they either require large amounts of hardware and are costly, or suffer from performance degradation due to packet blocking. These switching systems can be classified as input queueing or output queueing type. The input queueing type has a throughput performance which is approximately 58% that of the output queueing type [6]. The output queueing networks, however, are hampered by a need for more complex hardware. One of the first output queueing networks proposed is the Knockout switch [4]. The explosion in crosspoints of $O(N^2)$ makes large knockout switches impractical [7]. Recently, nonblocking $N \times N$ output queueing networks with only $N \log_2 N$ switching elements have been proposed, i.e., the input–output difference self-routing network [3]. These networks, although simple in hardware, require a large internal speed-up factor. In order to maintain the nonblocking property, the internal speed of the $N \times N$ switching network has to be N times higher than the maximum rate of the incoming packets. Internal speed is reduced by introducing parallel switching planes thereby increasing the hardware complexity. In this paper, we present a new switching network with the speed-up factor of 2. This network consists of simple 2×2 switching elements, and distributor stages. This self-routing switching network approaches a maximum throughput of 100%

Manuscript received May 15, 1989; revised November 10, 1989.
The authors are with the Department of Electrical Engineering, University of Toronto, Toronto, Ont. M5S 1A4.
IEEE Log Number 8934094.

and is capable of broadcasting. The structure and the operation of the switching network are described and the performance of the switching network is analyzed.

In Section II, we describe the overall structure of the new switching network. The basic elements of the switching network are the 2×2 switching element and distributor. The structure and the operation of the 2×2 switching element and the distributor are described in Section II. In Section III, the routing of the switching network is described. In Section IV, the performance of the switching network is analyzed. The conclusions are given in Section V.

II. THE SWITCHING NETWORK STRUCTURE

The $N \times N$ switching network is constructed with $\log_2 N$ stages of switching elements and $(\log_2 N - 1)$ stages of distributors as shown in Fig. 1. Each stage is labeled from 1 to $n = \log_2 N$, starting from the leftmost stage to the rightmost stage. The ith stage of $N \times N$ switching network consists of $N/2$ switching elements of size 2×2 and 2^i distributors of size $2^{n-i} \times 2^{n-i}$.

Fig. 1 shows block diagram of the 8×8 switching network. Each output line from $N/2$ switching elements in the first stage is connected to different distributors. The output lines of the distributors are in turn connected to the switching elements in the second stage. The switching elements in the second stage are connected to the distributors in the third stage in similar fashion, and so on up to the last stage. If the distributors are removed, the linking pattern is the same as that of the 8×8 banyan network. This linking pattern provides packets with a route through the switching network to reach their destinations.

The distributor takes the packets at its input and distributes them evenly among its outputs. In this fashion it provides a balanced load to the input ports of the switching elements in the next stage. The distributor uses buffers efficiently and prevents packet build-up in a particular buffer while other buffers are empty. This packet build-up phenomenon occurs in input queueing networks such as the Batcher-banyan network [8] due to head of line blocking.

Let T denote a clock cycle. Then, the switching elements and the distributor operate in subintervals of duration, $T_s = T/2$ s. The switching elements in the first stage receive packets every T seconds and output packets every

Reprinted from *IEEE J. Selected Areas Commun.*, vol. 8, no. 3, pp. 459–466, April 1990.

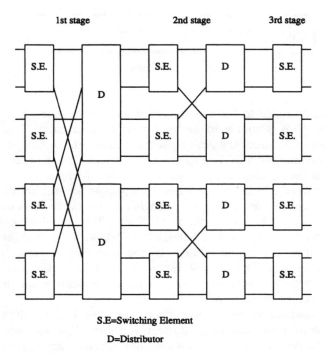

1st stage   2nd stage   3rd stage

S.E=Switching Element

D=Distributor

Fig. 1. Block diagram of 8 × 8 switching network.

T_s seconds. The distributors in the ith stage receive packets from the switching elements in ith stage every T_s seconds and the distributors output packets every T seconds. The switching elements in the last stage receive packets every T seconds from the distributor and output packets to the output buffers every T_s seconds.

The switching network is self-routing and no central control is required to route a packet through the switching network. The switching element in the ith stage needs only the ith bit of the destination address to route a packet. The switching element uses ith bit to route a packet to the proper distributor. The distributor distributes its input packets evenly across its output buffers and then the distributor forwards the packets to the switching element in the next stage. A packet reaches the destination after it has gone through $\log_2 N$ stages of switching elements and ($\log_2 N - 1$) stages of distributors. The linking pattern of the banyan network provides a unique path for each input–output pair. The introduction of the distributors in the switching network ensures that the packet can be delivered through different routes, even though it has the linking pattern of the banyan network. Therefore, the distributor provides multiple routes to the destination port for a packet in any input port. While the routing control is still distributed, the distributor effectively provides multiple routes for the input–output pair. There are other switching networks with alternate routings [9], [7]. However, these switching networks require central control to set up and find a route through the network. The central control can cause bottleneck in large switching network when fast switching is required.

If packets at the input ports of the switching elements in the ith stage have distinct destination addresses, the maximum number of the packets coming into a distributor in the ith stage is $N/2^i$. All packets then proceed to the switching elements in the $i + 1$th stage without blocking since the distributor is nonblocking as shown in a later section. If some of the packets at the input ports have the same destination addresses, then there might be more than $N/2^i$ packets coming into a distributor and only $N/2^i$ packets proceed to the next stage and the rest of the packets stay in the distributor's buffer until the next clock cycle. Therefore, the switching network can handle the multiple packets with the same destination addresses without packet loss if it has sufficient buffer space in the distributor. Now, we look at the structure and the operation of each of the building blocks.

A. Structure and Operation of 2 × 2 Switching Element

The 2 × 2 switching element is a nonblocking switch consisting of one stage of two switching subelements. (see Fig. 2) The upper switching subelement is labeled ''0'' and the lower switching subelement is labeled ''1.'' The label is used in the routing of a packet. The switching subelements are connected vertically in a circular manner and horizontally to the output port as shown in Fig. 2.

The operation of the switching element is synchronous. The clock cycle, T seconds, is divided into two subintervals of T_s seconds. Each input port accepts a packet at the beginning of each T seconds clock cycle. Every T_s seconds, each switching element shifts the packet either horizontally or vertically according to the destination address bit. In the ith stage, the ith bit of the destination address of a packet is compared to the label of the switching subelement. If the ith bit of the destination address is the same as the label, then the packet is shifted horizontally to the distributor. If the ith bit of the destination address and the label are not the same, then the packet is shifted vertically to the other switching subelement. In this manner any two input packets can be routed to its destination within two T_s subintervals. Thus the 2 × 2 switching element can route any two input packets to their destinations within one clock cycle.

B. Structure and Operation of Distributor

We now describe the structure and the operation of the distributor. The basic element of the distributor is the reverse banyan network, which is obtained from a mirror image of the banyan network (see Fig. 3). Buffers are connected to the output ports of the reversed banyan network. The distributor receives packets from the switching elements and distributes the packets in cyclic fashion across the output buffers. As a result, the number of packets in the output buffers of the distributor differs by at most one, thereby using the buffer space efficiently.

The structure of the $N' \times N'$ distributor is shown in Fig. 3. It has a running adder that routes packets to an output buffer through the $N' \times N'$ reverse banyan network. The running adder computes the address of the output buffer to which each incoming packet is to be routed. It assigns them in order so that the packets are distributed

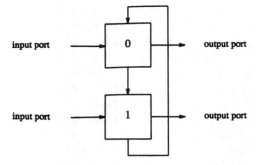

Fig. 2. 2 × 2 switching element.

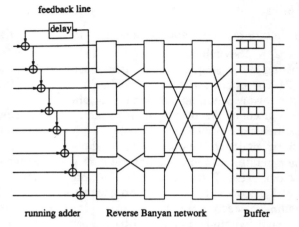

Fig. 3. Structure of 8 × 8 distributor.

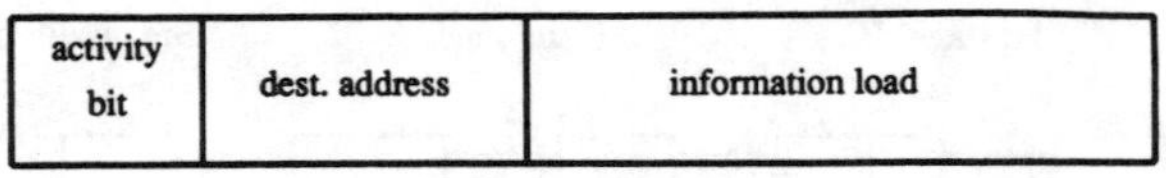

activity bit	dest. address	information load

Fig. 4. Packet format.

evenly among the output buffers. The running adder requires a feedback, delayed by one subinterval, from the lowest line as shown in Fig. 3. The complexity of the running adder is $O(N')$ and it mainly consists of exclusive-or gates to compute modulo N' additions.

The output buffers connected to the distributor are labeled from 0 to $N' - 1$. Let i, $0 \leq i \leq N' - 1$, be the current output buffer address. When the packets are coming into the empty buffers, the indicator i is 0. Then, the indicator i is the sum of the packets coming into the distributor obtained by modulo N' addition. Therefore, i indicates the label of the next buffer that is ready to accept the packet. The indicator i can be easily obtained from the running adder at the lowest line as explained below.

As shown in Fig. 4, the packet has an activity bit in the header indicating whether the packet is empty (0) or full (1). This is followed by the destination address and the information field. Now, the packet in the input line of the distributor is padded with the address of the output buffer that it has been assigned. The buffer address is obtained as follows. The indicator i is sent to the top line of the distributor to be added to the activity bit of the packet in the line with modulo N'. If the line has a packet, then the buffer address of that packet is the indicator i, the address of the buffer to be filled. Then, i is propagated to the line just below the first one and it is again added to the activity bit of the packet in the second line. Then, the sum is padded to the packet as the buffer address. Again the resulting

sum is added to the activity bit of the packet in the next line. This continues to the last line of the distributor. This operation forces the active packets to have the buffer address increased by 1 in modulo N' from i to $i + m$ where m is the number of active packets in the distributor.

After the buffer address has been padded into the active packets, the reverse banyan network routes the packets to the proper buffers without blocking. The nonblocking property of the reverse banyan network with a particular input packet pattern is stated by the following theorem.

Theorem: The $N' \times N'$ reverse banyan network is nonblocking when the input packets have consecutive output addresses, modulo N'.

Blocking in the reverse banyan network occurs when two packets need to go through the same link to reach different output ports. However, the reverse banyan network is nonblocking when input packets are arranged in a particular pattern described in the theorem. The banyan network has a unique route for every input–output port pair. Thus, if the banyan network is nonblocking for a particular input packet pattern, then the reverse banyan network is also nonblocking with the corresponding input packet pattern. Thus, the outline of the proof is as follows. We first assume that any two input–output port pair share the same link, and then contradict the assumption by using appropriate equations. This proof is an extension of Bechmann's proof [10] of the nonblocking property of the Batcher–banyan network. The details of the proof of the Theorem can be found in [11].

We now describe the routing of packets through the reverse banyan network. First, the running adder in the distributor computes the routing address of the packet for the reverse banyan network. Then, the routing addresses are appended to the incoming packets with an activity bit. The packets are routed through the reverse banyan network with the output buffer address in the following manner. Let $d_0 d_1 \cdots d_{n'-1}$ where $n' = \log_2 N'$, denote the binary output address computed in the running adder. The stage of the reverse banyan network is labeled from ($n' - 1$), for the first stage, to 0 for the last stage. Switching elements at the ith stage only require the d_ith bit of the output address in order to route it to its proper destination. If the d_i is "zero" or "one," switching elements route a packet to the upper output port or the lower output port, respectively. The particular interconnection pattern of reverse Banyan network allows this routing scheme to forward a packet to the proper destination. As stated in the theorem, there is no blocking in the routing of these packets. At every clock cycle, the packets in the front end of the buffers are advanced to the switching elements in the next stage. The delay incurred by the reverse banyan network is the processing time of the address bits.

III. ROUTING IN THE SWITCHING NETWORK

A. Point-to-Point Routing

The routing in the switching network does not require central control; it is done by each switching element by simply checking a bit of destination address. Let $d_1 d_2 d_3 \cdots d_i \cdots d_n$ be the binary destination address of the switching network of size $N \times N$ where $N = 2^n$. Each switching subelement in the ith stage has a label of "0" or "1" and it shifts the packet horizontally to the next stage if the ith bit of the destination address matches the label, or it shifts vertically if it does not match. Fig. 5 shows an example of the routing operation of the switching network. The input port 0 has a packet destined to the output port 6, so the destination address is "110" in binary. In the first stage, the packet is in the switching subelement labeled "0" so that the packet has to be shifted vertically and then to the next stage. The packet passes through the distributor and it could be placed in any of the switching elements in the next stage depending on the number of packets present in the distributor. We assume that the packet comes out to input port 5 of the second stage. Then, the packet is in the switching subelement labeled "1." Thus, the packet is shifted horizontally to the next stage. Again we assume that the packet is placed in the input port 7 of the third stage. Then, the packet is in the switching subelement labeled "1." Therefore, the packet is shifted vertically and then horizontally out of the switching subelement to reach the destination port. The routing is done in such a way that no two packets will be shifted to occupy the same switching subelement even for the packets with the destination address.

B. Point-to-Multipoint Routing

The structure of the switching element allows the routing of point-to-multipoint traffics where a single packet of an input port can be simultaneously destined to multiple output ports. This traffic might arise in broadcasting TV signals or in conference calls where information is sent to more than one site.

The routing of the point-to-multipoint traffic is done as follows. The header of the packet contains addresses of the desired output ports. Then the switching element shifts the packet through the switching element in the same fashion as before but now a packet can be shifted vertically and horizontally at the same time. Fig. 6 shows the routing operation of the point-to-multipoint traffic where the input port 0 sends packets to the output ports 0, 4, and 6.

IV. PERFORMANCE OF THE SWITCHING NETWORK

The performance of the switching network is studied in this section. We analyze the switching network under the uniform traffic pattern where an incoming packet is equally likely to go to any of the output ports.

A. Analysis of the Switching Network

The analysis is composed of three parts. The first part of the analysis deals with the buffers of the distributor in

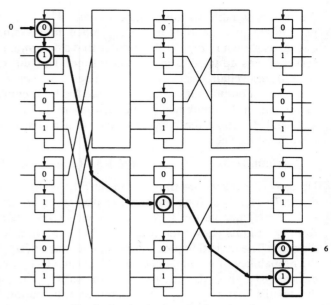

Fig. 5. Routing in the switching network.

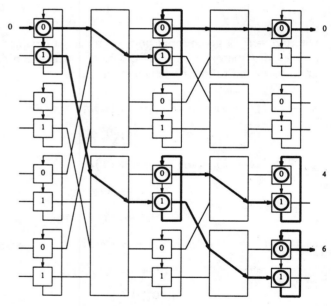

Fig. 6. Multicast routing in the switching network.

the first stage. The second part deals with the buffers of the distributor in the ith stage, $1 < i < n$ where $n = \log_2 N$. The last part deals with the output buffers of the switching network.

1) First Stage Analysis: First, we consider the output buffers of the distributors in the first stage. We assume that the arrival processes to the switching elements in the first stage are independent Bernoulli processes with $\lambda = P$ [a packet arrives at input port]. Let $p_{ij} = \lambda / N$ be the probability that there is a packet arrival to input port i destined to output port j.

Consider the output buffers of an arbitrary distributor. The output buffers in other distributors have the same statistics since the traffic is assumed to be uniform. We model the buffers as a discrete time Markov chain. The $N/2$ output buffers in a distributor are combined into a single queue, as shown in Fig. 7. We assume that a deterministic server serves up to $N/2$ packets at a time. This discrete time Markov chain has $(mN/2 + 1)$ states where m is the size of each output buffer. Let Q_t denote the total number of packets in the combined queue just before the tth time slot, and A_t denote the number of packet arrivals during tth time slot. Then,

$$Q_t = \min\left(\max\left(0, Q_{t-1} - N/2\right) + A_t, mN/2\right). \quad (1)$$

Up to $N/2$ packets are transmitted out of the distributor every T seconds as long as there are packets to be delivered. Let Q denote the steady-state queue size obtained from Q_t. Fig. 8 shows the Markov chain state transition diagram. The transition probabilities are in terms of $a_1(k)$, the probability of having k packet arrivals at a clock cycle directed to a given distributor where the subscript one indicates the first stage. $a_1(k)$ is obtained from following equation:

$$a_1(k) = \binom{N}{k} p^k (1-p)^{N-k} \quad \text{for } 0 \le k \le N \quad (2)$$

where $p = P$ [a given input port has a packet destined to a given distributor] $= \lambda/(2N)$. The state transition diagram gives the following balance equations:

$$q_0^1 \triangleq \Pr\left(Q = 0\right) = a_1(0)q_0^1 + a_1(0)q_1^1$$
$$+ \cdots + a_1(0)q_{N/2}^1 \quad (3)$$

$$q_1^1 \triangleq \Pr\left(Q = 1\right) = a_1(1)q_0^1 + a_1(1)q_1^1$$
$$+ \cdots + a_1(1)q_{N/2}^1$$
$$+ a_1(0)q_{N/2+1}^1 \quad (4)$$

$$q_2^1 \triangleq \Pr\left(Q = 2\right) = a_1(2)q_0^1 + a_1(2)q_1^1$$
$$+ \cdots + a_1(1)q_{N/2+1}^1$$
$$+ a_1(0)q_{N/2+2}^1 \quad (5)$$

$$\vdots$$

$$q_{mN/2}^1 \triangleq \Pr\left(Q = mN/2\right) = \left(\sum_{k=0}^{N/2} q_k^1\right) \sum_{i=mN/2}^{N} a_1(i)$$
$$+ \sum_{k=N/2+1}^{mN/2} q_k^1$$
$$\cdot \left[\sum_{\substack{i=mN/2 \\ -(k-N/2)}}^{N} a_1(i)\right]. \quad (6)$$

Replacing the last equation by $\sum_{k=0}^{mN/2} q_k^1 = 1$, we obtain the transition matrix for the Markov chain from above equations and invert the matrix to obtain the distribution of q_i^1, the steady-state probability of the queue occupancy

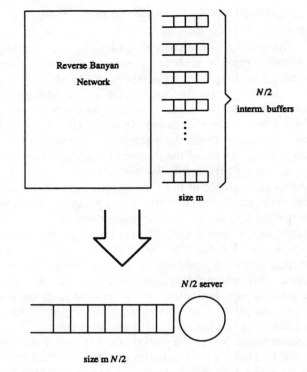

Fig. 7. $N/2$ buffers combined into a single queue.

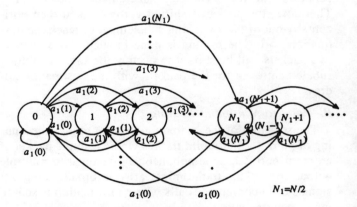

Fig. 8. Markov-state transition diagram of the first queue.

in the first stage, $\overline{W}_1$, the delay in the first queue, is computed using Little's formula.

$$\overline{W}_1 = \frac{\overline{Q}}{\lambda'} \quad (7)$$

where

$$\overline{Q} = \sum_{k=0}^{mN/2} k q_k^1 \quad (8)$$

$$\lambda' = \sum_{k=1}^{N/2} k q_k^1 + N/2 \sum_{k=N/2+1}^{mN/2} q_k^1. \quad (9)$$

2) Intermediate Stage Analysis: We now find the steady-state probability of the output buffers of the distributors in the intermediate stages. The incoming packet distribution is computed using q_i^1 from the first queue. First, we find $a_i(k)$, $1 < i < n$, probability that k packets in a given clock cycle are directed to the buffers of a distributor in the ith stage from the buffers in the $(i - 1)$th stage. For a given distributor let $P_{i-1}(k|l)$ be the probability that k packets are directed to it, given that there are l head-of-line packets in the corresponding distributors in the $(i - 1)$th stage. The $a_i(k)$ are then given by

$$a_i(k) = \sum_{l=k}^{N_{i-1}} P_{i-1}(k|l) q_l^{i-1} \quad \text{for } k \le l \le N_{i-1} \tag{10}$$

$$a_i(k) = P_{i-1}(k|N_{i-1}) \sum_{l=N_{i-1}+1}^{mN_{i-1}} q_l^{i-1} \quad \text{for } l > N_{i-1} \tag{11}$$

where m is the size of each buffer of the distributor and $N_{i-1} = 2^{n-(i-1)}$ is the maximum number of packets that can advance from the $(i - 1)$th stage distributors.

The head-of-line packets from each $(i - 1)$th stage distributor proceed to one of two ith stage distributors. Let p be the probability that a head-of-line packet in $(i - 1)$th stage proceeds to a given ith stage distributor, then

$$P_{i-1}(k|l) = \binom{l}{k} p^k (1-p)^{l-k} \quad \text{for } k \le l \le N_{i-1}. \tag{12}$$

The $a_i(k)$ are then obtained by substituting (12) into (10) and (11). Note that $p = 1/2$ since the traffic is uniform.

We model the buffers of the ith queue as a discrete time Markov chain, with the transition probabilities given in terms of probability $a_i(k)$'s. Let Q_t denote the total number of packets in the combined buffer queue in a intermediate distributor just before the tth time slot, and A_t denote the number of packet arrivals during tth time slot. Then,

$$Q_t = \min\left(\max\left(0, Q_{t-1} - 2^{n-i}\right) + A_t, m2^{n-i}\right) \tag{13}$$

where m denotes the buffer size of each output port in ith stage. The above equation uses the fact that up to 2^{n-i} packets are transmitted out of the output port at each time slot as long as there are packets to be delivered. Let Q_i denote the steady-state queue size obtained from Q_t where the subscript i indicates stage i. The Markov chain state transition diagram can be easily obtained from the Fig. 8 by replacing N_1 by N_i. The state transition diagram gives the following balance equations:

$$q_0^i \triangleq \Pr(Q_i = 0) = a_i(0)q_0^i + a_i(0)q_1^i$$
$$+ \cdots + a_i(0)q_{N_i}^i \tag{14}$$

$$q_1^i \triangleq \Pr(Q_i = 1) = a_i(1)q_0^i + a_i(1)q_1^i$$
$$+ \cdots + a_i(1)q_{N_i}^i$$
$$+ a_i(0)q_{N_i+1}^i \tag{15}$$

$$q_2^i \triangleq \Pr(Q_i = 2) = a_i(2)q_0^i + a_i(2)q_1^i$$
$$+ \cdots + a_i(1)q_{N_i+1}^i$$
$$+ a_i(0)q_{N_i+2}^i \tag{16}$$

$$\vdots$$

$$q_k^i \triangleq \Pr(Q_i = k) = a_i(k)q_0^i + a_i(k)q_1^i + a_i(k)q_{N_i}^i$$
$$+ a_i(k-1)q_{N_i+1}^i$$
$$+ \cdots + a_i(0)q_{N_i+k}^i \tag{17}$$

$$\vdots$$

$$q_{mN_i}^i \triangleq \Pr(Q_i = mN_i) = \sum_{j=mN_i}^{N_{i-1}} \left(a_i(j) \sum_{k=0}^{N_i} q_k^i\right)$$
$$+ \sum_{k=N_i+1}^{mN_i}$$
$$\cdot \left(q_k^i \sum_{j=mN_i-(k-N_i)}^{N_{i-1}} a_i(j)\right). \tag{18}$$

Equation (18) is replaced by $\sum_{k=0}^{mN_i} q_k^i = 1$ since the above set of equations is linearly dependent. The distribution of q_k^i, the steady-state probability of the intermediate queue occupancy, is then obtained by inverting the matrix obtained from the new set of equations. $\overline{W}_i$, the delay in the queue in the ith stage, is computed using Little's formula.

$$\overline{W}_i = \overline{Q}_i / \lambda' \tag{19}$$

where

$$\overline{Q}_i = \sum_{k=0}^{mN_i} k q_k^i \tag{20}$$

$$\lambda' = \sum_{k=1}^{N_i} k q_k^1 + N_i \sum_{k=N_i+1}^{mN_i} q_k^1. \tag{21}$$

3) Output Buffer Analysis: In the last part of the analysis, we model the output queue of buffer size m as a discrete time Markov chain of $(m + 1)$ states, with the transition probabilities given in terms of the probabilities $a_n(k)$'s. Let Q_t denote the number of packets in a given output queue just after the tth time slot, and A_t denote the number of packet arrivals during tth time slot, then,

$$Q_t = \min\left(\max\left(0, Q_{t-1} + A_t - 1\right), m\right). \tag{22}$$

In the above equation, we are assuming that the packet at the output port is removed immediately upon its arrival within the time slot [6]. Therefore, one packet is transmitted out of the output port at each time slot as long as there is a packet to be delivered. Let Q denote the steady-state queue size obtained from Q_t. Fig. 9 shows the Mar-

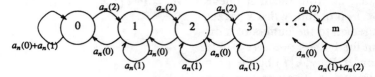

Fig. 9. Markov-state transition diagram of the output queue.

kov chain state transition diagram. From the state transition diagram, we obtain the following balance equations.

$$q_0^n \triangleq \Pr(Q = 0) = (a_n(0) + a_n(1))q_0^n + a_n(0)q_1^n \tag{23}$$

$$q_1^n \triangleq \Pr(Q = 1) = a_n(2)q_0^n + a_n(1)q_1^n + a_n(0)q_2^n \tag{24}$$

$$q_2^n \triangleq \Pr(Q = 2) = a_n(2)q_1^n + a_n(1)q_2^n + a_n(0)q_3^n \tag{25}$$

$$q_3^n \triangleq \Pr(Q = 3) = a_n(2)q_2^n + a_n(1)q_3^n + a_n(0)q_4^n \tag{26}$$

$$\vdots$$

$$q_m^n \triangleq \Pr(Q = m) = a_n(2)q_{m-1}^n + (a_n(1) + a_n(2))q_m^n. \tag{27}$$

Rather than using the Z transform of Q, we obtain the steady-state queue size probabilities directly by simple numerical method from the following balance equations.

$$q_1^n = \frac{(1 - a_n(0) - a_n(1))}{a_n(0)} q_0^n \tag{28}$$

$$\vdots$$

$$q_m^n = \frac{(1 - a_n(1))}{a_n(0)} q_{m-1}^n - \frac{a_n(2)}{a_n(0)} q_{m-2}^n \tag{29}$$

where

$$q_0^n = \frac{1}{1 + \Sigma \, q_i^n / q_0^n}. \tag{30}$$

The normalized throughput ρ is then

$$\rho = 1 - q_0^n a_n(0) \tag{31}$$

and the probability of the packet loss in the switching network is

$$\Pr[\text{packet loss}] = 1 - \frac{\rho}{\lambda} \tag{32}$$

where λ is the total load to an output port. The mean queue length is

$$\overline{Q} = \sum_{i=1}^{m} i q_i^n \tag{33}$$

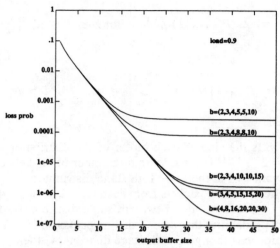

Fig. 10. Packet loss probability versus output buffer size for 128×128 network with $b = (a_1, a_2, \cdots, a_{n-1})$ where a_i is the size of buffer at ith stage.

and the mean waiting time in the queue is again obtained using Little's formula as follows:

$$\overline{W}_n = \frac{\sum_{i=1}^{m} i q_i^n}{\rho} = \frac{\sum_{i=1}^{m} i q_i^n}{1 - q_0^n a_n(0)}. \tag{34}$$

The total mean delay of a packet is then

$$\text{total mean delay} = 1 + \sum_{i=1}^{n} \overline{W}_i \tag{35}$$

where 1 is the normalized delay occurring in the first stage due to its transmission time using store-and-forward mode.

B. Performance Results

Here we present some results illustrating the performance of the switching network under the uniform traffic pattern. Fig. 10 shows the blocking probability versus the output buffer size curves with the load at 0.9. Several curves with different internal buffer sizes are plotted for 128×128 switching network. Let $b = (a_1, a_2, a_3, \cdots, a_{n-1})$ denote the set of number of buffers in different stages from the first stage to the $(n-1)$th stage; Then, Fig. 10 shows the packet loss probability versus the number of the output buffers for various internal buffer sizes. We can observe that the packet loss probability reaches a minimum and no further improvement is obtained by increasing the size of the output buffer. This saturation is due to the blocking occurring in the internal buffers. Thus we see that the saturation point decreases as we increase the number of buffers in the internal stages. With $b = (4, 8, 16, 20, 20, 30)$, the switching network's packet loss probability is in the order of 10^{-7} with the load of 0.9.

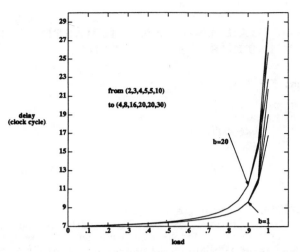

Fig. 11. Load versus delay for 128 × 128 network with different buffer sizes ($b = 1$ and b = 20).

Fig. 11 shows the load versus delay curve of the switching network with the same internal buffer parameters used in Fig. 10. Two sizes of the output buffer, 1 and 20, are used in the figure. The figure shows that the switching network has small delay up to a load of 0.9.

V. CONCLUSION

In this paper, we propose a new switching network that approaches a maximum throughput of 100% as buffering is increased. This self-routing switching network consists simple 2 × 2 switching elements, distributors, and buffers located between stages and in the output ports. The proposed switching network requires a speed-up factor of two. The switch has $\log_2 N$ stages that move packets in a store-and-forward fashion, thus incurring a latency of $\log_2 N$ time periods. The performance analysis of the switch under uniform traffic pattern shows that the additional delay is small and a maximum throughput of 100% is achieved as buffering is increased.

REFERENCES

[1] H. Ahmadi and W. Denzel, "A survey of modern high-performance switching techniques," *IEEE J. Select. Areas Commun.*, vol. 7, pp. 1091–1103, Sept. 1989.
[2] H. Suzuki, H. Nagano, T. Suzuki, T. Takeuchi, and S. Iwasaki, "Output-buffer switch architecture for asynchronous transfer mode," in *Proc. ICC '89*, Boston, MA, 1989, pp. 99–103.
[3] H. Imagawa, S. Urushidani, and K. Hagishima, "A new self-routing switch driven with input-to-input address difference," in *Proc. Globecom '88*, Hollywood Beach, FL, Dec. 1988, pp. 1607–1611.
[4] Y. Yeh, M. Hluchyj, and A. Acampora, "The knockout switch: A simple, modular architecture for high-performance packet switching," *IEEE J. Select. Areas Commun.*, vol. SAC-5, pp. 1274–1283, Oct. 1987.
[5] H. Ahmadi, W. Denzel, C. Murphy, and E. Port, "A high performance switch fabric for integrated circuit and packet switching," in *Proc. INFOCOM '88*, New Orleans, LA, Mar. 1988, pp. 9–18.
[6] M. Karol, M. Hluchyj, and S. Morgan, "Input versus output queueing on a space-division packet switch," *IEEE Trans. Commun.*, vol. COM-35, pp. 1347–1356, Dec. 1987.
[7] J. Degan, G. Luderer, and A. Vaidya, "Fast packet technology for future switches," *AT&T Tech. J.*, vol. 68, pp. 36–51, Mar./Apr. 1989.
[8] J. Hui and E. Arthurs, "A broadband packet switch for integrated transport," *IEEE J. Select. Areas Commun.*, vol. SAC-5, pp. 1264–1273, Oct. 1987.
[9] C. Lea, "The load-sharing banyan network," *IEEE Trans. Comput.*, vol. C-34, pp. 1025–1034, Dec. 1986.
[10] T. T. Lee, "Nonblocking copy networks for multicast packet switching," *IEEE J. Select. Areas Commun.*, vol. SAC-6, pp. 1455–1467, Dec. 1988.
[11] H. Kim and A. Leon-Garcia, "Non-blocking property of reverse banyan network," *IEEE Trans. Commun.*, to be published.

ANALYSIS, CONTROL AND DESIGN OF CROSSBAR AND BANYAN BASED BROADBAND PACKET SWITCHES FOR INTEGRATED TRAFFIC

Salman Z. Shaikh, Mischa Schwartz and Ted H. Szymanski

Department of Electrical Engineering, and
Center for Telecommunications Research
Columbia University, New York, NY 10027-6699
E-mail: salman@ctr.columbia.edu

ABSTRACT

Approximate Markov chain models are developed for Crossbar and Banyan packet switches built with nodes of arbitrary size, each node having output buffers of arbitrary size, operating in an environment of *multiclass* traffic with priority. A simple design change is incorporated to obtain a straightforward mechanism to 'control' the throughput of each traffic class. Effects of flexible priority disciplines are presented as a means for 'controlling' the average time delays of each traffic class. Techniques for obtaining the time delay distributions for packets of each traffic class are illustrated.

1. Introduction

Recently a variety of fast packet switching architectures have been proposed for broadband integrated networks of the future (e.g. [1,6,7,8,14,20,22]). Prominent among them are packet switches based on Multistage Banyans. The analysis of these networks reported in the literature has been confined to *single class* traffic (e.g. [4,9,12,13,16,19a,21]). In a recent paper [18], we analyzed the performance of Banyan packet networks with single buffered nodes, and an arbitrary number of traffic classes, under a non-preemptive priority scheme. In this paper, we extend the multiclass traffic analysis to 'preemptive' priority schemes and multibuffered nodes. We also consider ways of 'controlling' the throughputs and the time delays of the different classes.

Approximate Markov chains are developed for Banyan networks with nodes (basic switches) of arbitrary size, an arbitrary number of traffic classes, and each node having output buffers of arbitrary size. Since a single stage Banyan is just a crossbar, the analysis also applies to crossbar and other non blocking networks with output buffering for any number of traffic classes. (Input buffered crossbars (or non blocking networks) with two traffic types have been analyzed in [3]). We then present and model a straightforward mechanism for 'controlling' the division of the total throughput into the throughputs of the individual traffic classes. This mechanism just requires the addition of a single buffer and a simple controller at each of the input ports of the network. Flexible priority schemes are presented as a means for 'controlling' the average time delays of each of the traffic classes. We successively give priority to different traffic classes, in order to achieve the desired trade off in their time delays. With the help of the Markov chains, we also illustrate the calculation of the distribution of time delay of each traffic class. Simulations of small networks are used to validate the models.

Section 2 describes the Banyan network packet switch. Design considerations for incorporating a 'preemptive' priority scheme are discussed in section 3. The modeling assumptions and approach are discussed and used in section 4. Techniques for 'controlling' the throughputs and the time delays of the individual traffic classes are presented and modeled in section 5 and 6 respectively. Delay distribution calculations are illustrated in section 7, followed by some concluding remarks in section 8. The appendix reports on the analysis of some flexible priority schemes discussed in section 6.

2. The Banyan Packet Switch

Our approximate Markov chain models are applicable to a large number of topologically equivalent Banyan networks, including the SW-Banyan, the Omega, the indirect binary n-cube, delta and the baseline etc, which are often collectively termed as "Multistage Interconnection Networks" or "MINs". A square $k^n \times k^n$ SW-Banyan network of size $N \times N$ (N users or network input / output ports) consists of $n = \log_k N$ stages, where each stage consists of N/k nodes, each a $k \times k$ crossbar. (See fig. 1a). The cost of MINs is $O(N \log N)$ crosspoints; their path length is $\log_k N$ nodes; and most importantly, their routing scheme is extremely simple and distributed, the routing decisions being based on successive k-ary digits of the destination address [20]. In the multiclass traffic case, whenever there is a contention, the higher priority packet wins and moves ahead, while the other remains buffered where it was. If the two contending packets have equal priority, the winner is picked with the toss of a fair coin. In order for a packet to be able to move forward, either the desired buffer at the next stage should be empty or it should have a packet which is able to move forward. As in [9], we assume fixed size packets (e.g. with A.T.M.) and synchronized operation. In the first portion of a clock cycle, control signals are passed from the last stage towards the first stage, so that every packet knows whether it can move forward one stage or stay in the same buffer. In the second portion of a cycle the packets move in accordance with the control signals and the clock period ends.

Banyans with output buffered nodes have been shown to outperform those with input buffered nodes, for infinite and finite buffers in [4] and [19a] respectively. This is because of the inherent blocking with input queueing, shown for non blocking networks in [10]. (An output buffer can accept as many packets as it has room for, in a single cycle). Hence in this paper, we concentrate on Banyans with arbitrarily sized output buffered nodes only.

3. Preemptive-like priority

Consider the case of two traffic classes. For ease of discussion, call the higher priority class 'voice' and the lower priority one 'data'. In [18], we had reported models for single buffered Banyans under the *non-preemptive* priority discipline. Thus if the 'data' load increased, the average time delay for 'voice' also increased. We may desire that the high priority traffic should not be affected at all by the 'background' low priority packets. Clearly, for such a scheme, we have to separate the queues of the different classes inside the nodes. As shown in fig. 1b, in this paper, we assume separate buffers in the nodes for each traffic class. Once a 'data' packet reaches the head of a 'data' queue, we say its 'service' has begun. This 'service' may take one or more clock cycles, depending on how often the 'data' packet is blocked at this stage. By giving voice packets a priority at the output ports, a newly arrived 'voice' packet can interrupt the 'service' of this 'data' packet (bypass the data queue) and go first. This is true for every stage. Hence we achieve the effect of preemptive priority discipline.

Reprinted from *Proc. IEEE International Communications Conference 1990*, pp. 761–765, April 1990.

4. Modeling

Since we want the higher priority traffic to suffer smaller delays, at each network input port, we keep separate buffers for storing 'voice' and 'data' packets (instead of a common FIFO buffer). (See fig. 1b). In section 5, we consider the input design of fig. 1c, where in addition to the separate buffers for each class, there is also a common single buffer, which helps in 'controlling' the throughputs. An exact Markov chain model for the network would have an exponentially growing state space [9]. Hence as in [9,19a], we assume uniform traffic conditions: (1) At the beginning of each cycle every network input link receives either a 'voice' packet or a 'data' packet or none, with probabilities P_v, P_d and $(1 - P_v - P_d)$ respectively (from the external queues). (For more than two classes, we define a probability P_c for each class c). (Such uniform input may be a result of a distribution network, put before the Banyan network switching fabric [20]). (2) The packet destinations are randomly and uniformly distributed among the network output links. Then the state of each buffer in a stage is statistically identical. Further, assuming independence between the states of the buffers of the stage, we reduce the state of a stage to that of a single buffer. For small Banyans, these modeling assumptions give reasonable results, as verified by simulation. For crossbars the results are quite accurate (see section 4.2). For large fully loaded Banyans, the analytic results are optimistic. However, the models are reasonable for moderate or *stable* loads (packet arrival probability less than maximum possible throughput). Even for high loads, they faithfully convey the performance trends and the salient features (eg. equality of the 'voice' and 'data' throughputs etc.)

Each input to a Banyan based Broadband network would probably have multiplexed packetized traffic from several small users. San-qi Li [15] has shown that if more than 40 calls equally share a link, then the effect of input correlation can be neglected for crossbars. Thus the Bernoulli arrival assumption should be valid for Banyans which are multiple stages of crossbars. Instead of giving complete priority to 'voice' at the network inputs, we may accept packets from the 'voice' and 'data' buffers in any arbitrary fashion. This can help us 'control' P_v and P_d, which represent the effective 'voice' and 'data' input loads respectively. Our analysis is a generalization of the single traffic class analysis presented in [19a].

4.1 Markov chain analysis for an arbitrary number of traffic classes

Consider $k^n \times k^n$ Banyans. The classes are numbered 1 through m, smaller class number indicating a higher priority. The input design is similar to that in fig. 1b, with separate buffers for each traffic class. Let $Pr[e]$ denote the probability of event e, and let the time interval $[t, t+1]$ correspond to "cycle t". Stage 0 refers to the input of the network. We define the following variables:

bs_c = the size of the class c buffers at each node.

$q_{c,s,l,t}$ = Pr[class c output buffer at stage s has l packets at time t]

$q'_{c,s,l,t}$ = Pr[class c output buffer at stage s would have l packets at the end of cycle t, if it didn't receive any new packets in this cycle]

$u_{c,s,t}$ = Pr[class c output buffer at stage s has at least one packet at time t]

$u'_{c,s,t}$ = Pr[class c buffer at stage s is not empty at time t and there is no contention from any higher priority packet in the neighboring queues]

$rx_{c,s,t}$ = the expected number of arrivals at a class c buffer at stage s, during cycle t.

$block_{c,s,t}$ = Pr[class c packet (if any) at the head of a class c queue at stage s is not able to move forward during cycle t]

The relations between these variables are as follows:

$$q'_{c,s,0,t} = q_{c,s,0,t} + q_{c,s,1,t} \cdot (1 - block_{c,s,t}) \tag{1}$$

$$q'_{c,s,l,t} = q_{c,s,l,t} \cdot block_{c,s,t} + q_{c,s,l+1,t} \cdot (1 - block_{c,s,t})$$
$$\text{for } l = 1, 2, \ldots bs_c - 1 \tag{2}$$

$$q'_{c,s,bs_c,t} = q_{c,s,bs_c,t} \cdot block_{c,s,t} \tag{3}$$

$$u_{c,s,t} = 1 - q_{c,s,0,t} \tag{4}$$

$$u'_{c,s,t} = u_{c,s,t} \prod_{i=1}^{c-1} (1 - u_{i,s,t}(1 - block_{i,s,t})) \tag{5}$$

$$rx_{c,s,t} = \sum_{i=0}^{bs_c} \sum_{r=0}^{k} q'_{c,s,i,t} \binom{k}{r} (u'_{c,s-1,t}/k)^r (1 - u'_{c,s-1,t}/k)^{k-r}$$
$$\times \min(r, bs_c - i) \tag{6}$$

$$block_{c,s,t} = 1 - rx_{c,s+1,t}/u_{c,s,t} \tag{7}$$

$$q_{c,s,l,t+1} = \sum_{i=\max((l-k),0)}^{l} q'_{c,s,i,t} \binom{k}{l-i} (u'_{c,s-1,t}/k)^{l-i}$$
$$\times (1 - u'_{c,s-1,t}/k)^{k-(l-i)} \text{ for } l = 0, 1, \ldots bs_c - 1 \tag{8}$$

$$q_{c,s,bs_c,t+1} = 1 - \sum_{l=0}^{bs_c-1} q_{c,s,l,t+1} \tag{9}$$

$$block_{c,n,t} = 1 - \prod_{i=1}^{c-1} (1 - u_{i,n,t}) \tag{10}$$

$$u_{c,0,t} = u'_{c,0,t} = P_c \tag{11}$$

Equations (1) thorough (9) are valid for $s = 1, 2, \ldots n$, except equation (7), which does not hold for $s = n$. Equations (1), (2) and (3) are used to obtain the state of the queues considering only the possible transmission of a packet in the cycle, and not any packet receptions. For equation (6), note that from the k input ports of a node, r class c packets select a particular output port (out of k possible choices) at the node with binomial probability $\binom{k}{r}(u'_{c,s-1,t}/k)^r(1 - u'_{c,s-1,t}/k)^{k-r}$. Among all the packets at stage s which were available to move forward, the expected value of the fraction which do move forward in a cycle is $rx_{c,s+1,t}/u_{c,s,t}$. This gives equation (7). A class c buffer at stage s would have l class c packets at time $t+1$, if it had i packets after considering a possible packet transmission in cycle t, and $l - i$ class c packet requests arrive. Thus we get equation (8). Equation (10) assumes that after the last stage, there is no blocking, but there is at most one packet transmission per cycle per port.

Equations (1) through (11) converge on iteration to provide the stationary probabilities $q_{c,s,l}$, $q'_{c,s,l}$, $u_{c,s}$, $u'_{c,s}$, $rx_{c,s}$ and $block_{c,s}$, as a function of the input loads P_c. When solving these equations iteratively, the time subscripts are ignored. (We assume that the probabilities converge to a stationary distribution).

For class c, the throughput per user γ_c and the average time delay T_c (found by using Little's formula) are given by

$$\gamma_c = u_{c,s}(1 - block_{c,s}) \quad \text{packets/cycle for any stage } s \tag{12}$$

$$T_c = \sum_{s=1}^{n} \left(\sum_{l=1}^{bs_c} q_{c,s,l} \cdot l \right) \cdot \frac{1}{(1 - q_{c,s,0})(1 - block_{c,s})} \quad \text{cycles} \tag{13}$$

4.2 Results

Fig. 2 shows the performance of a 32×32 crossbar (which is a single stage Banyan), or any non blocking network with output buffers and two traffic classes ('voice' and 'data'). We keep $P_v = P_d$, both increasing from 0 to 0.5. The broken curves show the analysis results, while the solid curves indicate the simulation results with 95% confidence intervals, which were obtained by collecting the steady state data over six independent runs, and then using the student-t distribution. Note in fig. 2a that because of priority, the 'voice' throughput (per user) is always greater than the 'data' throughput. In fig. 2b, the time delay (inside the network) for 'voice' packets is much smaller than that for 'data' packets, due to the priority. Fig. 3 shows the performance of a $2^4 \times 2^4$ Banyan network. Again the 'voice' shows higher throughputs and lower time delays than 'data'. Observe that as $(P_v + P_d)$ exceeds the maximum possible total throughput for the network, the 'data' throughput drops due to increasing competition from 'voice' packets. In both fig. 2 and fig. 3, the 'voice' throughput curve and the 'voice' time delay curves are identical to the single class analysis curves given in [19a]. Thus in effect the 'voice' packets do not 'see' the 'data' packets.

5. 'Control' of throughputs

Observe in fig. 3a that for equal 'voice' and 'data' offered loads, the 'voice' throughput is greater than 'data' throughput. This happens because the priority of 'voice' packets inside the network results in their being able to move ahead faster than 'data' packets. In order to have easy control, we may prefer a 'fair' splitting of the throughputs (in proportion to the offered loads). In this way, if say we want the network to carry 60% 'voice' throughput and 40% 'data' throughput, we simply offer the 'voice' load P_v and the 'data' load P_d in the ratio 6 : 4 at the inputs of the network. We may easily maintain P_v and P_d to desired values by using a simple controller to appropriately accept packets from the 'voice' and 'data' buffers at the network inputs Hence we want the priority to affect only the time delays, and not the throughputs.

It is surprisingly easy to achieve this fair splitting of the throughputs, by simply adding a single input buffer for each user at the network inputs (common to all the traffic classes). This input design is illustrated in fig. 1c. Now the packets at each input port, regardless of their class, are fed into the single input buffer common to all the classes. Whether an arriving packet is blocked or not now simply depends on whether this single input buffer is full or empty, and not on the packet's class. Once a packet enters an input buffer, it must eventually come out. Hence the throughputs are divided in a 'fair' manner, and both 'voice' and 'data' throughputs exhibit the same performance with respect to their corresponding loads. Therefore by controlling the fraction of 'voice' and 'data' packets offered to the common input buffer, (i.e. P_v and P_d), the fraction of 'voice' and 'data' throughputs through the entire network can be controlled easily.

The Markov chain analysis for an arbitrary number of traffic classes, with this input design, changes as follows. The variables n, k, m, bs_c, $q_{c,s,l,t}$, $q'_{c,s,l,t}$, $u_{c,s,t}$, $u'_{c,s,t}$, $rx_{c,s,t}$, and $block_{c,s,t}$ remain the same as defined in section 4, except that some variables referring to stage 0 (network inputs) now refer to the common single buffer at each network input, introduced in this section. Thus we have

$u_{c,0,t} = Pr[$the single buffer at a network input has a class c packet at time $t]$

$block_{c,0,t} = Pr[$a class c packet (if any) in the single buffer at a network input is not able to move forward during cycle $t]$

$iwbe_t = Pr[$the single buffer at a network input will be free to receive a possible incoming packet during cycle $t]$

In addition to equations (1) through (10) of section 4, we have:

$$iwbe_t = (1 - \sum_{c=1}^{m} u_{c,0,t}) + \sum_{c=1}^{m} u_{c,0,t}(1 - block_{c,0,t}) \quad (14)$$

$$u'_{c,0,t+1} \equiv u_{c,0,t+1} = u_{c,0,t} \cdot block_{c,0,t} + iwbe_t \cdot P_c \quad (15)$$

The throughput and time delay expressions are still given by equations (12) and (13). (The time delay does not include the delay in the common single input buffer).

Concentrating again on the two traffic classes case, fig. 3c shows the analysis and simulation results for a $2^4 \times 2^4$ network with this input design, with ($P_v = P_d$). Observe that as expected, the 'voice' and 'data' throughput curves overlap. See [17] for the time delay curves, which show that the voice time delays remain small as before.

Results for the case of three traffic classes and for very large networks appear in [17]. Although the Markov chains may not remain sufficiently accurate for large fully loaded multistage Banyans, as noted in [19a], they do convey the trends of the performance measures accurately. Thus they are useful in comparing Banyan networks with different design variations.

6. 'Control' of Time delays

Observe in fig. 3b that the average time delay per stage for 'voice' packets is close to the minimum possible time delay of 1 cycle per stage, while that for 'data' is much higher. For some applications, we may want to reduce the time delay of 'data' packets when the 'data' load is small (and consequently tolerate some increase in 'voice' time delays). Consider a flexible priority scheme in which we have N_v successive cycles in which 'voice' is given priority followed by N_d cycles in which 'data' has priority. (N_v and N_d are small integers). (Note that these priorities apply only inside the network and not at the common single buffer at the network inputs. The throughputs are still shared 'fairly'). The analysis for this case appears in the appendix. Observe in table 1 how the relative delays per stage of 'voice' and 'data' packets get affected by changing N_v and N_d. For example, for equal 'voice' and 'data' loads, we can go in steps from equal 'voice' and 'data' time delays to a 'voice' delay of 1.196 and 'data' delay of 4.93. Thus by having suitable combinations of cycles in which we give priority to 'voice' packets, or 'data' packets, we can exercise some 'control' over the average time delays of the traffic classes. The effect of flexible priority is to divide as well as share the network between the traffic classes. For example, keeping $N_v = N_d = 1$ is like giving half the switch each to voice and data, and letting them use the other halves if they are free.

A note on the packet size: Equation (12) gives the class c throughput per user, γ_c, in packets per cycle. A cycle is the time required to transmit a packet between two stages, plus some overhead for control. If each packet consists of i information bits and h header bits, then the throughput is $\gamma_c \cdot (i/(i+h+o))$ information bits/unit time, where o corresponds to the overhead and unit time is the time to transmit a bit between two stages. Thus for $i \gg (h+o)$, the throughput in information bits per unit time is independent of the packet size. On the other hand, the average time delay for class c is T_c cycles or $T_c \cdot (i+h+o)$ units of time. Thus the time delay in units of time increases directly with i. Hence we can reduce the time delay while keeping the throughput almost a constant by reducing i, until i approaches $(h+o)$, at which point the effective throughput starts decreasing quickly. Hence the packet size should be kept small. This would also decrease the amount of memory needed for the buffers. Note that this analysis does not assume 'cut-through' [11]. Of course in a practical situation, other cost factors may become more important.

7. Distribution of Time delays

Consider the case of Banyan networks with nodes having arbitrarily sized output buffers for each class, as in fig. 1b. We follow the approach in [19a]. Assuming the blocking at stage s to be independent from cycle to cycle, the distribution of the delay of the packet at the head of the class c queue at stage s is given by:

$$p_{c,s}(k) = (1 - block_{c,s})(block_{c,s})^{k-1} \quad \text{for } k = 1, 2, \ldots; s = 1, 2, \ldots n \quad (16)$$

By summing over the cases that a packet arrives to find $l-1$ packets in the queue, the expected distribution of the total delay at stage s is given by

$$P_{c,s} = \sum_{l=1}^{bs_c} p_{c,s}^l \cdot \frac{q_{c,s,l}}{(1 - q_{c,s,0})} \quad \text{for } s = 1, 2, \ldots n; \quad (17)$$

where $p_{c,s}^l$ denotes an l fold convolution of $p_{c,s}$. Finally, the end to end delay in the n stage Banyan for class c packets, is given by

$$P_c = P_{c,1} * P_{c,2} * \ldots * P_{c,n} \quad (* \text{ denotes convolution}) \quad (18)$$

The results for a $2^4 \times 2^4$ Banyan with nodes having output buffers of size 4 for both 'voice' and for 'data', and $P_v = P_d = 0.4$, appear in fig. 4. The analysis and the simulation agree approximately, except for the tails where the probabilities are very small. See [17] for more plots. The delays in any external queues may be incorporated in a similar manner.

8. Conclusion

In this paper we reported approximate Markov chain models for Banyan based packet switches built with nodes of arbitrary size, an arbitrary number of traffic classes, and each node having output buffers of arbitrary sizes. While we analyzed the case of separate buffers for each traffic class to get the effect of preemptive priority, in practice, there may be some buffer sharing. Next we proposed simple design changes at the inputs of the network, in order to better 'control' the splitting of the total throughputs into the throughputs of the various traffic classes. We also reported on and modeled some flexible priority schemes, aimed at obtaining a better 'control' over the time delays of the individual traffic classes. We were able to incrementally decrease the time delay of one traffic class, while increasing that of the other traffic class. The time delay distribution for the different traffic classes was obtained. Since a single stage Banyan is simply a crossbar switch, our Markov chains also give the performance of arbitrarily sized crossbars (and other non blocking networks) having output buffers of arbitrary size, in a multiclass environment. The analyses presented explicitly modeled the time behavior, and may be used to evaluate transient responses. Quicker convergence may be easily obtained by using the steady state formulæ for birth death processes in the analyses, as done in [19b] for the single traffic class case.

9. Acknowledgements

The authors acknowledge useful discussions with Mark Karol and Aurel Lazar, and the financial support of the National Science Foundation (grant CDR-88-11111).

10. Appendix

Markov chain analysis for Banyan networks with flexible priority schemes

Consider the case of two traffic classes namely 'voice' (class 'v') and 'data' (class 'd'). Equations (1) through (10), (14) and (15) continue to hold for $c = v, d$, except that equation (5) is modified as:

$$u'_{d,s,t} = \begin{cases} u_{d,s,t}, & \text{if 'data' has priority in cycle } t \\ u_{d,s,t}((1 - u_{v,s,t}) + u_{v,s,t} \cdot block_{v,s,t}) & \text{otherwise} \end{cases} \quad (19)$$

$$u'_{v,s,t} = \begin{cases} u_{v,s,t}, & \text{if 'voice' has priority in cycle } t \\ u_{v,s,t}((1 - u_{d,s,t}) + u_{d,s,t} \cdot block_{d,s,t}) & \text{otherwise} \end{cases} \quad (20)$$

Let us consider the case of $N_v = 2, N_d = 1$ i.e., we consecutively have two cycles with 'voice' priority and one cycle with 'data' priority. In this case, when we iterate these equations, the probabilities would go through a cycle with a period of three ($N_v + N_d = 3$). Thus each of the probabilities 'converges' to three values, which we denote by the subscript cl. This subscript replaces the time subscript t, which is dropped to indicate stationary distribution. When solving these equations iteratively, we ignore the time subscripts and use only the cl subscript, $cl = 1, 2, \dots N_v + N_d$. Then $cl + 1$ implies $(cl + 1)$ mod $(N_v + N_d)$.

For class c, the throughput per user γ_c (in packets / cycle) and the average time delay T_c (in cycles) are given by

$$\gamma_c = \frac{1}{N_v + N_d} \sum_{cl=1}^{N_v + N_d} u_{c,s,cl}(1 - block_{c,s,cl}) \quad (21)$$

$$T_c = \sum_{s=1}^{n} \frac{\frac{1}{N_v + N_d} \sum_{cl=1}^{N_v + N_d} \sum_{l=1}^{bs_c} q_{c,s,l,cl} \cdot l}{\frac{1}{N_v + N_d} \sum_{cl=1}^{N_v + N_d} (1 - q_{c,s,0,cl})(1 - block_{c,s,cl})} \quad (22)$$

10. References

[1] H. Ahmadi and W. Denzel, "A survey of modern high-performance switching techniques", *IEEE JSAC*, Vol. SAC-7, No. 7, Sept. 1989, pp. 1091-1103.

[2] R.G. Bubenik and J.S. Turner, "Performance of a Broadcast Packet Switch", *IEEE Trans. Communications*, Vol. 37, No. 1, Jan. 1989, pp. 60-69.

[3] J.S.-C. Chen and R. Guerin, "Input Queueing of an Internally Non-Blocking Packet Switch with Two Priority Classes", *Proc. Infocom 1988*, pp. 529-537.

[4] D.M. Dias and J.R. Jump, "Analysis and Simulation of Buffered Delta Networks", *IEEE Trans. Comput.*, Vol. C-30, No. 4, April 1981, pp. 273-282.

[5] D.M. Dias and M. Kumar, "Packet switching in Log N Multistage Networks", *Proc. Globecom 1984*, pp. 5.2.1-5.2.7.

[6] K.Y. Eng, M.G. Hluchyj and Y.S. Yeh, "A Knockout switch for variable length packets", *Proc. ICC 1987*, pp. 22.6.1-22.6.6.

[7] A. Huang and S. Knauer, "Starlite: a Wideband Digital Switch", *Proc. Globecom 1984*.

[8] J. Hui and E. Arthurs, "A Broadband Packet Switch for Integrated Transport", *IEEE JSAC*, Vol. SAC-5, No. 8, Oct. 1987, pp. 1264-1273.

[9] Y-C. Jenq, "Performance Analysis of a Packet Switch Based on Single-Buffered Banyan Network", *IEEE JSAC*, Vol. SAC-1, No. 6, Dec. 1983.

[10] M.J. Karol, M.G. Hluchyj, and S.P. Morgan, "Input vs. Output Queueing In a Space Division Packet Switch", *Proc. Globecom 1986*.

[11] P. Kermani and L. Kleinrock, "Virtual Cut-Through: A New Computer Communication Switching Technique", *Computer Networks*, Vol. 3, 1979, pp. 267-286.

[12] H.S. Kim and A.L. Garcia, "Performance of Buffered Banyan Networks Under Nonuniform Traffic Patterns", *Proc. Infocom 1988*.

[13] C.P. Kruskal and M. Snir, "The Performance of Multistage Interconnection Networks for Multiprocessors", *IEEE Trans. Comput.*, Vol C-32, pp. 1091-1098, Dec. 1983.

[14] A.A. Lazar, R. Gidron and A. Temple, "A Switching Architecture for Asynchronous Time Sharing", *Proc. Globecom 1989*, pp. 1166-1172.

[15] San-qi Li, "Performance of a Non-Blocking Space-Division Packet Switch with Correlated Input Traffic", *Proc. Globecom 1989*, pp. 1754-1763.

[16] D. Mitra and R. A. Cieslak, "Randomized Parallel Communications on an extension of the Omega Network", *Journal of A.C.M.*, Vol. 34, No. 4, Oct. 1987, pp. 802-824.

[17] S.Z. Shaikh, *Ph. D. Thesis*, in preparation, Electrical Engr. Dept., Columbia Univ., New York.

[18] S.Z. Shaikh, M. Schwartz and T.H. Szymanski, "Performance Analysis and Design of Banyan Network Based Broadband Packet Switches for Integrated Traffic", *Proc. Globecom 1989*, pp. 1154-1158.

[19a] T.H. Szymanski and S.Z. Shaikh, "Markov chain analysis of Packet-Switched Banyans with arbitrary switch sizes, queue sizes, link multiplicities and speedups", Proc. *Infocom 1989*, pp. 960-971; [19b] extended version submitted for publication and also to appear as CTR Technical report.

[20] J.S. Turner, "Design of a Broadcast Packet Network", *IEEE Trans. Comput.*, Vol. C-36, No. 7, July 1987, pp. 810-822.

[21] L.T. Wu, "Mixing Traffic in a Buffered Banyan Network", *Proc. 9th Data Comm. Symp.*, Sept. 1985.

[22] Y.S. Yeh, M.G. Hluchyj and A.S. Acampora, "The Knockout switch: A simple modular architecture for high performance packet switching", *Proc. ISS 1987*, pp. B10.2.1-B10.2.8.

	$N_v=1$, $N_d=0$		$N_v=2$, $N_d=1$		$N_v=1$, $N_d=1$	
	T_v	T_d	T_v	T_d	T_v	T_d
$P_v=0.5$, $P_d=0.5$	1.196	4.933	2.058	4.315	3.360	3.360
$P_v=0.7$, $P_d=0.3$	1.426	7.715	2.963	3.724	3.306	2.200

Table 1: *Analysis of a $2^4 \times 2^4$ Banyan built with nodes having output buffers of size 4 each for 'voice' and 'data' packets, and input design of fig. 1c; 'Control' of Time Delays is illustrated.*

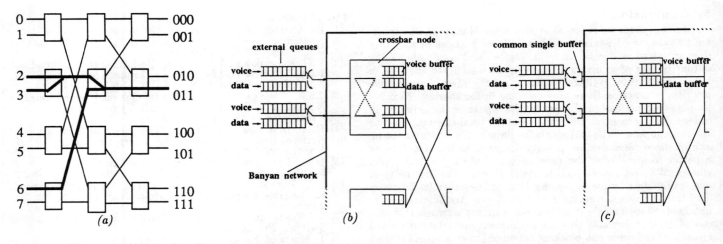

Figure 1: *(a) A $2^3 \times 2^3$ SW-Banyan network showing routing from inputs 2, 3 and 6 to output 3. (b,c) Two input designs for a $2^n \times 2^n$ Banyan with nodes having output buffers for 'voice' and 'data' at each output port.*

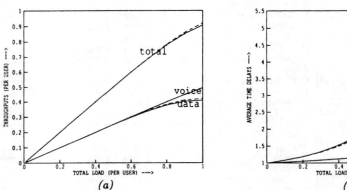

Figure 2: *Analysis (broken curves) and simulations (solid curves) for a 32×32 Crossbar with output buffers of size 4 each for both 'voice' and 'data' packets and $P_v = P_d$; (a) Throughputs per user and (b) Average time delays vs. $(P_v + P_d)$. For 'voice' throughputs and time delays, the analysis and simulation curves overlap.*

Figure 4: *Distribution of voice time delays for a $2^4 \times 2^4$ Banyan built with nodes having output buffers of size 4 each for both classes, and input design of fig. 1b; $P_v = P_d = 0.4$.*

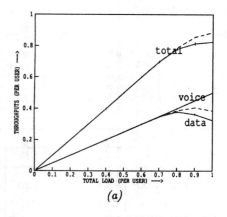
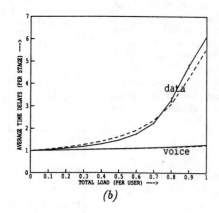
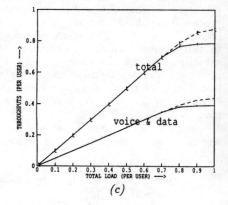

Figure 3: *Analysis (broken curves) and simulations (solid curves) for a $2^4 \times 2^4$ Banyan built with nodes having output buffers of size 4 each for both 'voice' and 'data' packets and $P_v = P_d$; (a) Throughputs per user and (b) Average time delays per stage vs. $(P_v + P_d)$, for input design of fig. 1b. For 'voice' throughputs and time delays, the analysis and simulation curves overlap. (c) Throughputs per user for input design of fig. 1c.*

AN ANALYSIS OF "HOT-POTATO" ROUTING IN A FIBER OPTIC
PACKET SWITCHED HYPERCUBE

Ted Szymanski

Department of Electrical Engineering and
Center for Telecommunications Research
Columbia University
New York, NY 10027
e-mail: teds@sirius.ctr.columbia.edu

Abstract: Two implementations of a fiber-optic packet-switched hypercube are proposed. In the first, each directed link is implemented with a fixed wavelength laser and photodetector, and all optical transmissions are wavelength multiplexed onto one or more fibers. In the second, the electronic crosspoint matrices within the nodes are eliminated by allowing each laser to tunable over a range of $logN$ wavelengths. Assume that a "hot-potato" or "deflection" routing algorithm is used; as soon as a packet is received at a node, a routing decision is made and the packet is sent out. The node attempts to send the packet towards its destination, but if this is not possible the packet is sent out away from its destination. Our analysis indicates that in a hypercube, hot-potato routing offers essentially optimal performance for random traffic, regardless of how large the hypercube grows, and it significantly outperforms traditional shortest path routing with buffering and flow control. A few variations, including an algorithm which gives priority to packets closer to their destinations and one which gives priority to various classes of traffic are also proposed and analysed. Our analytic methods for hot-potato routing and/or prioritized traffic are general and applicable to many other networks.

1. Introduction

The hypercube is very popular in commercial multiprocessors since it uses relatively simple distributed routing algorithms, it can simulate many other topologies including meshes, trees, rings, banyans and the Benes network, it is very fault tolerant and its cost is moderate. A fiber optic hypercube has a very high link bandwidth (in the Gigabit/sec range), it can be physically distributed, and its physical interconnection can be far simpler than that of an electronic hypercube. Such a network may be useful in future GaAs-based multiprocessors, in localized fast packet switches and as a distributed high-speed LAN/MAN.

Traditionally the hypercube is packet-switched and uses variations of shortest-path routing with buffering and flow-control. However, there are problems with flow control; (1) Flow control requires that the receiving node be able to inform the sending node when its buffers are full, requiring a backwards transmission mechanism. Such backwards status links can be easily implemented in electrical hypercubes by adding an extra wire to each directed link for this purpose. However, the creation of optical status links would double the number of lasers/photodetectors needed in any fiber-optic network. Also, in a distributed system where each link can dynamically buffer many packets flow control is not possible. (2) The existence of directed cycles in any "store-and-forward" network (i.e., with buffering and flow control) is a necessary and sufficient condition for deadlock to occur. Thus, both the hypercube and recirculating shuffle based networks with buffering and flow control can deadlock. (One can still estimate the performance of recirculating shuffles by assuming store-and-forward mode with large/infinite queues [10].)

This research was supported by CTR through NSF Grant CDR-88-11111.

The existence of directed cycles favours the use of the "hot-potato" routing in a hypercube; as soon as a packet is received at a node, a routing decision is made and the packet is sent out. Therefore, a node always has sufficient buffer space to receive any new packets and thus deadlock is impossible. This approach also eliminates the need for backwards status links to implement flow control. However, the routing algorithm occasionally sends packets out in the wrong direction, thereby increasing the amount of time and resources (links and buffers) that they use. It is necessary to quantify the effects of these deflections on the system's performance. This author is unaware of any previous analysis of hot-potato routing in the hypercube. (Deadlock can be avoided in hypercubes by constraining packets to traverse necessary dimensions in a fixed order and not allowing out-of-order traversals [5], but this 'LR' scheme eliminates most minimum distance alternate paths and can severely reduce the performance (see [1], pp. 1009)).

Two variations of the hot-potato algorithm are proposed. In the "basic" version, all packets have the same priority. In the "deluxe" version, packets closer to their destinations have higher priorities in order to remove them from the system as soon as possible. Our analysis and simulations indicate that even with the basic algorithm, the hypercube has essentially optimal performance when the traffic is randomly and uniformly distributed; i.e., in a hypercube hot-potato routing significantly out-performs shortest-path routing with buffering and flow control. In addition, hot-potato routing provides a powerful fault tolerance mechanism, by routing around faulty links (or nodes) whenever they are encountered.

Our analysis indicates that hypercubes with millions or billions of nodes still exhibit essentially optimal performance for random uniform traffic. Our simulations indicate that the hypercube with hot-potato routing is robust even with specific permutation traffic patterns with average path distances of $logN/2$; provided that each node receives no more than its fair share of bandwidth, it appears to be very difficult to "clog up" the network and cause a significant drop in throughput under these assumptions.

As a backbone LAN/MAN, the hypercube may carry voice, video and data traffic, each with different priorities and grades of service. An analysis of the network with an arbitrary number of prioritized traffic classes is also presented. It is shown that the assignment of priorities reduces the delays experienced by higher priority traffic, as expected.

In summary, our analytic models allow the designer to compute the average bandwidth, the average delay, delay distributions and the probability that an average packet (or a packet with a given initial distance) is delayed by more than an arbitrary threshold, for each traffic class. The analysis is also easily extendible to include buffering within the nodes. Our analysis assumes that all links are independent, and hence is mostly independent of the specific topology (the topology affects only a few equations). Thus the analysis

Reprinted from *Proc. IEEE INFOCOM '90*, pp. 918–925, June 1990.

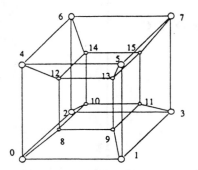

Figure 1: A binary 4-cube.

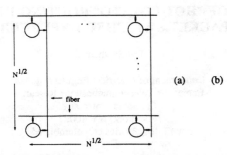

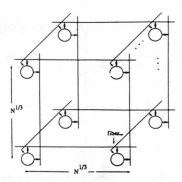

Figure 2: 2D (a) and 3D (b) constructions of fiber-optic hypercubes.

is easily adaptable to other topologies using deflection routing. (We have adapted our analysis to model recirculating shuffle based networks and the analytic results were reasonably accurate; those results will appear elsewhere). Our analysis of prioritized traffic is simple and applicable to other networks.

This paper is organized as follows. Section 3 describes the fiber optic hypercube implementations. Section 4 describes the hot-potato routing algorithm and presents the analyses. Section 5 presents both analytic and simulation results. Section 6 discusses a few extensions of the basic model. Section 7 contains concluding remarks.

2. Background

An n-dimensional binary hypercube consists of 2^n nodes, each of degree n ($logN = n$). † Each node has a unique binary address consisting of n bits, and all N nodes span the address range $0...N-1$. The hypercube has a recursive definition; to create an $n+1$ dimensional hypercube, two n dimensional hypercubes are arranged side-by-side. An extra address bit is prepended to all nodes; those in one half receive a leading 0 bit, and those in the other half receive a leading 1 bit. Finally, the degree of each node is increased by one, and each node in one half is connected to one node in the other half (the one with the same least significant n address bits) using this extra IO port. (Assume that each link is bidirectional.) A 4-cube of 16 nodes is shown in figure 1

3. Fiber Optic Hypercube Implementations

3.1. Implementation #1: Fixed Wavelength Lasers

The "multihop" technique described in [2] can be used to implement all $NlogN$ directed links in a hypercube. Each directed link is assigned a unique wavelength, and all links are wavelength multiplexed onto one (or more) optic fibers. The switching function in each node is still implemented with electronic crossbars. In principle, a single fiber can carry all the wavelengths and the fiber can be arranged to visit each node sequentially, much like a bus topology. Alternatively, the fiber can be arranged as a passive star where all nodes are at the leaves. All nodes transmit on their own wavelengths up to the root (a large star coupler), which combines the optical power and then broadcasts the sum to each node.

WDM offers the significant advantage of reducing the physical interconnection complexity of the hypercube considerably. The layouts shown in fig. 2 appear particularly attractive for both localized interconnection networks and distributed MANs. In fig. 2a, each axis has $\sqrt{N}$ nodes and each fiber along one axis carries $\sqrt{N} \cdot log(\sqrt{N}) = \sqrt{N} \cdot n/2$ wavelengths. In fig. 2b, each axis has $N^{1/3}$

---
† All logarithms are to base 2.

nodes and each fiber carries $N^{1/3} \cdot log(N^{1/3})$ wavelengths. (To change the number of wavelengths on each fiber, multiple fibers can be used, the number of nodes along any axis can be varied, and more than 3 axes can be used.)

3.2. Implementation #2: Tunable Lasers

It is possible to eliminate the need for the electronic crosspoint matrices in the above implementation by using tunable lasers. The implementation of a 2-cube ($logN = 2$) using tunable lasers is shown in fig. 3a, and the design of a node is shown in fig. 3b. Each node contains $logN$ fixed wavelength photodetectors, which perform the optical-to-electronic (OE) conversion. In every node, the photodetector associated with the incoming link on dimension j ($0 \le j < n$) is always tuned to receive wavelength λ_j. Every packet that arrives at a photodetector is fed horizontally into the corresponding GaAs shift register and stored there while the routing decisions are made. (It is necessary to buffer only the part of a packet that is received while a routing decision is made.)

While the crosspoint matrix is eliminated, the distributed routing algorithm must still be performed by GaAs logic. The logic determines whether each packet is destined for this node; if so, assume that a multiplexer (electrical or wavelength) can route the packet into a receive buffer. We assume that this buffer is big enough to avoid overflow, as is usually assumed in the analysis of hypercubes. (Modelling a finite size buffer is straight-forward.) If the packet is to be sent out, it is forwarded horizontally from its shift register into the corresponding tunable laser, which performs an electrical-to-optical (EO) conversion. In every node, if a packet is to be sent out over dimension j, ($0 \le j < n$), then its tunable laser is tuned to wavelength λ_j before the transmission starts. (Assume that the node initially submits its packet into an extra shift register which also has its its own tunable laser.)

Observe that within any node, at most one packet will be propagated out over dimension j, and therefore at most one laser in each node will be transmitting over wavelength λ_j. The outputs of all $logN$ lasers of any one node are combined in a passive star-coupler. For every node i in the hypercube, ($0 \le i < N$), one of its coupler outputs is fed to the jth photodetector of neighboring node $i' = i$ *exor* 2^j (exor is the "exclusive or" logic function). It is not difficult to verify that the connectivity of the hypercube is preserved with this WDM scheme.

Note that the methodology can be used to implement any network topology. In each case the electronic crosspoint matrices in the nodes are eliminated, although the controllers that perform the distributed routing are still required. If all of the directed links in an arbitrary network (or directed edges in a graph) can be coloured using k colors such that no two directed edges leaving the same node have the same colour, then the fiber optic implementation of this network will require that the lasers be tunable to at most k

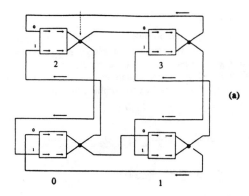

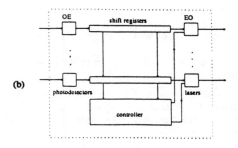

Figure 3: (a) WDM 2-cube using tunable lasers. (b) Node design.

wavelengths.

By including a tunable photodetector at each shift register and feeding a coupler output back to each input, it may be possible to simplify the GaAS controller; each input port with a packet first listens to the wavelength of a desired output port, and if it is idle then it attempts to claim it by initiating a transmission on that wavelength. In case of collisions, some input ports should backoff and search for other output ports. To minimize the probability of collisions, each input port i should search the necessary links in any prefered order (i.e., decreasing with wrap-around, starting at output link $i-1$).

Finally, the above implementation replaced the large central star coupler by many smaller couplers. One or more of the smaller couplers can be combined into one larger coupler, so that the fiber-efficient constructions shown in fig. 3 can be used, provided that the all wavelengths on any one coupler are distinct.

4. The Deflection Routing Algorithm

Assume the network is operated in a synchronous or discrete-time manner. At the beginning of a clock cycle (or time-slot), conceptually every node transmits all of its packets and simultaneously receives many packets from its neighbors. Assume that a node considers incoming packets, one at a time and in any order, when it makes routing decisions. Assume that the routing decisions made during time t will be used when the incoming packets are transmitted out in time $t+1$.

Each packet carries an n-bit destination tag, which identifies the destination node. Every time a packet is received, the node compares the packet's destination tag to its own address. If they are equal then the packet has reached its destination, and it is forwarded into the receive buffer and removed from the network. Otherwise the node must transmit it out in the next clock cycle to avoid deadlocking the system and to avoid overwriting the buffer when a new packet arrives.

Suppose a packet in a node is d hops away from its destination; then its destination tag differs from the node's binary address by exactly d bits (or dimensions). The packet can be sent out over any of these d dimensions and move closer to its destination, and all other $n-d$ dimensions take the packet away from its destination. When a packet has distance d, the node searches the d suitable outgoing links looking for one that will be free (in the next clock cycle). If such a link is found, then it is reserved for this packet (in the next clock cycle). Otherwise, the packet will be deflected (in the next clock cycle). There are two possible ways to handle packets that will be deflected. In the first way, a packet selects a free outgoing link from the unsuitable outgoing links at random, right away. It does not wait until all other packets have examined the set of outgoing links before doing so. Therefore, it may select an out-

going link that causes another packet to be deflected. In the second way, a packet that must be deflected waits for all other packets to examine and claim outgoing links; after all such packets have claimed links, only the deflected packets are "left over". These deflected packets then claim the remaining outgoing links in any order, since we assume that the direction in which a packet is deflected does not matter. Our analyses can model either of these variations.

The synchronous mode is assumed to simplify the analysis, but it is possible to operate the fiber-optic hypercube asynchronously. To insert a new packet the node finds a free suitable output link and starts the transmission. However, the transmission may have to be aborted if necessary in order to send out a packet that arrives from a neighbor. In this case, the shift registers should be able to buffer a complete packet, so that the successor will be able to detect an aborted packet. The asynchronous scheme is suitable for distributed hypercubes, which would probably have some output queueing within each node to minimize deflection probabilities.

4.1. Analysis for the Basic Deflection Routing Algorithm

Assume that any new packets inserted into the network must traverse any particular dimension with probability α, so that the average path distance (including self-destined traffic) is $\alpha \cdot logN$. Thus, a random uniform distribution has $\alpha = 0.5$ and an average path distance of $logN/2$, which is consist with other hypercube traffic models used in the literature. (One can easily remove self-destined traffic from our model, but the average distance is then slightly greater than $\alpha logN$.) Due to the traffic assumption and the symmetry of the hypercube, all nodes are statistically identical, and since there is no flow control then all nodes are also independent. Therefore the state of the entire hypercube can be characterized by the state of one node. To model a node further simplifying assumptions are made. Assume that all links (and therefore all input buffers) in the network are statistically identical and independent. In addition, simultaneous events that cause the state transitions of a node are considered one at a time, in a series of *conceptual time phases*, in order to simplify the analysis considerably.

Let n denote the number of dimensions in the hypercube, i.e., an n-dimensional hypercube contains 2^n nodes. Let Pr $[e]$ denote the probability that event e occurs, and let the phrase "cycle t" denote the time interval $[t, t+1)$. The phrase "time t" denotes the beginning of clock cycle t. Throughout the development, assume that variables without time subscripts correspond to steady state values. The following variables are defined.

$dst_{d,t}$ = Pr [a packet is distance d away from its destination at the beginning of time t]

ui_t = Pr [a packet will arrive on an incoming link during cycle t]

uo_t = Pr [a packet that arrived on an incoming link during cycle t did not reach its destination and must be propagated out]

409

$rxbw_t$ = average number of packets received by a node during cycle t

$txbw_t$ = average number of packets transmitted by a node during cycle t

λ = Pr [a node will be offer a packet to the network any cycle (the *offered load*)]

The mnemonics ui and ui denote the utilization of input and output links respectively. The equations (1)-(9) are not in the most compact form in order to make them self-explanatory. The arrays and variables are initialized as follows.

$$dst_{d,0} = \binom{n}{d}\alpha^d \cdot (1 - \alpha)^{n-d} \tag{1}$$

ui_0 is initialized to λ/n. The receive and transmit bandwidths are computed as follows.

$$rxbw_t = ui_t \cdot dst_{0,t} \cdot n \tag{2}$$

$$uo_t = ui_t(1.0 - dst_{0,t}) \tag{3}$$

$$txbw_t = \left[1 - \binom{n}{n}uo_t{}^n\right] \cdot \lambda \tag{4}$$

After any new packets are inserted into the network, each node will have on average $n \cdot uo_t + txbw_t$ packets. Assume that a node inserts at most one packet per clock cycle into the network. The new distance vector consists of the weighted average of the original vector and the current vector as follows;

$$dst'_{d,t} = \frac{uo_t \cdot n}{n \cdot uo_t + txbw_t} \cdot dst_{d,t} + \frac{txbw_t}{n \cdot uo_t + txbw_t} \cdot dst_{d,0} \tag{5}$$

for $0 \le d \le n$.

A node may transmit to itself with a small probability $txbw_t \cdot (1 - \alpha)^n$. The analysis must now consider self-destined packets and update the relevant variables as follows. (In reality, self destined traffic need not be inserted into the network before being removed, but the fraction of self-destined traffic should be fairly low and this assumption does not affect the results by much. In either case, it is straight forward to adapt the analysis to not include self-destined traffic.)

$$uo_t = uo_t + txbw_t \cdot (1 - (1-\alpha)^n / n) \tag{6}$$

$$rxbw_t = rxbw_t + txbw_t \cdot (1 - \alpha)^n \tag{7}$$

$$dst'_{0,t} = 0 \tag{8}$$

$$dst''_{i,t} = dst'_{i,t} / \sum_{j=0}^{n} dst'_{j,t} \tag{9}$$

During each clock cycle assume that the input buffers are serviced (in any order) one during each phase p ($1 \le p \le n$). (Assume that the input buffers are labelled $1...n$.) The following variables are defined for this aspect of the analysis, where p is used to denote the phase.

$cu_{p,t}$ = Pr [an outgoing link will carry a packet (in the next clock cycle) after considering the effects of input buffers $1..p$ during time t]

$pb_{p,d,t}$ = Pr [a packet in an input buffer whose current distance is d will not be able to move towards its destination over a suitable outgoing link, after considering the effects of input buffers $1...p-1$, during time t]

$amv_{p,t}$ = Pr [a packet will win exclusive access to a suitable outgoing link and hence will be able to move closer to its destination in the next cycle, averaged over all distances, during phase p of time t]

$mv_{p,t}$ = Pr [a packet in an input buffer will be able to move towards its destination over a suitable outgoing link in the next cycle, averaged over packets in input buffers $1...p-1$,

during time t]

The vectors are initialized to 0 initially, i.e., $pb_{0,t} = cu_{0,t} = 0$, for the phase 0 of time t, and assign $pb_{p,n,t} = 0$ and $pb_{p,d,t} = 0$ when $d \ge p$. Eqs. (10-12) are computed for each of the n phases labelled $p = 1..n$. These equations are approximations and can be replaced by more complicated combinatorial expressions.

$$pb_{p+1,d,t} = \binom{d}{d}cu_{p,t}^d \cdot (1 - cu_{p,t})^{d-d} \tag{10}$$

$$amv_{p+1,t} = \frac{1}{n} \cdot \sum_{d=1}^{n} (1 - pb_{p,d,t}) \tag{11}$$

$$cu_{p+1,t} = uo_t \cdot amv_{p+1,t} \cdot \frac{1}{n} + cu_{p,t} \tag{12}$$

As described earlier, the node operation could be changed to allow deflected packets to select outgoing links right away, which may cause other packets to be unnecessarily deflected. To model this mode of operation, eq. (12) is replaced by the eq. (13).

$$cu_{p+1,t} = uo_t \cdot p/n \tag{13}$$

Next, the probability that a packet whose distance is d at time t will not be deflected during time t is computed as follows:

$$mv_{d,t} = \frac{1}{n} \cdot \sum_{p=1}^{n} (1 - pb_{p,d,t}) \tag{14}$$

for $0 \le p < n$. (Set $mv_{n,t} = 1$.) The probability that a packet of distance d is deflected changes with each phase, and hence the average is determined over all phases. Finally, the probability that an incoming link will receive a packet in the next clock cycle is assigned;

$$ui_{t+1} = uo_t \tag{15}$$

The distance vector for the next clock cycle must be updated to reflect the fact that some packets move closer to their destinations and some are deflected and move further from their destinations, as follows;

$$dst_{d,t+1} = dst''_{d+1,t} \cdot mv_{d+1,t} + dst''_{d-1,t} \cdot (1.0 - mv_{d-1,t}) \tag{16}$$

$$dst_{n,t+1} = dst''_{n-1,t} \cdot (1.0 - mv_{n-1,t}) \tag{17}$$

The preceding equations express the dynamics of the Markov chain model and they are solved by iteration until they converge. In equilibrium, the average transmission and reception bandwidths ($txbw$ and $rxbw$) clearly must be equal; they represent the "carried load" as a function of the offered load λ. Note that this time-dependent analysis can also be used to compute the system response to periodic offered loads.

There does not appear to be a simple closed form expression using stationary probabilities for the average delay. An iterative approach appears to be the only tractable method. Let $ad_{d,t}$ be the average delay experienced by a packet with initial distance d to reach its destination. The t subscript here denotes the iteration number. The equilibrium values of the "move" probability mv are used in these equations, which are iteratively computed until ad converges. Initially, let $ad_{d,0} = 0$, for $0 \le d \le n$.

$$ad_{d,t+1} = mv_d \cdot (1.0 + ad_{d-1,t}) + (1.0 - mv_d) \cdot (1.0 + ad_{d+1,t}) \tag{18}$$

$$ad_{n,t+1} = mv_n \cdot (1.0 + ad_{n-1,t}) \tag{19}$$

Having computed the equilibrium values of ad_d ($0 \le d \le n$), the average delay nd of a packet leaving the network can be computed as follows.

$$nd = \sum_{d=0}^{n} ad_d \cdot \binom{n}{d}\alpha^d \cdot (1 - \alpha)^{n-d}. \tag{20}$$

Networks with large variances in the packet delay may slow down a computer system considerably, since often the entire system must wait for the slowest synchronizing event to occur before doing work. To compute the delay distribution of an average packet leaving the hypercube, define the following variables;

$dd_{i,j,t}$ = Pr[a packet with initial distance i requires j cycles before reaching its destination, at iteration t]

A packet with initial distance d may require an infinite number of cycles before reaching its destination, this case occurring when a packet is repeatedly deflected ad. infinitum. However, the probability that a packet requires greater than some finite number of clock cycles is very small. Therefore, we may place an upper bound on j, and a reasonable bound is $x = 10 \cdot n$. An iterative approach must be used to compute the equilibrium values of $dd_{i,j}$. In eq. (21) the value of i is first held constant while j varies from $0...x$. Initially, assign $dd_{i,j,0} = 0$ for all $i, j > 0$, and $dd_{0,0,0} = 1$. By truncating the delay distribution at x, the normalization in (23) is required.

$$dd'_{i,j,t+1} = mv_i \cdot dd_{i-1,j-1,t} + (1.0 - mv_i) \cdot dd_{i+1,j-1,t} \quad (21)$$

$$dd'_{n,j,t+1} = mv_n \cdot dd_{n-1,j-1,t} \quad (22)$$

$$dd_{i,j,t+1} = dd'_{i,j,t+1} / \sum_{j=0}^{x} dd'_{i,j,t+1} \quad (23)$$

The probability that a packet with initial distance i is deflected y times ($y \geq 0$) is given by $dd_{i,i+2y}$, and the average number of deflections for a packet with initial distance i is given by

$$\sum_{y=0}^{x/2} dd_{i,i+2y} \cdot y \quad (24)$$

The average delay for a packet with initial distance d is also given by

$$nd = \sum_{j=0}^{x} dd_{d,j} \cdot j \quad (25)$$

and therefore the average delay over all packets is given by

$$\sum_{d=0}^{n} \left[\binom{n}{d} \alpha^n \cdot (1 - \alpha)^{n-d} \cdot \sum_{j=0}^{x} dd_{d,j} \cdot j \right] \quad (26)$$

The probability that a packet with initial distance d experiences a delay greater than h time slots is given by

$$\sum_{j=h+1}^{x} dd_{d,j} \quad (27)$$

and the probability an average packet experiences a delay greater than h time slots is given by

$$\sum_{d=0}^{n} \left[dst_{d,0} \cdot \sum_{j=h+1}^{x} dd_{d,j} \right] \quad (28)$$

4.2. Analysis for the Deluxe Deflection Routing Algorithm

Packets closer to their destination have fewer minimum distance paths available. By allowing such packets to have higher priority, the probability that a packet is deflected may be reduced, and the entire system performance should be improved. To model this behavior, equations (10-12) are modified as follows. The following variables are defined:

$cu_{d,i,t}$ = Pr [an outgoing link carries a packet after considering the effects of packets with distances $1..d$ in input buffers $1..i$ during time t]

$pb_{d,i,t}$ = Pr [a packet with distance d in input buffer i cannot move closer to its destination, after considering the effects of packets with distances $1...d-1$ in all input buffers and packets with distance d in input buffers $1..i-1$, during time t]

$mv_{d,t}$ = Pr [a packet with distance d will move closer to its destination during time t]

We consider packets in order of increasing distance $1 \leq d \leq n$, and having selected a distance, we consider input buffers in a linear order $1 \leq i \leq n$. (The actual order for i is irrelevant, since by assumption all input buffers are identical and independent). Both eq. (29) and (30) are evaluated in this manner. Initially, $cu_{d,i,t} = 0$ for $0 \leq d \leq n$ and $0 \leq i \leq n$.

$$pb_{d,i,t} = \binom{d}{d} cu_{d,i-1,t}^d \cdot (1 - cu_{d,i-1,t})^0 \quad (29)$$

$$cu_{d,i,t} = \sum_{j=0}^{i-1} uo_t \cdot dst_{d,t} \cdot (1 - pb_{d,i,t}) \cdot \frac{1}{n} \quad (30)$$

The average probability that a packet of distance d moves forward during cycle t must be computed in eq. (31). Thereafter, the analysis remains the same as in the previous section.

$$mv_{d,t} = \frac{1}{n} \cdot \sum_{i=0}^{n} (1.0 - pb_{d,i,t}) \quad (31)$$

4.3. Prioritized Traffic

In integrated networks, the voice and video traffic typically receive higher priority than bulk data transfers, since voice packets are useless if their delay exceeds a certain threshold. Priority is especially important in networks using hot-potato routing, since it is possible (but not very probable in the hypercube) that some packets may be delayed indefinitely.

In every other analysis of integrated networks this author is aware of, the parameters of interest for each individual traffic class are explicitly modeled using variables for each traffic class (i.e., see [4][9]). Explicitly considering the effects of multiple traffic classes usually becomes complicated or intractable in complicated systems.

An analysis for a work-conserving priority queue was presented in [7][11]. This analysis first computes the average number of customers of each class within a single queue of infinite size, and then recursively computes the waiting time of each class, based on the average service time of each class. Our approach is also recursive, but it assumes that average service times for each class are identical (i.e., assume all packets are fixed sized ATM cells). It never computes the average number of customers of each class within a single queue, and in fact our approach applies to systems such as networks of queues and even to networks without buffers/queues. Our technique assumes the following; (1) all the system traffic is distinguishable only by its priority (i.e., all other parameters such as message/packet length, service times, etc. are identical), (2) the system is 'work-conserving', i.e., all service arbiters in the network are work-conserving, and (3) all resources in the system (i.e., buffers and links) are shared among all priority classes. In this case an analytic model for the system's performance with homogeneous traffic can be used to estimate the system's performance with an arbitrary number of prioritized traffic classes. (This technique is described in detail and applied to other systems including unbuffered or buffered space-division switches and unbuffered or buffered banyans in [13].)

Suppose there are k traffic classes constituting the offered load to the network, each with relative probabilities p_j, $0 \leq j < k$, ($p_0 + \cdots + p_k = 1.0$). Let the smallest subscripts denote the highest priority. To compute the statistics for the highest priority traffic class, simply let the offered load be $\lambda = p_0$ and evaluate the bandwidth, delay, delay distributions and probability that a packet is delayed by more than the relevant threshold(s). To compute the statistics for traffic class j, we first compute the statistics for the aggregate traffic classes $0..j-1$, and then for aggregate traffic classes $0..j$, and then compute the statistics for class j by computing the differences between the two. To compute the aggregate behavior of traffic classes $0...j$, $0 \le j \le k$, we define the offered load for aggregate classes $0..j$ as follows

$$\lambda_{0..j} = \sum_{i=0}^{j} p_j \qquad (32)$$

and then evaluate the network's performance with this offered load. The average bandwidth and average delay of traffic class j can be computed using the following recursion, where $bw_{0..j}$ denotes the network bandwidth when the offered load consists of traffic classes $0..j$, and where $ad_{0..j}$ denotes the average network delay when the offered load consists of traffic classes $0..j$.

$$bw_j = bw_{0..j} - bw_{0..j-1} \qquad (33)$$

$$ad_j = (ad_{0..j} \cdot \lambda_{0..j} - ad_{0..j-1} \cdot \lambda_{0..j-1}) \cdot \lambda_j^{-1} \qquad (34)$$

Let $dd_{j,x}$ be the probability that an average packet in class j has age x when it leaves the network. Given the delay distribution for aggregate traffic classes $0..j-1$ and $0..j$, the delay distribution for traffic class j is computed as follows.

$$dd_{j-1,x} = (dd_{0..j,x} \cdot \lambda_{0..j} - dd_{0..j-1,x} \cdot \lambda_{0..j-1}) \cdot \lambda_j^{-1} \qquad (35)$$

and the delay distribution for packets in a specific class with a specific starting distance can be computed similarly.

5. Numeric Results

5.1. Basic Algorithm

Graphs of the normalized bandwidth vs. $logN = n$ and the average delay vs. $logN$ are shown in fig. 4a,b. Throughout the paper, the solid lines represent simulations and the dotted lines represent analytic results. To obtain the 95% confidence intervals on the simulations, the method of "subruns" was used. The network was "started up" in the "cold" state (i.e., with no packets in the network), and it would be simulated until it was in equilibrium. Thereafter, one large simulation which required typically 5,000,000 packets to be transmitted would be performed. The run was divided into a number of subruns, and the statistics from each subrun were used to compute the 95% confidence intervals of the throughput and delay, which are very small and not easily visible.

First, observe that even at full offered load ($\lambda = 1.0$, $\alpha=0.5$), the analysis and simulations are in very close agreement. The average error in the normalized throughput and average delays is typically less than 2% of the simulated values. For $\lambda = 1$, $\alpha=0.5$ and $N=8$, the carried load from simulations is about 0.98, i.e., the network only accepts packets with probability 0.98. This slight drop is caused by the large fraction of packets which have distances of 1 or 2, and which therefore have either none or only 1 alternate paths available in each time slot, causing excessive link contention. As the node size increases, much less traffic has distances of 1 or 2 in the steady state, and therefore most packets have a large number of minimum distance alternate paths available, resulting in very high throughputs and very low delays. The only time there is a large drop is throughput is when $\lambda \approx 1.0$ and $\alpha > 0.5$, i.e., when the average path length is much larger than $logN/2$ and the network is heavily loaded. Even in this case the throughput is still very good and it approaches 1.0 as N increases, due to the increasing path multiplicity. For $\lambda \approx \alpha \approx 1.0$ the normalized bandwidth can be made essentially 1.0 by having nodes search the outgoing links in a prefered order (i.e., highest to lowest dimensions) rather than in a random order when attempting to propagate packets. Such routing also results in improvements to the deluxe algorithm when α is large.

Recently an analysis of a hypercube has been presented [1]. That model uses shortest-path routing with buffering but without flow control; when a packet could not be propagated towards its destination (or if it encountered a full output queue when output queues are used) it was simply lost forever, without the packet's sender being notified of the loss. For a fully-loaded 10-dimensional hypercube, [1] indicates that with a single buffer for each incoming link, the hypercube has a maximum possible bandwidth of 0.81. Due to the lack of flow control a significant amount of all traffic accepted into the network will simply be lost forever when congestion is encountered. Using their scheme, nodes must rely on time-outs to detect the lack of a packet's acknowledgement and initiate packet re-transmission, which may degrade system performance considerably. However, the results of [1] would appear to be an upper bound on the performance when flow control is used, since backpressure tends to lower the bandwidth.

Comparing the results of [1] to our own, it is clear that hot-potato routing is significantly better than shortest path routing with limited buffering. For ($\lambda = 1.0$, $\alpha=0.5$), in a 10-dimensional hypercube with hot-potato routing the normalized bandwidth observed from our simulations was $0.9993 \pm .003$, which is extremely high and much better than the maximum bandwidth of 0.81 reported in [1]. In addition, our scheme never loses packets once they are accepted into the network. (An external input queue can be added to each node to buffer packets that were not accepted into the network in any one slot. This case is easily analysed.) The average delay for our network was $5.88 \pm .014$, only slightly larger than the minimum of 5. (No average delays were reported in [1] for finite-buffered system as they had no provisions for the fraction of packets which were lost.)

The analytic delay distributions agree very closely with the simulations for the basic algorithm, but graphs are not included due to space limitations.

5.2. Deluxe Algorithm

Graphs of the normalized bandwidth vs. $logN = n$ and the average delay vs. n for the deluxe algorithm are shown in fig. 4c,d. Each graph contains both analytic and simulation results. Once again, even at full offered load ($\lambda = 1.0$), the analysis and simulations agree closely. The average error is typically less than 1% of the simulated values.

By comparing figures 4a,b and 4c,d, it is clear that the deluxe algorithm performs better than the basic algorithm, as expected. However, the improvements are not significant at light to moderate loads, since even the basic algorithm has nearly optimal performance and there is little room for improvement.

The delay distribution averaged over all packets leaving a network is shown in fig. 4e ($logN = 10$, $\lambda = 0.5, \alpha = 0.5$). Observe that the model is extremely accurate. The deluxe algorithm results in fewer deflections than the regular algorithm at heavy loads as expected, resulting in a slightly lower average delay. The delay distribution for packets with initial distance 6 are shown in fig. 4f. Observe that most packets exit the network after traversing the minimum number of hops, since the distribution has a peak at that age. When a packet with current distance d is deflected, it will require at least 2 hops just to get back to the same distance d.

Therefore, the delay distribution of a single packet with initial distance i is identically 0 for all ages less than d, and it is non-zero only for ages $d + 2 \cdot x$, for $x \ge 0$, where x is the number of times it has been deflected. The analysis models these effects with reasonable accuracy.

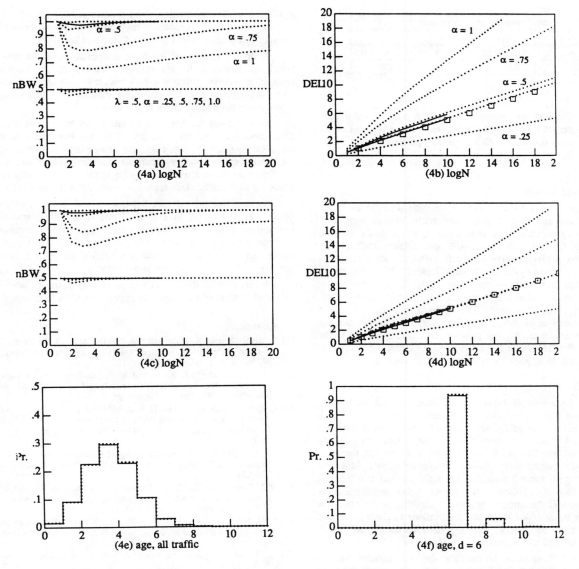

Figure 4: Normalized bandwidth (a) and delay (b) vs. *logN* for the basic algorithm. Simulations for λ = 1.0, α = 0.5 and λ = 0.5, α = 0.5. Analysis for offered load λ =   1.0 and locality α = 0.25, 0.5, 0.75, 1.0. (4c,d); same as 4a,b except for the deluxe algorithm. Boxes indicate minimum possible delays for α = 0.5. (4e,f); Delay distribution for deluxe algorithm, λ = 0.5, α = 0.5, N = 64.

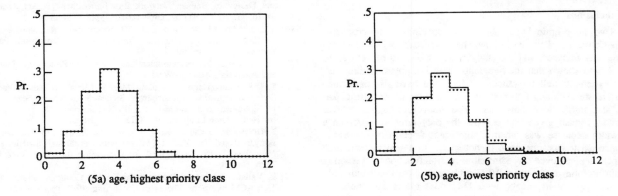

Figure 5: Delay distributions for basic algorithm, λ = 0.5, α = 0.5 and two traffic classes.

413

Table 1: analysis vs. simulations				
-	bw_0	nd_0	bw_1	nd_1
sim	$0.0499 \pm .01$	$3.012 \pm .02$	$0.450 \pm .01$	$3.257 \pm .01$
an	0.050	3.021	0.450	3.329

Table 1: Analytic and simulations results of a 6 dimensional hypercube, with aggregate offered load = 0.5, which was composed of 10 % high priority packets and 90 % low priority packets.

5.3. Prioritized Traffic

Simulations for a 6-dimensional hypercube with the basic hot potato algorithm and with 2 prioritized traffic classes were performed. The aggregate offered load was 0.5, which was composed of 10% high priority traffic and 90% low priority traffic. The bandwidth and average delays for each traffic classes were accurately given by the analysis, as shown in table 1. The delay distributions for both traffic classes are shown in fig. 5a,5b. Once again, the analyses are essentially superimposed on the simulations, indicating a very close agreement. The discrepancies are due to the small inaccuracies of the basic algorithm analysis. Observe that the delay distribution of the highest priority class is less skewed to the left than the distribution for the lowest priority class, indicating that the highest class undergoes fewer deflections on average.

6. Variations

In the preceding analyses it was assumed that each node inserts at most one packet into the network during any one clock cycle. The network is easily able to handle this offered load and the normalized throughput is essentially unity. Our analysis and simulations confirm that in the steady state, with $\alpha = 0.5$ the average link load is about 0.66. (Many simple arguments show that with shortest path routing with infinite buffering and $\alpha = 0.5$, the average link load is α. However, hot potato routing avoids queueing by placing heavier demands on the links, resulting in a higher link load.) It may be desirable in certain applications to allow a node to insert more than one packet into the network during one clock cycle. It is not difficult to model this variation.

It is not difficult to model the following variations of the hot-potato algorithm: Suppose each node has input or output queues of arbitrary sizes. With output queues, a packet is only deflected into another queue when every suitable queue is full. Our analysis can be modified using the techniques described in [12].

7. Conclusions

Two fiber optic hypercube implementations were proposed. The performance of the packet switched hypercube with hot potato routing was analysed, and the analysis was shown to be very accurate. It was shown that the hypercube with hot potato routing performs extremely well for random traffic. We have also performed extensive simulations of this algorithm for nonuniform traffic patterns and traffic with hot-spots. The hot-potato algorithm was extremely robust; given that $\alpha = 0.5$ the only time we observed a bandwidth decrease was when a significant fraction of the traffic was directed to just one node (a hot-spot). In a real system, such hot-spots are removed by admission control functions. Assuming random uniform traffic and a 10 Gigabit/sec optical link bandwidth, a 20 dimensional hypercube with 1M nodes offers each node 10 Gigabit/sec bandwidth on average, and it has an aggregate bandwidth of about 10K Terabits/sec. The network also has reason-

able 2D and 3D constructions, and appears to be extremely robust, attractive features for a localized multiprocessor interconnection network or fast packet switch or for a distributed LAN/MAN.

A deluxe version of the basic hot-potato algorithm was proposed and analysed. In this version, a packet that is closer to its destination receives priority over a packet further from its destination. Our analysis and simulations indicate that this algorithm offers fewer deflections under heavy loads.

Our analysis for hot potato routing can be adapted to model other topologies without difficulty. (We have extended it to model recirculating shuffles, with basic and deluxe hot potato routing, and the results are reasonably accurate.) Our approximate analysis for multiclass traffic in a work-conserving system with constant service times for each traffic class applies to many other networks [13].

Valiant has proved that hypercubes and d-way shuffles can deliver $O(N)$ packets in $O(logN)$ time slots with overwhelming probability [14]. Interestingly, our approximate probabilistic analysis indicates that the hypercube can deliver $O(N)$ uniformly distributed packets in each time slot, even if the average path distance is $logN$, and our simulations support this.

Extensions of the model and more graphs are included in an extended version of this paper, to appear as a CTR tech. report.

8. References

[1] S. Abraham and K. Padmanabhan, "Performance of the Direct Binary n-Cube Network for Multiprocessors", IEEE Trans. Comput., Vol. 38, No. 7, July 1989, pp. 1000-1011.

[2] A.S. Acampora, M.J. Karol and M.G. Hluchyj, "Terabit Lightwave Networks: The Multihop Approach", AT&T Technical Journal, Nov./Dec., 1987, Vol. 66, Issue 6, pp. 21-34

[3] E. Arthurs, M.S. Goodman, H. Kobrinski and V.P. Veechi, "HYPASS: An Optoelectronic Hybrid Package Switching System", IEEE Journal on Selected Areas in Comm., Dec. 1988, Vol. 6, No. 9, pp. 1500-1510

[4] J.S.-C. Chen and R. Guerin, "Input Queueing of an Internally Non-Blocking Packet Switch with Two Priority Classes", Infocom' 89, pp. 529-537

[5] W.J. Dally and C.L. Sietz, "Deadlock Free Message Routing in Multiprocessor Interconnection Networks", IEEE Trans. Comput., Vol. C-36, No. 5, May 1987, pp. 547-553

[6] M.G. Hluchyj and M.J. Karol, "Shufflenet: An application of generalized perfect shuffles to multihop lightwave networks", Infocom' 88, pp. 379-390

[7] L. Kleinrock, "Queueing Systems, Vol. 2: Computer Applications", Wiley 1976

[8] N.F. Maxemchuck, "Comparison of Deflection and Store-and-Forward Techniques in the Manhattan Street and Shuffle-Exchange Networks", IEEE Infocom' 89, pp. 800-8091

[9] S.Z. Shaikh, M. Schwartz and T.H. Szymanski, "Performance Analysis and Design of Banyan Network Based Broadband Packet Switches for Integrated Transport", Globecom' 89, pp. 1154-1158

[10] S.Z. Shaikh, M. Schwartz and T.H. Szymanski, "A Comparison of ShuffleNet and the Banyan Topologies for Broadband Packet Switches", Infocom' 90

[11] M. Schwartz, "Telecommunications Networks: Protocols, Modeling and Analysis", Addison-Wesley, 1987

[12] T.H. Szymanski and S. Shaikh, "Markov Chain Analysis of Packet-Switched banyans with Arbitrary Switch Sizes, Queue Sizes, Link Multiplicities and Speed-Ups", IEEE Infocom' 89, pp. 126-135 (revised version to appear as a CTR Tech. Report)

[13] T.H. Szymanski and C. Fang, "Design and Analysis of Crossbar and Banyan Based Fast Packet Switches with Cut-through Switching and Multiclass Traffic", submitted for publication, to appear as a CTR Tech. Report

[14] L. Valiant, "Universal Schemes for Parallel Communications", Proc. 13th ACM Symp. on Theory of Computing, 1981, pp.263-277

A Modular Architecture for Very Large Packet Switches

TONY T. LEE, SENIOR MEMBER, IEEE

Abstract—We propose a modular architecture for very large packet switches in this paper. Switch modules, the building blocks of our system, are independently operated packet switches. Each module consists of a Batcher sorting network, a stack of binary trees, and a bundle of banyan networks. The modular architecture is a unification of the Batcher–banyan switch and the knockout switch, and can be physically realized as an array of three-dimensional parallel processors. Switch modules are interconnected only at the outputs by multiplexers. The partitioned switch fabric provides a flexible distributed architecture, which is the key to simplify the operation and maintenance of the whole switching system. The modularity implies less stringent synchronization requirements and makes higher speed implementation possible. The proposed modular switch is intended to meet the needs of broadband telephone offices of all sizes. We estimate that a modular switch with Terabit capacity can be built using current VLSI technologies.

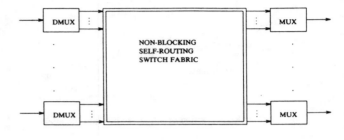

DMUX: DEMULTIPLEXER

MUX : MULTIPLEXER

Fig. 1. A very large packet switch with mixed input and output queueing.

I. INTRODUCTION

IN the future, it is desirable to have an integrated information transport network to support a wide range of communication services. Maturing technologies, such as VLSI and fiber optics, have brought this vision closer to reality, and have stimulated intensive research efforts on high-speed packet switches capable of handling real-time traffic [3], [6], [11], [12], [15], [18], [20], [23]. A well-known example is the combination of Batcher and banyan networks, forming a nonblocking, self-routing packet switch. The Batcher network performs the sorting [2], and the banyan network performs the routing of the input packets. Several systems, such as the Starlite of AT&T [11] and Sunshine of Bellcore [6], [15], have been prototyped based on this principle. In practice, it is estimated that a central office of a broadband network is expected to require switch fabrics with 10 000 high-speed ports and connections for 100 000 terminals [5]. A long-standing open problem is then how to scale research prototypes to these dimensions for practical applications. In this paper, we propose a modular architecture for very large packet switches. Specifically, we show that a large packet switch can be constructed from relatively small fixed-size Batcher and banyan networks, while at the same time preserving the nonblocking and self-routing properties.

Our modular architecture is motivated by several realistic concerns. First, the nonblocking property of the switch requires that the whole set of input packets are synchronized at every stage of the network. For a switch of 10 000 ports, this means synchronizing up to 10 000 packets over a network of about 100 stages. The second concern is the physical limitation of VLSI chips and the growing complexity of interconnection wiring between stages. The third is the reliability and maintainability considerations. It is clear that smaller switch fabrics are much easier to develop, to test and to replace. Finally, it is desirable to engineer the switch systematically to meet growing throughput and capacity requirements by incremental augmentation. All these issues are addressed by the modular architecture here.

As shown in Fig. 1, a switch fabric with both input demultiplexing and output multiplexing calls for a large termination capacity. Demultiplexing surge input-traffic to cope with high speed fiber optic trunk was originally suggested by Huang and Knauer [11]. It has been observed by many authors [10], [13], [15], [18], [19], [21], [23] that the switch throughput can be substantially improved by allowing more than one packet to be switched to the same output concurrently. At the same time, however, simultaneous arrivals will introduce unavoidable output queuing, waiting to be transmitted by the output trunk. For this reason, output multiplexers are essential for engineering the switch to satisfy the desirable throughput and performance.

Fig. 2 illustrates the basic idea of our modular approach to construct a switch fabric with a large number of ports. The set of all inputs is equally partitioned into K subsets. Each subset of the inputs are then interconnected to all N outputs, forming the basic building block of the system called the *switch module*. The number of inputs of a switch module, $M = N/K$, is called the *base dimension* of the switch. In this architecture, each module is an autonomous nonblocking, self-routing packet switch. The outputs with the same index, one from each switch module, are multiplexed together and fed into the output port bearing that index as its port address. The modular architecture proposed by Yeh, Hluchyj, and Acampora of AT&T [23], called the knockout switch, is equivalent to the special case when $M = 1$. Roughly speaking, our modular switch is a unification of the Batcher-banyan switch and the knockout switch. The details of this point will be discussed in Section II-C.

Physically, as shown in Fig. 3, the modular architecture can be realized as an array of three-dimensional parallel processors. Each processor is a switch module consisting of a Batcher sorting network and an expansion routing network. The expansion network is a set of binary trees cross interconnected with a set of banyan networks. The set of concurrent modules is interconnected at the outputs by multiplexers. Thus, no interferences between switch modules are possible, which is the key to simplify the operation and maintenance of the whole system. This architecture also allows independent clocking of modules which simplifies timing substantially. Within each module, the small physical size makes it fairly straightforward to synchro-

Paper approved by the Editor for Optical Switching of the IEEE Communications Society. Manuscript received June 16, 1989; revised November 13, 1989. This paper was presented at IEEE GLOBECOM '89, Dallas, TX, November 27–30, 1989.

The author is with Bellcore, Morristown, NJ 07960-1910.

IEEE Log Number 9036348.

Reprinted from *IEEE Trans. Commun.*, vol. 38, no. 7, pp. 1097–1106, July 1990.

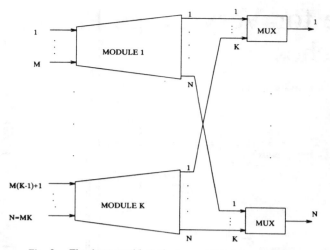

Fig. 2. The decomposition of a large switch into modules.

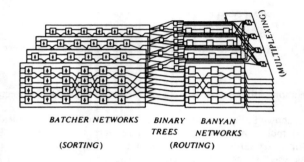

BATCHER NETWORKS   BINARY   BANYAN
                     TREES   NETWORKS

(SORTING)   (ROUTING)

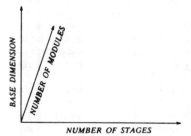

Fig. 3.   The 3-D modular architecture of Batcher-binary-banyan packet switch.

nize, and the simpler hardware makes higher speed implementation possible.

In summary, our modular architecture offers the following distinctive features.

1) The $N \times N$ switch fabric is partitioned into K independent modules. The dimension, either fan-in or fan-out, of every component involved in each module will not exceed $\max(K, M)$ where M is the base dimension.

2) The nonblocking and self-routing properties are preserved in the modular switch. There are no internal flow control or buffering problems. Adaptive path hunting procedures are completely embedded in VLSI chips, requiring no external control for connections. The switch performance is measured in terms of queueing behaviors of buffers at input and output ports, which are mainly determined by the link loading. This implies that the quality of services can be assured by flow control algorithms implemented at network level.

3) The required timing for synchronization can be decoupled. Only local clocks are needed for each of the $M \times M$ Batcher and banyan subnetworks.

4) Any fault will only disturb the local traffic carried by a single module, while the remaining switch modules can still be normally operated. Fault tolerance can be accomplished by providing a spare module, not a duplication of the entire switch.

5) The port capacity can be expanded incrementally, starting from small and then growing into large.

6) Because of the loosely interconnected architecture, the switch capacity can be distributed over a rural exchange area to reduce access cost.

7) A packet switch with Terabit capacity can be realized by this modular approach. It is estimated that a switch fabric with 8000 to 16 000 ports, each port is operated in the range of 150–200 Mbits/s, can be built by current VLSI technologies.

8) Contention resolution algorithms developed for Batcher-banyan switches remain valid. In particular, the ring reservation algorithm becomes an attractive scheme for switch modules. The K concurrent rings, one for each switch module, make this otherwise sequentially operated procedure a parallel algorithm.

9) The switch throughput can be engineered to meet any desired requirement. This can be achieved by adjusting the number of modules K and the output group size L, the maximum number of packets that can be received by an output port simultaneously.

The remainder of this paper is organized as follows. Section II describes the architecture and explains the properties of switch modules addressed above. Section III discusses the contention resolution and output space extension. We present an extended ring reservation algorithm to handle both problems in a unified way. Section IV describes the performance issues of the modular switch. Particularly, the effect of the modularity and the output group size to the switch throughput and packet contention probability is discussed in detail. Finally, Section V is the conclusion of the paper.

II. MODULAR BATCHER-BINARY-BANYAN SWITCH

With reference to the previous discussion, each of the K modules of an $N \times N$ switch is a cascade of an $M \times M$ Batcher network, a stack of M binary trees and a group of K banyan networks where $M = N/K$. We will use $\sigma(M, N, K)$ to represent such a switch module. The Batcher network sorts the input packets based on their destination addresses. The succeeding binary trees and banyan networks then route the packets to the correct outputs.

The principle of this modular architecture is based on a common approach called *divide-and-conquer*, which has been used extensively in designing efficient algorithms [1]. A well-known example is the fast Fourier transform. It simply splits a large problem into smaller parts, solve them, and then combine the solutions for the parts into a solution for the whole. This point is illustrated in Fig. 4. The set of inputs is first partitioned into K subsets. Each subset is sorted by a Batcher network. The sorted subset is then partitioned again by the binary trees into finer subsubsets. In each module, the ordered packets of these subsubsets are routed concurrently to their destinations by K parallel banyan subnetworks. Finally, the multiplexers will collect these packets and dispatch them to the output trunks.

The combination of binary trees and banyan networks forms an expansion network, a network has larger fan-out than fan-in. As such, the output space of an expansion network can be arbitrarily enlarged by adding more banyan networks. Thus, each module may even have more outlets than the number of output trunks. A group of outlets can then be multiplexed into a single output trunk to increase switch throughput and to reduce packet contention probability. The structure and property of the switch modules are detailed below.

A. Binary-Banyan Expansion Network

The use of a binary-banyan network as a broadcast network has been discussed in [17]. The same topology can also be employed as a routing network. An n-stage expansion network with $M = 2^m$ inputs and $N = 2^n$ outputs is a combination of a set of M binary trees and a set of $K = N/M$ banyan networks. Fig. 5 illustrates a 3-stage expansion network with 4 inputs and 8 outputs. Each $1 \times K$ binary tree has $k = \log K = \log N - \log M = n - m$ stages and each $M \times M$ banyan network has m stages. Every node of the network is

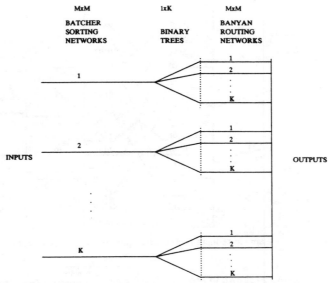

Fig. 4. The divide-and-conquer modular approach.

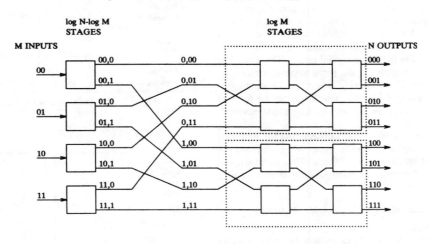

Fig. 5. A 4 × 8 expansion network.

either an 1×2 or a 2×2 switch element capable of performing the binary routing algorithm based on the n-bit destination address in the header of the packet. That is, a node at stage j sends the packet out on link 0 (up) or link 1 (down) according to the jth bit of the header [6], [11], [12], [15], [17].

The cross interconnection of M binary trees and K banyan networks is similar to the link system of a multistage crossbar switch. The outputs of a binary tree can be labeled by two binary numbers $(x_1 \cdots x_m, y_1 \cdots y_{n-m=k})$ where $x_1 \cdots x_m$ is the top-down numbering of the binary tree and $y_1 \cdots y_k$ is the local address of each input within the binary tree. Similarly, the inputs of the banyan network can also be identified by two binary numbers $(a_1 \cdots a_{n-m=k}, b_1 \cdots b_m)$ where $a_1 \cdots a_{n-m}$ is the top-down numbering of the banyan network and $b_1 \cdots b_m$ is the local address of the input. The binary trees and banyan networks are transpose interconnected according to this numbering scheme; the two ports are interconnected if $(x_1 \cdots x_m, y_1 \cdots y_{n-m}) = (b_1 \cdots b_m, a_1 \cdots a_{n-m})$. This transpose cross interconnection results in a three-dimensional realization of switch modules as portrayed in Fig. 3.

The binary trees of the expansion network consist of 1×2 elements, which only allow one input packet at any instant of time. Packets will never collide in this stage of the network, but this may occur in the subsequent banyan subnetworks. In the following, we will discuss the condition on inputs that prevents any possible packet collisions in an expansion network.

B. Nonblocking Property of Expansion Networks

A interconnection network is nonblocking if the packet routing is not store-and-forward, and internal buffers are not needed within the switch nodes. It is known that if the incoming packets with distinct destination addresses are arranged in an ascending or a descending order, then the banyan network is internally nonblocking (see [17, Appendix]).

The cross interconnection defined by the above numbering scheme guarantees that the routing performed by the set of binary trees preserves the ordering of the destination addresses in input packet headers. Since a binary tree is actually an $1 \times K$ demultiplexer, at most one packet will be routed by a tree during any time slot. Packet collisions may only occur in the subsequent banyan subnetworks. This implies that the same nonblocking condition for banyan networks can still be applied to expansion networks. Formally, this property states the following.

[NB] *If the set of destination addresses of input packets to the expansion network is monotone and concentrated, then so is every subset of input packets to each banyan subnetwork of the expansion network.*

This nonblocking condition can be best demonstrated by the following example. Consider two packets:

A: from 01 to 1001

and

417

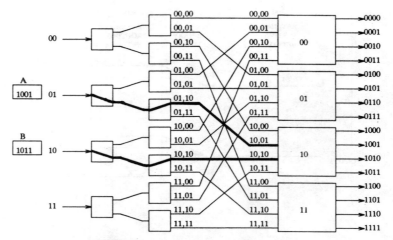

Fig. 6.   An example of input packets order preserving by binary trees.

B: from 10 to 1011

input adjacently to the 4×16 expansion network as shown in Fig. 6. According to the self-routing algorithm, they will emerge at outputs labeled 01, 10 and 10, 10 of the binary trees 01, and 10, respectively. Following the transpose interconnection, packets *A* and *B* will be routed to the inputs 10, 01 and 10, 10, respectively, of the banyan subnetwork 10. The two packets are still neighboring contiguously, no gap has been created, and in the same spatial order. The assertion in [*NB*] can be established by applying the same argument to every adjacent pair of input packets. Consequently, a rectangular nonblocking self-routing switch with larger fan-out than fan-in can be formed by combining a Batcher sorting network and an expansion network with arbitrary input/output ratio.

It should be noticed that the perfect shuffle proceeding the inputs of banyan subnetworks is topologically equivalent to the perfect shuffle of inputs proceeded to the stack of binary trees. This point can be demonstrated by the three dimensional configuration shown in Fig. 3.

C. Complexity and Maintainability of Modular Switches

The aim of most conventional modular approaches, such as the well-known Clos network [4], are solely to reduce its complexity in terms of number of switch nodes. In contrast, our primary goals are focused on relaxing the limitation of VLSI implementation, simplifying interstage wiring and synchronization. As far as switch operation is concerned, preserving nonblocking and self-routing properties are certainly more important than reducing the switch hardware. The statistics of a switch module $\sigma(M, N, K)$ is tabulated in Fig. 7. The modularity only cuts down the complexity of Batcher networks. The total number of nodes is increasing with respect to K, the number of modules. These extra nodes are not entirely overhead. The modularity can improve switch throughput and performance. Intuitively, this is simply because there are fewer input packets competing for outputs in each module. The analysis of this point will be elaborated in Section IV.

Theoretically, there is virtually no limit on the dimension of the modular switch that can be built from fixed-size Batcher and banyan networks. However, there is a tradeoff between the base dimension M and the number of modules K for a given switch with dimension N, since $N = MK$. There are two notable special cases worth mentioning here. Namely, the Batcher–banyan switch for $K = 1$ [3], [6], [11], [12], [15], and the knockout switch for $K = N$ [23]. Fig. 8 illustrates the basic idea of knockout switch, each input is connected to every output by a broadcast bus. The bus interface at each output provides packet filters to allow packets addressed to the output to pass and block all others, and a concentrator to regulate the maximum number of concurrent packet arrivals at the output during a time slot. It is clear that if every module has only one input, $M = 1$, then our modular switch becomes a knockout switch. The three-dimensional modular architecture is actually the unification of Batcher–banyan and

	Number of Networks	Dimension	Number of Stages	Number of Nodes
Batcher Network	1	$M \times M$	$\dfrac{\log M (1 + \log M)}{2}$	$\dfrac{M \log M (1 + \log M)}{4}$
Binary Tree	M	$1 \times K$	$\log K$	$K - 1$
Banyan Network	K	$M \times M$	$\log M$	$\dfrac{M}{2} \log M$

Fig. 7.   The statistics of a switch module $\sigma(M, N, K)$.

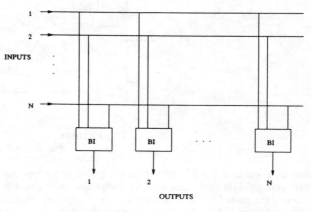

BI:BUS INTERFACE

Fig. 8.   Knockout switch.

knockout switches. This point is illustrated in Fig. 9. Both switch architectures cannot be scaled to very large size N because either the base dimension $M = N$ or the number of modules $K = N$. That is, the decomposition of N into smaller factors does not apply. A choice for balancing these factors is $M = K = \sqrt{N}$, from which we can claim that this approach has at least the power to square the performable dimensions of these interconnection networks.

An important objective of our modular approach is to simplify the timing requirement. Both sorting and routing demand packet synchronization at every stage of the network. If we take a close look at these paths from inputs to outputs, the following can be observed.

• Each binary tree carries at most one packet during a time slot. Therefore, synchronization is unnecessary in this demultiplexing stage.

• The statistical multiplexers at the output ports, operated asynchronously, are able to collect packets coming from different banyan subnetworks at different instants of time. Thus, each banyan subnetwork can be clocked independently without global synchronization.

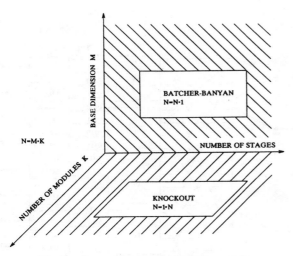

Fig. 9. Two extreme cases of modular switches.

• There are at most M packets to be synchronized over $\log M(\log M + 1)/2$ stages in each sorting network and $\log M$ stages in each banyan routing network.

Based on the above observation, we assert that the clocks needed for Batcher and banyan networks can be localized to each plane of the three-dimensional structure. Indeed, the decoupled timing requirement is the crucial point that enables us to build very large switches realistically.

Higher reliability, easier growth, and maintainability are few other advantages that can also be attributed to the modularity. Any fault on a switch module will disturb local traffic only. All other switch modules can still be operated normally. Fault tolerance can be achieved easily by providing a spare switch module, while most other switches require complete duplication to satisfy reliability requirement. The table in Fig. 7 suggests that the modular switch can grow incrementally from $N \times N$ to $2N \times 2N$ by using the same sized Batcher and banyan networks. The fan-out of binary trees and fanin of the multiplexers should be doubled, which can be realized by combining two binary trees or two multiplexers together.

Since every switch module is connected to every multiplexer, the interconnections between the K switch modules and the N multiplexers form a complete bipartite graph [9] with KN links. An example of this interconnection pattern is shown in Fig. 10. Current advances in optical technologies may provide a cost-effective solution for this stage of interconnection to avoid mutual interference effects caused by electronics [8].

A goal of building a large packet switch is to achieve Tbit/s (10^{12}) capacity in order to cope with the corresponding capacity of fiber optic trunks, which is on the order of Gbit/s (10^9). To reach this target, we estimated that a modular switch with 16 384 ports, operated at 150–200 Mbits/s per port, can be easily constructed using present-day VLSI technologies. The complexity of such a modular switch is tabulated in Fig. 11. If we use the "square root rule" mentioned before, this switch, achieving a total capacity of 2.2–3.2 Tbit/s, only requires 128×128 Batcher and banyan networks as its building blocks.

III. CONTENTION RESOLUTION AND EXTENDED OUTPUT SPACE

An inherent problem in the design of a space-division-based packet switch is output port conflicts, which occur when multiple packets simultaneously request the same output. The procedure to arbitrate input packets for the same destination is called a *contention resolution algorithm*. Conceptually, the packet contention probability can be reduced by increasing the number of packets allowed for the same output concurrently, an output space extension. There are several existing solutions for contention resolution [3], [6], [11], [12] and output space extension [15], [19], [21]. In this section, we will discuss a unified scheme to handle both problems together.

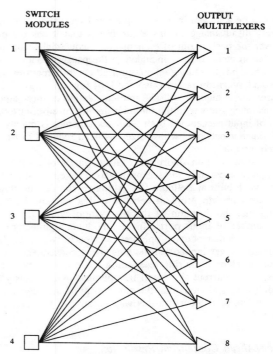

Fig. 10. The cross interconnections between switch modules and multiplexers.

M	K	$M \times M$ Batcher $(N/M) = K$	$M \times M$ Banyan $(N/M)^2$	$1 \times K$ Binary tree (N)	$K \times 1$ Multiplexer (N)
64	256	256	65,536	16,384	16,384
128	128	128	16,384	16,384	16,384
256	64	64	4,096	16,384	16,384
512	32	32	1,024	16,384	16,384

Fig. 11. The complexity of a switch with Terabit capability (assuming each port is operated in the range of 150–200 Mbits/s).

A. Contention Resolution

Three algorithms have been devised for resolving output conflicts of a Batcher-banyan packet switch.

1) *Recirculation Algorithm* [6], [11], [15]: This mechanism was originally proposed by Huang and Knauer in their Starlite system. It simply fed the blocked packets back to the inputs for reentry at the next time slot.

2) *Three Phase Algorithm* [12]: This algorithm is also a feedback scheme. It requires each input port sending a probing header for arbitration in phase 1. The inputs then receive a positive or negative acknowledgment back from the outputs in phase 2. The actual transmission of the winning packets, these who received positive acknowledgments, takes place in phase 3.

3) *Ring Reservation Algorithm* [3]: Unlike the previous two feedback algorithms, the ring reservation is a token passing scheme. At the beginning of a time slot, a cleared token is issued by a token generator. The token has an N-bit field to indicate the availability of N output ports. The token is circulated around a ring connecting the input ports. When the token comes by, the packet at the head of each input buffer will make the reservation. If the intended output is successfully reserved, the packet is transmitted at the next time slot. The reservation cycle and transmission cycle can be overlapped to minimize the overhead.

Certainly, all three resolution algorithms remain valid for individual switch modules. The advantage of the first two schemes is that they both are parallel algorithms. The recirculation algorithm uses a trap to catch excess packets and sends them back to an auxiliary set of input ports. Sharing common buffers and smoothing bursty traffic can be achieved by this trapping and recirculating procedure. The only disadvantage of this algorithm is that it has the potential to deliver the packets out of their original sequential order, unless a recirculation count field is included in the packet header. The three phase algorithm is an elegant solution although it requires stringent timing and synchronization (also see [16] for some modification of this three phase algorithm). The implementation of these algorithms becomes much simpler in the modular architecture because the arbitration of input packets for both feedback schemes is mainly carried out by Batcher network, whose size has been reduced to $M \times M$ in a switch module.

The ring reservation algorithm is fairly straightforward to operate, and this was apparently the reason it was selected for the prototype packet switch built at Bellcore [3]. Although simple, the ring reservation has a growth problem because it is a sequential procedure, meaning that the duration of the reservation cycle is proportional to the number of input ports. In our modular architecture, switch modules are independently operated and each switch module has substantially fewer inputs than outputs. As a consequence, the serialization problem of the ring reservation algorithm can be alleviated. That is, K concurrent rings will be operated independently for the K modules. It is obvious that the parallelism of the ring reservation algorithm can be achieved in this way because of the modularity of the switch.

B. Look-Ahead Contention Resolution

The throughput of a switch is limited mainly due to the head of line blocking phenomena. This limitation can be relaxed and throughput can be increased if the service discipline of the input buffers is not restricted to first-in first-out. The look-ahead contention resolution scheme described in [10] provides a simple solution in this regard. The contention resolution process during a time slot is divided up into w phases where w is called *window size*. The packets at the heads of the input buffers content for access to the desired outputs. In the second phase, the blocked inputs then contend with their second packets for access to those outputs that are not seized in this time slot. This process is repeated up to w times and at most one packet will be allowed per time slot from each input. Since any cleared packets can only access to those outputs bequeathed to them by the packets blocked previously. The look-ahead contention resolution process will not introduce out-of-sequence problem. The throughput of the switch is monotonically increasing with respect to increased window size. The analysis of this point will be carried out in Section IV-A.

C. Output Space Extension

As we mentioned previously, the packet contention probabilities can be substantially reduced by providing more output spaces. Several schemes have been proposed for this purpose. For example, the duplex switch [18], [21] is a simple and effective method to expand both input and output spaces to increase the switch throughput. Others include the trunk grouping suggested by Pattavina [19], and the multiple banyan networks employed in the Sunshine switch devised by Littlewood [15]. They are all based on the principle of output address extension.

Considering an $M \times NL$ expansion network as a routing network, which has only N output addresses. That is, each output port can have L outlets which allow up to L packets, destining for the same output address, to be switched concurrently. Such a switch module will be denoted by $\sigma_L(M, N, K)$. Since the network has $\log LN = \log L + \log N$ stages. It requires $\log N + \log L$ address bits to perform the self-routing algorithm. Therefore, additional $\log L = l$ bits, called *group index*, should be determined and appended to the $\log N$-bit destination address in the header of a winning packets, after contending for one of the L outlets, to form the required routing address. This point is illustrated in Fig. 12. An algorithm which performs both contention resolution and output address extension is described in the remainder of this section.

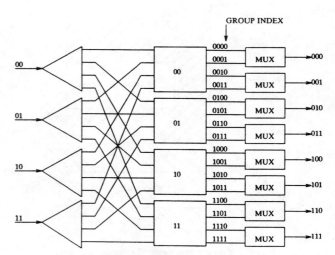

Fig. 12. Output space extension by expansion network.

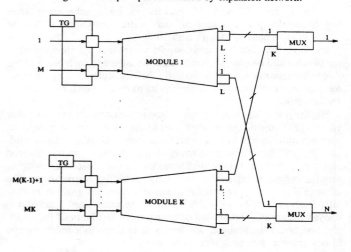

TG : TOKEN GENERATOR

Fig. 13. A modular switch with concurrent rings for resolution of output contention.

D. The Extended Ring Reservation Algorithm

Fig. 13 illustrates an $M \times NL$ switch module whose input ports are sequentially interconnected by a ring. The token generator on the ring issues a cleared token at the beginning of each time slot. The tokens are used by the input ports to reserve the outputs of the module requested by their head of line packets. The formats of packet headers and tokens are shown at the top of Fig. 14. The routing address in the packet header consists of an n-bit destination address and an l-bit group index to be determined by the extended ring reservation algorithm. The priority field represents the class of the packet. The tokens also consist of two fields, a group index field and a priority field. Initially, all the fields in the token are cleared, and set to zero. The reservation cycle is divided into P subcycles; the token will circulate around the ring P times where P is the number of priority classes. At the ith subcycle, only packets in the ith priority class can make reservations. It should be noticed that the length of this token is still proportional to N, the total number of outputs. The rules governing the reservation process are described below.

1) Suppose the destination address D in the packet header is $d_n \cdots d_1 = (i)_{10}$. If the priority class in the packet header agrees with the priority class in the token and the ith subfield $G_i = g_{l+1} g_l \cdots g_1$ of the group index is strictly less than L, then the reservation is allowed. In this case, the least significant l bits of G_i are attached to the destination address to form the routing address $d_n \cdots d_1 g_l \cdots g_1$ for the winning packet.

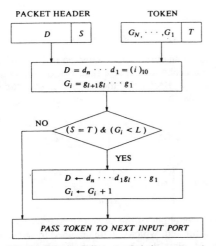

Fig. 14. The flow chart of the extended ring reservation algorithm.

2) The contents of the token should be modified each time after a successful reservation has been made. In the case of 1) above, the ith subfield of the group index G_i should be stepped up by one. The most significant bit of G_i represents the status of output i. If space is still available then $g_{l+1} = 0$, otherwise $g_{l+1} = 1$.

3) If the reservation attempt is rejected, the failed packet has to wait for the next cycle, possibly with higher priority class by modifying its priority field.

4) When the token returns to the starting point, the token generator will increase its priority field by one, and reissue the token to commence the next subcycle.

The flow chart of the above control logic is given in Fig. 14. Functionally, this reservation scheme can also be used to protect the reserved packets for circuit emulated calls [22]. Assigning highest priority class to reserved packets will guarantee that they are switched immediately without experiencing any interference.

IV. Switch Performance Analysis

In this section, we analyze the throughput and performance of a switch consisting of $\sigma_L(M, N, K)$ modules. First, we will determine the maximum throughput of the switch module $\sigma_1(M, N, K)$ with look-ahead contention resolution mechanism described before in Section III-B. In this analysis, we assume that those packets lost contention will be retransmitted. Next, the switch modules with grouped outputs will be studied. We will be interested in finding the relationship between the number of modules K, the output group size L, the switch throughput and the packet delay. We will assume that the input traffic is homogeneous and packets are uniformly destined for the N outputs. Our investigation will be concentrated on the behavior of a single switch module because of the symmetry of the structure and the independent operation of each module.

A. Maximum Throughput with Look-Ahead Contention Resolution

The maximum throughput of a switch consisting of $\sigma_1(M, N, K)$ modules with look-ahead contention resolution is approximated in this section. The formula is recursive on the window size of the contention resolution process. To achieve maximal possible loading, we assume that there are always w packets waiting at each input buffer for window size w.

Suppose λ_t is the maximum throughput of an $M \times N$ switch with look-ahead window t. At the end of tth phase, let M_0 be the expected number of inputs being blocked, and N_0 be the expected number of outputs not being seized so far in this time slot. Since $\lambda_t M = M - M_0 = N - N_0$, it follows that

$$\frac{M_0}{M} = 1 - \lambda_t, \tag{1}$$

and

$$\frac{N_0}{N} = 1 - \frac{\lambda_t}{K} \tag{2}$$

where $K = N/M$ is the number of modules. Without loss of generality, we assume that outputs 1 through N_0 are those outputs not yet seized at the end of tth phase in this time slot. In the $(t + 1)$th phase, the contention resolution process will be focusing only on those blocked inputs that their $(t + 1)$th packets destined for outputs 1 through N_0. Thus, we can approximate the situation by considering a "virtual" switch with $M_0(N_0/N)$ inputs and N_0 outputs. The first step of our analysis is to determine the number of input packets, say F, of this virtual switch that can be cleared in a contention resolution cycle. Following the analysis in [10], we define B^i as the number of packets destined for output i but not selected to pass the switch, and A^i as the number of fresh packets destined for output i to replace those being cleared. By definition, we should have

$$F = \sum_{i=1}^{N_0} A^i, \tag{3}$$

and

$$F = M_0 \left(\frac{N_0}{N} \right) - \sum_{i=1}^{N_0} B^i. \tag{4}$$

It follows immediately that

$$E[A^i] = \frac{E[F]}{N_0} \triangleq \rho, \tag{5}$$

and

$$E[B^i] = \frac{M_0}{M} - \frac{E[F]}{N_0}$$
$$= \frac{M_0}{M} \cdot \frac{M}{N} - \frac{E[F]}{N_0}$$
$$= \left(\frac{1 - \lambda_t}{K} \right) - \rho. \tag{6}$$

It has been shown in [10] that $E[A^i] = \rho$ and $E[B^i]$ are related by the mean steady-state $M/D/1$ queue size formula as follows:

$$E[B^i] = \frac{\rho^2}{2(1 - \rho)}. \tag{7}$$

Solving (6) and (7) simultaneously, we obtain

$$\rho = (K + 1 - \lambda_t - \sqrt{K^2 + (1 - \lambda_t)^2})/K. \tag{8}$$

The maximum throughput λ_{t+1} can be interpreted as the probability that an input has a packet being selected to pass the switch at the end of $(t + 1)$th phase. Therefore, it follows from (2), (5), and (8) that

$$\lambda_{t+1} = \lambda_t + \frac{E[F]}{M}$$
$$= \lambda_t + \left(\frac{E[F]}{N_0} \right) \left(\frac{N_0}{N} \right) \left(\frac{N}{M} \right)$$
$$= \lambda_t + \rho \left(1 - \frac{\lambda_t}{K} \right) K$$
$$= \lambda_t + (K + 1 - \lambda_t - \sqrt{K^2 + (1 - \lambda_t)^2})$$
$$\cdot \left(1 - \frac{\lambda_t}{K} \right), \tag{9}$$

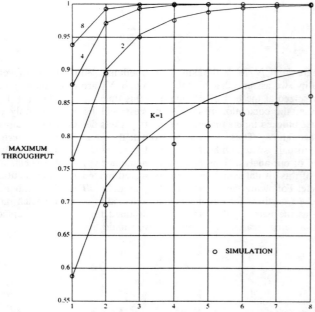

Fig. 15. Maximum throughput of modular switches with look-ahead contention resolution.

with initial condition $\lambda_0 = 0$. It is easy to verify that the following limiting cases are valid

$$\lim_{t \to \infty} \lambda_t(K) = \lim_{K \to \infty} \lambda_t(K) = 1.$$

Fig. 15 shows that this recursive formula fits well with simulation results. We find that maximum throughput can be easily greater than 90% when the window size $w = 2$ for $K \geq 2$.

B. Analysis of Modules with Grouped Outputs

The throughput and delay analysis of a switch module $\sigma_L(M, N, K)$ with grouped outputs is briefly discussed in the sequel. We presume that the contention probabilities can be arbitrary small by properly designed output group size. Contention resolution, either packet retransmission or output port reservation, is therefore not considered in the following analysis for the sake of simplicity. We assume that if there are $S(> L)$ packets waiting at the heads of input queues addressed to the same output, the selection of L to pass through the switch is done at random; and the $S - L$ which lose the contention will be ignored. Under these assumptions, the switch throughput is defined to be the carried load on an output line when all input lines are fully loaded. Packets will not be delayed at inputs and the time for packets traversing the switch is a constant. All queueing is done at the output multiplexing stage.

Suppose the input line loading, or the probability that a slot contains a packet, is λ. Let M_i be the number of packets addressed to output i, we have

$$\Pr\{M_i = j\} = \binom{M}{j} \left(\frac{\lambda}{N}\right)^j \left(1 - \frac{\lambda}{N}\right)^{M-j},$$

$$j = 0, \cdots, M. \tag{10}$$

Suppose $A_i = \min(M_i, L)$ is the number of packets that actually arrived at output i in a time slot. For sufficiently large N, we can use Poisson approximation with parameter $M\lambda/N = \lambda/K$ to obtain

$$a_j = \Pr\{A_i = j\} = \left(\frac{\lambda}{K}\right)^j \frac{e^{-\lambda/K}}{j!}$$

$$j = 0, \cdots, L - 1 \tag{11}$$

and

$$a_L = 1 - a_0 - \cdots - a_{L-1}.$$

The *carried load* on an output trunk, denoted by λ', can be intuitively interpreted as the probability that a time slot on the output trunk is occupied by a packet. Because of the symmetricity of the K modules, we immediately obtain

$$\lambda' = K \cdot E[A_i] = K \cdot \left[L - \sum_{j=0}^{L-1} (L-j)a_j\right]$$

$$= K \cdot \left[L - e^{-\lambda/K} \sum_{j=0}^{L-1} \left(\frac{\lambda}{K}\right)^j \frac{(L-j)}{j!}\right]. \tag{12}$$

The maximum throughput of the switch is the carried load when all input lines are saturated, which can be obtained by substituting $\lambda = 1$ into (12). Fig. 16 plots the maximum throughput versus number of modules K with varying output group size L. For $K \geq 2$, maximum throughput $\lambda = 0.95$ can be achieved with group size $L = 2$.

The output trunk is considered as a single server with constant service time. For a fixed output i, the packets are input from K identical independently operated modules. Suppose A is the total number of packets switched to output port i during a time slot. The generating function of A is then given by

$$A(Z) = E[Z^A] = E[Z^{A_i}]^K = \left(\sum_{j=0}^{L} a_j Z^j\right)^K$$

$$= \left[Z^L - e^{-\lambda/K} \sum_{j=0}^{L-1} \frac{(Z^L - Z^j)}{j!} \left(\frac{\lambda}{K}\right)^j\right]^K. \tag{13}$$

Let Q be the number of packets in the output queue i at the end of a time slot. In the next time slot, the number of packets in this queue must be

$$Q' = Q - U(Q) + A \tag{14}$$

where $U(X) = 1$ if $X > 0$ and $U(X) = 0$ if $X = 0$ is the unit step function. It is a routine exercise to determine the generating function of Q [14], which is given as follows:

$$Q(Z) = E[Z^Q] = \frac{(1 - \lambda')(1 - Z)A(Z)}{A(Z) - Z}. \tag{15}$$

From Little's Law, the average packet delay at output queue i is given by

$$E[W] = \frac{\dot{Q}(1)}{\lambda'} = 1 + \frac{(K-1)\lambda'^2 - K\lambda' + K^2 E[A_i^2]}{2K\lambda'(1 - \lambda')} \tag{16}$$

where

$$E[A_i^2] = L^2 - e^{-\lambda/K} \sum_{j=0}^{L-1} \frac{(L^2 - j^2)}{j!} \left(\frac{\lambda}{K}\right)^j. \tag{17}$$

The average packet delay is monotonically increasing with respect to K and L, as shown in Fig. 17 for offered load $\lambda = 0.75$ per input line. It should be noticed that although this average delay caused by batch arrivals is bounded, but not the delay W itself. For voice or video packets, additional control to bound the delay may be needed. For large number of modules, each output trunk behaves like an $M/D/1$ queue. It is easy to show from (16) that

$$\lim_{K \to \infty} E[W] = \lim_{L \to \infty} E[W] = 1 + \frac{\lambda}{2(1 - \rho)}. \tag{18}$$

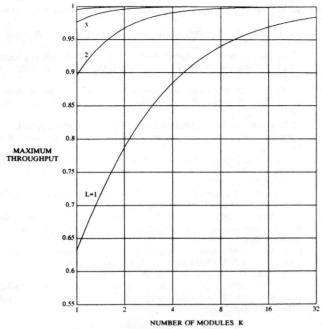

Fig. 16. Maximum throughput of modular switches with grouped outputs.

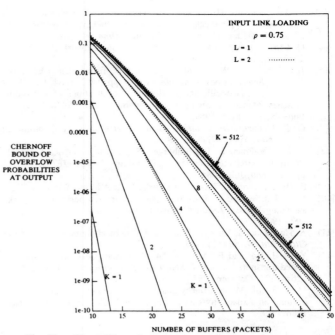

Fig. 18. Chernoff bound of overflow probabilities at output port.

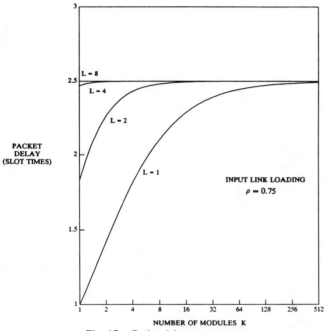

Fig. 17. Packet delay at output port.

Another performance measurement of interest is the probability of packet overflow at each output port. For the buffer of size B, the overflow probability P_o at output port can be estimated by the Chernoff bounds [7] as follows:

$$P_o \leq \Pr\{Q > B\} \leq e^{-sB} Q(e^s) \tag{19}$$

where s is determined by solving the following equation:

$$\frac{dQ(e^s)}{ds} - BQ(e^s) = 0, \tag{20}$$

which provides a tightest upper bound of P_o that can be obtained from (19). It should be noted that Chernoff bound is only a conservative estimate and the operation of the multiplexers at the outputs may be far more complicated than a single server FIFO queue. Still, the

above bound can be used to project the memory space needed at the outputs. Fig. 18 plots the Chernoff bound for offered load $\lambda = 0.75$ per input line with varying K and L. Again, we observed that the family of these curves converges quickly as K becomes large, from which we estimate that a buffer of about 50 packets is sufficient for any sized switch. The analysis in this section shows that the performances of nonblocking modular switches, measured by the input and output queuing behaviors, are mainly determined by the link loading. This implies that the quality of communication services can be assured by network management procedures such as congestion or flow control algorithms.

V. Conclusion

We have presented a modular architecture for very large packet switches. This new modular approach enables one to construct packet switches of arbitrary size from relatively small fixed-size interconnection networks. Two remarkable features of Batcher–banyan switches, namely nonblocking and self-routing properties, are sustained in our design to simplify the switch operation and to avoid internal flow control problem. This paper has discussed the structure of the switch module, the basic building block of our system, its performance and complexity. We also described an extended ring reservation algorithm for contention resolution and output space extension.

We estimated that a packet switch with 16 384 ports can be constructed using 128×128 Batcher and banyan networks. Such a switch can reach a total capacity of 2 to 3 Tbit/s provided that each input/output is operated in the range of 150–200 Mbits/s. These dimension and speed are certainly achievable by current VLSI technologies. What the proposed modular approach can achieve is to simply square the performable dimensions of interconnection networks.

Acknowledgment

I thank S. C. Liew for his insightful comments which improve this paper significantly. I also thank two anonymous reviewers for their useful comments and corrections.

References

[1] A. V. Aho, J. E. Hopcroft, and J. D. Ullman, *The Design and Analysis of Computer Algorithms.* Reading, MA: Addison-Wesley, 1974.
[2] K. E. Batcher, "Sorting networks and their applications," *AFIPS Proc. Spring Joint Comput. Conf.*, 1968, pp. 307–314.

[3] B. Bingham and H. Bussey, "Reservation-based contention resolution mechanism for Batcher-banyan packet switches," *Electron. Lett.*, 23rd, vol. 24, no. 13, pp. 772–773, June 1988.

[4] C. Clos, "A study of non-blocking switching networks," *Bell Syst. Tech. J.*, vol. 32, pp. 406–424, 1953.

[5] J. P. Coudreuse, W. D. Sincoskie, and J. S. Turner, Guest Editorials, *IEEE J. Select. Areas Commun.*, vol. 6, pp. 1452–1454, Dec. 1988.

[6] C. Day, J. Giacopelli, and J. Hickey, "Applications of self-routing switches to LATA fiber optic networks," *Int. Switch. Symp.*, Phoenix, AZ, Mar. 1987.

[7] G. R. Gallager, *Information Theory and Reliability Communication*. New York: Wiley, 1968.

[8] J. W. Goodman, F. I. Leonberger, S.-Y. Kung, and R. A. Athale, "Optical interconnections for VLSI systems," *Proc. IEEE*, vol. 72, pp. 850–866, July 1984.

[9] F. Harary, *Graph Theory*. Reading, MA: Addison-Wesley, 1972.

[10] M. G. Hluchyj and M. J. Karol, "Queueing in high-performance packet switching," *IEEE J. Select. Areas Commun.*, vol. 6, pp. 1587–1597, Dec. 1988.

[11] A. Huang and S. Knauer, "Starlite: A wideband digital switch," *Proc. Globecom 84*, pp. 121–125.

[12] Y. N. J. Hui and E. Arthurs, "A broadband packet switch for integrated transport," *IEEE J. Select. Areas Commun.*, vol. SAC-5, pp. 1264–1273, October 1987.

[13] M. J. Karol, M. G. Hluchyj, and S. P. Morgan, "Input vs. output queuing on a space division packet switch," *GLOBECOM 86*, Nov. 1986.

[14] L. Kleinrock, *Queuing Systems, Vol. 1: Theory*. New York: Wiley, 1975.

[15] M. Littlewood, "Sunshine: A broadband packet switch," *Bell Commun. Res. Tech. Rep.*, Sept. 28, 1987.

[16] T. T. Lee, R. Boorstyn, and E. Arthurs, "The architecture of a multicast broadband packet switch," *Proc. INFOCOM'88*, Mar. 1988, pp. 1–8.

[17] T. T. Lee, "Non-blocking copy networks for multicast packet switching," *IEEE J. Select. Areas Commun.*, vol. 6, pp. 1455–1467, Dec. 1988.

[18] P. Newman, "A fast packet switch for the integrated services backbone network," *IEEE J. Select. Areas Commun.*, vol. 6, pp. 1468–1479, Dec. 1988.

[19] A. Pattavina, "Multichannel bandwidth allocation in a broadband packet switch, *IEEE J. Select. Areas Commun.*, vol. 6, pp. 1489–1499, Dec. 1988.

[20] J. S. Turner, "Design of an integrated service packet network," *IEEE J. Select. Areas Commun.*, vol. SAC-4, Nov. 1986.

[21] L. T. Wu, E. Arthurs, and W. D. Sincoskie, "A packet network for BISDN applications," *Proc. 1988 Zurich Sem. Digit. Commun.*, Zurich, 1988, pp. 191–197.

[22] L. T. Wu and M. Kerner, "Emulating circuits on a broadband packet network," *GLOBECOM'88*, Miami, FL, Nov. 1988.

[23] Y.-S. Yeh, M. G. Hluchyj, and A. S. Acampora, "The knockout switch: A simple, modular architecture for high-performance packet switching," *IEEE J. Select. Areas Commun.*, vol. SAC-5, pp. 1274–1283, Oct. 1987.

Stochastic Models for ATM Switching Networks

A. Descloux

Abstract—It is widely recognized that the input streams to cell switching networks cannot, in general, be adequately modeled as Poisson point processes. The fact that the traffic passing through such systems often displays marked burstiness must be taken into account, and many contributions are concerned with the practical and analytical consequences of the resulting dependencies between the cell interarrival times. The models considered in this paper include yet another feature that is usually ignored, namely that the switching speed may be appreciably greater than the speed at which some sources generate their respective cells. It is therefore important, as we do here, to investigate the situation where the cells making up individual bursts are spaced according to some prescribed probability distribution.

The purpose of this paper is to show that the switch performance is strongly dependent on the input parameters and, at the same time, to demonstrate that analytical approaches provide useful alternatives to the lengthy simulation runs needed to evaluate probabilities of rare events such as cell losses.

I. INTRODUCTION

THIS paper deals with the traffic handling capacity of non-blocking cell switching networks, the emphasis here being on the contention probabilities and the cell losses.

A switching network is said to be nonblocking if, irrespectively of momentary connections, there is always a free path from any free inlet to any free outlet. Hence, in the case of circuit switching, contentions can only occur at call setup times when two or more requests for the same idle inlet–outlet pair are placed simultaneously. However, these events are quite infrequent and can usually be ignored. By contrast, in a synchronous cell switching network, all the inputs and all the outputs become reusable at the beginning of each and every time slot and any subgroup of n distinct inlets can try to send n cells to n (not necessarily) distinct outlets. Since contentions (collisions) will occur if two or more inputs attempt to reach the same output during the same time slot, there are now many opportunities for such events to occur and their effects can no longer be ignored. In case of collision, only one cell can go through the network and the others must be stored and wait in buffers for later delivery.

Buffering can be done in many different ways. One possibility is to store the delayed cells at queues associated with the input ports and serve the cells at the head of the line first. With this approach, the access to the output will be "limited," as the cells at the head of the waiting line may delay cells to other destinations that are or have become free (see [1], [6], [7]). Here, however, we shall only be concerned with the case where all the delayed cells destined to the same outlet wait in output buffers. With this arrangement, delayed cells can always reach the output side of the network without any further interference from cells with other addresses, and maximum throughput (up to saturation) can then be achieved. Under these conditions, the input ports are "transparent" and their presence can be ig-

nored. All we need to know is the joint input of the sources assumed here to be evenly spread among destinations. In first approximation, the outputs can then be modeled as single-servers with constant service time (equal to the fixed cell length) and, in general, special features. Here, however, the delayed cells that cannot immediately reach their respective output queues may have to wait in shared buffers and the queues are therefore not independent (a second-order effect that can usually be ignored). There is considerable freedom for the physical location of the buffers without infringing on the nonblocking property of the network. They can all be placed on the output side of the network or at some intermediate points. For instance, in one switch design involving a nonblocking self-routing Batcher–banyan network, the delayed cells are intercepted at the output of the banyan network and resubmitted to the input side of that network via additional Batcher sorting networks which prevent conflicts between cells with different destinations (see [3]). The purpose of these remarks is to stress that the model investigated here covers a wide variety of options, some new and some that have been proposed in the literature, none of which involve throughput losses. Incidentally, we also mention that full throughput could also be achieved by queueing on the input side of the network. This could be accomplished by means of a token-passing ring that would fully poll all the input ports during each cell time slot and thus reduce head-of-the-line delays but not eliminate them.

In the systems now being developed, the switching speed is relatively high (up to 150 Mb/s) and the ratio between the latter and the user's input speeds may vary over a wide range and strongly affect the characteristics of the traffic that reaches the switches. Cells generated by slow terminals will appear at widely spaced intervals at the switches as if they had been generated randomly. By contrast, if the input and switching speeds are of commensurate magnitudes, then the successive interarrival times of the cells sent by a given source (to a given destination) will no longer be statistically independent and the cells will appear to arrive in bunches of varying lengths. As in several published papers, we can still view the output queues as single servers with constant service (cell) length, but the classical queueing models do not apply since the cells from a given burst do not arrive independently but are *strung* over not necessarily adjacent cell time slots. Accordingly, the switch performance was determined under the following assumptions:

i) the leading cells of successive cell-bursts arrive according to a Poisson process,

ii) the cells of a given burst arrive over consecutive or randomly spaced cell time-slots,

iii) the number of cells in a single burst is a random variable with a geometric distribution.

The particular case where the cells of a given burst arrive over consecutive cell time-slots is easy to handle since the underlying equations can then be solved recurrently. A far more complicated situation arises when the cells making up a burst

Manuscript received April 23, 1990; revised November 15, 1990.
The author is with Bellcore, Morristown, NJ 07962.
IEEE Log Number 9042047.

Reprinted from *IEEE J. Selected Areas Commun.*, vol. 9, no. 3, pp. 450–457, April 1991.

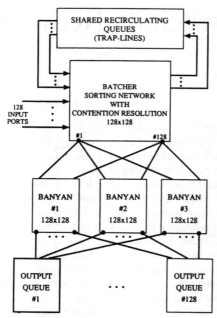

Fig. 1. Conceptual description of a *Sunshine* network with three banyan networks.

are randomly spaced because the number of active bursts may then exceed the number of cells present in the system. In that case, solutions were obtained by matrix exponentiation, a process whose convergence speed decreases as the occupancy and/or the cell separation increases.

The impact of the preceding assumptions was determined for two switch designs, each with 128 inputs and 128 outputs. The first one, referred to as *multiply-reentrant* switch (see [3]), consists of a nonblocking network made up of a Batcher sorting network followed by a self-routing banyan network. Here, contentions are resolved by means of a fixed number of output buffers that are assigned to each of the output ports, together with an additional set of shared buffers used to handle the cells overflowing from the individual buffers. The second model, called *Sunshine* in the literature (see [5] and Fig. 1), has an arbitrary number, b say, of banyan networks which can transmit simultaneously as many as b cells to any output with the result that the cells may arrive at the output buffers in *batches* of size b or less. (For $b = 1$, the *Sunshine* network transfers at most one cell to each of the output queues during any cell time slot, but there is no such restriction for the system first mentioned above.)

The quantitative results of the paper are presented in the form of graphs showing cell loss probabilities and some queue-length averages in terms of the average burst size (n) and the average spacing (s). Clearly, the burst size and the spacing have complementary effects: an increase of the former may be compensated by an increase of the latter. Numerous examples show that the cell loss probability increases very rapidly as the burstiness increases and, in the absence of controls, even apparently harmless traces of burstiness can degrade the system performance below acceptable levels. In practice, highly bursty traffic as well as long nonrandom trains of cells may have to be segregated and circuit switched or be given lower priorities in order to protect nondelayable services such as voice.

The present paper deals only with one class of arrival processes. Many others will undoubtedly gain in importance in the

near future. In this connection, we note that with the predicted introduction of new services (such as home video) the cells from some sources may arrive in periodic streams at the switch. A similar situation may also arise in systems involving minipacketization, where constant spacing between the cells could occur either naturally or as the result of load or flow control measures.

Although the results presented here were at first obtained to evaluate specific switches, it is clear that the analytical approach and conclusions of this paper should remain valid for a broad class of switch architectures with output queueing.

II. PROBABILISTIC MODELS

In this section, we describe the conditions under which the analysis is carried out. For reasons stated earlier, the cell generation mechanism at the input ports can be ignored and, 'bypassing' the switch, we shall assume that the cells arrive at the outputs in bursts of various sizes. The first cell of such bursts is referred to as the leading cell. The successive cells of a given burst may not be submitted over consecutive cell time slots and the spacing between any two consecutive cells is defined as the number of cell time-slots that are skipped over. For each destination, the cell arrival process is as follows.

1) All the cells are of unit length and arrive synchronously at the switch.

2) The leading cells arrive according to a discretized Poisson process of intensity λ.

3) The cells of a given burst arrive over randomly spaced cell time-slots, i.e., the spacing between two consecutive cells from the same burst is equal to n with probability $\beta^n (1 - \beta)$, $n \geq 0$. (A spacing of n between two cells of a burst means that they are separated by n time slots, none of them being occupied by another cell from that burst.)

4) The number of cells in a given burst has a modified geometric distribution with mean $(1 - \alpha)^{-1}$ (empty bursts are excluded, i.e., the probability that a burst contains n cells is equal to $\alpha^{n-1}(1 - \alpha)$, $n \geq 1$).

We now list the features of the service (transfer) mechanism.

1) A fixed number of trap-lines act as common buffers for all the cells that cannot be immediately transferred to the outputs for lack of an available banyan network. New cells are trapped, if and only if the number of cells (old and new) destined to the same output, momentarily exceeds the number of banyan networks.

2) The banyan networks transfer to the output buffers as many cells as possible during each cell time-slot, irrespectively of the state of the output queues, and thus create truncated *batch* arrival processes at the output ports which means greater output buffer requirements.

3) The trap-lines become instantly and simultaneously available as soon as the cells that occupied them are switched by the banyans.

4) A cell is lost if it cannot be immediately switched to its output port while all the trap-lines are being used, or if it is transmitted to an output without available buffer.

In the next two sections, we shall deal with two cases. First, we consider the situations where there is a single banyan and the cells belonging to the same burst arrive in consecutive cells slots (zero-spacing) and then we will analyze the multibanyan case with randomly spaced cells. In the first instance, we are led to a set of difference equations that can be quickly solved recurrently whereas, in the second, the solution is obtained iteratively.

III. State Equations for Zero-Spacing and a Single Banyan

To facilitate the exposition, we introduce some definitions. We shall say that a burst is active at the beginning of a cell time-slot if it may still generate a cell. At that time, each active burst either comes to an end with probability $1 - \alpha$ or submits another cell with probability α.

At the beginning of each cell time-slot, a series of events occur simultaneously: a cell begins service, bursts do not terminate and immediately submit new cells, new bursts generate their leading cells. (Of course, depending on the state of the system, some or even none of these events may occur.) The instantaneous intermediate "states" the system goes through during such a sequence of events are called here pseudo-states (illustrations are given in Figs. 2 and 3).

An output queue will be said to be in state (i, j) at the beginning of a time slot when (i) there are i active bursts and (ii) j cells are waiting. Let $P(i, j)$ be the equilibrium probability of state (i, j). Since there is no spacing between the cells of the active bursts, $P(i, j) = 0$ for $i > j$.

Let

$$\pi(n) \equiv e^{-\lambda} \cdot \frac{\lambda^n}{n!}, \qquad n = 0, 1, 2, \cdots. \qquad (1)$$

Consider first the case where $i = j$. This situation can arise if and only if the system was in one of three states, namely $(0, 0)$, $(0, 1)$, or $(1, 1)$ at the beginning of the previous cell time-slot and i new cells arrived during that cell time-slot. Hence, we have:

$$P(0, 0) = \pi(0) \cdot [P(0, 0) + P(0, 1) + P(1, 1)(1 - \alpha)], \qquad (2)$$

$$P(i, i) = \pi(i) \cdot [P(0, 0) + P(0, 1) + P(1, 1)(1 - \alpha)] + P(1, 1)\alpha\pi(i - 1). \qquad (3)$$

Equations (2) and (3) yield the simple relations from which all the $P(i, i)$'s are readily obtained:

$$P(i, i) = \frac{\pi(i)}{\pi(0)[1 - \alpha\pi(i - 1)]} \cdot P(0, 0). \qquad (4)$$

Furthermore $P(0, 1)$ can now be determined from (2):

$$P(0, 1) = \frac{[P(0, 0)[1 - \pi(0)] - \pi(0)P(1, 1)(1 - \alpha)]}{\pi(0)}. \qquad (5)$$

Next consider the case where $j = i + 1$. Now the state $(i, i + 1)$ can only be reached if, at the beginning of the current time slot, none of the active bursts submits a new cell and i new cells arrive *or* if, at that same instant, one and only one of the active bursts submit another cell and $i - 1$ new cells arrive. Note that the state $(i, i + 1)$ can only be reached from three states, viz. $(0, 2)$, $(1, 2)$, and $(2, 2)$. Summarizing, it follows that

$$P(i, i + 1)$$
$$= [P(0, 2) + P(1, 2)(1 - \alpha) + P(2, 2)(1 - \alpha)^2]\pi(i)$$
$$+ \left[P(1, 2)\alpha + \binom{2}{1} \cdot P(2, 2)\alpha(1 - \alpha)\right]\pi(i - 1). \qquad (6)$$

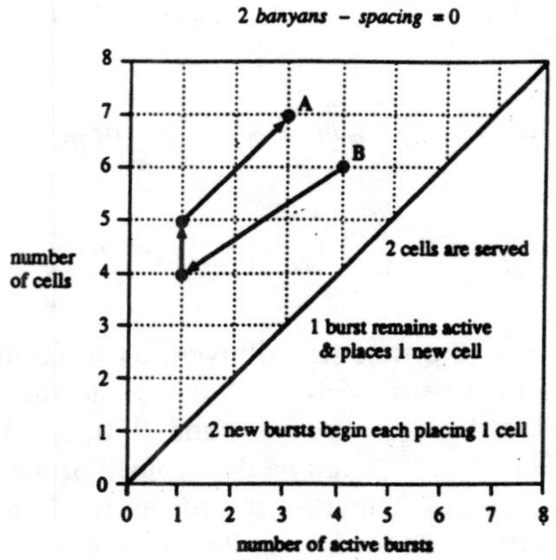

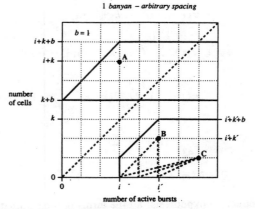

Fig. 2. Pseudo-states for one-step transitions. The graph shows the three components of a one-step transition from **B** to **A**.

Fig. 3. Transition domains. The graph shows the two (framed) sets of states from which **A** and **B** can be reached in one (composite) step. Also drawn are the three possible one-step transitions from **C** to **B**. (When the spacing is equal to zero, all the states below the diagonal have probability 0 and the transition domains reduce to horizontal segments.)

With the convention that $\pi(-1) \equiv 0$, the preceding relations are satisfied for all i's.

Since $P(2, 2)$ is known from the previous step, we can now determine $P(0, 2)$ and $P(1, 2)$ from (6) for $i = 0$ and $i = 1$.

We now consider the general case. Suppose that a transition from state (p, q) to state $(i, i + k)$ takes place during a single cell time-slot. Let $j (\leq p)$ and r be, respectively, the number of bursts that terminate and the number of new bursts that begin service at the start of the cell time-slot. Then:

$$i = p - j + r$$

and

$$i + k = (q - 1) + (p - j) + r.$$

Subtracting the first from the second of these two relations shows that the transition under consideration is possible only if $q = k + 1$.

Let $k^* \equiv \min(i, k + 1)$. Then, for any given value of $p (0 \leq p \leq k + 1)$, the state $(i, i + k)$ can be reached from state $(p, k + 1)$ in $k^* + 1$ ways, each one of them being uniquely determined by the number, $n (n \leq p)$, of bursts that submit new cells. Indeed, for a given n, state $(i, i + k)$ is reached via the following sequence of pseudo-states $(p, k), (n, k), (n, n + k)$. Finally, with the arrival of $i - n$, new bursts state $(i, i + k)$ is

427

reached. The $P(i, i + k)$'s therefore satisfy the following equilibrium equations:

$$P(i, i + k) = \sum_{n=0}^{k^*} \pi(i - n) \cdot \sum_{m=n}^{k+1} P(m, k + 1)$$

$$\cdot \binom{m}{n} \alpha^n (1 - \alpha)^{m-n}. \tag{7}$$

As we have seen, the first step allowed us to compute all the probabilities $P(i, i)$'s as well as $P(0, 1)$ and the second step yielded $P(i, i + 1)$, $i = 1, \cdots$, and $P(0, 2)$. It is easy to show that all the $P(i, k)$'s can be determined at the kth step. In view of the preceding remarks, it suffices to show that if the previous property holds for $k - 1$, then it is necessarily true for k. To this end, we rewrite (7) as follows:

$$P(i, i + k) = \pi(i)P(0, k + 1) + \pi(i)P(1, k + 1)(1 - \alpha)$$

$$+ \pi(i - 1)P(1, k + 1)\alpha$$

$$+ \pi(i) \sum_{m=2}^{k+1} P(m, k + 1)(1 - \alpha)$$

$$+ \pi(i - 1) \sum_{m=2}^{k+1} P(m, k + 1)\alpha(1 - \alpha)$$

$$+ \sum_{n=2}^{k^*} \pi(i - 1) \sum_{m=n}^{k+1}$$

$$\cdot P(m, k + 1)\alpha^n (1 - \alpha)^{m-n}. \tag{8}$$

By the induction hypothesis, all the terms in the summations of (8) are known and $P(0, k + 1)$ and $P(1, k + 1)$ can be computed from the two equations obtained from (8) by setting i equal to 0 and to 1. Hence, the kth step determines all the $P(i, i + k)$'s as well as $P(0, k + 1)$ and the proof is complete. (Normalization yields $P(0, 0)$.)

The computation of the $P(i, j)$'s is only an intermediate step. What we really need here are the probabilities, $Q(j, k)$, of having j unserved cells at the beginning of a cell time-slot and k unserved at the start of the next slot. To compute the cell loss probability, we need the distribution of the number of *new* cells that arrive during a single cell time-slot. These new *one-step* joint probabilities can be obtained from the $P(i, j)$'s. We can first determine the probability, $Q^*(j, r)$ that there are j unserved cells at the beginning of a cell time-slot and that the active bursts generate r new cells.

$$Q^*(j, r) = \sum_{i=0}^{j} P(i, j) \binom{i}{r} (1 - \alpha)^{i-r} \cdot \alpha^r. \tag{9}$$

The $Q(i, j)$'s can be expressed in terms of the $Q^*(i, r)$. Indeed, we have:

$$Q(0, k) = \pi(k) \cdot P(0, 0), \quad k = 0, \cdots, \infty,$$

$$Q(i, k) = \sum_{j=0}^{\min(k+1-i, i)} \pi(k - i - j + 1) \cdot Q^*(i, j), i > 0. \tag{10}$$

In principle, the equilibrium probabilities that j cells are simultaneously present could be obtained by solving the equations

associated with the *one-step* transition probabilities

$$Q(i, j) / \sum_{k=0}^{i} P(k, i).$$

However, the complexity of the transition probabilities seems to preclude an explicit solution and we shall proceed numerically.

Let

$$Q_1 \equiv [Q(i, j)] \tag{11}$$

be the matrix of these transitions probabilities. It is well known that the successive powers of this matrix tend to a matrix with identical rows whose elements are the equilibrium probabilities. Accordingly, let

$$Q_2 \equiv Q_1 \times Q_1, \quad Q_4 \equiv Q_2 \times Q_2, \quad Q_8 \equiv Q_4 \times Q_4, \cdots,$$

$$Q_{2^n} \equiv Q_{2^{n-1}} \times Q_{2^{n-1}}.$$

The elements of Q_{2^n}, which are obtained after n matrix multiplications, are the transition probabilities over a sequence of 2^n consecutive cell time-slots. Extensive computations have shown that eight or even fewer exponentiations sufficed to determine the equilibrium state probabilities for the individual output queues with the accuracy needed (eight significant figures or more) in the next steps.

A. Cell Loss Probability (Zero-Spacing-1 Banyan)

Thus far, we have only dealt with single output ports. Of practical interest here are the buffer requirements for networks with arbitrary numbers of inputs and outputs, set equal to 128 throughout the computations. The cell loss probability (defined as the ratio of the number of cells that cannot be accommodated by the common trap-lines to the total number of submitted cells) is obtained by cumulating the contributions of the output ports to the shared trap-lines (or reentry loops in [3]). Strictly speaking, the output queues are not statistically independent since they compete at times for common facilities. However, in practical applications, the cell loss probability is always very small by design and the dependence between the output queues can be ignored. Under this condition, the cell loss probability can be obtained by convolving the overflow distributions of the individual output queues which for 128 ($=2^7$) output ports can be performed in seven steps.

IV. State Probabilities for Random Spacing and Multiple Banyans

In this section, we deal with the case where the cells from a given burst need not arrive at successive cell time-slots but may be randomly spaced over nonadjacent cell time-slots. This new feature complicates the analysis because the number of cells simultaneously present in the system are no longer necessarily larger than or equal to the number of active bursts: for instance, an active burst may submit cells at times when *all* the cells it had placed earlier have been served.

The number of (one-step) transition paths from state (n, m) to state $(i, i + k)$ is equal to $\min(i, n) - [\max(0, m - b) - k] + 1$, and all these paths involve distinct pseudo-states. For example (see Fig. 3), from the state with coordinates $n' = 7$ and $m' = 1$ there are three paths to the state $(5, 2)$: i) one cell is served, five bursts remain active and two of the latter submit a new cell, ii) one cell is served, four bursts remain active, one of them places a new cell and a new burst enters the system, and iii) one cell is served, three bursts remain active but do not submit new cells and two new bursts arrive.

State $(i, i + k)$ can be reached from state (n, m) if and only if that state belongs to the *transition domain* defined by the inequalities (see Fig. 3):

    i. $n \geq \max(0, -k)$,

    ii. $m \leq \min(n, i) + k + b, n \geq 0$,

    iii. $m \geq \begin{cases} \max(k + b, 0) & \text{if } k > 0 \\ 0 & \text{if } k \leq 0. \end{cases}$

    iv. The number of bursts that remain active, r, is at most equal to $\min(i, n)$,

    v. Among the r bursts that remain active, $r + k - \max(0, m - b)$ of them submit exactly $r + k - \max(0, m - b)$ new cells,

    vi. $i - r$ new bursts enter the system.

We shall say that an active burst is eligible at the beginning of a cell time-slot if it is allowed access to the system at that time. (Eligibility is used to modeled cell spacing.) Let θ be the average spacing between successive cell time-slots. If n is the number of active bursts at the beginning of a cell time-slot, then, in accordance with our assumptions, the probability that r of these bursts become eligible and place r cells is equal to

$$\binom{n}{r} \beta^{n-r} (1 - \beta)^r$$

where $\beta = \theta/(1 + Q)$ and the probability that j of these bursts then terminate is equal to

$$\binom{r}{j} \alpha^{r-j} (1 - \alpha)^j$$

Upon becoming eligible, a burst must necessarily place a new cell and only then can it terminate. Under these circumstances, the probability of a transition from state (n, m) to state $(i, i + k)$ is equal to

$$\sum_{r=0}^{i+k-[(m-b)]^+} \binom{n}{r} \beta^{n-r}(1 - \beta)^r \times \binom{r}{m - i} \alpha^{r-(m-i)}$$
$$(1 - \alpha)^{m-i} \cdot \pi(i - r) \qquad (12)$$

with the proviso that sums whose lower limits exceed their upper limits are set equal to zero.

Our main purpose here is to determine the equilibrium distribution of the number of cells in a single output queue. As shown later, these probabilities are then used to determine queue-length distributions and cell loss probabilities.

We now briefly describe how the state probabilities associated with the individual output ports were obtained. First, we now have to determine the equilibrium state probabilities $P(i, j)$ defined (and determined recurrently) in the preceding section. Let P_1 be the matrix of the transition probabilities given in (12) above. Then, we have

$$\lim_{n \to \infty} P_1^n \equiv P, \qquad (13)$$

where P is a matrix with identical rows whose elements are the equilibrium probabilities, $P(i, j)$, that there are i active bursts and j cells in an output queue at the beginning of a randomly chosen cell slot. As before P can be approximated as closely as needed by raising P_1 to a sufficiently high power (itself preferably a power of 2 to speed up the computations).

In the case of multiple banyans, practical considerations have

led us to carry out the computations for the case of zero-spacing only. This case, however, is of particular interest since it provides upper bounds on the cell loss probabilities and other traffic characteristics. Under this condition, we recall that the state (i, j) has a non-vanishing probability if and only if $i \leq j$. Accordingly (and conveniently), we can reindex the states as follows:

$$(i, j) \leftrightarrow (j \cdot (j + 1))/2 + i + 1.$$

For any given value of the right-hand side in the preceding correspondence, i and j are uniquely determined. The advantage of this new indexing is that each state is now specified by a single integer (rather than 2) and this simplifies some programming steps.

Computations have shown that, for practical applications, the range of (i, j) can be restricted to $0 \leq j \leq 32$. With this additional restriction, the "new" index will vary from 1 to 561 ($= 33 \times (33 + 1)/2$) and the transition matrices have more than 300 000 entries! We note here that a method due to Winograd (see [8]) greatly speeds up taking powers of matrices of such magnitudes. According to this approach, the number of multiplications is reduced, for large dimensions, by nearly one half while the total number of operations remains about the same.

Let $X = [x_{i,j}]$ be an $m \times n$ matrix and let $Y = [y_{j,k}]$ be an $n \times r$ matrix. Let $Z = [z_{i,k}]$ be the product of X and Y in that order. By definition,

$$z_{i,k} = \sum_{j=1}^{n} x_{i,j} y_{j,k}$$

and the Winograd matrix multiplication scheme is based on the easily verified identity:

$$\sum_{j=1}^{n} x_{i,j} y_{j,k} \equiv \sum_{1 \leq j \leq n/2} (x_{i,2j} + y_{2j-1,k})$$
$$\cdot (x_{i,2j-1} + y_{2j,k}) - a_i - b_k + c_{i,k}$$

where

$$a_i = \sum_{1 \leq j \leq n/2} x_{i,2j} \cdot x_{i,2j-1}$$

$$b_k = \sum_{1 \leq j \leq n/2} y_{2j-1,k} \cdot y_{2j,k}$$

and $c_{i,k} = x_{i,n} \cdot y_{n,k}$ when n is odd and equal to 0 when n is even.

V. Cell Loss Probabilities

The determination of the cell loss probability is straightforward in the case of the *multiply-reentrant* network (see [3] and [4]) and we shall only be concerned here with the losses in the *Sunshine* network. In that case, a cell can be lost in two ways: i) if it arrives when no banyan network is available for its immediate transfer to the output queue and all the trap-lines are being used or ii) if it is transferred to a busy output without free buffers.

Let b be the number of banyan networks. Two extreme cases provide upper bounds.

1) For $b = 1$, all the delayed cells wait in the trap-lines, the output-queue loss probability is equal to 0, at most one output buffer is ever needed and, for a given number of trap-lines, the trap-line loss probability takes its maximum value. (We assume tacitly that the output ports discharge the cells one at a time and without further congestion delays.)

2) When $b = \infty$, then all the delayed cells must wait in the output buffers, the trap-line loss probability is equal to 0, no trap-line is ever needed, and all losses occur at the outputs.

429

The two components of the overall cell loss probability are dealt with separately in the next two subsections.

A. Trap-Loss Probability

We begin by focusing attention on the cells destined to a particular output port and ignore losses due to shortages of output buffers. Then, during any time-slot, the only cells that can be lost are those that arrived during that time-slot, did not have immediate access to a banyan network, and could not find idle trap-lines (delayed cells assigned to trap-lines are never lost while waiting for a banyan network.) To compute the trap-loss probability, we have, therefore, to determine the distribution of the number of new cells that must use trap-lines. We do this first for the individual destinations (in two steps) and then, by convolving and truncating, obtain the overall trap-loss probability.

Let $\Pi_b(i)$ be the equilibrium probability that there are i cells for a given destination and let $\Pi_b^*(j, i)$ be the joint probability that i trap-lines with (i) unserved cells are occupied at the beginning of a cell time-slot and that there are j new requests for trap-lines by the end of the time-slot. Let $p(j|i)$ be the conditional probability that j new cells arrive given i, with i as defined above. Since

$$\Pi_b^*(j, i) \equiv \Pi_b(i) \cdot p(j|i)$$
$$\equiv \sum_{m,n} P(m, i) \cdot P_1[(n, j)|(m, i)],$$

the following relations describe the cell assignment mechanism to the traps:

$$\Pi_b^*(0, i) = \Pi_b(i) \cdot \sum_{j=0}^{b} p(j|i), \quad 0 \le i < b,$$

$$\Pi_b^*(0, i) = \Pi_b(i) \cdot \sum_{j=0}^{2b-i} p(j|i), \quad b < i \le 2b,$$

$$\Pi_b^*(0, i) = \Pi_b(i) \cdot p(0|i), \quad i > 2b,$$

$$\Pi_b^*(j, i) = \Pi_b(i) \cdot p(b + j|i), \quad 0 < j, 0 \le i \le b,$$

$$\Pi_b^*(j, i) = \Pi_b(i) \cdot p(2b - i + j|i), \quad 0 < j, b < i \le 2b,$$

$$\Pi_b^*(j, i) = \Pi_b(i) \cdot p(j|i), \quad 0 < j, i > 2b.$$

For completeness sake, we note that $p(j|i)$ can be obtained from the one-step transition probabilities. Indeed, we have:

$$p(j - \max(0, m - \max[0, m - b])|n)$$
$$= \sum_{i,m} P_1[(i, j)|(m, n)], \quad i, j \ge 0.$$

The joint probability that there are j trapped cells at a given instant and that i of these cells had been previously delayed is readily obtained from the expressions for the Π_b^*'s. All we have to do is to take into account that $\max(0, i - b)$ trapped cells remain trapped. Writing Π_b^+ for that distribution, we have:

$$\Pi_b^+(j, 0) = \sum_{i=0}^{b} \Pi_b^*(j, i), \quad 0 \le j, 0 \le i \le b,$$

$$\Pi_b^+(j + i - b, i - b) = \Pi_b^*(j, i), \quad i > b.$$

In all applications, the cell loss probability is quite small, which means that there is very little interaction between the output queues and the latter are practically independent of each other.

Let N be the number of output ports. Then, under the stated assumption, the probability $T_b^+(m, n)$ that the total number of

busy trap-lines at the beginning of a cell time-slot is n and that m among the new cells that will arrive during the ensuing cell time-slot will be delayed in getting to their destination or even be lost, can be obtained by taking the Nth fold convolution of the bivariate distribution $\Pi_b^+(\cdot, \cdot)$ with itself evaluated at (m, n).

Let τ be the number of trap-lines. The expected number of lost cells, L_τ, during a single cell time slot is then given by the following formula:

$$L_\tau = \sum_{m=\tau}^{\infty} (m - \tau) \sum_{n=0}^{\tau-1} T_k^+(m, n)$$
$$+ \gamma_b(\tau) \sum_{m=\tau}^{\infty} (m - \tau) \cdot T_b^+(m - \tau, \tau)$$

where

$$\gamma_b(\tau) \equiv \frac{\sum_{n=\tau}^{\infty} \sum_{m=0}^{\infty} T_b^+(m, n)}{\sum_{m=0}^{\infty} T_b^+(m, \tau)}.$$

The factor $\gamma_b(\tau)$ takes into account the truncation of the distribution T_b^+ at $n = \tau$. Finally, with ρ the cell arrival rate per destination, the trap-loss probability Λ_τ, can be expressed as follows:

$$\Lambda_\tau = \frac{L_\tau}{N \cdot \rho}.$$

B. Output-Queue Loss Probability

Let T and B be, respectively, the number of busy trap-lines and the number of output buffers occupied, at a given instant, by cells destined to a specific output port and let τ and κ be, respectively, the upper limits of these two variables. Clearly,

$$\Pr[T + B = m] = \sum_{n=0}^{m} \Pr[T = n] \cdot \Pr[B = m - n | T = n]$$

where $0 \le n \le \tau, 0 \le m - n \le \kappa$. So long as the trap losses are small and under the assumption that the cells at the outputs are served one at a time, we have, approximately,

$$\Pr[T + B = m] = \Pi_1(m).$$

Since

$$\Pr[T = n] = \Pi_b(n),$$

the first equations of this subsection yield the following recursive system of linear equations for the conditional probabilities $\Pr[B = i | T = j]$:

$$\Pi_1(m) = \sum_{i+j=m} \Pi_b(j) \cdot \Pr[B = i | T = j],$$
$$0 \le i \le \kappa, 0 \le j \le \tau.$$

The output-queue cell loss probability, Λ_κ, can then be expressed as follows:

$$\Lambda_\kappa = \sum_{i=0}^{b-1} \Pi_b(i) \cdot \Pr[B \ge \kappa | T = i].$$

Note that equating the distribution of $T + B$ to Π_1 yields an upper bound for the output-queue loss probability Λ_κ. For low trap-loss probabilities, this upper bound can be expected to be quite close to the true output-queue loss. By using a truncated version of Π_1, an exact formula for the trap-loss probability can be obtained. Such a refinement does not appear to be warranted.

VI. Numerical Results

We have performed extensive computations to evaluate the performance changes induced by varying the load levels, the burstiness, and the cell spacing. A by-product of this work was a check of a simulation study of the *Sunshine* switch (see [5]) which, in turn, led to refinements of the analytic model. The data obtained by these two approaches are in very good agreement (see Figs. 6–9) and establish confidence in both of them. Clearly, the analytic method does not have the flexibility of simulation but was far more efficient than the latter at low losses and also yielded, at no extra cost, the higher moments of the equilibrium distribution of the buffer requirements.

The curves of Figs. 4 and 5 pertain to a switch with a single banyan and an arbitrary number of unshared (dedicated) buffers at each output. All the cells that cannot be accommodated in these buffers are placed in a common buffer that can hold up to 128 cells. For a given occupancy (0.6), Fig. 4 shows the dependence of the cell loss probability on the unshared buffers and the cell spacing. Fig. 5 illustrates how the cell loss probability varies as a function of the load. Jointly, these two figures demonstrate that even minimal cell spacing and very marginal load variations strongly affect performance.

Figs. 6 and 8 give the expected number of trap-lines that are requested (needed to completely satisfy the momentary demand) in terms of the actual number of trap-lines (cells unable to get a trap-line are lost). The first of these two figures is for Poisson input while the second is for an average burst size of 2. (There are two banyans in both instances). Under the same sets of conditions, Figs. 7 and 9 show the cell loss probabilities. Here again, we see that an apparently minor change of burst size brings about a surprisingly large change of buffer requirements: the ratio between the average asymptotic number of busy trap-lines for the (minimally) bursty traffic (about 41) and the Poisson traffic (about 16) exceeds 2.5. The impact of the same change on the trap-loss probability is even more spectacular as can be seen by comparing the steadily dropping curves of Figs. 7 and 9. These observed effects have been closely duplicated by a simulation whose results appear as stars (✳). (The independence assumption between output queues fails to be satisfied for small numbers of trap-lines and explains the small discrepancies between the analytical and simulation results visible in Fig. 9.)

VII. Summary and Conclusions

In this paper, we have described simple tools for the evaluation of a broad class of switch architectures with the property that their throughputs are not curtailed by access limitations. Although we have focused attention on the *Sunshine* and the *multiply-reentrant* switches, the same approach (possibly with some tuning) should apply to many other configurations.

The *Sunshine* switch with multiple Banyans stands out because it allows simultaneous cell transfers to the same destination. Its study requires a more general model that must take into account this *batch* arrival process, as well as the fact that losses can occur in two places: at the trap-lines and/or at the output ports.

Trap-line and reentry-loop sharing also implies that the contributions of the output ports must be integrated, and our results therefore depend on the switch size. The analytic approach is then best suited to investigate the asymptotic behavior of the cell loss probability as the switch size gets large (in our exam-

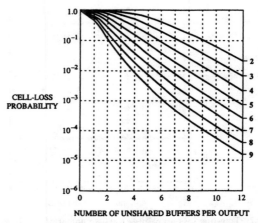

Fig. 4. Cell loss probability versus spacing. The numbers of the right-hand side of the graph stand for the average spacings between cells.

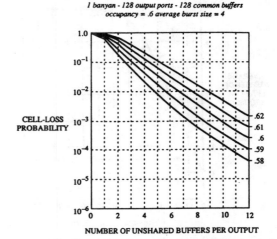

Fig. 5. Cell loss probability versus occupancy. The numbers of the right-hand side of the graph stand for the output occupancies.

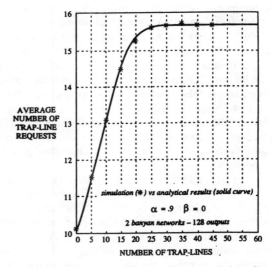

Fig. 6. Average number of trap-line requests versus number of trap-lines.

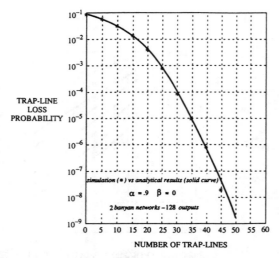

Fig. 7. Trap-line loss probability versus number of trap-lines.

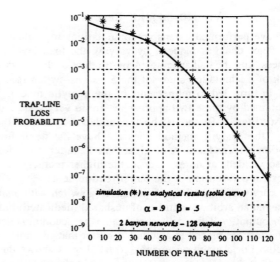

Fig. 9. Trap-line loss probability versus number of trap-lines.

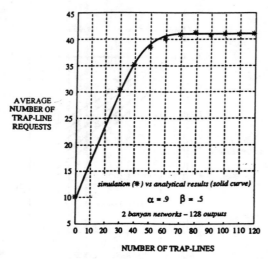

Fig. 8. Average number of trap-lines requests versus number of trap-lines.

that this may indeed happen even if there is some contraction of cell spacing (see [2]).

It is questionable whether models that would include the refinements alluded to above could ever serve as manageable engineering tools. The results of the present and similar performance studies appear to be, it seems, mainly useful warning flags while the most urgent practical need remains the development of robust engineering procedures whose application does not require very detailed descriptions of traffic patterns.

Acknowledgment

The author would like to thank C. M. Day, J. N. Giacopelli, and M. Littlewood for their useful discussions and comments.

ples, the computed cell loss probabilities can be equated to their asymptotic values, which are also upper bounds).

The analysis was carried out under assumptions that are rather stringent but still general enough to reveal the salient characteristics of networks handling bursty traffic. Yet, if needed, the model can be modified to account for more general input processes. For instance, the assumption that the leading cells arrive according to a Poisson distribution can be easily relaxed. Similarly, the introduction of *phase-type* distributions (see [9]) for the burst sizes and spacing, for example, would only involve simple modifications of the analysis. Many other generalizations are possible.

We have observed that a network's performance is strongly affected by even very minor perturbations of its input process. Because of this sensitivity, it becomes important to determine how burstiness and cell spacing vary as one proceeds from the periphery to the core of a multihop network. As cell streams from several distinct sources progress though a set of switching nodes and are intermingled, the traffic reaching *inner* switching points may tend to be less bursty. A simulation study indicates

References

[1] F. Baskett and A. J. Smith, "Interference in multiprocessor computer systems with interleaved memory," *Commun. ACM*, vol. 19, no. 6, pp. 327–334, June 1976.

[2] B. S. Davie, "ATM network modeling and design for bursty traffic," *Proc. ISS'90*, May–June, 1990.

[3] C. M. Day, J. N. Giacopelli, and J. Hickey, "Applications of self-routing switches to LATA fiber optic networks," *Proc. ISS'87*, Mar. 1987.

[4] A. Descloux, "Contention probabilities in packet switching networks with strung input processes," in *Proc. 12th Int. Teletraffic Cong.*, Torino, Italy, June 1–8, 1988.

[5] J. N. Giacopelli, M. Littlewood, and W. D. Sincoskie, "The sunshine switch," *Proc. ISS'90*, May–June 1990.

[6] J. Hui and E. Arthurs, "A broadband packet switch for integrated transport," *IEEE J. Select. Areas Commun.*, vol. SAC-5, no. 8, Oct. 1987.

[7] M. J. Karol, M. G. Hluchyj, and S. P. Morgan, "Input vs. output queuing on a space-division packet switch," in *Proc. IEEE Global Commun. Conf.*, Dec. 1–4, 1986, vol. 2, pp. 19.4.1–19.4.7.

[8] D. E. Knuth, *The Art of Computer Programming (Seminumerical Algorithms), Vol. 2, Second Edition.* New York: Addison-Wesley, 1981.

[9] M. F. Neuts, *Matrix-Geometric Solutions in Stochastic Models.* Baltimore, MD: Johns Hopkins University Press, 1981.

On nonuniform packet switched delta networks and the hot-spot effect

P.G. Harrison

Indexing terms: Networks, Computer applications, Stochastic modelling

Abstract: We analyse the performance of a multi-stage interconnection network (MIN) with a packet-switching protocol, embedded in a closed network of processors. First, the expected value of transmission time through a delta-2 type of MIN is determined. We then obtain, for the first time by an analytical method, a formula for its probability density function from which the variance and higher moments follow. Previously densities have only been estimated by simulation which is expensive and can be unreliable, especially in the often crucial tail region. Numerical results reveal new insights into the hot-spot phenomenon which occurs when one output address is selected more frequently than the others. We first show how mean transmission time on hot paths increases with the hot-spot intensity and compare this with cooler paths. We also plot the density functions for these transmission times. Hence it is possible to determine precisely transmission time variability and to obtain reliability measures from their tails. The approach can handle arbitrary routing frequencies to the MIN output addresses and suggests new approximation techniques with wider applicability.

1 Introduction

Interest in multistage interconnection networks (MINs) has increased dramatically in recent years because of their crucial role in parallel computers; see for example [15, 16, 20]. In general it is necessary for any processing element of one type, e.g. parallel processors, to be able to communicate at a high data-rate with any of another, e.g. distributed shared memory modules. In some cases it may be that any processing element (of any type) must be able to communicate with any other, e.g. in a loosely coupled architecture. The most efficient means of communication is provided by a full crossbar switch which provides concurrent connection between source-destination pairs provided all of the sources and all of the destinations are distinct. Thus, if there are n sources and n destinations, it would be possible for n transmissions to take place simultaneously provided each source wished to transmit to a different destination. The switch described here is called an $n \times n$ crossbar and has inter-

nal complexity of order n^2. Unfortunately, however, there may be very many processing elements that must be so connected and such crossbars soon become prohibitively expensive and then impossible to fabricate as n increases.

The delta network is a type of regular MIN that provides a cost-effective alternative to the full crossbar. It is constructed from relatively small crossbars and provides the same connectivity, i.e. any input can be connected to any output. In the case of n-way communication, where n is a power of b, a *delta-b* MIN comprises $\log_b n$ stages of n/b parallel $b \times b$ crossbars. Its complexity is therefore of order $n \log_b n$, i.e. close to linear for even quite large n. In the present paper, we will take b to be 2, which is common in optical networks, but the analysis generalises immediately to any value of b.

There are a number of different ways of connecting the outputs of one stage of crossbars to the inputs of the next so as to obtain the desired connectivity. We will use the partial shuffle topology [15] which is defined below. However, any topology would be equally appropriate for our analysis since it merely serves to provide an enumeration for the queues in our queueing network model.

(i) a one-stage network, Δ_1 is the single 2-way crossbar

(ii) an s-stage network, Δ_s ($s > 0$) consists of one stage (numbered s) of 2^{s-1} switches connected to the right of 2 sub-networks of type Δ_{s-1}. The ith switch in stage s takes its top input and bottom input from the ith pin of the upper sub-network and the ith pin of the lower sub-network, respectively.

A 3-stage network with the partial shuffle topology is shown in Fig. 1.

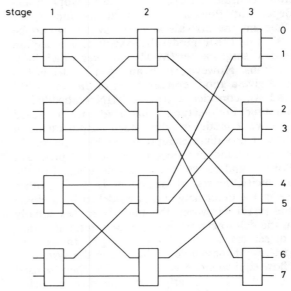

Fig. 1 *Three-stage (8-way) partial shuffle delta network*

Paper 7961E (C3), first received 16th February and in revised form 25th October 1990

The author is with the Department of Computing, Imperial College of Science, Technology and Medicine, 180 Queen's Gate, London SW7 2BZ, United Kingdom

The price to pay for this cheaper solution to the communication problem is, of course, performance. The use of several stages introduces the possibility of contention for internal links in the MIN; it is easy to find examples where two different sources wishing to transmit to two different destinations require a common internal link. This effect is compounded when the network traffic is nonuniform, for example in the presence of a hot-spot where one output address is selected more often than the others which are selected uniformly.

The protocol used in a MIN may be circuit-switching, packet-switching or some kind of hybrid, as well as more exotic alternatives such as packet-switching with no buffers. Circuit-switched delta networks are analysed in depth in Reference 10, although transmission time densities are not derived. For packet switching, paths need not be held throughout a message's transmission. Hence buffered internal crossbars can be modelled as conventional servers in a queueing network since their transmission times are non-negligible because of the intelligence required to provide the buffering. However, similar problems arise to those in circuit-switching if the buffers become exhausted and cannot reasonably be assumed infinite. This is because blocking occurs, i.e. a crossbar output pin's transmission rate is reduced to zero when the buffer to which it wishes to transmit is full. Blocking therefore propagates backwards from the hot-spot as network traffic increases and more buffers fill, eventually blocking whole paths.

The present paper considers a packet-switched delta network with sufficient buffer size that blocking effects can be neglected. We assume that message transmission times through any crossbar or other processor are exponentially distributed and that all queueing disciplines are first-come-first-served. We can therefore model the delta network as a Markovian queueing network [5, 12] which provides a benchmark against which to assess precisely more generally applicable approximate models such as [7, 9]. Moreover, we will see that our analysis suggests techniques for developing new approximations.

2   Queueing network model

If we were to consider an open queueing network model with independent Poisson arrivals at its inputs, by Burke's theorem the arrivals at every switch would be Poisson. We could then model the system as a collection of independent single server (M/M/1) queues which have simple solutions. However, such an open model is too simplistic to give a good representation of the real world and instead we consider a delta-2 MIN embedded in some computer architecture. This leads to a closed queueing network model in which each output queue in the constituent crossbars is represented by a server. In this closed model, independence is lost, arrival processes are not Poisson and so the 'open' approach fails. For the sake of simplicity, the rest of the system, i.e. servers other than those in the delta network, is modelled by a single exponential server. However, the analysis applies equally well when the delta network is embedded in an arbitrary network of servers of the BCMP type [2]. In fact, we could short-circuit the whole of the rest of the system to derive a 'flow equivalent server'. This could then be used to replace the MIN in any encompassing queueing network in a decomposition-based analysis, [10].

We therefore consider a closed Markovian queueing network of M servers with population N, [5, 12]. Each

server has a 'first-come first-served' (FCFS) queueing discipline and exponentially distributed service times with load-independent mean, i.e. we have constant rate servers. If only mean transmission times are required, milder assumptions suffice. Then, each server's rate may vary with its queue length, other disciplines can be modelled and multiple customer classes are allowed, see for example Reference 2. The results of Section 3 (where we consider means) generalise immediately to the more general cases but we will need the stronger assumptions in Section 4 when we consider probability densities of transmission times.

2.1   Notation and basic results

We will use the following notation:

(i) μ_i is the constant service rate specified for server i $(1 \leqslant i \leqslant M)$

(ii) e_i is the visitation rate of server i $(1 \leqslant i \leqslant M)$. The vector e is any nonzero solution of the equations

$$e_i = \sum_{j=1}^{M} e_j p_{ji} \quad (1 \leqslant i \leqslant M)$$

where p_{ji} is the routing probability between servers j and i, i.e. the constant probability that, on completing service at server j, a customer next visits server i. Either the routing probabilities or the visitation rates directly are normally specified for a queueing network model

(iii) $x_i = \dfrac{e_i}{\mu_i}$

(iv) $S(N) = \left\{ \underline{n} \middle| \sum_{i=1}^{M} n_i = N ; n_i \geqslant 0, 1 \leqslant i \leqslant M \right\}$ is the state space of the queueing network

(v) $P(\underline{n}) = \dfrac{1}{G(N)} \prod_{i=1}^{M} x_i^{n_i}$ is the equilibrium probability distribution of the state $\underline{n} \in S(N)$ by the result in Reference 5, where $G(N)$ is the normalising constant defined by:

(vi) $G(N) = \sum_{\underline{n} \in S(N)} \prod_{i=1}^{M} x_i^{n_i}$

By a result in Reference 3, the normalising constant can be computed by the following simple recurrence:

$$G(N) = g(M, N)$$

where

$$g(m, n) = g(m - 1, n) + x_m g(m, n - 1) \quad (m, n \geqslant 1)$$
$$g(m, 0) = 1 \quad (m \geqslant 0)$$
$$g(0, n) = 0 \quad (n > 0)$$

Finally, we give expressions for the mean queue length and throughput of a server in terms of the normalising function g which we will find invaluable in the following Section. Let $G_j(n)$ be the normalising constant for the network with server j removed and population n, i.e.

$$G_j(N) = \sum_{n \in S_j(N)} \prod_{\substack{i=1 \\ i \neq j}}^{M} x_i^{n_i}$$

where

$$S_j(N) = \left\{ (n_i) \middle| \sum_{i \neq j} n_i = N ; n_i \geqslant 0, 1 \leqslant i \neq j \leqslant M \right\}$$

Then we have the following:

2.1.1 Proposition:
The equilibrium probability that the queue length at server j is k is $(G_j(N-k)/G(N))x_j^k$, $(1 \leqslant j \leqslant M, 0 \leqslant k \leqslant N)$.

Proof: The required probability is equal to

$$\frac{1}{G(N)} \sum_{\substack{n \in S(N) \\ n_j = k}} \prod_{i=1}^{M} x_i^{n_i} = \frac{x_j^k}{G(N)} \sum_{n \in S_j(N-k)} \prod_{\substack{i=1 \\ i \neq j}}^{M} x_i^{n_i}$$

2.1.2 Proposition:
If server j has a fixed service rate, its equilibrium throughput is $(G(N-1)/G(N))e_j$, $(1 \leqslant j \leqslant M)$.

Proof: The required throughput is equal to μ_j multiplied by the equilibrium probability that the queue length at server j is nonzero, i.e.

$$\frac{\mu_j}{G(N)} \sum_{\substack{n \in S(N) \\ n_j \geqslant 1}} \prod_{i=1}^{M} x_i^{n_i} = \frac{e_j}{G(N)} \sum_{n \in S(N-1)} \prod_{i=1}^{M} x_i^{n_i}$$

2.2 Model parameterisation

The packet-switched MIN is modelled as a Markovian network of stochastically identical queues with a first-come-first-served queueing discipline, constant service rates and adequate buffer sizes to avoid blocking, i.e. at least as large as the population of the whole closed system. The queues are associated with the output pins of each switch, and logically this is where the buffers are placed. Thus, a delta network with m $p \times p$ switches in all has a total of mp servers in its queueing model representation. We assume that the rest of the system is represented by a single additional exponential server with arbitrary fixed rate λ. In practice, there would typically be one buffer in each crossbar into which incoming messages from any input pin would be inserted as they arrived, assuming no overflow, and tagged with their required output pin number. Each output pin, when it became free after transmitting to the next crossbar, could then search the buffer for the first message addressed to it. This sharing of a single buffer delays the onset of blocking for as long as possible for a given total buffer space and is cheap to construct.

In our case the MIN is constructed from 2×2 crossbars and we assume that all inputs are used uniformly, i.e. have stochastically identical arrival processes. We number the pins in any stage of the network consecutively, starting at zero for the top pin. Now, the visitation rates of the output pins in the final stage are proportional to their selection probabilities which are specified. The rate for a pin in another stage is simply half of the sum of the rates of the output pins reachable in the next stage from the said pin. These rates are determined from the previous iteration using the interconnection topology. The 'half' factor arises since the inputs to each switch in stages after the first come from corresponding outputs in identical subnetworks; this is an immediate consequence of the partial shuffle topology and the uniformity of the inputs to the whole network.

In an S-stage MIN, let the visitation rate for pin number i in stage s be denoted by e_{si} ($1 \leqslant s \leqslant S$, $0 \leqslant i < 2^S$). Thus $\{e_{si} | 0 \leqslant i < 2^S\}$ is given and

$$e_{si} = \frac{e_{s+1, 2j} + e_{s+1, 2j+1}}{2} \quad \text{for } i = (k-1)2^s + j$$

where $0 \leqslant j \leqslant 2^s - 1$, $1 \leqslant k \leqslant 2^{S-s}$, $1 \leqslant s < S$. The visitation rate of the other server, e_M say, is of course

$\sum_{i=0}^{2^S-1} e_{Si}$. We assume without loss of generality that all output pins have rate 1, i.e. the mean message transmission time between switches in adjacent stages is unity.

First, we consider a uniform network in which all outputs are selected with the same probability and so all visitation rates e_{sj} are equal, say to 1. All paths (of length S) through the MIN are then stochastically identical and so we have $M = 1 + S2^S$, $e_i = 1$ for $1 \leqslant i \leqslant M - 1$ and $e_M = 2^S$. We therefore have

$$G(N) = \sum_{k=0}^{N} \left[\frac{2^S}{\lambda}\right]^k \binom{M + N - k - 1}{M - 1}$$

although we will not use this result, preferring the recurrence of Section 2.1.

When there is a hot-spot, we set $e_{S0} = \rho$, the hot-spot selection probability and $e_{Si} = (1 - \rho)/(2^S - 1)$ for $1 \leqslant i \leqslant 2^S - 1$. The remainder of the visitation rates are then computed successively as described above and the e_{sj} are mapped onto the e_i ($1 \leqslant i \leqslant M - 1$) according to some algorithm such as $e_{(s-1)2^s + j + 1} = e_{sj}$ ($0 \leqslant j \leqslant 2^s - 1, 1 \leqslant s \leqslant S$).

In this way, the embedded MIN is mapped into a conventional closed queueing network model. So far, the structure of the physical network has been used only to determine the visitation rates of the servers. When we consider transmission time probability densities, we will find that we will need the feed-forward property of the delta network which ensures that a message cannot be overtaken by another message on any given path through it. However, the recursive structure of the partial shuffle topology is not exploited which contrasts with the analysis given for the circuit switching protocol in Reference 10.

3 Mean transmission times

A message's transmission time along a given path in a network is the sum of its sojourn times at the servers comprising that path. Therefore, mean transmission time in the sum of the mean sojourn times at each server. This is true whether or not the sojourn times are independent. Here, of course, we have a closed network in which the sojourn times are dependent.

Now, in the steady state, Little's result [13] implies that the mean sojourn time of a message at any server is the ratio of the server's mean queue length and throughput. Therefore, using propositions 2.1.1 and 2.1.2, and considering a valid path of FCFS servers numbered, arbitrarily, $1, \ldots, m$ in a network of M servers with population N, we see that mean transmission time along the path can be written as:

$$\sum_{j=1}^{m} \sum_{k=0}^{N-1} \frac{G_j(N-k)kx_j^k}{G(N-1)e_j} = \sum_{j=1}^{m} \frac{1}{\mu_j}$$
$$+ \sum_{j=1}^{m} \sum_{k=0}^{N-1} \frac{G_j(N-1-k)kx_j^k}{G(N-1)\mu_j}$$

Now, this involves the calculation of m separate normalising constant functions for population $N-1$, m mean queue lengths and their sum. We can number the servers so that servers $1, \ldots, m$ come last, hence, Buzen's algorithm does not need to recompute $g(h, k)$ for $h \leqslant M - m$ (see Section 2.1). However, we can do better if all the μ_i are the same. The summation over j sums mean queue lengths in the network with population $N-1$. It is therefore equal to the mean number of messages in all of the queues $1, \ldots, m$ in the steady state. We therefore have the following:

3.1 Proposition
Mean transmission time on path $1, \ldots, m$ in the closed Markovian queueing network under discussion, with service rates $\mu_i = \mu$ for $1 \leqslant i \leqslant m$, is

$$\frac{m}{\mu} + \frac{\sum\limits_{p=0}^{N-1} p G_m(N - p - 1) G_{\bar{m}}(p)}{G(N-1)\mu}$$

where for $k \geqslant 0$, $G_{\bar{m}}(k)$ is the normalising constant for the subnetwork comprising servers $1, \ldots, m$ with population k, defined by

$$G_{\bar{m}}(k) = \sum_{\substack{\sum\limits_{i=1}^{m} n_i = k \\ n_i \geqslant 0}} \prod_{i=1}^{m} \left[\frac{e_i}{\mu_i} \right]^{n_i}$$

and $G_m(k)$ is the normalising constant of the whole network with servers $1, \ldots, m$ removed and population k, defined by

$$G_m(k) = \sum_{\substack{\sum\limits_{i=m+1}^{M} n_i = k \\ n_i \geqslant 0}} \prod_{i=m+1}^{M} \left[\frac{e_i}{\mu_i} \right]^{n_i}$$

Proof: It is sufficient to prove that

$$\sum_{j=1}^{m} \sum_{k=0}^{n} G_j(n-k) k x_j^k = \sum_{k=0}^{n} G_{m+1, \ldots, M}(n-k) G_{1, \ldots, m}(k) k$$

for all $n \geqslant 0$ and $m \geqslant 1$, where for $1 \leqslant p \leqslant q \leqslant M$,

$$G_{p, \ldots, q}(n) = \sum_{\underline{n} \in S_{p, \ldots, q}(n)} \prod_{i=p}^{q} x_i^{n_i}$$

and

$$S_{p, \ldots, q}(n) = \left\{ (n_i) \, \middle| \, \sum_{i=p}^{q} n_i = n; \, n_i \geqslant 0, \, p \leqslant i \leqslant q \right\}$$

We prove this by induction on m. For $m = 1$ the result is obvious since $G_{2, \ldots, M} \equiv G_1$ and $G_{11}(k) \equiv x_1^k$.

Now assume the result holds for $m \geqslant 1$. Then we have

$$\sum_{j=1}^{m+1} \sum_{k=0}^{n} G_j(n-k) k x_j^k = \sum_{k=0}^{n} G_{m+1, \ldots, M}(n-k) G_{1, \ldots, m}(k) k$$
$$+ \sum_{k=0}^{n} G_{m+1}(n-k) k x_{m+1}^k$$

But it follows directly from the definition of a normalising constant in Section 2.1 that

$$G(n) = \sum_{p=0}^{n} G_{1, \ldots, m}(p) G_{m+1, \ldots, M}(n-p)$$

with a similar result for $G_{m+1}(n-k)$. Thus the right hand side is equal to

$$\sum_{k=0}^{n} G_{m+1, \ldots, M}(n-k) G_{1, \ldots, m}(k) k$$
$$+ \sum_{k=0}^{n} \sum_{p=0}^{n-k} G_{m+2, \ldots, M}(n-k-p) G_{1, \ldots, m}(p) k x_{m+1}^k$$

Now, in the same way,

$$G_{m+1, \ldots, M}(n-k) = \sum_{p=0}^{n-k} G_{m+2, \ldots, M}(n-k-p) x_{m+1}^p$$

Since both summation domains are $0 \leqslant k + p \leqslant n$, we may interchange p and k in the first sum and then combine it with the second to get

$$\sum_{0 \leqslant p+k \leqslant n} G_{m+2, \ldots, M}(n-k-p) G_{1, \ldots, m}(p)(p+k) x_{m+1}^k$$

$$= \sum_{v=0}^{n} \sum_{k=0}^{v} G_{m+2, \ldots, M}(n-v) G_{1, \ldots, m}(v-k) x_{m+1}^k v$$

The result then follows since

$$\sum_{k=0}^{v} G_{1, \ldots, m}(v-k) x_{m+1}^k = G_{1, \ldots, m+1}(v)$$

Notice that a simple probabilistic alternative proof is the following. The left hand side of the equation to be proved is the sum of the means of the queue length random variables at servers $1, \ldots, m$. This is equal to the mean of the sum of these same random variables, whether they are independent or not (as in a closed network). This is the expression on the right hand side.

It is therefore simple to compute mean transmission times via the appropriate normalising constants. In Section 5, we study the variation of mean transmission time along a number of paths through a delta network as the intensity of a hot-spot increases. However, to obtain the variance and higher moments, and certainly densities, requires a deeper analysis. This we give in the following Section.

4   Transmission time densities

The problem of finding passage time densities in closed queueing networks has proved a difficult problem which can be solved in closed form only for a small class of Markovian networks. These must possess the non-overtaking property, i.e. be such that no customer behind the customer being timed can influence this customer's progress through the network in any way. The delta networks considered here have a feed forward interconnection topology and crossbar switches with fixed transmission rates. They therefore satisfy the non-overtaking property and so we can apply the following result of Reference 4.

4.1 Proposition
The Laplace transform of the passage time density for the overtake-free path of servers numbered (arbitrarily) $1, \ldots, m$ in a closed Markovian network of M fixed-rate servers with FCFS queueing disciplines and population N is

$$\frac{1}{G(N-1)} \sum_{\underline{n} \in S(N-1)} \prod_{i=1}^{M} \left(\frac{e_i}{\mu_i} \right)^{n_i} \prod_{j=1}^{m} \left\{ \frac{\mu_j}{s + \mu_j} \right\}^{n_j + 1}$$

Now, for paths in a homogeneous delta network such as ours, all the rates μ_i are the same and the Laplace transform is a mixed sum of terms of the form $[\mu/(s+\mu)]^n$ which can be inverted by inspection to give a corresponding mixture of Erlangians for the transmission time density. We therefore have

4.2 Proposition
If the centres in overtake-free path $(1, 2, \ldots, m)$ in the Markovian network of proposition 4.1 all have service rate μ, the path's transmission time density function is

$$\frac{\mu^m e^{-\mu t}}{G(N-1)} \sum_{p=0}^{N-1} G_m(N - p - 1) G_{\bar{m}}(p) \mu^p \frac{t^{p+m-1}}{(p+m-1)!}$$

Proof: By proposition 4.1, the Laplace transform of the density may be written as

$$L(s) = \frac{1}{G(N-1)} \sum_{p=0}^{N-1} \sum_{\substack{\sum_{i=m+1}^{M} n_i = N-p-1 \\ n_i \geqslant 0}} \prod_{i=m+1}^{M} \left[\frac{e_i}{\mu}\right]^{n_i}$$

$$\times \sum_{\substack{\sum_{i=1}^{m} n_i = p \\ n_i \geqslant 0}} \prod_{i=1}^{m} \left[\frac{e_i}{\mu}\right]^{n_i} \prod_{i=1}^{m} \left[\frac{\mu}{s+\mu}\right]^{n_i+1}$$

$$= \frac{1}{G(N-1)} \sum_{p=0}^{N-1} \left[\frac{\mu}{s+\mu}\right]^{p+m}$$

$$\times \sum_{\substack{\sum_{i=m+1}^{M} n_i = N-p-1 \\ n_i \geqslant 0}} \prod_{i=m+1}^{M} \left[\frac{e_i}{\mu}\right]^{n_i} \sum_{\substack{\sum_{i=1}^{m} n_i = p \\ n_i \geqslant 0}} \prod_{i=1}^{m} \left[\frac{e_i}{\mu}\right]^{n_i}$$

$$= \frac{1}{G(N-1)} \sum_{p=0}^{N-1} \left[\frac{\mu}{s+\mu}\right]^{p+m} G_m(N-p-1)G_{\bar{m}}(p)$$

From this result we can immediately obtain formulae for moments higher than the mean of transmission time. In particular, we use the first two moments to find its variance in our case study in the following Section.

4.3 Corollary

For a path of equal rate servers, message transmission time has kth moment equal to

$$\frac{1}{\mu^k G(N-1)} \sum_{p=0}^{N-1} G_m(N-p-1)$$

$$\times G_{\bar{m}}(p)(p+m) \cdots (p+m-k+1)$$

5 The hot-spot effect quantitatively

Our queueing network model can be parameterised to represent any utilisation pattern of the output addresses provided all inputs are selected with equal probability by arriving tasks. The output selection probabilities are in direct proportion to the visitation rates of the queues corresponding to them from which all visitation rates follow. However, for the purposes of a quantitative demonstration, we consider a system in which there is a single hot-spot. All servers representing buffers in the delta network have unit service rate, with no loss of generality, and we take the other server's rate λ to be twice the total rate of a bank of crossbars; the same for all banks in the steady state. This means that the delta network will saturate relatively quickly as the network population increases. Our model is also parameterised by the hot-spot ratio which is equal to the hot pin selection probability divided by the selection probability of any other output pin.

We consider a 16-way network, i.e. one having 4 stages of 8 crossbars. Thus our queueing network model has 65 servers in total and we set the population to 100. Notice that this by no means represents saturation since, on average, each queue would contain about 1.5 messages. Numerical results were computed for hot-spot ratios in the range 1 (corresponding to a uniform delta network) to 8 and for four different types of path. These path types lead to the hot-spot, to the pin adjacent to the hot-spot, to the pin adjacent to that pin (two away from the hot-spot) and to the coldest pin (furthest from the hot-spot). Of course it is immaterial from which input pin the paths originate since the inputs are selected uniformly. Each path type will experience a different level of contention:

(i) the first type will be most congested

(ii) the second type will compete for internal links with the first type at all stages, deviating only in the selection of the final output pin

(iii) the third type will compete for internal links with the first type at all but the last stage, i.e. at two out of three

(iv) the fourth type will compete for no internal links with the first type since it will select a different decode tree in its first stage crossbar

All of our results were obtained by first computing the normalising constants $G(i)$, $G_m(i)$ and $G_{\bar{m}}(i)$ for $0 \leqslant i \leqslant 100$ where servers are enumerated so that the path of interest comprises servers numbered 62, ..., 65. This enumeration allows, in the notation of Section 2.1, $G_m(i) \equiv g(61, i)$ to be used in the computation of $G(i)$ which corresponds to a network with just 4 more servers. To avoid both underflow and overflow in the calculation of normalising constants we choose the visitation rates e_i to be such that the maximum ratio $x_i = e_i/\mu_i$ is exactly 1. This is always possible in view of the uniqueness of the e_i only up to an arbitrary multiplicative constant. By construction, the server i with maximum x_i must be the hot-spot. The programs that compute the moments and density of transmission time are now very simple. They were actually implemented using Mathematica [21] which produced the graphs shown in Figs. 2–7.

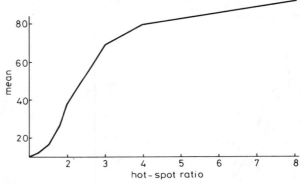

Fig. 2 *Mean transmission time for hot paths*

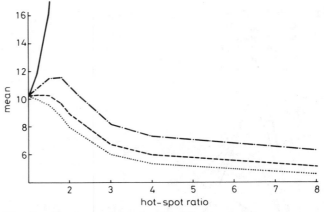

Fig. 3 *Mean transmission times for different output pins*

——— hot-spot
– – – two from hot-spot
—·—· next to hot-spot
········· bottom pin

5.1 Mean and standard deviation of transmission times

We compared the performance of the four pins identified at the beginning of this Section in terms of the mean and standard deviation of their steady state message transmission times. Fig. 2 shows how the mean transmission time on paths to the hot-spot rises quickly as the hot-spot ratio increases above 1 (the uniform case), before levelling off at higher ratios.

The mean and standard deviation of transmission time are plotted in Figs. 3 and 4 for hot-spot ratios 1, 1.1, 1.2, 1.5, 1.8, 2, 3, 4, 8. Fig. 4 shows that the standard deviation on hot paths has a maximum near the ratio 2, a result which is not obvious *a priori*. The suggestion is that as the traffic in the network increases, the hot pin begins to dominate and the network's overall behaviour

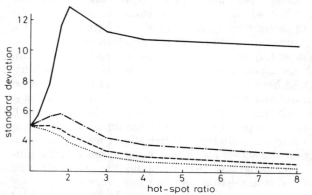

Fig. 4 *Standard deviations of transmission times*

——— hot-spot
- - - two from hot-spot
-·-· next to hot-spot
······ bottom pin

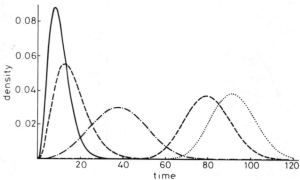

Fig. 5 *Probability density function for hot paths*

——— hot-spot ratio = 1.0
- - - hot-spot ratio = 1.5
-·-· hot-spot ratio = 2.0
- - - - hot-spot ratio = 4.0
······ hot-spot ratio = 8.0

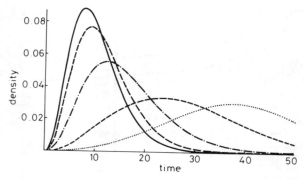

Fig. 6 *Probability density function for hot paths at low ratio*

——— hot-spot ratio = 1.0
- - - hot-spot ratio = 1.2
-·-· hot-spot ratio = 1.5
- - - - hot-spot ratio = 1.8
······ hot-spot ratio = 2.0

becomes more predictable. However, a deeper explanation is given below when we consider the hot pin's buffer as a bottleneck server.

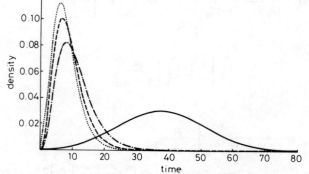

Fig. 7 *Transmission time densities at hot-spot ratio 2*

——— hot-spot
- - - two from hot-spot
-·-· next to hot-spot
······ bottom pin

In the first stage, paths to the coldest pin enter a different subnetwork from paths to the hot pin by selecting a different decode tree. As a result they only experience increased contention in the first stage and subsequently there will be less traffic to compete with as the hot-spot ratio increases. Thus we would expect to see both the mean and standard deviation decrease for the coldest pin, as indeed we do. However, for the other two cooler pins the results are more surprising. Applying a similar argument, we observe that a path to the pin adjacent to the hot-spot shares all internal links with hot paths and only diverges at the output stage. Therefore we might expect message transmission time on such a path to increase, although less rapidly than for a hot path. The same would also apply, to a lesser degree, to the next output pin which shares only one less internal link. But this is not what is observed. Instead, mean message transmission time decreases as the hot-spot ratio increases, after small initial increases for the pins adjacent to the hot-spot; barely noticable for the second of these.

The reason for this is that we are considering a network with unbounded buffers in which the hot pin's server in our model is the single bottleneck whenever the ratio is greater than 1. Thus for large populations, the model approaches an open system with Poisson arrivals at unit rate from the always busy hot pin to the 'other server' (outside the delta network). Messages that select the hot pin in the final stage now depart the network. The interesting implication of this is that even paths to the hot-spot do not saturate until the very last stage, the hot-spot itself. This is because the arrival rates to all internal buffers are less than their service rates and will actually decrease as the hot-spot ratio increases — since more network output pins will become idle. Asymptotically, the arrival rates will approach those of the open model above and may be computed easily from its traffic equations. Notice that these arrivals are non-Poisson in view of the feedback which still exists for non-hot paths.

It is the decrease in the utilisation of all internal links which eventually causes message transmission times to decrease on all non-hot paths. In the final stage, of course, the hot output buffer saturates and the increasing message transmission time on hot paths is due to the increase in this queue length. Comparing a hot path with a path to the pin adjacent to the hot-spot, we see that, stochastically, each has the same partial transmission time (up to the final output buffer). However, the hot path experiences an increasing delay in the final stage whereas the adjacent path (and all other cooler paths) experience a decreasing delay. In the case of a hot-spot ratio only slightly bigger than 1 we saw in Fig. 3 that the mean transmission time of adjacent paths increases. This is explained by the fact that the increase in partial transmission time outweighs the decrease in sojourn time in the final stage. This is not the case for the other cooler paths, nor for any non-hot path at higher hot-spot ratios.

5.2 Transmission time density

We first computed the density function for paths to a hot-spot (hot paths) with seven different ratios: 1, 1.2, 1.5, 1.8, 2, 4, 8. These were plotted in the two graphs shown in Figs. 5 and 6. Fig. 5 shows the density in the uniform network with a relatively sharp peak which has decreased considerably at ratio 1.5, further at ratio 2 and then increases at higher ratios, suggesting that transmission time is becoming less variable as the intensity of the hot-spot increases. This is entirely consistent with the results of the previous Section showing an increasing mean transmission time and a peak in its variance near a hot-spot ratio of 2.

The most interesting region here is at ratios immediately above 1 where the density is changing rapidly, with a decaying peak, widening spread and increasing mean. Fig. 6 shows the densities for ratios 1, 1.2, 1.5, 1.8, 2, which demonstrate the hot-spot effect from a new point of view; before it has only been considered in terms of throughput, e.g. Reference 17.

In Fig. 7 we compare the drastically degraded hot paths with paths to the three cooler pins at hot-spot ratio 2. This gives a more precise picture, showing how the density narrows for cooler paths. The corresponding distributions, easily derived by integration but not shown here, allow the probability of tranmission time exceeding some given value to be read off from the appropriate quantile.

5.3 Comparison with other protocols

The hot-spot behaviour observed here is very different from that seen in circuit-switched networks or packet switched networks with small buffers. In these, saturation propagates back from the final stage since complete paths are held throughout data transmission in the former case and blocking occurs in the latter. However, we noted above that only the hot buffer in the final stage saturates in our idealised model. This suggests that only one very large buffer would be necessary in a real system to improve performance in the presence of a hot-spot. More generally, the same would be true if there were more than one hot-spot: a large buffer need only be given to each of the equal hottest network output pins.

6 Conclusion

We have analysed transmission times in a class of non-uniform, packet-switched delta networks and studied the hot-spot effect quantitatively from a new angle. In particular, an exact expression for the probability density function of transmission time has been obtained for the first time. This is an advance over previous work which has either used simulation, which can be unreliable and is expensive to run, or produced only Laplace transforms. Although not all of the model's assumptions are realistic, the result provides a standard against which to assess approximations, and itself suggests new approximate

techniques such as the one described below. Of more immediate use, the asymptotic analysis of Section 5.3 suggested an inexpensive way of reducing the degradation introduced by hot-spots, by assigning large buffers to the hottest output pin(s) only.

The results obtained in Section 4 rely on the degeneracy of the rates of the servers in the path in question. An immediate generalisation is therefore to consider what happens when some of the rates are distinct. In fact such a result has been derived by the author, but only for the other extreme in which all rates are distinct, i.e. when there is no degeneracy, [8]. We may achieve the result for arbitrary degeneracy as follows. First, Reference 8 inverts the Laplace transform

$$L(\underline{n}, s) = \prod_{i=1}^{M} \left[\frac{\mu_i}{s + \mu_i} \right]^{n_i}$$

where $\underline{n} = (n_1, \ldots, n_M)$, $n_i \geqslant 1$ and the μ_i's are distinct to obtain

$$f(\underline{n}, t) = \left[\prod_{i=1}^{M} \mu_i^{n_i} \right] \sum_{j=1}^{M} D_j(\underline{n}, t)$$

where the $D_j(\underline{n}, t)$ are given by a recurrence on $\underline{n}$.

Next, given real numbers $a_1, \ldots, a_M$ let

$$H_{jm}(\underline{z}) = \sum_{\underline{n} \in S(M+m)} D_j(\underline{n}, t) \prod_{i=1}^{M} (a_i z_i)^{n_i - 1}$$

so that, by proposition 4.1, transmission time density is obtained from the $H_{jm}(1, \ldots, 1)$ with $a_i = x_i$. The key result is then that

$$H_{jm}(\underline{z}) = \frac{e^{-\mu_j t}}{\prod_{1 \leqslant i \neq j \leqslant M} (\mu_i - \mu_j)} \left\{ \sum_{i=0}^{m} \frac{(a_j z_j t)^{m-i}}{(m-i)!} \right.$$

$$\left. \times \sum_{\substack{\underline{n} \in S_{M+i} \\ n_j = 1}} \prod_{\substack{k=1 \\ k \neq j}}^{M} \left[\frac{(a_k z_k - a_j z_j)}{\mu_k - \mu_j} \right]^{n_k - 1} \right\}$$

Now, given this result, suppose first that $\mu_{M-1} = \mu_M$ and that the values $\mu_1, \ldots, \mu_{M-1}$ are distinct. Then if we define $\underline{n}' = (n_1, \ldots, n_{M-2}, n_{M-1} + n_M)$ and use S', D'_j to denote the respective functions S, D_j when M is replaced by $M - 1$, we require, for $1 \leqslant j \leqslant M - 1$, $m \geqslant 0$,

$$H_{jm}(\underline{z}) = \sum_{\underline{k} \in S'(M+m)} D'_j(\underline{k}, t) \prod_{i=1}^{M-2} (a_i z_i)^{k_i - 1}$$

$$\times \sum_{p+q=k_{M-1}} (a_{M-1} z_{M-1})^p (a_M z_M)^q$$

$$= M a_M z_M H'_{jm}(z_1, \ldots, z_{M-1})$$

$$\text{if } a_{M-1} z_{M-1} = a_M z_M$$

$$= \frac{a_{M-1} z_{M-1} a_M z_M}{a_M z_M - a_{M-1} z_{M-1}}$$

$$\times \left\{ H'_{jm}\left(z_1, \ldots, z_{M-2}, \frac{a_M z_M}{a_{M-1}} \right) \right.$$

$$\left. - H'_{jm}(z_1, \ldots, z_{M-1}) \right\} \quad \text{otherwise}$$

where H'_{jm} is defined as H_{jm} in the $(M - 1)$-dimensional case. Further degeneracy can be handled by repeating the above procedure.

Finally, the form of proposition 4.2 suggests an approximate approach for obtaining transmission time densities in non-overtake-free networks with FCFS discipline, and a similar observation applies when the degeneracy is not total. The result states that transmission time

439

density is a weighted sum of Erlang-$(m + p)$ densities where the weights are the probabilities that there are p tasks in the whole path in a network with its population reduced by one. Now, formulae exist for these probabilities in much more general networks: exact for BCMP networks or approximate for many others including ones with blocking for example. Similarly, the Erlang-$(m + p)$ density is the convolution of $m + p$ exponentials. The suggestion is therefore that the more general probabilities and convolutions of $(m + p)$ service time densities be used in place of their counterparts in proposition 4.2. This is currently being investigated.

7 References

1 ABATE, T., and WHITT, W.: 'Seeing through the Laplace curtain: numerical and approximate methods for Laplace Transform inversion'. Tutorial at SIGMETRICS 1988, Santa Fe, May, 1988
2 BASKETT, F., CHANDY, K.M., MUNTZ, R.R., and PALACIOS, F.G.: 'Open, closed and mixed networks of queues with different classes of customers', *J. ACM*, 1975, **22**, (2)
3 BUZEN, J.P.: 'Computational algorithms for closed queueing networks with exponential servers', *C. ACM*, 1973, **16**, (9), pp. 527–531
4 DADUNA, H.: 'Passage times for overtake-free paths in Gordon Newell networks', *Adv. Appl. Prob.*, 1982, **14**, pp. 672–686
5 GORDON, W.J., and NEWELL, G.F.: 'Closed queueing systems with exponential servers', *Oper. Res.*, 1967, **15**, pp. 254–265
6 HARRISON, P.G.: 'The distribution of cycle times in tree-like networks of queues', *Comp. J.*, 1984, **27**, (1), pp. 27–36
7 HARRISON, P.G.: 'An enhanced approximation by pair-wise analysis of servers for time delay distributions in queueing networks', *IEEE Trans. Comp.*, 1986, **35-1**, pp. 54–61
8 HARRISON, P.G.: 'Passage time distributions and Laplace transform inversion', *J. Appl. Prob.*, 1990, **27**, pp. 74–87
9 HOHL, S.D., and KUEHN, P.J.: 'Approximate analysis of flow and cycle times in queueing networks'. Proc. 3rd Int. Conf. on Data Communication Systems and their Performance, Rio de Janeiro, N. Holland, 1987
10 HARRISON, P.G., and PATEL, N.M.: 'The representation of multi-stage interconnection networks in queueing models of parallel systems', *J. ACM*, 1990, **37**, (4), pp. 863–898
11 HARRISON, P.G., and PATEL, N.M.: 'Performance modelling of communication networks' (Addison-Wesley), to appear
12 JACKSON, J.: 'Jobshop-like queueing systems', *Man. Sci.*, 1963
13 LITTLE, J.D.C.: 'A proof of the queueing formula $L = \lambda W$', *Operations Research*, 1961, **9**, (3), pp. 383–387
14 LAVENBERG, S., and REISER, M.: 'Stationary state probabilities at arrival instants for closed queueing networks with multiple types of customers', *J. Appl. Prob.*, 1980, **17**, pp. 1048–1061
15 PATEL, J.H.: 'Performance of processor-memory interconnections for multiprocessors', *IEEE Trans. Computers*, October 1981, pp. 771–780
16 POMBORTSIS, A., and HALATSIS, C.: 'Performance of crossbar interconnection networks in presence of "Hot Spots"', *Electron. Lett.*, 1988, **24**, (3), pp. 182–184
17 PFISTER, G.F., and NORTON, V.A.: 'Hot-spot contention and combining in multi-stage interconnection networks'. Proc. 1985 Int. Conf. on Parallel Processing, August 1985, pp. 790–797
18 SCHASSBERGER, R., and DADUNA, H.: 'The time for a round trip in a cycle of exponential queues', *J. ACM*, 1983, **30**, (1), pp. 146–150
19 SEVCIK, K.C., and MITRANI, I.: 'The distribution of queueing network states at input and output instants', *J. ACM*, 1981, **28**, (2), pp. 358–371
20 SIEGEL, H.J.: 'Interconnection networks for large-scale parallel processing: theory and case studies' (Lexington Books, 1985)
21 WOLFRAM, S.: 'Mathematica' (Addison-Wesley, 1988)

PERFORMANCE OF STATISTICAL MULTIPLEXERS

WITH FINITE NUMBER OF INPUTS AND TRAIN ARRIVALS

Yijun Xiong   and   Herwig Bruneel

Research Center   Laboratory for Communications Engineering
Alcatel Bell Telephone Mfg. Co.   University of Ghent
Francis Wellesplein 1   Sint—Pietersnieuwstraat 41
B—2018 Antwerp, Belgium   B—9000 Gent, Belgium

ABSTRACT — A slotted statistical multiplexer with a finite number of input links is considered in this paper. Messages arriving on each input link contain a fixed number of fixed—length packets and are carried to the multiplexer in the form of a packet "train" at the rate of one packet per slot. Several messages may arrive contiguously on an input link; idle periods are geometrically distributed. The multiplexer buffer is modeled as a discrete—time single—server queueing system with train arrivals. By means of a generating function approach, a technique to derive the moments of the buffer occupancy is developed and an explicit expression for the mean buffer occupancy is given. Furthermore, an approximate method is presented to get a tight upper bound for the tail distribution of the buffer occupancy, especially for large traffic load.

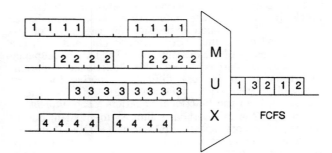

Fig.1   A statistical multiplexer with 4 input links and train length m = 4.

arrivals (or new *message arrivals*) in the consecutive slots are not independent. In fact, the above *packet arrival* process includes a two—fold correlation : (1) if the first packet of a message arrives in a given slot, the probability is one that the remaining $(m-1)$ packets of this message will arrive consecutively in the next $(m-1)$ slots; (2) the arrival of leading packets in a given slot is dependent on the arrival of leading packets in the previous $(m-1)$ slots.

The motivation for studying the above train arrival model is as follows. For the large ATM switching network described in [1], an internal transfer mode is used. Externally, the switching network is synchronized on slot which is the time unit to carry one ATM cell of fixed size, while internally, all the components (e.g., switching element) of the switching network operate in terms of a smaller time unit called a "minislot", which is the time interval for transmission of one "minicell" of fixed size. At the edge of the switching network, each incoming ATM cell is converted into a fixed number (e.g., m) of minicells which are transferred through the switching network consecutively. Since all the components within the network are synchronized on a minislot basis, the minicell arrival process on each internal link is just the train arrival process described above. A Bernoulli arrival process is a special case of this train arrival model when m=1. The advantages of this internal transfer mode are : (1) to reduce the required buffering capacity for the switching element and to reduce the total network cell delay, (2) to facilitate the adaptation to different external cell sizes, and most importantly, (3) to optimize the internal characteristics for optimum usage of technologies [1]. In order to investigate the impact of the train arrival process on the performance of multiplexers (or switching elements), a

I. INTRODUCTION

In this paper, we consider a statistical multiplexer with N input links and one output link. All these links have the same transmission rate and are slotted in time. One slot is the fixed—length time unit required to carry exactly one "data—unit" or "packet". The information to be transmitted is generated in the form of "messages", where each message contains a constant number, say m, of packets. Owing to the equal link rates, the m packets of a message on an input link look like a m—length packet train which enters the multiplexer buffer at the rate of one packet per slot. This arrival process is called a "fixed—length packet—train arrival process" or simply a "train arrival process" in the paper. Traffic on different input links is assumed to be independent and to have the same statistical characteristics. Packets are sent out from the buffer according to a first—come—first—served (FCFS) rule (see Fig. 1). The multiplexer is assumed to have infinite buffer capacity.

We further assume that on each input link, after the previous message, a new message may arrive with probability q and no new message starts with probability 1—q. In other words, if we model the packet stream on an input link by a two—state pattern called "active" state and "idle" state (or "on" and "off" states) which alternates in time, then the number of messages in an active period and the length of an idle period are both geometrically distributed with mean values 1/(1—q) and 1/q respectively. Since N is finite, the numbers of leading—packet

Reprinted from *Proc. IEEE INFOCOM '92*, pp. 2036—2044,
May 1992.

statistical multiplexer with train arrivals is considered in the paper. For simplicity, we use the terms message and packet instead of cell and minicell, respectively.

Statistical multiplexers with correlated arrivals have been studied on several occasions. For instance, the case of a fixed–length train arrival process, assuming the generation of new messages (or trains) independent from slot to slot, has been thoroughly treated in [2–5]. In [6–7], a similar model for geometrically distributed message lengths was investigated, whereas in [8–9] the combination of geometric message lengths and nonindependent message generation (as a consequence of the finiteness of the number of input links) was considered. In this paper, we will study the buffer behavior with the correlated arrival process described above. By means of a generating function approach, a functional equation is established which includes all the information about the buffer behavior. The explicit expression for the mean of the distribution of the buffer occupancy is obtained. All the higher moments of the distribution of the buffer occupancy can also be derived from this functional equation. Unfortunately, it is very difficult to derive the exact probability generation function (pgf) of the buffer occupancy from the functional equation. The implementation of numerical method is limited to small m due to the huge memory space requirement ($\approx (N+1)^{m-1}(S+1)$, where S is the buffer size to be reached). So we develop an approximate method to obtain an upper bound for the tail distribution of the buffer occupancy. Comparing with simulation results, we found this upper bound to be easy to evaluate and very tight, especially for large traffic load.

The rest of the paper is organized as follows. In section II, the fundamental relationships between the random variables and parameters are described, and a functional equation for the m–dimensional joint pgf of the state vector is established. In section III, the packet arrival process in steady state is analyzed. In section IV, an explicit expression for the mean buffer occupancy is derived. In section V, a method to obtain an upper–bound tail distribution of the buffer occupancy is presented. The upper–bound tail distribution and simulation results are also compared there. Conclusions are given in section VI.

II. FUNDAMENTAL RELATIONSHIPS AND FUNCTIONAL EQUATION

Consider a discrete–time single–server queue with N input links and fixed–length (equal to m) packet–train arrivals. On each input link, the probability that the first packet of a message will enter the buffer in a given slot is q if the first packet of the previous message on this link did not enter the buffer during one of the previous (m−1) slots; otherwise the probability is zero. Let $\{c_j, j \geq 1\}$ be a series of i.i.d. Bernoulli random variables with pgf

$$C(z) = 1 - q + qz .\qquad(1)$$

Let $a_{k,n}$ ($1 \leq k \leq m$) be the random variable denoting the number of input links having sent the kth packet of a message to the buffer in slot n. It is obvious that

$$a_{k,n+1} = a_{k-1,n} , \qquad 2 \leq k \leq m \qquad(2)$$

and

$$a_{1,n+1} = \sum_{j=1}^{N-I_n} c_j , \qquad(3)$$

where $\qquad I_n = \sum_{k=1}^{m-1} a_{k,n} .$

Let v_n ($n \geq 0$) be the random variable representing the number of packets in the buffer just after slot n, we have

$$v_{n+1} = (v_n - 1)^+ + \sum_{k=1}^{m} a_{k,n+1} . \qquad(4)$$

As $a_{1,n+1}$ is dependent on $a_{k,n}$ ($1 \leq k \leq m-1$) (see equation (3)), it can be seen from the above equation that v_{n+1} also depends on these $a_{k,n}$'s. It is thus clear that $\{a_{1,n}, ..., a_{m-1,n}, v_n\}$ forms an m–dimensional Markov process which can completely characterize the queueing system with m–length packet–train arrivals.

Now let us introduce the m–dimensional joint pgf of the random variables $a_{1,n}, ..., a_{m-1,n}, v_n$:

$$P_n(x_1, ..., x_{m-1}, z) \triangleq E\left[x_1^{a_{1,n}} ... x_{m-1}^{a_{m-1,n}} z^{v_n} \right] . \qquad(5)$$

From this definition, $P_{n+1}(x_1, ..., x_{m-1}, z)$ can be expressed as follows, by using equations (1)–(4) and averaging over the distribution of the c_j's,

$$P_{n+1}(x_1, ..., x_{m-1}, z) = \left[C(x_1 z) \right]^N E\left[\left[\frac{x_2 z}{C(x_1 z)} \right]^{a_{1,n}} ... \left[\frac{x_{m-1} z}{C(x_1 z)} \right]^{a_{m-2,n}} \left[\frac{z}{C(x_1 z)} \right]^{a_{m-1,n}} z^{(v_n-1)^+} \right] ,$$

where the expectation is over the joint distribution of $\{a_{1,n}, ..., a_{m-1,n}, v_n\}$. Note that $v_n = 0$ also implies $a_{k,n} = 0$ for $1 \leq k \leq m$. It follows that

$$P_{n+1}(x_1, ..., x_{m-1}, z) = \left[C(x_1 z) \right]^N \left\{ \frac{1}{z} P_n\left[\frac{x_2 z}{C(x_1 z)}, ..., \frac{x_{m-1} z}{C(x_1 z)}, \frac{z}{C(x_1 z)}, z \right] + \text{Prob}[v_n = 0](1 - \frac{1}{z}) \right\} .$$

When n approaches infinity, we assume that the buffer system can reach a steady state. Let $P(x_1, ..., x_{m-1}, z)$ be the steady–state pgf, it is clear from the above equation that the P–function satisfies

$$P(x_1, ..., x_{m-1}, z) = \left[C(x_1z)\right]^N \left\{ \frac{1}{z}P\left[\frac{x_2z}{C(x_1z)}, ..., \frac{x_{m-1}z}{C(x_1z)}, \frac{z}{C(x_1z)}, z\right] + p_0(1-\frac{1}{z}) \right\}, \quad (6)$$

where p_0 is the steady–state probability that the buffer is empty.

In principle, equation (6) contains all the information on the buffer behavior. In order to obtain more information about the buffer behavior imbedded in this functional equation, e.g., the moments of the buffer occupancy, as in [8], let the arguments of the P–function on the left hand side and the right hand side of equation (6) be chosen equal to each other, i.e.,

$$\begin{cases} x_k = \dfrac{x_{k+1}z}{C(x_1z)}, & 1 \le k \le m-2, \\ x_{m-1} = \dfrac{z}{C(x_1z)}. \end{cases} \quad (7)$$

From the above equation, x_k ($1 \le k \le m-1$) can be solved in terms of z. It turns out that there is more than one set of solutions; we only select a set of solutions which satisfies $x_k=1$ for $z=1$. Let $x_k = r_k(z)$ ($1 \le k \le m-1$) denote this set of solutions; it is easy to show that it is completely determined by

$$\begin{cases} r_k(z) = \dfrac{z^{m-k}}{[C(zr_1(z))]^{m-k}}, \\ r_k(1) = 1, \end{cases} \quad 1 \le k \le m-1. \quad (8)$$

Choosing $x_k=r_k(z)$ ($1\le k \le m-1$) in equation (6) then yields the following normalized form

$$P(r_1(z), ..., r_{m-1}(z), z) = \frac{[1-H'(1)](z-1)H(z)}{z - H(z)}, \quad (9)$$

where

$$H(z) \triangleq [C(zr_1(z))]^N = [1-q + qzr_1(z)]^N. \quad (10)$$

The analysis carried out in the remainder of this paper is mainly based on equation (9). A method to obtain the moments of the buffer occupancy from this equation is described in section IV. An analytical approach to obtain the upper bound tail distribution of the buffer occupancy from equation (9) is described in section V. In the next section, we first consider the packet arrival process in the steady state.

III. PACKET ARRIVAL PROCESS IN STEADY STATE

Let a_k ($1 \le k \le m$) be the number of input links having sent the kth packet of a message to the buffer in an arbitrary slot in the steady state, with marginal pgf $A_k(x_k)$. $A(x_1, ..., x_m)$, the m–dimensional joint pgf of the steady–state random variables $a_1, ..., a_m$ is defined as

$$A(x_1, ..., x_m) \triangleq E\left[x_1^{a_1} ... x_m^{a_m}\right].$$

In order to derive $A(x_1, ..., x_m)$, let us consider one input link j ($1 \le j \le N$). If p denotes the average traffic load, then in an arbitrary slot in the steady state, it is clear that a packet arrives on this input link with probability p/N and no packet arrives with probability $1-p/N$. Since each message consists of m packets, the probability of having the arrival of a kth packet of a message ($1 \le k \le m$) is given by p/Nm. Formally, if $b_k(j)$ denotes the number of kth packets sent by input link j to the buffer in an arbitrary slot in steady state, we have

$$\text{Prob}[b_1(j)=i_1, ..., b_m(j)=i_m]$$
$$\begin{cases} p/Nm, & i_k=1, \; i_n=0, \; 1 \le n \le m \text{ and } n \ne k, \\ 1- p/N, & i_n=0, \; 1 \le n \le m, \\ 0, & \text{otherwise}. \end{cases}$$

The m–dimensional joint pgf of the random variables $b_1(j)$, ..., $b_m(j)$ can be easily obtained from this as

$$B_j(x_1, ..., x_m) = 1 - \frac{p}{N} + \frac{p}{mN}\left(\sum_{k=1}^{m} x_k\right),$$

where $1 \le j \le N$. Since all N input links are independent and have the same statistical characteristics, we have

$$\begin{aligned} A(x_1, ..., x_m) &= \prod_{j=1}^{N} B_j(x_1, ..., x_m) \\ &= \left[1 - \frac{p}{N} + \frac{p}{mN}\left(\sum_{k=1}^{m} x_k\right)\right]^N. \end{aligned} \quad (11)$$

Equation (11) explicitly describes the packet arrival process in the steady state. From equation (11), the marginal pgf $A_k(x_k)$ of a_k ($1 \le k \le m$) can be derived by setting $x_n=1$ for $1 \le n \le m$ and $n \ne k$, i.e.,

$$A_k(x_k) = \left[1 - \frac{p}{mN} + \frac{p}{mN}x_k\right]^N, \quad (12)$$

which is intuitively clear.

IV. MEAN BUFFER OCCUPANCY

Let v denote the number of packets stored in the buffer just after a slot in the steady state, while a_k $(1 \leq k \leq m)$ denotes the number of input links having sent the kth packet of a message to the buffer during that slot. From the definition of $P(x_1, ..., x_{m-1}, z)$, the pgf $V(z)$ of v can be expressed as $V(z) = P(1, ..., 1, z)$. In the following, a method to obtain the mean buffer occupancy is described.

Taking the first derivative of equation (9) and letting z=1, after some algebra, we have

$$\left.\frac{dP}{dz}\right|_{z=1} = H'(1) + \frac{H''(1)}{2[1-H'(1)]}, \qquad (13)$$

where

$$\frac{dP}{dz} = \frac{\partial P}{\partial x_1}\frac{dx_1}{dz} + ... + \frac{\partial P}{\partial x_{m-1}}\frac{dx_{m-1}}{dz} + \frac{\partial P}{\partial z}.$$

Since $x_k = r_k(z)$ and $r_k(1) = 1$ for $1 \leq k \leq m-1$, $\left.\frac{\partial P}{\partial z}\right|_{z=1} = V'(1)$, which is the mean buffer occupancy. Also, from the definition in the previous section it is clear that $\left.\frac{\partial P}{\partial x_k}\right|_{z=1} = A_k'(1)$. Thus, equation (13) can be written as

$$V'(1) = H'(1) + \frac{H''(1)}{2[1-H'(1)]} - \sum_{k=1}^{m-1} A_k'(1)r_k'(1).$$

Here, $H'(1)$ and $H''(1)$, the first and the second derivative of $H(z)$ at z=1, can be derived from equation (10). The first derivative of $r_k(z)$ $(1 \leq k \leq m-1)$ and the second derivative of $r_1(z)$ can be obtained from equation (8). These derivations are summarized in appendix A. Also, $A_k'(1)$ can be easily obtained from equation (12). Finally, the above equation leads to the following explicit expression for the mean buffer occupancy:

$$\bar{v} = V'(1) = p + \frac{N-1}{N}\frac{mp^2}{2(1-p)} - \frac{N-1}{N^2}\frac{(m-1)p^3}{2(1-p)}. \qquad (14)$$

Higher moments of the buffer occupancy can also be obtained in a similar way, although the mathematical derivations become more and more complicated. From the above equation it is easy to see that, for a given mean load p, the mean buffer occupancy will get larger as m, the number of packets in a message, increases. Note that when m = 1, the train arrivals model is deduced to a simple Bernoulli arrival process. The mean buffer occupancy for N superposed Bernoulli arrival processes can be found in many literatures (e.g., in [12]). Fig. 2 shows some plots of the mean buffer occupancy versus the load p for N=4 and various values of m.

V. UPPER–BOUND TAIL DISTRIBUTION OF THE BUFFER OCCUPANCY

Although the functional equation (6) implicitly contains all the information about the buffer behavior, it appears to

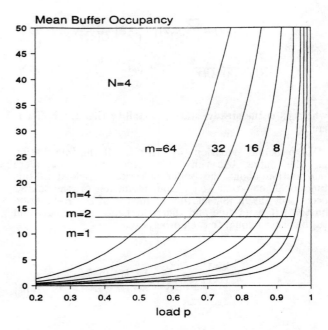

Fig.2 Mean buffer occupancy versus load p, for N = 4, and m = 1, 2, 4, 8, 16, 32 and 64.

be very difficult to obtain the distribution of the buffer occupancy from this functional equation. In this section, we shall present a method to get an upper bound for the tail distribution of the buffer occupancy from equation (9). As mentioned in [13], for a wide range of queueing systems including G/G/c, the tail distribution of the buffer occupancy has geometric form. That is, when the buffer occupancy is sufficiently large (e.g., > B), the probability of the buffer occupancy can be approximated as

$$\text{Prob}[v = n] \cong K \gamma^n, \qquad n > B. \qquad (15)$$

This is also valid for our train arrivals case. In the following, we shall describe an analytical approach to obtain the geometric decay rate γ and an upper bound for the coefficient K. In this way, we can get the upper–bound tail distribution of the buffer occupancy.

A. Geometric Decay Rate γ

Let us define the steady–state partial generating function

$$Q_{(i_1, ..., i_m)}(z) \triangleq \sum_{n=0}^{\infty} \text{Prob}[a_1=i_1, ..., a_m=i_m, v=n] z^n.$$

Then the pgf $V(z)$ of v can be expressed as

$$V(z) = \sum_{i_1=0}^{N} ... \sum_{i_m=0}^{N} Q_{(i_1, ..., i_m)}(z) \qquad (16)$$

and the P–function defined in equation (5) (when n→∞) can be rewritten as

$$P(x_1, ..., x_{m-1}, z)$$

$$= \sum_{i_1=0}^{N} ... \sum_{i_m=0}^{N} x_1^{i_1} ... x_{m-1}^{i_{m-1}} Q_{(i_1, ..., i_m)}(z) .$$

Since N is finite, the above equation shows that as long as the $Q_{(i_1, ..., i_m)}(z)$'s are finite for a given value of z, $P(x_1, ..., x_{m-1}, z)$ exists for any finite value of x_k ($1 \leq k \leq m-1$). This means that, if z_j is a pole of $P(x_1, ..., x_{m-1}, z)$ for any given set of finite x_k's, it must also be a pole of at least one of the partial generating functions $Q_{(i_1, ..., i_m)}(z)$. From equation (16), z_j should then also be a pole of $V(z)$. The inverse implication also holds. Thus we have more or less proved that $V(z)$ has completely the same poles as $P(x_1, ..., x_{m-1}, z)$. As x_k ($1 \leq k \leq m-1$) can be any set of finite values, we can select $x_k = r_k(z)$ which is defined in equation (8). From equation (8), it is easy to see that the $r_k(z)$'s exist for any finite value of z; therefore $V(z)$ has completely the same poles as $P(r_1(z), ..., r_{m-1}(z), z)$ which is given in equation (9). Note that a similar reasoning can be done if the $r_k(z)$'s are replaced by another set of solutions $x_k(z)$ of equations (7), for which $x_k(1)$ is not necessarily equal to 1, but, in appendix B, we show that it suffices to consider the $r_k(z)$'s.

Let $z_0, ..., z_u$ denote all the possible poles of $V(z)$. The inversion formula for z–transforms [14], together with the residue theorem from complex analysis, leads to an expression for Prob[v=n] which is a weighted sum of negative powers of the poles of $V(z)$. Since the modulus of all the poles is larger than one, i.e., $|z_j| > 1$ ($0 \leq j \leq u$), it is obvious that for sufficiently large n (i.e., n > B), Prob[v=n] is dominated by the contribution of the pole having the smallest modulus. We denote this dominating pole by z_0. As explained in [10–11], z_0 must necessarily be real and positive to ensure that Prob[v=n] is nonnegative anywhere for large n. In appendix B, we prove that there really exists such a real pole z_0 and that its multiplicity is one. So, with respect to the asymptotic behavior of the buffer occupancy, $V(z)$ can be approximated as

$$V(z) \cong \frac{\beta}{z - z_0} , \tag{17}$$

where β is an unknown parameter. Taking the inverse z–transform of equation (17), we thus obtain

$$Prob[v = n] \cong -\frac{\beta}{z_0}\left(\frac{1}{z_0}\right)^n , \quad n > B . \tag{18}$$

Comparing to equation (15), we have

$$\gamma = \frac{1}{z_0} . \tag{19}$$

As stated above, z_0 is also a pole of $P(r_1(z), ..., r_{m-1}(z), z)$ in equation (9), i.e., z_0 is a real root of $z - H(z) = 0$. In the real positive domain ($z \geq 0$) it can be proved (see appendix B) that $r_1(z) > 0$ and hence, $1-q+qzr_1(z) > 0$. From equation (8) and $z_0 - [1-q+qz_0r_1(z_0)]^N = 0$, it gives

$$z_0 = \left[1-q + qz_0^{\frac{(N-1)m+1}{N}} \right]^N . \tag{20}$$

By using a repeated substitution algorithm, z_0 can easily be obtained numerically. So, γ can be gotten from equation (19).

B. Upper Bound for the Coefficient K of Geometric Form

If the number of packets in the buffer is sufficiently large (>> N) just after a given slot, it is reasonable to think that the number of packets arriving in that slot ($\leq$ N) has nearly no influence on the total number of packets in the buffer, i.e., when n is sufficiently large (n > B), we may assume that the arrival process is independent of the state of the buffer system. Define the limiting arrival probability function $\omega(i_1, ..., i_m)$ as

$$\omega(i_1, ..., i_m) \triangleq \lim_{n \to \infty} Prob[a_1=i_1, ..., a_m=i_m \mid v=n] ,$$

with pgf $\Omega(x_1, ..., x_m)$. When n is very large (e.g., > B), it is sure that $Prob[a_1=i_1, ..., a_m=i_m \mid v=n] \cong \omega(i_1, ..., i_m)$.

Let $p(i_1, ..., i_m, n) = Prob[a_1=i_1, ..., a_m=i_m, v=n]$ and $v(n) = Prob[v=n]$, then, the joint pgf $P(x_1, ..., x_{m-1}, z)$ can be approximately expressed as

$$P(x_1, ..., x_{m-1}, z) \cong \sum_{\underline{i}=0}^{N} \sum_{n=0}^{B} p(i_1, ..., i_m, n) x_1^{i_1} ... x_{m-1}^{i_{m-1}} z^n$$

$$+ \sum_{\underline{i}=0}^{N} \omega(i_1, ..., i_m) x_1^{i_1} ... x_{m-1}^{i_{m-1}} \left[V(z) - \sum_{n=0}^{B} v(n) z^n \right] ,$$

where

$$\sum_{\underline{i}=0}^{N} \triangleq \sum_{i_1=0}^{N} ... \sum_{i_m=0}^{N} .$$

Setting $x_k = r_k(z)$ for $1 \leq k \leq m-1$ in the above equation, we know from previous subsection that z_0 is the pole of $V(z)$ and $P(r_1(z), ..., r_{m-1}(z), z)$. Since $r_k(z_0)$ is finite and B is extremely large, but as long as B < ∞, the other terms in the above equation are finite. Thus, multiplying by $(z-z_0)$ on both sides of the above equation and taking the z → z_0 limit, leads to

$$\beta = \frac{\alpha}{\Omega(r_1(z_0), ..., r_{m-1}(z_0), 1)} ,$$

where $\Omega(x_1, ..., x_m)$ is the pgf of the limiting arrival probability $\omega(i_1, ..., i_m)$ and α is given by equation (9) as

$$\alpha = \frac{[1-H'(1)](z_0-1)H(z_0)}{1-H'(z_0)}.$$

Comparing equations (15) and (18) gives that $K = -(\beta/z_0)$. As $z_0 = H(z_0)$, from the above two equations, we get

$$K = \frac{[1-H'(1)](z_0-1)}{[H'(z_0)-1]\Omega(r_1(z_0), ..., r_{m-1}(z_0), 1)}. \quad (21)$$

It is difficult to derive $\Omega(x_1, ..., x_m)$ directly. In the following, a method to obtain an upper bound for K is described.

Let us consider two different cases : (1) the arrival process is investigated when the buffer occupancy is extremely large, and (2) the arrival process is randomly observed. It is intuitive that the probability of observing more packet arrivals in the former case is larger than the later case. That is, the limiting arrival probability $\omega(i_1, ..., i_m)$ is larger than the unconditional arrival probability $a(i_1,..., i_m)$ (= Prob$[a_1=i_1, ..., a_m=i_m]$) for large i_k's ($1 \leq k \leq m$). It is shown in appendix B that $r_k'(z) > 0$ for $z > 0$. As $r_k(1) = 1$ and $z_0 > 1$, so $r_k(z_0) > 1$ for $1 \leq k \leq m-1$. Due to the normalization condition, We thus have

$$\Omega(r_1(z_0), ..., r_{m-1}(z_0), 1) > A(r_1(z_0), ..., r_{m-1}(z_0), 1),$$

where $A(x_1, ..., x_m)$ is the pgf of $a(i_1,..., i_m)$ and is given in equation (11). Using this inequality, an upper bound for the coefficient K in equation (21) can be found as

$$K < K_u = \frac{[1-H'(1)](z_0-1)}{[H'(z_0)-1]A(r_1(z_0), ..., r_{m-1}(z_0), 1)}, \quad (22)$$

where $r_1(z_0)$ can be obtained from $z_0 - H(z_0) = 0$ and $r_k(z_0)$ ($2 \leq k \leq m-1$) is given by equation (8).

To conclude this section, the tail distribution of the buffer occupancy can be approximated by a geometric form for which an explicit upper bound can be obtained, i.e., for sufficiently large n,

$$\text{Prob}[v=n] \cong K \left(\frac{1}{z_0}\right)^n < K_u \left(\frac{1}{z_0}\right)^n, \quad (23)$$

where z_0 is determined by equation (20) and K_u is given by equation (22).

C. Comparison with Simulation Results

The above analysis is based on the assumption that the buffer size is infinite, which means that no packet can get lost. The probability of the buffer occupancy exceeding a proposed buffer size S (i.e., Prob$[v > S]$) is often used to

estimate the packet loss ratio due to the finite buffer size in a real system. From equation (23), we have

$$\text{Prob}[v > S] \cong \frac{K}{z_0-1}\left(\frac{1}{z_0}\right)^S < \frac{K_u}{z_0-1}\left(\frac{1}{z_0}\right)^S. \quad (24)$$

In Figs. 3–8, we compare the upper–bound tail distribution of the buffer occupancy given by equation (24) with simulation results. The simulation results are not stable in the low probability area (roughly starting from 10^{-5}) due to the limited simulation time.

The curves show that the approximate method described above gives a very tight upper–bound tail distribution, especially in the interesting region of high traffic loads p (see Figs 3, 5 and 7). For low traffic load, the difference between the upper–bound tail distribution and the simulation results is more pronouced, but the difference is still small (see Figs 4, 6 and 8). The reason is that the limiting arrival probability function $\omega(i_1, ..., i_m)$ has less bias from the unconditional arrival probability function $a(i_1, ..., i_m)$ for high traffic load than for low traffic load. Since the m–length train arrival process is rather bursty, for high load, large buffer occupancy is easily to be reached. This means that the difference between $\omega(i_1, ..., i_m)$ and $a(i_1, ..., i_m)$ is small for high load. This explains why the upper bound is very tight for high traffic load. To investigate the characteristics of the limiting arrival probability function $\omega(i_1, ..., i_m)$ and to obtain the exact coefficient K of the geometric form of the buffer occupancy distribution remain for further study.

Furthermore, the simulation results confirm that the analytical approach used in this section to obtain z_0, the pole of $V(z)$ with smallest modulus outside the unit circle, is correct. This pole is imbedded in the m–dimensional joint pgf $P(x_1, ..., x_{m-1}, z)$ and can be found by choosing $x_k = r_k(z)$ which has the property that $r_k(1) = 1$ for $1 \leq k \leq m-1$. Fig. 3 also tells us that, for a given traffic load, the probability of the buffer occupancy exceeding a certain buffer size becomes larger if the correlation of the packet arrival process gets stronger (i.e., if m, the number of packets in a message, gets larger). Note that m=1 corresponds to an uncorrelated packet arrival process.

VI. CONCLUSIONS

In this paper, we have analyzed the performance of a statistical multiplexer with a finite number of input links and train arrivals. By means of a generating function approach, we have indicated a method to obtain the moments of the buffer occupancy in the form of closed form expressions. We have also given an approximate method to obtain an upper bound for the tail distribution of the buffer occupancy. Comparing with simulation results, we found this upper bound to be very tight, especially for high traffic load.

In general, it is very difficult to analytically derive the exact distribution of the buffer occupancy in case of a correlated arrival process. Since for most discrete–time single–server queueing systems, the tail distribution of the buffer occupancy has a geometric form, the approximate method presented in this paper could be used to obtain upper–bound performance of a discrete–time single–server queueing system with any correlated arrival process.

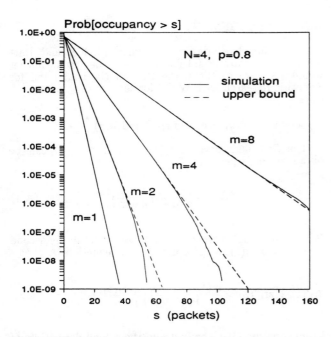

Fig.3 Complementary distribution of the buffer occupancy versus buffer size, for N = 4, p = 0.8 Erlang, and m = 1, 2, 4 and 8.

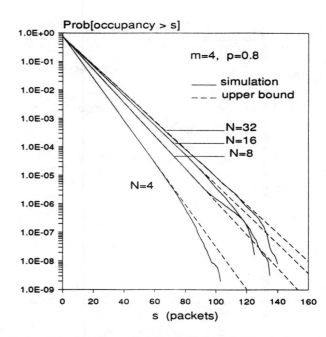

Fig.5 Complementary distribution of the buffer occupancy versus buffer size, for m = 4, p = 0.8 Erlang, and N = 4, 8, 16 and 32.

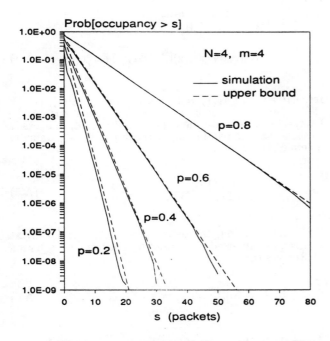

Fig.4 Complementary distribution of the buffer occupancy versus buffer size, for N = 4, m = 4, and p = 0.2, 0.4, 0.6 and 0.8 Erlang.

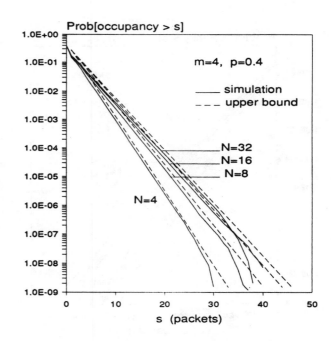

Fig.6 Complementary distribution of the buffer occupancy versus buffer size, for m = 4, p = 0.4 Erlang, and N = 4, 8, 16 and 32.

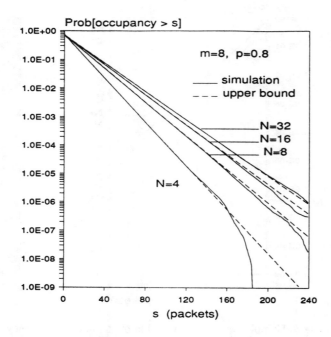

Fig.7 Complementary distribution of the buffer occupancy versus buffer size, for m = 8, p = 0.8 Erlang, and N = 4, 8, 16 and 32.

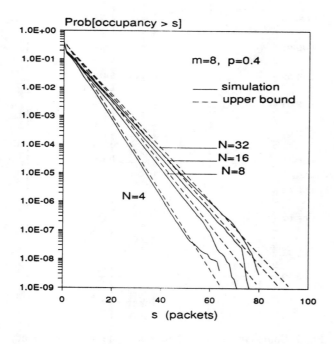

Fig.8 Complementary distribution of the buffer occupancy versus buffer size, for m = 8, p = 0.4 Erlang, and N = 4, 8, 16 and 32.

ACKNOWLEDGEMENT

We would like to thank Herman Michiel and Bart Steyaert for many useful discussions. The second author would also like to thank the Belgian National Fund for Scientific Research (N.F.W.O.) for support of this work.

Appendix A. Derivatives of $r_k(z)$ and H(z)

The first derivative of $r_k(z)$ $(1 \leq k \leq m-1)$ can be obtained from equation (8) as

$$r_k'(z) = \frac{(m-k)(1-q)r_k(z)}{z[1-q+mqzr_1(z)]} . \qquad (A.1)$$

It is a bit complicated but can be shown that the second derivative of $r_1(z)$ can be obtained from equation (A.1) as

$$2r_1'(z) + zr_1''(z) = \frac{m(1-q)[1-q+qzr_1(z)]}{[1-q+mqzr_1(z)]^2} r_1'(z) . \qquad (A.2)$$

From equation (10), the first and the second derivatives of H(z) can be derived as

$$H'(z) = Nq[1-q+qzr_1(z)]^{N-1}[r_1(z)+zr_1'(z)] \qquad (A.3)$$

and

$$H''(z) = \frac{N-1}{N} \frac{[H'(z)]^2}{H(z)} +$$

$$Nq[1-q+qzr_1(z)]^{N-1}[2r_1'(z)+zr_1''(z)] , \qquad (A.4)$$

respectively. As $r_k(1) = 1$ $(1 \leq k \leq m-1)$, $r_k'(1)$, $r_1''(1)$, $H'(1)$ and $H''(1)$ can be obtained from equations (A.1)–(A.4). Let p indicate the average traffic load, it is easily to show that the relationship between p and the parameter q defined in section I is

$$\frac{p}{N} = \frac{m/(1-q)}{m/(1-q)+1/q} = \frac{mq}{1+(m-1)q} . \qquad (A.5)$$

Using this relation, the derivatives of $r_k(z)$ and H(z) at z=1 can be expressed in terms of p. For instance, equation (A.3) gives that

$$H'(1) = Nq[1+r_1'(1)] = p . \qquad (A.6)$$

Appendix B. Properties of Smallest Real Pole z_0 of V(z)

Consider $f(z) \triangleq [1-q+qzx_1(z)]^N - z$ in the real positive domain $(z \geq 0)$ where $x_1(z)$ is implicitly defined by (see equation (7))

$$x_1(z)[1-q+qzx_1(z)]^{m-1} = z^{m-1} . \qquad (B.1)$$

There are m solutions for $x_1(z)$ in the above equation (m "branches") and it is obvious that one of them has the property that $x_1(1) = 1$, which is denoted by $r_1(z)$ (see equation (8)). In the following, we will prove that the smallest real zero (denoted by z_0) of $f(z)$ for $z > 1$, is completely determined by the branch $x_1(z) = r_1(z)$, and its multiplicity is one.

First, assume that $f(z)$ has real zeros for $z > 1$. Let z^* (>1) be such a zero of $f(z)$. Based on equation (B.1) and $f(z^*) = 0$, one can easily prove that $x_1(z^*)$ has two properties: (i) $x_1(z^*)$ must be real; (ii) $|x_1(z^*)| > 1$. Hence, either $-\infty < x_1(z^*) < -1$ or $1 < x_1(z^*) < \infty$. Let z_0^* and z_1^* be zeros of $f(z)$ and $1 < x_1(z_0^*) < \infty$, $-\infty < x_1(z_1^*) < -1$. Equation (B.1) shows that z_1^* exists only when m and N are even. In this case, $1-q+qz_1^*x_1(z_1^*) < 0$. Using equation (B.1) and $f(z_1^*) = 0$, it can be shown that

$$(z_1^*)^{\frac{1}{N}} \left[q\,(z_1^*)^{\frac{m(N-1)}{N}} - 1 \right] = (1-q) .$$

Similarly z_0^* can be expressed as

$$(z_0^*)^{\frac{1}{N}} \left[1 - q\,(z_0^*)^{\frac{m(N-1)}{N}} \right] = (1-q) .$$

From these two equations it is clear that $z_1^* > z_0^*$.

Summarizing our results so far, we have shown that if $f(z)$ has a zero $z_0^* > 1$ for which $x_1(z_0^*) > 1$, then no other zero ($z_1^* > 1$) of $f(z)$ for which $x_1(z_1^*) < -1$ is smaller than z_0^*. We next prove that there actually always exists exactly one such zero z_0^* (> 1) of $f(z)$ which therefore is the smallest pole of $V(z)$, i.e., $z_0^* = z_0$. In order to do so, we first prove that there always exists exactly one real positive solution of $x_1(z)$ in equation (B.1) which is larger than one for $z > 1$ and this solution is $r_1(z)$, and, next, that $f(z)$ has indeed exactly one real positive zero (z_0^*) > 1 for $x_1(z) = r_1(z)$.

Let

$$g(y) \triangleq y\,[1-q+qzy]^{m-1} - z^{m-1} , \quad z > 0 . \quad (B.2)$$

The first derivative of $g(y)$ is

$$g'(y) = [1-q+qzy]^{m-1} + (m-1)qzy[1-q+qzy]^{m-2} . \quad (B.3)$$

Since $g(0) = -z^{m-1} < 0$, $g(+\infty) = +\infty$ and $g'(y) > 0$ for $y > 0$, $g(y)$ crosses the y–axis in the positive region one time, which means that there always and only exists one real positive zero of $g(y)$ for $z > 0$. It is obvious that this zero is a continuous function of z. As $r_1(1) = 1$ is a zero of $g(y)$ when $z=1$, so $r_1(z)$ is the real positive zero of $g(y)$, i.e., the only real positive solution of $x_1(z)$ in equation (B.1). As $r_1(z) > 0$, from equations (A.1) and (8), we have $r_k'(z) > 0$ ($1 \leq k \leq m-1$), so $r_k(z) > 1$ for $z > 1$.

When $x_1(z) = r_1(z)$, it is obvious that $z = 1$ is a zero of $f(z)$ and $f'(1) = H'(1)-1 = p-1 < 0$ (see equation (A.6)). Since $r_1(z) > 1$ for $z > 1$, it is clear that $f(z) > 0$ for large z. As $f''(z) = H''(z)$, it is easy to see from equations (A.2)

and (A.4) that $f''(z) > 0$ for $z > 0$, i.e., $f(z)$ is a concave function of z. So, there really exists another zero of $f(z)$, i.e., z_0, which is larger than one and has multiplicity one.

REFERENCES

[1] M.A. Henrion, K.J. Schrodi, D. Boettle, M.De Somer and M. Dieudonne, "Switching network architecture for ATM based broadband communications", Proceedings of ISS'90 (Stockholm, 1990), vol. V, pp. 1–8.

[2] B. Gopinath and J.A. Morrison, "Discrete–time single server queues with correlated inputs", BSTJ, vol. 56, 1977, pp. 1743–1768.

[3] A.G. Fraser, B. Gopinath and J.A. Morrison, "Buffering of slow terminals", BSTJ, vol.57, 1978, pp. 2865–2885.

[4] W.A. Massey and J.A. Morrison, "Calculation of steady–state probabilities for content of buffer with correlated inputs", BSTJ, vol. 57, 1978, pp. 3097–3117.

[5] Y. Xiong and H. Bruneel, "Buffer contents and delay for statistical multiplexers with fixed–length packet–train arrivals", submitted for publication (1991).

[6] H. Bruneel and I. Bruyland, "Performance study of statistical multiplexing in case of slow message generation", Proceedings of ICC'89 (Boston, 1989) pp. 951–955.

[7] H. Bruneel, "Message waiting times and delays in ATDM switching elements", Proceedings of GLOBECOM'90 (San Diego, 1990), pp. 1450–1454.

[8] H. Bruneel, "Queueing behavior of statistical multiplexers with correlated inputs", IEEE Trans. Commun., vol. COM–36, Dec., 1988, pp. 1339–1341.

[9] B. Steyaert and H. Bruneel, "An effective algorithm to calculate the distribution of the buffer contents and the packet delay in a multiplexer with bursty sources", Proceedings of GLOBECOM'91 (Phoenix, 1991).

[10] C.M. Woodside and E.D.S. Ho, "Engineering calculation of overflow probabilities in buffers with Markov–interrupted service", IEEE Trans. Commun., vol. COM–35, Dec., 1987, pp. 1272–1277.

[11] E. Desmet, B. Steyaert, H. Bruneel and G. Petit, "Tail distributions of queue length and delay in discrete–time multiserver queueing models, applicable in ATM networks", Proceedings of ITC'13 (Copenhagen, 1991).

[12] M.J. Karol, M.G. Hluchyj and S.P. Morgan, "Input versus output queueing on a space–division packet switch", IEEE Trans. Commun., vol. COM–35, Dec., 1987, pp. 1347–1356.

[13] J.N. Daigle, Y. Lee and M.N. Magalhaes, "Discrete time queues with phase dependent arrivals", Proceedings of INFOCOM'90 (San Francisco, 1990), pp. 728–732.

[14] L. Kleinrock, Queueing Systems, Volume I: Theory (Wiley, New York, 1975).

Epilogue to the Volume

A few observations are warranted at the conclusion of this reprint volume on the performance evaluation of high-speed switching networks. One is that all the switching networks in this volume are essentially parallel processors. Using a multiplicity of switching elements, they boost throughput by allowing traffic to flow in parallel over a variety of simultaneous paths.

In reading through these papers it is also interesting to note the need for statistical models to evaluate their performance. We now have the capability to build sophisticated switching systems, where the basic building blocks are $m \times n$ switching elements. With an interconnection of large numbers of such switching elements and associated buffers, a statistical analysis is crucial to understand the overall switch behavior.

Of course, performance evaluation doesn't exist in a vacuum. It is motivated by a desire to build new communications systems, particularly (at this time) ATM-based networks. Part and parcel of the design process is the need to evaluate different architectural configurations. To this end, statistical performance evaluation is a cost-effective design tool.

Index

N

Editor's Biography

Thomas G. Robertazzi received a B.E.E. from the Cooper Union in 1977 and a Ph.D. in electrical engineering from Princeton University in 1981. During 1982–1983 he was an assistant professor of electrical engineering at Manhattan College, Riverdale, N.Y. Since 1983 he has taught and conducted research at the electrical engineering department of the State University of New York at Stony Brook, where he is presently an associate professor.

Prof. Robertazzi is a senior member of the IEEE. During the fall of 1990 he was a Visiting Research Scientist at Columbia University's Center for Telecommunications Research. He is a reviewer for a number of engineering journals and has been a member of the INFOCOM Technical Program Committee since 1989.

Prof. Robertazzi's research interests are in the area of the performance evaluation of communication and computer systems.